P9-CNC-799

WITHDRAWN

STANDARD NORMAL
DISTRIBUTION

Please remember that this is a library book, and that it belongs only temporarily to each person who uses it. Be considerate. Do not write in this, or any, library book.

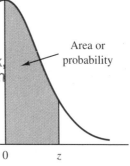

Area or probability

0   z

Entries in the table give the area under the curve between the mean and z standard deviations above the mean. For example, for $z = 1.25$ the area under the curve between the mean and z is .3944.

z	.00	.01	.02	.03	.04	.05	.06	.07	.08	.09
.0	.0000	.0040	.0080	.0120	.0160	.0199	.0239	.0279	.0319	.0359
.1	.0398	.0438	.0478	.0517	.0557	.0596	.0636	.0675	.0714	.0753
.2	.0793	.0832	.0871	.0910	.0948	.0987	.1026	.1064	.1103	.1141
.3	.1179	.1217	.1255	.1293	.1331	.1368	.1406	.1443	.1480	.1517
.4	.1554	.1591	.1628	.1664	.1700	.1736	.1772	.1808	.1844	.1879
.5	.1915	.1950	.1985	.2019	.2054	.2088	.2123	.2157	.2190	.2224
.6	.2257	.2291	.2324	.2357	.2389	.2422	.2454	.2486	.2518	.2549
.7	.2580	.2612	.2642	.2673	.2704	.2734	.2764	.2794	.2823	.2852
.8	.2881	.2910	.2939	.2967	.2995	.3023	.3051	.3078	.3106	.3133
.9	.3159	.3186	.3212	.3238	.3264	.3289	.3315	.3340	.3365	.3389
1.0	.3413	.3438	.3461	.3485	.3508	.3531	.3554	.3577	.3599	.3621
1.1	.3643	.3665	.3686	.3708	.3729	.3749	.3770	.3790	.3810	.3830
1.2	.3849	.3869	.3888	.3907	.3925	.3944	.3962	.3980	.3997	.4015
1.3	.4032	.4049	.4066	.4082	.4099	.4115	.4131	.4147	.4162	.4177
1.4	.4192	.4207	.4222	.4236	.4251	.4265	.4279	.4292	.4306	.4319
1.5	.4332	.4345	.4357	.4370	.4382	.4394	.4406	.4418	.4429	.4441
1.6	.4452	.4463	.4474	.4484	.4495	.4505	.4515	.4525	.4535	.4545
1.7	.4554	.4564	.4573	.4582	.4591	.4599	.4608	.4616	.4625	.4633
1.8	.4641	.4649	.4656	.4664	.4671	.4678	.4686	.4693	.4699	.4706
1.9	.4713	.4719	.4726	.4732	.4738	.4744	.4750	.4756	.4761	.4767
2.0	.4772	.4778	.4783	.4788	.4793	.4798	.4803	.4808	.4812	.4817
2.1	.4821	.4826	.4830	.4834	.4838	.4842	.4846	.4850	.4854	.4857
2.2	.4861	.4864	.4868	.4871	.4875	.4878	.4881	.4884	.4887	.4890
2.3	.4893	.4896	.4898	.4901	.4904	.4906	.4909	.4911	.4913	.4916
2.4	.4918	.4920	.4922	.4925	.4927	.4929	.4931	.4932	.4934	.4936
2.5	.4938	.4940	.4941	.4943	.4945	.4946	.4948	.4949	.4951	.4952
2.6	.4953	.4955	.4956	.4957	.4959	.4960	.4961	.4962	.4963	.4964
2.7	.4965	.4966	.4967	.4968	.4969	.4970	.4971	.4972	.4973	.4974
2.8	.4974	.4975	.4976	.4977	.4977	.4978	.4979	.4979	.4980	.4981
2.9	.4981	.4982	.4982	.4983	.4984	.4984	.4985	.4985	.4986	.4986
3.0	.4986	.4987	.4987	.4988	.4988	.4989	.4989	.4989	.4990	.4990

WITHDRAWN

AAV-2489
VC- LibStudies

68,50

WITHDRAWN

AAV-2489

Introduction to
STATISTICS
Concepts and Applications

Third Edition

WITHDRAWN

INTRODUCTION TO
STATISTICS
CONCEPTS AND APPLICATIONS

THIRD EDITION

DAVID R. ANDERSON
University of Cincinnati

DENNIS J. SWEENEY
University of Cincinnati

THOMAS A. WILLIAMS
Rochester Institute of Technology

WEST PUBLISHING COMPANY
Minneapolis/St. Paul   New York   Los Angeles   San Francisco

Copyediting   Luana Richards
Design   John Rokusek
Artwork   Rolin Graphics
Composition   Carlisle Communications
Proofreading   Jerrold Moore
Cover Design   John Rokusek
Cover Image   The Stock Market/Peter Saloutos © 1991
Indexing   Sandi Schroeder
Prepress, Printing, and Binding by West Publishing Company

Minitab is a registered trademark of Minitab, Inc., 3081 Enterprise Drive, State College, PA, 16801 (telephone 814/238–3280; telex 881612; fax 814/238–4383).

West's Commitment to the Environment

In 1906, West Publishing Company began recycling materials left over from the production of books. This began a tradition of efficient and responsible use of resources. Today, up to 95 percent of our legal books and 70 percent of our college texts are printed on recycled, acid-free stock. West also recycles nearly 22 million pounds of scrap paper annually—the equivalent of 181,717 trees. Since the 1960s, West has devised ways to capture and recycle waste inks, solvents, oils, and vapors created in the printing process. We also recycle plastics of all kinds, wood, glass, corrugated cardboard, and batteries, and have eliminated the use of styrofoam book packaging. We at West are proud of the longevity and the scope of our commitment to our environment.

COPYRIGHT © 1986, 1991   By WEST PUBLISHING COMPANY
COPYRIGHT © 1994   By WEST PUBLISHING COMPANY
                         610 Opperman Drive
                         P.O. Box 64526
                         St. Paul, MN 55164-0526

All rights reserved
Printed in the United States of America
00 99 98 97 96 95 94   8 7 6 5 4 3 2 1

Library of Congress Cataloging-in-Publication Data

Anderson, David Ray, 1941–
   Introduction to statistics : concepts and applications / David R.
  Anderson, Dennis J. Sweeney, Thomas A. Williams.—3rd ed.
     p.   cm.
  Includes bibliographical references and index.
  ISBN 0-314-02813-7
   1. Statistics. I. Sweeney, Dennis J. II. Williams, Thomas
Arthur, 1944–   . III. Title.
QA276.12.A44 1993
519.5—dc20

                              93-30031

 CIP

CONTENTS

CHAPTER 3   DESCRIPTIVE STATISTICS II: MEASURES OF LOCATION AND DISPERSION, 60

CHAPTER 4   INTRODUCTION TO PROBABILITY, 110

CHAPTER 9   INFERENCES ABOUT A POPULATION PROPORTION,   340

CHAPTER 12 ANALYSIS OF VARIANCE AND EXPERIMENTAL DESIGN, 434

CHAPTER 13 LINEAR REGRESSION AND CORRELATION, 500

CHAPTER 16 TESTS OF GOODNESS OF FIT AND INDEPENDENCE, 694

CHAPTER 17 NONPARAMETRIC METHODS, 714

CHAPTER 18 SAMPLE SURVEY, 756

APPENDIXES

LIST OF STATISTICS IN PRACTICE

PREFACE

The purpose of this book is to provide a comprehensive treatment of introductory statistics for students from a wide variety of academic backgrounds. The text is applications oriented and has been written with the needs of the nonmathematician in mind. The mathematical prerequisite is a course in college algebra.

APPLICATIONS AND METHODOLOGY INTEGRATED

Applications of the statistical methodology are an integral part of the organization and presentation of the material. Each chapter begins by motivating the student with a general interest writeup involving statistics in practice that demonstrates a use of the statistical procedures that will be introduced in the chapter. Statistical techniques are then introduced using examples where the techniques have been successfully applied. The discussion and development of each technique is centered around an application setting, with the statistical results providing information helpful in solving the underlying problem.

In addition, we have taken care to provide a sound methodological development. Throughout the text we have utilized notation that is generally accepted for the topic being covered. Students will thus find that the text provides good preparation for the study of more advanced statistical material. A bibliography that should prove useful as a guide to further study has been included as an appendix.

CHANGES IN THE THIRD EDITION

In making modifications for this new edition, we have maintained the presentation style of the previous edition. However, this edition includes a more complete and detailed topical coverage than the previous editions, including the addition of counting rules for combinations and permutations and expanded binomial probability tables. We have also extended our treatment of experimental design, revised the three-chapter sequence on regression analysis, and included a new chapter on sample survey methods. The more significant changes in the third edition are summarized below.

MORE EXAMPLES AND PROBLEMS BASED ON REAL DATA

We have continued our emphasis on helping students understand the wide range of statistical applications by expanding the number of examples and problems based on real statistical studies. Sources such as *The New York Times, USA Today, Psychology Today, New England Journal of Medicine,* and *The Wall Street Journal* provide referenced applications and problems that demonstrate wide-ranging uses of statistics. The use of real data means that students not only learn about the statistical methodology but also learn about the content of applications encountered in practice. Approximately 300 new problems have been added in this edition.

METHODS EXERCISES AND APPLICATIONS EXERCISES

The end-of-section exercises have been split into two parts. With the Methods exercises, students are required to make sure they can handle the formulas and make the necessary computations. In the Applications exercises, students are given problems that require using the chapter material in real-world situations. With this approach, students first focus on the computational "nuts and bolts," then move on to the subtleties of statistical application and interpretation.

SELF-TEST EXERCISES

Certain exercises are identified as self-test exercises. Completely worked-out solutions for these exercises are provided in an appendix at the end of the text. Students can attempt the self-test exercises and immediately check the solution to evaluate their understanding of the concepts presented in the chapter.

COMPUTER CASES

Many chapters have computer cases, which contain problem scenarios accompanied by modest-sized data sets. Computer solution by Minitab, The Data Analyst, or another statistical software package is required. Each case outlines a report that the student prepares to summarize statistical results as well as present interpretations and recommendations. The data sets for all computer cases are available on data disks formatted for Minitab, The Data Analyst, or MYSTAT.

A NEW CHAPTER ON SAMPLING AND SURVEY METHODOLOGY

Chapter 18 treats sampling and survey methodology. A discussion of issues involved in designing and conducting survey research is presented. The probability sampling methods of simple random sampling, stratified simple random sampling, cluster sampling, and systematic sampling are covered. The nature and control of sampling and nonsampling error are treated.

ANALYSIS OF VARIANCE AND EXPERIMENTAL DESIGN

The presentation of analysis of variance and experimental design (Chapter 12) has been revised with a new *Statistics in Practice* used to introduce students to the concepts. The chapter begins with analysis of variance and its use in testing for the equality of means with three or more populations. Procedures for making multiple comparisons have been expanded and now appear immediately following the material on testing for the equality of means. Experimental design concepts are presented after the introduction to analysis of variance. Completely randomized, randomized block, and factorial experimental designs complete the coverage. Formulas for the analysis of variance computations appear in the chapter appendix.

REGRESSION ANALYSIS

The three-chapter sequence on regression analysis (Chapters 13, 14, and 15) has been enhanced with an expanded discussion of the use and interpretation of dummy variables and interaction terms. The best-subset selection technique and the use of normal probability plots are now included, and the discussion of residual analysis has been expanded.

FEATURES

We have continued many of the features that appeared in earlier editions. Some of the more important ones include the following.

LEARNING OBJECTIVES

Each chapter begins with a statement of learning objectives which falls under the heading of "What You Will Learn in This Chapter." This list contains the concepts that the student will be expected to master and should help guide the student's study of the material.

STATISTICS IN PRACTICE

Each chapter opens with a general interest news article that demonstrates a use of the statistical procedures about to be introduced. These *Statistics in Practice* applications are based on actual articles appearing in journals, magazines, and newspapers such as *The New England Journal of Medicine, Time, USA Today, The New York Times,* and so on. The *Statistics in Practice* is a condensed version of the original article and specifically focuses on the use of statistics as reported in the publication. Topical selections include the cost of a college education, marriage statistics, pay differentials for men and women, and professional sports; these selections should capture the student's interest and show them that the statistical procedures that he or she is about to learn have some interesting applications.

CHAPTER PEDAGOGY

Each chapter introduces statistical methodology in the context of examples that demonstrate the use of the methodology in a wide variety of general interest applications. Problems are provided after each section to enable the student to check his or her progress. Answers to the even-numbered exercises and self-test exercises are provided at the back of the book. Each chapter concludes with a review of key concepts and topics introduced therein. A glossary of statistical terms found in the chapter follows, and a key formula section itemizes the important equations that the student should know how to apply. A review quiz is then included to reinforce the key concepts presented. Supplementary exercises, based on the material throughout the chapter, then provide additional opportunities to practice applying the methodology presented. Where appropriate, computer printouts are included; these print-outs demonstrate how computer packages can be used in statistical computation and summaries.

REVIEW QUIZZES

The review quizzes consist of true-false and multiple choice questions. Each review quiz provides the student with an opportunity to evaluate his or her progress after the chapter material has been covered. Answers for the review quiz questions are included at the back of the book.

NOTES AND COMMENTS

At the end of many sections, we have provided *Notes and Comments* designed to give the student additional insights about the statistical methodology presented in the section. They include warnings and/or limitations about the methodology, recommendations for application, brief descriptions of additional technical considerations, and so on. It is hoped that this feature will expand the students' understanding of statistics and their ability to use the material.

COMPUTER SOFTWARE

The text contains numerous examples and discussions of the important role statistical software packages play in the computation and presentation of statistical results. More examples and problems based on the Minitab statistical package have been added. Use and interpretation of the information provided by Minitab printouts are emphasized. Available to adopters is The Data Analyst version 2.0, an IBM-compatible microcomputer software package developed by the authors. The Data Analyst or MYSTAT may be ordered shrink-wrapped with the text. The Data Analyst, MYSTAT, Minitab, or other statistical packages can be used to solve problems and computer cases appearing in the text.

DATA DISK

Data sets for text examples, problems, and computer cases are available on special Data Disks, that can be ordered with the text. The Data Disks contain the data sets in a format acceptable to Minitab, MYSTAT, or The Data Analyst software packages.

ANCILLARIES

Accompanying the text is a complete package of support materials. A solutions manual, prepared by the authors, contains complete solutions to all exercises in the text. A test bank contains multiple choice questions followed by problems, and WESTTEST Computerized Testing also accompanies the package. Transparency masters display worked-out demonstration problems, and a student workbook (prepared by Meredith Many and Charlotte Lewis, University of New Orleans) provides chapter summaries and glossaries, formula references, and additional practice problems with solutions. The Data Analyst microcomputer software package, developed by the authors, is available free to adopters, and students may purchase the software shrinkwrapped with the text for a small additional charge. In addition, the MYSTAT statistical software package may also be purchased shrinkwrapped with the text for a small additional charge.

ACKNOWLEDGMENTS

We owe a debt to many of our colleagues and friends for their helpful comments and suggestions during the development of this manuscript. Among these are Paul B. Berger, Ben P. Bockstege, John M. Burns, Curtis K. Church, Louis J. Cote, Robert H. Cranford, Henry Crouch, Carl Cuneo, Shirley Dowdy, Ruby A. Evans, David Gillman, Penelope Greene, Terry H. Hughes, Peter Kerwin, Basil P. Korin, Robert L. Lacher, Stanley M. Lukawecki, David R. Lund, Daniel E. McNamara, Robert Mee, Jeff Mock, Alex Papadopoulos, Charles D. Reinauer, Franklin D. Rich, Arnold L. Schroeder, A.K. Shah, Robert K. Smidt, W. Robert Stephenson, Bill Stines, Glen H. Swindle, Chin-Chyuan Tai, Barbara Treadwell, David L. Turner, and Vasant B. Waikar.

We are also indebted to our editor, Mary Schiller, for her counsel and support during the preparation of this text. We would like to express our appreciation to Tom Hilt, our production editor, for the outstanding job he performed in pulling all the pieces together to produce this book. Finally, a sincere thank you to everyone else at West Publishing Company that played a part in helping us complete this project.

David R. Anderson

Dennis J. Sweeney

Thomas A. Williams

CHAPTER 1

DATA, MEASUREMENT, AND STATISTICS

WHAT YOU WILL LEARN IN THIS CHAPTER

- The meaning of the term *statistics*
- The meanings of the terms *data, elements, variables,* and *observations*
- The nominal, ordinal, interval, and ratio measurement scales
- The distinction between qualitative and quantitative data
- Sources of data for statistical analysis
- How errors can arise in data
- The difference between a population and a sample
- The advantages of summarizing data
- The process of statistical inference

CONTENTS

Statistics Provide Facts About Life

The *Statistical Abstract of the United States,* published annually since 1878, provides summaries of statistics on the social, political, and economic organization of the United States. Containing almost 1000 pages and over 1400 tables of statistical facts, the *Abstract* is designed to serve as a convenient statistical reference and also as a guide to other statistical publications and sources. Many tables in the annual report allow readers to track trends in spending, manufacturing, recreation, and countless other activities from the 1970s and even from the 1920s. Statistical facts such as the following are included in the *Abstract:*

- 11% of the 1970 population of persons 25 and older completed college. By 1991, this percentage had risen to 21%.
- 33.2% of women 20–24 years of age are married, while only 19.2% of men 20–24 years of age are married.
- 22% of Americans aged 18 and over report sleeping 6 hours or less per night.
- Expenditures for drugs and other nondurable medical goods soared from $8.8 billion in 1970 to $54.6 billion in 1990.
- Americans consume 111 pounds of fresh vegetables per person per year.
- 27.6% of persons 20–24 years of age smoke cigarettes.

The *Abstract* provides statistics on a variety of social, political, and economic topics.

- The life expectancy of current 22 year olds is 74 years for males and 80 years for females.

This *Statistics in Practice* is based on the *Statistical Abstract of the United States,* 1992.

Frequently, we see news articles and reports with statements such as the following:

- The average fuel economy of new cars sold in the United States is 27.5 miles per gallon.
- The average starting salary for college graduates with degrees in the social sciences is $21,375 per year.
- Employment in the 500 largest U.S. industrial corporations dropped 3.9% last year.
- Spending on health care has replaced the military as the largest outlay in the federal budget at an average of $3160 per person per year.
- 38% of the voters approve of the job the president is doing.
- The median family income is $34,788.

The numerical facts in the preceding statements (27.5, $21,375, 3.9%, $3160, 38%, and $34,788) are called statistics. Thus, in everyday usage, the term *statistics* refers to numerical facts. However, the field, or subject, of statistics involves much more than simply the calculation and presentation of numerical facts.

The subject of statistics involves the study of how data are collected, how they are analyzed, how they are presented, and how they are interpreted. Chapter 1 begins with a discussion of how *data,* the raw material of statistics, are acquired and used. The two uses of data, referred to as descriptive statistics and statistical inference, are discussed in Sections 1.4 and 1.5.

1.1 ▽ DATA

Data are the facts and figures that are collected, analyzed, and summarized. All the data collected in a particular study are referred to as the *data set* for the study. Table 1.1 shows a data set for 40 colleges and universities.

ELEMENTS, VARIABLES, AND OBSERVATIONS

The *elements* are the entities on which data are collected. For the data set in Table 1.1, each individual college or university is an element. With 40 colleges and universities, there are 40 elements in the data set. A *variable* is a characteristic of interest for the elements. The data set in Table 1.1 consists of six variables, which are defined as follows.

Variable	Definition
Enrollment	The number of students enrolled.
Type	A variable that specifies whether the school is primarily for women, primarily for men, or coed.
Control	A variable that specifies whether the school is public or private.
Tuition, resident	The tuition charged for students who are residents of the state where the school is located.
Tuition, nonresident	The tuition charged for students who are not residents of the state where the school is located.
Room and Board	The annual room-and-board cost.

Data are obtained by collecting measurements on each variable for every element in the study. The set of measurements collected for a particular element is called an *observation.* Referring to Table 1.1, we see that the first element (University of Alabama) has the observation 15,943, coed, public, 2008, 5016, and 3288. With 40 elements, there are 40 observations in the data set. Note that data for some variables are numeric (enrollment, tuition resident, tuition nonresident, and room and board), and data for other variables are nonnumeric (type and control).

1.2 ▽ SCALES OF MEASUREMENT

The type of statistical analysis that is appropriate for a particular variable depends on the scale of measurement used for the variable. There are four scales of measurement: nominal, ordinal, interval, and ratio. The scale of measurement determines the amount of

TABLE 1.1   A DATA SET FOR 40 COLLEGES AND UNIVERSITIES

College/University	Enrollment	Type	Control	Tuition ($) Resident	Tuition ($) Nonresident	Room and Board ($)
University of Alabama	15,943	Coed	Public	2,008	5,016	3,288
Amherst College	1,579	Coed	Private	17,900	17,900	4,800
Barnard College	2,200	Women	Private	16,854	16,854	7,316
Boise State University	10,725	Coed	Public	1,378	3,578	2,923
Boston College	8,806	Coed	Private	15,002	15,002	6,470
Cameron University	4,894	Coed	Public	1,481	3,645	2,732
Cazenoiva College	1,078	Coed	Private	8,592	8,592	4,388
Clarkson University	2,848	Coed	Private	14,190	14,190	5,077
Cleveland State University	13,408	Coed	Public	2,682	5,364	3,525
University of Colorado	20,407	Coed	Public	1,970	9,900	3,540
Columbia College	1,235	Women	Private	9,190	9,190	3,460
Delaware State College	2,624	Coed	Public	1,295	3,170	4,182
Drake University	4,273	Coed	Private	11,780	11,780	4,215
East Carolina University	13,883	Coed	Public	1,254	6,308	3,100
Florida Atlantic University	10,273	Coed	Public	1,350	4,850	3,450
Fordham University	7,089	Coed	Private	12,105	12,105	6,840
Glassboro State College	5,427	Coed	Public	2,543	3,532	4,515
Hollins College	852	Women	Private	12,200	12,200	4,950
Indiana State University	10,245	Coed	Public	2,272	5,490	3,211
Jackson State University	6,051	Coed	Public	1,786	2,968	2,424
Kenyon College	1,486	Coed	Private	17,610	17,610	3,570
Louisiana State University	21,243	Coed	Public	2,040	5,240	2,710
Morehouse College	2,720	Men	Private	7,430	7,430	4,980
Meredith College	2,159	Women	Private	6,020	6,020	2,970
University of Montana	8,752	Coed	Public	1,800	5,400	3,600
SUNY-Buffalo	17,263	Coed	Public	2,520	6,120	4,376
University of Oregon	13,074	Coed	Public	4,402	9,188	3,670
Pomona College	11,000	Coed	Private	2,413	2,413	2,452
University of Richmond	2,883	Coed	Private	12,620	12,620	3,040
Russell Sage College	961	Women	Private	11,270	11,270	4,720
St. John's University	1,971	Men	Private	10,543	10,543	3,670
Shepherd College	3,503	Coed	Public	1,894	4,314	3,690
Smith College	2,607	Women	Private	16,850	16,850	6,100
Temple University	22,859	Coed	Public	4,756	8,696	4,681
Tufts University	4,393	Coed	Private	18,344	18,344	5,443
Union College	1,957	Coed	Private	18,009	18,009	5,722
University of Vermont	7,992	Coed	Public	6,140	14,740	4,588
Wake Forest University	3,764	Coed	Private	12,000	12,000	4,100
Wayne State University	21,085	Coed	Public	2,460	5,460	6,300
Yale University	5,150	Coed	Private	17,500	17,500	6,200

Source: "Accredited U.S. Senior Colleges and Universities," *The 1993 Information Please Almanac,* 46th Ed. Boston: Houghton Mifflin.

information contained in the data and indicates the data summarization and statistical analyses that are most appropriate.

NOMINAL SCALE

The scale of measurement for a variable is *nominal* when the data are simply labels used to identify an attribute of the element. For example, referring to the data set in Table 1.1, we see that the second variable, type, is measured on a nominal scale. This is so because women, men, and coed are labels used to identify whether the school is primarily for women, primarily for men, or coed. The third variable, control, is also measured on a nominal scale with the labels public and private indicating whether the school is public or private.

In cases where the scale of measurement is nominal, a numeric code as well as a nonnumeric symbol may be used. For example, the data for the type variable could be 1 for women, 2 for men, and 3 for coed. Such a code would facilitate recording and computer processing of the data. However, it is important to remember that the numeric values of 1, 2, and 3 are simply labels used to identify the type of school. Thus, the scale of measurement is nominal.

Other examples of variables where data have a nominal scale are as follows:

- Sex (male, female)
- Marital status (single, married, widowed, divorced)
- Religious affiliation (many possibilities)
- Parts identification code (A13622, 12B63)
- Employment status (employed, unemployed)
- House number (5654, 2712, 624)
- Occupation (many possibilities)

As the preceding examples show, the key feature of the nominal scale is that the data obtained are labels used to identify an attribute of the element.

Finally, it is important to note that arithmetic operations such as addition, subtraction, multiplication, and division *do not* make sense for nominal data. Thus, even if the nominal data are numeric, arithmetic operations including averaging are inappropriate.

ORDINAL SCALE

The scale of measurement for a variable is *ordinal* when

1. The data have the properties of nominal data, and
2. The order or rank of the data is meaningful.

Let us use the following example to illustrate the ordinal scale of measurement. Some restaurants place questionnaires on the tables to solicit customer opinions concerning the restaurant's performance in terms of food, service, atmosphere, and so on. A questionnaire used by the Lobster Pot Restaurant in Redington Shores, Florida, is shown in Figure 1.1. Note that the customers completing the questionnaire are asked to provide ratings for six different variables: food, drinks, service, waiter, captain, and hostess. The response categories are excellent, good, and poor for each variable. The observations for each variable possess characteristics of nominal data in that each response rating is a label for excellent, good, or poor quality. In addition, the data can be ranked, or ordered, with respect to quality. For example, consider the food quality variable. After collecting the

FIGURE 1.1   CUSTOMER OPINION QUESTIONNAIRE USED BY THE LOBSTER POT RESTAURANT, REDINGTON SHORES, FLORIDA (USED WITH PERMISSION)

The
LOBSTER
Pot
RESTAURANT

	Excellent	Good	Poor	Comments
Food				
Drinks				
Service				
Waiter				
Captain				
Hostess				

Waiter's Name _____

Captain's Name _____

Other Comments _____

data, we can rank the data in order of food quality by beginning with excellent, followed by good, and, finally, followed by poor. With only three response categories, we can expect to see many data with tied rankings. Nonetheless, the data can be ranked in terms of food quality.

Like data obtained from a nominal scale, data obtained from an ordinal scale may be either nonnumeric or numeric. For the Lobster Pot questionnaire, the nonnumeric letters of E for excellent, G for good, and P for poor could be used to record the data. Or, a numeric code with values of 1 for excellent, 2 for good, and 3 for poor could be used equally well. Finally, as with nominal data, it is important to remember that arithmetic operations *do not* make sense for ordinal data. Thus, even if the ordinal data are numeric, the arithmetic operations such as addition, subtraction, multiplication, division, and averaging are inappropriate.

INTERVAL SCALE

The scale of measurement for a variable is *interval* when

1. The data have the properties of ordinal data, and
2. The difference or interval between data values indicates how much more or less of a variable one element possesses when compared to another element.

Interval data are always *numeric*. Subtracting the interval data value for one element from the interval data value for another element provides the difference between the two elements in terms of the variable being measured. Temperature is a good example of a variable that is measured on an interval scale. The high temperatures on April 16, 1993 (*USA Today*), for Phoenix, Arizona; Charlotte, North Carolina; and Minneapolis, Minnesota, are shown in Table 1.2. The interval Fahrenheit-scale temperatures are numeric with the property of ordinal data in that Phoenix, Charlotte, and Minneapolis can be ranked in terms of temperature with Phoenix being the warmest (highest temperature) and Minneapolis being the coldest (lowest temperature). In addition, the difference between data values is meaningful with Phoenix being $85 - 71 = 14$ degrees warmer than Charlotte and $85 - 44 = 41$ degrees warmer than Minneapolis. Charlotte was $71 - 44 = 27$ degrees warmer than Minneapolis.

Using the Celsius scale for temperature, the above high temperatures on April 16, 1993, would have been Phoenix (29.4), Charlotte (21.7), and Minneapolis (6.7). These interval-scaled data can again be ordered or ranked from highest to lowest with the differences meaningful in that Phoenix was $29.4 - 6.7 = 22.7$ degrees warmer than Minneapolis on the Celsius scale.

Scholastic Aptitude Test (SAT) scores are another example of interval-scaled data. Three students with SAT scores of 1120, 1050, and 970 can be ranked or ordered in terms of best to poorest performance on the SAT. In addition, the differences are meaningful, showing student 1 scored $1120 - 1050 = 70$ points more than student 2, while student 2 scored $1050 - 970 = 80$ points more than student 3.

Finally, with interval-scaled data, the arithmetic operations of addition, subtraction, multiplication, division, and averaging are meaningful. As a result, data obtained using this scale lend themselves to more alternatives for statistical analysis than do data obtained from nominal or ordinal scales.

RATIO SCALE

The scale of measurement for a variable is a *ratio* scale when

1. The data have all the properties of interval data, and
2. The ratio of two data values is meaningful.

Variables such as distance, height, weight, and time use the ratio scale of measurement. A requirement of this scale is that a zero value is inherently defined on the scale. Specifically, the zero value indicates nothing exists for the variable at the zero point. Whenever we collect data on cost, we use a ratio scale of measurement. Consider a variable indicating the cost of an automobile. The zero point is inherently defined in that a zero cost indicates that the automobile is free (no cost). Then, comparing the $20,000 cost of one automobile with the $10,000 cost of a second automobile, the ratio property of the data shows that the first automobile is $20,000/10,000 = 2$ times, or twice, the cost of the second automobile.

TABLE 1.2	HIGH TEMPERATURES FOR APRIL 16, 1993

City	High Temperature
Phoenix	85
Charlotte	71
Minneapolis	44

Since ratio data have all of the properties of interval data, ratio data are always *numeric*. Arithmetic operations such as addition, subtraction, multiplication, division, and averaging are possible with ratio data. Thus, as with interval data, ratio data lend themselves to more alternatives for statistical analysis than do data obtained from nominal or ordinal scales.

Table 1.3 provides a summary of the relationship between nonnumeric and numeric data and the four scales of measurement. We note that nominal and ordinal scales can generate both nonnumeric and numeric data, but that interval and ratio scales generate only numeric data.

The amount of information in the data varies with the scale of measurement. Nominal data contain the least amount of information, followed by ordinal, interval, and then ratio data. Since arithmetic operations are meaningful only for interval and ratio data, it is important to know the measurement scale used in order to employ the most appropriate statistical procedures. There are many statistical procedures that can be used with interval and ratio data that are not meaningful with nominal and ordinal data.

QUALITATIVE AND QUANTITATIVE DATA

Data can also be classified as being either qualitative or quantitative. *Qualitative data* provide labels or names for categories of like items. The categories for qualitative data may be identified by either nonnumeric descriptions or by numeric codes. Qualitative data are obtained from either a nominal or an ordinal scale of measurement. On the other hand, *quantitative data* indicate either "how much" or "how many" of something. Quantitative data are always numeric and are obtained from either an interval or a ratio scale of measurement.

In terms of the statistical methods used for summarizing data, qualitative data provided by nominal and ordinal scales employ similar methods, whereas quantitative data provided by interval and ratio scales employ similar methods. Tabular and graphical methods for summarizing qualitative data are presented in Section 2.1. Tabular and graphical methods for summarizing quantitative data are presented in Section 2.2.

TABLE 1.3	THE RELATIONSHIP BETWEEN NONNUMERIC AND NUMERIC DATA AND SCALES OF MEASUREMENT	

Scale of Measurement	Nonnumeric Data	Numeric Data
Nominal	Description indicates the category for the element	Numeric value indicates the category for the element
Ordinal	Description permits ranking or ordering of the data	Numeric value permits ranking or ordering of data
Interval		Numeric values* are defined such that the interval between data values is meaningful
Ratio		Numeric values* have an inherently defined zero, and the ratio of data values is meaningful

*Arithmetic operations are meaningful for these kinds of data.

NOTES AND
COMMENTS

1. An element is an entity on which measurements are obtained. It could, for example, be a company, a person, an automobile, and so on. An observation is the set of measurements obtained for each element. Thus, the number of observations and the number of elements in a data set will always be the same. The number of measurements obtained on each element is the number of variables. Thus, the total number of data values in a data set is the number of elements times the number of variables.

2. For purposes of statistical analysis, the most important distinguishing characteristic of the types of data is that ordinary arithmetic operations are meaningful *only* with quantitative (interval- and ratio-scaled) data.

1.3   DATA ACQUISITION

We have introduced the concepts of elements, variables, and observations and of how data collected in a particular study form a data set. In this section we describe the sources of data that are available and how the data can be acquired. The data needed for statistical analysis may be obtained through already-existing sources of data or through specially designed statistical studies that are undertaken to obtain new data.

EXISTING SOURCES OF DATA

In some cases, data needed for a particular application are available from existing sources. For example, many organizations maintain a variety of data on their operations. Data on salaries, ages, and years of experience can usually be obtained relatively easily from internal personnel records. Data on sales, advertising costs, distribution costs, inventory levels, and production quantities are available from other internal information and record-keeping systems.

Governmental agencies provide another important source of existing data. For instance, the U.S. Department of Labor maintains considerable data on characteristics such as employment rates, wage rates, size of the labor force, and union memberships. Data are also available from a variety of professional associations and special-interest organizations. The Travel Industry Association of America maintains travel-related information such as the number of tourists and travel expenditures by state. Such data would be of interest to firms and individuals participating in the travel industry.

STATISTICAL STUDIES

Sometimes data are not available from existing sources. If the data are considered essential, a statistical study will have to be conducted to obtain the data. Such statistical studies can be classified as being either *experimental* or *observational.*

In an experimental study, variables of interest are identified. Then one or more factors in the study are controlled so that data may be obtained about how those factors influence the variables. For example, a medical researcher might be interested in conducting an experiment designed to learn about how a new drug affects blood pressure. To obtain data about the effect of the new drug, a sample of individuals will be selected. The dosage of the new drug will be controlled, with different groups of individuals being given different dosages. Data on blood pressure will be collected for each group. Statistical analysis of the experimental data will help determine how the new drug affects blood pressure.

In observational, or nonexperimental, statistical studies, no attempt is made to control or influence the variables of interest. A survey is perhaps the most common type of observational study. In a survey, research questions are identified. A questionnaire is then designed and administered to a sample of elements. In this way, data are obtained about the research variables but no attempt is made to control the factors that influence the variables.

Individuals wishing to use data and statistical analyses as an aid to decision making must always be aware of the time and cost required to obtain the data. The use of existing data sources is desirable when data must be obtained in a relatively short period of time. If the data are not available from an existing source, the additional time and cost involved in collecting the data must be taken into account. In all cases, it is desirable for the decision maker to consider the contribution of the statistical analysis to the decision-making process. The cost of data acquisition and the subsequent statistical analysis should not be more than the savings generated by using the statistical information to make a better decision.

POSSIBLE DATA-ACQUISITION ERRORS

Always be aware of the possibility of data errors in the use of statistical studies. Using erroneous data and the misleading statistical analyses generated from such data would be worse than not using the data and statistical information at all. An error in data acquisition occurs whenever the data value obtained is not equal to the true or actual value that would have been obtained with a correct procedure. Such errors can occur in a number of ways. For example, an interviewer might mistakenly record the age of a 24-year-old person as 42. Or, the person answering an interview question might misinterpret the question and make an incorrect response.

Care must be exercised in collecting and recording data to ensure that errors are not made. Special procedures may be used to check for internal consistency of the data. Such procedures would indicate the need to review the data for a respondent who is shown to be 22 years of age but who reports 20 years of work experience. Statisticians also review data for unusually large and small values, called *outliers,* which are candidates for possible data errors. In Chapter 3 we present some of the methods used to identify outliers.

The point of this discussion is to alert the users of statistical information of the possibility of errors occurring during data acquisition. Blindly using any data that happen to be available or using data that were acquired with little care can lead to poor and misleading information.

1.4 ▽ DESCRIPTIVE STATISTICS

Most of the statistical information in newspapers, magazines, reports, and other publications comes from data that have been summarized and presented in a form that is easy for the reader to understand. These summaries of data, which may be tabular, graphical, or numerical, are referred to as *descriptive statistics.*

Refer to the data set in Table 1.1, where data for 40 colleges and universities are listed. Methods of descriptive statistics can be used to summarize these data. For example, a tabular summary of the data for the type variable is shown in Table 1.4. A graphical

TABLE 1.4   FREQUENCIES AND PERCENTAGES FOR TYPE OF SCHOOL

Type	Frequency	Percent
Women	6	15
Men	2	5
Coed	32	80
Totals	40	100

FIGURE 1.2   BAR GRAPH FOR TYPE OF SCHOOL

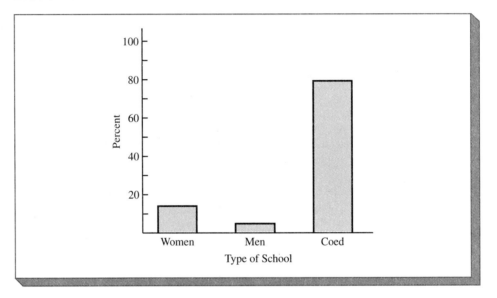

summary of the same data is shown in Figure 1.2. The purpose of tabular and graphical summaries such as these is to make it easier to interpret the data. Referring to Table 1.4 and Figure 1.2, we can easily see that the majority of the colleges and universities in the data set are coed. On a percentage basis, 15% of the schools are primarily for women, 5% are primarily for men, and 80% are coed.

The data on annual room-and-board cost in Table 1.1 is reorganized and summarized in Table 1.5 and presented graphically by the histogram in Figure 1.3. From the histogram, it is easy to see how the room-and-board costs are distributed over the range of values shown.

In addition to tabular and graphical displays, numerical descriptive statistics are often used to summarize data. The most common numerical descriptive statistic is the *average,* or *mean.* Using the data on room-and-board cost in Table 1.1, we can compute the average room-and-board cost by adding the room-and-board costs for all 40 colleges and universities and dividing the sum by 40. Doing so tells us that the average annual room-and-board cost for the 40 schools is $4275. This average is taken as a measure of the central value, or central location, of the data.

In recent years there has been a growing interest in statistical methods that can be used for developing and presenting descriptive statistics for data sets. Chapters 2 and 3 are devoted to the tabular, graphical, and numerical methods of descriptive statistics.

TABLE 1.5   **TABULAR SUMMARY OF THE ANNUAL ROOM-AND-BOARD COST FOR A SAMPLE OF 40 U.S. COLLEGES AND UNIVERSITIES**

Annual Room-and-Board Cost ($)	Number of Schools	Percent
2000–2499	2	5.0
2500–2999	4	10.0
3000–3499	6	15.0
3500–3999	7	17.5
4000–4499	5	12.5
4500–4999	7	17.5
5000–5499	2	5.0
5500–5999	1	2.5
6000–6499	4	10.0
6500–6999	1	2.5
7000–7499	1	2.5
Totals	40	100.0

FIGURE 1.3   **HISTOGRAM OF ANNUAL ROOM-AND-BOARD COST**

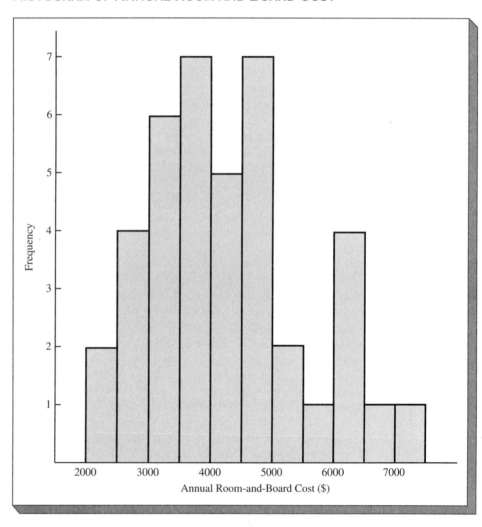

1.5 ▽ STATISTICAL INFERENCE

In many situations, there exists a large group of elements (individuals, voters, households, products, customers, etc.) about which data are sought. Because of time, cost, and other considerations, data are collected from only a small portion of the group. The larger group of elements in a particular study is called the *population,* and the smaller group is called the *sample.* Formally, we will use the following definitions.

Population

A *population* is the collection of all elements of interest in a particular study.

Sample

A *sample* is a subset of the population.

A major contribution of statistics is that it enables us to use data from a sample to make estimates and test claims or hypotheses about the characteristics of a population. This process is referred to as *statistical inference.*

As an example of statistical inference, let us consider the study conducted by Norris Electronics. Norris manufactures a high-intensity light bulb that is used in a variety of electrical products. In an attempt to increase the useful life of the light bulb, the product design department has developed a new light-bulb filament. To evaluate the advantages of the new filament, a sample of 200 new-filament bulbs was manufactured and tested. Data were collected on the number of hours each bulb operated before the filament burned out; the data are shown in Table 1.6.

The corresponding population is the collection of all bulbs that could be produced using the new filament. Suppose that Norris is interested in using the data from the sample to make an inference about the average or mean lifetime for the population of all bulbs that could be produced using the new filament. Adding the hours of useful life for the 200 bulbs and dividing the total by 200 provides the sample average, or sample mean, lifetime. Doing so for the data in Table 1.6 provides a sample average lifetime of 76 hours. Thus, we would use this sample result to estimate that the average lifetime for the bulbs in the population would be 76 hours.

Whenever statisticians estimate a characteristic of a population based on a sample, as in the Norris Electronics example, they usually provide a statement of the precision associated with the estimate. For the Norris example, the statistician might state that the estimate of the average lifetime for the population of new light bulbs is 76 hours with a margin of error of ±4 hours. Thus, 72–80 hours would be the interval estimate for the average lifetime of all bulbs that could be produced using the new filament. Using concepts from probability theory, a statistician can also state how confident he or she is that the interval of 72–80 hours contains the true average lifetime for the population. Figure 1.4 summarizes the statistical inference process used in the Norris Electronics example.

TABLE 1.6	HOURS UNTIL BURNOUT FOR A SAMPLE OF 200 BULBS FOR NORRIS ELECTRONICS								
107	73	68	97	76	79	94	59	98	57
54	65	71	70	84	88	62	61	79	98
66	62	79	86	68	74	61	82	65	98
62	116	65	88	64	79	78	79	77	86
74	85	73	80	68	78	89	72	58	69
92	78	88	77	103	88	63	68	88	81
75	90	62	89	71	71	74	70	74	70
65	81	75	62	94	71	85	84	83	63
81	62	79	83	93	61	65	62	92	65
83	70	70	81	77	72	84	67	59	58
78	66	66	94	77	63	66	75	68	76
90	78	71	101	78	43	59	67	61	71
96	75	64	76	72	77	74	65	82	86
66	86	96	89	81	71	85	99	59	92
68	72	77	60	87	84	75	77	51	45
85	67	87	80	84	93	69	76	89	75
83	68	72	67	92	89	82	96	77	102
74	91	76	83	66	68	61	73	72	76
73	77	79	94	63	59	62	71	81	65
73	63	63	89	82	64	85	92	64	73

FIGURE 1.4	THE PROCESS OF STATISTICAL INFERENCE FOR THE NORRIS ELECTRONICS EXAMPLE

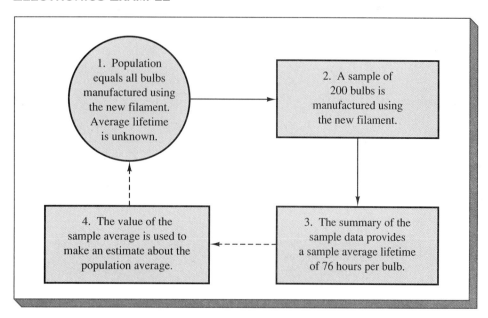

SUMMARY

In everyday usage, the term *statistics* refers to numerical facts. However, the field of statistics requires a broader definition, such as the processes of collecting, analyzing, presenting, and interpreting data.

Data were described as being the raw material of statistics. Thus, data are the facts and figures that are collected, analyzed, presented, and interpreted. A significant portion of the chapter was devoted to data, including how data are measured and how data are acquired.

Four scales of measurement are available for obtaining data on a particular variable: nominal, ordinal, interval, and ratio. We showed that nominal and ordinal data are classified as qualitative data and may be recorded by either a nonnumeric description or a numeric code. Interval and ratio data are classified as quantitative data. Quantitative data are always numeric and indicate how much or how many for the variable of interest. Ordinary arithmetic operations are meaningful only if the data are *quantitative*.

We introduced the topics of descriptive statistics and statistical inference in Sections 1.4 and 1.5. Descriptive statistics are the tabular, graphical, and numerical methods used to summarize data. Statistical inference is the process of using data obtained from a sample to make inferences about the characteristics of a population.

GLOSSARY

Data The facts and figures that are collected, analyzed, presented, and interpreted. Data may be numeric or nonnumeric.

Data set All the data collected in a particular study.

Elements The entities on which data are collected.

Variable A characteristic of interest for the elements.

Observation The set of measurements or data obtained for a single element.

Nominal scale A scale of measurement that uses a label to define an attribute of an element. Nominal data may be nonnumeric or numeric.

Ordinal scale A scale of measurement that has the properties of a nominal scale and can be used to rank or order the data. Ordinal data may be nonnumeric or numeric.

Interval scale A scale of measurement that has the properties of an ordinal scale with the interval between data values indicating how much more or less of a variable one element possesses when compared to another element. Interval data are always numeric.

Ratio scale A scale of measurement that has the properties of an interval scale and is meaningful in terms of the ratio of two data values. Ratio data are always numeric.

Qualitative data Data obtained with a nominal or ordinal scale of measurement. Qualitative data may be nonnumeric or numeric.

Quantitative data Data obtained with an interval or ratio scale of measurement. Quantitative data are always numeric and indicate how much or how many for the variable of interest.

Outliers Unusually large and small data values that are candidates for possible data errors.

Descriptive statistics Tabular, graphical, and numerical methods used to summarize data.

Population The collection of all elements of interest in a particular study.

Sample A subset of the population.

Statistical inference The process of using data obtained from a sample to make inferences about the characteristics of a population.

REVIEW QUIZ

TRUE/FALSE

1. The speed of an airplane in miles per hour is ordinal data.
2. A football player's uniform number is nominal data.
3. Room numbers in a building are examples of interval data.
4. The place a person finishes in a golf tournament (first, second, third, etc.) is an example of ordinal data.
5. Interval and ratio data must be numeric data.
6. Nominal and ordinal data must be numeric data.
7. A population is the set of all elements of interest in a particular study.
8. A sample may be larger than the population.
9. Methods for data summarization are referred to as prescriptive statistics.
10. The use of probability enables the statistician to make statements concerning the precision of statistical inferences.

MULTIPLE CHOICE

11. Which of the following measures involve nominal data?
    a. the test score on an exam
    b. the number on an automobile license plate
    c. the speed of an automobile
    d. the class rank of a college student
12. Which of the following measures involve ordinal data?
    a. the score in a baseball game
    b. the place of a baseball team in the league standings
    c. the height of a flagpole
    d. the weight of a fish
13. When data are collected for only a subset of the elements of interest, we are using a
    a. population
    b. sample
    c. statistical inference
    d. summary
14. Descriptive statistics is concerned with
    a. arriving at a conclusion for the population based on sample information
    b. the summarization and presentation of data
    c. statistical inference
15. Statistical inference is concerned with
    a. arriving at a conclusion about the population based on sample information
    b. the summarization and presentation of data
16. In a recent study based on an inspection of 100 homes in Central City, 60 homes were found to violate one or more city codes. Using this information, the city manager released a statement that 60% of all Central City's homes were in violation of city codes. The manager's statement is an example of
    a. descriptive statistics
    b. statistical inference
17. Refer to Question 16. The manager's statement that 60% of all Central City's homes are in violation of city codes is
    a. exactly correct
    b. only an estimate, since it is based on sample information
    c. very misleading, since it is based on a study of only 100 homes

18. The statement "Based on sample data, we expect to sell $50,000 worth of snow removal equipment this winter" is an example of
   a. descriptive statistics
   b. statistical inference

EXERCISES

1. Discuss the difference between the concept of statistics as numerical facts and the concept of statistics as a subject or field of study.

2. Table 1.7 shows the executive compensation, sales, return on equity, and profit rating for seven firms in the aerospace industry (*Business Week,* May 1, 1989).
   a. How many elements are in this data set?
   b. How many variables are in this data set?
   c. How many observations are in this data set?
   d. Which of the variables are qualitative and which are quantitative?
   e. What measurement scale is being used for each of the variables?

SELF TEST
3. Consider the data set on the sample of Fortune 500 companies shown in Table 1.8 (Fortune, April 20, 1992).
   a. How many variables are in the data set?
   b. Which of the variables are qualitative and which are quantitative?
   c. What type of measurement scale is being used for each of the variables?

4. Columbia House provides CDs, tapes, and records to its mail-order club members. A Columbia House Music Survey conducted in 1992 asked new club members to complete an 11-question survey. Some questions asked were as follows:
   a. How many albums (CDs, tapes, or records) have you bought in the last 12 months?
   b. Are you currently a member of a national mail-order book club? (Yes or No)
   c. What is your age?
   d. Including yourself, how many people (adults and children) are in your household?
   e. What kind of music are you interested in buying? (There were 15 categories listed, including hard rock, soft rock, adult contemporary, heavy metal, rap, and country.)

   Comment on each question as to whether it provides qualitative or quantitative data.

TABLE 1.7   **EXECUTIVE COMPENSATION AND PROFITABILITY FOR SEVEN FIRMS IN THE AEROSPACE INDUSTRY**

Company	CEO Salary ($1000s)	Sales ($1,000,000s)	Return on Equity (%)	Pay Versus Corporation Profit Rating*
Boeing	846	16,962.0	11.4	3
General Dynamics	1,041	9,551.0	19.7	3
Lockheed	1,146	10,590.0	17.9	3
Martin Marietta	839	5,727.5	26.6	2
McDonnell Douglas	681	15,069.0	11.0	3
Parker Hannifin	765	2,397.3	12.4	3
United Technology	1,148	18,000.1	13.7	4

*CEOs are assigned to five groupings according to pay versus corporate profitability. Those in group 1 are given a rating of 1, those in group 2 a rating of 2, and so on. A rating of 1 is the best rating; a rating of 5 is the worst.

Source: Business Week, May 1, 1989.

TABLE 1.8   A SAMPLE OF 10 OF THE FORTUNE 500 LARGEST U.S. INDUSTRIAL
CORPORATIONS

Company	Sales ($1,000,000s)	Rank	Profit ($1,000,000s)	Assets ($1,000,000s)	Industry Code
Raytheon	9356	51	592	6087	7
Texas Instruments	6812	77	−409	5009	7
Hershey Foods	2902	160	220	2342	8
Polaroid	2096	203	684	1889	22
Zenith Electronics	1322	282	−52	687	7
Brown-Foreman	1126	315	145	1083	3
Cooper Tire	1002	338	79	671	21
Quaker State	814	392	23	752	18
Standard Register	694	432	33	464	20
La-Z-Boy Chair	610	464	23	363	10

Source: Fortune, April 20, 1992

5. A California state agency classifies worker occupations as professional, white collar, or blue collar.
   a. The variable is worker occupation. Is it a qualitative or quantitative variable?
   b. What type of measurement scale is being used for this variable?

6. A Bruskin-Goldring research poll asked 1000 people, Where is the best place to meet a "suitable companion"? (*USA Today,* February 13, 1992) Response categories were religious gathering, friend's home, evening class, work, nightspot, or other.
   a. What was the sample size in this survey?
   b. What measurement scale is being used?
   c. Would it make more sense to use averages or percentages as a summary of the data in this survey?
   d. Nightspot was the least preferred choice with only 170 people selecting it as the best place to meet a "suitable companion." What percentage was reported for the nightspot category?

7. A study of last-minute holiday shoppers conducted by Best Products, Inc. (*USA Today,* December 22, 1988) asked individuals to indicate when they completed their holiday shopping. The response alternatives were as follows:

   Before Halloween   A few days before Christmas
   Before Thanksgiving   Christmas Eve

   a. The variable of interest is the date by which holiday shopping is completed. Is this a qualitative or quantitative variable?
   b. What measurement scale is being used?
   c. Would it make more sense to use averages or percentages as a summary of the data in this study? Explain.

8. State whether each of the following variables is qualitative or quantitative, and indicate the measurement scale that is appropriate for each.
   a. Age   b. Sex   c. Class rank   d. Make of automobile
   e. Number of people favoring the death penalty

9. State whether each of the following variables is qualitative or quantitative, and indicate the measurement scale being used.
   a. Annual sales
   b. Soft-drink size (small, medium, or large)
   c. Employee classification (GS1 through GS18)
   d. Earnings per share
   e. Method of payment (cash, check, credit card)

10. The Hawaii Visitors Bureau collects data on visitors to Hawaii. The following questions were among the 16 questions asked in a questionnaire handed out to passengers during incoming airline flights in July 1992.

    1. This trip to Hawaii is my: 1st, 2nd, 3rd, 4th, etc.
    2. The primary reason for this trip is: (10 categories including vacation, convention, honeymoon)
    3. Where I plan to stay: (11 categories including hotel, apartment, relatives, camping)
    4. Total days in Hawaii

    a. What is the population being studied?
    b. Is the use of questionnaires for passengers during incoming airline flights a good way to reach this population?
    c. Comment on each of the four questions above in terms of whether it will provide qualitative or quantitative data.

11. A manager of a large corporation has recommended a $10,000 raise be given in order to keep a valued subordinate from moving to another company. What internal and external sources of data might be used to decide whether such a salary increase is appropriate?

12. The marketing group at your company has come up with a new diet soft drink that it claims will capture a large share of the young adult market.
    a. What data would you want to see before deciding to invest substantial funds to introduce the new product into the marketplace?
    b. How would you expect the data mentioned in part (a) to be obtained?

SELF TEST

13. A 1988 *Newsweek/Gallup* poll investigated whether adults preferred staying home or going out as their favorite way of spending time in the evening. The poll of 1500 adults concluded that the majority of adults (70%) indicated that "staying at home with family" was the favorite evening activity.
    a. What is the population of interest in this study?
    b. What is the variable being studied?
    c. Is the variable being studied qualitative or quantitative?
    d. What was the size of the sample used?
    e. Where was a descriptive statistic used in this study?
    f. Describe the process of statistical inference in this study.

14. In a recent study of causes of death in males 60 years of age and older, a sample of 120 men indicated that 48 died due to some form of heart disease.
    a. Develop a descriptive statistic that can be used as an estimate of the percentage of males 60 years of age or older who die from some form of heart disease.
    b. Are the data on cause of death qualitative or quantitative?
    c. Discuss the role of statistical inference in this type of medical research.

15. The 25th Annual Report on Shoplifting in Supermarkets *(Commercial Service Systems, Inc., 1987)* used 391 supermarkets in southern California to compile the following statistics on supermarket shoplifters caught in the act:

    ■ Most items stolen were valued between $1 and $5
    ■ Most shoplifters were male (56%)
    ■ Most frequent age category of shoplifters was "under 30" (51%)

Answer the following questions assuming that the purpose of the study was to present statistical data on the national trend and impact of shoplifting in supermarkets.
    a. Cite two descriptive statistics reported.
    b. What warning would you issue if the results of the study were to be used to make a statistical inference about the national trend and impact of shoplifting in supermarkets?

16. Select a recent copy of the newspaper *USA Today.*
    a. Note four examples of statistical information.
    b. For each example, indicate the descriptive statistics used and discuss any statistical inferences made.

17. A 7-year medical research study *(Journal of the American Medical Association,* December 1984) reported that women whose mothers took the drug DES during pregnancy were *twice* as likely to develop tissue abnormalities that might lead to cancer as women whose mothers did not take the drug.
   a. This study involved the comparison of two populations. What were the populations involved?
   b. Do you suppose the data obtained here were the result of a survey or an experiment?
   c. For the population of women whose mothers took the drug DES during pregnancy, a sample of 3980 women showed 63 developed tissue abnormalities that might lead to cancer. Provide a descriptive statistic that could be used to estimate the number of women out of 1000 in this population who have tissue abnormalities.
   d. For the population of women whose mothers did not take the drug DES during pregnancy, what is the estimate of the number of women out of 1000 who would be expected to have tissue abnormalities?
   e. Medical studies of diseases and disease occurrence often use a relatively large sample (in this case, 3980). Why is this done?

18. A firm is interested in testing the advertising effectiveness of a new television commercial. As part of the test, the commercial is shown on a 6:30 P.M. local news program in Denver, Colorado. Two days later a market research firm conducts a telephone survey to obtain information on recall rates (percentage of viewers who recall seeing the commercial) and impressions of the commercial.
   a. What is the population for this study?
   b. What is the sample for this study?
   c. Why would a sample be used in this situation? Explain.

19. The Nielsen organization conducts weekly surveys of television viewing throughout the United States. The Nielsen statistical ratings indicate the size of the viewing audience for each major network television program. Rankings of the television programs and of the viewing-audience market shares for each network are published each week.
   a. What is the Nielsen organization attempting to measure?
   b. What is the population?
   c. Why would a sample be used for this situation?
   d. What kinds of decisions or actions are taken based on the Nielsen studies?

20. A sample of midterm grades for five students showed the following results: 72, 65, 82, 90, 76. Which of the following statements are correct, and which should be challenged as being too generalized?
   a. The average midterm grade for the sample of five students is 77.
   b. The average midterm grade for all students who took the exam is 77.
   c. An estimate of the average midterm grade for all students who took the exam is 77.
   d. More than half of the students who take this exam will score between 70 and 85.
   e. If five other students are included in the sample, their grades will be between 65 and 90.

CHAPTER 2

DESCRIPTIVE STATISTICS I:

TABULAR AND

GRAPHICAL METHODS

WHAT YOU WILL LEARN IN THIS CHAPTER

- How to construct and interpret summarization procedures for qualitative data using a:

  frequency distribution

  relative frequency distribution

  bar graph

  pie chart

- How to construct and interpret summarization procedures for quantitative data using a

  frequency distribution

  relative frequency distribution

  cumulative frequency distribution

  cumulative relative frequency distribution

  dot plot

  histogram

  ogive

- The role of the computer in summarizing data
- How to construct and interpret a stem-and-leaf display
- How to construct and interpret summarization procedures for bivariate data

CONTENTS

2.1 Summarizing Qualitative Data

2.2 Summarizing Quantitative Data

2.3 The Role of the Computer

2.4 Exploratory Data Analysis (Optional)

2.5 Summarizing Bivariate Data (Optional)

What Is America's Favorite Beverage?

Based on annual consumption statistics, Americans prefer soft drinks over all other beverages including beer, milk, fruit juice, coffee, and bottled water. An average American drinks approximately 10–12 cans or bottles of soft drinks per week, resulting in an average consumption of about 50 gallons per person per year. According to *Beverage Digest,* these soft-drink lovers spend over $47 billion per year on their favorite beverage.

Competition for a share of the soft-drink market is intense; new products are proliferating and much money is being spent on advertising campaigns. Including diet and caffeine-free soft drinks, consumers may now choose from approximately 250 different brands. Coca-Cola and Pepsi continue to be the industry leaders with their brands accounting for approximately 90% of the market.

Given the importance of this billion-dollar market and the intense competition, the crucial question is, What is America's favorite soft drink? The battle between Coca-Cola and Pepsi is close, but according to market share data compiled by *Beverage Digest,* Coke Classic is number 1. Pepsi is running a close second, followed by Diet Coke, Diet Pepsi, Dr Pepper, Mountain Dew, Sprite, and 7 Up. A bar graph showing market shares of the leading soft drinks is provided.

This *Statistics in Practice* is based on "Soft Drinks: Coke Classic No. 1," *Associated Press,* February 1993.

In America, soft drinks are number 1.

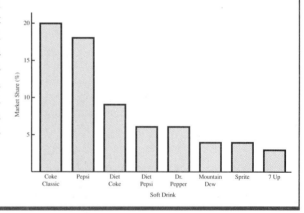

As indicated in Chapter 1, data may be classified as being either *qualitative* or *quantitative.* Qualitative data provide labels or names for categories of like items. The categories for qualitative data may be identified by nonnumeric descriptions or by numeric codes. Qualitative data are measured on nominal or ordinal scales. Quantitative data indicate "how much" or "how many." Quantitative data are always numeric and are measured on interval or ratio scales.

The purpose of this chapter is to introduce several tabular and graphical procedures commonly used to summarize both qualitative and quantitative data. Tabular and graphical summaries of data can be found in reports, newspaper articles, and research studies. Everyone is exposed to these types of presentations. Hence, it is important to understand how tabular and graphical summaries are prepared and also to know how they can be interpreted.

2.1 SUMMARIZING QUALITATIVE DATA

FREQUENCY DISTRIBUTION

We begin the discussion of how tabular and graphical methods can be used to summarize qualitative data with the definition of a *frequency distribution.*

Frequency Distribution

A *frequency distribution* is a tabular summary of a set of data showing the frequency (or number) of items in each of several nonoverlapping classes.

RELATIVE FREQUENCY DISTRIBUTION

We are often interested in knowing the fraction, or proportion, of the data items that fall within each class. The *relative frequency* of a class is simply the fraction, or proportion, of the total number of data items belonging to the class. For a data set with a total of n observations, or items, the relative frequency of each class is given by the following definition.

Relative Frequency

$$\text{Relative Frequency of a Class} = \frac{\text{Frequency of the Class}}{n} \qquad \textbf{(2.1)}$$

The definition of a *relative frequency distribution* is as follows.

Relative Frequency Distribution

A *relative frequency distribution* is a tabular summary of a set of data showing the relative frequency of items in each of several nonoverlapping classes.

The objective of both frequency distributions and relative frequency distributions is to provide insights about the data that cannot be easily obtained by looking only at the original data.

EXAMPLE 2.1 ▶ *USA Today* (December 13, 1988) reported that, based on 1988 automobile sales data, the Chevrolet Cavalier, Ford Escort, Fort Taurus, Honda Accord, and Hyundai Excel were the top choices among women buying cars. The *USA Today* article summarized the purchase choices of several thousand women. Assume that the data shown in Table 2.1 were collected from a sample of 50 women who made a recent purchase of one of the top five selling automobiles.

To develop a frequency distribution for the qualitative data in Table 2.1, we simply *count* the number of times each of the five automobiles appears in the data set. The Chevrolet Cavalier appears 9 times, the Ford Escort appears 14 times, the Ford Taurus

TABLE 2.1   DATA FROM A SAMPLE OF 50 NEW-CAR PURCHASES BY WOMEN

Honda Accord	Ford Escort	Ford Taurus
Ford Taurus	Chevrolet Cavalier	Honda Accord
Honda Accord	Ford Escort	Ford Taurus
Honda Accord	Hyundai Excel	Hyundai Excel
Ford Escort	Hyundai Excel	Chevrolet Cavalier
Ford Taurus	Ford Escort	Ford Escort
Honda Accord	Chevrolet Cavalier	Chevrolet Cavalier
Ford Escort	Ford Escort	Chevrolet Cavalier
Honda Accord	Honda Accord	Hyundai Excel
Ford Taurus	Chevrolet Cavalier	Ford Escort
Honda Accord	Hyundai Excel	Hyundai Excel
Honda Accord	Ford Escort	Ford Escort
Ford Escort	Honda Accord	Hyundai Excel
Chevrolet Cavalier	Chevrolet Cavalier	Ford Taurus
Hyundai Excel	Ford Escort	Ford Escort
Chevrolet Cavalier	Ford Escort	Honda Accord
Ford Taurus	Ford Taurus	

appears 8 times, the Honda Accord appears 11 times, and the Hyundai Excel appears 8 times. These counts are summarized in the frequency distribution shown in Table 2.2.

The advantage of the frequency distribution is that it provides a better understanding of the preferences of women in the sample than does the original data shown in Table 2.1. Using Table 2.2, we can see at a glance that the Ford Escort and the Honda Accord were the first and second choices of the women in the sample. Ford models did very well, with Escort and Taurus accounting for $14 + 8 = 22$ of the 50 new-car purchases. The information about the women's new-car purchases contained in Table 2.1 was much easier to grasp after the data had been systematically summarized in the frequency distribution of Table 2.2.

Using equation (2.1) we can develop a relative frequency distribution for the data presented in Table 2.2. For example, with the sample size $n = 50$, the relative frequency for the Chevrolet Cavalier is $9/50 = .18$. Computing the relative frequency for each automobile provides the relative frequency distribution shown in Table 2.3. The relative frequency of .28 for the Ford Escort shows this automobile was selected by 28% of the women in the sample. Similarly, .22, or 22%, of the women selected the Honda Accord; .16, or 16%, selected the Hyundai Excel; and so on. ◀

TABLE 2.2   FREQUENCY DISTRIBUTION OF NEW-CAR PURCHASES BASED ON A SAMPLE OF 50 WOMEN

Automobile Purchased	Frequency
Chevrolet Cavalier	9
Ford Escort	14
Ford Taurus	8
Honda Accord	11
Hyundai Excel	8
Total	50

TABLE 2.3	RELATIVE FREQUENCY DISTRIBUTION OF NEW-CAR PURCHASES BASED ON A SAMPLE OF 50 WOMEN

Automobile Purchased	Relative Frequency
Chevrolet Cavalier	.18
Ford Escort	.28
Ford Taurus	.16
Honda Accord	.22
Hyundai Excel	.16
Total	1.00

BAR GRAPHS AND PIE CHARTS

A *bar graph* is a graphical device for depicting qualitative data that have been summarized in a frequency distribution or a relative frequency distribution. On the horizontal axis of the graph, we specify the labels that are used for each of the classes. Either a frequency scale or a relative frequency scale can be used for the vertical axis of the graph. Then, using a bar of fixed width drawn above each class label, we extend the height of the bar until we reach the frequency or relative frequency of the class as indicated by the vertical axis. The bars are separated to emphasize the fact that each class is a separate category. A bar graph of the frequency distribution for the 50 new-car purchases by women is shown in Figure 2.1. Note how the graphical presentation shows the Ford Escort and the Honda Accord to be the two most preferred models.

The *pie chart* is a commonly used graphical device for presenting relative frequency distributions for qualitative data. To draw a pie chart, first draw a circle; then use the relative frequencies to subdivide the circle into sectors, or parts, that correspond to the relative frequency for each class. For example, since there are 360 degrees in a circle and since the Ford Escort has a relative frequency of .28, the sector of the pie chart labeled Ford Escort should consist of .28 × 360 = 100.8 degrees. Similar calculations for the other classes yield the pie chart shown in Figure 2.2. The numerical values shown in each sector may be frequencies, relative frequencies, or percentages.

EXAMPLE 2.2 ▶ Course evaluations at a major university are obtained by student responses to a questionnaire that is filled out on the last day of class. There are a variety of questions, which use a five-category response scale. One question is as follows:

> Compared to other courses that you have taken, what is the overall quality of the course you are now completing?

_____	_____	_____	_____	_____
Poor	Fair	Good	Very Good	Excellent

Data from the course evaluations measure the quality of the course on an ordinal scale using the five labels shown above.

Tabular and graphical methods can be used to summarize the teaching evaluation data set. Table 2.4 and Figure 2.3 provide these summaries for the overall quality of the course question shown. Rating data were provided by 125 students. Since the scale is ordinal, we have ordered the labels for the five categories in both the table and the bar graph. A review of these summaries shows that the quality of the course is generally very good.

◀

FIGURE 2.1   BAR GRAPH OF NEW-CAR PURCHASES BASED ON A SAMPLE OF 50 WOMEN

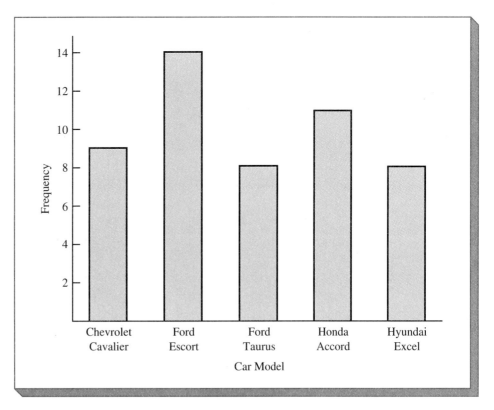

FIGURE 2.2   PIE CHART OF NEW-CAR PURCHASES BASED ON A SAMPLE OF 50 WOMEN

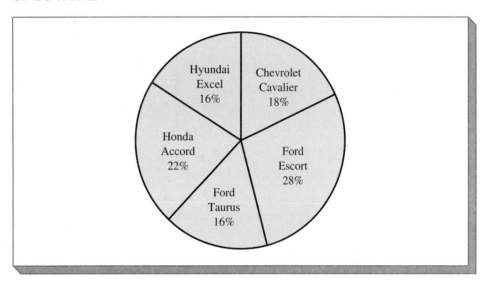

FIGURE 2.3   BAR GRAPH FOR 125 COURSE EVALUATIONS

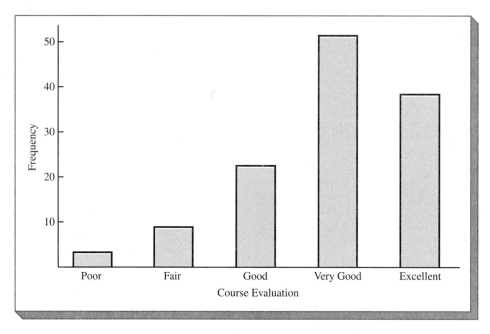

TABLE 2.4   FREQUENCY DISTRIBUTION AND RELATIVE FREQUENCY DISTRIBUTION
FOR 125 COURSE EVALUATIONS

Rating	Frequency	Relative Frequency
Poor	3	.024
Fair	8	.064
Good	24	.192
Very Good	52	.416
Excellent	38	.304
Totals	125	1.000

NOTES AND
COMMENTS

1. Often the number of classes in a frequency distribution for qualitative data
    will be the same as the number of categories found in the data. This was the
    case for the automobile-purchase and course-evaluation data presented in this
    section. Since the automobile-purchase data considered only five
    automobiles, or categories, a separate frequency distribution class was
    defined for each automobile. If the data set had included purchases of all
    models of automobiles, there would have been too many models, or
    categories, to develop a separate class for each. In such a situation, the lower
    frequency categories may be grouped together to form an aggregate class.
    With the automobile-purchase data, models such as the Honda Civic,
    Chevrolet Corsica, and Toyota Camry could have been summarized in an
    aggregate class identified as "other models." As a general guideline, most

    —*Continues on next page*

—Continued from previous page

statisticians recommend that from 5 to 20 classes be used in a frequency distribution. Whenever there are too many data categories to present a separate class for each, good judgment must be exercised in deciding which categories to group together.

2. The sum of the frequencies in any frequency distribution always equals the total number of elements in the data set. The sum of the relative frequencies in any relative frequency distribution always equals 1.

EXERCISES

METHODS

1. The response to a question has three alternatives: A, B, and C. A sample of 120 responses provides 60 A's, 24 B's, and 36 C's. Show the frequency and relative frequency distributions.

2. A partial relative frequency distribution is shown below.

Class	Relative Frequency
A	.22
B	.18
C	.40
D	

   a. What is the relative frequency of class D?
   b. The total sample size was 200. What is the frequency of class D?
   c. Show the frequency distribution for the four classes.

SELF TEST

3. A questionnaire provides 58 Yes answers, 42 No answers, and 20 No-Opinion answers.
   a. In the construction of a pie chart, how many degrees would be in the section of the pie showing the Yes answers?
   b. How many degrees would be in the section of the pie showing the No answers?
   c. Construct a pie chart.
   d. Construct a bar graph.

APPLICATIONS

4. *Psychology Today* (October, 1988) provided data on the consumption of beverage products including milk (M), fruit juice (F), soft drinks (S), beer (B), and bottled water (W). The following data show the results of a sample of 30 individuals who were asked to select their most frequently consumed beverage.

   | M | F | S | S | B | S | B | M | W | S | S | S | F | B | B |
   | S | B | S | W | S | M | F | S | S | B | B | S | F | S | B |

   a. Comment on why these are qualitative data. Is the scale of measurement nominal or ordinal?
   b. Provide a frequency distribution and a relative frequency distribution summary of the data.
   c. Provide a bar graph and a pie chart summary of the data.
   d. Based on this sample, what is the favorite beverage?

5. Freshmen entering Eastern University were asked to indicate their preferred major, and the data shown below were obtained.

Major	Number
Psychology	55
Biology	51
History	28
Communications	82

Summarize the data by constructing
a. a relative frequency distribution
b. a bar graph
c. a pie chart

6. What are the best-selling home videos of all time? Sales of the top-selling videos were reported by *Video Store Magazine* (*U.S. News & World Report,* February 24, 1992). Assume that sample data for the top six best-selling home videos are summarized with the following letter codes.

Code	Video
B	Bambi
E	ET The Extra-Terrestrial
F	Fantasia
H	Home Alone
L	The Little Mermaid
T	Batman

The following sample data are available:

```
F  T  H  E  H  H  L  B  E  H  B  F  B  T  T
B  F  F  F  H  B  F  F  H  F  F  T  F  B  F
L  F  B  T  F  F  H  T  E  T  E  L  E  E  L
T  E  E  L  E  B  T  H  B  E  E  E  T  E  L
```

a. Prepare a frequency distribution and a relative frequency distribution for the data set.
b. Rank order the top six best-selling home videos. Which video has been the most successful?

SELF TEST **7.** Leverock's Waterfront Steakhouse in Maderia Beach, Florida, uses a questionnaire to ask customers how they rate the server, food quality, cocktails, prices, and atmosphere at the restaurant. Each characteristic is rated on a scale from Outstanding (O), Very Good (V), Good (G), Average (A), and Poor (P). Use descriptive statistics to summarize the following data collected on food quality. What is your feeling about the food-quality ratings at the restaurant?

```
G  O  V  G  A  O  V  O  V  G  O  V  A
V  O  P  V  O  G  A  O  O  O  G  O  V
V  A  G  O  V  P  V  O  O  G  O  O  V
O  G  A  O  V  O  O  G  V  A  G
```

8. Data on position played for a sample of 55 members of the Baseball Hall of Fame in Cooperstown, New York, is shown below (*Sports Illustrated,* April 6, 1992). Each data item indicates the primary position played by the Hall of Famers with pitcher (P), catcher (H), first base (1), second base (2), third base (3), shortstop (S), left field (L), center field (C), and right field (R).

L	P	C	H	2	P	R	1	S	S	1	L	P	R	P
P	P	P	R	C	S	L	R	P	C	C	P	P	R	P
2	3	P	H	L	P	1	C	P	P	P	S	1	L	R
R	1	2	H	S	3	H	2	L	P					

   a. Use frequency and relative frequency distributions to summarize the data for the nine positions.
   b. What position provides the most Hall of Famers?
   c. What position provides the fewest Hall of Famers?
   d. What outfield position (L, C, or R) provides the most Hall of Famers?
   e. Compare infielders (1, 2, 3, and S) to outfielders (L, C, and R).

9. Employees at Electronics Associates are on a flextime system; under this system, the employees can begin their working day at 7:00, 7:30, 8:00, 8:30, or 9:00 A.M. The following data represent a sample of the starting times selected by the employees.

7:00	8:30	9:00	8:00	7:30	7:30	8:30	8:30	7:30	7:00
8:30	8:30	8:00	8:00	7:30	8:30	7:00	9:00	8:30	8:00

   Summarize the data by constructing
   a. a frequency distribution
   b. a relative frequency distribution
   c. a bar graph
   d. a pie chart
   e. What do the summaries tell you about employee preferences concerning the flextime system?

10. Students at the University of Cincinnati are asked to fill out a course-evaluation questionnaire upon completion of their courses. One of the questions is as follows:

   Compared to other courses that you have taken, what is the overall quality of the course you are now completing?

Poor	Fair	Good	Very Good	Excellent

   A sample of 60 students completing a course during the spring quarter of 1993 provided the following responses. To aid in computer processing of the questionnaire results, a numeric scale was used with 1 = Poor, 2 = Fair, 3 = Good, 4 = Very Good, and 5 = Excellent.

3	4	4	5	1	5	3	4	5	2	4	5	3	4	4
4	5	5	4	1	4	5	4	2	5	4	2	4	4	4
5	5	3	4	5	5	2	4	3	4	5	4	3	5	4
4	3	5	4	5	4	3	5	3	4	4	3	5	3	3

   a. Comment on why these are qualitative data. Is the scale of measurement nominal or ordinal?
   b. Provide a frequency distribution and a relative frequency distribution summary of the data.
   c. Provide a bar graph and a pie chart summary of the data.
   d. Based on your summaries, comment on the students' overall evaluation of the course.

2.2 SUMMARIZING QUANTITATIVE DATA

FREQUENCY DISTRIBUTION

A frequency distribution also provides a tabular method for summarizing quantitative data. We defined a frequency distribution in Section 2.1 to be a tabular summary of a set of data showing the number of items in each of several nonoverlapping classes. This definition holds for quantitative data as well as for qualitative data. However, with quantitative data, we have to be more careful in defining the classes. We do not have natural separate categories for the values of the data items. Three steps are necessary to define the classes for a frequency distribution for quantitative data: determine the number of classes, determine the width of each class, and determine the class limits for each class. These three steps are discussed next.

NUMBER OF CLASSES   As a general guideline, we recommend using between 5 and 20 classes. Large data sets usually require more classes; small data sets usually require fewer. The goal is to use enough classes to show the variation in the data but not so many that there are only a few items in many of the classes.

WIDTH OF CLASSES   The second step in constructing a frequency distribution for quantitative data is to choose a width for each class. Choosing the number of classes and the width of the classes are not independent decisions. Since the smallest data value must be in the lowest class and the largest data value must be in the highest class, the fewer classes used, the wider the classes must be. The relationship between the number of classes and the class width can be written as follows:

$$\text{Approximate Class Width} = \frac{\text{Largest Data Value} - \text{Smallest Data Value}}{\text{Number of Classes}} \qquad (2.2)$$

Once the number of classes has been chosen, the approximate class width is determined. Experimenting with a few different numbers of classes and class widths is usually helpful in constructing a frequency distribution.

CLASS LIMITS   The last step in constructing a frequency distribution for quantitative data is to choose *class limits*. The class limits determine the range of data values that are grouped into each class. Care must be taken to be sure the limits are stated such that all data values fall into *one and only one class*.

For example, suppose we had integer data and were interested in defining class limits for class widths of 10. The class limits of 60–70, 70–80, 80–90, and so on would not be acceptable due to the fact that data values of 70 and 80 appear to belong to two classes. There is, however, flexibility in defining class limits. In this case, the class limits may be written as 60–69, 70–79, 80–89, because with integer data, there are no items with values in the intervals from 69 to 70 and 79 to 80. Another alternative for writing class limits for quantitative data is 60–under 70, 70–under 80, and 80–under 90. An alternative such as this would be necessary if data values between 69 and 70, 79 and 80, and so on were possible. The final specification of the class limits is left to the user's discretion. However, the overriding consideration is that the limits must be defined so that all data values belong to one and only one class.

RELATIVE FREQUENCY DISTRIBUTION

We define the relative frequency distribution for quantitative data in the same manner as for qualitative data. Recall that the relative frequency is simply the fraction or proportion of the total number of items belonging to a class. For a data set having n items,

$$\text{Relative Frequency of Class} = \frac{\text{Frequency of the Class}}{n}$$

EXAMPLE 2.3 ▶ A mathematics achievement test consisting of 100 questions was given to 50 sixth-grade students at Maple Elementary School. The data shown in Table 2.5 indicate the number of questions each of the 50 students answered correctly.

To develop a frequency distribution for the mathematics achievement test scores, we first observe that the largest data value in Table 2.5 is 98 and the smallest data value is 41. At this point, we made the choice or preference to try six classes for the frequency distribution. Using equation (2.2), the approximate class width is computed to be

$$\text{Approximate Class Width} = \frac{98 - 41}{6} = 9.5$$

Rounding up, we selected 10 as the class width.

The final decision of determining the class limits for the six classes requires some judgment and preference. Our choice of class limits for this example is 40–49, 50–59, 60–69, 70–79, 80–89, and 90–99. However, we emphasize that the choices of the number of classes, the class width, and the class limits were simply our preferences for the frequency distribution. Your preferences for the frequency distribution might differ, and you would construct a different, but equally valid, frequency distribution for the data set.

The frequency distribution is constructed by counting the number of mathematics achievement test scores in each of the six classes. The relative frequency distribution is constructed by dividing the frequency for each class by the total of 50 students who took the achievement test. Both the frequency distribution and the relative frequency distribution for the mathematics achievement test scores are presented in Table 2.6.

Referring to Table 2.6, we can make some observations about the performance of the sixth graders on the achievement test. For example, note the following:

1. The most frequently occurring test scores are in the 60–69 interval, with 15 students and a relative frequency of .30, or 30%, of the students answering this many questions correctly.

TABLE 2.5 **MATHEMATICS ACHIEVEMENT TEST SCORES FOR A SAMPLE OF 50 SIXTH-GRADE STUDENTS**

75	48	46	65	71
49	61	51	57	49
84	85	79	85	83
55	69	88	89	55
61	72	64	67	60
77	51	61	68	54
63	98	54	53	71
84	79	75	65	50
41	65	77	71	63
67	57	63	71	77

TABLE 2.6

FREQUENCY AND RELATIVE FREQUENCY DISTRIBUTIONS FOR 50 SIXTH-GRADE MATHEMATICS ACHIEVEMENT TEST SCORES

Test Scores	Frequency	Relative Frequency
40–49	5	.10
50–59	10	.20
60–69	15	.30
70–79	12	.24
80–89	7	.14
90–99	1	.02
Totals	50	1.00

2. Only 5 students, or 10%, answered fewer than 50 questions correctly.

3. Only 8 students, or 16%, answered 80 or more questions correctly.

Frequency and relative frequency distributions are valuable in that they provide insights about the entire data set that cannot be easily obtained by viewing only the individual test scores as presented in Table 2.5.

CUMULATIVE FREQUENCY AND CUMULATIVE RELATIVE FREQUENCY DISTRIBUTIONS

Variations of the frequency and relative frequency distributions are the *cumulative frequency* and *cumulative relative frequency distributions.* The cumulative forms of these distributions provide additional information and insight about a data set. They contain the same number of classes as the frequency and relative frequency distributions; however, the cumulative forms show the frequency and the relative frequency of items with values *less than or equal to the upper limit* of the class.

EXAMPLE 2.3 ▶
(CONTINUED)

The frequency distribution and the cumulative frequency distribution for the mathematics achievement test score data presented in Example 2.3 are shown together in Table 2.7.

To see how the cumulative frequency distribution is constructed, consider the interval 60–69 in the frequency distribution. The upper limit for this class is 69. To determine the number of items with values less than or equal to 69, we simply sum the frequencies of the intervals 40–49, 50–59, and 60–69; doing so, we obtain $5 + 10 + 15 = 30$. Hence,

TABLE 2.7

FREQUENCY AND CUMULATIVE FREQUENCY DISTRIBUTIONS FOR THE 50 SIXTH-GRADE MATHEMATICS ACHIEVEMENT TEST SCORES

Frequency Distribution		Cumulative Frequency Distribution	
Test Scores	Frequency	Test Scores	Cumulative Frequency
40–49	5	Less than or equal to 49	5
50–59	10	Less than or equal to 59	15
60–69	15	Less than or equal to 69	30
70–79	12	Less than or equal to 79	42
80–89	7	Less than or equal to 89	49
90–99	1	Less than or equal to 99	50

the value that we enter in the cumulative frequency distribution corresponding to a score of less than or equal to 69 is 30.

The cumulative relative frequency distribution is constructed in a similar fashion and is shown in Table 2.8.

Several interpretations and insights are available from the cumulative forms of the distributions. For example, it is easy to see that 30 students, or 60%, scored less than or equal to 69 on the achievement test. Other observations concerning cumulative frequencies are possible. ◀

NOTES AND COMMENTS

1. The description of the classes for a cumulative frequency or cumulative relative frequency distribution will vary, depending on how the class limits are defined for the frequency distribution. In Example 2.3, we used 40–49, 50–59, 60–69, and so on for the class limits. Using the data value 59 as an example, we see that 59 belongs to the second class. In the cumulative distribution, the description use for this class was "less than or equal to 59." If we had defined the class limits in the original frequency distribution as being 40–under 50, 50–under 60, 60–under 70, and so on, the description used for the classes of the cumulative distribution would have been "less than 50," "less than 60," "less than 70," and so on. In this case, the less-than-or-equal-to form would not have been used.

2. Given class limits for a frequency distribution, we may occasionally want to know the *class midpoints*. Each class midpoint is simply one-half of the distance between the class limits. For example, with a class limit denoted by 60–69, the class midpoint is 64.5. If the class limits are denoted as 60–under 70, the midpoint is one-half of the distance between 60 and 70, thus making the midpoint 65.

DOT PLOTS

One of the simplest graphical summaries of quantitative data is a *dot plot*. A horizontal axis shows the range of values for the data. Then each data value is represented by a dot

TABLE 2.8   RELATIVE FREQUENCY AND CUMULATIVE RELATIVE FREQUENCY DISTRIBUTIONS FOR THE 50 SIXTH-GRADE MATHEMATICS ACHIEVEMENT TEST SCORES

Relative Frequency Distribution		Cumulative Relative Frequency Distribution	
Test Scores	Relative Frequency	Test Scores	Cumulative Relative Frequency
40–49	.10	Less than or equal to 49	.10
50–59	.20	Less than or equal to 59	.30
60–69	.30	Less than or equal to 69	.60
70–79	.24	Less than or equal to 79	.84
80–89	.14	Less than or equal to 89	.98
90–99	.02	Less than or equal to 99	1.00

FIGURE 2.4   DOT PLOT FOR THE 50 MATHEMATICS ACHIEVEMENT TEST SCORES

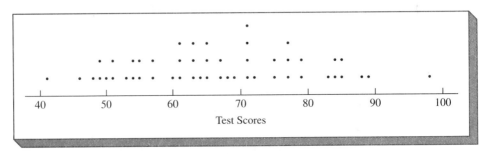

placed above the horizontal axis. One dot is plotted for each data value. The dot plot for the 50 mathematics achievement test scores presented in Table 2.5 is shown in Figure 2.4. The four dots located at the value of 71 indicates that 71 occurs four times in the data set. Dot plots show the details of the data and are very useful for comparing data for two or more variables.

HISTOGRAMS

Perhaps the most often used graphical presentation of quantitative data is a *histogram*. A histogram can be prepared for data that have previously been summarized in either a frequency distribution or a relative frequency distribution. A histogram is constructed by placing the variable of interest on the horizontal axis and the frequency or relative frequency values on the vertical axis. The frequency, or relative frequency, is shown by drawing a rectangle whose base is the class interval on the horizontal axis and whose height is the corresponding frequency or relative frequency.

EXAMPLE 2.3 ▶
(CONTINUED)

The frequency distribution of test scores for the 50 sixth-grade students who completed the mathematics achievement test was shown in Table 2.6. A histogram of this data set is shown in Figure 2.5. Note that the highest frequency is shown by the rectangle appearing above the class interval 60–69. The height of the rectangle shows that the frequency of this class is 15. A histogram for the relative frequency distribution of this data set would look the same as the histogram in Figure 2.5 with the exception that the vertical axis would be labeled in terms of relative frequency rather than frequency.   ◀

As Figure 2.5 shows, the rectangles of a histogram are drawn contiguous to one another—that is, there is no space between the rectangles of adjacent classes. This is the usual convention. Since the class limits of the frequency distribution were stated as 40–49, 50–59, 60–69, and so on, there appear to be intervals between classes corresponding to the one-unit intervals of 49 to 50, 59 to 60, 69 to 70, and so on. The spaces are eliminated in the histogram by drawing the vertical lines halfway between the class limits. For example, the vertical lines for the class 50–59 are drawn above the values 49.5 and 59.5 on the horizontal axis. Similarly, the vertical lines for the class 60–69 are shown above the values 59.5 and 69.5. This minor adjustment causes no interpretation problems, since it is clear in Figure 2.5 that the largest rectangle corresponds to scores in the 60s. In most histograms, this slight adjustment used to eliminate the spaces between classes goes unnoticed. If the class limits in the frequency distribution had been defined as 40–under 50, 50–under 60, and so on, the vertical lines would have been drawn above 40, 50, 60, and so on, and the one-half unit adjustment would have been unnecessary.

FIGURE 2.5   HISTOGRAM FOR 50 SIXTH-GRADE MATHEMATICS ACHIEVEMENT
                  TEST SCORES

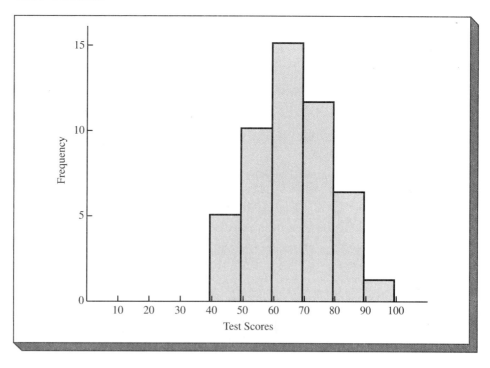

OGIVES

A graphical method used to present a cumulative frequency or a cumulative relative frequency distribution is called an *ogive*. The values for the variable of interest are again placed on the horizontal axis, with the cumulative frequencies or the cumulative relative frequencies on the vertical axis. A point is plotted for each class. The point is plotted above the upper limit of the class at a height corresponding to the cumulative frequency or cumulative relative frequency of the class. One additional point is then plotted above the lower limit of the first class at a height of zero. These points are then connected by straight-line segments to form the ogive. Given any value, the ogive can be used to estimate the number of items or percentage of items with values less than or equal to the value being considered.

EXAMPLE 2.3 ▶
(CONTINUED)

The cumulative frequency distribution for the 50 mathematics achievement test scores is shown in Table 2.7. Figure 2.6 shows the ogive for this cumulative frequency distribution. The points are plotted above the upper class limit values of 49, 59, 69, and so on. Using the ogive, we could estimate that approximately 25 data values are less than or equal to 65. ◀

EXERCISES

METHODS

11. Consider the following data:

3, 5, 12, 14, 8, 9, 2, 10, 18, 7.

Develop a frequency distribution and relative frequency distribution using class limits of 0–4, 5–9, 10–14, and 15–19.

SELF TEST ▷ 12. Consider the following frequency distribution.

Class	Frequency
10–19	8
20–29	12
30–39	15
40–49	5

Construct a cumulative frequency distribution and a cumulative relative frequency distribution.

13. Construct a histogram and an ogive for the data in Exercise 12.

14. Consider the following data.

8.9	10.2	11.5	7.8	10.0	12.2	13.5	14.1	10.0	12.2
6.8	9.5	11.5	11.2	14.9	7.5	10.0	6.0	15.8	11.5

a. Construct a dot plot.
b. Construct a frequency distribution.
c. Construct a relative frequency distribution.

FIGURE 2.6   OGIVE FOR 50 SIXTH-GRADE MATHEMATICS ACHIEVEMENT TEST SCORES

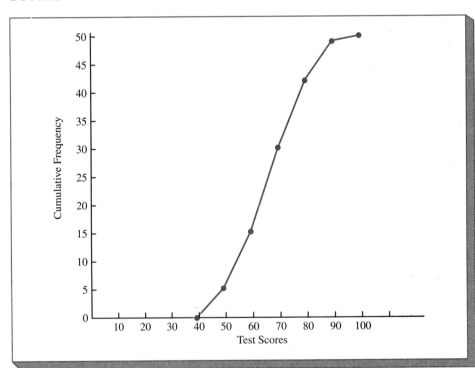

APPLICATIONS

SELF TEST

15. A doctor's office staff has studied the waiting times for patients who arrive at the office with a request for emergency service. The following data were collected over a 1-month period (the waiting times are in minutes):

2, 5, 10, 12, 4, 4, 5, 17, 11, 8, 9, 8, 12, 21, 6, 8, 7, 13, 18, 3.

Use classes of 0–4, 5–9, and so on.
  a. Show the frequency distribution.
  b. Show the relative frequency distribution.
  c. Show the cumulative frequency distribution.
  d. Show the cumulative relative frequency distribution.
  e. What proportion of patients needing emergency service has a waiting time of 9 minutes or less?

 AUTOCOST

16. Data on selling prices of new automobiles and used automobiles were provided in *U.S. News & World Report* (September 9, 1991). Assume the following sample data apply with values stated in thousands of dollars.

New Automobiles:	10.3	24.7	14.7	10.8	14.4	9.8	16.3
	17.8	20.8	17.6	17.1	20.1	16.5	13.8
Used Automobiles:	5.0	4.4	5.6	10.4	9.4	7.0	
	9.3	4.5	4.2	7.7	11.7	11.2	

Use frequency distributions, relative frequency distributions, and histograms to summarize the data. For comparative purposes, use classes of 0.0–4.9, 5.0–9.9, 10.0–14.9, and so on for both data sets. What comparisons can you make about the selling prices of new and used automobiles?

17. National Airlines accepts flight reservations by phone. The data below show the call durations (in minutes) for a sample of 20 phone reservations. Construct the frequency and relative frequency distributions for the data. Also provide a histogram.

2.1	4.8	5.5	10.4	3.3	3.5	4.8	5.8	5.3	5.5
2.8	3.6	5.9	6.6	7.8	10.5	7.5	6.0	4.5	4.8

 EXSALARY

18. How much do executives of some of the largest corporations get paid? *Business Week* (May 1, 1989) reported executive compensation for 1988, including salary and bonus. The data for 25 chief executive officers were reported in thousands of dollars and are as follows.

Company	Compensation	Company	Compensation
Boeing	846	Delta Airlines	457
Whirlpool	563	Chrysler	1466
Bank of Boston	1200	Coca-Cola	2164
Sherwin-Williams	746	Du Pont	1611
Bristol-Myers	824	Motorola	824
General Mills	1310	Marriott	1007
Sara Lee	1367	Honeywell	575
Eastman Kodak	1252	Exxon	1354
Apple Computer	2479	Scott Paper	1238
Bausch & Lomb	927	CBS	1253
K Mart	925	AT&T	1284
Goodyear	1279	Philip Morris	1660
Teledyne	860		

Summarize the data by constructing
a. a frequency distribution
b. a relative frequency distribution
c. a histogram
d. a cumulative frequency distribution
e. Using these summaries, comment on what you learned
    about executive salaries from the sample data.

19. The data shown below are the number of units produced by a production employee for the
    most recent 20 days.

| 160 | 170 | 181 | 156 | 176 | 148 | 198 | 179 | 162 | 150 |
| 162 | 156 | 179 | 178 | 151 | 157 | 154 | 179 | 148 | 156 |

Summarize the data by constructing
a. a frequency distribution
b. a relative frequency distribution
c. a cumulative frequency distribution
d. a cumulative relative frequency distribution
e. an ogive

 MPGDATA

20. *Car and Driver* magazine (January 1989) selected the Honda Civic as one of the 10 best
    cars of the year. Data provided about the Civic included fuel-economy information stated
    in miles per gallon. Assume that the following miles-per-gallon data were obtained from a
    sample of mileage tests with the Civic.

30.2	29.0	27.5	28.3	29.2	32.1	33.8	25.2	34.3	30.6
30.5	28.3	26.0	28.5	29.4	30.3	30.8	29.2	25.9	26.4
27.7	33.9	30.4	29.4	29.4	30.2	28.8	27.5	30.8	30.0

Summarize the data by constructing
a. a frequency distribution
b. a relative frequency distribution
c. a histogram
d. *Car and Driver* magazine reported the fuel economy of the Honda Accord as being in
    the range of 22–27 miles per gallon. Which of the two, the Accord or the Civic,
    appears to have better fuel economy?

 COMPUTER

21. The personal computer has brought computer convenience and power into the home
    environment. But just how many hours a week are people actually using their home
    computers? A study designed to determine the usage of personal computers at home (*U.S.
    News & World Report,* December 26, 1988) provided the following data in hours per week.

.5	1.2	4.8	10.3	7.0	13.1	16.0	12.7	11.6	5.1
2.2	8.2	.7	9.0	7.8	2.2	1.8	12.8	12.5	14.1
15.5	13.6	12.2	12.5	12.8	13.5	1.3	5.5	5.0	10.8
2.5	3.9	6.5	4.2	8.8	2.8	2.5	14.4	16.0	12.4
2.8	9.5	1.5	10.5	2.2	7.5	10.5	14.1	14.9	.3

Summarize the data using
a. a frequency distribution (use a class width of 3 hours)
b. a relative frequency distribution
c. a histogram
d. an ogive
e. Comment on what the data indicate about personal computer usage at home.

2.3 ▽ THE ROLE OF THE COMPUTER

Computers play an important role in providing statistical summaries of data. Prior to 1970, there were relatively few statistical packages for analyzing data using a computer. Since that time, however, the situation has changed dramatically. Today, the user of statistics has a choice of packages such as Minitab, SAS (Statistical Analysis System), SPSS (Statistical Package for the Social Sciences), and BMDP (UCLA Biomedical statistical package), to name a few. In this and future chapters we illustrate various ways in which statistical computing can assist in the analysis and interpretation of data.

Throughout the text, we will use the Minitab system to illustrate the application of statistical computing packages. Minitab is a widely used, general-purpose system that can be installed on a variety of mainframe and personal computers. It has been designed for users who have had little or no previous computer experience. Although easy to use, it still has a large capacity for data summarization and analysis.

Minitab provides a worksheet of columns and rows that is used to store the data set. Each column of the worksheet corresponds to a variable. The columns are identified by labels of C1 for column 1, C2 for column 2, and so on. However, the user may select the NAME command of Minitab to rename the columns (variables) so that the variables can be easily identified. Each row of the worksheet corresponds to an element of the data set. The data in each row provide the observation for the corresponding element.

Let us illustrate the use of the Minitab system by using Minitab to summarize the mathematics achievement test scores data shown in Table 2.5. Specifically, we will show how the Minitab commands of DOTPLOT and HISTOGRAM can be used to quickly, easily, and accurately develop a dot plot, frequency distribution, and histogram for the data. Refer to Figure 2.7. We will assume that the user has loaded the Minitab system on the computer. The MTB > symbol appears on the computer screen, which indicates that the Minitab system is waiting for a command from the user. The user begins with the command READ C1, which indicates that the system should take the data from the input lines that follow and store the data in column 1 of the worksheet. Note that the next line shows Minitab's response, DATA >, indicating that the user is to enter the first data value. The data input process continues with one data value being entered per line. When all the data have been input, the user enters the command END, which terminates the data input process. At this point, the data reside in column 1 of the Minitab worksheet.

With the data in the worksheet, the user can continue with a variety of statistical summary and analysis commands. Let us assume that the user would like to view a dot plot of the data. The Minitab command DOTPLOT C1 is all that is needed. This command instructs the Minitab system to construct a dot plot for the data in column C1. As Figure 2.7 shows, the dot plot is displayed immediately following the command.

The next three commands in Figure 2.7 show that the user has instructed the computer to develop a frequency distribution and histogram for the data. The user would like a frequency distribution with a class width of 10 and class limits of 40–49, 50–59, 60–69, 70–79, 80–89, and 90–99. Minitab will develop such a frequency distribution and its histogram after receiving the following user commands:

HISTOGRAM C1;
START 44.5;
INCREMENT 10.

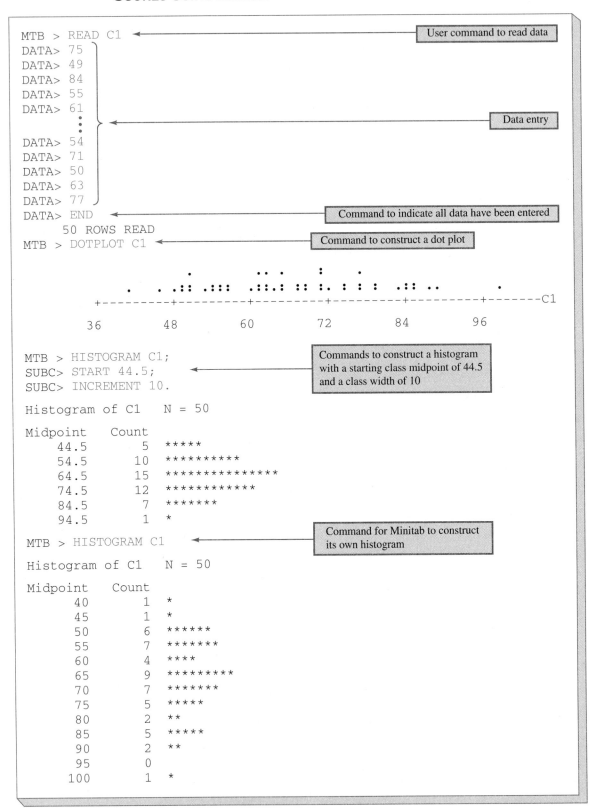

The command HISTOGRAM C1 requests a frequency distribution and a histogram for the data stored in column 1. The semicolon following C1 is used to indicate that user instructions about the format of the frequency distribution and histogram follow. The next two lines are *subcommands* and are requested following the Minitab prompt SUBC >. Minitab uses the *midpoint* of each class, the value halfway between the class limits, to identify the class. Accordingly, the first class of 40–49 has a midpoint of $(40 + 49)/2 = 44.5$; the second class of 50–59 has a midpoint of $(50 + 59)/2 = 54.5$; and so on. The Minitab subcommand START 44.5 tells the system that the first class of the frequency distribution will have a midpoint at 44.5. The semicolon following the 44.5 indicates that another subcommand will follow. The last command of INCREMENT 10 indicates the desired class width. The period after the 10 is used to indicate that the subcommands have been completed.

When the user enters the above three commands, Minitab responds with the frequency distribution and histogram shown in Figure 2.7. The label for the output reminds the user that the histogram command has been requested for the variable in column 1 and that this variable has N = 50 observations. Note that the class midpoints are used to identify the requested classes of 40–49, 50–59, 60–69, 70–79, 80–89, and 90–99. The frequencies of each class are provided in the column labeled Count. The histogram is shown by the pattern of asterisks appearing to the right of the counts. Note that the results are identical to the frequency distribution and histogram provided earlier in this section except that the histogram is on its side. Relative frequencies, cumulative frequencies, and cumulative relative frequencies can be easily computed from the count and histogram information in Figure 2.7.

NOTES AND COMMENTS

This section has focused on the use of the computer to summarize quantitative data. Statistical software packages often handle qualitative data by using numeric codes to represent the data. For example, the data set for new-car purchases by women in Table 2.1 could have been represented by 1 for Chevrolet Cavalier, 2 for Ford Escort, 3 for Ford Taurus, 4 for Honda Accord, and 5 for Hyundai Excel. Using this coding, the Minitab command HISTOGRAM develops the following frequency distribution and bar graph information.

```
Histogram of C1   N = 50

Midpoint    Count
       1        9    *********
       2       14    **************
       3        8    ********
       4       11    ***********
       5        8    ********
```

Interpretation of the midpoint column requires the user to be aware of the numeric codes used and to recall that 1 corresponds to Chevrolet Cavalier, 2 corresponds to Ford Escort, and so on.

If the user had input the command HISTOGRAM C1 and had omitted the START and INCREMENT subcommands, Minitab would have selected its own format for the frequency distribution and histogram. Such a command is shown next in Figure 2.7. Note that Minitab has selected 13 classes with a class width of 5. Since the output is produced directly after the command is given, the user can quickly look at the output, form some initial judgments about the data, enter another command to modify the format for the frequency distribution and histogram, and continue. Data analysis done interactively provides quick computer response and is an important reason why statistical packages are so valuable to the data analyst.

2.4 ▽ EXPLORATORY DATA ANALYSIS (OPTIONAL)

The techniques of *exploratory data analysis* focus on how simple arithmetic and easy-to-draw pictures can be used to summarize data quickly. In this section we study how one of these techniques—referred to as a *stem-and-leaf display*—can be used to rank order data and provide an idea of the shape of the distribution of a set of quantitative data.

One simple method of displaying data involves arranging the data in ascending or descending order. This process, referred to as rank ordering data, provides some degree of organization. However, such an approach provides little insight concerning the shape of the distribution of data values. A stem-and-leaf display is a device that provides a display of both rank order and shape simultaneously.

EXAMPLE 2.3 (CONTINUED) ▶ Let us return to the mathematics achievement test scores for 50 sixth-grade students. The data set is shown below.

75	48	46	65	71
49	61	51	57	49
84	85	79	85	83
55	69	88	89	55
61	72	64	67	60
77	51	61	68	54
63	98	54	53	71
84	79	75	65	50
41	65	77	71	63
67	57	63	71	77

To create a stem-and-leaf display of these data, we will place the first digit of each data value to the left of a vertical line and the second digit to the right of the vertical line. The first digit forms the "stem" and the second digit is the "leaf." For the first item in the data set, 75, the stem-and-leaf numbers are arranged as follows:

Stem	*Leaf*
7	5

Scanning the data set, we find that the smallest value is 41, whereas the largest value is 98. Thus, the stems must range from 4 (for data in the 40s) to 9 (for data in the 90s). The complete set of stems for this data set is

4 |
5 |
6 |

```
7 |
8 |
9 |
```

The stem-and-leaf display is completed by writing the second digit for each item as a leaf in the row containing the first digit. Thus, the first value of 75 is noted by writing a 5 as a leaf in the row headed by 7. The second item, with a value of 48, is noted by writing an 8 as a leaf in the row headed by 4. Each value in the data set can be added to the display by entering the second digit in the appropriate row. Adding the leaves for the entire data set provides the following:

```
4 | 8  6  9  9  1
5 | 1  7  5  5  1  4  4  3  0  7
6 | 5  1  9  1  4  7  0  1  8  3  5  5  3  7  3
7 | 5  1  9  2  7  1  9  5  7  1  1  7
8 | 4  5  5  3  8  9  4
9 | 8
```

Given the organization of the data, it is a simple matter to arrange the leaf values in each row in ascending order from smallest to largest. Doing so provides the following stem-and-leaf display:

```
4 | 1  6  8  9  9
5 | 0  1  1  3  4  4  5  5  7  7
6 | 0  1  1  1  3  3  3  4  5  5  5  7  7  8  9
7 | 1  1  1  1  2  5  5  7  7  7  9  9
8 | 3  4  4  5  5  8  9
9 | 8
```

The stem-and-leaf display is now complete. Using the first line of the display, we see that the ordered data set contains the data values 41, 46, 48, and two 49s. In addition, notice how the stem-and-leaf display provides information about the pattern, or distribution, of the test scores. For example, we see that the largest number of test scores are in the 60s and the second-largest number of test scores are in the 70s. Five students scored in the 40s, and one student scored above 90. Other observations and interpretations are up to the user. ◀

The stem-and-leaf display of quantitative data provides visual insights concerning the pattern of the data as well as a means of arranging the data in ascending order. We note here that there is no one right way to develop and present a stem-and-leaf display. Personal preferences can lead to different displays for the same data set. For instance, if we believe that the six stems shown for the data set have condensed the data set too much, it is a simple matter to stretch the display by using two or more stems for each first digit. For example, to use two lines for each first digit, we place all data values ending in 0, 1, 2, 3, or 4 on one line and all values ending in 5, 6, 7, 8 and 9 on a second line.

EXAMPLE 2.3
(CONTINUED) ▶ Using the preceding convention, the stems of the display for the 50 mathematics achievement test scores would be written as follows:

```
4 |
4 |
5 |
5 |
6 |
6 |
7 |
7 |
8 |
8 |
9 |
9 |
```

Using this format, the first stem row for 6 contains data values from 60 to 64 and the second stem row for 6 contains data values from 65 to 69. The complete stretched stem-and-leaf display appears as follows:

```
4 | 1
4 | 6  8  9  9
5 | 0  1  1  3  4  4
5 | 5  5  7  7
6 | 0  1  1  1  3  3  3  4
6 | 5  5  5  7  7  8  9
7 | 1  1  1  1  2
7 | 5  5  7  7  7  9  9
8 | 3  4  4
8 | 5  5  8  9
9 |
9 | 8
```

Since there is no one best way to set up a stem-and-leaf display, we are free to use any part of the number as the stem and the remaining part of the number as the leaf. For example, with data values in the thousands (such as 1644, 1765, 1852, etc.), stems could be expressed in terms of the first two digits: 16, 17, and 18. The next digits 4, 6, and 5, would be written as the leaves. Data of the form 22.75, 24.63, 25.30, and so on could be best summarized by using 22, 23, 24, and 25 as the stems with the next decimal digits 7, 6, and 3, appearing as the leaves.

Note that the above convention uses only single-digit leaves. Thus, when we have four digit numbers and stems involving two digits, the last digit is truncated. The numbers in the stem-and-leaf display will approximate the actual numbers to within 10 units. For instance, the stem and leaf for the number 1644 would be 16 | 4. The stem and leaf for the number 22.75 would be 22 | 7. When digits have been truncated, the exact numerical values of the data do not appear in the display, but the display still provides a convenient summary for the data set.

EXERCISES

METHODS

22. Construct a stem-and-leaf display for the following data.

70	72	75	64	58	83	80	82
76	75	68	65	57	78	85	72

SELF TEST ▷ **23.** Construct a stem-and-leaf display for the following data.

11.3	9.6	10.4	7.5	8.3	10.5	10.0
9.3	8.1	7.7	7.5	8.4	6.3	8.8

24. Construct a stem-and-leaf display for the following data. Use the first two digits as the stem and the third digit as the leaf.

1161	1206	1478	1300	1604	1725	1361	1422
1221	1378	1623	1426	1557	1730	1706	1689

APPLICATIONS

SELF TEST ▷ **25.** A psychologist developed a new test of adult intelligence. The test was administered to 20 individuals, and the following scores were obtained:

114	99	131	124	117	102	106	127	119	115
98	104	144	151	132	106	125	122	118	118

Construct a stem-and-leaf display for these data.

26. The earnings-per-share data for a sample of 20 companies from the Fortune 500 largest U.S. industrial corporations are as follows (*Fortune,* April 20, 1992).

Company	Earnings per Share	Company	Earnings per Share
Procter & Gamble	4.92	Sara Lee	2.15
Goodyear	1.61	General Dynamics	12.06
Ralston Purina	3.34	Eli Lilly	4.50
Chiquita Brands	2.55	Compaq Computer	1.49
Hershey Foods	2.43	Sunstrand	3.02
Data General	2.62	Briggs & Stratton	2.52
Helene Curtis	.18	Interlake	−1.31
Huffy	1.52	Dell Computer	1.36
Quaker State	.84	Harley-Davidson	2.08
Anchor Glass	3.30	Zenith Electronics	−1.79

Develop a stem-and-leaf display for these data. Comment on what you learned about the earnings per share for these companies.

 JOBSAT   **27.** In a study of job satisfaction, a series of tests were administered to 50 subjects. The following data were obtained; higher scores represent greater satisfaction.

87	76	67	58	92	59	41	50	90	75	80	81	70
73	69	61	88	46	85	97	50	47	81	87	75	60
65	92	77	71	70	74	53	43	61	89	84	83	70
46	84	76	78	64	69	76	78	67	74	64		

Construct a stem-and-leaf display for these data.

28. The net profit margins for oil companies were reported in the *Forbes* 41st Annual Report on American Industry (*Forbes,* January 9, 1989). Data on net profit margins are as follows.

Company	Net Profit Margin	Company	Net Profit Margin
Exxon	6.8	Phillips Petroleum	4.8
AMOCO	9.8	Quaker State	1.6
Du Pont	6.6	Ashland	2.2
Chevron	7.1	Union Pacific	8.9
Mobil	4.1	Kerr McGee	3.4
Occidental	1.8	Crown Central	6.5
Getty	2.1	Pacific Resources	1.9
Union Texas	9.2	American Petrofina	4.6
Atlantic Richfield	8.9	Coastal Corp	1.5

a. Develop a stem-and-leaf display for these data.
b. Use the results of the stem-and-leaf display to develop a frequency distribution and relative frequency distribution for the data.

2.5 ▽ SUMMARIZING BIVARIATE DATA (OPTIONAL)

In some statistical applications, data are collected on two attributes of the elements in a study. Such a data set contains *bivariate data.* By providing a data summary referred to as a *cross-tabulation,* we are often able to gain some insight about the relationship between two variables. In Example 2.4 we consider the case of bivariate data when both variables are qualitative. In Example 2.5 we consider the case of bivariate data when one variable is qualitative and one variable is quantitative.

EXAMPLE 2.4 ▶ Alber's Brewery of Phoenix, Arizona, manufactures and distributes three types of beers: a low-calorie light beer, a regular beer, and a dark beer. The market research group has raised questions concerning differences in preferences for the three beers among male and female beer drinkers. A sample of 150 beer drinkers has been selected. After taste-testing each beer, the individuals in the sample are asked to state their preference or first choice. The two qualitative variables of interest in this study are gender of the beer drinker and beer preference. A partial listing of the data is shown next:

Individual	Gender	Beer Preference
1	Female	Dark
2	Male	Light
3	Male	Regular
.	.	.
.	.	.
.	.	.
149	Female	Regular
150	Female	Light

TABLE 2.9   FORMAT OF A CONTINGENCY TABLE FOR ALBER'S BREWERY

		Variable 2: Beer Preference		
		Light	Regular	Dark
Variable 1:	Male	Cell (1, 1)	Cell (1, 2)	Cell (1, 3)
Gender	Female	Cell (2, 1)	Cell (2, 2)	Cell (2, 3)

Organizing the data for two qualitative variables into a table often provides valuable insights. Suppose we utilize the format of Table 2.9. Every individual in the sample can be classified as belonging to one of the six cells in the table. In general, cell (i, j) corresponds to the cell in row i and column j. For example, an individual may be a male preferring regular beer (cell 1, 2), a female preferring light beer (cell 2, 1), a female preferring dark beer (cell 2, 3), and so on. Since we have included all possible combinations of beer preferences and gender—or, in other words, listed all possible contingencies—Table 2.9 is called a *contingency table*.

A summary of the data in the Alber's Brewery study in the form of a contingency table is shown next with percentages indicated in parentheses.

		Beer Preference			
		Light	*Regular*	*Dark*	**Total**
Gender	*Male*	20 (13.3)	40 (26.7)	20 (13.3)	80 (53.3)
	Female	30 (20.0)	30 (20.0)	10 (6.7)	70 (46.7)
Total		50 (33.3)	70 (46.7)	30 (20.0)	150 (100.0)

Percentage → (46.7)   Number of individuals ← 70

In reviewing the contingency table or cross-tabulation summary, we see that of the 150 beer drinkers in the sample, 20, or 13.3%, were men who favored light beer; 30, or 20.0%, were women who favored light beer; and so on. The cross-tabulation presentation facilitates inferences about the population of beer drinkers and their preferences. For example, the largest cell frequency of 40 (26.7%) suggests that the largest segment of beer drinkers consists of men who prefer regular beer. The smallest cell frequency of 10 (6.7%) suggests the smallest segment of beer drinkers consists of women who prefer dark beer.

Additional inferences are possible from the information in the margins of the contingency table. The entries in the bottom row show that 33.3% of beer drinkers prefer light beer, 46.7% prefer regular beer, and 20% prefer dark beer. Information in the total column (the right margin) shows that 53.3% of beer drinkers in the study are male and 46.7% are female.   ◀

EXAMPLE 2.5 ▶   Let us now consider a bivariate data set in which one variable is quantitative and one variable is qualitative. In Table 1.1, we presented data from a sample of 40 accredited senior colleges and universities in the United States. Six variables were recorded from each college and university in the sample. One variable noted whether the school was public or private and another variable provided the school's annual room-and-board cost. Assume that we are interested in learning about the similarities or differences in the annual room-and-board cost at public as compared to private colleges and universities. In this case, we have bivariate data with each school in the sample providing data for a public/private variable and an annual room-and-board cost variable.

We will use a contingency table cross-tabulation similar to the one shown in Table 2.9. The qualitative variable indicates whether the school is public or private. The quantitative variable of annual room-and-board cost will have to be summarized with classes much like we developed for a frequency distribution in Section 2.2. We will use five classes for the annual room-and-board cost with class limits of 2000–2999, 3000–3999, 4000–4999, 5000–5999, and 6000 or more. The cross-tabulation of the bivariate data simply counts the number of data items for each combination of private/public and annual room-and-board cost categories. The cross-tabulation summary including both frequencies and relative frequencies is shown in Table 2.10.

Many interpretations are possible with the cross-tabulation summary in Table 2.10. Generally, such a tabulation provides insights that are difficult to obtain by viewing the original data set. Some interpretations based on Table 2.10 are as follows:

1. Overall, the sample is roughly evenly split with 19 (47.5%) public institutions and 21 (52.5%) private institutions.
2. The most frequently occurring category is public colleges and universities with annual room-and-board costs in the $3000–3999 range.
3. Most noticeable is the fact that annual room-and-board costs are greater in the private institutions. Only one public school reported annual room-and-board costs

---

TABLE 2.10   CROSS-TABULATION OF ANNUAL COLLEGE ROOM-AND-BOARD COSTS BY PUBLIC/PRIVATE COLLEGES AND UNIVERSITIES

Frequency

		Annual Room-and-Board Cost ($)					
		2000–2999	*3000–3999*	*4000–4999*	*5000–5999*	*6000+*	**Total**
Type	*Public*	4	9	5	0	1	19
	Private	2	4	7	3	5	21
Total		6	13	12	3	6	40

Relative Frequency

		Annual Room-and-Board Cost ($)					
		2000–2999	*3000–3999*	*4000–4999*	*5000–5999*	*6000+*	**Total**
Type	*Public*	.100	.225	.125	.000	.025	.475
	Private	.050	.100	.175	.075	.125	.525
Total		.150	.325	.300	.075	.150	1.000

of $5000 or more, while 8 of the 21 private schools reported annual room-and-board costs of $5000 or more.

Other interpretations are possible. The information in the bottom rows provides the frequency distribution and relative frequency distribution for the annual room-and-board cost regardless of type of school. The information in the rightmost column provides the frequency and relative frequency distributions for the public/private school variable. ◀

EXERCISES

APPLICATIONS

SELF TEST

29. A large amusement park surveyed park visitors at the end of the day to investigate what effect (if any) the distance traveled had on how satisfied visitors are with regard to the food-service facilities at the park. Responses were coded as S for satisfied and NS for not satisfied. The classes for distance traveled were summarized as follows: less than 25 miles, 25–100 miles, and over 100 miles. The following data were collected.

Opinion	Distance (Miles)	Opinion	Distance (Miles)
S	50	NS	15
S	15	S	5
NS	10	NS	175
NS	120	S	120
S	115	S	35
NS	12	S	10
S	20	NS	70
S	18	S	14
S	140	NS	120
NS	40	S	20

Summarize the distance traveled using the classes 0–24, 25–99, and 100 or more. Construct a contingency table for the data, and develop whatever conclusions appear to be appropriate.

30. Eastern Pharmaceutical Corporation is testing a new drug intended to help relieve the symptoms associated with hay fever. One hundred patients were given different levels of the drug (A, B, or C) and then observed for any possible side effects. The following results were obtained: 60 patients experienced no side effects, and of this group 25 had been given level A of the drug and 30 had been given level B; 20 of the patients who were given level B and 15 of the patients who were given level C experienced some side effects. Construct a contingency table for these data, and develop whatever conclusions appear to be appropriate.

SUMMARY

A set of data, even if modest in size, is often difficult to interpret directly in the form in which it is gathered. Tabular and graphical procedures provide means of organizing and summarizing the data so that patterns are revealed and the data are more easily interpreted. Frequency distributions,

FIGURE 2.8   TABULAR AND GRAPHICAL PROCEDURES FOR SUMMARIZING DATA

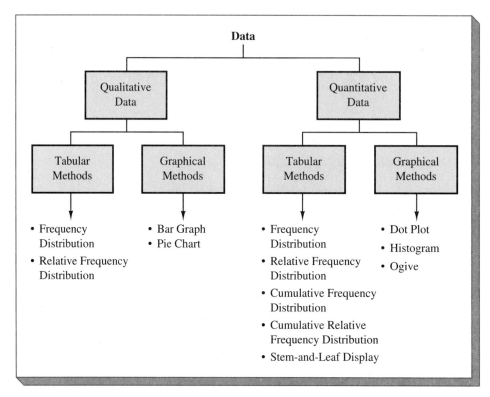

relative frequency distributions, bar graphs, and pie charts were presented as tabular and graphical procedures for summarizing qualitative data. Frequency distributions, relative frequency distributions, cumulative frequency distributions, cumulative relative frequency distributions, dot plots, histograms, and ogives were presented as ways of summarizing quantitative data. A stem-and-leaf display was presented as an exploratory data analysis technique that can be used to summarize quantitative data. The chapter concluded with tabular summaries of bivariate data. Figure 2.8 provides a summary of the tabular and graphical methods presented in this chapter.

GLOSSARY

Qualitative data Data that provide labels or names for categories of like items. These data use either a nominal or an ordinal scale of measurement.

Quantitative data Data that indicate how much or how many. These data use either an interval or a ratio scale of measurement.

Frequency distribution A tabular summary of a set of data showing the frequency (or number) of items in each of several nonoverlapping classes.

Relative frequency distribution A tabular summary of a set of data showing the relative frequency—that is, the fraction or proportion—of the total number of items in each of several nonoverlapping classes.

Bar graph A graphical device for depicting the information presented in a frequency distribution or relative frequency distribution of qualitative data constructed by placing the class labels on the horizontal axis and the frequency or relative frequency scale on the vertical axis. The rectangles or bars are separated to emphasize the fact that each class is a separate category.

Pie chart A graphical device for presenting qualitative data summaries based on subdividing a circle into sectors that correspond to the relative frequency for each class.

Dot plot A graphical device for presenting quantitative data where dots correspond to the data values.

Cumulative frequency distribution A tabular summary of a set of quantitative data showing the number of items having values less than or equal to the upper class limit of each class.

Cumulative relative frequency distribution A tabular summary of a set of quantitative data showing the fraction or proportion of the items having values less than or equal to the upper class limit of each class.

Class midpoint The point in each class that is halfway between the lower and upper class limits.

Histogram A graphical presentation of a frequency distribution or relative frequency distribution of quantitative data constructed by placing the class intervals on the horizontal axis and the frequencies or relative frequencies on the vertical axis. The rectangles are contiguous to one another to show that there is no gap between the class intervals.

Ogive A graphical presentation of a cumulative frequency distribution or a cumulative relative frequency distribution.

Exploratory data analysis The use of simple arithmetic and easy-to-draw pictures to present data more effectively.

Stem-and-leaf display An exploratory data analysis technique that simultaneously rank orders quantitative data and provides insight into the shape of the distribution.

Bivariate data Measurements of two variables are collected for each element.

Cross-tabulation A table that summarizes the frequency and relative frequency for the various categories of bivariate data.

Contingency table A table that provides cells or categories for each possible combination of bivariate data.

KEY FORMULAS

Relative Frequency

$$\frac{\text{Frequency of the Class}}{n} \tag{2.1}$$

Approximate Class Width

$$\frac{\text{Largest Data Value} - \text{Smallest Data Value}}{\text{Number of Classes}} \tag{2.2}$$

REVIEW QUIZ

TRUE/FALSE

1. A frequency distribution is a graphical summary of a set of data.
2. A relative frequency distribution has twice as many classes as a frequency distribution.
3. In a bar graph the width of the bar is proportional to the number of items in the class.
4. The class midpoint is halfway between the class limits.
5. A histogram is a graphical presentation of a frequency distribution for quantitative data.
6. A stem-and-leaf display provides insight into the shape of the distribution of a set of data.

MULTIPLE CHOICE

7. A group of union members indicated what they felt was the most important issue in the upcoming labor–management negotiations; the responses are summarized in the accompanying frequency distribution.

Issue	Frequency
Wages	22
Medical benefits	10
Retirement	15
Working conditions	13
Total	60

The relative frequency of the wages issue is closest to

a. .22

b. .15

c. .40

d. .60

8. Refer to the frequency distribution in Question 7. Fringe benefits include both medical benefits and retirement. What percentage of the union membership feels that fringe benefits are the most important issue?

a. 16.7%

b. 25%

c. 63%

d. 41.7%

9. Consider the following frequency distribution with the completion time given in minutes.

Completion Time (minutes)	Frequency
8–10	5
11–13	8
14–16	15
17–19	12
20–22	7
23–25	3
Total	50

The width of the class intervals is

a. 3

b. 6

c. 2

d. 2.5

10. Refer to the frequency distribution in Question 9. The midpoint for the second class is

a. 11

b. 11.5

c. 12

d. 12.5

11. Consider the following stem-and-leaf display.

```
6 | 2  2  4
7 | 3  6  6  7  8  9
8 | 1  5  7
9 | 0  1
```

The data set consists of how many items?

a. 14   c.   6

b. 18   d. 20

SUPPLEMENTARY EXERCISES

31. The Gallup Poll News Service selected a random sample of adults to learn what sports fans select as their favorite sport to watch, in person or on television. (*USA Today,* December 12, 1990). The sample results shown below are consistent with the findings of the Gallup poll. In the data, B is baseball, K is basketball, F is football, I is ice hockey, T is tennis, and O is other sports.

```
O   F   B   O   B   F   F   K   O   K   F   F   O   F   O
T   F   F   B   K   F   F   O   F   B   F   O   O   B   K
I   F   O   B   F   K   B   K   O   O
```

a. Show a frequency distribution.

b. Show a relative frequency distribution. What is the favorite spectator sport?

c. Show a pie chart summary of the data.

32. Each of the Fortune 500 companies is classified as belonging to one of several industries (*Fortune,* April 20, 1992). Shown below is a sample of 20 companies with their corresponding industry classification.

Company	Industry Classification	Company	Industry Classification
Coca-Cola	(Beverage)	McDonnell Douglas	(Aerospace)
Union Carbide	(Chemicals)	Morton Thiokol	(Chemicals)
General Electric	(Electronics)	Quaker Oats	(Food)
Motorola	(Electronics)	Pepsico	(Beverage)
Beatrice	(Food)	Maytag	(Electronics)
Kellogg	(Food)	Pillsbury	(Food)
Dow Chemical	(Chemicals)	Lockheed	(Aerospace)
Campbell Soup	(Food)	RJR Nabisco	(Food)
Square D	(Electronics)	Westinghouse	(Electronics)
Ralston Purina	(Food)	TRW	(Electronics)

a. Provide a frequency distribution showing the number of companies in each industry.

b. Provide a relative frequency distribution.

c. Provide a bar graph for these data.

33. Airline travelers were asked to indicate the airline they believed offered the best overall service. The four choices were American Air (A), East Coast Air (E), Suncoast (S), and Great Western (W). The following data were obtained.

```
E   A   E   S   W   W   E   S   W   E   W   E   E   A   S
S   W   E   A   W   W   S   E   E   A   E   E   S   W   A
S   E   A   W   A   A   W   E   S   W
```

Summarize the data by constructing

a. a frequency distribution   b. a relative frequency distribution

c. a bar graph   d. a pie chart

e. Which airline appears to offer the best service?

34. Voters participating in a recent election exit poll in Michigan were asked to state their political party affiliation. Coding the data 1 for Democrat, 2 for Republican, and 3 for Independent, the following data were collected.

```
1   2   2   1   3   1   2   2   2   1   2   3   2   3   2   1   1   2   1   2
2   1   1   1   2   1   2   3   1   1   2   1   3   1   1   2   1   2   3   2
```

a. Show a frequency distribution and a relative frequency distribution for the data.
b. Show a bar graph for the data.
c. Comment on what the data suggest about the strengths of the political parties in this voting area.

35. What were the favorite movies of 1988? The biggest box-office successes were listed in *U.S. News & World Report,* December 26, 1988. Assume that sample data collected on those movie preferences were summarized with the letter codes shown below.

Code	Movie
A	Coming to America
B	Big
D	Crocodile Dundee II
R	Who Framed Roger Rabbit
V	Good Morning, Vietnam
O	Other motion picture preferred

The following sample data are available.

```
V   R   O   R   B   A   D   V   V   R   R   D   A   B   V
R   R   A   B   V   A   B   R   R   V   R   A   V   B   V
D   R   A   V   B   A   O   R   R   B   R   A   D   R   R
B   A   R   O   A   A   V   A   D   A   B   R   B   R   B
```

a. Prepare a frequency distribution and a relative frequency distribution for the data set.
b. Prepare a bar graph for the data set.
c. Rank order the top five motion pictures for 1988. What motion picture appears to have been the most successful?

 LAWAGES

36. An international survey was conducted by the Union Bank of Switzerland in order to obtain data on the hourly wages of blue- and white-collar workers throughout the world (*Newsweek,* February 17, 1992). Workers in Los Angeles ranked seventh in the world in terms of highest hourly wage. Assume that the following 25 values indicate hourly wages for workers in Los Angeles.

```
11.50   8.40   11.75   10.05   10.25    8.00   13.65    7.05   9.05
11.90   9.90    6.85   15.35   11.10   14.70   13.15   13.10   6.65
13.10   9.20    9.15   12.05    8.45    5.85    9.80
```

a. Construct a frequency distribution using classes of 4.00–5.99, 6.00–7.99, and so on.
b. Construct a relative frequency distribution.
c. Construct cumulative frequency and cumulative relative frequency distributions.
d. Use these distributions to comment on what you have learned about the hourly wages of workers in Los Angeles.

37. The data below represent sales in millions of dollars for 17 companies in the health care services industry (*The 1992 Business Week 1000*).

```
6099   1709   847   3973    604   2104   166   282   170   233
 868    230   225    491   2452   2301   393
```

a. Construct a frequency distribution to summarize these data. Use a class width of 1000.
b. Develop a relative frequency distribution for the data.
c. Construct a cumulative frequency distribution for the data.
d. Construct a cumulative relative frequency distribution for the data.
e. Construct a histogram as a graphical representation of the data.

 COMSTOCK   **38.** Given below are the closing prices of 40 common stocks (*Investor's Daily*, April 25, 1992).

29⅝	34	43¼	8¾	37⅞	8⅝	7⅝	30⅜	35¼	19⅜
9¼	16½	38	53⅜	16⅝	1¼	48⅜	18	9⅜	9¼
10	37	18	8	28½	24¼	21⅝	18½	33⅝	31⅛
32¼	29⅝	79⅜	11⅜	38⅞	11½	52	14	9	33½

a. Construct frequency and relative frequency distributions for these data.
b. Construct cumulative frequency and cumulative relative frequency distributions for these data.
c. Construct a histogram for the data.
d. Using your summaries, make comments and observations about the price of common stock.

 GRADEAVE   **39.** The grade point averages for 30 students majoring in economics are given below.

2.21	3.01	2.68	2.68	2.74	2.60	1.76	2.77	2.46	2.49
2.89	2.19	3.11	2.93	2.38	2.76	2.93	2.55	2.10	2.41
3.53	3.22	2.34	3.30	2.59	2.18	2.87	2.71	2.80	2.63

a. Construct a relative frequency distribution for the data.
b. Construct a cumulative relative frequency distribution for the data.
c. Construct a histogram for the data.

 ROOMCOST   **40.** The annual cost of room and board at a sample of 50 U.S. colleges and universities is given in the following table (America's Best Colleges, *U.S. News & World Report*, June 5, 1992).

College	Annual Cost of Room and Board ($)	College	Annual Cost of Room and Board ($)
Auburn University	3167	University of Missouri	3004
University of Alaska, Fairbanks	2860	Montana State University	3278
Northern Arizona University	2800	University of Nebraska	2800
University of Arkansas	3150	University of Nevada at Las Vegas	4850
California State University, Fullerton	3249	University of New Hampshire	3600
Colorado State University	3462	Fairleigh Dickinson	5166
University of Connecticut	4522	University of New Mexico	3274
University of Delaware	3540	Syracuse University	5860
Georgetown University	5732	Duke University	4960
Howard University	4040	North Dakota State University	2436
University of Florida	3790	Ohio State University	3639
University of Georgia	2988	University of Oklahoma	3000
University of Hawaii at Manoa	3072	Oregon State University	2950
DePaul University	4333	University of Pittsburgh	3514
Indiana University	3730	Providence College	5000
Iowa State University	2850	Clemson University	3153
University of Kansas	2684	University of South Dakota	2302
University of Kentucky	3734	University of Tennessee	3166

—Table continued on next page

—Table continued from previous page

College	Annual Cost of Room and Board ($)	College	Annual Cost of Room and Board ($)
Tulane University	5505	University of Texas	3300
University of Maine	4241	University of Vermont	4358
University of Maryland	4712	University of Virginia	3312
Amherst College	4400	University of Washington	3684
University of Michigan	3853	West Virginia University	3846
University of Minnesota	3400	University of Wisconsin	3721
University of Mississippi	3004	University of Wyoming	3262

Develop the following summaries:

a. a frequency distribution   **b.** a relative frequency distribution   **c.** a histogram

d. Use your summaries to make some comments about the annual room-and-board costs associated with attending college.

41. There were 79 new shadow stocks reported by the American Association of Individual Investors, *AAII Journal,* April 1992. The term *shadow* is used to indicate that the stocks are for small- to medium-sized firms that are not followed closely by the major brokerage houses. The stock exchange listing the stock—New York Stock Exchange (NYSE), American Stock Exchange (AMEX), and Over the Counter Exchange (OTC)—the earnings per share, and the price-earnings ratio are provided for the following sample of 15 shadow stocks.

Stock	Exchange	Earnings per share	Price-Earnings Ratio
Selas Corp of America	AMEX	1.61	7.1
CE Software Holdings	OTC	4.13	7.5
Shult Homes Corp	AMEX	1.05	11.0
Basic American Medical	OTC	1.06	12.0
Titan Corporation	NYSE	.24	18.3
First Team Sports	OTC	.51	18.6
Cooker Restaurant Corp	OTC	.76	43.1
Chempower	OTC	.22	20.5
Benchmark Electronics	AMEX	.62	25.0
Mylex Corp	OTC	.11	37.5
Pharmacy Mgmt Service	OTC	.22	50.0
Arrow Automotive Indus.	AMEX	.23	38.6
U.S. Filter Corp	AMEX	.23	73.4
Sun Sportswear	OTC	.21	28.6
Village Supermarket	OTC	.61	13.5

a. Provide frequency and relative frequency distributions for the exchange data. What exchange carries the most shadow stocks?

b. Provide frequency and relative frequency distributions for the earnings-per-share and price-earnings ratio data. Use class limits of .00–.49, .50–.99, etc., for the earnings-per-share data and class limits of 0.0–9.9, 10.0–19.9, and so on for the price-earnings ratio data. What observations and comments can you make about the shadow stocks?

STATES

42. A state-by-state listing of per capita incomes for 1991 is shown below (*The Wall Street Journal,* April 23, 1992). Develop a frequency distribution, a relative frequency distribution, and a histogram for these data.

State	Income ($)	State	Income ($)	State	Income ($)
Ala.	15,567	Ky.	15,539	N.D.	16,088
Alaska	21,932	La.	15,143	Ohio	17,916
Ariz.	16,401	Maine	17,306	Okla.	15,827
Ark.	14,753	Md.	22,080	Ore.	17,592
Calif.	20,952	Mass.	22,897	Pa.	19,128
Colo.	19,440	Mich.	18,679	R.I.	18,840
Conn.	25,881	Minn.	19,107	S.C.	15,420
Del.	20,349	Miss.	13,343	S.D.	16,392
D.C.	24,439	Mo.	17,842	Tenn.	16,325
Fla.	18,880	Mont.	16,043	Texas	17,305
Ga.	17,364	Neb.	17,852	Utah	14,529
Hawaii	21,306	Nev.	19,175	Vt.	17,747
Idaho	15,401	N.H.	20,951	Va.	19,976
Ill.	20,824	N.J.	25,372	Wash.	19,442
Ind.	17,217	N.M.	14,844	W.Va.	14,174
Iowa	17,505	N.Y.	22,456	Wis.	18,046
Kan.	18,511	N.C.	16,642	Wyo.	17,118

43. The conclusion from a 40-state poll conducted by the Joint Council on Economic Education (*Time,* January 9, 1989) is that students do not learn enough economics. The findings were based on test results from 11th- and 12th-grade students who took a 46-question, multiple-choice test on basic economic concepts such as profit and the law of supply and demand. The sample data shown below represent data on the number of questions answered correctly.

```
12   16   22   17   18   23   31   18   24   20
24   28   24   25   19   26   18   18   22   16
13   33   19   14    8   15   16   14   22   19
10   21   15    9   16   14   30   12   12   17
```

Summarize these data using

a. a stem-and-leaf display   **b.** a frequency distribution
c. a relative frequency distribution   **d.** a cumulative frequency distribution
e. Based on these data, do you agree with the claim that students are not learning enough economics? Explain.

 CITIES

44. The daily high and low temperatures for 24 cities follow (*USA Today,* April 6, 1992).

City	High	Low	City	High	Low
Tampa	80	58	Birmingham	68	32
Kansas City	69	40	Minneapolis	62	39
Boise	58	35	Portland	50	41
Los Angeles	71	57	Memphis	67	41
Philadelphia	56	35	Buffalo	44	28
Milwaukee	47	29	Cincinnati	55	29
Chicago	52	25	Charlotte	61	37

—Table continues on next page

—Table continued from previous page

City	High	Low	City	High	Low
Albany	50	28	Boston	50	35
Houston	63	50	Tulsa	73	50
Salt Lake City	61	49	Washington, D.C.	56	35
Miami	79	56	Las Vegas	80	53
Cheyenne	66	35	Detroit	52	29

a. Prepare a stem-and-leaf display for the high temperatures.
b. Prepare a stem-and-leaf display for the low temperatures.
c. Compare the stem-and-leaf display from parts (a) and (b), and make some comments about the differences between daily high and low temperatures.
d. Use the stem-and-leaf display from part (b) to determine the number of cities having a low temperature of freezing (32°F) or below.
e. Provide frequency distributions for both the high- and low-temperature data.

COMPUTER CASE: CONSOLIDATED FOODS, INC.

Consolidated Foods, Inc. operates a chain of supermarkets located in New Mexico, Arizona, and California. A recent promotional campaign has advertised the chain's offering of a new credit-card

TABLE 2.11   PURCHASE AMOUNT ($) AND METHOD OF PAYMENT* FOR A RANDOM SAMPLE OF 100 CONSOLIDATED FOODS' CUSTOMERS

Cash	Personal Check	Credit Card	Cash	Personal Check	Credit Card
7.40	27.60	50.30	5.08	52.87	69.77
5.15	30.60	33.76	20.48	78.16	48.11
4.75	41.58	25.57	16.28	25.96	
15.10	36.09	46.24	15.57	31.07	
8.81	2.67	46.13	6.93	35.38	
1.85	34.67	14.44	7.17	58.11	
7.41	58.64	43.79	11.54	49.21	
11.77	57.59	19.78	13.09	31.74	
12.07	43.14	52.35	16.69	50.58	
9.00	21.11	52.63	7.02	59.78	
5.98	52.04	57.55	18.09	72.46	
7.88	18.77	27.66	2.44	37.94	
5.91	42.83	44.53	1.09	42.69	
3.65	55.40	26.91	2.96	41.10	
14.28	48.95	55.21	11.17	40.51	
1.27	36.48	54.19	16.38	37.20	
2.87	51.66	22.59	8.85	54.84	
4.34	28.58	53.32	7.22	58.75	
3.31	35.89	26.57		17.87	
15.07	39.55	27.89		69.22	

*The above data are based on actual bills and types of payments reported for grocery purchases (*The Wall Street Journal*, April 9, 1992).

policy in which Consolidated Foods' customers have the option of paying for their purchases with credit cards such as Visa and MasterCard in addition to the usual options of cash or personal check. The new policy is being implemented on a trial basis with the hope that the credit-card option will encourage customers to make larger purchases.

After the first month of operation, a random sample of 100 customers was selected over a 1-week period. Data were collected on the method of payment and how much was spent by each of the 100 customers. The sample data are shown in Table 2.11. Prior to the new credit-card policy, approximately 50% of Consolidated Foods' customers paid in cash, and approximately 50% paid by personal check.

REPORT

Use the tabular and graphical methods of descriptive statistics presented in Chapter 2 to summarize the sample data in Table 2.11. Your report should contain summaries such as the following:

1. a frequency and relative frequency distribution for the method of payment
2. a bar graph or pie chart for the method of payment
3. frequency and relative frequency distributions for the amount spent using each method of payment
4. histograms and/or stem-and-leaf displays for the amount spent using each method of payment

 CONSOLID

What preliminary insights do you have about the amounts spent and method of payment at Consolidated Foods? The data set for this computer case is available in the data file CONSOLID (see Appendix D).

CHAPTER 3

DESCRIPTIVE STATISTICS II: MEASURES OF LOCATION AND DISPERSION

WHAT YOU WILL LEARN IN THIS CHAPTER

- How to compute and interpret the mean, median, and mode
- How to compute and interpret percentiles and quartiles
- How to compute and interpret the range, interquartile range, variance, standard deviation, and coefficient of variation
- Some practical uses of the mean and standard deviation, including the use of z-scores, Chebyshev's theorem, and the empirical rule
- What outliers are and how to identify them
- The use of exploratory data analysis techniques such as five-number summaries, box plots, and fences
- The role of computer software packages for data summarization and analysis
- How to compute the mean, variance, and standard deviation from grouped data

CONTENTS

The American Dream: Is it Still Affordable?

Home ownership is said to be part of the American dream. It is the symbol for a way of life sought by most Americans. However, with the continued rise in the cost of housing, the dream of home ownership is beginning to fade for a significant portion of the population. In 1980, 64.4% of all households were homeowners; since then the percentage has declined steadily.

Statistics summarizing the trend in the increasing cost of home ownership point to the fact that in 1970 the median price of a house in the United States was $23,400. In 1993 the median price of a house had risen to $107,200. The higher housing costs are forcing individuals in the lower- and middle-income brackets out of the housing market. If the trend continues, home ownership will be out of reach for all but the wealthiest of American families.

The upward trend in housing costs affects not only families, builders, construction workers, building materials suppliers, and realtors but, as some suggest, our democracy itself. Anything can happen when a human dream goes unfulfilled. And the evidence is that the American dream of home ownership may be unfulfilled in the future.

Median resale prices in some major metropolitan areas are as follows:

Honolulu	$361,700
Newark	$201,400
Boston	$183,800
Washington, DC	$164,700
Bakersfield, CA	$161,800
Seattle	$152,200
Chicago	$147,700
Dallas/Ft. Worth	$ 98,500
Milwaukee	$ 98,100

Minneapolis/St. Paul	$97,300
Portland	$97,300
Denver	$96,000
Phoenix	$92,700
St. Louis	$88,600
Charleston, SC	$85,000
Kansas City	$85,000
Buffalo	$84,800
Syracuse	$84,400
Indianapolis	$84,000
Detroit	$83,000
Houston	$81,000
Pittsburgh	$79,300
Tampa	$77,400
Louisville	$70,300

This *Statistics in Practice* is based on "What's Your Home Worth?" *U.S. News & World Report,* April 5, 1993.

In this chapter we continue the discussion of descriptive statistics by introducing several numerical measures of location and dispersion such as the mean, the median, the variance, and the standard deviation. Numerical measures that are computed for a population are called *population parameters;* when they are computed for a sample, they are called *sample statistics.* As we showed in the *Statistics in Practice* article on housing costs, numerical measures are an important part of many statistical presentations; in the article, for example, median resale prices are reported as a measure of housing costs in various cities.

3.1 ▽ MEASURES OF LOCATION

MEAN

Perhaps the most important numerical measure is the *mean,* or average value, of the data. The mean provides a measure of central location for a data set. It is obtained by adding all the data values and dividing by the number of items. If the data set is a sample, the sample mean is denoted by $\bar{x}$ (pronounced *x* bar); if the data set is a population, the population mean is denoted by the Greek letter μ (pronounced mū).

EXAMPLE 3.1 ▶ A sample of five class sections in the College of Arts and Science provided the following data on the number of students in each class section:

$$46, 54, 42, 46, 32.$$

The sample mean for these data is computed as follows:

$$\bar{x} = \frac{46 + 54 + 42 + 46 + 32}{5} = \frac{220}{5} = 44$$

Thus for the five sections sampled, the mean is 44 students per section.   ◀

In specifying general statistical formulas, it is customary to denote the value of the first data item by x_1, the value of the second data item by x_2, and so on. Using this notation, the general formula for the sample mean is as follows.

Sample Mean

$$\bar{x} = \frac{x_1 + x_2 + \cdots + x_n}{n} = \frac{\sum\limits_{i=1}^{n} x_i}{n} \tag{3.1}$$

where n = number of items in the sample.

In this formula, the numerator denotes the sum of the data values starting with $i = 1$ and ending with $i = n$. That is,

$$\sum_{i=1}^{n} x_i = x_1 + x_2 + \cdots + x_n$$

The uppercase Greek letter Σ (sigma) is used as a summation sign.

EXAMPLE 3.1 ▶
(CONTINUED)

Using the notation x_1, x_2, x_3, x_4, x_5 to represent the number of students in each of the five sections sampled, we have:

$$x_1 = 46$$
$$x_2 = 54$$
$$x_3 = 42$$
$$x_4 = 46$$
$$x_5 = 32$$

Thus, to compute the sample mean, we can write

$$\overline{x} = \frac{\sum\limits_{i=1}^{n} x_i}{n} = \frac{\sum\limits_{i=1}^{5} x_i}{5} = \frac{x_1 + x_2 + x_3 + x_4 + x_5}{5}$$

$$= \frac{46 + 54 + 42 + 46 + 32}{5} = 44$$

When we want to sum all the values in a data set, we use the following abbreviated summation notation:

$$\sum x_i = \sum_{i=1}^{n} x_i$$

The starting and ending points for the summation, shown below and above the Σ, respectively, are dropped in the abbreviated notation. It is understood that all the data values are to be included in the sum. A summary of the summation notation and operations used in this text is contained in Appendix C.

EXAMPLE 3.2 ▶ The Nielsen organization provides data on television viewing in the United States. A study (*USA Today*, December 13, 1988) reported that the mean number of hours of television viewing per week was increasing. Suppose that the following data provide the hours of television viewing per week for a sample of 16 college students:

14, 9, 12, 4, 20, 26, 17, 15, 18, 15, 10, 6, 16, 15, 8, 5.

Since there are 16 sample observations, $n = 16$. The sample mean is given by

$$\overline{x} = \frac{\Sigma x_i}{n} = \frac{14 + 9 + \cdots + 5}{16} = \frac{210}{16} = 13.125$$

Hence, the mean or average time spent viewing television for the 16 students is 13.125 hours per week. ◀

Equation (3.1) shows how the mean is computed for a sample of n items. The formula for computing the mean of a population is the same, but we use different notation to indicate that we are dealing with the entire population. The number of items in the population is denoted by N, and, as we mentioned previously, the symbol for the population mean is μ.

Population Mean

$$\mu = \frac{\Sigma x_i}{N} \tag{3.2}$$

TRIMMED MEAN

Occasionally, a variable will have one or more unusually small and/or unusually large data values that significantly influence the value of the mean. In this situation, the mean may provide a poor description of the central location of the data. To remove the effect of the unusually small or large data values, we eliminate, or trim, a percentage of small and large data values from the data set. The mean of the remaining data is called the *trimmed mean*. The intent is for the trimmed mean to be a better indicator of the central location of the data. For example, a 5% trimmed mean removes the smallest 5% of the data values *and* the largest 5% of the data values. The 5% trimmed mean is then computed as the mean of the middle 90% of the data. In general, an α percent trimmed mean is obtained by trimming α percent of the items from each end of the data and computing the mean for the remaining items.

EXAMPLE 3.2 ▶
(CONTINUED)

Let us consider the 5% trimmed mean for the data on television-viewing times for 16 students. The 5% trimmed mean indicates that the smallest 5% and the largest 5% of the data values are to be trimmed from the data set. Here, 5% of 16 is .05(16) = .80 data values. In computing the trimmed mean, we simply round the number of data values trimmed to its nearest integer value; thus, in this example, .80 is rounded to 1, indicating that one data value will be removed from each end of the data set. After eliminating the smallest television-viewing time (4) and the largest television-viewing time (26), the 5% trimmed mean based on the 14 remaining data values is 180/14 = 12.857.   ◀

Another measure of central location that is not influenced by extreme data values is the median. Let us show how the median is computed.

MEDIAN

The *median* is another numerical measure of central location for a set of data. The median is the value falling in the middle when the data items are arranged in ascending order (rank ordered from smallest to largest). If there is an odd number of data items, the median is the middle item. If there is an even number of data items, there is no single middle value. In this case, we follow the convention of defining the median to be the average of the middle two data values. For convenience, this definition is restated below.

Median

- If there are an odd number of items in the data set, the median is the value of the middle item when all items are arranged in ascending order.
- If there are an even number of items in the data set, the median is the average value of the two middle items when all items are arranged in ascending order.

EXAMPLE 3.1 ▶
(CONTINUED)

Let us apply the above definition to compute the median for the sample of five class sections from the College of Arts and Science. Arranging the five data values in ascending order provides the following rank-ordered list:

32, 42, 46, 46, 54.

Since $n = 5$ is odd, the median is the middle item in the above rank-ordered list. That is, the median is the value of the third item, 46. Even though there are two values of 46 for this data set, each value is treated as a separate item when we rank order the data.   ◀

EXAMPLE 3.2
(CONTINUED) ▶
A rank ordering of the television-viewing times for the 16 students in Example 3.2 produces the following list:

$$4, 5, 6, 8, 9, 10, 12, 14, 15, 15, 15, 16, 17, 18, 20, 26.$$

$$\uparrow$$
$$\text{Median} = 14.5$$

Since $n = 16$ is even, the median is the average value of the two middle items—that is, the eighth and ninth items. Hence, the median is $(14 + 15)/2 = 14.5$. ◀

Although the mean is the most commonly used measure of central location, there are a number of situations in which the median is preferred. The mean is influenced by extreme values in a data set, but the median is not. The following example demonstrates this situation.

EXAMPLE 3.3 ▶
A sample of five families in Herrold, Iowa, showed the following rank-ordered annual incomes:

$$\$17,500 \quad \$23,000 \quad \$24,000 \quad \$26,000 \quad \$320,000$$

The median annual income is $24,000 (the third item). The mean is given by

$$\bar{x} = \frac{\$17,500 + \$23,000 + \$24,000 + \$26,000 + \$320,000}{5} = \$82,100$$

In this case, the median is a better indication of central annual incomes than the mean. The mean has been substantially influenced by the single family with a very large income. Indeed, the mean annual income for the sample is over three times the annual income for four of the five families. ◀

MODE

A third numerical measure, the *mode,* provides the location of the most frequently occurring value in the data set. The mode is defined as follows.

Mode

The mode of a set of data is the value that occurs with greatest frequency.

EXAMPLE 3.1
(CONTINUED) ▶
Referring to the sample of five class section sizes with data values 46, 54, 42, 46, and 32, we see that the only value that occurs more than once is 46. Since this value, occurring with a frequency of 2, has the greatest frequency in the data set, it is the mode. ◀

EXAMPLE 3.2
(CONTINUED) ▶
Referring to the 16 weekly television-viewing times for college students, we see that 15 hours of viewing time occurs with the greatest frequency. Thus 15 (occurring with a frequency of 3) is the mode. ◀

Although there is a single value that occurs with the greatest frequency in both of these examples, situations can arise for which the greatest frequency occurs at two or more different values. If a data set has exactly two data values that occur with the same greatest frequency, we say that the data set has two modes and is *bimodal*. If a data set has more than two modes, it is said to be *multimodal;* in such cases, the mode is almost never reported. In the extreme case, every data value can be different, and an argument can be made that every observation is a mode. In such cases, a mode would not be reported.

The mode can be a good measure of location for nominal data. For example, the nominal data set introduced in Example 2.1 of Chapter 2 resulted in the following frequency distribution for automobile purchases by women.

Car Model	Frequency
Chevrolet Cavalier	9
Ford Escort	14
Ford Taurus	8
Honda Accord	11
Hyundai Excel	8

The mode, or most frequently preferred automobile, is the Ford Escort. For this type of data, it obviously makes no sense to speak of the mean or median automobile purchase. But the mode does provide a good indicator of what we are interested in, the automobile preferred by the greatest number of women.

PERCENTILES

A *percentile* is a numerical measure that also locates values of interest in a data set. A percentile provides information regarding how the data items are spread over the interval from the lowest value to the highest value. In large data sets that do not have numerous repeated values, the pth percentile is a value that divides the data set into two parts. Approximately p percent of the items take on values less than the pth percentile; approximately $(100 - p)$ percent of the items take on values greater than the pth percentile. The definition of the pth percentile is as follows.

> **Percentile**
>
> The pth percentile of a data set is a value such that *at least p* percent of the items take on this value or less and *at least* $(100 - p)$ percent of the items take on this value or more.

Admission test scores for colleges and universities are frequently reported in terms of percentiles. For instance, suppose an applicant has a raw score of 530 on the verbal portion of an admissions test. It may not be readily apparent how this student performed relative to other students taking the same test. However, if the raw score of 530 corresponds to the 82nd percentile, then we know that approximately 82% of the students had scores less than this individual and approximately 18% scored better. The following three-step procedure can be used to calculate the pth percentile of a data set.

Calculating the p th Percentile

Step 1 Arrange the data values in ascending order.

Step 2 Compute an index i as follows:

$$i = \left(\frac{p}{100}\right)n$$

where p is the percentile of interest and n is the number of data values.

Step 3 (a) If i *is not an integer*, the next integer *greater than i* denotes the position of the pth percentile.

(b) If i *is an integer*, the pth percentile is the average of the data values in positions i and $i + 1$.

EXAMPLE 3.2
(CONTINUED)

As an illustration, let us determine the 90th percentile for the television-viewing times of Example 3.2.

Step 1 Arrange the 16 data values in ascending order:

$$4, 5, 6, 8, 9, 10, 12, 14, 15, 15, 15, 16, 17, 18, 20, 26.$$

Step 2 Compute i as follows:

$$i = \left(\frac{90}{100}\right)16 = 14.4$$

Step 3 Since i is not an integer, the position of the 90th percentile is the next integer value greater than 14.4, the 15th position. Thus, the 90th percentile, in the 15th position for the data shown in step 1, is 20.

As another illustration, let us compute the 50th percentile for the data set. Applying step 2, we obtain

$$i = \left(\frac{50}{100}\right)16 = 8$$

Since i is an integer, step 3(b) indicates that the 50th percentile is the average of the eighth and ninth data values; thus, the 50th percentile is $(14 + 15)/2 = 14.5$. ◀

The 50th percentile divides the data set into two equal parts; 50% of the data values are less and 50% of the data values are greater. This is exactly what is accomplished by finding the median of a data set. In other words, *the median and the 50th percentile are the same.* Refer to the section on the median to see that the median for the television-viewing data set was also shown to be 14.5.

QUARTILES AND HINGES

It is often desirable to divide a data set into four parts with each part containing one-fourth, or 25%, of the data values. Figure 3.1 shows a data set divided into four parts. The division points are referred to as the *quartiles* and are defined as follows:

$$Q_1 = \text{first quartile, or 25th percentile}$$
$$Q_2 = \text{second quartile, or 50th percentile (also the median)}$$
$$Q_3 = \text{third quartile, or 75th percentile}$$

FIGURE 3.1   LOCATION OF THE QUARTILES

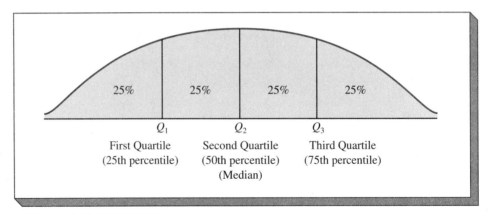

We have defined the quartiles as the 25th, 50th, and 75th percentiles. Thus, we

EXAMPLE 3.2 (CONTINUED)

The television-viewing data are again arranged in ascending order, and Q_2, the second quartile or median, has already been identified as 14.5:

$$4, 5, 6, 8, 9, 10, 12, 14, 15, 15, 15, 16, 17, 18, 20, 26.$$

The computations of Q_1 and Q_3 require the use of the rule for finding the 25th and 75th percentiles of a set of data. These calculations are as follows:

For Q_1,

$$i = \left(\frac{25}{100}\right)16 = 4$$

Since i is an integer, step 3(b) indicates that the first quartile, or 25th percentile, is the average of the fourth and fifth data values; thus, $Q_1 = (8 + 9)/2 = 8.5$.

For Q_3,

$$i = \left(\frac{75}{100}\right)16 = 12$$

Again, since i is an integer, step 3(b) indicates that the third quartile, or 75th percentile, is the average of the 12th and 13th data values; thus, $Q_3 = (16 + 17)/2 = 16.5$.

As shown below, the first quartile, the second quartile, and the third quartile have divided the 16 data values into four parts, with each part consisting of 25% of the data values.

$$4 \quad 5 \quad 6 \quad 8 \mid 9 \quad 10 \quad 12 \quad 14 \mid 15 \quad 15 \quad 15 \quad 16 \mid 17 \quad 18 \quad 20 \quad 26$$
$$Q_1 = 8.5 \qquad\qquad Q_2 = 14.5 \qquad\qquad Q_3 = 16.5 \qquad \blacktriangleleft$$

We have defined the quartiles as the 25th, 50th, and 75th percentiles. Thus, we computed their values in the same fashion as for that of other percentiles. However, there is some variation in the conventions followed in computing quartiles, and the actual value computed may vary slightly depending on the approach used. Nevertheless, the objective of all procedures for computing quartiles is to divide a data set into roughly four equal parts.

Another approach to dividing a data set into four equal parts has been developed by proponents of exploratory data analysis. A *lower hinge* (lower 25%) and an *upper hinge* (upper 25%) are computed. To find these hinges, the median position is found for the data set. Then the data set is divided into two equal parts: data in positions less than or equal to the median position and data in positions greater than or equal to the median position. The median for the data *less than or equal to the median position* is the lower hinge. The median for the data *greater than or equal to the median position* is the upper hinge.

EXAMPLE 3.2 ▶
(CONTINUED)

Referring to the television-viewing data with 16 items, we find that the median position is halfway between the eighth and ninth items. The data values in positions less than or equal to the median position are

$$4, 5, 6, 8, 9, 10, 12, 14.$$

The median of these data values is 8.5, and the lower hinge is thus 8.5. The data values in positions greater than or equal to the median position are

$$15, 15, 15, 16, 17, 18, 20, 26.$$

The median of these data values is 16.5, and the upper hinge is 16.5.   ◀

For the television-viewing time data, the lower hinge is equal to the first quartile and the upper hinge is equal to the third quartile. However, this should not be expected to hold for every data set. In some cases, the hinges and the quartiles take on slightly different values because they are based on slightly different computational procedures.

NOTES AND
COMMENTS

In computing the hinges for data with an odd number of items, the median position is included in the computation of the lower hinge *and* in the computation of the upper hinge. For example, with nine elements, the median position is 5. The median of the data in positions 1–5, the data value in position 3, is the lower hinge, and the median of the data in positions 5–9, the data value in position 7, is the upper hinge. The median position 5 is used in both computations.

EXERCISES

METHODS

1. Consider a sample of size 5 with data values as follows:

$$10, 20, 12, 17, 16.$$

Compute the mean and median.

2. Consider a sample of size 6 with data values as follows:

$$10, 20, 21, 17, 16, 12.$$

Compute the mean and median.

SELF TEST

3. Consider a sample of size 8 with data values as follows:

27, 25, 20, 15, 30, 34, 28, 25.

Compute the 20th, 25th, 65th, and 75th percentiles.

4. Given a sample of 36 items, how many items should be trimmed from each end of the data set to compute the 5% trimmed mean?

APPLICATIONS

5. A *College Placement Council* survey of starting salaries for college graduates (March 1992) found that 1992 starting salaries were up approximately 3% compared with the preceding year. A sample of typical starting salaries for graduates is shown here. Data are shown in thousands of dollars.

STARTSAL

24.8	26.2	29.8	30.1	23.5	22.5	19.7
22.5	24.0	31.2	28.1	19.3	23.5	24.5
20.8	20.0	22.6	24.2	24.0	21.0	24.0
23.5	24.3	26.6	28.5			

    a. What is the mean annual starting salary of the graduates?
    b. What is the median annual starting salary?
    c. What is the mode?
    d. What is the first quartile?
    e. What is the third quartile?

6. Manufacturers of Japanese automobiles established export quotas for automobiles to be shipped to the United States. While the quotas pleased the Detroit automobile manufacturers, the quotas meant Japanese cars would be in short supply and more expensive for the U.S. consumer. *The Wall Street Journal* (January 11, 1989) listed the Japanese export quotas for each of the 12 months of 1988. The data shown below are in terms of thousands of automobiles:

145, 135, 100, 220, 170, 145, 190, 155, 210, 200, 205, 180.

    a. What are the mean, median, and mode for the automobile quota data?
    b. Compute and interpret the first and third quartiles for this data set.

7. A quality-control inspector found the following number of defective parts on 16 different days:

11, 14, 18, 14, 21, 17, 13, 21, 25, 19, 17, 13, 28, 13, 17, 18.

Compute the mean, median, mode, and 90th percentile.

SELF TEST

8. *American Demographics* (December 1988) reported that 25 million Americans get up each morning and go to work in their offices at home. The growing use of personal computers is suggested as one of the reasons more people can operate at-home businesses. The article presented data on the ages of individuals who work at home. Assume the following is a sample of age data for these individuals.

| 22 | 58 | 24 | 50 | 29 | 52 | 57 | 31 | 30 | 41 |
| 44 | 40 | 46 | 29 | 31 | 37 | 32 | 44 | 49 | 29 |

    a. Compute the mean and mode.
    b. Compute a 5% and a 10% trimmed mean.
    c. The median age of the population of all adults is 40.5 years. Use the median age of the preceding data to comment on whether the at-home workers tend to be younger or older than the population of all adults.
    d. Compute the first and third quartiles.
    e. Compute and interpret the 32nd percentile.

9. The American Association of Advertising Agencies records data on nonprogramming minutes per half-hour of prime-time television programming (*U.S. News & World Report,* April 13, 1992). Representative data are shown here for a sample of prime-time programs on major networks at 8:30 P.M.

| 6.0 | 6.6 | 5.8 | 7.0 | 6.3 | 6.2 | 7.2 | 5.7 | 6.4 | 7.0 |
| 6.5 | 6.2 | 6.0 | 6.5 | 7.2 | 7.3 | 7.6 | 6.8 | 6.0 | 6.2 |

a. Compute the mean and median.
b. Compute the first and third quartiles.
c. Using the sample mean, what percentage of viewing time is spent on prime-time advertisements, promotions, and credits? What percentage of viewing time is spent on the programs themselves?

10. A bowler has the following scores for six games:

$$182, 168, 184, 190, 170, 174.$$

Using these data as a sample, compute the following descriptive statistics.
a. mean
b. median
c. mode
d. 75th percentile

11. Monthly sales data for car telephone units for the RC Radio Corporation are as follows:

$$80, 115, 82, 102, 94, 90, 88, 91, 89, 95, 105, 108.$$

Compute the mean, median, and mode for monthly sales.

12. The *Los Angeles Times* regularly reports the air-quality index for various areas of Southern California. Index ratings of 0–50 are considered good, 51–100 are considered moderate, 101–200 unhealthy, 201–275 very unhealthy, and over 275, hazardous. Recent air-quality indexes for Pomona were 28, 42, 58, 48, 45, 55, 60, 49, and 50.
a. Compute the mean, median, and mode for the data. Could the Pomona air-quality index be considered good?
b. Compute the 25th percentile and 75th percentile for the Pomona air-quality data.
c. Compute the lower and upper hinges. Compare your result with the answers to part (b).

13. The following data show the number of automobiles arriving at a toll booth during 20 intervals, each of 10 minutes' duration. Compute the mean, median, mode, first quartile, and third quartile for the data.

| 26 | 26 | 58 | 24 | 22 | 22 | 15 | 33 | 19 | 27 |
| 21 | 18 | 16 | 20 | 34 | 24 | 27 | 30 | 31 | 33 |

14. In automobile mileage and gasoline-consumption testing, 13 automobiles were road tested for 300 miles in both city and country driving conditions. The following data were recorded for miles-per-gallon performance:

| *City:* | 16.2 | 16.7 | 15.9 | 14.4 | 13.2 | 15.3 | 16.8 | 16.0 | 16.1 | 15.3 | 15.2 | 15.3 | 16.2 |
| *Country:* | 19.4 | 20.6 | 18.3 | 18.6 | 19.2 | 17.4 | 17.2 | 18.6 | 19.0 | 21.1 | 19.4 | 18.5 | 18.7 |

Use the mean, median, and mode to make a statement about the difference in performance for city and country driving.

15. A sample of 15 college seniors showed the following credit hours taken during the final term of the senior year:

$$15, 21, 18, 16, 18, 21, 19, 15, 14, 18, 17, 20, 18, 15, 16.$$

a. What are the mean, median, and mode for credit hours taken? Compute and interpret.
b. Compute the first and third quartiles.

   c. Compute the lower and upper hinges. Compare your answers to part (b).

   d. Compute and interpret the 70th percentile.

16. Parents and others have expressed concern over the number of acts of violence occurring in prime time television shows. The following data show the number of acts of violence per hour for ten television shows studied during the months of September, October and November, 1992 (*The Cincinnati Enquirer,* July 18, 1993).

Angel Street (CBS)	41
Covington Cross (ABC)	45
The Edge (Fox)	33
FBI: The Untold Stories (ABC)	28
Final Appeal (NBC)	27
The Hat Squad (CBS)	42
Raven (CBS)	42
Secret Service (NBC)	24
Top Cops (CBS)	38
The Young Indiana Jones (ABC)	60

   a. Compute the mean, median and mode.

   b. Compute the 15th and 80th percentiles.

   c. Compute the quartiles and hinges.

3.2 ▽ MEASURES OF DISPERSION

It is often desirable to consider the dispersion or variability in the values of a data set. For example, assume that an individual has the choice of using public transportation or driving an automobile to work. A major consideration would likely be the travel time associated with the two methods of transportation. Suppose that over a period of several months the individual used public transportation and an automobile the same number of times and found that the mean time to get to work was 32 minutes for both alternatives. At first glance, then, it would appear that both alternatives offer comparable service. However, before reaching a final conclusion, consider the histograms shown in Figure 3.2. Although the mean travel time is 32 minutes for both methods of transportation, do both alternatives possess the same degree of reliability in terms of getting to work on time? Note the dispersion, or variability, in the data. Which method of transportation would you prefer?

For many people, the greater variability exhibited in the times for the public transportation system would be a major concern. That is, to protect against arriving late, one would have to allow for the maximum possible travel times of approximately 40 minutes for public transportation. With a car, one would only need to allow for maximum times up to approximately 34 minutes. Of even more concern, however, are the wide extremes and variation that must be expected when using public transportation. This illustration shows that although the average, or mean, travel time is an important consideration, the dispersion, or variability, in the travel times is also important and might be an overriding consideration. We turn now to a discussion of some commonly used numerical measures of the dispersion, or variability, in a set of data.

FIGURE 3.2   HISTOGRAMS OF TRAVEL TIMES USING PUBLIC TRANSPORTATION AND AN AUTOMOBILE

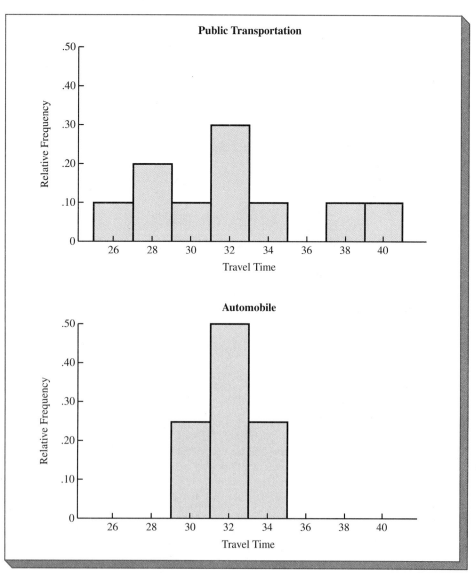

RANGE

The range is the simplest measure of dispersion to compute.

Range

The *range* for a set of data is the difference between the largest and smallest values.

EXAMPLE 3.1
(CONTINUED)

The data on class section size from Example 3.1 are 46, 54, 42, 46, and 32. The largest section size is 54 and the smallest section size is 32. Thus, the range is $54 - 32 = 22$.

Although the range is easy to compute, it is not widely used because it is based on only two of the items in the data set and is influenced by extreme or unusual data values.

EXAMPLE 3.3
(CONTINUED)

The annual incomes for the sample of five families in Herrold, Iowa, are as follows:

$17,500, $23,000, $24,000, $26,000, $320,000.

The range for this data set is $320,000 - 17,500 = $302,500, showing a large dispersion in the data set. In this case, the extreme value of $320,000 leads to the large value for the range.

INTERQUARTILE RANGE

A form of the range that avoids the dependence on extreme values in the data set is the *interquartile range* (IQR).* This descriptive measure of dispersion is simply the difference between the third quartile (Q_3) and the first quartile (Q_1). In effect, it is showing the range for the middle 50% percent of the data and, as such, is not affected by the extreme values in the data set.

Interquartile Range

$$IQR = Q_3 - Q_1 \qquad\qquad \textbf{(3.3)}$$

EXAMPLE 3.2
(CONTINUED)

Previously, we considered the data showing television-viewing time for a sample of 16 college students. The largest viewing time was 26 hours per week, whereas the smallest viewing time was 4 hours per week. In the discussion of quartiles, we found the first quartile for this data set was $Q_1 = 8.5$ and the third quartile was $Q_3 = 16.5$. As measures of dispersion in the data set, we could provide the following:

$$\text{Range} = 26 - 4 = 22 \text{ hours}$$

$$\text{IQR} = Q_3 - Q_1$$

$$= 16.5 - 8.5 = 8 \text{ hours}$$

Thus, we know that the entire data set range is 22 hours, whereas the middle 50% of the data values range is 8 hours.

VARIANCE

A key step in computing the *variance* measure of dispersion involves the computation of the difference between each data value and the mean for the data set. The difference between each data value x_i and the mean ($\bar{x}$ for a sample, μ for a population) is called a *deviation about the mean*.

---

*Another term used for the interquartile range is the *Q-spread* of the data set.

EXAMPLE 3.1
(CONTINUED) ▶

The sample mean for the data set of Example 3.1 is a section size of 44 students. A summary of the data, including the computation of the deviations about the sample mean, is as follows.

Number of Students in Section (x_i)	Mean Section Size $(\bar{x})$	Deviation about Sample Mean $(x_i - \bar{x})$
46	44	2
54	44	10
42	44	−2
46	44	2
32	44	−12
	Total	0

Note: Sum of deviations about the mean is zero.   ◀

We might first think of summarizing the dispersion in a data set by computing the average deviation about the mean. However, a little reflection based on the above table would lead us to discard that idea—the sum of the deviations about the mean for the five sections sampled is equal to zero. This is true for any data set; that is

$$\Sigma(x_i - \bar{x}) = 0$$

The positive and negative deviations cancel each other, causing the average deviation to equal zero. Thus, the average deviation cannot measure the variability in a data set since it will always be equal to $0/n = 0$.

One approach to preventing the positive and negative deviations from canceling out is to take the absolute value of each deviation. The average absolute deviation can then be computed as a measure of variability. While this measure is sometimes used, the most common approach to preventing the cancellation of deviations about this mean is to square them.

EXAMPLE 3.1
(CONTINUED) ▶

The squared deviations and their sum are shown below.

Number of Students in Section (x_i)	Deviation about Sample Mean $(x_i - \bar{x})$	Squared Deviation $(x_i - \bar{x})^2$
46	2	4
54	10	100
42	−2	4
46	2	4
32	−12	144
Totals	0	256
	$\Sigma(x_i - \bar{x})$	$\Sigma(x_i - \bar{x})^2$

◀

A measure of variability based on the squared deviations is the *average squared deviation*. If the data set involved is a population, the average of the squared deviations is called the *population variance*. The population variance is denoted by the Greek symbol σ^2 (pronounced sigma squared). Given a population of N items and using μ to represent the population mean, the definition of the population variance is given by equation (3.4).

Population Variance

$$\sigma^2 = \frac{\Sigma(x_i - \mu)^2}{N}$$

(3.4)

In most statistical applications, the data being analyzed is a sample. When we compute the variance for a sample, we are often interested in using the sample variance as an estimate of the population variance σ^2. At this point, it might seem that the average of the squared deviations of the sample values from $\bar{x}$ would provide a good estimate of the population variance. However, statisticians have found that the average squared deviation for the sample has the undesirable feature of providing a biased estimate of the population variance σ^2; specifically, it tends to underestimate the population variance.

Although it is beyond the scope of this text, it can be shown that if the sum of the squared deviations in a sample is divided by $n - 1$, and not n, then the resulting sample statistic provides an unbiased estimate of the population variance. For this reason, the *sample variance*, denoted by s^2, is defined as follows.

Sample Variance

$$s^2 = \frac{\Sigma(x_i - \bar{x})^2}{n - 1}$$

(3.5)

EXAMPLE 3.1
(CONTINUED) ▶ Let us now compute the sample variance for the data on the number of students in the College of Arts and Science classes. Recall that for this data set, $\Sigma(x_i - \bar{x})^2 = 256$. Hence with $n - 1 = 4$, we obtain

$$s^2 = \frac{\Sigma(x_i - \bar{x})^2}{n - 1} = \frac{256}{4} = 64$$ ◀

While it is admittedly difficult to obtain an intuitive feel for the meaning of the variance, we can note that larger variances could be obtained only from data sets with larger deviations about the mean and, therefore, more dispersion.

EXAMPLE 3.2
(CONTINUED) ▶ Recall that the sample mean for average weekly television-viewing time of a sample of college students is $\bar{x} = 13.125$. Let us use equation (3.5) to compute the sample variance for this data set. The data set and computations are as follows:

x_i	$(x_i - \bar{x})$	$(x_i - \bar{x})^2$
14	.875	.765625
9	−4.125	17.015625
12	−1.125	1.265625
4	−9.125	83.265625
20	6.875	47.265625
26	12.875	165.765625
17	3.875	15.015625
15	1.875	3.515625
18	4.875	23.765625
15	1.875	3.515625
10	−3.125	9.765625
6	−7.125	50.765625
16	2.875	8.265625
15	1.875	3.515625
8	−5.125	26.265625
5	−8.125	66.015625
Totals 210	0.000	525.750000
Σx_i	$\Sigma(x_i - \bar{x})$	$\Sigma(x_i - \bar{x})^2$

Using equation (3.5), we obtain a sample variance of

$$s^2 = \frac{\Sigma(x_i - \bar{x})^2}{n - 1} = \frac{525.75}{15} = 35.05$$

◀

STANDARD DEVIATION

The *standard deviation* of a data set is defined to be the positive square root of the variance. Following the notation we adopted for a sample variance and a population variance, we use s to denote the sample standard deviation and σ to denote the population standard deviation. The standard deviation is derived from the variance in the following manner.

Standard Deviation

Sample Standard Deviation $= s = \sqrt{s^2}$ (3.6)

Population Standard Deviation $= \sigma = \sqrt{\sigma^2}$ (3.7)

EXAMPLE 3.1
(CONTINUED) ▶

Recall that the sample variance for the sample of section sizes in the College of Arts and Science is $s^2 = 64$. Thus, the sample standard deviation is $s = \sqrt{64} = 8$. ◀

Obviously, the standard deviation is also a measure of dispersion, since the square root of a larger variance will provide a larger standard deviation. However, the standard deviation is more often used as a measure of dispersion because it is in the same units as the data. For instance, in the example just considered, the variance of section sizes is 64 "students squared." The standard deviation is 8 students. The fact that variance is reported in units of the original data squared makes it difficult to obtain an intuitive feel for it as a measure of variability.

EXAMPLE 3.2
(CONTINUED) ▶

Referring to the data on television-viewing time in Example 3.2, we see that with $s^2 = 35.05$, the sample standard deviation measure of dispersion is $s = \sqrt{35.05} = 5.92$ hours. ◀

COEFFICIENT OF VARIATION

Sometimes, we are interested in a relative measure of the variability in a data set. For example, a standard deviation of 1 inch would be considered very large for a batch of motor-mount bolts used in automobiles. However, a standard deviation of 1 inch would be considered small for the length of a telephone pole. When the means for data sets differ greatly, we do not get an accurate picture of the relative variability in the two data sets by comparing the standard deviations. A measure of variability that can be used for such comparisons is the *coefficient of variation*. The formula for computing the coefficient of variation is given as follows:

Coefficient of Variation

$$CV = \frac{\text{Standard Deviation}}{\text{Mean}} \times 100 \qquad (3.8)$$

As shown above, the coefficient of variation, CV, is the ratio of the standard deviation to the mean, expressed as a percentage.

EXAMPLE 3.1
(CONTINUED) ▶

Recall that for the sample of section sizes in the College of Arts and Science, the sample mean and sample standard deviation are 44 and 8, respectively. Thus, for this data set, the coefficient of variation is $(8/44) \times 100 = 18.18$. In other words, we could say that the standard deviation of the sample is 18.18% of the value of the sample mean. ◀

NOTES AND COMMENTS
▼

1. Rounding the value of the sample mean $\bar{x}$ and the values of the squared deviations $(x_i - \bar{x})^2$ may introduce rounding errors in the values of the variance and standard deviation. To reduce rounding errors, we recommend carrying at least six significant digits during intermediate calculations. The resulting variance or standard deviation may then be rounded to fewer digits.

2. An alternative formula for the computation of the sample variance is

$$s^2 = \frac{\Sigma x_i^2 - n\bar{x}^2}{n - 1}$$

where $\Sigma x_i^2 = x_1^2 + x_2^2 + \cdots + x_n^2$. Using this formula eases the computational burden slightly and helps reduce rounding errors. Exercise 23 requires this alternative formula to compute the sample variance.

EXERCISES

METHODS

17. Consider a sample of size 5 with data values as follows:

    10, 20, 12, 17, 16.

    Compute the range and interquartile range.

18. Consider a sample of size 5 with data values as follows:

    10, 20, 12, 17, 16.

    Compute the variance and standard deviation.

SELF TEST   19. Consider a sample of size 8 with data values as follows:

    27, 25, 20, 15, 30, 34, 28, 25.

    Compute the range, interquartile range, variance, and standard deviation.

APPLICATIONS

20. Data on selling prices of new and used automobiles were provided in *U.S. News & World Report* (September 9, 1991). Assume the following sample data apply, with data in thousands of dollars.

New Automobiles:	10.3	24.7	14.7	10.8	14.4	9.8	16.3
	17.8	20.8	17.6	17.1	20.1	16.5	13.8
Used Automobiles:	5.0	4.4	5.6	10.4	9.4	7.0	9.3
	4.5	4.2	7.7	11.7	11.2		

    a. Compute the mean and median selling prices for the new and used automobiles.
    b. Compute the range, interquartile range, and standard deviation in the selling prices for new and used automobiles.
    c. What comparisons can you make about the selling prices of new and used cars?

21. The *Washington Post* (January 7, 1989) reported on overcrowding in the Virginia prison system. Use the following data as the population of capacities of the five Virginia state prisons:

    233, 164, 587, 52, 175.

    Compute the range, variance, and standard deviation for this population.

22. The *Los Angeles Times* regularly reports the air-quality index for various areas of Southern California. A sample of air-quality index values for Pomona provided the following data:

    28, 42, 58, 48, 45, 55, 60, 49, and 50.

    a. Compute the range and interquartile range.
    b. Compute the sample variance and sample standard deviation.
    c. A sample of air-quality index readings for Anaheim provided a sample mean of 48.5, a sample variance of 136, and a sample standard deviation of 11.66. What comparisons can you make between the air quality in Pomona and Anaheim based on these descriptive statistics?

23. The Davis Manufacturing Company has just completed five weeks of operation using a new process that is supposed to increase productivity. The numbers of parts produced each week are

    410, 420, 390, 400, 380.

Compute the sample variance and sample standard deviation using the definition of sample variance equation (3.5) as well as the alternative formula provided in the Notes and Comments.

24. A firm has two suppliers who provide an important raw material. Data on the number of days required to fill orders for Dawson Supply, Inc. and J. C. Clark Distributors are as follows:

| *Dawson Supply:* | 11 | 10 | 9 | 10 | 11 | 11 | 10 | 11 | 10 | 10 |
| *Clark Distributors:* | 8 | 10 | 13 | 7 | 10 | 11 | 10 | 7 | 15 | 12 |

Use the range and standard deviation to support the position that Dawson Supply provides the more consistent and reliable delivery times.

SELF TEST ▷ **25.** A bowler's scores for six games were as follows:

$$182, 168, 184, 190, 170, 174.$$

Using these data as a sample, compute the following descriptive statistics:
a. range
b. variance
c. standard deviation
d. coefficient of variation

LAWAGES **26.** The Union Bank of Switzerland conducted a survey to obtain data on the hourly wages of blue- and white-collar workers throughout the world (*Newsweek,* February 1992). Assume a sample of 25 workers from the Los Angeles area provided the data shown here.

11.50	8.40	11.75	10.05	10.25	8.00	13.65	7.05	9.05
11.90	9.90	6.85	15.35	11.10	14.70	13.15	13.10	6.65
13.10	9.20	9.15	12.05	8.45	5.85	9.80		

Provide the following descriptive statistics:
a. mean
b. median
c. range
d. interquartile range
e. variance
f. standard deviation

27. A production department uses a sampling procedure to test the quality of newly produced items. The department employs the following decision rule at an inspection station: If a sample of 14 items has a sample variance of more than .005, the production line must be shut down for repairs. Suppose that the following data have just been collected:

| 3.43 | 3.45 | 3.43 | 3.48 | 3.52 | 3.50 | 3.39 |
| 3.48 | 3.41 | 3.38 | 3.49 | 3.45 | 3.51 | 3.50 |

Should the production line be shut down? Why or why not?

28. The following times were recorded by the quarter-mile and mile runners of a university track team (times are in minutes):

| *Quarter-mile Times:* | .92 | .98 | 1.04 | .90 | .99 |
| *Mile Times:* | 4.52 | 4.35 | 4.60 | 4.70 | 4.50 |

After viewing this sample of running times, one of the coaches commented that the quarter-milers turned in the more consistent times. Use the standard deviation and the coefficient of variation to summarize the variability in the data. Does the use of the coefficient of variation measure indicate that the coach's statement should be qualified?

▽

3.3   SOME USES OF THE MEAN AND STANDARD DEVIATION

We have now developed several measures of location and variability for a set of data. You will find that the mean is the most widely used measure of location, whereas the standard deviation and variance are the most widely used measures of variability. Using only the mean and the standard deviation, we can learn much about a data set.

Z-SCORES

Using the mean and standard deviation together enables us to make statements about the relative location of any item in a data set. Suppose we have a sample of n items, with the values denoted by $x_1, x_2, \ldots, x_n$, and that we have already computed the sample mean $\bar{x}$ and the sample standard deviation s. Associated with each item is a value, called a *z-score*. Equation (3.9) shows how the *z*-score is computed for the value x_i.

---

z-Score

$$z_i = \frac{x_i - \bar{x}}{s} \qquad (3.9)$$

where

z_i is the *z*-score for item i
$\bar{x}$ is the sample mean
s is the sample standard deviation

The *z*-score for an item is often referred to as its standardized value. For instance, the standardized value z_1 can be interpreted as the *number of standard deviations x_1 is from the mean $\bar{x}$*. For example, $z_1 = 1.2$ would indicate x_1 is 1.2 standard deviations above, or larger than, the sample mean. Similarly, $z_2 = -.5$ would indicate data value x_2 is .5, or ½, standard deviation below, or less than, the sample mean. As can be seen from equation (3.9), *z*-scores greater than zero occur for items with values larger than the sample mean, and *z*-scores less than zero occur for items with values smaller than the mean. A *z*-score of zero indicates that the value of the item is equal to the mean.

CHEBYSHEV'S THEOREM

Chebyshev's theorem permits us to make statements about the percentage of items that must be within a specified number of standard deviations from the mean. The statement of Chebyshev's theorem is as follows.

Chebyshev's Theorem

At least $(1 - 1/k^2)$ of the items in any data set must be within k standard deviations of the mean, where k is any value greater than 1.

Some of the implications of this theorem, using $k = 2$, 3, and 4 standard deviations, are as follows:

- At least .75, or 75%, of the items must be within $k = 2$ standard deviations of the mean.
- At least .89, or 89%, of the items must be within $k = 3$ standard deviations of the mean.

■   At least .94, or 94%, of the items must be within $k = 4$ standard deviations of the mean.

EXAMPLE 3.4 ▶   For an example using Chebyshev's theorem, assume that the midterm test scores for 100 students in a college business statistics course had a mean of 70 and a standard deviation of 5. How many students had test scores between 60 and 80? How many students had test scores between 58 and 82?

For the test scores between 60 and 80, we note that the value of 60 is two standard deviations below the mean and the value of 80 is two standard deviations above the mean. Using Chebyshev's theorem with $k = 2$, we see that at least 75% of the items must have values within two standard deviations of the mean. Thus, at least 75 of the 100 students must have scored between 60 and 80.

For the test scores between 58 and 82, we see that $(58 - 70)/5 = -2.4$ indicates 58 is 2.4 standard deviations below the mean and that $(82 - 70)/5 = +2.4$ indicates 82 is 2.4 standard deviations above the mean. Applying Chebyshev's theorem with $k = 2.4$, we have

$$1 - \frac{1}{k^2} = 1 - \frac{1}{(2.4)^2} = .826$$

At least 82.6% of the 100 students must have test scores between 58 and 82.

In the next subsection we will see that when the distribution of data is known to be mound-shaped, larger percentages of the data can be said to lie within two and three standard deviations of the mean. Indeed, we will see that then it is even possible to estimate that a significant portion of the data lies within one standard deviation of the mean.

THE EMPIRICAL RULE

One of the advantages of Chebyshev's theorem is that it applies to any data set, regardless of the shape of the distribution of the data. In practical applications, however, it has been found that many data sets have a mound-shaped, or bell-shaped, distribution like the one shown in Figure 3.3. When it is believed that the data set approximates this pattern, the *empirical rule* can be used to estimate the percentage of items that fall within a specified number of standard deviations of the mean.*

Empirical Rule

For a data set having a bell-shaped distribution similar to the one shown in Figure 3.3, the following statements apply:

■   Approximately 68% of the data items will fall within one standard deviation of the mean.

■   Approximately 95% of the data items will fall within two standard deviations of the mean.

■   Almost all the data will fall within three standard deviations of the mean.

---

*The empirical rule is based on the normal probability distribution, which is presented in detail in Chapter 6.

FIGURE 3.3   A MOUND-SHAPED, OR BELL-SHAPED, DISTRIBUTION

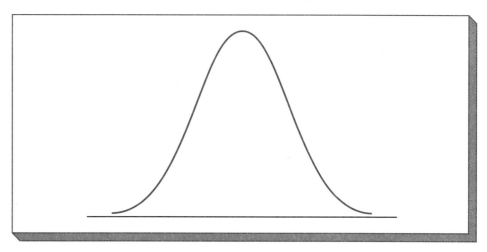

EXAMPLE 3.5 ▶   Liquid detergent cartons are filled automatically on a production line. Filling weights frequently have a bell-shaped distribution. If the mean filling weight is 16 ounces and the standard deviation is .25 ounces, we can use the empirical rule to make the following conclusions.

Approximately 68% of the filled items will have weights between 15.75 and 16.25 ounces (i.e., within one standard deviation of the mean).
Approximately 95% of the filled items will have weights between 15.50 and 16.50 ounces (i.e., within two standard deviations of the mean).
Almost all filled items will have weights between 15.25 and 16.75 ounces (i.e., within three standard deviations of the mean).   ◀

DETECTING OUTLIERS

Sometimes a set of data will have one or more items with unusually large or unusually small values. Extreme values such as these are called *outliers*. Experienced statisticians take steps to identify and review each outlier carefully. An outlier may be an item for which the value has been incorrectly recorded. If so, the value can be corrected before proceeding with further analysis. An outlier may also be an item that was incorrectly included in the data set; if so, it can be removed. Finally, an outlier may just be an unusual item that has been correctly recorded and does belong in the data set. In such cases, the item should remain in the data set, but care should be exercised to ensure that it does not cause misleading inferences to be made.

Standardized data values (z-scores) can be used to detect outliers. The farther a standardized data value is from zero, the more likely the item is an outlier. Recall that the empirical rule allows us to conclude that for many data sets encountered in practice, almost all the items lie within three standard deviations of the mean. Thus, when using z-scores to identify outliers, we recommend treating any item with a z-score less than -3 or greater than 3 as an outlier. Such items can then be reviewed for accuracy and for whether or not they belong in the data set.

EXAMPLE 3.1
(CONTINUED) ▶

Let us return to the sample of five class sections in the College of Arts and Sciences; the class sizes were 46, 54, 42, 46, and 32. The sample mean is $\bar{x} = 44$ and the sample standard deviation is $s = 8$. The calculations shown below provide the corresponding standardized values, or z-scores.

x_i	$x_i - \bar{x}$	$z_i = (x_i - \bar{x})/s$
46	2	$2/8 = +.25$
54	10	$10/8 = +1.25$
42	-2	$-2/8 = -.25$
46	2	$2/8 = +.25$
32	-12	$-12/8 = -1.50$

The class size of 32 provides the most extreme value, occurring at $z = -1.50$, or 1.50 standard deviations below the mean. However, this standardized value is within the -3 to $+3$ guideline, and thus, the class size of 32 is not identified as an outlier. ◀

NOTES AND COMMENTS

1. Before analyzing a data set, statisticians usually make a variety of checks to ensure the validity of the data. In a large study, it is not uncommon for errors to be made recording data values or inputting the values at a computer workstation. Identifying outliers is one tool used to catch such errors and to ensure the validity of data.

2. Chebyshev's theorem is applicable for any data set; it makes a statement about the minimum number of items that lie within a certain number of standard deviations of the mean. If the data set is known to be approximately mound-shaped, more can be said. For instance, the empirical rule allows us to say that *approximately* 95% of the items lie within two standard deviations of the mean; Chebyshev's theorem allows us to conclude only that at least 75% of the items lie in the same interval.

EXERCISES

METHODS

29. Consider a sample of size 5 with data values as follows:

$$10, 20, 12, 17, 16.$$

Compute the z-score for each of the five data values.

30. Consider a sample with a mean of 500 and a standard deviation of 100. What is the z-score for each of the following data values: 520, 650, 500, 450, and 280?

SELF TEST ▶

31. Consider a sample with a mean of 30 and a standard deviation of 5. Use Chebyshev's theorem to determine the proportion, or percentage, of the data within each of the following ranges:
    a. 20 to 40   b. 15 to 45   c. 22 to 38   d. 18 to 42   e. 12 to 48

32. Data with a bell-shaped distribution have a mean of 30 and a standard deviation of 5. Use the empirical rule to determine the proportion, or percentage, of data within each of the following ranges:

a. 20 to 40   **b.** 15 to 45   **c.** 25 to 35

APPLICATIONS

33. The average household income in Tucson, Arizona, is $41,747 (*U.S. News & World Report,* April 6, 1992). If the standard deviation is $12,200, what is the z-score for a household with an annual income of $25,000? What is the z-score for a household with an annual income of $125,000? Interpret these z-scores, and use the empirical rule to comment on whether either of these income values should be considered an outlier.

SELF TEST ▷ **34.** A sample of 10 NCAA men's college basketball scores (*USA Today,* January 5, 1989) provided the following winning teams and the number of points scored:

Winner	Score	Winner	Score
Rutgers	87	Tulsa	70
Niagara	79	Texas—El Paso	82
Mississippi	80	Stanford	83
Western Kentucky	64	Iowa	93
Purdue	75	Montana	62

a. Compute the sample mean and sample standard deviation for these data.
b. In another game, Penn State beat Massachusetts by a score of 110 to 79. Use the z-score to determine if the Penn State score should be considered an outlier. Explain.
c. Assume that the distribution of the points scored by winning teams has a mound-shaped distribution. Estimate the percentage of all NCAA basketball games in which the winning team will score 87 or more points. Estimate the percentage of all NCAA basketball games in which the winning team will score 58 or less points.

35. A 1989 salary survey (*Working Woman,* January 1989) listed the average salary of elementary and secondary school teachers as $28,085. Assume that the standard deviation of salaries is $4500.

a. Janice Herbranson was identified as a teacher in a one-room schoolhouse in McLeod, North Dakota. Her salary was reported to be $8100 per year. What is the z-score associated with $8100? Comment on whether this salary figure is an outlier.
b. Compute the z-score for each of the following salaries: $33,500, $25,200, $28,985, and $39,000. Should any of these be reviewed as possible outliers?

36. Use the salary data in Exercise 35 and Chebyshev's theorem to find the percentage of elementary school teachers that must have salaries in the following ranges:

a. $19,085 to $37,085   **b.** $14,585 to $41,585
c. Repeat parts (a) and (b) if it can be assumed that the distribution of teacher salaries is approximately bell-shaped.

37. Birth rates and IQ scores were discussed in an article in the *Atlantic Monthly,* May 1989. The IQ scores have a bell-shaped distribution with a mean of 100 and a standard deviation of 15.

a. What percentage of the population should have an IQ score between 85 and 115?
b. What percentage of the population should have an IQ score between 70 and 130?
c. What percentage of the population should have an IQ score of more than 130?
d. A person with an IQ score of more than 145 is considered a genius. Does the empirical rule support this statement? Explain.

38. The average fuel economy of new cars sold in the United States is 27.5 miles per gallon (*The Wall Street Journal*, April 8, 1992). Assume that the standard deviation is 3.5 miles per gallon.

    a. Use Chebyshev's theorem to calculate the percentage of new cars sold with miles-per-gallon ratings between 20.5 and 34.5 miles per gallon; between 18.75 and 36.25 miles per gallon; and between 17 and 38 miles per gallon.

    b. If it were reasonable to assume miles-per-gallon ratings for new cars followed a bell-shaped distribution, what can be said about the percentage of new cars sold with miles-per-gallon ratings between 20.5 and 34.5 miles per gallon? Between 17 and 38 miles per gallon?

39. Cruise ship inspections (*St. Petersburg Times,* December 16, 1990) are performed by the Center for Environmental Health and Injury Control. General areas of inspection include potable water, food preparation and holding, general cleanliness, and storage. Ships scoring 85 or less are reinspected more quickly than those with higher ratings. Inspection scores for 20 cruise ships are as follows:

Ship	Score	Ship	Score
Americana	98	Regent Sun	91
Costa Riviera	89	Royal Princess	87
Crown Princess	87	Seaward	94
Daphne	78	Song of America	96
Dolphin IV	95	Song of Norway	91
Fair Princess	76	Starship Atlantic	93
Jubilee	93	Starship Oceanic	97
Meridian	92	Sun Viking	88
Nordic Prince	93	Tropicana	91
Pegasus	62	Viking Princess	86

    a. Compute the mean and median.

    b. Compute the first and third quartiles.

    c. Compute the standard deviation.

    d. What are the *z*-scores associated with the Fair Princess and Pegasus? What is your interpretation of these values?

    e. Are there any outliers? Explain.

3.4 ▽ EXPLORATORY DATA ANALYSIS

In Chapter 2 we introduced exploratory data analysis with the use of stem-and-leaf displays. Exploratory data analysis focuses on using simple arithmetic and easy-to-draw pictures to summarize data. In this section we continue our introduction of exploratory data analysis by considering five-number summaries and box plots.

FIVE-NUMBER SUMMARIES

In a *five-number summary,* the following five numbers are used to summarize a data set:

1. Smallest value in the data set

2. First quartile (Q_1)

3. Median (second quartile)

4. Third quartile (Q_3)
5. Largest value in the data set

EXAMPLE 3.2
(CONTINUED)

Referring to the data and computations we have made with the sample of 16 television-viewing times, the smallest value in the data set is 4 and the largest value is 26. Previous analysis provided $Q_1 = 8.5$, median $= 14.5$, and $Q_3 = 16.5$. Thus the five-number summary of the television-viewing time data is 4, 8.5, 14.5, 16.5, and 26. One-fourth, or 25%, of the data values fall between adjacent numbers in the five-number summary.   ◀

BOX PLOTS

Figure 3.4 shows the box plot for the television-viewing times from Example 3.2. The dashed lines extend from Q_1 to the smallest value in the data set and from Q_3 to the largest value in the data set. Each dashed line contains 25% of the data. The *box* extends from Q_1 to Q_3 and includes the middle 50% of the data. The vertical line drawn within the box identifies the median of the data set. Basically, the box plot provides an easy-to-interpret graphical presentation of the information contained in the five-number summary. When outliers are present, the construction of the box plot is modified somewhat. To see how, we must first discuss the notion of fences.

FENCES AND THE DETECTION OF OUTLIERS

Previously, we defined outliers as unusually small or large values in a data set that should be reviewed to be sure that they are correctly recorded or that they belong in the data set. We showed how z-scores could be used to identify outliers. In exploratory data analysis, fences are often used to identify outliers.

In Figure 3.5 we have added four vertical lines to the box plot for the television-viewing time data. Two of the lines are identified as *inner fences,* and two are identified as *outer fences.* The location of the fences depends on the location of the first and third quartiles, Q_1 and Q_3, as well as the size of the interquartile range (IQR) for the data set. For the television-viewing time data, IQR $= Q_3 - Q_1 = 16.5 - 8.5 = 8$. The inner fences are located at a distance of 1.5(IQR) below Q_1 and at a distance of 1.5(IQR) above Q_3. With 1.5(IQR) = 1.5(8) = 12, the inner fences are located at $Q_1 - 12 = 8.5 - 12 = -3.5$ and $Q_3 + 12 = 16.5 + 12 = 28.5$. The outer fences are located 3(IQR) below Q_1 and 3(IQR) above Q_3. With 3(IQR) = 3(8) = 24, the outer fences are located at $Q_1 - 24 = 8.5 - 24 = -15.5$ and $Q_3 + 24 = 16.5 + 24 = 40.5$.

FIGURE 3.4   **BOX PLOT OF THE TELEVISION-VIEWING TIME DATA**

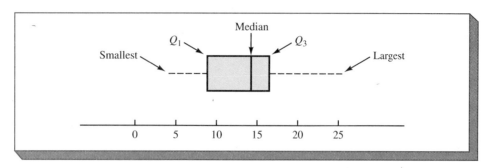

FIGURE 3.5   **FENCES AND OUTLIER INFORMATION FOR THE TELEVISION-VIEWING TIME DATA**

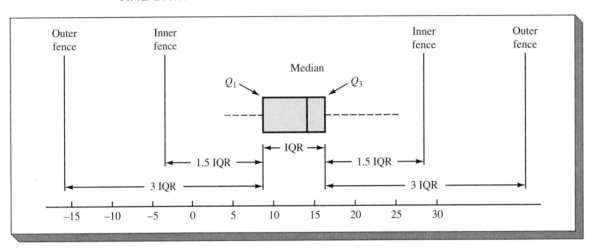

Fences are used to determine how far data values must be from the first and third quartiles before they should be considered outliers. Data values falling between the inner and outer fences are considered *mild outliers*. Data values falling outside the outer fences are considered *extreme outliers*. Both mild and extreme outliers should be reviewed to ensure validity. As Figure 3.5 shows, the television-viewing time data all fall within the inner fences. As such, the fences provide no indication of outliers being present in the data set.

When outliers are present, a box plot is somewhat differently constructed. Figure 3.6 shows what a box plot could look like for a data set with outliers. When outliers exist, the dashed lines are extended to the smallest and largest data values *within the inner fences*. Mild outliers are then denoted by *, and extreme outliers are denoted by ○. The data set in Figure 3.6 can be seen to have two mild outliers and one extreme outlier.

FIGURE 3.6   **A BOX PLOT OF A DATA SET WITH THREE OUTLIERS**

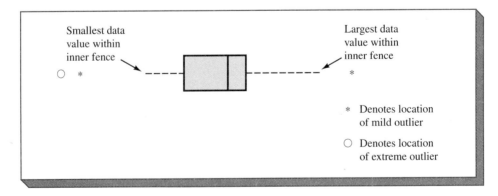

NOTES AND
COMMENTS

1. Using the fences to identify outliers, one will not always identify the same items as when using the z-score approach. However, the objective of both approaches is simply to identify items that should be reviewed to ensure the validity of a data set. Thus, items identified using either procedure should be reviewed.

2. An advantage of the exploratory data analysis procedures is that they are easy to use; few numerical calculations are necessary. One simply needs to put the items into ascending order to construct the five-number summary, the fences, and the box plot. It is not necessary to compute the mean and standard deviation.

EXERCISES

METHODS

40. Consider a sample of size 8 with data values as follows:

$$27, 25, 20, 15, 30, 34, 28, 25.$$

Provide the five-number summary for this data.

41. Show the box plot for the data in Exercise 40.

 SELF TEST   42. Show the five-number summary and the box plot for the following data:

$$5, 15, 18, 10, 8, 12, 16, 10, 6.$$

43. A data set has a first quartile of 42 and a third quartile of 50. Compute the inner and outer fences. Should a data value of 65 be considered an outlier?

APPLICATIONS

 GROWTH   44. *Fortune* (April 20, 1992) published its annual report on the 500 largest U.S. industrial corporations. Data on the percent growth in sales for the past 12 months for a sample of 26 companies are as follows:

16.1	49.9	23.7	15.6	1.9	10.8	20.4
12.2	22.4	4.9	13.4	6.8	15.8	12.1
19.3	10.0	46.1	27.0	7.0	12.5	6.1
15.6	6.3	16.7	10.1	55.9		

a. Provide a five-number summary.
b. Compute the inner and outer fences.
c. Do there appear to be outliers? How would this information be helpful to a financial analyst?
d. Show a box plot.

SELF TEST   45. Annual sales in millions of dollars for 17 companies in the chemical industry are as follows (*Forbes,* January 9, 1989):

484	2,731	598	2,472	3,261	1,220
4,514	32,249	15,980	8,030	3,122	8,258
1,061	2,188	2,366	1,049	636	

    a. Provide a five-number summary of these data.

    b. Compute the inner and outer fences.

    c. Do there appear to be outliers? What does outlier information tell you in this case? *Note:* Companies associated with some of the preceding data are as follows: Valspar (484), Dow Chemical (15,980), and Du Pont (32,249).

    d. Show a box plot.

46. *Consumer Reports* provides performance and quality ratings for numerous consumer products. The overall ratings were provided for a sample of 16 midpriced VCRs in the *Consumer Reports 1992 Buying Guide.* The manufacturer brands and overall scores are as follows:

Manufacturer	Score	Manufacturer	Score
Fisher	77	Panasonic	77
General Electric	81	Phillips	73
Hitachi	89	Quasar	72
J. C. Penney	78	Radio Shack	76
JVC	79	RCA	79
Magnavox	80	Sanyo	75
Montgomery Ward	78	Sony	86
Mitsubishi	90	Toshiba	79

    a. Provide the mean and median overall rating.

    b. Compute the first and third quartiles.

    c. Provide the five-number summary.

    d. Similar ratings for camcorders showed a mean of 82.56, standard deviation of 6.39, and a five-number summary of 75, 77, 82, 86, and 93. Compare the *Consumer Reports* ratings data for VCRs and camcorders. Show the box plots for both.

    e. Are there any outliers in the VCR data? Explain.

 INJURY

47. The Highway Loss Data Institute "Injury and Collision Loss Experience" (September 1988) rates car models based on the number of insurance claims filed after accidents. Index ratings near 100 are considered average. Lower ratings are better, and the car model is considered safer. Shown are ratings for 20 midsize cars and 20 small cars.

Midsize Cars:	81	91	93	127	68	81	60	51	58	75
	100	103	119	82	128	76	68	81	91	82
Small Cars:	73	100	127	100	124	103	119	108	109	113
	108	118	103	120	102	122	96	133	80	140

Summarize the data for the midsize and small cars separately.

    a. Provide a five-number summary for midsize cars and for small cars.

    b. Show the box plots.

    c. Make a statement about what your summaries indicate about the safety of midsize cars compared to small cars.

WORLD

48. Morgan Stanley Capital International in Geneva, Switzerland, provided percent changes in stock markets around the world (*Barrons,* March 30, 1992). Data shown at the top of the next page are percentage changes over the preceding 1-year period.

    a. What are the mean and median percent changes among these world markets?

    b. What are the first and third quartiles?

    c. Are there any outliers? Show a box plot.

    d. What percentile would you report for the United States?

Country	Percent Change	Country	Percent Change
Australia	29.1	Japan	8.3
Austria	−13.4	Luxembourg	6.7
Belgium	9.3	Netherlands	13.9
Canada	8.3	New Zealand	12.6
Denmark	15.3	Norway	−15.4
Europe	10.0	Pacific	10.3
Finland	−16.7	Portugal	−7.2
France	15.8	Singapore	22.7
Germany	6.3	Spain	11.6
Hong Kong	42.8	Sweden	9.1
Italy	−4.1	United Kingdom	11.6
Ireland	9.6	United States	27.2

3.5 ▽ THE ROLE OF THE COMPUTER

We have introduced more than 10 numerical measures of descriptive statistics that can be used to summarize and interpret data. When the size of a data set increases, the computational task of computing these numerical measures can become overwhelming. Fortunately, statistical software packages are available to do the computations. After the user has entered the data, a few simple commands instruct the computer to generate the desired descriptive statistics. In Chapter 2 we described a session with the Minitab computer system, showing how frequency distributions, histograms, and dot plots could be quickly, easily, and accurately developed. In this section we show how Minitab can be used to develop additional descriptive statistics, including a box plot.

To provide an illustration of how Minitab works, we use the data set originally presented in Example 2.3. Recall that these data showed the number of questions answered correctly by each of 50 students who took a 100-question mathematical achievement test. For convenience, this data set is shown again in Table 3.1.

Consider the Minitab session shown by the computer printout in Figure 3.7. At the Minitab prompt, MTB >, the user enters the command RETRIEVE 'MATH', which retrieves the data set of 50 mathematical aptitude test scores previously saved in the Minitab session discussed in Chapter 2. After this command, Minitab responds by indicating that the data set worksheet has been retrieved and is available for analysis.

TABLE 3.1   **MATHEMATICS ACHIEVEMENT TEST SCORES**

75	48	46	65	71
49	61	51	57	49
84	85	79	85	83
55	69	88	89	55
61	72	64	67	60
77	51	61	68	54
63	98	54	53	71
84	79	75	65	50
41	65	77	71	63
67	57	63	71	77

The command DESCRIBE C1 instructs Minitab to develop the numerical measures of descriptive statistics as shown in Figure 3.7. These measures are defined as follows:

Measure	Definition
N	number of data values
MEAN	mean
MEDIAN	median
TRMEAN	5% trimmed mean
STDEV	standard deviation
MIN	minimum data value
MAX	maximum data value
Q1	first quartile
Q3	third quartile

We have not discussed the numerical measure labeled SEMEAN, which refers to the standard error of the mean. This measure is computed by dividing the standard deviation by the square root of N. The interpretation and use of this measure are discussed in Chapter 7 when we introduce the topics of sampling and sampling distributions.

Finally, although the numerical measures of range, interquartile range, variance, and coefficient of variation do not appear on the Minitab output, these values, if desired, can be easily computed from the results in Figure 3.7 using the following:

$$Range = MAX - MIN$$

$$IQR = Q3 - Q1$$

$$Variance = (STDEV)^2$$

$$Coefficient\ of\ Variation = (STDEV/MEAN) \times 100$$

The Minitab session in Figure 3.7 continues with the user command BOXPLOT C1. The Minitab box plot output uses the symbol + to locate the median. The "I" symbols locate the first and third quartiles and identify the ends of the box containing the middle 50% of the data.

A NOTE ON THE COMPUTATION OF QUARTILES

Occasionally, the slightly different computational procedures employed by various computer packages will provide slightly different values for the quartiles Q_1 and Q_3. The reason for any differences in computed quartiles is due to the fact that there is more than one convention that can be used to identify the quartiles of a data set. However, any differences in computed quartiles should be slight, and the interpretation of the quartiles as dividing the data set into quarters is appropriate.

To see why differences in the values of quartiles can occur, consider the first quartile for the mathematics achievement test scores in Table 3.1. Arranging the data in ascending order will show that a test score of 55 is in position 12 and a test score of 56 is in position 13. Following the logic we used in Section 3.1, the first quartile Q_1, the 25th percentile of the 50 test scores, can be found by computing $i = (p/100)n = (25/100)50 = 12.5$. Since

FIGURE 3.7   DESCRIPTIVE STATISTICS FOR MATHEMATICAL ACHIEVEMENT TEST SCORES USING MINITAB

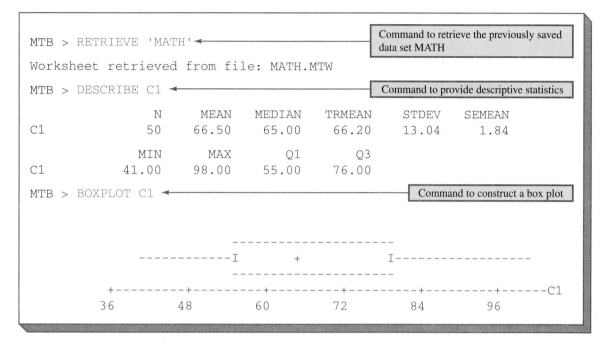

i is not an integer, we round up to position 13. As a result, we would report the first quartile as being $Q_1 = 56$.

The Minitab procedure for computing the first quartile uses the following position formula:

$$\frac{p}{100}(n + 1) = \frac{25}{100}(50 + 1) = 12.75$$

To find the value associated with position 12.75, Minitab interpolates between the data value in position 12 (55) and the data value in position 13 (56). Using the .75 portion of 12.75 to indicate the interpolated value is .75, or ¾, of the way between position 12 (55) and position 13 (56), Minitab reports $Q_1 = 55.75$. Thus, we see the difference between the computed quartiles of 56 and 55.75 is slight. The user would make the correct interpretation by concluding that approximately 25% of the data are below and approximately 75% of the data are above these values.

3.6   MEASURES OF LOCATION AND DISPERSION FOR GROUPED DATA (OPTIONAL)

In most cases, measures of location and dispersion for a data set are computed using the individual data values. However, sometimes we are presented with data in grouped, or frequency distribution, form. This section describes how approximations of the mean, variance, and standard deviation can be obtained from a frequency distribution.

MEAN

Recall that in order to compute the sample mean using the individual data items, we simply sum all the values and divide by n, the sample size. If the data are available only in frequency distribution form, we can approximate the sum of the values by first finding an approximation for the sum of the values for each class and then adding these for all classes.

To do this, we treat the midpoint of each class as if it were the mean of the items in the class. Let M_i denote the midpoint for class i and f_i denote the frequency of the class. Then an approximation to the sum of the items in class i is given by $f_i M_i$. Summing these approximations over all classes, we obtain $\Sigma f_i M_i$, which approximates the sum of all the data values.

Once this approximation of the sum of all the data values is obtained, an approximation of the mean is computed by dividing this sum by the total number of data items. The following formula is thus used to compute the sample mean for *grouped data*.

Sample Mean for Grouped Data

$$\bar{x} = \frac{\Sigma f_i M_i}{n} \tag{3.10}$$

As we indicated at the beginning of this section, we do not expect the calculations based on the individual data values and grouped data to provide exactly the same numerical result. Thus, $\bar{x}$ calculated using equation (3.10) is an approximation of $\bar{x}$ calculated when all data values are known.

EXAMPLE 3.6 ▶ In Example 2.3 of Chapter 2 we analyzed a data set involving the scores obtained by each of 50 sixth graders on a mathematics achievement test. We developed a frequency distribution using the six class intervals shown in Table 3.2. Note that we have added two columns to the frequency distribution. One column is for the class midpoints and the other is for the approximation of the sum of data values in each class, $f_i M_i$. The sum of the items in the last column provides the approximation to the sum of the 50 data values. Using this sum, we compute the mean for the grouped data as follows:

$$\bar{x} = \frac{\Sigma f_i M_i}{50} = \frac{3315}{50} = 66.3$$

TABLE 3.2   **COMPUTATION OF THE SAMPLE MEAN FOR MATHEMATICS ACHIEVEMENT TEST SCORES USING GROUPED DATA**

Class	Frequency (f_i)	Class Midpoint (M_i)	$f_i M_i$
40–49	5	44.5	222.5
50–59	10	54.5	545.0
60–69	15	64.5	967.5
70–79	12	74.5	894.0
80–89	7	84.5	591.5
90–99	1	94.5	94.5
Total	50		$\Sigma f_i M_i = 3315.0$

Referring to the original data as presented in Example 2.3 and computing the sample mean using the ungrouped data, we find $\bar{x} = 66.46$. Thus, the approximation error using the grouped data computation is only $66.46 - 66.3 = .16$. In this case, the grouped data approximation of the sample mean is very good.   ◀

VARIANCE

The approach to computing the sample variance for a set of grouped data is to use a slightly altered form of the formula for the sample variance as provided in equation (3.5). Since we no longer have the individual data values, we treat the class midpoint as a representative value for the data items in each class. Then, we weight the squared deviation from each midpoint by the frequency of its corresponding class. Using this approach, equation (3.5) is modified as follows.

> **Sample Variance for Grouped Data**
>
> $$s^2 = \frac{\Sigma f_i (M_i - \bar{x})^2}{n - 1}$$ (3.11)

EXAMPLE 3.6
(CONTINUED) ▶

The calculation of the sample variance for the grouped data of Example 3.6 is shown in Table 3.3. Using the grouped data sample mean of $\bar{x} = 66.3$, we see that the sample variance is 157.92.   ◀

STANDARD DEVIATION

The standard deviation computed from grouped data is simply the square root of the variance computed from grouped data. For the data of Example 3.6, the sample standard deviation computed from grouped data is $s = \sqrt{157.92} = 12.57$.

TABLE 3.3   **COMPUTATION OF THE SAMPLE VARIANCE FOR MATHEMATICS ACHIEVEMENT TEST SCORES USING GROUPED DATA**

Class	Frequency (f_i)	Class Midpoint (M_i)	$(M_i - \bar{x})$	$(M_i - \bar{x})^2$	$f_i(M_i - \bar{x})^2$
40–49	5	44.5	−21.8	475.24	2376.20
50–59	10	54.5	−11.8	139.24	1392.40
60–69	15	64.5	−1.8	3.24	48.60
70–79	12	74.5	8.2	67.24	806.88
80–89	7	84.5	18.2	331.24	2318.68
90–99	1	94.5	28.2	795.24	795.24
Total	50				$\Sigma f_i(M_i - \bar{x})^2 = 7738.00$

$$s^2 = \frac{\Sigma f_i(M_i - \bar{x})^2}{n - 1} = \frac{7738.00}{49} = 157.92$$

POPULATION SUMMARIES FOR GROUPED DATA

Before closing this section on computing measures of location and dispersion for grouped data, we note that the formulas in this section were presented only for data sets constituting a sample. Population summary measures are computed in a similar manner. The grouped data formulas for a population mean and variance are as follows.

Population Mean for Grouped Data

$$\mu = \frac{\Sigma f_i M_i}{N}$$

(3.12)

Population Variance for Grouped Data

$$\sigma^2 = \frac{\Sigma f_i (M_i - \mu)^2}{N}$$

(3.13)

NOTES AND COMMENTS

An alternative formula for the computation of the sample variance for grouped data is as follows:

$$s^2 = \frac{\Sigma f_i M_i^2 - n\bar{x}^2}{n - 1}$$

where $\Sigma f_i M_i^2 = f_1 M_1^2 + f_2 M_2^2 + \cdots + f_k M_k^2$ and k is the number of classes used to group the data. Using this formula may ease the computations slightly.

EXERCISES

METHODS

SELF TEST

49. Consider the sample data in the following frequency distribution:

Class	Midpoint	Frequency
3–7	5	4
8–12	10	7
13–17	15	9
18–22	20	5

Compute the sample mean.

SELF TEST

50. Compute the sample variance and sample standard deviation for the grouped data in Exercise 49.

APPLICATIONS

51. The *Journal of Personal Selling and Sales Management* (August 1988) reported a study investigating the use of a persuasion technique in selling. A sample of 56 participants was used in the study. Assume that the following frequency distribution shows the number of sales presentations made per week by the 56 participants.

Number of Presentations	Frequency
10–12	5
13–15	9
16–18	22
19–21	12
22–24	8
Total	56

a. What is the mean number of sales presentations made per week for participants in this study?

b. What are the variance and the standard deviation?

52. The following frequency distribution for the first examination in operations management was posted on the department bulletin board.

Examination Grade	Frequency
40–49	3
50–59	5
60–69	11
70–79	22
80–89	15
90–99	6
Total	62

Treating these data as a sample, compute the mean, variance, and standard deviation.

53. *Psychology Today* (November 1988) reported on the characteristics of individuals caught in the act of shoplifting in a supermarket. Data from 9832 cases showed 56% of the shoplifters were male and 44% were female. The value of most items stolen was in the $1–5 range. The following frequency distribution shows the age of the shoplifters.

Age of Shoplifter	Frequency
6–11	364
12–17	1249
18–29	3392
30–59	3962
60–80	865
Total	9832

a. What is the sample mean age of supermarket shoplifters?

b. What are the variance and standard deviation of the ages?

54. A service station has recorded the following frequency distribution for the number of gallons of gasoline sold per car in a sample of 680 cars.

Gasoline (gallons)	Frequency
0–4	74
5–9	192
10–14	280
15–19	105
20–24	23
25–29	6
Total	680

Compute the mean, variance, and standard deviation for these grouped data. If the service station expects to service about 120 cars on a given day, what is an estimate of the total number of gallons of gasoline that will be sold?

55. Test scores obtained by a sample of patients on their depression levels are summarized in the following frequency distribution.

Depression Level Score	Frequency
25–34	3
35–44	1
45–54	2
55–64	6
65–74	4
75–84	6
85–94	2
95–104	1
Total	25

Using the given grouped data, compute the following:

a. mean

b. variance

c. standard deviation

SUMMARY

In this chapter we introduced several statistical measures that can be used to describe the location and dispersion of data. When the numerical values obtained are for a sample, they are called sample statistics. When the numerical values obtained are for a population, they are called population parameters. Some of the notation used for sample statistics and population parameters are summarized as follows.

Measure	Sample Statistic	Population Parameter
Mean	$\bar{x}$	μ
Variance	s^2	σ^2
Standard deviation	s	σ

As measures of location, we defined the mean, median, and mode for both sample and population data. Then the concept of a percentile was used to describe the location of other values in the data. Next, we presented the range, interquartile range, variance, standard deviation, and coefficient of variation as statistical measures of variability or dispersion.

A discussion of two exploratory data analysis techniques that can be used to summarize data more effectively was included in Section 3.4. Specifically, we showed how to develop a five-number summary and a box plot in order to provide simultaneous information about the location, dispersion, and shape of the distribution. A session with the software package Minitab was used to illustrate how statistical computing systems can support the analysis and interpretation of data. Finally, we described how the mean, variance, and standard deviation could be computed for grouped data. However, we recommend using the individual data values unless the grouped format is the only manner in which the data are available.

GLOSSARY

Population parameter A numerical value used as a summary measure for a population of data (e.g., the population mean μ, the population variance σ^2, and the population standard deviation σ).

Sample statistic A numerical value used as a summary measure for a sample (e.g., the sample mean $\bar{x}$, the sample variance s^2, and the sample standard deviation s).

Mean A measure of central location for a data set. It is computed by summing all the data values and dividing by the number of items.

Trimmed mean The mean of the data remaining after α percent of the smallest and α percent of the largest items have been removed. The purpose of a trimmed mean is to provide a measure of central location that has eliminated the effect of any extremely large and/or extremely small data values.

Median A measure of central location. It is the value that splits the data into two equal groups—one with values greater than or equal to the median and one with values less than or equal to the median.

Mode A measure of location, defined as the most frequently occurring data value.

Percentile A value such that at least p percent of the items are less than or equal to this value and at least $(100 - p)$ percent of the items are greater than or equal to this value. The 50th percentile is the median.

Quartiles The 25th, 50th, and 75th percentiles referred to as the first quartile, the second quartile (median), and third quartile, respectively. The quartiles can be used to divide the data set into four parts, with each part containing approximately 25% of the data.

Hinges The value of the lower hinge is approximately the first quartile, or 25th percentile. The value of the upper hinge is approximately the third quartile, or 75th percentile. The values of the hinges and quartiles may differ slightly due to differing computational conventions.

Range A measure of dispersion, defined to be the difference between the largest and smallest data values.

Interquartile range (IQR) A measure of dispersion, defined to be the difference between the third and first quartiles.

Variance A measure of dispersion for a data set based on the squared deviations of the data values about the mean.

Standard deviation A measure of dispersion for a data set, found by taking the positive square root of the variance.

Coefficient of variation A measure of relative dispersion for a data set, found by dividing the standard deviation by the mean and multiplying by 100.

z-Score For each data item, a value found by dividing the deviation about the mean $(x_i - \bar{x})$ by the standard deviation s. A z-score is referred to as a standardized value and denotes the number of standard deviations a data value x_i is from the mean.

Chebyshev's theorem A theorem applying to any data set that can be used to make statements about the percentage of items that must be within a specified number of standard deviations of the mean.

Empirical rule A rule that states the percentages of items that are within one, two, and three standard deviations from the mean for mound-shaped, or bell-shaped, distributions.

Outlier An unusually small or unusually large data value.

Five-number summary An exploratory data analysis technique that uses the following five numbers to summarize the data set: smallest value, first quartile, median, third quartile and largest value.

Box plot A graphical summary of data. A box, drawn from the first to the third quartiles, shows the location of the middle 50% of the data. Dashed lines, extending from the ends of the box, show the location of data greater than the third quartile and data less than the first quartile. The locations of any outliers are also noted.

Fences Values used to identify outliers. Inner fences are located 1.5(IQR) below the first quartile and 1.5(IQR) above the third quartile. Outer fences are located 3(IQR) below the first quartile and 3(IQR) above the third quartile. Data falling between the inner and outer fences are considered mild outliers. Data falling outside the outer fences are considered extreme outliers.

Grouped data Data available in class intervals as summarized by a frequency distribution. Individual values of the original data are not recorded.

KEY FORMULAS

Sample Mean

$$\bar{x} = \frac{\sum_{i=1}^{n} x_i}{n} \tag{3.1}$$

Population Mean

$$\mu = \frac{\Sigma x_i}{N} \tag{3.2}$$

Interquartile Range

$$\text{IQR} = Q_3 - Q_1 \tag{3.3}$$

Population Variance

$$\sigma^2 = \frac{\Sigma(x_i - \mu)^2}{N} \tag{3.4}$$

Sample Variance

$$s^2 = \frac{\Sigma(x_i - \bar{x})^2}{n - 1} \tag{3.5}$$

Standard Deviation

$$\text{Sample Standard Deviation} = s = \sqrt{s^2} \qquad (3.6)$$

$$\text{Population Standard Deviation} = \sigma = \sqrt{\sigma^2} \qquad (3.7)$$

Coefficient of Variation

$$\text{CV} = \frac{\text{Standard Deviation}}{\text{Mean}} \times 100 \qquad (3.8)$$

z-Score

$$z_i = \frac{x_i - \bar{x}}{s} \qquad (3.9)$$

Sample Mean for Grouped Data

$$\bar{x} = \frac{\Sigma f_i M_i}{n} \qquad (3.10)$$

Sample Variance for Grouped Data

$$s^2 = \frac{\Sigma f_i (M_i - \bar{x})^2}{n - 1} \qquad (3.11)$$

Population Mean for Grouped Data

$$\mu = \frac{\Sigma f_i M_i}{N} \qquad (3.12)$$

Population Variance for Grouped Data

$$\sigma^2 = \frac{\Sigma f_i (M_i - \mu)^2}{N} \qquad (3.13)$$

REVIEW QUIZ

TRUE/FALSE

1. The mean for a set of data is found by adding all the data values and dividing by the number of items minus 1.
2. The mean and median can never be equal.
3. The median and the 50th percentile are the same.
4. The mode of a set of data is the value that occurs with greatest frequency.
5. The standard deviation is the average of the differences between the data values and the mean.
6. The variance is the square root of the standard deviation.
7. The interquartile range is the difference between the fourth and second quartiles.
8. The sample variance is the average of the deviations in a sample.
9. The coefficient of variation is a measure of the relative variability in a data set.
10. Fences are used to detect outliers in a data set.

MULTIPLE CHOICE

11. The mean for the sample data 6, 9, 10, 12, 13 is closest to
    a. 9.5
    b. 9.9

    c. 10.5

    d. 11.0

12. The median for the data set in Question 11 is

    a. 9.5

    b. 10

    c. 11

    d. 12

13. The standard deviation for the data set in Question 11 is closest to

    a. 3

    b. 5

    c. 8

    d. 10

14. The range for the data set in Question 11 is closest to

    a. 3

    b. 4

    c. 5

    d. 6

15. The coefficient of variation for the data set in Question 11 is closest to

    a. 27

    b. 2

    c. 5

    d. 20

16. The interquartile range for the data set in Question 11 is closest to

    a. 4

    b. 3

    c. 1

    d. 7

SUPPLEMENTARY EXERCISES

 PRICES

56. A sample of economists predicted what would happen to consumer prices during 1992 (*Business Week,* December 30, 1991). Their predictions about the percentage increase in consumer prices were as follows:

4.6	3.4	2.3	2.9	2.7	3.2	2.9
3.7	3.7	3.3	3.7	2.9	3.3	2.9
3.6	3.7	3.3	3.5	3.4	2.7	3.3
2.9	3.0	3.0	1.8			

    a. Compute the mean, median, and mode.

    b. Compute the first and third quartiles.

    c. Compute the range and interquartile range.

    d. Compute the variance and standard deviation.

    e. Are there any outliers?

57. In 1989 a sample of six home mortgage loans showed the following interest rates:

12.5, 13.2, 11.2, 13.0, 12.0, 12.5.

Compute the following descriptive statistics for the data set.

a. mean	**b.** median	**c.** mode
d. 25th percentile	**e.** range	**f.** interquartile range
g. variance	**h.** standard deviation	**i.** coefficient of variation

58. *Time* (January 9, 1989) published an article on the academic ability of college athletes. The article noted that some of the most successful athletic programs (citing the University of Notre Dame and Duke University) have athletes with very good college board scores.

Assume that the following sample data are typical of college board scores for Notre Dame football players:

1100, 970, 1000, 1250, 880, 790, 1300, 1050, 900, 950, 1120.

a. Compute the mean, median, and mode.
b. Compute the range and interquartile range.
c. Compute the variance and standard deviation.
d. Using z-scores, state whether there are any outliers in this data set.

 MORTGAGE **59.** The following data show home mortgage loan amounts handled by one loan officer at the Westwood Savings and Loan Association. Data are in thousands of dollars.

20.0	38.5	33.0	27.5	34.0	12.5	25.9	43.2	37.5	36.2
25.2	30.9	23.8	28.4	13.0	31.0	33.5	25.4	33.5	20.2
39.0	38.1	30.5	45.5	30.5	52.0	40.5	51.6	42.5	44.8

a. Find the mean, median, and mode.
b. Find the first and third quartiles.

60. *Newsweek* (January 9, 1989) reported statistics on the number of visits a couple makes to a therapist while undergoing marriage counseling. With data based on this article, assume that the following show the number of visits for a sample of nine couples:

12, 8, 3, 13, 18, 20, 10, 9, 18.

Compute the following descriptive statistics for these data.
a. mean   **b.** median   **c.** mode   **d.** 40th percentile
e. range **f.** variance **g.** standard deviation

61. A sample of 10 stocks on the New York Stock Exchange (May 22, 1992) shows the following price-earnings ratios:

9, 4, 6, 7, 3, 11, 4, 6, 4, 7.

Using these data, compute the mean, median, mode, range, variance, and standard deviation.

62. The Hawaii Visitors' Bureau reports visitors coming to Hawaii from the U.S. mainland spend an average of $102 per day (*St. Petersburg Times,* December 11, 1988). Assume the standard deviation is $27 and the distribution of expenditures is approximately bell-shaped. Answer the following:
a. What percentage of visitors will have an average daily expenditure of between $75 and $129 per day?
b. What percentage of visitors will have an average daily expenditure of between $48 and $156 per day?
c. Should an average daily expenditure of $225 per day be considered an outlier?

63. The cost of new homes in selected cities throughout the United States was reported in *U.S. News & World Report* (April 5, 1993). Assume the cost of homes in Richmond, Virginia, had a mean of $100,000 and a standard deviation of $40,000.
a. Should a home selling for $200,000 be considered an outlier? Explain.
b. Use Chebyshev's theorem to determine the percentage of homes selling between $40,000 and $160,000.

64. Public transportation and an automobile are two methods an employee has of getting to work each day. Samples of times recorded for each method are shown. Times are in minutes.

| *Public Transportation:* | 28 | 29 | 32 | 37 | 33 | 25 | 29 | 32 | 41 | 34 |
| *Automobile:* | 29 | 31 | 33 | 32 | 34 | 30 | 31 | 32 | 35 | 33 |

a. Compute the sample mean time to get to work for each method.
b. Compute the sample standard deviation for each method.

c. Based on your results from parts (a) and (b), which method of transportation should be preferred? Explain.

d. Develop a box plot for each method. Does a comparison of the box plots support your conclusion in part (c)?

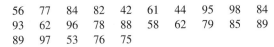 EXAM

65. Final examination scores for 25 statistics students are as follows:

56	77	84	82	42	61	44	95	98	84
93	62	96	78	88	58	62	79	85	89
89	97	53	76	75					

a. Provide a five-number summary.

b. Provide a box plot.

66. The following data show the total yardage accumulated during the NCAA college football season for a sample of 20 receivers.

744	652	576	1112	971	451	1023	852	809	596
941	975	400	711	1174	1278	820	511	907	1251

a. Provide a five-number summary.

b. Provide a box plot.

c. Identify any outliers.

67. A frequency distribution for the duration of 20 long-distance telephone calls (rounded to the nearest minute) is as follows:

Call Duration	Frequency
4–7	4
8–11	5
12–15	7
16–19	2
20–23	1
24–27	1
	Total 20

Compute the mean, variance, and standard deviation for the data.

68. Dinner check amounts at La Maison French Restaurant have the following frequency distribution:

Dinner Check (dollars)	Frequency
25–34	2
35–44	6
45–54	4
55–64	4
65–74	2
75–84	2
	Total 20

Compute the mean, variance, and standard deviation for the given data.

69. Automobiles traveling on the New York State Thruway are checked for speed by a state police radar system. A frequency distribution of speeds is shown:

Speed (miles per hour)	Frequency
45–49	10
50–54	40
55–59	150
60–64	175
65–69	75
70–74	15
75–79	10
Total	475

a. What is the mean speed of the automobiles traveling on the New York State Thruway?

b. Compute the variance and the standard deviation.

 DUKE

70. In the 1992 NCAA Division I Basketball Championships, the Duke Blue Devils became the first team since the 1973 UCLA Bruins to win back-to-back national championships. Duke's season record was 34–2, with its only losses coming at North Carolina and Wake Forest. The scores of the 36 games, with Duke's score listed first, follow (*NCAA Final Four Program,* April 1992).

Opponent	Score	Opponent	Score
East Carolina	103–75	Louisiana State	77–67
Harvard	118–65	Georgia Tech	71–62
St. John's	91–81	North Carolina State	71–63
Canisius	96–60	Maryland	91–89
Michigan	88–85	Wake Forest	68–72
William & Mary	97–61	Virginia	76–67
Virginia	68–62	UCLA	75–65
Florida State	86–70	Clemson	98–97
Maryland	83–66	North Carolina	89–77
Georgia Tech	97–84	Maryland	94–87
North Carolina State	110–75	Georgia Tech	89–76
N.C.—Charlotte	104–82	North Carolina	94–74
Boston University	95–85	Campbell	82–56
Wake Forest	84–68	Iowa	75–62
Clemson	112–73	Seton Hall	81–69
Florida State	75–62	Kentucky	104–103
Notre Dame	100–71	Indiana	81–78
North Carolina	73–75	Michigan	71–51

a. Compute the mean and median scores for Duke and its opponents.

b. Compute the range and interquartile range for Duke and its opponents.

c. Provide box plots for Duke and its opponents.

d. Comment on what you learned.

COMPUTER CASE 1: CONSOLIDATED FOODS, INC.

Consolidated Foods, Inc., operates a chain of supermarkets located in New Mexico, Arizona, and California. (See Computer Case, Chapter 2). Data in Table 3.4 show the dollar amounts and method of payment for a sample of 100 customers. Consolidated's management requested the sample be taken in an attempt to learn about payment practices for the store's customers. In particular, management was interested in learning about how a new credit-card payment option was related to the customers' purchase amounts.

REPORT

Use the methods of descriptive statistics presented in Chapter 3 to summarize the sample data. Provide summaries of the dollar purchase amounts for cash customers, personal-check customers, and credit-card customers separately. Your report should contain summaries and discussions such as the following:

1. a comparison and interpretation of means and medians
2. a comparison and interpretation of measures of dispersion such as the range and standard deviation
3. the identification and interpretation of the five-number summaries for each method of payment
4. box plots for each method of payment

 CONSOLID   Use the summary section of your report to discuss what you have learned about the method of payment and the amounts of payments for Consolidated Foods' customers. The data set for this computer case is in the data file CONSOLID.

TABLE 3.4   PURCHASE AMOUNT AND METHOD OF PAYMENT FOR A RANDOM SAMPLE OF 100 CONSOLIDATED FOODS' CUSTOMERS

Cash	Personal Check	Credit Card	Cash	Personal Check	Credit Card
7.40	27.60	50.30	5.08	52.87	69.77
5.15	30.60	33.76	20.48	78.16	48.11
4.75	41.58	25.57	16.28	25.96	
15.10	36.09	46.24	15.57	31.07	
8.81	2.67	46.13	6.93	35.38	
1.85	34.67	14.44	7.17	58.11	
7.41	58.64	43.79	11.54	49.21	
11.77	57.59	19.78	13.09	31.74	
12.07	43.14	52.35	16.69	50.58	
9.00	21.11	52.63	7.02	59.78	
5.98	52.04	57.55	18.09	72.46	
7.88	18.77	27.66	2.44	37.94	
5.91	42.83	44.53	1.09	42.69	
3.65	55.40	26.91	2.96	41.10	
14.28	48.95	55.21	11.17	40.51	
1.27	36.48	54.19	16.38	37.20	
2.87	51.66	22.59	8.85	54.84	
4.34	28.58	53.32	7.22	58.75	
3.31	35.89	26.57		17.87	
15.07	39.55	27.89		69.22	

COMPUTER CASE 2: NATIONAL HEALTH CARE ASSOCIATION

HEALTH-1

The National Health Care Association is concerned about the shortage of nurses projected for the future. To assess the current degree of job satisfaction among nurses, the association has sponsored a study of hospital nurses throughout the country. As part of this study, a sample of 50 nurses were asked to indicate their degree of satisfaction in their work, their pay, and their opportunities for promotion. Each of the three categories was measured on a scale from 0 to 100, with larger values indicating higher degrees of satisfaction. The data in Table 3.5 were collected and are available in the data set HEALTH-1.

HEALTH-2

In addition, the sample data were broken down by the types of hospitals employing the nurses and are given in Table 3.6. The types of hospitals considered were private (P), Veterans Administration (VA), and university (U). The data are also available in the data set HEALTH-2.

REPORT

Use methods of descriptive statistics to summarize these data. Present the summaries that will be beneficial in communicating the results to others. Discuss your findings. Specifically, comment on the following:

1. Using the entire data set and the three job-satisfaction variables, what aspect of the job is most satisfying for the nurses? What appears to be the least satisfying? In what area(s), if any, do you feel improvements should be made? Discuss.

TABLE 3.5   **WORK, PAY, AND PROMOTION JOB-SATISFACTION SCORES FOR A SAMPLE OF 50 NURSES**

Work	Pay	Promotion	Work	Pay	Promotion
71	49	58	72	76	37
84	53	63	71	25	74
84	74	37	69	47	16
87	66	49	90	56	23
72	59	79	84	28	62
72	37	86	86	37	59
72	57	40	70	38	54
63	48	78	86	72	72
84	60	29	87	51	57
90	62	66	77	90	51
73	56	55	71	36	55
94	60	52	75	53	92
84	42	66	74	59	82
85	56	64	76	51	54
88	55	52	95	66	52
74	70	51	89	66	62
71	45	68	85	57	67
88	49	42	65	42	68
90	27	67	82	37	54
85	89	46	82	60	56
79	59	41	89	80	64
72	60	45	74	47	63
88	36	47	82	49	91
77	60	75	90	76	70
64	43	61	78	52	72

TABLE 3.6   WORK, PAY, AND PROMOTION JOB-SATISFACTION SCORES FOR NURSES IN PRIVATE, VA, AND UNIVERSITY HOSPITALS

Private Hospitals			VA Hospitals			University Hospitals		
Work	Pay	Promotion	Work	Pay	Promotion	Work	Pay	Promotion
72	57	40	71	49	58	84	53	63
90	62	66	84	74	37	87	66	49
84	42	66	72	37	86	72	59	79
85	56	64	63	48	78	88	55	52
71	45	68	84	60	29	74	70	51
88	49	42	73	56	55	85	89	46
72	60	45	94	60	52	79	59	41
88	36	47	90	27	67	69	47	16
77	60	75	72	76	37	90	56	23
64	43	61	86	37	59	77	90	51
71	25	74	86	72	72	71	36	55
84	28	62	95	66	52	75	53	92
70	38	54	65	42	68	76	51	54
87	51	57	82	37	54	89	80	64
74	59	82	82	60	56			
89	66	62	90	76	70			
85	57	67	78	52	72			
74	47	63						
82	49	91						

2. Using descriptive measures of dispersion, what measure of job satisfaction appears to generate the greatest difference of opinion among the nurses? Explain.
3. What can be learned about the types of hospitals? Does any particular type of hospital seem to have better levels of job satisfaction than the other types of hospitals? Do your results suggest any recommendations for learning about and/or improving job satisfaction? Discuss.
4. What additional descriptive statistics and insights can you point to in terms of learning about and possibly improving job satisfaction?

CHAPTER 4

INTRODUCTION
TO PROBABILITY

WHAT YOU WILL LEARN IN THIS CHAPTER

- The meaning of probability
- Three methods used for assigning probabilities to experimental outcomes
- How to compute the probability of an event from the probabilities of experimental outcomes
- How to use the rules of probability to compute the probability of an event from known probabilities of related events
- The purpose and use of Bayes' theorem to revise probabilities
- The meaning of terms such as experiment, sample space, event, mutually exclusive event, independent event, and conditional probability

CONTENTS

Probability: What It Tells About the Typical American

Americans are the most widely researched people on earth. Every day, between 20,000 and 30,000 of us participate in some sort of survey. Based on these surveys, estimates are made of the probability that a "typical" American will engage in some event or activity; have certain likes, dislikes, habits, and beliefs; and so on. For example, the probability that a "typical" American believes he or she exercises enough is .43, and the probability that a "typical" American watches TV almost every evening is .63. Some other common events and their probability estimates are shown below.

Event	Probability
Goes to McDonald's every day	.05
Water skis	.07
Has high blood pressure	.12
Jogs	.13
Is underweight	.18
Never exercises at all	.25
Has never smoked	.46
Daydreams about being rich	.52
Has experienced ESP	.67
Lives in a metropolitan area	.77
Considers themselves happy	.90

Are any of these people "typical" Americans?

This *Statistics in Practice* is from *100% American* by Daniel Evan Weiss, New York: Poseidon Press, 1988.

Life is filled with uncertainties such as the following:

1. What is the chance that your car will be towed away if you park illegally in the visitors' parking lot?
2. What is the likelihood of receiving an A on the first exam if you wait until the night before to begin studying?
3. How likely is completing your college program on time if you take a part-time job that requires you to work 20 hours per week?
4. What are the chances you will get a 7 on the first roll of the dice at a casino in Las Vegas?

Probabilities are quite effective in dealing with such uncertainties. In everyday terminology, *probability* can be thought of as a numerical measure of the chance, or likelihood, that a particular event will occur. The probabilities for several events involving typical Americans are presented in the chapter-opening *Statistics in Practice*.

Probability values are always assigned on a scale of 0 to 1. A probability near 0 indicates that an event is very unlikely to occur; a probability near 1 indicates that an event is almost certain to occur. Other probabilities between 0 and 1 represent varying

FIGURE 4.1   **PROBABILITY AS A NUMERICAL MEASURE OF THE LIKELIHOOD OF AN EVENT'S OCCURRENCE**

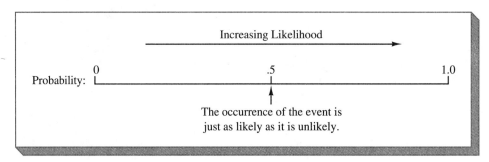

degrees of likelihood that an event will occur. Figure 4.1 depicts this view of probability as a numerical measure of the likelihood an event will occur.

EXAMPLE 4.1 ▶   If we consider the event "rain tomorrow," we understand that when the television weather report indicates a near-zero probability of rain, there is almost no chance of rain. However, if a .90 probability of rain is reported, we know that it is very likely or almost certain that rain will occur. A .50 probability indicates that rain is just as likely to occur as not.   ◀

4.1   EXPERIMENTS, COUNTING RULES, AND ASSIGNING PROBABILITIES

Using the terminology of probability, we define an *experiment* to be any process that generates well-defined outcomes. On any single repetition of an experiment, one and only one of the possible experimental outcomes will occur. Several examples of experiments and their associated outcomes are as follows:

Experiment	Experimental Outcomes
Toss a coin	Head, tail
Apply for a job	Hired, not hired
Roll a die	1, 2, 3, 4, 5, 6
Play a football game	Win, lose, tie

When we have specified all possible experimental outcomes, we have identified the *sample space* for the experiment.

> **Sample Space**
>
> The sample space for an experiment is the set of all experimental outcomes.

EXAMPLE 4.2 ▶ Consider the experiment of tossing a coin. The experimental outcomes are determined by the upward face of the coin—a head or a tail. If we let S denote the sample space, we can use the following notation to describe the sample space:

$$S = \{\text{Head, Tail}\}$$ ◀

EXAMPLE 4.3 ▶ Consider the experiment of rolling a die, with the experimental outcomes defined as the number of dots appearing on the upward face of the die. In this experiment, the numerical values 1, 2, 3, 4, 5, and 6 represent the possible experimental outcomes. We can denote the sample space for this experiment as follows:

$$S = \{1, 2, 3, 4, 5, 6\}$$ ◀

COUNTING RULES, COMBINATIONS, AND PERMUTATIONS

For some experiments, it is not easy to determine the number of experimental outcomes. But, being able to identify and count the experimental outcomes is a necessary step in assigning probabilities. We now discuss three counting rules that are useful.

EXAMPLE 4.4 ▶ Consider the experiment of tossing two coins. Let the experimental outcomes be defined in terms of the pattern of heads and tails appearing on the upward faces of the two coins. How many experimental outcomes are possible for this experiment? This can be thought of as a two-step experiment in which step 1 is the tossing of the first coin and step 2 is the tossing of the second coin. If we use H to denote a head and T to denote a tail, (H, H) indicates the experimental outcome with a head on the first coin and a head on the second coin. Continuing this notation, we can describe the sample space (S) for this coin-tossing experiment as follows:

$$S = \{(H, H), (H, T), (T, H), (T, T)\}$$

Thus, we see that there are four outcomes for this experiment. ◀

Example 4.4 describes an experiment consisting of two steps, in which there are two possible outcomes (head or tail) on the first step and two possible outcomes (head or tail) on the second step. Let us introduce a counting rule that is helpful in determining the number of outcomes for an experiment consisting of multiple steps.

A Counting Rule for Multiple-Step Experiments

If an experiment can be described as a sequence of k steps in which there are n_1 possible outcomes on the first step, n_2 possible outcomes on the second step, and so on, then the total number of experimental outcomes is given by $(n_1)(n_2) \ldots (n_k)$.

EXAMPLE 4.4
(CONTINUED) ▶ Looking at the experiment of tossing two coins as a sequence of first tossing one coin ($n_1 = 2$) and then tossing the other coin ($n_2 = 2$), we can see from the counting rule that there must be $(2)(2) = 4$ distinct experimental outcomes. As shown above, they are $S = \{(H, H), (H, T), (T, H), (T, T)\}$ ◀

A graphical device that is helpful in visualizing an experiment and enumerating outcomes in a multiple-step experiment is a *tree diagram.* Figure 4.2 shows a tree diagram for the coin-tossing experiment described in Example 4.4. The sequence of steps is depicted by moving from left to right through the tree. Step 1 corresponds to tossing the first coin, and there are two branches corresponding to the two possible outcomes. Step 2 corresponds to tossing the second coin, and for each possible outcome at step 1, there are two branches corresponding to the two possible outcomes at step 2. Finally, each of the points on the right-hand end of the tree corresponds to an experimental outcome. Each path through the tree from the leftmost node to one of the nodes at the right-hand side of the tree corresponds to a unique sequence of outcomes for each step.

A second counting rule that is often useful allows one to count the number of experimental outcomes when n objects are to be selected from a set of N objects. It is called the counting rule for combinations.

Counting Rule for Combinations

The number of combinations of N objects taken n at a time is as follows:

$$C_n^N = \binom{N}{n} = \frac{N!}{n!(N-n)!} \tag{4.1}$$

where

$$N! = N(N-1)(N-2)\ldots(2)(1)$$
$$n! = n(n-1)(n-2)\ldots(2)(1)$$

and

$$0! = 1$$

The notation ! means *factorial;* for example, 5 factorial is $5! = (5)(4)(3)(2)(1) = 120$. By definition, 0! is equal to 1.

FIGURE 4.2   **TREE DIAGRAM FOR THE EXPERIMENT OF TOSSING TWO COINS**

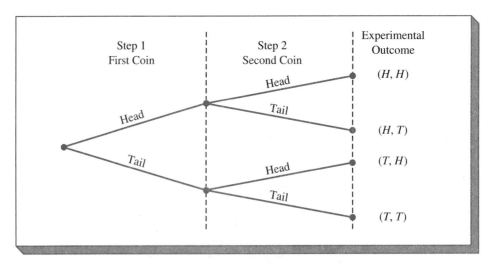

EXAMPLE 4.5 ▶

In a quality-control procedure, an inspector randomly selects two of five parts to test for defects. In a group of five parts, how many combinations of two parts may be selected? The counting rule in equation (4.1) shows that with $N = 5$ and $n = 2$, we have

$$\binom{5}{2} = \frac{5!}{2!(5-2)!} = \frac{(5)(4)(3)(2)(1)}{(2)(1)(3)(2)(1)} = \frac{120}{12} = 10$$

Thus, there are 10 outcomes for the experiment of randomly selecting two parts from a group of five. If we label the five parts as A, B, C, D, and E, the 10 combinations, or experimental outcomes, can be identified as AB, AC, AD, AE, BC, BD, BE, CD, CE, and DE. ◀

EXAMPLE 4.6 ▶

The State of Ohio lottery system uses the random selection of 6 numbers from a group of 47 numbers to determine the weekly lottery winner. The counting rule for combinations, equation (4.1), can be used to determine the number of ways 6 different numbers can be selected from a group of 47 numbers.

$$\binom{47}{6} = \frac{47!}{6!(47-6)!} = \frac{47!}{6!41!} = \frac{(47)(46)(45)(44)(43)(42)}{(6)(5)(4)(3)(2)(1)} = 10,737,573$$

The counting rule for combinations has told us that there are over 10 million experimental outcomes in the lottery drawing. If an individual buys a lottery ticket, there is 1 chance in 10,737,573 that the individual will win. ◀

A third counting rule that is sometimes useful is the counting rule for permutations. It allows one to compute the number of experimental outcomes when n objects are to be selected from a set of N objects where the order of selection is important. The same n objects selected in a different order is considered a different experimental outcome.

Counting Rule for Permutations

The number of permutations of N objects taken n at a time is given by

$$P_n^N = n!\binom{N}{n} = \frac{N!}{(N-n)!} \tag{4.2}$$

The counting rule for permutations is closely related to the one for combinations; however, there are more permutations than combinations for the same number of objects. This is because for every selection of n objects there are $n!$ different ways to order them.

EXAMPLE 4.5 ▶
(CONTINUED)

The quality-control inspector selects two of five parts to inspect for defects. How many permutations may be selected? The counting rule in equation (4.2) shows that with $N = 5$ and $n = 2$, we have

$$P_2^5 = \frac{5!}{(5-2)!} = \frac{5!}{3!} = \frac{(5)(4)(3)(2)(1)}{(3)(2)(1)} = (5)(4) = 20$$

Thus, there are 20 outcomes for the experiment of randomly selecting two parts from a group of five when the order of selection must be taken into account. If we label the parts A, B, C, D, and E, the 20 permutations are AB, BA, AC, CA, AD, DA, AE, EA, BC, CB,

BD, DB, BE, EB, CD, DC, CE, EC, DE, ED. Note that this is twice as many as the number of combinations because each distinct selection of two parts can be done in two different orders. ◀

EXAMPLE 4.6 ▶
(CONTINUED)

Suppose the State of Ohio changes its rules so that to win the lottery you must not only select the correct 6 numbers out of 47, but you must also select them in the correct order. The counting rule for permutations can be used to compute the number of experimental outcomes (permutations of 6 numbers out of 47).

$$P_6^{47} = \frac{47!}{(47 - 6)!} = (47)(46)(45)(44)(43)(42) = 7,731,052,560$$

There is now only one chance in over 7 billion to win. The odds are stacked much more in favor of the state when the numbers must be selected in the proper order. ◀

ASSIGNING PROBABILITIES

Now let us see how probabilities can be assigned to experimental outcomes. The three approaches most frequently used are the classical, relative frequency, and subjective methods. Regardless of the method used, the probabilities assigned must satisfy two basic requirements.

Basic Requirements for Assigning Probabilities

1. The probability assigned to each experimental outcome must be between 0 and 1, inclusively. If we let E_i denote the ith experimental outcome and $P(E_i)$ be its probability, then this requirement can be written as

$$0 \le P(E_i) \le 1 \text{ for all } i \tag{4.3}$$

2. The sum of the probabilities for all the experimental outcomes must equal 1. If there are n experimental outcomes, this requirement can be written as

$$P(E_1) + P(E_2) + \cdots + P(E_n) = 1 \tag{4.4}$$

The *classical method* of assigning probabilities is appropriate when all the experimental outcomes are equally likely. If there are n experimental outcomes, a probability of $1/n$ is assigned to each experimental outcome. When using this approach, the two basic requirements for assigning probabilities are automatically satisfied.

EXAMPLE 4.2 ▶
(CONTINUED)

For the experiment of tossing a fair coin, the two experimental outcomes—head and tail—are equally likely. Therefore, since one of the two equally likely outcomes is a head, the probability of observing a head is ½, or .50. Similarly, the probability of observing a tail is also ½, or .50. ◀

EXAMPLE 4.3 ▶
(CONTINUED)

For the experiment of rolling a die, it would also seem reasonable to conclude that the six possible outcomes are equally likely, and hence each outcome is assigned a probability of ⅙. If $P(1)$ denotes the probability that one dot appears on the upward face of the die, then $P(1) = ⅙$. Similarly, $P(2) = ⅙$, $P(3) = ⅙$, $P(4) = ⅙$, $P(5) = ⅙$, and $P(6) = ⅙$. ◀

EXAMPLE 4.4
(CONTINUED) ▶

Recall that there are four experimental outcomes for the experiment involving the tossing of two coins. Since it seems reasonable to assume that the four possible experimental outcomes are equally likely, we assign a probability of ¼ to each; thus, the probability of experimental outcome (*H, H*) is ¼, the probability of experimental outcome (*H, T*) is ¼, the probability of experimental outcome (*T, H*) is ¼, and the probability of experimental outcome (*T, T*) is ¼. ◀

The *relative frequency method* of assigning probabilities is appropriate when data are available to estimate the proportion of the time the experimental outcome will occur when the experiment is repeated a large number of times.

EXAMPLE 4.7 ▶

As part of a study of the X-ray department for a local hospital, the number of patients waiting for service at 9:00 A.M. was recorded for 20 successive days. The following results were obtained.

Number Waiting	Number of Days Outcome Occurred
0	2
1	5
2	6
3	4
4	3
	Total 20

These data show that on 2 of the 20 days, zero patients were waiting for service; on 5 of the days, one patient was waiting for service; and so on. Using the relative frequency method, we would assign a probability of $2/20 = .10$ to the experimental outcome of zero patients waiting for service, $5/20 = .25$ to the experimental outcome of one patient waiting, $6/20 = .30$ to two patients waiting, $4/20 = .20$ to three patients waiting, and $3/20 = .15$ to four patients waiting. ◀

EXAMPLE 4.8 ▶

In the test market evaluation of a new product, 400 potential customers were contacted; 100 actually purchased the product, but 300 did not. In effect, then, we have repeated an experiment of contacting a customer 400 times and have found that the product was purchased 100 times. Thus, using the relative frequency approach, we assign a probability of $100/400 = .25$ to the experimental outcome of a potential customer purchasing the product. Similarly, $300/400 = .75$ is assigned to the experimental outcome of a potential customer not purchasing the product. ◀

As examples 4.7 and 4.8 illustrate, the two basic requirements of probability are also satisfied automatically when the relative frequency approach is used.

The *subjective method* of assigning probabilities is most appropriate when it is unrealistic to assume that the experimental outcomes are equally likely and when little relevant data are available. When the subjective method is used to assign probabilities to the experimental outcomes, we may use any information available, such as our experience or intuition. After considering all available information, a probability value that expresses our *degree of belief* that the experimental outcome will occur is specified. Since subjective probability expresses a person's degree of belief, it is personal. Using the subjective method, different people can be expected to assign different probabilities to the same experimental outcome.

When using the subjective probability assignment method, extra care must be taken to ensure that requirements (4.3) and (4.4) are satisfied. Regardless of a person's degree of belief, the probability value assigned to each experimental outcome must be between 0 and 1, inclusive, and the sum of all the experimental outcome probabilities must equal 1.

EXAMPLE 4.9 ▶ Consider the next football game that the Pittsburgh Steelers will play. What is the probability that the Steelers will win? The experimental outcomes of win, lose, and tie are obviously not equally likely. Also, since the teams involved will not have played several times previously in the same year, there are no relative frequency data available relevant to the game. Thus, if we want an estimate of the probability of the Steelers' winning, we must use the subjective method and state a value that expresses our degree of belief that they will win.   ◀

EXAMPLE 4.10 ▶ Tom and Judy Elsbernd have just made an offer to purchase a house. Two outcomes are possible:

$$E_1 = \text{their offer is accepted}$$

$$E_2 = \text{their offer is rejected}$$

Judy believes that the probability their offer will be accepted is .8; thus, Judy would set $P(E_1) = .8$ and $P(E_2) = .2$. Tom, however, believes that the probability that their offer will be accepted is .6; hence, Tom would set $P(E_1) = .6$ and $P(E_2) = .4$. Note that Tom's probability estimate for E_1 reflects the fact that he is a bit more pessimistic than Judy is about their offer being accepted.

Both Judy and Tom have assigned probabilities that satisfy the two requirements for assigning probabilities. The fact that their probability estimates are different emphasizes the personal nature of the subjective method.   ◀

NOTES AND COMMENTS ▼

1. In statistics, the notion of an experiment is somewhat different than the notion of an experiment in the physical sciences. In the physical sciences, an experiment is usually conducted in a laboratory or a controlled environment in order to learn about a scientific occurrence. In statistical experiments, the outcomes are determined by probability. Even though the experiment is repeated in exactly the same way, an entirely different outcome may occur. Because of this influence of probability on the outcome, the experiments of statistics are sometimes called *random experiments*.

2. When drawing a random sample without replacement from a population of size N, the counting rule for combinations is used to find the number of different samples of size n that can be selected.

EXERCISES

METHODS

1. Consider the experiment of drawing 1 card from a deck of 52 playing cards and observing whether the card is red or black.
   a. What are the experimental outcomes?
   b. Assign probabilities to the experimental outcomes.

2. Consider the experiment of drawing 1 card from a deck of 52 playing cards and observing whether the card is a spade.
   a. What is the probability of drawing a spade?
   b. What is the probability of not drawing a spade?

3. Consider the experiment of tossing a coin three times.
   a. Develop a tree diagram for the experiment.
   b. List the experimental outcomes.
   c. What is the probability for each outcome?

4. An experiment has three steps with three outcomes possible for the first step, two outcomes possible for the second step, and four outcomes possible for the third step. How many experimental outcomes exist for the entire experiment?

SELF TEST

5. How many ways can three items be selected from a group of six items? Use the letters A, B, C, D, E, and F to identify the items, and list each of the different combinations of three items.

6. How many permutations of three items can be selected from a group of three?

7. How many permutations of three items can be selected from a group of six? Use the letters A, B, C, D, E, and F to identify the items, and list each of the permutations when the three items (B, D, F) are selected.

8. Consider the experiment of rolling a pair of dice. Each die has six possible results (the number of dots on its face).
   a. How many outcomes are possible for this experiment?
   b. Show a tree diagram for the experiment.
   c. How many experimental outcomes provide a sum of 7 for the dots on the dice?

APPLICATIONS

9. Consider the experiment of administering a true–false exam consisting of 10 questions. Each different sequence of answers is an experimental outcome.
   a. How many experimental outcomes are there?
   b. If a student guesses on every question, what is the probability of any particular experimental outcome?

10. A major resort in Florida is concerned about weather conditions in the Northeast as well as in Florida. In characterizing the temperature in both areas, the following three categories are used: below average, average, or above average. A combination of below-average temperatures in the Northeast with above-average temperatures in Florida means an increased volume of business for the resort. For the experiment of observing the weather conditions on a particular day, answer the following questions.
    a. How many experimental outcomes are possible?
    b. Develop a tree diagram for this experiment.

11. In the city of Milford, applications for zoning changes go through a two-step process: a review by the planning commission and a final decision by the city council. At step 1 the planning commission will review the zoning change request and make a positive or negative recommendation concerning the change. At step 2 the city council will review the

planning commission's recommendation and then vote to approve or to disapprove the zoning change. An application for a zoning change has just been submitted by the developer of an apartment complex. Consider the application process as an experiment.

   a. How many sample points are there for this experiment? List the sample points.

   b. Construct a tree diagram for the experiment.

12. An investor has two stocks: stock A and stock B. Each stock may increase in value, decrease in value, or remain unchanged. Consider the experiment of investing in the two stocks and observing the change (if any) in value.

   a. How many experimental outcomes are possible?

   b. Show a tree diagram for the experiment.

   c. How many of the experimental outcomes result in an increase in value for at least one of the two stocks?

   d. How many of the experimental outcomes result in an increase in value for both of the stocks?

13. Many states design their automobile license plates such that space is available for up to six letters or numbers.

   a. If a state decides to use only numerical values for the license plates, how many different license plate numbers are possible? Assume that 000000 is an acceptable license plate number, although it will be used only for display purposes at the license bureau. (*Hint:* Use the counting rule for multiple-step experiments.)

   b. If the state decides to use two letters followed by four numbers, how many different license plate numbers are possible? Assume that the letters I and O will not be used because of their similarity to numbers 1 and 0.

   c. Would larger states, such as New York or California tend to use more or fewer letters in license plates? Explain.

SELF TEST ▷ **14.** Simple random sampling uses a sample of size n from a population of size N to obtain data that can be used to make inferences about the characteristics of a population. Suppose we have a population of 50 bank accounts and wish to take a random sample of 4 accounts in order to learn about the population. How many different random samples of 4 accounts are possible?

15. A company that manufactures toothpaste has five different package designs they want to study. Assuming that one design is just as likely to be preferred by a consumer as any other design, what probability would you assign to a randomly selected consumer preferring each of the package designs? In an actual experiment, 100 consumers were asked to pick the design they preferred. The following data were obtained.

Design	1	2	3	4	5
Total	5	15	30	40	10

Do the data appear to confirm the belief that one design is just as likely to be selected as another? Explain.

SELF TEST ▷ **16.** In a survey of new matriculants to MBA programs ("School Selection by Students," *GMAC Occasional Papers,* March 1988), the following data were obtained on the marital status of the students.

Marital status	Frequency
Never married	1106
Married	826
Other (separated, widowed, divorced)	106
Total	2038

Consider the experiment of interviewing a new MBA student and recording her or his marital status. Show your probability assignments.

17. The final exam in a course in contemporary science resulted in the following grades.

Grade	A	B	C	D	F
Number	7	12	16	5	3

    a. What is the probability that a randomly selected student received an A?
    b. What is the probability that a randomly selected student received a C?

18. A small-appliance store in Madeira has collected the following data on refrigerator sales for the last 50 weeks.

Number of Refrigerators Sold	Number of Weeks
0	6
1	12
2	15
3	10
4	5
5	2
	Total 50

Suppose that we are interested in the experiment of observing the number of refrigerators sold in 1 week of store operations.
    a. How many experimental outcomes are there?
    b. Which approach would you recommend for assigning probabilities to the experimental outcomes?
    c. Assign probabilities to the experimental outcomes, and verify that your assignments satisfy the two basic requirements.

19. Strom Construction has made a bid on two contracts. The owner has identified the possible outcomes and subjectively assigned probabilities as follows.

Experimental Outcome	Obtain Contract 1	Obtain Contract 2	Probability
1	Yes	Yes	.15
2	Yes	No	.15
3	No	Yes	.30
4	No	No	.25

    a. Are these valid probability assignments? Why or why not?
    b. If not, what would have to be done to make the probability assignments valid?

20. Faced with the question of determining the probability of obtaining either 0 heads, 1 head, or 2 heads when flipping a coin twice, an individual argued that it seems reasonable to treat the outcomes as equally likely and that the probability of each event is $\frac{1}{3}$. Do you agree? Explain.

4.2   EVENTS AND THEIR PROBABILITIES

In the introduction to this chapter we used the term *event* much as it would be used in everyday conversation. We now introduce the formal definition of an *event* as it relates to probability.

Event

An *event* is a subset of the sample space.

This means that an event is a collection of experimental outcomes. The sample space consists of all the experimental outcomes and is itself an event. Note also that a subset of the sample space can consist of just one experimental outcome; in such cases, we refer to the resulting event as a *simple event.*

EXAMPLE 4.3
(CONTINUED)

Recall that for the experiment of rolling a die, the sample space was $S = \{1, 2, 3, 4, 5, 6\}$. There are six simple events, each of which corresponds to one of the experimental outcomes.

$$E_1 = \{1\} \quad E_4 = \{4\}$$

$$E_2 = \{2\} \quad E_5 = \{5\}$$

$$E_3 = \{3\} \quad E_6 = \{6\}$$

If we define A to be the event that an even number of dots appears on the upward face of the die, then we can describe event A as follows:

$$A = \{2, 4, 6\}$$

Similarly, if B is the event that an odd number of dots appears on the upward face of the die, then

$$B = \{1, 3, 5\}$$

We see that events A and B are simply different subsets of the sample space.   ◀

EXAMPLE 4.11

A cab company has analyzed its operating records for the past 20 days. On 8 of these days, no vehicle breakdowns were observed; on 6 of the days, one cab had a breakdown; on 3 days, there were two breakdowns; on 2 days, there were three breakdowns; and on 1 of the days, four cabs had breakdowns. Let

$$S = \{0, 1, 2, 3, 4\}$$

denote the sample space for the experiment of observing the number of cab breakdowns on a day. The numerical values 0, 1, 2, 3, and 4 denote the number of breakdowns (the experimental outcomes). If A is defined as the event that two or more vehicle breakdowns are observed on a typical day, then

$$A = \{2, 3, 4\}$$

If B is defined as the event that less than two breakdowns are observed, then

$$B = \{0, 1\}$$

Similarly, if C is defined as the event that no breakdowns are observed, then

$$C = \{0\}$$

We see that the event C consists of just one experimental outcome; thus, C is a simple event. ◀

Given the probabilities of the experimental outcomes, we can use the following definition to compute the probability of any event.

Probability of an Event

The probability of an event is equal to the sum of the probabilities of the experimental outcomes in the event.

Using this definition, we calculate the probability of a particular event by adding the probabilities of the experimental outcomes that make up the event.

EXAMPLE 4.3
(CONTINUED)

Recall that for the experiment of rolling a die, we used the classical approach to assign probabilities of $P(1) = \frac{1}{6}, P(2) = \frac{1}{6}, \ldots, P(6) = \frac{1}{6}$. Thus to compute the probability of an even number, event $A = \{2, 4, 6\}$, we sum the probabilities of the experimental outcomes 2, 4, and 6:

$$P(A) = P(2) + P(4) + P(6)$$

$$= \frac{1}{6} + \frac{1}{6} + \frac{1}{6} = \frac{3}{6} = \frac{1}{2}$$

Similarly, for an odd number, $B = \{1, 3, 5\}$,

$$P(B) = P(1) + P(3) + P(5)$$

$$= \frac{1}{6} + \frac{1}{6} + \frac{1}{6} = \frac{3}{6} = \frac{1}{2}$$ ◀

EXAMPLE 4.11
(CONTINUED)

Using the relative frequency method, we can use the data on taxicab breakdowns provided to estimate the probability of a specific number of breakdowns on a day selected at random. The probabilities assigned to the experimental outcomes are shown below.

Number of Breakdowns	Number of Occurrences	Probability
0	8	8/20 = .40
1	6	6/20 = .30
2	3	3/20 = .15
3	2	2/20 = .10
4	1	1/20 = .05
Totals	20	1.00

Thus, the probability of two or more breakdowns, event $A = \{2, 3, 4\}$, is given by

$$P(A) = P(2) + P(3) + P(4)$$
$$= .15 + .10 + .05 = .30$$

Similarly, the probability of less than two breakdowns, event $B = \{0, 1\}$, is

$$P(B) = P(0) + P(1)$$
$$= .40 + .30 = .70$$

Finally, we see that for no breakdowns, event $C = \{0\}$, $P(C) = P(0) = .40$.

NOTES AND COMMENTS

1. As noted, the sample space is itself an event. Since it contains all the experimental outcomes, it has a probability of 1; that is, $P(S) = 1$.
2. When the classical method is used to assign probabilities, the assumption is that the experimental outcomes are equally likely. In such cases, the probability of any event of interest can be computed by dividing the number of experimental outcomes in the event by the total number of experimental outcomes. For instance, recall Example 4.3 where $A = \{2, 4, 6\}$. Since we assumed that the experimental outcomes are equally likely and since the total number of experimental outcomes is 6, $P(A) = \frac{3}{6} = .5$.

EXERCISES

METHODS

21. An experiment has three outcomes with $P(E_1) = .35$, $P(E_2) = .40$, $P(E_3) = .25$.
    a. What is the probability that E_1 or E_2 occurs?
    b. What is the probability that E_1 or E_3 occurs?
    c. What is the probability that E_1, E_2, or E_3 occurs?

22. An experiment has four equally likely outcomes (E_1, E_2, E_3, and E_4).
    a. What is the probability that E_2 occurs?
    b. What is the probability that any two of the outcomes occurs (e.g., E_1 or E_3)?
    c. What is the probability that any three of the outcomes occurs (e.g., E_1 or E_2 or E_4)?

SELF TEST

23. Consider the experiment of selecting a card from a deck of 52 cards. Each card corresponds to an experimental outcome with a 1/52 probability.
    a. List the experimental outcomes in the event an ace is selected.
    b. List the experimental outcomes in the event a club is selected.
    c. List the experimental outcomes in the event a face card (jack, queen, or king) is selected.
    d. Find the probabilities associated with each of the events in parts (a), (b), and (c).

24. Consider the experiment of rolling a pair of dice. Suppose that we are interested in the sum of the face values showing on the dice.
    a. How many experimental outcomes are possible? (*Hint:* Use the counting rule for multiple-step experiments.)
    b. List the experimental outcomes.
    c. What is the probability of obtaining a value of 7?
    d. What is the probability of obtaining a value of 9 or greater?
    e. Since there are six possible even values (2, 4, 6, 8, 10, and 12) and only five possible odd values (3, 5, 7, 9, and 11), the dice should show even values more often than odd values. Do you agree with this statement? Explain.
    f. What method did you use to assign the probabilities requested above?

APPLICATIONS

25. Suppose that a manager of a large apartment complex provides the subjective probability estimates shown below concerning the number of vacancies that will exist next month.

Vacancies	Probability
0	.05
1	.15
2	.35
3	.25
4	.10
5	.10

List the experimental outcomes in each of the following events and provide the probability of the event:

a. no vacancies   **b.** at least four vacancies   **c.** two or fewer vacancies

26. The manager of a furniture store sells from 0 to 4 china hutches each week. Based on past experience, the following probabilities are assigned to sales of 0, 1, 2, 3, or 4 hutches:

$$P(0) = .08$$
$$P(1) = .18$$
$$P(2) = .32$$
$$P(3) = .30$$
$$P(4) = \underline{.12}$$
$$1.00$$

a. Are these valid probability assignments? Why or why not?
b. Let A be the event that two or fewer are sold in one week. Find $P(A)$.
c. Let B be the event that four or more are sold in one week. Find $P(B)$.

27. A sample of 100 customers of Montana Gas and Electric resulted in the following frequency distribution of monthly charges.

Amount ($)	0–49	50–99	100–149	150–199	200–249
Number	13	22	34	26	5

a. Let A be the event that monthly charges are $150 or more. Find $P(A)$.
b. Let B be the event that monthly charges are less than $150. Find $P(B)$.

28. A survey of 50 students at Tarpon Springs College regarding the number of extracurricular activities resulted in the data shown below.

Number of activities	0	1	2	3	4	5
Frequency	8	20	12	6	3	1

a. Let A be the event that a student participates in at least one activity. Find $P(A)$.
b. Let B be the event that a student participates in three or more activities. Find $P(B)$.
c. What is the probability a student participates in exactly two activities?

29. A telephone survey was used to determine viewer response to a new television show. The following data were obtained.

Rating	Poor	Below average	Average	Above average	Excellent
Frequency	4	8	11	14	13

    a. What is the probability that a randomly selected viewer rates the new show as average or better?

    b. What is the probability that a randomly selected viewer rates the new show below average or worse?

30. A bank has observed that credit-card account balances have been growing over the past year. A sample of 200 customer accounts resulted in the following data.

Amount Owed (\$)	0–99	100–199	200–299	300–399	400–499	500–599
Frequency	62	46	24	30	26	12

    a. Let A be the event that a customer's balance is less than \$200. Find $P(A)$.

    b. Let B be the event that a customer's balance is \$300 or more. Find $P(B)$.

4.3 ▽ RULES FOR COMPUTING EVENT PROBABILITIES

Any time we can identify all the experimental outcomes and assign the corresponding probabilities, we can use the definition of the previous section to compute the probability of an event of interest. However, in many experiments the number of experimental outcomes is large, and their identification—as well as the determination of their associated probabilities—becomes extremely cumbersome, if not impossible. In this and the remaining sections of this chapter, we present rules that can often be used to compute the probability of an event without knowledge of the probability of each experimental outcome. These probability relationships require a knowledge of the probabilities for related events. Probabilities of events are then computed directly from the related event probabilities using one or more of the probability rules.

COMPLEMENT OF AN EVENT

Given an event A, the *complement* of A is defined to be the event consisting of all experimental outcomes that are *not* in A. The complement of A is denoted by A^c. Figure 4.3 provides a diagram known as a *Venn diagram*, which illustrates the concept of an event and its complement. The rectangular area represents the sample space for the experiment and, as such, contains all possible experimental outcomes. The circle

---

FIGURE 4.3   COMPLEMENT OF EVENT A

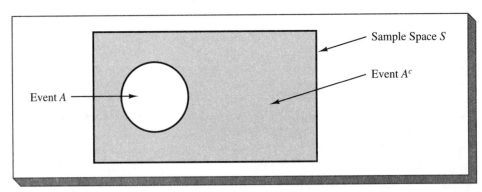

represents event A and contains only the experimental outcomes in A. The shaded region of the diagram contains all experimental outcomes not in event A, which is the definition of event A^c, the complement of A.

In any probability application, event A and its complement A^c must satisfy

$$P(A) + P(A^c) = 1 \tag{4.5}$$

Solving for $P(A)$, we obtain the following result.

Computing Probability Using the Complement

$$P(A) = 1 - P(A^c) \tag{4.6}$$

Equation (4.6) shows that the probability of an event A can easily be computed if the probability of its complement, $P(A^c)$, is known.

EXAMPLE 4.12 ▶ Based on an analysis of student records, the placement director at a university states that 98% of the students who interview with a particular firm are not given a job offer. Letting A denote the event of a job offer and A^c denote the event of no job offer, the placement director is stating that $P(A^c) = .98$. Using equation (4.6), we see that

$$P(A) = 1 - P(A^c) = 1 - .98 = .02$$

This shows that there is a .02 probability that a student who interviews with the firm will receive a job offer.   ◀

EXAMPLE 4.13 ▶ A purchasing agent states that there is a .90 probability that a supplier will send a shipment that is free of defective parts. Using the complement, we can conclude that there is a $1 - .90 = .10$ probability that the shipment will contain at least one defective part.   ◀

UNION AND INTERSECTION OF EVENTS

We are often interested in the events obtained by combining two or more other events. To begin, let us consider the *union* of two events.

Union of Two Events

Given two events A and B, the *union* of A and B is the event containing all experimental outcomes belonging to A *or* B *or both*. The union is denoted by $A \cup B$.

The Venn diagram shown in Figure 4.4 depicts the union of events A and B. Note that the shaded region contains all experimental outcomes in event A as well as all experimental outcomes in event B. The fact that the circles overlap indicates that there are some experimental outcomes contained in both A and B.

FIGURE 4.4   **UNION OF EVENTS A AND B**

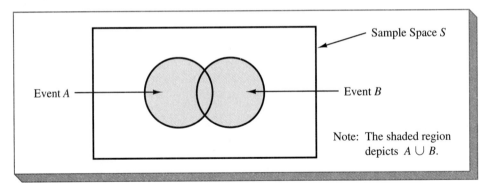

The *intersection* of two events is defined as follows.

Intersection of Two Events

Given two events A and B, the *intersection of A and B* is the event containing the experimental outcomes belonging to *both A and B*. The intersection is denoted by $A \cap B$.

A Venn diagram depicting the intersection of the events A and B is shown in Figure 4.5. The area where the two circles overlap is the intersection; it is an event containing the experimental outcomes that are in both A and B.

ADDITION RULE

The *addition rule* is used to compute the probability of the union of two events.

Addition Rule

$$P(A \cup B) = P(A) + P(B) - P(A \cap B) \qquad \textbf{(4.7)}$$

FIGURE 4.5   **INTERSECTION OF EVENTS A AND B**

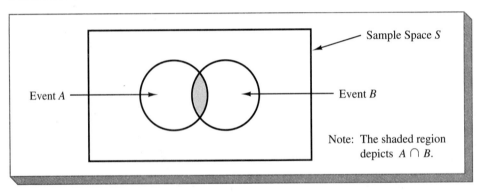

To obtain an intuitive understanding of the addition rule, refer to Figure 4.4; note that the first two terms in the addition rule, $P(A) + P(B)$, account for all the experimental outcomes in A and all the experimental outcomes in B. However, since the experimental outcomes in the intersection, $A \cap B$, are in both A and B, when we compute $P(A) + P(B)$ we are in effect counting each of the experimental outcomes in $A \cap B$ twice. The addition rule corrects for this by subtracting $P(A \cap B)$.

EXAMPLE 4.14 ▶ As an illustration of the addition rule, consider the following grades obtained in an introductory psychology course. Of 200 students taking the course, 160 passed the midterm exam and 140 passed the final exam; 124 students passed both exams. Letting

$$M = \text{event of passing the midterm exam}$$
$$F = \text{event of passing the final exam}$$
$$M \cap F = \text{event of passing both exams}$$

the given relative frequency information leads to the following event probabilities:

$$P(M) = \frac{160}{200} = .80$$

$$P(F) = \frac{140}{200} = .70$$

$$P(M \cap F) = \frac{124}{200} = .62$$

After reviewing the grades, the professor of the course decides to give a passing grade to any student who passed at least one of the two exams. That is, a passing grade will be given to any student who passes the midterm, to any student who passes the final, and to any student who passes both exams. What is the probability of receiving a passing grade in this course?

While your first reaction may be to try to count how many of the 200 students passed at least one exam, that information is not available; even if it were, the counting process would be tedious. However, note that the question concerns the union of the events M and F. That is, we want to know the probability a student passes the midterm (M), passes the final (F), or both. Thus, we want to know $P(M \cup F)$. Here is where the addition rule can be helpful.

$$P(M \cup F) = P(M) + P(F) - P(M \cap F)$$

Knowing the three probabilities on the right-hand side of the above equation, we can write

$$P(M \cup F) = .80 + .70 - .62 = .88$$

Thus, there is a .88 probability of passing the course because there is a .88 probability of passing at least one of the exams.   ◀

EXAMPLE 4.15 ▶ A study involving the television-viewing habits of married couples found that 30% of the husbands and 20% of the wives were regular viewers of a particular Friday evening program. Both husband and wife were regular viewers in 12% of the households. What is the probability that at least one of the spouses is a regular viewer of the program?

Letting

$$H = \text{husband is a regular viewer}$$
$$W = \text{wife is a regular viewer}$$

we have $P(H) = .30$, $P(W) = .20$, and $P(H \cap W) = .12$. Using the addition rule, we have

$$P(H \cup W) = P(H) + P(W) - P(H \cap W) = .30 + .20 - .12 = .38$$

This shows there is a .38 probability that at least one of the spouses is a regular viewer of the program.

MUTUALLY EXCLUSIVE EVENTS

Let us now see how the addition rule is applied to *mutually exclusive events*. First, we define mutually exclusive events.

> ### Mutually Exclusive Events
>
> Two or more events are said to be *mutually exclusive* if the events do not have any experimental outcomes in common.

Another term for mutually exclusive events is *disjoint* events. Events A and B are said to be mutually exclusive (disjoint) if when one event occurs the other cannot occur. Thus, if A and B are mutually exclusive, then their intersection contains no experimental outcome. A Venn diagram for the mutually exclusive events A and B is shown in Figure 4.6. Since $P(A \cap B) = 0$, the addition rule for mutually exclusive events can be shortened as follows.

> ### Addition Rule for Mutually Exclusive Events
>
> $$P(A \cup B) = P(A) + P(B) \qquad\qquad (4.8)$$

EXAMPLE 4.16 ▶ If A denotes the event that you get a grade of A for a term paper and B denotes the event that you get a grade of B, then A and B are mutually exclusive events. Thus if $P(A) = .20$ and $P(B) = .50$, then

$$P(A \cup B) = P(A) + P(B)$$

$$= .20 + .50 = .70 \qquad ◀$$

FIGURE 4.6   MUTUALLY EXCLUSIVE EVENTS

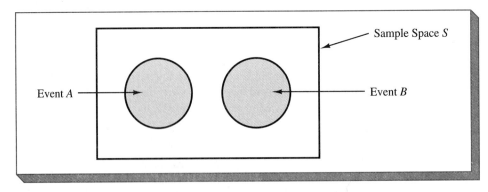

NOTES AND
COMMENTS

1. When working probability problems, some students have difficulty in determining whether the problem involves the computation of the union of two events or the intersection of two events. As a general rule of thumb, if the problem statement uses the word "or" or the expression "at least one," the problem is asking you to compute the union; if the problem statement uses the word "and," the problem is asking you to compute the intersection.
2. The addition rule for mutually exclusive events follows directly from the definition of the probability of an event. Since for mutually exclusive events none of the experimental outcomes are in more than one event, we are summing the probabilities of the experimental outcomes when we sum the event probabilities.
3. Any event and its complement are mutually exclusive events.

EXERCISES

METHODS

31. Suppose $P(A) = .13$. Find $P(A^c)$.

32. Suppose $P(A) = .70$, $P(B) = .20$, and $P(A \cap B) = .15$.
a. Find $P(A \cup B)$.
b. Find $P(B^c)$.
c. Find $P(A \cap B^c)$.

SELF TEST

33. Suppose that we have a sample space $S = \{E_1, E_2, E_3, E_4, E_5, E_6, E_7\}$, where $E_1, E_2, \ldots, E_7$ denote the experimental outcomes. The following probability assignments apply:

$$
\begin{aligned}
P(E_1) &= .05 \\
P(E_2) &= .20 \\
P(E_3) &= .20 \\
P(E_4) &= .25 \\
P(E_5) &= .15 \\
P(E_6) &= .10 \\
P(E_7) &= \underline{.05} \\
& 1.00
\end{aligned}
$$

Let

$$
\begin{aligned}
A &= \{E_1, E_4, E_6\} \\
B &= \{E_2, E_4, E_7\} \\
C &= \{E_2, E_3, E_5, E_7\}
\end{aligned}
$$

a. Find $P(A)$, $P(B)$, and $P(C)$.
b. Find $A \cup B$ and $P(A \cup B)$.
c. Find $A \cap B$ and $P(A \cap B)$.
d. Are events A and C mutually exclusive?
e. Find B^c and $P(B^c)$.

34. Let

$$
\begin{aligned}
M &= \text{person interviewed is male} \\
F &= \text{person interviewed is female} \\
D &= \text{person believes the man should always pay on first date} \\
W &= \text{person believes the one who requests the date should pay} \\
E &= \text{person believes the one who earns more money should pay on first date}
\end{aligned}
$$

As reported in *Money* magazine (November 1988), $P(M) = .61$, $P(F) = .39$, $P(D) = .41$, $P(W) = .41$, $P(E) = .02$, and $P(D \cap M) = .29$.
a. Are M and D mutually exclusive events?
b. Find $P(D \cup M)$.
c. Find $P(D \cup W)$.
d. Find $P(D \cap F)$.
e. Find $P(D \cup F)$.

35. Consider an experiment where eight possible outcomes exist. We will denote the experimental outcomes as $E_1, E_2, \ldots, E_8$. Suppose the following events are defined:

$$A = (E_1, E_2, E_3, E_5)$$
$$B = (E_2, E_4, E_5, E_8)$$

Note that experimental outcomes E_6 and E_7 are in neither event A nor event B. List the experimental outcomes making up the following events:
a. $A \cup B$
b. $A \cap B$
c. A^c
d. Are A and B mutually exclusive events? Explain.

APPLICATIONS

36. The public accounting firm Grant Thornton conducted a survey to see how executives felt about the 1992 recession and the potential for recovery (*Journal of Accountancy,* February 1992). The probability that an executive indicated a recession existed was .74. If we were to choose one of the executives, what is the probability that she or he would indicate a recession does not exist?

37. A pharmaceutical company conducted a study to evaluate the effectiveness of an allergy relief medicine; 250 patients with symptoms that included itchy eyes and a skin rash were given the new drug. The results of the study were as follows: 90 of the patients treated experienced eye relief, 135 had their skin rash clear up, and 45 experienced relief both from itchy eyes and the skin rash. What is the probability that a patient who takes the drug will experience relief for at least one of the two symptoms?

38. In a study of 100 students who had been awarded university scholarships, it was found that 40 had part-time jobs, 25 had made the dean's list the previous semester, and 15 had both a part-time job and had made the dean's list. What was the probability that a student had a part-time job or was on the dean's list?

39. A survey of the subscribers to *Fortune* magazine showed that 46% have mutual funds, 63% have money market funds, and 74% have mutual funds and/or money market funds (*Fortune Subscriber Portrait,* 1988). What is the probability a subscriber will have investments in both money market and mutual funds? What is the probability a subscriber will not have investments in either type of fund?

 SELF TEST

40. The survey of subscribers to *Fortune* magazine referred to in Exercise 39 (*Fortune Subscriber Portrait,* 1988) showed 54% rented a car in the past 12 months for business reasons, 51% rented a car for personal reasons, and 72% rented a car for either business or personal reasons.
a. What is the probability a subscriber rented a car in the past 12 months for business reasons and for personal reasons?
b. What is the probability a subscriber did not rent a car in the past 12 months?
c. What is the probability a subscriber rented a car for business reasons only during the past 12 months?

41. Let

A = the event that a person runs 5 miles or more per week
B = the event that a person dies of heart disease
C = the event that a person dies of cancer

Further, suppose that $P(A) = .01$, $P(B) = .25$, and $P(C) = .20$.

a. Are events A and B mutually exclusive? Can you find $P(A \cap B)$?

b. Are events B and C mutually exclusive? Find the probability that a person dies of heart disease or cancer.

c. Find the probability that a person dies from causes other than cancer.

42. During winter in Cincinnati, Mr. Krebs experiences difficulty in starting his two cars. The probability that the first car starts is .80, and the probability that the second car starts is .40. There is a probability of .30 that both cars start.

a. Define the events involved, and use probability notation to show the probability information given above.

b. What is the probability that at least one car starts?

c. What is the probability that Mr. Krebs cannot start either of the two cars?

43. Let A be an event that a person's primary method of commuting to work is an automobile and B be an event that a person's primary method of commuting is a bus. Suppose that in a large city we find $P(A) = .45$ and $P(B) = .35$.

a. Are events A and B mutually exclusive? What is the probability that a person uses an automobile or a bus in going to and from work?

b. Find the probability that a person's primary method of commuting is something other than a bus.

44. From *Statistics in Practice* at the beginning of the chapter, we can define the following events and probabilities.

$$H = \text{person has high blood pressure}$$
$$E = \text{person never exercises at all}$$
$$S = \text{person has never smoked}$$
$$P(H) = .12, P(E) = .25, P(S) = .46$$

a. Are any of these events mutually exclusive?

b. What is the probability a randomly selected person has smoked?

c. Is the probability that a person never exercises at all or has high blood pressure greater than .40? Explain.

4.4 CONDITIONAL PROBABILITY, INDEPENDENCE, AND THE MULTIPLICATION RULE

CONDITIONAL PROBABILITY

Often, the probability of an event is influenced by whether or not a related event has occurred. For instance, most people believe that the probability of an accident increases if the driver has been drinking. In such cases, the concept of *conditional probability* can be used to compute the probability of one event given that another related event is known to have occurred.

EXAMPLE 4.17 ▶ A major metropolitan police force in the eastern United States consists of 1200 officers—960 men and 240 women. Over the past 2 years, 324 officers on the police force have been awarded promotions. The breakdown of promotions for male and female officers is shown in Table 4.1.

After reviewing the data in Table 4.1, a committee of female officers charged discrimination on the basis that 288 male officers had received promotions, whereas only 36 female officers had received promotions. The police administration countered with the

TABLE 4.1   **PROMOTIONAL STATUS OF POLICE OFFICERS OVER THE PAST 2 YEARS**

	Male	**Female**	**Totals**
Promoted	288	36	324
Not Promoted	672	204	876
Totals	960	240	1200

argument that the relatively low number of promotions for female officers was not due to discrimination but due to the fact that there are fewer female officers on the police force.

After reflecting on this situation, we see that the real issue involves not the number of promotions but the probability of promotion given that the officer is male or that the officer is female.   ◀

In dealing with conditional probabilities, our interest is in computing probabilities that can be used to answer questions such as those raised in Example 4.17. Suppose that we have an event A with probability denoted by $P(A)$. If we should learn that a related event, B, has occurred, we would want to take advantage of this additional information in computing the probability for event A.

The probability of event A *given* that another event B is known to have occurred is written $P(A \mid B)$. The vertical line, |, between A and B is used to denote the fact that we are considering the probability of event A *given* the condition that event B has occurred. The notation $P(A \mid B)$ is read "the probability of A *given* B." Shown below are the mathematical definitions for the conditional probabilities of A given B and of B given A.

Conditional Probability

$$P(A \mid B) = \frac{P(A \cap B)}{P(B)} \qquad (4.9)$$

or

$$P(B \mid A) = \frac{P(A \cap B)}{P(A)} \qquad (4.10)$$

EXAMPLE 4.17
(CONTINUED)   ▶   Returning to the police department discrimination case, we let

M = event that a randomly selected officer is male
F = event that a randomly selected officer is female
A = event that a randomly selected officer is promoted

Dividing the data values in Table 4.1 by the total of 1200 officers permits us to summarize the available information in the following probability values:

$$P(M \cap A) = \frac{288}{1200} = .24 = \text{probability that an officer is male } and \text{ is promoted}$$

$$P(M \cap A^c) = \frac{672}{1200} = .56 = \text{probability that an officer is male } and \text{ is not promoted}$$

$$P(F \cap A) = \frac{36}{1200} = .03 = \text{probability that an officer is female } and \text{ is promoted}$$

$$P(F \cap A^c) = \frac{204}{1200} = .17 = \text{probability that an officer is female } and \text{ is not promoted}$$

Since each of these values gives the probability of the intersection of two events, the probabilities are given the name of *joint probabilities*. Table 4.2, which provides a summary of the probability information for the police officer promotion situation, is referred to as a *joint probability table*.

The values in the margins of the joint probability table provide the probabilities of each event separately. That is, $P(M) = .80$, $P(F) = .20$, $P(A) = .27$, and $P(A^c) = .73$. Thus, we see that 80% of the force is male, 20% of the force is female, 27% of all officers received promotions, and 73% were not promoted. These probabilities are referred to as *marginal probabilities* because of their location in the margins of the joint probability table.

The conditional probabilities of relevance in the discrimination charge are $P(A \mid M)$, the probability of an officer being promoted given the officer is male, and $P(A \mid F)$, the probability that an officer is promoted given the officer is female. Using the joint and marginal probabilities, we can apply the definition of conditional probability to find the probability of promotion given a male officer and the probability of promotion given a female officer.

$$P(A \mid M) = \frac{P(A \cap M)}{P(M)} = \frac{.24}{.80} = .30$$

$$P(A \mid F) = \frac{P(A \cap F)}{P(F)} = \frac{.03}{.20} = .15$$

We see that the probability of promotion is twice as great for male officers. These conditional probabilities do not necessarily prove that the female officers have been discriminated against, but they do support the female officers' argument. ◀

EXAMPLE 4.18 ▶ A research study concerning the relationship between smoking and heart disease in men over 50 years old led to the finding that 10% of the men smoked and had experienced heart disease. Furthermore, it was known that 30% of the men in the study were smokers. Let

$$H = \text{has experienced heart disease}$$
$$S = \text{smoker}$$

TABLE 4.2   **JOINT PROBABILITY TABLE FOR PROMOTION OF POLICE OFFICERS**

	Male (*M*)	Female (*F*)	Totals
Promoted (*A*)	.24	.03	.27
Not Promoted (*A^c*)	.56	.17	.73
Totals	.80	.20	1.00

Using equation (4.9), we can compute the conditional probability of a man over 50 experiencing heart disease given that he smokes.

$$P(H \mid S) = \frac{P(H \cap S)}{P(S)} = \frac{.10}{.30} = .33$$

◀

INDEPENDENT EVENTS

In Example 4.17 involving promotional practices for male and female police officers, we saw that $P(A) = .27$, $P(A \mid M) = .30$, and $P(A \mid F) = .15$. As you will recall, the events were defined as follows: A = promotion, M = male officer, and F = female officer. These data show that the probability of a promotion (event A) is affected, or influenced by, whether the officer is male or female. In particular, since $P(A \mid M) \neq P(A)$, we say events A and M are *dependent* events. That is, the probability of event A (promotion) is influenced by knowing whether or not M (the officer is male) occurs. Similarly, with $P(A \mid F) \neq P(A)$, we say events A and F are *dependent* events. On the other hand, if the probability of event A is not affected by the occurrence of event M—that is, $P(A \mid M) = P(A)$—we say events A and M are *independent* events. This leads us to the following definition of the independence of two events.

Independent Events

Two events A and B are independent if

$$P(A \mid B) = P(A) \qquad (4.11)$$

or

$$P(B \mid A) = P(B) \qquad (4.12)$$

Otherwise, the events are dependent.

EXAMPLE 4.19 ▶ Suppose that $P(A) = .30$, $P(B) = .25$, and $P(A \mid B) = .20$. Are events A and B independent? Since $P(A \mid B) \neq P(A)$, the events are dependent. ◀

MULTIPLICATION RULE

Recall that the addition rule is used to compute the probability of the union of two events. We now show how the *multiplication rule* can be used to find the probability of the intersection of two events (i.e., the joint probability of the two events). The multiplication rule follows from the definition of conditional probability. Using equations (4.9) and (4.10) and solving for $P(A \cap B)$, we obtain the multiplication rule.

Multiplication Rule

$$P(A \cap B) = P(B)P(A \mid B) \qquad (4.13)$$

or

$$P(A \cap B) = P(A)P(B \mid A) \qquad (4.14)$$

EXAMPLE 4.20 ▶ A newspaper circulation department knows that 84% of its customers subscribe to the daily edition of the paper. Letting D denote the event that a customer subscribes to the daily edition, we set $P(D) = .84$. In addition, it is known that the probability that a customer already holding a daily subscription also subscribes to the Sunday edition (event S) is .75; that is, $P(S \mid D) = .75$. What is the probability that a customer subscribes to both the daily and Sunday editions of the newspaper? Using the multiplication rule, we compute the desired probability, $P(D \cap S)$, as follows:

$$P(D \cap S) = P(D)P(S \mid D) = (.84)(.75) = .63$$

This tells us that 63% of the newspaper's customers take both the daily and Sunday editions.   ◀

Before concluding this section, let us consider the special case of the multiplication rule for independent events. Recall that events A and B are independent whenever $P(A \mid B) = P(A)$ or $P(B \mid A) = P(B)$. Applying equations (4.13) and (4.14) for the special case of independent events, we obtain equation (4.15).

Multiplication Rule for Independent Events

$$P(A \cap B) = P(A)P(B) \qquad\qquad \textbf{(4.15)}$$

Thus to compute the probability of the intersection of two independent events, we simply multiply the probabilities of the two events. The multiplication rule for independent events provides another means of determining whether two events are independent. For instance, if $P(A \cap B) = P(A)P(B)$, then A and B are independent; if $P(A \cap B) \neq P(A)P(B)$, then A and B are dependent.

EXAMPLE 4.21 ▶ A service station manager knows from past experience that 80% of the customers use a credit card when purchasing gasoline. What is the probability that the next two customers purchasing gasoline will both use credit cards? If we let

A = event that the first customer uses a credit card
B = event that the second customer uses a credit card

then the event of interest is $A \cap B$. It seems reasonable to assume that A and B are independent events. Thus, $P(A \cap B) = P(A)P(B) = (.80)(.80) = .64$.   ◀

NOTES AND COMMENTS

Do not confuse the notion of mutually exclusive events with that of independent events. Two events with nonzero probabilities cannot be both mutually exclusive and independent. If one mutually exclusive event is known to occur, the probability of the other occurring is reduced to zero. They are therefore dependent.

▷

EXERCISES

METHODS

SELF TEST ▷ 45. Suppose that we have two events A and B with $P(A) = .5$, $P(B) = .60$, and $P(A \cap B) = .40$.
   a. Find $P(A \mid B)$.   **b.** Find $P(B \mid A)$.   **c.** Are A and B independent? Why or why not?

46. Let $P(A) = .60$, $P(B) = .45$, and $P(A \cap B) = .30$.
  a. Find $P(A \cup B)$.
  b. Find $P(A \mid B)$.
  c. Find $P(B \mid A)$.
  d. Are events A and B independent?
  e. Are events A and B mutually exclusive?

47. Assume that we have two events, A and B, that are mutually exclusive. Assume further that it is known that $P(A) = .30$ and $P(B) = .40$.
  a. What is $P(A \cap B)$?
  b. What is $P(A \mid B)$?
  c. A student in statistics argues that the concepts of mutually exclusive events and independent events are really the same and that, if events are mutually exclusive, they must be independent. Do you agree with this statement? Use the probability information in this problem to justify your answer.
  d. What general conclusion would you make about mutually exclusive and independent events given the results of this problem?

APPLICATIONS

48. A Daytona Beach nightclub has the following data on the age and marital status of 140 customers.

		Marital Status	
		Single	*Married*
Age	*Under 30*	77	14
	30 or Over	28	21

  a. Develop a joint probability table using these data.
  b. Use the marginal probabilities to comment on the age of customers attending the club.
  c. Use the marginal probabilities to comment on the marital status of customers attending the club.
  d. What is the probability of finding a customer who is single and under the age of 30?
  e. If a customer is under 30, what is the probability that he or she is single?
  f. Is marital status independent of age? Explain, using probabilities.

SELF TEST **49.** In a survey of MBA students, the following data were obtained on "Students' first reason for application to the school to which they matriculated" ("School Selection by Students," *GMAC Occasional Papers,* Stolzenberg and Giarrusso, March 1988).

		Reason for Application			
		School Quality	*School Cost or Convenience*	*Other*	**Totals**
Enrollment	*Full time*	421	393	76	890
Status	*Part time*	400	593	46	1039
Totals		821	986	122	1929

  a. Develop a joint probability table using these data.
  b. Use the marginal probabilities of school quality, cost/convenience, and other to comment on the most important reason for choosing a school.

  c. If a student goes full time, what is the probability school quality will be the first reason for choosing a school?

  d. If a student goes part time, what is the probability school quality will be the first reason for choosing a school?

  e. Let A be the event that a student is full-time and let B be the event that the student lists school quality as the first reason for applying. Are events A and B independent? Justify your answer.

50. A survey of automobile ownership was conducted for 200 families in Houston. The results of the study showing ownership of automobiles of United States and foreign manufacture are summarized as follows.

		Do You Own a U.S. Car?		
		Yes	*No*	**Totals**
Do you own	*Yes*	30	10	40
a foreign car?	*No*	150	10	160
Totals		180	20	200

  a. Show the joint probability table for the above data.

  b. Use the marginal probabilities to compare U.S. and foreign car ownership.

  c. What is the probability that a family will own both a U.S. car and a foreign car?

  d. What is the probability that a family owns a car, U.S. or foreign?

  e. If a family owns a U.S. car, what is the probability that it also owns a foreign car?

  f. If a family owns a foreign car, what is the probability that it also owns a U.S. car?

  g. Are U.S. and foreign car ownership independent events? Explain.

51. The probability that Ms. Smith will get an offer on the first job she applies for is .5, and the probability that she will get an offer on the second job she applies for is .6. She thinks that the probability that she will get an offer on both jobs is .15.

  a. Define the events involved, and use probability notation to state the probability information given above.

  b. What is the probability that Ms. Smith gets an offer on the second job given that she receives an offer for the first job?

  c. What is the probability that Ms. Smith gets an offer on at least one of the jobs she applies for?

  d. What is the probability that Ms. Smith does not get an offer on either of the two jobs she applies for?

  e. Are the job offers independent? Explain.

52. Shown below are data from a sample of 80 families in a midwestern city. The data shows the record of college attendance by fathers and their oldest sons.

		Son	
		Attended College	*Did Not Attend College*
Father	*Attended College*	18	7
	Did Not Attend College	22	33

  a. Show the joint probability table.

  b. Use the marginal probabilities to comment on the comparison between fathers and sons in terms of attending college.

    c. What is the probability that a son attends college given that his father attended college?

    d. What is the probability that a son attends college given that his father did not attend college?

    e. Is attending college by the son independent of whether or not his father attended college? Explain, using probability values.

53. The Texas Oil Company provides a limited partnership arrangement whereby small investors can pool resources in order to invest in large-scale oil exploration programs. In the exploratory drilling phase, locations for new wells are selected based on the geologic structure of the proposed drilling sites. Experience shows that there is a .40 probability of a type A structure present at the site given a productive well. It is also known that 50% of all wells are drilled in locations with type A structure. Finally, 30% of all wells drilled are productive.

    a. What is the probability of a well being drilled in a type A structure *and* being productive?

    b. If the drilling process begins in a location with a type A structure, what is the probability of having a productive well at the location?

    c. Is finding a productive well independent of the type A geologic structure? Explain.

54. The Grant Thornton public accounting firm conducted a survey to see how executives felt about the 1992 recession and recovery potential (*Journal of Accountancy,* February 1992). Results showed that the probability of an executive indicating the existence of a recession was .74. The probability of an executive stating that both a recession existed and a recovery would occur within 6 months was .41.

    a. Given that an executive indicated the existence of a recession, what is the probability that the executive felt that a recovery would occur within 6 months?

    b. Assume that if an executive denied the existence of a recession, he or she also believed that a recovery had already begun. Construct a joint probability table for "recession," "no recession," and "recovery," "no recovery." Those who feel a recovery is already underway and those who feel one will occur in 6 months should be put in the same category for this table.

    c. Use the joint probability table in part (b) to find the marginal probability of a recovery.

55. A purchasing agent has placed a rush order for a particular raw material with two different suppliers, A and B. If neither order arrives in 4 days, the production process must be shut down until at least one of the orders arrives. The probability that supplier A can deliver the material in 4 days is .55. The probability that supplier B can deliver the material in 4 days is .35.

    a. What is the probability that both suppliers deliver the material in 4 days? Since two separate suppliers are involved, we are willing to assume independence.

    b. What is the probability that at least one supplier delivers the material in 4 days?

    c. What is the probability the production process is shut down in 4 days because of a shortage in raw material (i.e., both orders are late)?

56. In a 1992 study of the consumer's view of the economy, the probability that a consumer would buy a house during the year was .033 and the probability that a consumer would buy a car during the year was .168 (*U.S. News & World Report,* April 13, 1992). Assume that there was only a .004 probability that a consumer would buy a house and a car during the year.

    a. What is the probability that a consumer would buy either a car or a house during the year?

    b. What is the probability that a consumer would buy a car during the year given that the consumer purchased a house during the year?

    c. Are buying a car and buying a house independent events? Explain.

4.5 ▽ BAYES' THEOREM

The discussion of conditional probability suggests that it is possible to revise or update probabilities given new information. Often, we begin a probability analysis with *prior probability* estimates for specific events of interest. Then, from sources such as a sample, a special report, a product test, and so on, we obtain additional information affecting the probability of the events. Given this new information, we want to revise or update the prior probability values. The updated, or revised, probabilities for the events are referred to as *posterior* probabilities. *Bayes' theorem* provides a means for computing posterior probabilities. The steps of the Bayesian probability revision process are shown in Figure 4.7.

EXAMPLE 4.22 ▶ A manufacturing firm receives 65% of its parts from one supplier and 35% from a second supplier. The quality of the purchased parts varies with the supplier. Table 4.3 shows the percentages of good and defective parts received from the two suppliers. Let A_1 denote the event a part comes from supplier 1 and A_2 the event that a part comes from supplier 2. If we let G denote the event that a part is good and D denote the event that a part is defective, the information in Table 4.3 leads to the following conditional probability values:

$$P(G \mid A_1) = .98 \qquad P(D \mid A_1) = .02$$

$$P(G \mid A_2) = .95 \qquad P(D \mid A_2) = .05$$

Furthermore, given the percentage of parts received from each supplier, if a part is selected at random, the probability it came from supplier 1 is $P(A_1) = .65$ and the probability it came from supplier 2 is $P(A_2) = .35$.

FIGURE 4.7   PROBABILITY REVISION USING BAYES' THEOREM

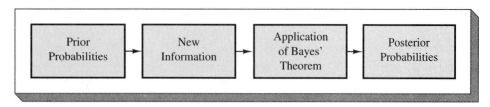

TABLE 4.3   PERCENTAGES OF GOOD AND DEFECTIVE PARTS RECEIVED FROM SUPPLIERS

Supplier	Good Parts (%)	Defective Parts (%)
1	98	2
2	95	5

Suppose that the manufacturing firm has just had a machine breakdown due to a defective part and wants to determine the probability that the part came from supplier 1 and the probability that the part came from supplier 2. Thus, it is desired to compute the posterior probabilities $P(A_1 \mid D)$ and $P(A_2 \mid D)$ given the new information, D, that the part is defective. Since $P(A_1 \cap D)$ and $P(A_2 \cap D)$ are not known, we cannot use the conditional probability rule for these calculations. This is a problem in which Bayes' theorem is needed; we return to make the calculations shortly. ◀

Bayes' theorem can only be used to compute posterior (revised) probabilities when the events of interest are mutually exclusive and *collectively exhaustive*. Two events are mutually exclusive and collectively exhaustive if (1) when one event occurs, the other cannot and (2) the sum of probabilities for the events equals 1. This means that exactly one of the events must occur. An event and its complement will always be mutually exclusive and collectively exhaustive.

Suppose we have two mutually exclusive and collectively exhaustive events, A_1 and A_2, and know the prior probabilities, $P(A_1)$ and $P(A_2)$. If it becomes known that a related event B has occurred, the following formulas, known as Bayes' theorem, can be used to compute the posterior probabilities $P(A_1 \mid B)$ and $P(A_2 \mid B)$.

Bayes' Theorem

$$P(A_1 \mid B) = \frac{P(B \mid A_1)P(A_1)}{P(B \mid A_1)P(A_1) + P(B \mid A_2)P(A_2)} \qquad (4.16)$$

and

$$P(A_2 \mid B) = \frac{P(B \mid A_2)P(A_2)}{P(B \mid A_1)P(A_1) + P(B \mid A_2)P(A_2)} \qquad (4.17)$$

EXAMPLE 4.22
(CONTINUED)

Recall that we are interested in finding the posterior probabilities that the defective part came from supplier 1, $P(A_1 \mid B)$, and from supplier 2, $P(A_2 \mid B)$. Since the part came from exactly one of the suppliers, the events A_1 and A_2 are mutually exclusive and collectively exhaustive. Using the prior probabilities $P(A_1) = .65$ and $P(A_2) = .35$ and the conditional probabilities $P(D \mid A_1) = .02$ and $P(D \mid A_2) = .05$, we can utilize Bayes' theorem to compute the probabilities in question. Using equation (4.16) and using event D as the related event that has occurred, we find

$$P(A_1 \mid D) = \frac{P(D \mid A_1)P(A_1)}{P(D \mid A_1)P(A_1) + P(D \mid A_2)P(A_2)}$$

$$= \frac{(.02)(.65)}{(.02)(.65) + (.05)(.35)} = \frac{.0130}{.0130 + .0175}$$

$$= \frac{.0130}{.0305} = .426$$

Using equation (4.17), we find $P(A_2 \mid D)$ as follows:

$$P(A_2 \mid D) = \frac{(.05)(.35)}{(.02)(.65) + (.05)(.35)}$$

$$= \frac{.0175}{.0130 + .0175} = \frac{.0175}{.0305} = .574$$

We initially had a probability of .65 that a part selected at random was from supplier 1. However, given the information that the part is defective, the probability that the part is from supplier 1 drops to .426. Thus, if the part is defective, there is a better than 50–50 chance that the part came from supplier 2; that is, $P(A_2 \mid D) = .574$. ◀

THE TABULAR APPROACH

A tabular approach helpful in organizing and conducting the Bayes' theorem calculations is shown in Table 4.4 for the data presented in Example 4.22. The computations shown in that table are conducted in the following manner.

Step 1 Prepare the following three columns.

■ Column 1: The list of the mutually exclusive and collectively exhaustive events that can occur in the problem.
■ Column 2: The prior probabilities for the events. Note that since the events are mutually exclusive and collectively exhaustive, the probabilities in column 2 must sum to 1.
■ Column 3: The conditional probabilities of the new information (event D in example 4.22) *given* each of the mutually exclusive events in column 1.

Step 2 In column 4, compute the joint probabilities for each mutually exclusive event and the event providing new information. These joint probabilities are found by multiplying the values in column 2 by the corresponding values in column 3. For Example 4.22, we obtain $P(A_1 \cap D) = P(A_1)P(D \mid A_1)$ and $P(A_2 \cap D) = P(A_2)P(D \mid A_2)$.

Step 3 Add the joint probability column (column 4) to find the probability of the event representing new information. We see that in Example 4.22 there is a .0130 probability of a defective part from supplier 1 and a .0175 probability of a defective part from supplier 2. Since these are the only two ways a defective part can be obtained, the sum .0130 + .0175 = .0305 shows there is an overall probability of .0305 of finding a defective part from the combined shipments of both suppliers.

Step 4 In column 5, compute the posterior probabilities using the basic relationship of conditional probability. For Example 4.22, this is

TABLE 4.4 **TABULAR APPROACH TO BAYES' THEOREM CALCULATIONS FOR THE TWO-SUPPLIER PROBLEM**

Column 1 A_i	Column 2 $P(A_i)$	Column 3 $P(D \mid A_i)$	Column 4 $P(A_i \cap D)$	Column 5 $P(A_i \mid D)$
A_1	.65	.02	.0130	$\frac{.0130}{.0305} = .426$
A_2	.35	.05	.0175	$\frac{.0175}{.0305} = .574$
	1.00		$P(D) = .0305$	1.000

$$P(A_1 \mid D) = \frac{P(A_1 \cap D)}{P(D)} = \frac{.0130}{.0305} = .426$$

$$P(A_2 \mid D) = \frac{P(A_2 \cap D)}{P(D)} = \frac{.0175}{.0305} = .574$$

The joint probabilities $P(A_i \cap D)$ are found in column 4, whereas the probability $P(D)$ appears as the sum of column 4.

NOTES AND COMMENTS

1. The Reverend Thomas Bayes (1702–1761), a Presbyterian minister, is credited with the original work leading to the version of Bayes' theorem in use today.
2. An event and its complement are mutually exclusive and collectively exhaustive since their union is the entire sample space. Thus, Bayes' theorem is always applicable for computing posterior probabilities of an event and its complement.
3. Bayes' theorem can be extended to the case of n events and is used widely in the field of decision analysis. These topics are beyond the scope of this text.

EXERCISES

METHODS

SELF TEST

57. The prior probabilities for events A_1 and A_2 are $P(A_1) = .40$ and $P(A_2) = .60$. It is also known that $P(A_1 \cap A_2) = 0$. Suppose $P(B \mid A_1) = .20$ and $P(B \mid A_2) = .05$.
    a. Are A_1 and A_2 mutually exclusive? Why or why not?
    b. Compute $P(A_1 \cap B)$ and $P(A_2 \cap B)$.
    c. Compute $P(B)$.
    d. Apply Bayes' theorem to compute $P(A_1 \mid B)$ and $P(A_2 \mid B)$.

58. The prior probability for event A is $P(A) = .3$. Event B is related to event A. It is known that $P(B \mid A) = .7$ and $P(B \mid A^c) = .5$.
    a. Find $P(A^c)$.
    b. Find $P(A \cup A^c)$.
    c. Find $P(A \mid B)$.
    d. Find $P(A^c \mid B)$.
    e. Find $P(B)$.

APPLICATIONS

59. A consulting firm has submitted a bid for a large research project. The firm's management initially felt there was a 50–50 chance of getting the bid. However, the agency to which the bid was submitted has subsequently requested additional information on the bid. Past experience indicates that on 75% of the successful bids and 40% of the unsuccessful bids the agency requested additional information.
    a. What is your prior probability the bid will be successful (i.e., prior to receiving the request for additional information)?
    b. What is the conditional probability of a request for additional information given that the bid will ultimately be successful?
    c. Compute a posterior probability that the bid will be successful given that a request for additional information has been received.

SELF TEST ▷ **60.** A local bank is reviewing its credit-card policy with a view toward recalling some of its credit cards. In the past, approximately 5% of cardholders have defaulted, and the bank has been unable to collect the outstanding balance. Thus management has established a prior probability of .05 that any particular cardholder will default. The bank has further found that the probability of missing one or more monthly payments for those customers who do not default is .20. Of course, the probability of missing one or more payments for those who default is 1.

    a. Given that a customer has missed a monthly payment, compute the posterior probability that the customer will default.

    b. The bank would like to recall its card if the probability that a customer will default is greater than .20. Should the bank recall its card if the customer misses a monthly payment? Why or why not?

61. In a major eastern city, 60% of the automobile drivers are 30 years of age or older, and 40% of the drivers are under 30 years of age. Of all drivers 30 years of age or older, 4% will have a traffic violation in a 12-month period. Of all drivers under 30 years of age, 10% will have a traffic violation in a 12-month period. Assume that a driver has just been charged with a traffic violation; what is the probability that the driver is under 30 years of age?

62. A certain college football team plays 55% of its games at home and 45% of its games away. Given that the team has a home game, there is a .80 probability that it will win. Given that the team has an away game, there is a .65 probability that it will win. If the team wins on a particular Saturday, what is the probability that the game was played at home?

63. *M.D. Computing* (Vol. 8, No. 5, 1991) describes the use of Bayes' theorem and the use of conditional probability in medical diagnosis. Prior probabilities of diseases are based on the physician's assessment of such things as geographical location, seasonal influence, occurrence of epidemics, and so forth. Assume that a patient is believed to have one of two diseases, denoted D_1 and D_2, with $P(D_1) = .60$ and $P(D_2) = .40$ and that medical research has shown there is a probability associated with each symptom that may accompany the diseases. Suppose that, given diseases D_1 and D_2, the probabilities a patient will have symptoms S_1, S_2, or S_3 are as follows.

		Symptoms		
		S_1	S_2	S_3
Disease	D_1	.15	.10	.15 ◀——— $P(S_3 \mid D_1)$
	D_2	.80	.15	.03

After finding a certain symptom is present, the medical diagnosis may be aided by finding the revised probabilities the patient has each particular disease. Compute the posterior probabilities of each disease given the following medical findings:

    a. The patient has symptom S_1.

    b. The patient has symptom S_2.

    c. The patient has symptom S_3.

    d. For the patient with symptom S_1 in part (a), suppose we also find symptom S_2 present? What are the revised probabilities of D_1 and D_2?

SUMMARY

In this chapter we introduced probability. We described how probability can be interpreted as a numerical measure of the likelihood that an event will occur. In addition, we showed that the probability of an event could be computed directly by summing the probabilities of the experimental outcomes comprising the event or from related events utilizing the addition, conditional probability, and multiplication rules of probability. For cases where additional information is available, we demonstrated how Bayes' theorem could be used to obtain revised, or posterior, probabilities.

GLOSSARY

Probability A numerical measure of the likelihood that an event will occur.

Experiment Any process that generates well-defined outcomes.

Sample space The set of all possible experimental outcomes.

Tree diagram A graphical device helpful in determining the experimental outcomes for an experiment involving multiple steps.

Basic requirements of probability Two requirements that restrict the manner in which probability assignments can be made:
1. For each experimental outcome E_i, we must have $0 \le P(E_i) \le 1$.
2. If n is the number of experimental outcomes, then $P(E_1) + P(E_2) + \cdots + P(E_n) = 1$.

Classical method A method of assigning probabilities that assumes the experimental outcomes are equally likely.

Relative frequency method A method of assigning probabilities based on experimentation or historical data.

Subjective method A method of assigning probabilities based on judgment.

Event A subset of the sample space.

Simple event An event consisting of a single experimental outcome.

Complement of event A The event containing all experimental outcomes that are not in A.

Venn diagram A graphical device for symbolically representing the sample space and operations involving events.

Union of events A and B The event containing all experimental outcomes that are in A, in B, or in both.

Intersection of A and B The event containing all experimental outcomes that are in both A and B.

Mutually exclusive events Events that have no experimental outcome in common; that is, $A \cap B$ is empty and $P(A \cap B) = 0$.

Addition rule A probability law used to compute the probability of a union, $P(A \cup B)$. It is $P(A \cup B) = P(A) + P(B) - P(A \cap B)$ in general. For mutually exclusive events, since $P(A \cap B) = 0$, it reduces to $P(A \cup B) = P(A) + P(B)$.

Conditional probability The probability of an event given that another event has occurred. The conditional probability of A given B is $P(A \mid B) = P(A \cap B)/P(B)$.

Joint probability The probability that two events both occur.

Independent events Two events A and B where $P(A \mid B) = P(A)$ or $P(B \mid A) = P(B)$; that is, the events have no influence on each other.

Multiplication rule A probability rule used to compute the probability of an intersection, $P(A \cap B)$. It is $P(A \cap B) = P(A)P(B \mid A)$, or $P(A \cap B) = P(B)P(A \mid B)$. For independent events, reduces to $P(A \cap B) = P(A)P(B)$.

Collectively exhaustive events A set of events that represent all the possible outcomes. The probability for the union of these events must equal 1.

Prior probabilities Probabilities for a set of mutually exclusive and collectively exhaustive events prior to being updated by Bayes' theorem.

Posterior probabilities The revised probabilities for events resulting from the application of Bayes' theorem.

Bayes' theorem A formula for revising prior probabilities concerning mutually exclusive and collectively exhaustive events. The revised probabilities are called posterior probabilities.

▼ KEY FORMULAS

Computing Probability Using the Complement

$$P(A) = 1 - P(A^c) \tag{4.6}$$

Addition Rule

$$P(A \cup B) = P(A) + P(B) - P(A \cap B) \tag{4.7}$$

Addition Rule for Mutually Exclusive Events

$$P(A \cup B) = P(A) + P(B) \tag{4.8}$$

Conditional Probability

$$P(A \mid B) = \frac{P(A \cap B)}{P(B)} \tag{4.9}$$

or

$$P(B \mid A) = \frac{P(A \cap B)}{P(A)} \tag{4.10}$$

Multiplication Rule

$$P(A \cap B) = P(B)P(A \mid B) \tag{4.13}$$

or

$$P(A \cap B) = P(A)P(B \mid A) \tag{4.14}$$

Multiplication Rule for Independent Events

$$P(A \cap B) = P(A)P(B) \tag{4.15}$$

Bayes' Theorem

$$P(A_1 \mid B) = \frac{P(B \mid A_1)P(A_1)}{P(B \mid A_1)P(A_1) + P(B \mid A_2)P(A_2)} \tag{4.16}$$

and

$$P(A_2 \mid B) = \frac{P(B \mid A_2)P(A_2)}{P(B \mid A_1)P(A_1) + P(B \mid A_2)P(A_2)} \tag{4.17}$$

▼ REVIEW QUIZ

TRUE/FALSE

1. Probabilities can never be greater than 1 or less than 0.
2. The sum of probabilities for all experimental outcomes equals 1.
3. The sum of the probabilities for all experimental outcomes may be any number between 0 and 1, inclusive.
4. The subjective method of assigning probabilities is one of the two methods permitting probabilities less than 0.

5. The sum of the probabilities of the experimental outcomes in an event must equal 1.
6. If we know the probability of an event, then the probability of the complement of the event can also be computed.
7. To use the addition rule to compute the probability of the union of two events, we must know the probability of the intersection of the events.
8. If two events are independent, they must be mutually exclusive.
9. If two events are independent, we need only know each event's probability to compute the probability of the intersection of the events.
10. Posterior probabilities must be known before Bayes' theorem can be applied.

MULTIPLE CHOICE

The following event probabilities for a statistical experiment are utilized in Questions 11–13.

$$P(A) = .60 \qquad P(B) = .40$$

$$P(A \cap B) = .25$$

11. $P(A \cup B)$ is closest to
    a. .65
    b. .72
    c. .79
    d. .82

12. $P(A \mid B)$ is closest to
    a. .60
    b. .67
    c. .74
    d. .81

13. $P(A^c)$ is closest to
    a. .50
    b. .60
    c. .70
    d. .80

In Questions 14–16, assume events A and B are independent, $P(A \mid B) = .70$, and $P(A \cap B) = .21$.

14. $P(A)$ is closest to
    a. .30
    b. .50
    c. .70
    d. .90

15. $P(B)$ is closest to
    a. .20
    b. .30
    c. .50
    d. .70

16. $P(A \cup B)$ is closest to
    a. .50
    b. .60
    c. .70
    d. .80

17. If A and B are independent events with $P(A) = .3$ and $P(B) = .5$, then $P(A \mid B) =$
    a. 0
    b. .15
    c. .20
    d. .30

18. If J and K are mutually exclusive events with $P(J) = .4$ and $P(K) = .5$, then $P(J \cap K) =$
    a. 0   **b.** .2
    c. .7   **d.** .9

19. If J and K are mutually exclusive events with $P(J) = .4$ and $P(K) = .5$, then $P(J \cup K) =$
    a. 0   **b.** .2
    c. .7   **d.** .9

SUPPLEMENTARY EXERCISES

64. The Food and Drug Administration (FDA) places new drug applications in one of three categories:

$$A = \text{Potential breakthrough}$$
$$B = \text{Improvement over existing product}$$
$$C = \text{Me-too drug}$$

During 1985, the FDA approved 3 A's, 15 B's, and 12 C's (*Financial World,* January 24, 1989).
    a. Consider the experiment of observing the category to which a new FDA-approved drug is assigned. How many experimental outcomes are there?
    b. Using the data for 1985, assign probabilities to the experimental outcomes.

65. A financial manager has just made two new investments—one in the oil industry and one in municipal bonds. After a 1-year period, each of the investments will be classified as either successful or unsuccessful. Consider the making of the two investments as an experiment.
    a. How many experimental outcomes exist for this experiment?
    b. Show a tree diagram and list the experimental outcomes.
    c. Let $O =$ the event that the oil investment is successful and $M =$ the event that the municipal bond investment is successful. List the experimental outcomes in O and in M.
    d. List the experimental outcomes in the union of the events $(O \cup M)$.
    e. List the experimental outcomes in the intersection of the events $(O \cap M)$.
    f. Are events O and M mutually exclusive? Explain.

66. Consider an experiment where eight experimental outcomes exist. We will denote the experimental outcomes as $E_1, E_2, \ldots, E_8$. Suppose that the following events are identified:

$$A = \{E_1, E_2, E_3\}$$
$$B = \{E_2, E_4\}$$
$$C = \{E_1, E_7, E_8\}$$
$$D = \{E_5, E_6, E_7, E_8\}$$

Determine the experimental outcomes making up the following events:
    a. $A \cup B$ **b.** $C \cup D$ **c.** $A \cap B$
    d. $C \cap D$ **e.** $B \cap C$ **f.** A^c
    g. D^c **h.** $A \cup D^c$ **i.** $A \cap D^c$
    j. Are A and B mutually exclusive?
    k. Are B and C mutually exclusive?

67. Referring to Exercise 66 and assuming that the classical method is an appropriate way of establishing probabilities, find the following probabilities:
    a. $P(A)$, $P(B)$, $P(C)$, and $P(D)$
    b. $P(A \cap B)$ **c.** $P(A \cup B)$ **d.** $P(A|B)$
    e. $P(B|A)$ **f.** $P(B \cap C)$ **g.** $P(B|C)$
    h. Are B and C independent events?

68. A survey of 2125 subscribers to *Fortune* magazine indicated the following with respect to the number of subscribers owning various credit cards (*Fortune Subscriber Portrait*, 1988).

Type of Card	Number Holding
American Express	1360
Diners Club	234
Telephone Credit Card	1530
Gasoline Credit Card	1424
MasterCard	1466
VISA	1679
Discover	425

a. If 2083 subscribers indicated they have at least one credit card, what is the probability a *Fortune* subscriber does not have any credit cards?

b. What is the probability a subscriber holds an American Express card?

c. Suppose it were known that 55% have both a MasterCard and a VISA card. What is the probability an individual holds one or the other or both?

d. After studying the data, an analyst concluded that *at least* 956, or about 45%, of the subscribers must hold both an American Express and VISA card. Does this make sense? Why or why not?

69. A survey of new matriculants to MBA programs was conducted by the Graduate Management Admissions Council during 1985 ("School Selection by Students," *GMAC Occasional Papers,* March 1988). Shown below are the number of schools to which students applied.

Number of Schools	Number of Students
1	1230
2	304
3	184
4	118
5	78
6	51
7	25
8	13
9	20
10	8
11	9
12	6
Total	2046

a. Use these data to assign probabilities to the number of schools to which an MBA student applies.

b. What is the probability a student will apply to only one school?

c. What is the probability a student will apply to three or more schools?

d. What is the probability a student will apply to more than six schools?

70. In September 1988 the House of Representatives voted on an amendment requiring life imprisonment for drug-related murders. Results of the vote were reported as shown at the top of page 151.

	Yea	Nay	Did Not Vote	Totals
Democrat	153	83	19	255
Republican	169	0	8	177
Totals	322	83	27	432

a. What is the probability that a randomly selected representative voted for the amendment?
b. What is the probability that a randomly selected representative is both a Democrat and voted against the amendment?
c. What is the probability that a representative known to have voted for the amendment is a Democrat?
d. If someone is known not to have voted, is it more likely this person is a Democrat or Republican?
e. What is the joint probability of a randomly selected representative being a Republican and voting for the amendment?
f. Are the events "voted for the amendment" and "is a Democrat" independent?

71. A bank has observed that credit-card account balances have been growing over the past year. A sample of 200 customer accounts resulted in the data shown below.

Amount Owed ($)	Frequency
0–99	62
100–199	46
200–299	24
300–399	30
400–499	26
500 and over	12

a. Let A be the event that a customer's balance is less than $200. Find $P(A)$.
b. Let B be the event that a customer's balance is $300 or more. Find $P(B)$.

72. The GMAC MBA survey of 2018 new-matriculants (*GMAC Occasional Papers*, March 1988) gives the following data.

		Applied to More Than One School	
		Yes	*No*
Age Group	*23 and under*	207	201
	24–26	299	379
	27–30	185	268
	31–35	66	193
	36 and over	51	169

a. Prepare a joint probability table for the experiment consisting of observing the age and number of schools to which a randomly selected MBA student applies.
b. What is the probability an applicant will be 23 or under?
c. What is the probability an applicant will be older than 26?
d. What is the probability an applicant applies to more than one school?

73. Refer to the data from the GMAC new-matriculants survey in Exercise 72.
  a. Given that a person applied to more than one school, what is the probability the person is 24–26 years old?
  b. Given that a person is in the 36-and-over age group, what is the probability the person applies to more than one school?
  c. What is the probability a person is 24–26 years old *or* applies to more than one school?
  d. Suppose a person is known to have applied to only one school. What is the probability the person is 31 or more years old?
  e. Is the number of schools applied to independent of age? Explain.

74. Suppose that $P(A) = .60$, $P(B) = .30$, and events A and B are mutually exclusive.
  a. Find $P(A \cup B)$ and $P(A \cap B)$.
  b. Are events A and B independent?
  c. Can you make a general statement about whether mutually exclusive events can be independent?

75. A market survey of 800 people found the following facts about the ability to recall a television commercial for a particular product and the actual purchase of the product.

	Could Recall Television Commercial	Could Not Recall Television Commercial	Totals
Purchased	160	80	240
Did Not Purchase	240	320	560
Totals	400	400	800

Let T be the event of the person recalling the television commercial and B the event of buying or purchasing the product.
  a. Find $P(T)$, $P(B)$, and $P(T \cap B)$.
  b. Are T and B mutually exclusive events? Use probability values to explain.
  c. What is the probability that a person who could recall seeing the television commercial has actually purchased the product?
  d. Are T and B independent events? Use probability values to explain.
  e. Comment on the value of the commercial in terms of its relationship to purchasing the product.

76. A research study investigating the relationship between smoking and heart disease in a sample of 1000 men over 50 years of age provided the following data.

	Smoker	Nonsmoker	Totals
Record of Heart Disease	100	80	180
No Record of Heart Disease	200	620	820
Totals	300	700	1000

  a. Show a joint probability table that summarizes the results of this study.
  b. What is the probability a man over 50 years of age is a smoker and has a record of heart disease?
  c. Compute and interpret the marginal probabilities.
  d. Given that a man over 50 years of age is a smoker, what is the probability that he has a record of heart disease?

e. Given that a man over 50 years of age is a nonsmoker, what is the probability that he has a record of heart disease?

f. Does the research show that heart disease and smoking are independent events? Use probability to justify your answer.

g. What conclusion would you draw about the relationship between smoking and heart disease?

77. Cooper Realty is a small real estate company located in Albany, New York, specializing primarily in residential listings. They have recently become interested in determining the likelihood of one of their listings being sold within a certain number of days. An analysis of company sales of 800 homes for the previous years produced the accompanying data.

| | | Days Listed Until Sold | | | |
		Under 30	31–90	Over 90	Totals
Initial Asking Price	Under $50,000	50	40	10	100
	$50,000–99,999	20	150	80	250
	$100,000–150,000	20	280	100	400
	Over $150,000	10	30	10	50
Totals		100	500	200	800

a. If A is defined as the event that a home is listed for over 90 days before being sold, estimate the probability of A.

b. If B is defined as the event that the initial asking price is under $50,000, estimate the probability of B.

c. What is the probability of $A \cap B$?

d. Assuming that a contract has just been signed to list a home that has an initial asking price of less than $50,000, what is the probability the home will take Cooper Realty more than 90 days to sell?

e. Are events A and B independent?

78. In the evaluation of a sales training program, a firm found that of 50 salespersons making a bonus last year, 20 had attended a special sales training program. The firm has 200 salespersons. Let B = the event that a salesperson makes a bonus and S = the event a salesperson attends the sales training program.

a. Find $P(B)$, $P(S \mid B)$, and $P(S \cap B)$.

b. Assume that 40% of the salespersons have attended the training program. What is the probability that a salesperson makes a bonus given that the salesperson attended the sales training program, $P(B \mid S)$?

c. If the firm evaluates the training program in terms of the effect it has on the probability of a salesperson's making a bonus, what is your evaluation of the training program? Comment on whether B and S are dependent or independent events.

79. A company has studied the number of lost-time accidents occurring at its Brownsville, Texas, plant. Historical records show that 6% of the employees had lost-time accidents last year. Management believes that a special safety program will reduce such accidents to 5% during the current year. In addition, it is estimated that 15% of those employees who had lost-time accidents last year will have a lost-time accident during the current year.

a. What percentage of the employees will have lost-time accidents in both years?

b. What percentage of the employees will have at least one lost-time accident over the 2-year period?

80. In a study of television-viewing habits among married couples, a researcher found that, for a popular Saturday night program, 25% of the husbands viewed the program regularly and

30% of the wives viewed the program regularly. The study found that, for couples where the husband watches the program regularly, 80% of the wives also watch regularly.
   a. What is the probability that both the husband and wife watch the program regularly?
   b. What is the probability that at least one—husband or wife—watches the program regularly?
   c. What percentage of married couples do not have at least one regular viewer of the program?

81. A statistics professor has noted from past experience that students who do the homework for the course have a .90 probability of passing the course. On the other hand, students who do not do the homework for the course have a .25 probability of passing the course. The professor estimates that 75% of the students in the course do the homework. Given a student passes the course, what is the probability that she or he completed the homework?

82. A salesperson for Business Communication Systems, Inc., sells automatic envelope-addressing equipment to medium- and small-size businesses. The probability of making a sale to a new customer is .10. During the initial contact with a customer, sometimes the salesperson will be asked to call back later. Of the 30 most recent sales, 12 were made to customers who initially told the salesperson to call back later. Of 100 customers who did not make a purchase, 17 had initially asked the salesperson to call back later. If a customer asks the salesperson to call back later, should the salesperson do so? What is the probability of making a sale to a customer who has asked the salesperson to call back later?

83. Migliori Industries, Inc., manufactures a gas-saving device for use on natural gas forced-air residential furnaces. The company is currently trying to determine the probability that sales of this product will exceed 25,000 units during the next year's winter sales period. The company believes that sales of the product depend to a large extent on the winter conditions. Management's best estimate is that the probability of sales exceeding 25,000 units if the winter is severe is .8. This probability drops to .5 if the winter conditions are moderate. If the weather forecast is .7 for a severe winter and .3 for moderate conditions, what is Migliori's best estimate that sales will exceed 25,000 units?

84. The Dallas IRS auditing staff is concerned with identifying potential fraudulent tax returns. From past experience, they believe that the probability of finding a fraudulent return given that the return contains deductions for contributions exceeding the IRS standard is .20. Given that the deductions for contributions do not exceed the IRS standard, the probability of a fraudulent return decreases to .02. If 8% of all returns exceed the IRS standard for deductions due to contributions, what is the best estimate of the percentage of fraudulent returns?

85. Carstab Corporation,* a subsidiary of Morton International, makes an expensive catalyst used in chemical processing. One customer has very stringent requirements for the catalyst and, as a result, has typically returned 40% of Carstab's shipments. In order to cut down on returned shipments, Carstab has developed a test that, it is hoped, will help identify whether or not a lot will be acceptable to the customer.
   A sample of production lots was tested using Carstab's new procedure and by the customer. Of the lots tested, 55% passed Carstab's test and 50% passed both the customer's and Carstab's tests.
   a. Find the probability a lot will be acceptable to the customer given that it has passed Carstab's new test.
   b. Would you recommend that Carstab implement the new testing procedure? Why or why not?

86. In the setup of a manufacturing process, a machine is either correctly or incorrectly adjusted. The probability of a correct adjustment is .90. When correctly adjusted, the machine operates with a 5% defective rate. However, if it is incorrectly adjusted, a 75% defective rate occurs.

---

*Information provided by Michael Haskell of Morton International's Carstab subsidiary.

a. After the machine starts a production run, what is the probability that a defect is observed when one part is tested?

b. Suppose that the one part selected by an inspector is found to be defective. What is the probability that the machine is incorrectly adjusted? What action would you recommend?

c. Before your recommendation in part (b) was followed, a second part is tested and found to be good. Using your revised probabilities from part (b) as the most recent prior probabilities, compute the revised probability of an incorrect adjustment given that the second part is good. What action would you recommend now?

87. Western Airlines has done an analysis of a price promotion it is offering to frequent air travelers in order to increase the number of people on their New York to San Francisco route. Some 20% of the people in a large sample of individuals identified as frequent travelers from New York to San Francisco were aware of the Western promotion and elected to fly with Western on their next trip. It was further found that 80% were aware of the promotion and that prior to the promotion, 25% of these travelers flew with Western.

a. What is the probability that a person will fly with Western on their next trip given that he or she is aware of the price promotion?

b. For a randomly selected traveler, are the events "fly with Western" and "aware of the price promotion" independent? Why or why not?

88. A Bayesian approach can be used to revise probabilities that a prospect field will produce oil (*Oil & Gas Journal,* January 11, 1988). In one case, geological assessment indicates a 25% chance the field will produce oil. Further, there is an 80% chance that a particular well will strike oil given that oil is present on the prospect field.

a. Suppose that one well is drilled on the field and it comes up dry. What is the probability the prospect field will produce oil?

b. If two wells come up dry, what is the probability the field will produce oil?

c. The oil company would like to keep looking as long as the chances of finding oil are greater than 1%. How many dry wells must be drilled before the field will be abandoned?

CHAPTER 5

RANDOM VARIABLES AND DISCRETE PROBABILITY DISTRIBUTIONS

CONTENTS

Pete Rose Versus Joe DiMaggio: Baseball Greats

Many baseball enthusiasts will agree that the most difficult record for modern baseball players to challenge is Joe DiMaggio's 56-game hitting streak in 1941. In baseball circles, the prevailing view is that Joe's performance set an unbreakable major league standard. However, for 6 weeks of the 1978 season (June 14 to August 1) Pete Rose of the Cincinnati Reds was unstoppable as he knocked down consecutive-game hitting records. In game 38 of the hitting streak, Rose passed Tommy Holmes's 1945 modern National League record. In game 44, he became the National League's all-time consecutive-game hitting record-holder when he matched Willie Keeler's 1897 streak. Joe DiMaggio's 56-game record was next.

However, on August 1, Pete Rose's hitting streak of 44 consecutive games came to an end when he went 0 for 4 in a game with the Atlanta Braves in Atlanta. During the streak, Pete hit .376 with 56 singles, 14 doubles, and 13 walks. He provided plenty of excitement when on six occasions he saved the streak in his last at-bat. On four occasions during the streak, Rose's only hit was a bunt. Reds manager Sparky Anderson exclaimed, "Watching Pete break the National League record is the biggest thrill I've had as a manager, but Joe's 56-game record is an impossibility."

Pete Rose, baseball's all-time hit leader, challenged DiMaggio's record with a 44 consecutive-game hitting streak in 1978.

When Pete set the 44-game record, Las Vegas odds makers were still strong in their belief that he could not continue the hitting streak to DiMaggio's 56-game record. Probability specialists say that a special probability distribution, known as the binomial distribution, can be used to estimate the probability, or odds, of continuing a hitting streak for 12 more games. Using Pete's .376 batting average as the probability of a hit on each at-bat and assuming Pete would come to bat four times in a game, the probability of Pete having at least one hit in the four at-bats can be computed to be .8484. The probability of hitting in 12 successive games can be computed from this to be only .1391. The probability specialists say there was a .1391 probability of Pete reaching Joe's record and a .8609 probability that he would not. Thus, the odds were still better than 6 to 1 that Pete could not match DiMaggio's performance.

While Pete Rose's 44-game hitting streak provided much interest and excitement, Joe DiMaggio's 56-game record stands as baseball's "most unbreakable" record.

Joe DiMaggio, the "Yankee Clipper," hit safely in 56 consecutive games during the 1941 season.

This *Statistics in Practice* is based on "Doing Much," *Sports Illustrated* (August 7, 1978).

In this chapter we introduce the concepts of random variables and probability distributions. We will see that random variables provide a means of assigning numerical values to experimental outcomes. Probability distributions provide a means of determining the probability of the different values of the random variable occurring. They are used in statistics to provide a shorthand model, or description, of a population of data. Some probability distributions are used so extensively in statistical analysis that special formulas and/or tables have been developed for computing the probabilities associated with them. Three of these probability distributions, the binomial, the Poisson, and the hypergeometric are introduced in this chapter.

5.1 ▽ RANDOM VARIABLES

In Chapter 4 we studied the role of an experiment and the associated experimental outcomes in statistics. Random variables provide a means of assigning numerical values to experimental outcomes. These numerical values are used in computing means, variances, and other measures. The definition of a random variable is as follows.

Random Variable

A *random variable* is a numerical description of the outcome of an experiment.

For any experiment, a random variable can be defined such that each possible experimental outcome generates one and only, one numerical value for the random variable. The particular numerical value that the random variable takes on depends on the outcome of the experiment. That is, the value of the random variable is not known until the experimental outcome is observed.

EXAMPLE 5.1 ▶ Consider the experiment that consists of tossing a coin twice. With H indicating a head and T indicating a tail, the sample space for this experiment is

$$S = \{(H, H), (H, T), (T, H), (T, T)\}$$

Suppose we let x = number of heads occurring on the two coin tosses. Then x is a random variable (it provides a numerical description of the experimental outcome) that can assume the values 0, 1, and 2. ◀

EXAMPLE 5.2 ▶ To receive state certification as medical lab technicians, candidates must pass a series of three examinations. If we define the random variable x as the number of examinations any one candidate passes, then x can assume the values 0, 1, 2, and 3. ◀

EXAMPLE 5.3 ▶ The construction of a new library has just gotten underway at Lakeland Community College. If we define a random variable x as the percentage of the project that is completed after 6 months, the possible values of x range from 0 to 100. In other words, $0 \le x \le 100$. ◀

A random variable is classified as either discrete or continuous depending on the numerical values it can assume. A random variable that may assume either a finite number of values or an infinite sequence (e.g., 1, 2, 3, . . .) of values is referred to as a *discrete random variable*. The number of units sold, the number of defects observed, and the number of customers that enter a bank during 1 day of operation are examples of discrete random variables. Examples 5.1 and 5.2 involve discrete random variables assuming a finite number of values. The number of people entering a bank during 1 day of operation (0, 1, 2, 3, . . .) is an example of a discrete random variable assuming an infinite sequence of values. The distinguishing feature of a discrete random variable is the separation between the successive values it may assume.

Random variables such as weight, time, and temperature, which may take on all values in a certain interval or collection of intervals, are referred to as *continuous random variables*. For instance, the random variable in Example 5.3 (percentage of project completed after 6 months) is a continuous random variable because it may take on any value in the interval from 0 to 100 (e.g., 56.33 or 64.227). The feature that distinguishes a continuous random variable is the lack of separation between the successive values the random variable may assume.

NOTES AND COMMENTS

One way to determine whether a random variable is discrete or continuous is to think of the values of the random variable as points on a line segment. If the entire line segment between any two of these points also represents values the random variable may assume, the random variable is continuous.

EXERCISES

METHODS

SELF TEST

1. Consider the experiment of tossing a coin three times.
   a. List the experimental outcomes.
   b. Define a random variable that represents the number of heads occurring on the three tosses.
   c. Show what value the random variable would assume for each of the experimental outcomes.
   d. Is this random variable discrete or continuous?

2. For each of the following random variables state whether it is discrete or continuous:
   a. Number of heads on 5 tosses of a coin
   b. Number of heads on 500 tosses of a coin
   c. Number of people entering a drugstore in a 2-hour period
   d. Time from beginning of year until first baby is born
   e. Weight of a person
   f. Number of customers calling an airline reservation service in 1 minute

APPLICATIONS

SELF TEST

3. Three students have interviews scheduled for summer employment at the Brookwood Institute. In each case, the result of the interview will either be that a position is offered or not offered. Experimental outcomes are defined in terms of the results of the three interviews.
   a. List the experimental outcomes.
   b. Define a random variable that represents the number of offers made. Is this a discrete or continuous random variable?
   c. Show the value of the random variable for each of the experimental outcomes.

4. Subscribers to the Turner Broadcasting System may select one or more of the following services: Cable News Network, Superstation WTBS, Headline News, and Turner Network Television (*Business Week,* April 17, 1989). Suppose an experiment is designed to take a sample of five subscribers' selections. The random variable x is defined as the number in the sample who receive the Cable News Network service. What values may the random variable take on?

5. To perform a certain type of blood analysis, lab technicians use two procedures. The first procedure requires either one or two separate steps, and the second procedure requires either one, two, or three steps.

   a. List the experimental outcomes associated with performing an analysis.

   b. If the random variable of interest is the total number of steps required to do the complete analysis, show what value the random variable will assume for each of the experimental outcomes.

6. Listed below is a series of experiments and associated random variables. In each case, identify the values that the random variable can take on and state whether the random variable is discrete or continuous.

Experiment	Random Variable (x)
a. Take a 20-question examination	Number of questions answered correctly
b. Observe cars arriving at a tollbooth for 1 hour	Number of cars arriving at tollbooth
c. Audit 50 tax returns	Number of returns containing errors
d. Observe an employee's work	Number of nonproductive hours in an 8-hour work day
e. Weigh a shipment of goods	Number of pounds

5.2   DISCRETE PROBABILITY DISTRIBUTIONS

For any discrete random variable x, the *probability function,* denoted by $f(x)$, provides the probabilities associated with the values the random variable may assume. A *discrete probability distribution* is a table, graph, or mathematical formula that shows all possible values of the random variable x and the associated probability function $f(x)$.

EXAMPLE 5.4 ▶

As part of a study of 300 households in a village on the coast of Maine, a sociologist collected data showing the number of children in each household. The following data were obtained: 54 of the households had no children, 117 had one child, 72 had two children, 42 had three children, 12 had four children, and 3 had five children.

Suppose we consider the experiment of randomly selecting one of these households to participate in a follow-up study. If we let x be a random variable denoting the number of children in the household selected, possible values of x are 0, 1, 2, 3, 4, and 5. Thus, $f(0)$ provides the probability that a randomly selected household has no children, $f(1)$ provides the probability that a randomly selected household has one child, and so on. Since 54 of the 300 households have no children, we assign the value $54/300 = .18$ to $f(0)$. Similarly, since 117 of the 300 households have one child, we assign the value $117/300 = .39$ to $f(1)$. Continuing in this fashion for the other values the random variable x may assume, we obtain Table 5.1. This table, showing the values

TABLE 5.1 PROBABILITY DISTRIBUTION FOR THE NUMBER OF CHILDREN PER HOUSEHOLD

x	$f(x)$
0	.18
1	.39
2	.24
3	.14
4	.04
5	.01
Total	1.00

the random variable may assume and the associated probabilities $f(x)$, is the probability distribution for the random variable x.

We can also present the probability distribution of x graphically. In Figure 5.1 the values of the random variable x from Example 5.4 are shown on the horizontal axis and

FIGURE 5.1 GRAPHICAL PRESENTATION OF THE PROBABILITY DISTRIBUTION FOR NUMBER OF CHILDREN PER HOUSEHOLD

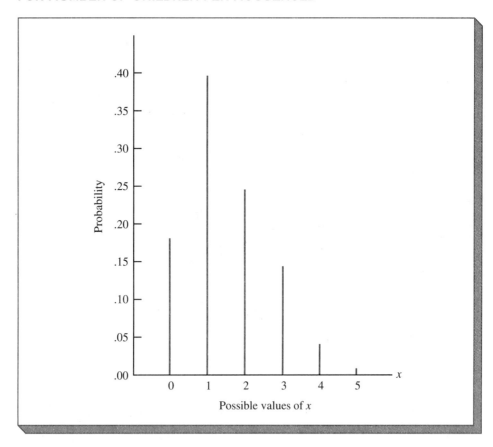

the probability that x assumes these values is shown on the vertical axis. For many discrete random variables, the probability can also be given as a formula that yields $f(x)$ for every possible value of x.

EXAMPLE 5.5 ▶

A multiple choice exam provides 4 possible answers for each question: a, b, c, and d. The exam has been designed so that each response is correct for $\frac{1}{4}$ of the questions. Let

$$x = \begin{cases} 1 \text{ if a is selected} \\ 2 \text{ if b is selected} \\ 3 \text{ if c is selected} \\ 4 \text{ if d is selected} \end{cases}$$

If $f(x)$ provides the probability that x is the correct answer, then the probability distribution for x can be given by the formula

$$f(x) = \frac{1}{4} \qquad \text{for } x = 1, 2, 3, 4$$

A discrete probability distribution such as this, where every value of the random variable has the same probability, is called a *discrete uniform probability distribution*. ◀

    More complex examples of discrete random variables with probability distributions given by formulas are the binomial, Poisson, and hypergeometric probability distributions discussed later in the chapter.

    In the development of the probability function for a discrete random variable, the following two conditions must always be satisfied.

Required Conditions for a Discrete Probability Function

$$f(x) \geq 0 \tag{5.1}$$

$$\Sigma f(x) = 1 \tag{5.2}$$

The symbol Σ in equation (5.2) is used to indicate that the summation is over all the values x may assume.

EXAMPLE 5.4
(CONTINUED) ▶

Table 5.1 shows that the probabilities for the random variable x, number of children per household, are all greater than or equal to zero, $f(x) \geq 0$. In addition, we note that

$$\Sigma f(x) = f(0) + f(1) + f(2) + f(3) + f(4) + f(5)$$

$$= .18 + .39 + .24 + .14 + .04 + .01 = 1.00$$

Since equations (5.1) and (5.2) are satisfied, the probability function developed by the sociologist is a valid discrete probability function. ◀

▷

EXERCISES

METHODS

7. State whether each of the following is a valid probability distribution. Why or why not?

a.			b.			c.	
x	f(x)		x	f(x)		x	f(x)
0	.60		4.5	.05		40	.20
1	.15		6.0	.35		80	.70
2	.20		7.33	.40		100	.10
			8.00	.20			

SELF TEST **8.** The table below shows the probability distribution of the random variable x.

x	f(x)
20	.20
25	.15
30	.25
35	.40
Total	1.00

a. Is this a proper probability distribution? Check to see that equations (5.1) and (5.2) are satisfied.
b. What is the probability $x = 30$?
c. What is the probability x is less than or equal to 25?
d. What is the probability x is greater than 30?

APPLICATIONS

SELF TEST **9.** The following data were collected by counting the number of operating rooms in use at Tampa General Hospital over a 20-day period: On 3 of the days, only one operating room was used; on 5 of the days two were used; on 8 of the days, three were used; and on 4 days, all four of the hospital's operating rooms were used.
a. Use the relative frequency approach to construct a probability distribution for the number of operating rooms in use on any given day.
b. Draw a graph of the probability distribution.
c. Show that your probability distribution satisfies the required conditions for a valid discrete probability distribution.

10. According to *Dataquest*, the personal computer market during 1986 was as follows.

Company	Company Identification Code	Market Share (%)
IBM	1	29.8
Apple	2	9.2
COMPAQ	3	6.6
Zenith	4	3.5
Tandy	5	6.6
Commodore	6	1.9
Other	7	42.4

Let x be a random variable based on the company identification code.
a. Use these data to develop a probability distribution. Specify the values for the random variable and the corresponding values for the probability function $f(x)$.
b. Draw a graph of the probability distribution.
c. Show that the probability distribution satisfies equations (5.1) and (5.2).

11. QA Properties is considering making an offer to purchase an apartment building. Management has subjectively assessed a probability distribution for x, the purchase price, as shown in the table below.

x	$148,000	$150,000	$152,000
$f(x)$	.20	.40	.40

a. Determine whether this is a proper probability distribution by checking equations (5.1) and (5.2).
b. What is the probability that the apartment house can be purchased for $150,000 or less?

12. The cleaning and changeover operation for a production system requires 1, 2, 3, or 4 hours, depending on the specific product that will begin production. Let x be a random variable indicating the time in hours required to make the changeover. The following probability function can be used to compute the probability associated with any changeover time x:

$$f(x) = \frac{x}{10} \qquad \text{for } x = 1, 2, 3, \text{ or } 4$$

a. Show that the probability function meets the required conditions of equations (5.1) and (5.2).
b. What is the probability that the changeover will take 2 hours?
c. What is the probability that the changeover will take more than 2 hours?
d. Graph the probability distribution for the changeover times.

13. The director of admissions at Lakeville Community College has subjectively assessed a probability distribution for x, the number of entering students, as shown in the table below.

x	1000	1100	1200	1300	1400
$f(x)$	.15	.20	.30	.25	.10

a. Is this a valid probability distribution?
b. What is the probability there will be 1200 or fewer entering students?

14. A psychologist has determined that the number of treatment hours required to obtain the trust of a new patient is either 1, 2, or 3. Let x be a random variable indicating the time in hours required to gain the patient's trust. The following probability function has been proposed.

$$f(x) = \frac{x}{6} \qquad \text{for } x = 1, 2, \text{ or } 3$$

a. Is this a valid probability function? Explain.
b. What is the probability that it takes exactly 2 hours to gain the patient's trust?
c. What is the probability that it takes at least 2 hours to gain the patient's trust?

15. The table below shows a partial probability distribution for the MRA Company's projected profits (x = profits in $000s) for the first year of operation (the negative value denotes a loss).

x	−100	0	50	100	150	200
$f(x)$	.10	.20	.30	.25	.10	

    a.   What is the missing value of $f(200)$? What is your interpretation of this value?

    b.   What is the probability that MRA will be profitable?

    c.   What is the probability that MRA will make at least $100,000?

5.3   EXPECTED VALUE AND VARIANCE

Random variables and probability distributions are used to describe populations of data. Just as we want to know the mean (μ) and the variance (σ^2) for a population of data, we also want to know the corresponding measures for the probability distribution being used to describe that population. These measures for the probability distribution are the expected value and variance.

EXPECTED VALUE

The *expected value* of a random variable provides the mean, or average, value for the random variable. The mathematical expression for the expected value of a discrete random variable x is as follows.

> **Expected Value of a Discrete Random Variable**
>
> $$E(x) = \mu = \Sigma xf(x) \qquad\qquad \textbf{(5.3)}$$

Both the notations $E(x)$ and μ are used to refer to the expected value of a random variable. Technically, μ is the mean of the population being modeled (described) by the probability distribution. However, for that population, $\mu = E(x)$, so the notations μ and $E(x)$ are used interchangeably.

Equation (5.3) shows that to compute the expected value of a discrete random variable, we must multiply each value of the random variable by the corresponding value of its probability function, and then add the resulting products.

EXAMPLE 5.4
(CONTINUED)

Table 5.2 shows the calculation of the expected value of the random variable x, which is the number of children in a randomly selected household. As shown, $E(x) = 1.50$. Since it is impossible for any household to have 1.5 children, we see that the expected value of a

TABLE 5.2   **EXPECTED VALUE OF RANDOM VARIABLE FOR EXAMPLE 5.4**

x	$f(x)$	$xf(x)$
0	.18	.00
1	.39	.39
2	.24	.48
3	.14	.42
4	.04	.16
5	.01	.05
		1.50

$$E(x) = \mu = \Sigma xf(x) = 1.50$$

random variable does not have to be one of the values the random variable can assume. The expected value is thought of as the mean or average value and not necessarily some value we expect the random variable to assume. ◀

VARIANCE

While the expected value provides the mean value of the random variable, we often need a measure of the dispersion, or variability, of the random variable. Just as we used variance in Chapter 3 to summarize the dispersion in a data set, we now use the *variance* measure to summarize the variability in the values of a random variable. The mathematical expression for the variance of a discrete random variable is as follows.

Variance of a Discrete Random Variable

$$\text{Var}(x) = \sigma^2 = \Sigma(x - \mu)^2 f(x) \tag{5.4}$$

Both the notations $\text{Var}(x)$ and σ^2 are used for the variance of a random variable. Technically, σ^2 is the variance of the population being modeled by the probability distribution, but the notations $\text{Var}(x)$ and σ^2 are used interchangeably.

As equation (5.4) shows, an essential part of the variance formula is the deviation, $x - \mu$, which measures how far a particular value of the random variable is from the expected value or mean, μ. In computing the variance of a random variable, the deviations are squared and then weighted by the corresponding value of the probability function. The sum of these weighted squared deviations from the mean for all values of the random variable is referred to as the variance. The *standard deviation* σ is defined as the positive square root of the variance.

EXAMPLE 5.4 ▶
(CONTINUED)

The calculation of the variance for the random variable representing the number of children per household is summarized in Table 5.3. We see that the variance for the number of children per household is 1.25. The standard deviation of the number of children per household is

$$\sigma = \sqrt{1.25} = 1.118$$ ◀

TABLE 5.3   **CALCULATION OF VARIANCE FOR EXAMPLE 5.4**

x	$x - \mu$	$(x - \mu)^2$	$f(x)$	$(x - \mu)^2 f(x)$
0	−1.50	2.25	.18	.4050
1	− .50	.25	.39	.0975
2	.50	.25	.24	.0600
3	1.50	2.25	.14	.3150
4	2.50	6.25	.04	.2500
5	3.50	12.25	.01	.1225
				1.2500

$$\sigma^2 = \Sigma(x - \mu)^2 f(x)$$
$$= 1.25$$

The standard deviation is measured in the same units as the random variable ($\sigma = 1.118$ children per household in Example 5.4); for this reason, σ is often preferred in describing the variability of a random variable. The variance (σ^2) is measured in squared units and is thus more difficult to interpret.

An alternate formula for the variance of a random variable is

$$\text{Var}(x) = \sigma^2 = \Sigma x^2 f(x) - \mu^2 \qquad (5.5)$$

When computing the variance using a calculator, this formula is often preferred because it does not require the computation of the deviations about the mean ($x - \mu$). To illustrate the use of this formula, we recompute the variance for Example 5.4.

EXAMPLE 5.4
(CONTINUED)

The calculations necessary to use equation (5.5) to compute the variance of the number of children per household are summarized below.

x	x^2	$f(x)$	$x^2 f(x)$
0	0	.18	.00
1	1	.39	.39
2	4	.24	.96
3	9	.14	1.26
4	16	.04	.64
5	25	.01	.25
			3.50

$$\mu = 1.50 \text{ (see Table 5.2)}$$

$$\text{Var}(x) = \sigma^2 = \Sigma x^2 f(x) - \mu^2 = 3.50 - (1.50)^2 = 1.25$$

As we should expect, this is the same answer we obtained previously using equation (5.4).

EXERCISES

METHODS

16. The table below shows a probability distribution for the random variable x.

x	$f(x)$
3	.25
6	.50
9	.25
Total	1.00

    a. Compute $E(x)$, the expected value of x.
    b. Compute σ^2, the variance of x.
    c. Compute σ, the standard deviation of x.

SELF TEST **17.** Shown below is a probability distribution for the random variable *x*.

x	*f(x)*
2	.20
4	.30
7	.40
8	.10
Total	1.00

   a. Compute $E(x)$.
   b. Compute $Var(x)$ and σ.

APPLICATIONS

18. Glazer's Winton Woods apartment building has 20 two-bedroom apartments. The number of apartment air-conditioner units that must be replaced during the summer season has the probability distribution shown.

Air Conditioners Replaced	0	1	2	3	4
Probability	.30	.35	.20	.10	.05

   a. What is the expected number of air-conditioner units that will be replaced during a summer season?
   b. What is the variance in the number of air-conditioner replacements?
   c. What is the standard deviation?

19. A volunteer ambulance service handles from zero to five service calls on any given day. Assume the probability distribution for the number of service calls is as shown in the table below.

Number of Service Calls	0	1	2	3	4	5
Probability	.10	.15	.30	.20	.15	.10

   a. What is the expected number of service calls?
   b. What is the variance in the number of service calls? What is the standard deviation?

SELF TEST **20.** Individual investors allocated their funds among four different investment categories in 1988 (*Money*, January 1989). The annual return for the categories and the proportion of investments in each category are as follows.

Investment Category	Annual Return	Proportion of Total Investments
Stocks	26.0%	.31
Bonds	8.6%	.23
CDs and money funds	7.7%	.45
Real estate and gold	−2.9%	.01

Let x be a random variable indicating the annual return percentage for a $1 investment. Assume the probabilities of each investment category $f(x)$ are provided by the proportion of total investments data.
   a. What is the expected annual return percentage on a $1 investment?
   b. What are the variance and standard deviation?
   c. Suppose the annual return on stocks drops to 5%. Recompute the expected return and the standard deviation of the return on a $1 investment.

21. The actual shooting records of the 1992 NCAA championship final four teams (*NCAA Final Four Program,* April 1992) showed the probability of making a 2-point basket was .50 while the probability of making a 3-point basket was .39.
   a. What is the expected value of a 2-point shot for these teams?
   b. What is the expected value of a 3-point shot for these teams?
   c. Since the probability of making a 2-point basket is greater than the probability of making a 3-point basket, why do coaches allow some players to shoot the 3-point shot if they have the opportunity? Use expected value to explain your answer.

22. The probability distribution for damage claims paid by the Newton Automobile Insurance Company on collision insurance is shown in the table below.

Payment ($)	0	400	1000	2000	4000	6000
Probability	.90	.04	.03	.01	.01	.01

   a. Use the expected collision payment to determine the collision insurance premium that would allow the company to break even.
   b. The insurance company charges an annual rate of $260 for the collision coverage. What is the expected value of the collision policy for a policyholder? (*Hint:* It is the expected payments from the company minus the cost of coverage.) Why does the policyholder purchase a collision policy with this expected value?

23. The number of dots observed on the upward face of a die has the following probability function:

$$f(x) = \frac{1}{6} \qquad \text{for } x = 1, 2, 3, 4, 5, 6$$

   a. Show that this probability function possesses the properties necessary for probability distributions.
   b. Draw a graph of the probability distribution.
   c. What is the expected value? What is the interpretation of this value?
   d. What are the variance and the standard deviation for the number of dots?

24. The demand for a product of Carolina Industries varies greatly from month to month. Based on the past 2 years of data, the probability distribution in the table below shows the company's monthly demand.

Unit demand	300	400	500	600
Probability	.20	.30	.35	.15

   a. If the company places monthly orders based on the expected value of the monthly demand, what should Carolina's monthly order quantity be for this product?
   b. Assume that each unit demanded generates $70 in revenue and that each unit ordered costs $50. How much will the company gain or lose in a month if it places an order based on your answer to part (a) and the actual demand for the item is 300 units?

25. What are the variance and the standard deviation for the number of units demanded in Exercise 24?

26. The J. R. Ryland Computer Company is considering a plant expansion that will enable the company to begin production of a new computer product. The company's president must determine whether to make the expansion a medium- or large-scale project. An uncertainty involves the demand for the new product, which for planning purposes may be low, medium, or high demand. The probability estimates for the demands are .20, .50, and .30, respectively. Letting x indicate the annual profit in $000s, the firm's planners have developed profit forecasts for the medium- and large-scale expansion projects.

Demand	Medium-Scale Expansion Profits x	$f(x)$	Large-Scale Expansion Profits x	$f(x)$
Low	50	.20	0	.20
Medium	150	.50	100	.50
High	200	.30	300	.30

a. Compute the expected value for the profit associated with the two expansion alternatives. Which decision is preferred for the objective of maximizing the expected profit?

b. Compute the variance for the profit associated with the two expansion alternatives. Which decision is preferred for the objective of minimizing the risk or uncertainty?

5.4 ▽ THE BINOMIAL PROBABILITY DISTRIBUTION

The binomial probability distribution is a discrete probability distribution that has many applications. It is associated with a multiple-step experiment that we call the binomial experiment. For a probability experiment to be classified as a *binomial experiment,* it must have the following four properties.

Properties of a Binomial Experiment

1. The experiment consists of a sequence of n identical trials.
2. Two outcomes are possible on each trial. We refer to one as a *success* and the other as a *failure.*
3. The probability of a success, denoted by p, does not change from trial to trial. Consequently, the probability of failure, denoted by $1 - p$, does not change from trial to trial. Also, since there are only two possible outcomes on each trial, the probability of a success plus the probability of a failure must equal 1.
4. The trials are independent.

Figure 5.2 depicts a sequence of possible outcomes for a binomial experiment involving eight trials. Properties 1 and 2 are illustrated; there are $n = 8$ trials and each trial results in either success or failure.

In a binomial experiment, our interest is in the *number of successes occurring in the n trials*. If we let x denote the number of successes occurring in the n trials, we see that x is a random variable that can assume the values of 0, 1, 2, 3, . . ., n. Since the number of values is finite, x is a *discrete* random variable. The probability distribution associated with this random variable is called the *binomial probability distribution*.

EXAMPLE 5.6 ▶ Consider the experiment of tossing a coin five times and on each toss observing whether the coin lands with a head or a tail on its upward face. Suppose we are interested in counting the number of heads appearing during the five tosses. Does this experiment have the properties of a binomial experiment? What is the random variable of interest? Note the following:

1. The experiment consists of five identical trials, where each trial involves the tossing of one coin.
2. There are two outcomes possible for each trial. The possible outcomes are a head and a tail. We can designate head as success and tail as failure.
3. The probability of a head and the probability of a tail are the same for each trial, with $p = .5$ and $1 - p = .5$.
4. The trials or tosses are independent, since the outcome on any one trial is not affected by what happens on other trials or tosses.

Thus, the properties of a binomial experiment are satisfied. The random variable of interest is $x =$ the number of heads appearing in the five trials. In this case, x can assume the values of 0, 1, 2, 3, 4, or 5. ◀

Property 2, two outcomes possible for each trial, is not as restrictive as it may first appear. For instance, consider rolling a die and observing whether a 5 comes up. Defining a success to be "5 comes up" and a failure to be "5 does not come up" we have defined the experiment in such a fashion that there are exactly two outcomes possible on each trial, 5 or not 5. This is true even though there are actually six numbers that may appear on the upward face of a die.

FIGURE 5.2   **DIAGRAM OF AN EIGHT-TRIAL BINOMIAL EXPERIMENT**

Property 1: The experiment consists of $n = 8$ identical trials.

Property 2: Each trial results in either success (S) or failure (F).

Trials ⟶ 1 2 3 4 5 6 7 8

Outcomes ⟶ S F F S S F S S

EXAMPLE 5.7 ▶

An insurance salesperson pays a visit to 10 randomly selected families. An outcome associated with a visit is classified as a success if the family purchases an insurance policy and a failure if the family does not. From past experience, the salesperson knows the probability that a randomly selected family purchases an insurance policy is .10. Show that the process of the salesperson contacting the 10 families and recording the number of families that purchase an insurance policy is a binomial experiment.

Checking the properties of a binomial experiment, we observe the following:

1. The experiment consists of 10 identical trials, where each trial involves contacting one family.
2. There are two outcomes possible on each trial: the family purchases a policy or the family does not purchase a policy.
3. The probabilities of a purchase and a nonpurchase are assumed to be the same for each family, with $p = .10$ and $1 - p = .90$.
4. The trials are independent since the families are randomly selected.

Since the four assumptions are satisfied, this is a binomial experiment. The random variable of interest is the number of sales obtained in contacting the 10 families. In this case, x can assume the values of 0, 1, 2, 3, 4, 5, 6, 7, 8, 9, and 10. ◀

Property 3 of the binomial experiment is often called the *stationarity assumption* and is sometimes confused with Property 4, independence of trials. To see how they differ, consider again the case of the salesperson in Example 5.7 calling on families to sell insurance policies. If, as the day wore on the salesperson got tired and lost enthusiasm, then the probability of success (selling a policy) might drop to .05 by the 10th call. In such a case, Property 3 (stationarity) would not be satisfied, and we would not have a binomial experiment. This would be true even if Property 4 held—that is, the purchase decisions of each family were made independently.

In applications involving binomial experiments, a special mathematical formula, called the *binomial probability function,* can be used to compute the probability of x successes in the n trials. We develop the binomial probability function by considering a situation that can be analyzed using the methods from Chapter 4. We then show that if the properties of a binomial experiment are satisfied, the binomial probability function can be used to compute the desired probabilities.

EXAMPLE 5.8 ▶

A moving target at a police academy target range can be hit 80% of the time by a particular individual. Suppose this person takes three shots at the target. What is the probability of exactly two hits?

We can see from the tree diagram in Figure 5.3 that the experiment of taking three shots at the target has eight possible outcomes. Using S to denote success (hitting the target) and F to denote failure (missing the target), we are interested in outcomes having two successes in the three trials, or shots.

Next, let us verify that the experiment of taking three shots at the target has the properties of a binomial experiment. Check to see if you agree with the following conclusions:

1. The experiment consists of three identical trials or shots at the target.
2. The two outcomes per trial are a hit (S) or a miss (F).
3. The probability of a hit and the probability of a miss are the same for each trial, with $p = .80$ and $1 - p = 1 - .80 = .20$.
4. The trials, or shots, are independent. ◀

FIGURE 5.3   TREE DIAGRAM OF EXPERIMENT INVOLVING TAKING THREE SHOTS AT A TARGET

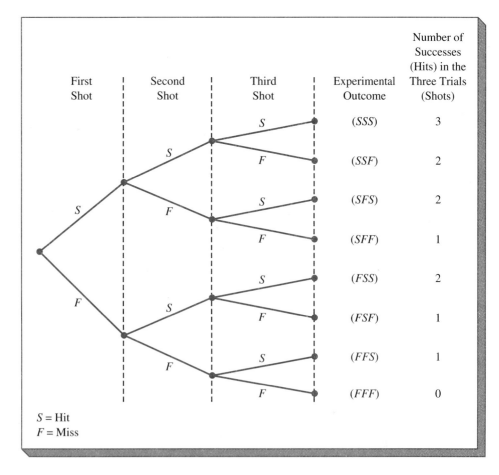

First Shot	Second Shot	Third Shot	Experimental Outcome	Number of Successes (Hits) in the Three Trials (Shots)
		S	(SSS)	3
	S	F	(SSF)	2
		S	(SFS)	2
S	F	F	(SFF)	1
		S	(FSS)	2
	S	F	(FSF)	1
F		S	(FFS)	1
	F	F	(FFF)	0

S = Hit
F = Miss

The number of outcomes of a binomial experiment that result in exactly x successes in n trials can be computed from the counting rule for combinations.*

$$\text{Number of Experimental Outcomes Providing Exactly } x \text{ Successes in } n \text{ Trials} = \binom{n}{x} = \frac{n!}{x!(n-x)!} \qquad (5.6)$$

where

$$n! = n(n-1)(n-2)\cdots(2)(1) \qquad (5.7)$$

and

$$0! = 1$$

The term $n!$ is called *n factorial*. For example, $5! = (5)(4)(3)(2)(1) = 120$.

---

*This is the counting rule for combinations introduced in Chapter 4. It is commonly used to determine the number of combinations of n objects selected x at a time. For the binomial experiment, this combinatorial formula provides the number of experimental outcomes having x successes in n trials.

EXAMPLE 5.8
(CONTINUED)

Now let us return to the experiment of taking three shots at a target. Equation (5.6) can be used to determine the number of experimental outcomes involving two hits; that is, the number of ways of obtaining $x = 2$ successes in the $n = 3$ trials. From equation (5.6) we have

$$\binom{n}{x} = \binom{3}{2} = \frac{3!}{2!(3-2)!} = \frac{(3)(2)(1)}{[(2)(1)](1)} = \frac{6}{2} = 3$$

Equation (5.6) shows that three of the outcomes yield two successes. From Figure 5.3 we see these three outcomes are denoted by *SSF, SFS,* and *FSS.*

Using equation (5.6) to determine how many experimental outcomes have three successes (hits) in the three trials, we obtain:

$$\binom{n}{x} = \binom{3}{3} = \frac{3!}{3!(3-3)!} = \frac{3!}{3!0!} = \frac{(3)(2)(1)}{[(3)(2)(1)](1)} = \frac{6}{6} = 1$$

From Figure 5.3 we see that the one experimental outcome with three successes is identified by *SSS.*   ◀

We know that the counting rule for combinations, equation (5.6), can be used to determine the number of experimental outcomes that result in x successes. But, if we are to determine the probability of x successes in n trials, we must also know the probability associated with each experimental outcome. Since the trials of a binomial experiment are independent, we can simply multiply the probabilities associated with each trial outcome to find the probability of a particular sequence of outcomes.

EXAMPLE 5.8
(CONTINUED)

With three shots at a target, the probability of hitting on the first and second shots but missing on the third is given by

$$pp(1 - p)$$

With a .80 probability of hitting the target on any one shot, the probability of hitting the target with the first two shots and missing the target with the third shot is given by

$$(.80)(.80)(.20) = (.80)^2(.20) = .128$$

There are two other sequences of outcomes resulting in two successes and one failure. The probabilities for all three sequences involving two successes are as shown.

Trial Outcomes				
1st Shot	2nd Shot	3rd Shot	Success/Failure Notation	Probability of Outcome
Hit	Hit	Miss	SSF	$pp(1 - p) = p^2(1 - p) = (.80)^2(.20) = .128$
Hit	Miss	Hit	SFS	$p(1 - p)p = p^2(1 - p) = (.80)^2(.20) = .128$
Miss	Hit	Hit	FSS	$(1 - p)pp = p^2(1 - p) = (.80)^2(.20) = .128$

Observe that all three outcomes with two successes have exactly the same probability.   ◀

In any binomial experiment, each sequence of trial outcomes yielding x successes in n trials has the *same probability* of occurrence. The probability of each sequence of trials yielding x successes in n trials is as follows:

$$\text{Probability of a Particular}$$
$$\text{Sequence of Trial Outcomes} = p^x(1 - p)^{(n-x)} \qquad \textbf{(5.8)}$$
$$\text{with } x \text{ Successes in } n \text{ Trials}$$

For the target practice situation of Example 5.8, this formula shows that any outcome with two successes has a probability of $p^2(1 - p)^{(3-2)} = p^2(1 - p)^1 = (.80)^2(.20)^1 = .128$, as shown.

Since equation (5.6) shows the number of outcomes in a binomial experiment with x successes and since equation (5.8) gives the probability for each sequence involving x successes, we combine the two to obtain the following *binomial probability function.*

Binomial Probability Function

$$f(x) = \binom{n}{x} p^x(1 - p)^{(n-x)} \qquad \textbf{(5.9)}$$

where

$$f(x) = \text{the probability of } x \text{ successes}$$

$$n = \text{the number of trials}$$

$$\binom{n}{x} = \frac{n!}{x!(n - x)!}$$

$$p = \text{the probability of a success on any one trial}$$

$$(1 - p) = \text{the probability of a failure on any one trial}$$

EXAMPLE 5.8
(CONTINUED)

In the experiment of taking three shots at the target, what is the probability for each of the following: hits on all three shots, hits on exactly two shots, a hit on exactly one shot, and misses on all three shots?

The binomial probability function can be used to answer these questions.

$$f(3) = \frac{3!}{3!(3 - 3)!}(.80)^3(.20)^0 = \frac{6}{6}(.512)(1) = .512$$

$$f(2) = \frac{3!}{2!(3 - 2)!}(.80)^2(.20)^1 = \frac{6}{2}(.64)(.20) = .384$$

$$f(1) = \frac{3!}{1!(3 - 1)!}(.80)^1(.20)^2 = \frac{6}{2}(.80)(.04) = .096$$

$$f(0) = \frac{3!}{0!(3 - 0)!}(.80)^0(.20)^3 = \frac{6}{6}(1)(.008) = .008$$

Summarizing these calculations in a tabular form provides the following probability distribution for the number of hits in three shots at the target.

x	$f(x)$
3	.512
2	.384
1	.096
0	.008
	1.000

A graphical representation of this probability distribution is shown in Figure 5.4. ◄

The binomial probability function can be applied to *any* binomial experiment. If we are satisfied that a situation has the properties of a binomial experiment and if we know the values of n and p, equation (5.9) can be used to compute the probability of x successes in the n trials.

EXAMPLE 5.9 ▶ For the typical American,* the probability of being hospitalized during a 1-year period is .12. Assume that four randomly selected individuals have been interviewed as part of a health-care study.

1. What is the probability that none of the individuals have been hospitalized during the past year?

FIGURE 5.4   **BINOMIAL PROBABILITY DISTRIBUTION FOR NUMBER OF HITS IN THREE TRIALS**

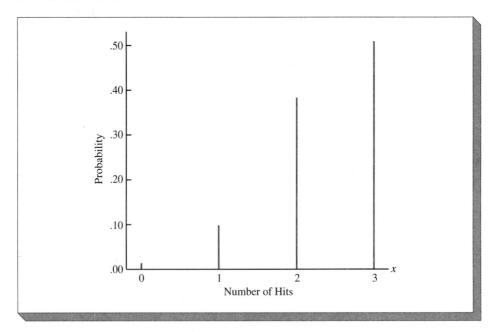

*From *How You Rate* by Tom Biracree, New York: Dell Publishing Co., 1984.

We need to compute $f(0)$ for the binomial probability function with $n = 4$ trials, $p = .12$, and $1 - p = .88$. We obtain

$$f(0) = \frac{4!}{0!(4-0)!}(.12)^0(.88)^4 = .5997$$

2. What is the probability that two or more of the individuals have been hospitalized during the past year?

We need to determine the probability of 2, 3, or 4 individuals being hospitalized. This is $f(2) + f(3) + f(4)$. But the same probability is given by $1 - f(0) - f(1)$. Since we already know $f(0)$, the second approach involves less computation.

$$f(1) = \frac{4!}{1!(4-1)!}(.12)^1(.88)^3 = .3271$$

Therefore, the probability that two or more of the individuals have been hospitalized is $1 - f(0) - f(1) = 1 - .5997 - .3271 = .0732$. ◀

EXAMPLE 5.10 ▶ Eight customers enter a clothing store during a 1-hour period. From past experience, it is known that approximately 30% of the people entering the store make a purchase. Answer the following questions.

1. What is the probability that exactly three of the eight customers make a purchase?

Assuming the properties of the binomial experiment apply, we have $n = 8$ trials with $p = .30$ and $1 - p = 1 - .30 = .70$.

$$f(3) = \frac{8!}{3!(8-3)!}(.30)^3(.70)^{(8-3)}$$

$$= \frac{40,320}{720}(.027)(.16807) = .2541$$

2. What is the probability at least one customer makes a purchase?

The probability of *at least* one customer purchase is the sum of the probabilities of 1, 2, 3, 4, 5, 6, 7, and 8 customer purchases. Although we could compute each of these probabilities separately and add them together, it is helpful to note that the probability of at least one person making a purchase is 1 minus the probability of no customer purchases. Computing the probability of no successes in the eight trials, we have:

$$f(0) = \frac{8!}{0!(8-0)!}(.30)^0(.70)^{(8-0)}$$

$$= \frac{8!}{1(8!)}(1)(.0576) = .0576$$

Therefore, the probability of at least one success must be:

$$P(\text{at least one success}) = 1 - f(0)$$

$$= 1 - .0576 = .9424$$ ◀

USING TABLES OF BINOMIAL PROBABILITIES

Tables have been developed that give the probability of x successes in n trials for a binomial experiment. These tables are generally easy to use and quicker than applying equation (5.9). A table of binomial probabilities is provided as Table 5 of Appendix B. A portion of this table is given in Table 5.4. To use this table, it is necessary to specify the values of n, p, and x for the binomial experiment of interest. In the example at the top of Table 5.4, we see that the probability of $x = 3$ successes in a binomial experiment with $n = 10$ and $p = .40$ is .2150. You might want to use equation (5.9) to verify that this is the answer you would obtain using the binomial probability function directly.

TABLE 5.4

SELECTED VALUES OF THE BINOMIAL PROBABILITY TABLE
EXAMPLE: $n = 10$, $x = 3$, $p = .40$; $P(x = 3$ SUCCESSES) = .2150

n	x	p									
		.05	.10	.15	.20	.25	.30	.35	.40	.45	.50
9	0	.6302	.3874	.2316	.1342	.0751	.0404	.0207	.0101	.0046	.0020
	1	.2985	.3874	.3679	.3020	.2253	.1556	.1004	.0605	.0339	.0176
	2	.0629	.1722	.2597	.3020	.3003	.2668	.2162	.1612	.1110	.0703
	3	.0077	.0446	.1069	.1762	.2336	.2668	.2716	.2508	.2119	.1641
	4	.0006	.0074	.0283	.0661	.1168	.1715	.2194	.2508	.2600	.2461
	5	.0000	.0008	.0050	.0165	.0389	.0735	.1181	.1672	.2128	.2461
	6	.0000	.0001	.0006	.0028	.0087	.0210	.0424	.0743	.1160	.1641
	7	.0000	.0000	.0000	.0003	.0012	.0039	.0098	.0212	.0407	.0703
	8	.0000	.0000	.0000	.0000	.0001	.0004	.0013	.0035	.0083	.0176
	9	.0000	.0000	.0000	.0000	.0000	.0000	.0001	.0003	.0008	.0020
10	0	.5987	.3487	.1969	.1074	.0563	.0282	.0135	.0060	.0025	.0010
	1	.3151	.3874	.3474	.2684	.1877	.1211	.0725	.0403	.0207	.0098
	2	.0746	.1937	.2759	.3020	.2816	.2335	.1757	.1209	.0763	.0439
	3	.0105	.0574	.1298	.2013	.2503	.2668	.2522	**.2150**	.1665	.1172
	4	.0010	.0112	.0401	.0881	.1460	.2001	.2377	.2508	.2384	.2051
	5	.0001	.0015	.0085	.0264	.0584	.1029	.1536	.2007	.2340	.2461
	6	.0000	.0001	.0012	.0055	.0162	.0368	.0689	.1115	.1596	.2051
	7	.0000	.0000	.0001	.0008	.0031	.0090	.0212	.0425	.0746	.1172
	8	.0000	.0000	.0000	.0001	.0004	.0014	.0043	.0106	.0229	.0439
	9	.0000	.0000	.0000	.0000	.0000	.0001	.0005	.0016	.0042	.0098
	10	.0000	.0000	.0000	.0000	.0000	.0000	.0000	.0001	.0003	.0010
11	0	.5688	.3138	.1673	.0859	.0422	.0198	.0088	.0036	.0014	.0005
	1	.3293	.3835	.3248	.2362	.1549	.0932	.0518	.0266	.0125	.0054
	2	.0867	.2131	.2866	.2953	.2581	.1998	.1395	.0887	.0531	.0269
	3	.0137	.0710	.1517	.2215	.2581	.2568	.2254	.1774	.1259	.0806
	4	.0014	.0158	.0536	.1107	.1721	.2201	.2428	.2365	.2060	.1611
	5	.0001	.0025	.0132	.0388	.0803	.1321	.1830	.2207	.2360	.2256
	6	.0000	.0003	.0023	.0097	.0268	.0566	.0985	.1471	.1931	.2256
	7	.0000	.0000	.0003	.0017	.0064	.0173	.0379	.0701	.1128	.1611
	8	.0000	.0000	.0000	.0002	.0011	.0037	.0102	.0234	.0462	.0806
	9	.0000	.0000	.0000	.0000	.0001	.0005	.0018	.0052	.0126	.0269
	10	.0000	.0000	.0000	.0000	.0000	.0000	.0002	.0007	.0021	.0054
	11	.0000	.0000	.0000	.0000	.0000	.0000	.0000	.0000	.0002	.0005

EXAMPLE 5.10 ▶
(CONTINUED)

Assume that 10 customers enter the clothing store during a 1-hour period and that the probability of a customer purchase is .30. Use the table of binomial probabilities to answer the following questions.

1. What is the probability that exactly 3 of the 10 customers make a purchase? From Table 5.4, $f(3) = .2668$.
2. What is the probability that at least 1 customer makes a purchase? From Table 5.4, $P(\text{at least 1 success}) = 1 - f(0) = 1 - .0282 = .9718$.
3. What is the probability that 3 or fewer customers make a purchase? From Table 5.4, $P(3 \text{ or fewer successes}) = f(3) + f(2) + f(1) + f(0) = .2668 + .2335 + .1211 + .0282 = .6496$.

Using all the probabilities corresponding to $n = 10$ trials with $p = .30$, Figure 5.5 provides a graphical representation of the binomial probability distribution for the number of customer purchases. ◀

While Example 5.10 demonstrates the relative ease of using the tables of binomial probabilities, it is impossible to have tables that show all possible values of n and p that might be encountered in a binomial experiment. In cases where the appropriate probabilities are not available, one can interpolate to arrive at an approximation. For instance, suppose we had $p = .275$, $n = 10$, and $x = 3$. Using Table 5.4 for $p = .25$, $n = 10$, and $x = 3$, we find that $f(3) = .2503$. Similarly for $p = .30$, $n = 10$, and $x = 3$ we find that $f(3) = .2668$. Interpolating halfway between these values, we obtain $f(3) = (.2503 + .2668)/2 = .2586$ for the case of $p = .275$. Alternatively, with today's calculators, it is not too difficult to calculate the desired probability using equation (5.9), especially if the number of trials is not too large. In the exercises, you should practice using equation (5.9) to compute the binomial probabilities unless the problem specifically requests that you use the binomial probability table.

FIGURE 5.5   **BINOMIAL PROBABILITY DISTRIBUTION FOR NUMBER OF CUSTOMER PURCHASES IN 10 TRIALS**

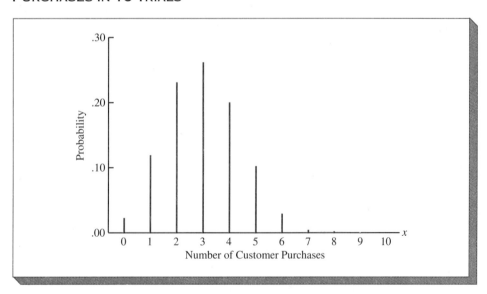

THE EXPECTED VALUE AND VARIANCE FOR THE BINOMIAL PROBABILITY DISTRIBUTION

In Section 5.3 we provided formulas for computing the expected value and variance of a discrete random variable. In the special case where the random variable has a binomial probability distribution with a known number of trials (n) and known probability, the general formulas for the expected value and variance, equations (5.3) and (5.4), can be simplified; the results are as follows.

Expected Value and Variance for the Binomial Probability Distribution

$$E(x) = \mu = np \qquad\qquad (5.10)$$

$$\text{Var}(x) = \sigma^2 = np(1 - p) \qquad\qquad (5.11)$$

EXAMPLE 5.11 ▶ A statistics class has 25 students. From past experience, it is known that 20% of all students who take this course withdraw before the end of the term. What is the expected number of withdrawals, and what is the variance in the number of withdrawals?

Assuming each student makes a decision independently, the process can be viewed as a binomial experiment with $n = 25$, $p = .20$, and $(1 - p) = 1 - .20 = .80$. The random variable of interest is the number of withdrawals during the term.

Using equations (5.10) and (5.11), the expected value and variance of the number of withdrawals are as follows:

$$\mu = np = (25)(.20) = 5$$

$$\sigma^2 = np(1 - p) = (25)(.20)(.80) = 4$$

The corresponding standard deviation is $\sigma = \sqrt{4} = 2$.   ◀

EXAMPLE 5.12 ▶ A shipment of 500 parts is received from a supplier. From past experience, it is known that the probability that any particular part is defective is .03. What is the expected number of defective parts in the shipment? What are the variance and standard deviation in the number of defective parts?

Viewing this as a binomial experiment with $n = 500$, $p = .03$, and $1 - p = 1 - .03 = .97$, we have

$$\mu = (500)(.03) = 15$$

$$\sigma^2 = (500)(.03)(.97) = 14.55$$

$$\sigma = \sqrt{14.55} = 3.81$$ ◀

To obtain an intuitive understanding of the expected value and variance for the binomial distribution, think of taking a sample of size n from an infinite population where p is the proportion of successes. Clearly np would be the expected number of successes in the sample. Note also that if $p = 0$ or 1, then $\text{Var}(x) = 0$. This is as it should be. If $p = 0$, then all items in the population represent failure and there is no variance. If $p = 1$, then all items in the population represent success and there is again no variance.

NOTES AND COMMENTS

1. Some binomial tables only show values of p up to and including $p = .50$. Thus, it would appear these tables cannot be used when the probability of success exceeds $p = .50$. However, such tables can be used by noting that the probability of $n - x$ failures is also the probability of x successes. So when the probability of success is greater than $p = .50$, one can compute the probability of $n - x$ failures instead. The probability of failure, $1 - p$, will then be less than .50 when $p > .50$. The tables in this text include all values of p.

2. Some sources present binomial tables in a cumulative form. Using these, one must subtract to find the probability of x successes in n trials. For example, $f(2) = P(x \leq 2) - P(x \leq 1)$. The tables in this text provide these probabilities directly. To compute cumulative probabilities using our tables, one simply sums the individual probabilities. For example, to compute $P(x \leq 2)$, we sum $f(0) + f(1) + f(2)$.

EXERCISES

METHODS

SELF TEST

27. Consider a binomial experiment with two trials and $p = .4$.
   a. Draw a tree diagram showing this as a two-trial experiment (see Figure 5.3).
   b. Compute the probability of one success, $f(1)$.
   c. Compute $f(0)$.
   d. Compute $f(2)$.
   e. Find the probability of at least one success.
   f. Find the expected value, variance, and standard deviation.

28. Suppose in a binomial experiment we have $n = 4$ and $p = .15$.
   a. What is the probability of no successes?
   b. What is the probability of one success?
   c. What is the probability of two or more successes?

29. For $n = 15$ and $p = .35$, use the binomial probability tables to compute the following:
   a. $f(3)$
   b. $f(5)$
   c. $f(10)$
   d. $1 - f(0)$

30. For $n = 10$ and $p = .125$, use the binomial tables and interpolate to find the following:
   a. $f(2)$
   b. $f(3)$
   c. $f(8)$

31. Consider a binomial experiment with $n = 12$ and $p = .20$.
   a. Find $E(x)$.
   b. Find $\text{Var}(x)$.

32. Consider a binomial experiment with $n = 10$ and $p = .10$. Use the binomial tables (Table 5 of Appendix B) to answer parts (a) through (d).
   a. Find $f(0)$.   d. Find $P(x \geq 1)$.
   b. Find $f(2)$.   e. Find $E(x)$.
   c. Find $P(x \leq 2)$.   f. Find $\text{Var}(x)$ and σ.

33. Consider a binomial experiment with $n = 20$ and $p = .70$. Use the binomial tables (Table 5 of Appendix B) to answer parts (a) through (d).

    a. Find $f(12)$.   **d.** Find $P(x \leq 15)$.

    b. Find $f(16)$.   **e.** Find $E(x)$.

    c. Find $P(x \geq 16)$.   **f.** Find $Var(x)$ and σ.

APPLICATIONS

34. The greatest number of complaints by owners of two-year-old automobiles is in the area of electrical system performance (*Consumer Reports 1992 Buying Guide*). In an annual questionnaire of owners of over 300 makes and models of automobiles, *Consumer Reports* found that 10% of the owners of two-year-old automobiles found trouble spots in the electrical system that included the starter, alternator, battery, switch controls, instruments, wiring, lights, and radio.

    a. What is the probability that a sample of 12 owners of two-year-old automobiles will find exactly 2 owners with electrical system problems?

    b. What is the probability that a sample of 12 owners of two-year-old automobiles will find at least 2 owners with electrical system problems?

    c. What is the probability that a sample of 20 owners of two-year-old automobiles will find at least one electrical system problem?

35. The New York State Bar Examination is the basis for admitting law school graduates into the law profession. Historically, 30% of the individuals taking the examination pass on their first attempt. Suppose that a group of 15 individuals will be taking the examination for the first time and that we are interested in the number of individuals in this group who will pass the exam. Describe the conditions necessary for this situation to be a binomial experiment.

SELF TEST   ▷ **36.** When a new machine is functioning properly, only 3% of the items produced are defective. Assume that we will randomly select two parts produced on the machine and that we are interested in the number of defective parts found.

    a. Describe the conditions under which this situation would be a binomial experiment.

    b. Draw a tree diagram similar to Figure 5.3 showing this as a two-trial experiment.

    c. How many experimental outcomes result in exactly one defect being found?

    d. Compute the probabilities associated with finding no defects, exactly one defect, and two defects.

37. Ninety-five percent of licensed practical nurses are women (*Statistical Abstract of the United States,* 1992). Suppose 10 licensed practical nurses are randomly selected to be interviewed about the quality of work conditions.

    a. Is the selection of the 10 licensed practical nurses a binomial experiment? Explain.

    b. What is the probability that three of the licensed practical nurses will be women?

    c. What is the probability that none will be women?

    d. What is the probability that at least one will be a woman?

38. The book *100% American* by Daniel Evan Weiss reports over 1000 statistical facts about the United States and its people. One fact reported is that 64% of the people live in the state where they were born.

    a. What is the probability that in a random sample of 10 people at least 8 people will be living in the state where they were born?

    b. What is the probability that in a random sample of 3 people exactly 1 person will be living in the state where she or he was born?

39. National Oil Company conducts exploratory oil drilling operations in the southwestern United States. To fund the operation, investors form partnerships, which provide the financial support necessary to drill a fixed number of oil wells. Each well drilled is classified as a producer well or a dry well. Past experience shows that this type of exploratory operation provides producer wells for 15% of all wells drilled. A newly formed partnership has provided the financial support for drilling at 12 exploratory locations.

a. What is the probability that all 12 wells will be producer wells?

b. What is the probability that all 12 wells will be dry wells?

c. What is the probability that exactly 1 well will be a producer well?

d. To make the partnership venture profitable, at least 3 of the exploratory wells must be producer wells. What is the probability that the venture will be profitable?

40. Military radar and missile detection systems are designed to warn a country against enemy attacks. A reliability question deals with the ability of the detection system to identify an attack and issue a warning. Assume that a particular detection system has a .90 probability of detecting a missile attack. Answer the following questions using the binomial probability distribution.

a. What is the probability that a single detection system will detect an attack?

b. If two detection systems are installed in the same area and operate independently, what is the probability that at least one of the systems will detect the attack?

c. If three systems are installed, what is the probability that at least one of the systems will detect the attack?

d. Would you recommend that multiple detection systems be used? Explain.

41. Assume that the binomial experiment applies for the case of a college basketball player shooting free throws. Late in a basketball game, a team will sometimes foul intentionally in the hope that the player shooting the free throw will miss and the team committing the foul will get the ball. Assume that the best player on the opposing team has a .82 probability of making a free throw and that the worst player has a .56 probability of making a free throw.

a. What are the probabilities that the best player makes 0, 1, and 2 points if fouled and given two free throws?

b. What are the probabilities that the worst player makes 0, 1, and 2 points if fouled and given two free throws?

c. Does it make sense for a coach to have a preset plan about which player to intentionally foul late in a basketball game? Explain.

42. A firm estimates the probability of having employee disciplinary problems on any day to be .10.

a. What is the probability that the company experiences 5 days without a disciplinary problem?

b. What is the probability of exactly 2 days with disciplinary problems in a 10-day period?

c. What is the probability of at least 2 days with disciplinary problems in a 20-day period?

43. At a particular university, it has been found that 20% of the students withdraw without completing the calculus course. Assume that 20 students have registered for the course this quarter.

a. What is the probability that two or fewer will withdraw?

b. What is the probability that exactly four will withdraw?

c. What is the probability that more than three will withdraw?

d. What is the expected number of withdrawals?

44. Suppose that a newly married couple is planning to have three children and that the couple is interested in knowing the probabilities of having no girls, one girl, two girls, and three girls. Assume that the probability of having a girl is .50 on any one birth.

a. What are the trials of the experiment in this application? How many trials are there?

b. How many outcomes are possible on each trial and what are they?

c. What are the probabilities associated with the outcomes for each trial? Are these probabilities the same for each trial?

    d. What additional assumption must be made about the trials for this to be a binomial experiment?

    e. What is the random variable of interest in this problem? Is it a discrete or a continuous random variable, and what values can it assume?

45. A study found that for 60% of the couples who have been married 10 years or less, both spouses work. A sample of 20 couples who have been married 10 years or less will be selected from marital records available at a local courthouse. We will be interested in the number of couples in the sample in which both spouses work. Describe the conditions necessary for this sampling process to be viewed as a binomial experiment.

46. Forty-five percent of the residents in a township who are of voting age are not registered to vote.

    a. In a sample of 10 people, what is the probability 5 are not registered to vote?

    b. In a sample of 10 people, what is the probability 2 or fewer are not registered to vote?

47. Suppose a salesperson makes a sale on 20% of customer contacts. A normal work week will enable the salesperson to contact 25 customers. What is the expected number of sales for the week? What is the variance for the number of sales for the week? What is the standard deviation for the number of sales for the week?

48. Of the next-day express mailings handled by the U.S. Postal Service, 85% are actually received by the addressee 1 day after the mailing. What is the expected value and variance for the number of 1-day deliveries in a group of 250 express mailings?

49. Betting on the color red in the game of roulette has an $18/38$ chance of winning. What is the expected value and variance for the number of wins in a series of 100 bets on red?

5.5 ▽ THE POISSON PROBABILITY DISTRIBUTION (OPTIONAL)

In this section we consider a discrete random variable that is often useful when dealing with the number of occurrences over a specified interval of time or space. For example, the random variable of interest might be the number of arrivals at a car wash in 1 hour, the number of repairs needed in 10 miles of highway, or the number of leaks in 100 miles of pipeline. If two assumptions are satisfied, the number of occurrences is a random variable described by the *Poisson probability function*. The two assumptions are

1. The probability of an occurrence is the same for any two intervals of equal length.

2. The occurrence or nonoccurrence in any interval is independent of the occurrence or nonoccurrence in any other interval.

The Poisson probability function is given by equation (5.12).

Poisson Probability Function

$$f(x) = \frac{\mu^x e^{-\mu}}{x!} \qquad \text{for } x = 0, 1, 2, \ldots \qquad (5.12)$$

where

$f(x) = $ the probability of x occurrences in an interval

$\mu = $ expected value or average number of occurrences in an interval

$e = 2.71828$

Before we consider a specific example to see how the Poisson distribution can be applied, note that there is no upper limit on x, the number of occurrences. It is a discrete random variable that may assume an infinite sequence of values ($x = 0, 1, 2, \ldots$). The Poisson random variable has no upper limit; any nonnegative integer value is permissible.

EXAMPLE 5.13 ▶ Suppose we are interested in the number of arrivals at the drive-up window of a fast-food restaurant during a 5-minute period. If we can assume that the probability of a car arriving is the same for any two periods of equal length and that the arrival or nonarrival of a car in any period is independent of the arrival or nonarrival in any other period, the Poisson probability function is applicable. Suppose that we are interested only in the busy lunch hour period, and these assumptions are satisfied. The average number of cars arriving in a 5-minute period is 10; thus, the following probability function applies:

$$f(x) = \frac{10^x e^{-10}}{x!} \qquad \text{for } x = 0, 1, 2, \ldots$$

The random variable x represents the number of cars arriving during a 5-minute period.

If we wanted to know the probability of exactly five arrivals in a 5-minute period, we would set $x = 5$ and obtain*

$$\begin{array}{c}\text{Probability of Exactly}\\ \text{5 Arrivals in 5 Minutes}\end{array} = f(5) = \frac{10^5 e^{-10}}{5!} = .0378 \qquad ◀$$

The probability in Example 5.13 is determined by evaluating the Poisson probability function with $\mu = 10$ and $x = 5$. It is often easier to refer to tables to find Poisson probabilities. Poisson probability tables can be used when μ and x are known. Such a table is included as Table 7 of Appendix B. For convenience, we have reproduced a portion of this table as Table 5.5 where we see that the probability of five arrivals in a period with $\mu = 10$ is found by locating the value in the row of the table corresponding to $x = 5$ and the column of the table corresponding to $\mu = 10$. It is $f(5) = .0378$.

Example 5.13 involves computing the probability of five arrivals in a 5-minute period. To compute Poisson probabilities for time periods of different length, one must first determine the expected number of occurrences during the period of interest. Then equation (5.12), or the Poisson probability tables, can be used.

EXAMPLE 5.13 ▶
(CONTINUED) Suppose we want to compute the probability of three arrivals during a 2-minute period. Since 10 is the expected number of arrivals in a 5-minute period, $\mu = 4$ is the expected number of arrivals in a 2-minute period. Thus, the probability of x arrivals in a 2-minute period is given by

$$f(x) = \frac{4^x e^{-4}}{x!} \qquad \text{for } x = 0, 1, 2, \ldots$$

To find the probability of three arrivals during a 2-minute period, we can either use the above formula or Table 7 in Appendix B. Using the above formula, we obtain

$$f(3) = \frac{4^3 e^{-4}}{3!} = .1954 \qquad ◀$$

---

*Values of $e^{-\mu}$ can be found in Table 6 of Appendix B.

TABLE 5.5   **SELECTED VALUES FROM THE POISSON PROBABILITY TABLES**
EXAMPLE: $\mu = 10$, $x = 5$; $f(5) = .0378$

					μ					
x	9.1	9.2	9.3	9.4	9.5	9.6	9.7	9.8	9.9	10
0	.0001	.0001	.0001	.0001	.0001	.0001	.0001	.0001	.0001	.0000
1	.0010	.0009	.0009	.0008	.0007	.0007	.0006	.0005	.0005	.0005
2	.0046	.0043	.0040	.0037	.0034	.0031	.0029	.0027	.0025	.0023
3	.0140	.0131	.0123	.0115	.0107	.0100	.0093	.0087	.0081	.0076
4	.0319	.0302	.0285	.0269	.0254	.0240	.0226	.0213	.0201	.0189
5	.0581	.0555	.0530	.0506	.0483	.0460	.0439	.0418	.0398	**.0378**
6	.0881	.0851	.0822	.0793	.0764	.0736	.0709	.0682	.0656	.0631
7	.1145	.1118	.1091	.1064	.1037	.1010	.0982	.0955	.0928	.0901
8	.1302	.1286	.1269	.1251	.1232	.1212	.1191	.1170	.1148	.1126
9	.1317	.1315	.1311	.1306	.1300	.1293	.1284	.1274	.1263	.1251
10	.1198	.1210	.1219	.1228	.1235	.1241	.1245	.1249	.1250	.1251
11	.0991	.1012	.1031	.1049	.1067	.1083	.1098	.1112	.1125	.1137
12	.0752	.0776	.0799	.0822	.0844	.0866	.0888	.0908	.0928	.0948
13	.0526	.0549	.0572	.0594	.0617	.0640	.0662	.0685	.0707	.0729
14	.0342	.0361	.0380	.0399	.0419	.0439	.0459	.0479	.0500	.0521
15	.0208	.0221	.0235	.0250	.0265	.0281	.0297	.0313	.0330	.0347
16	.0118	.0127	.0137	.0147	.0157	.0168	.0180	.0192	.0204	.0217
17	.0063	.0069	.0075	.0081	.0088	.0095	.0103	.0111	.0119	.0128
18	.0032	.0035	.0039	.0042	.0046	.0051	.0055	.0060	.0065	.0071
19	.0015	.0017	.0019	.0021	.0023	.0026	.0028	.0031	.0034	.0037
20	.0007	.0008	.0009	.0010	.0011	.0012	.0014	.0015	.0017	.0019
21	.0003	.0003	.0004	.0004	.0005	.0006	.0006	.0007	.0008	.0009
22	.0001	.0001	.0002	.0002	.0002	.0002	.0003	.0003	.0004	.0004
23	.0000	.0001	.0001	.0001	.0001	.0001	.0001	.0001	.0002	.0002
24	.0000	.0000	.0000	.0000	.0000	.0000	.0000	.0001	.0001	.0001

The Poisson probability distribution is not limited to computing probabilities for occurrences over time periods. Example 5.14 illustrates another application.

EXAMPLE 5.14 ▶ Suppose we are interested in the number of defects in a section of highway shortly after it is resurfaced. It seems reasonable to assume that the probability of a defect is the same for any two intervals of equal length and that the occurrence of a defect in one interval is independent of a defect in any other interval. Assume that after resurfacing, defects can be expected to occur at the mean rate of two per mile.

Let us determine the probability that there will be no defects in a particular 3-mile stretch of the highway. Since we are interested in an interval with a length of 3 miles, $\mu = (2 \text{ defects/mile})(3 \text{ miles}) = 6$ represents the expected number of defects over the 3-mile stretch. Using either equation (5.12), or Table 7 in Appendix B, we find that the probability of no defects is .0025. Thus, it is very unlikely that there will be no defects in the 3-mile section. In fact, there is a $1 - .0025 = .9975$ probability of at least one defect.

◀

EXERCISES

METHODS

50. Consider a Poisson probability distribution with $\mu = 3$.
   a. Write the appropriate Poisson probability function.
   b. Find $f(2)$.   **c.** Find $f(1)$.   **d.** Find $P(x \geq 2)$.

SELF TEST

51. Consider a Poisson probability distribution with an average number of occurrences per time period of 2.
   a. Write the appropriate Poisson probability function.
   b. What is the average number of occurrences in three time periods?
   c. Write the appropriate Poisson probability function to determine the probability of x occurrences in three periods.
   d. Find the probability of two occurrences in one period.
   e. Find the probability of six occurrences in three periods.
   f. Find the probability of five occurrences in two periods.

52. Consider a Poisson random variable with $\mu = 4$. Compute the following:
   a. $f(0)$
   b. $f(1)$
   c. $f(2)$

53. Suppose we have a Poisson random variable with three occurrences per minute.
   a. Find $f(1)$.
   b. What is the probability of four or more occurrences in 1 minute?
   c. What is the probability of four or more occurrences in 2 minutes?
   d. What is the probability of 10 occurrences in 5 minutes?

APPLICATIONS

SELF TEST

54. Phone calls arrive at the rate of 48 per hour at the reservation desk for Regional Airways.
   a. Find the probability of receiving three calls in a 5-minute interval.
   b. Find the probability of receiving exactly 10 calls in 15 minutes.
   c. Suppose no calls are currently on hold. If it takes the agent 5 minutes to complete processing the current call, how many callers do you expect to be waiting by that time? What is the probability none will be waiting?
   d. If there are no calls currently being processed, what is the probability the agent can take 3 minutes for personal time without being interrupted?

55. During the period of time phone-in reservations are being taken at a local university, calls come in at the rate of one every 2 minutes.
   a. What is the expected number of calls in 1 hour?
   b. What is the probability of three calls in 5 minutes?
   c. What is the probability of no calls in a 5-minute period?

56. A certain restaurant has a reputation for good food. Restaurant management boasts that on a Saturday night, groups of customers arrive at the rate of 15 groups every half-hour.
   a. What is the probability that 5 minutes will pass with no groups of customers arriving?
   b. What is the probability that eight groups of customers will arrive in 10 minutes?
   c. What is the probability that more than five groups will arrive in a 10-minute period?

57. During rush hours, accidents occur in a particular metropolitan area at the rate of two per hour. The morning rush period lasts for 1 hour 30 minutes and the evening rush period lasts for 2 hours.
   a. On a particular day, what is the probability that there will be no accidents during the morning rush period?
   b. What is the probability of two accidents during the evening rush period?
   c. What is the probability of four or more accidents during the morning rush period?

d. On a particular day, what is the probability there will be no accidents during bot48h the morning and evening rush periods?

58. Airline passengers arrive randomly and independently at the passenger-screening facility at a major international airport. The mean arrival rate is 10 passengers per minute.
   a. What is the probability of no arrivals in a 1-minute period?
   b. What is the probability that three or fewer passengers arrive in a 1-minute period?
   c. What is the probability of no arrivals in a 15-second period?
   d. What is the probability of at least one arrival in a 15-second period?

59. Williams Company has observed that calculators fail and need to be replaced at the rate of three every 25 days.
   a. What is the expected number of calculators that will fail in 30 days?
   b. What is the probability that at least two will fail in 50 days?
   c. What is the probability that exactly three will fail in 10 days?

60. Cars arrive at a car wash at the average rate of 15 cars per hour. If the number of arrivals per hour follows a Poisson distribution, what is the probability of 20 or more arrivals during any given hour of operation? Use the Poisson probability table.

61. A new automated production process has been experiencing an average of 1.5 breakdowns per day. Because of the cost associated with a breakdown, management is concerned about the possibility of having three or more breakdowns during a given day. Assume that the number of breakdowns per day follows a Poisson distribution. What is the probability of observing three or more breakdowns?

62. A regional director responsible for business development in Pennsylvania is concerned about the number of businesses that end as failures. If the average number of failures per month is 10, what is the probability that exactly 4 businesses will fail during a given month? Assume that the number of businesses failing per month follows a Poisson distribution.

63. During the registration period at a local university, students consult advisors with questions about course selection. A particular advisor noted that, during the registration period, an average of eight students per hour ask questions, although the exact arrival times of the students are random in nature. Use the Poisson distribution to answer the following questions:
   a. What is the probability that exactly eight students come in for consultation during a particular 1-hour period?
   b. What is the probability that three students come in for consultation during a particular ½-hour period?

5.6 ▽ THE HYPERGEOMETRIC PROBABILITY DISTRIBUTION (OPTIONAL)

The *hypergeometric probability distribution* is closely related to the binomial probability distribution. It also provides the probability of obtaining x successes in n trials when there are two possible outcomes (success and failure) on each trial. The key difference between the two probability distributions is that with the hypergeometric distribution, the probability of success changes from trial to trial.

EXAMPLE 5.15 ▶ A five-member committee consists of three women and two men. Two of the committee members are expected to represent the group at a meeting in Las Vegas. The committee has decided to randomly choose the two members that will attend the meeting. What is the probability that both persons chosen will be women?

Since three of the five committee members are women, the probability that the first person randomly selected is a woman is $\frac{3}{5} = .60$. However, if the first person chosen is a woman, then the probability of randomly selecting another woman from the four remaining committee members drops to $\frac{2}{4} = .50$. Therefore, the probability that two women will be selected to attend the meeting is $(.60)(.50) = .30$. ◀

One of the most important applications of the hypergeometric probability distribution involves sampling without replacement from a finite population. The objective is to choose a random sample of n items out of a population of N items, under the condition that once an item has been selected it is not returned to the population. Thus, on the next selection, the probability of selecting an item of that type goes down.

The usual notation in applications of the hypergeometric probability distribution is to let r denote the number of items in the population that are labeled success and $N - r$ denote the number of items in the population that are labeled failure. The hypergeometric probability function is used to compute the probability that in a random sample of n items, selected without replacement, we will obtain x items labeled success and $n - x$ items labeled failure. Note that for this to occur, we must obtain x successes from the r successes in the population and $n - x$ failures from the $N - r$ failures in the population. The hypergeometric probability function provides $f(x)$, the probability of obtaining x successes in a sample of size n.

Hypergeometric Probability Function

$$f(x) = \frac{\binom{r}{x}\binom{N - r}{n - x}}{\binom{N}{n}} \qquad x = 0, 1, \ldots, r \qquad (5.13)$$

where

$f(x)$ = probability of x successes
N = number of items in the population
r = number of items in the population labeled success
n = number of items in the sample

The term $\binom{N}{n}$ in the denominator of equation (5.13) simply represents the number of ways a sample of size n can be selected from a population of size N. In the numerator, $\binom{r}{x}$ represents the number of ways x successes can be selected from a total of r successes in the population and $\binom{N - r}{n - x}$ represents the number of ways $n - x$ failures can be selected from a total of $N - r$ failures in the population. The product in the numerator is thus the number of ways x successes and $n - x$ failures may be obtained in a sample of size n. The ratio of the numerator to the denominator is the probability of x successes and $n - x$ failures in the sample. To illustrate the computations involved in using equation (5.13), let us reconsider Example 5.15.

EXAMPLE 5.15
(CONTINUED)

Recall that the objective is to select two members from the five-member committee consisting of three women and two men. To determine the probability of obtaining a sample that consists of two women, we can use equation (5.13) with $N = 5$, $r = 3$, and $x = 2$.

$$f(2) = \frac{\binom{3}{2}\binom{2}{0}}{\binom{5}{2}} = \frac{\frac{3!}{2!1!}\frac{2!}{2!0!}}{\frac{5!}{3!2!}} = \frac{3}{10} = .30$$

Note that this is the same answer that we obtained previously. Suppose, however, that we now learn that three committee members will be allowed to make the trip. The probability that two of the three members will be women is

$$f(2) = \frac{\binom{3}{2}\binom{2}{1}}{\binom{5}{3}} = \frac{\frac{3!}{1!2!}\frac{2!}{1!1!}}{\frac{5!}{3!2!}} = \frac{6}{10} = .60$$

EXAMPLE 5.16

Suppose a population consists of 10 items, 6 of which are classified as acceptable and 4 of which are classified as defective. What is the probability that a random sample of size 3 contains two defective items?

For this problem, we can think of obtaining a defective item as a "success." Thus, $N = 10$, $r = 4$, $n = 3$, and $x = 2$. Using equation (5.13) we can compute $f(2)$, as shown below:

$$f(2) = \frac{\binom{4}{2}\binom{6}{1}}{\binom{10}{3}} = \frac{\frac{4!}{2!2!}\frac{6!}{1!5!}}{\frac{10!}{3!7!}} = \frac{36}{120} = .30$$

NOTES AND COMMENTS

The hypergeometric distribution is used to determine the probability of obtaining a certain sample outcome when sampling without replacement. When sampling with replacement, each item that is selected is returned to the population before another item is selected. The binomial probability distribution is then used to compute the probability of x successes.

EXERCISES

METHODS

SELF TEST

64. Suppose $N = 10$ and $r = 3$. Compute the hypergeometric probabilities for the following values of n and x.
    a. $n = 4, x = 1$ c. $n = 2, x = 0$
    b. $n = 2, x = 2$ d. $n = 4, x = 2$

65. Suppose $N = 15$ and $r = 4$. What is the probability of $x = 3$ for $n = 10$?

66. What is the probability of being dealt three aces in a seven-card poker hand?

APPLICATIONS

67. There are 25 students (14 boys and 11 girls) in the sixth-grade class at St. Andrew School. Five students were absent Thursday.
  a. What is the probability that two of the absent students were girls?
  b. What is the probability that two of the absent students were boys?
  c. What is the probability that all were boys?
  d. What is the probability that none were boys?

SELF TEST   **68.** Axline Computers manufactures personal computers at two plants; one is in Las Vegas, the other in Hawaii. There are 40 employees at the Las Vegas plant and 20 in Hawaii. A random sample of 10 different employees is to be asked to fill out a benefits' questionnaire.
  a. What is the probability that none will be from the plant in Hawaii?
  b. What is the probability that one will be from the plant in Hawaii?
  c. What is the probability that two or more will be from the plant in Hawaii?
  d. What is the probability that nine will be from the plant in Las Vegas?

SUMMARY

A random variable provides a numerical description of the outcome of an experiment. We saw that the probability distribution for a random variable describes how the probabilities are distributed over the values the random variable can take on. For any discrete random variable x, the probability distribution is defined by a probability function, denoted by $f(x)$, which provides for each value of the random variable its corresponding probability. Once the probability function has been defined, we can then compute the expected value and the variance for the random variable.

The special probability distributions that were discussed in this chapter are the binomial, Poisson, and hypergeometric distributions. The binomial probability distribution can be used to determine the probability of x successes in n trials whenever the experiment has the following properties:

1. The experiment consists of a sequence of n identical trials.
2. Two outcomes are possible on each trial, one called success and the other failure.
3. The probability of a success, p, does not change from trial to trial. Consequently, the probability of failure, $1 - p$, does not change from trial to trial.
4. The trials are independent.

When the above conditions hold, a binomial probability function, or a table of binomial probabilities, can be used to determine the probability of x successes in n trials. Formulas were also presented for the mean and variance of the binomial random variable.

The Poisson probability distribution is used when it is desired to determine the probability of obtaining x occurrences over an interval of time or space. The following assumptions are required for the Poisson distribution to be applicable:

1. The probability of an occurrence of the event is the same for any two intervals of equal length.
2. The occurrence or nonoccurrence of the event in any interval is independent of the occurrence or nonoccurrence in any other interval.

Both the binomial and Poisson are discrete probability distributions. The binomial random variable may assume a finite number of values $(0, 1, \ldots, n)$; the Poisson random variable may assume an infinite sequence of values $(0, 1, 2, \ldots)$.

A third discrete probability distribution, the hypergeometric, was introduced in Section 5.6. Like the binomial, it is used to compute the probability of x successes in n trials. But, unlike with the binomial, the probability of success changes from trial to trial.

GLOSSARY

Random variable A numerical description of the outcome of an experiment.

Discrete random variable A random variable that can assume only a finite or infinite sequence of values.

Continuous random variable A random variable that may assume all values in an interval or collection of intervals.

Probability function A function, denoted $f(x)$, that gives the probability that the discrete random variable x assumes a particular value.

Discrete probability distribution A table, graph, or equation showing the values of a discrete random variable and the associated probabilities.

Expected value A measure of the average, or mean, value of a random variable.

Variance A measure of the dispersion, or variability, of a random variable.

Standard deviation The positive square root of the variance.

Binomial experiment A probability experiment possessing the four properties stated in Section 5.4.

Binomial probability function The function used to compute the probability of x successes in n trials for a binomial experiment.

Poisson probability function The function used to compute the probability of x occurrences during one interval for a Poisson random variable.

Hypergeometric probability function The function used to compute the probability of x successes in n trials when the trials are dependent.

KEY FORMULAS

Expected Value of a Discrete Random Variable

$$E(x) = \mu = \Sigma x f(x) \tag{5.3}$$

Variance of a Discrete Random Variable

$$\text{Var}(x) = \sigma^2 = \Sigma (x - \mu)^2 f(x) \tag{5.4}$$

Computational Formula for Variance of a Discrete Random Variable

$$\text{Var}(x) = \sigma^2 = \Sigma x^2 f(x) - \mu^2 \tag{5.5}$$

Number of Experimental Outcomes Providing Exactly x Successes in n Trials

$$\binom{n}{x} = \frac{n!}{x!(n-x)!} \tag{5.6}$$

Binomial Probability Function

$$f(x) = \binom{n}{x} p^x (1-p)^{(n-x)} \tag{5.9}$$

Expected Value for the Binomial Probability Distribution

$$E(x) = \mu = np \tag{5.10}$$

Variance for the Binomial Probability Distribution

$$\text{Var}(x) = \sigma^2 = np(1-p) \tag{5.11}$$

Poisson Probability Function

$$f(x) = \frac{\mu^x e^{-\mu}}{x!} \qquad \text{for } x = 0, 1, 2, \ldots \tag{5.12}$$

Hypergeometric Probability Function

$$f(x) = \frac{\binom{r}{x}\binom{N-r}{n-x}}{\binom{N}{n}} \qquad x = 1, 2, \ldots, r \tag{5.13}$$

REVIEW QUIZ

TRUE/FALSE

1. A random variable may assume only numerical values.
2. A random variable that may assume any value between 5 and 6 is a discrete random variable.
3. The sum of the probabilities for all values that a discrete random variable may assume cannot be greater than 1.
4. The expected value for a random variable must be a value the random variable can assume.
5. The variance of a random variable is the sum of the squared deviations from the mean.
6. In a binomial experiment, the probability of success is 1 minus the probability of failure.
7. The binomial random variable is a discrete random variable.
8. In a five-trial binomial experiment, there are six possible values for the random variable.
9. In a binomial experiment involving three trials with a success probability of .3, the probability of one success is .441.
10. In a binomial experiment involving 100 trials with a success probability of .22, the expected number of successes is less than or equal to 20.
11. The Poisson random variable is a continuous random variable.
12. The probability of one success in n trials is the same for the binomial random variable as it is for the hypergeometric random variable.

MULTIPLE CHOICE

For Questions 13–15, consider the random variable x, which gives the number of successes in six identical trials, each of which has a probability of success of .3.

13. The probability of one success is
    a. .0000
    b. .1176
    c. .3025
    d. .1780

14. The probability of at least four successes is
    a. .0704
    b. .0595
    c. .0109
    d. not able to be computed from the information given

15. The expected value of x is closest to
    a. 1.50
    b. 1.75
    c. 2.00
    d. 3.00

16. A random variable assumes the values 1, 2, and 3 with probabilities .10, .60, .30, respectively. The expected value of this random variable is
   a. 1.80
   b. 2.00
   c. 2.20
   d. 3.00

17. The variance of the random variable in Question 16 is
   a. 2.00
   b. 2.20
   c. 8.22
   d. .680

18. Consider a binomial random variable with $p = .4$ and $n = 18$. The standard deviation is
   a. 7.20
   b. 4.32
   c. 2.08
   d. none of the above

SUPPLEMENTARY EXERCISES

69. Which of the following random variables are discrete, and which are continuous?
   a. x = amount of time until the first foul is called in a basketball game
   b. x = number of leaks in 10 miles of sewer
   c. x = distance traveled between meeting oncoming cars
   d. x = number of credit hours carried by a randomly selected college student

70. For the given values of n, x, and p, compute $f(x)$ for a binomial random variable.
   a. $n = 10$, $x = 3$, $p = .2$
   b. $n = 20$, $x = 14$, $p = .6$
   c. $n = 5$, $x = 2$, $p = .7$
   d. $n = 8$, $x = 3$, $p = .3$

71. Which of the following are, and which are not, probability distributions? Explain.

x	$f(x)$	x	$f(x)$	x	$f(x)$
0	.20	0	.25	−1	.20
1	.30	2	.05	0	.50
2	.25	4	.10	1	−.10
3	.35	6	.60	2	.40

72. An automobile agency located in Beverly Hills specializes in the rental of luxury automobiles. Assume that the probability distribution of daily demand at the agency is as shown below.

x	0	1	2	3	4
$f(x)$	.15	.30	.40	.10	.05

   a. Compute the expected value of daily demand.
   b. If the daily rental cost for an automobile is $75, what is the expected value of daily automobile rental?

73. At a large university, the number of student problems handled by the dean for student affairs varies from semester to semester. Assume that the number of student problems (x) handled by the dean has the following probability distribution.

x	0	1	2	3	4	5
$f(x)$	.10	.15	.30	.25	.10	.10

What are the mean and variance of the number of student problems handled by the dean each semester?

74. The number of weekly lost-time injuries at a particular plant (x) has the probability distribution shown below.

x	0	1	2	3	4
$f(x)$	.05	.20	.40	.20	.15

  a. Compute the expected value.
  b. Compute the variance.

75. Assume that the plant in Exercise 74 initiated a safety training program and that the number of lost-time injuries during the 20 weeks following the training program was as follows.

Number of Injuries	Number of Weeks
0	2
1	8
2	6
3	3
4	1
	Total 20

  a. Construct a probability distribution for weekly lost-time injuries based on these data.
  b. Compute the expected value and the variance, and use both to evaluate the effectiveness of the safety training program.

76. The Hub Real Estate Investment stock is currently selling for $16 per share. An investor plans to buy shares and hold the stock for 1 year. Let x be the random variable indicating the price of the stock after 1 year. The probability distribution for x is shown below.

Price of Stock (x)	16	17	18	19	20
$f(x)$	.35	.25	.25	.10	.05

  a. Show that the above probability distribution possesses the properties of all probability distributions.
  b. What is the expected price of the stock after 1 year?
  c. What is the expected gain per share of the stock over the 1-year period? What percent return on the investment is reflected by this expected value?
  d. What is the variance in the price of the stock over the 1-year period?
  e. Another stock with a similar expected return has a variance of 3. Which stock appears to be the better investment in terms of minimizing risk or uncertainty associated with the investment? Explain.

77. The budgeting process for a midwestern college resulted in expense forecasts for the coming year (in 000,000s) of $9, $10, $11, $12, and $13. Since the actual expenses are unknown, the following respective probabilities are assigned: .3, .2, .25, .05, and .2.
   a. Show the probability distribution for the expense forecast.
   b. What is the expected value of the expenses for the coming year?
   c. What is the variance in the expenses for the coming year?
   d. If income projections for the year are estimated at $12 million, comment on the financial position of the college.

78. The probability function for x, the hours required to change over a production system, is as follows:

$$f(x) = \frac{x}{10} \qquad \text{for } x = 1, 2, 3, \text{ or } 4$$

   a. What is the expected value of the changeover time?
   b. What is the variance of the changeover time?

79. A study conducted at the University of Southern California investigated the use of music in television commercial messages (*New York*, March 23, 1992). The study found 42% of commercials make use of music. Consider a sample of 12 commercials.
   a. What is the probability that exactly six of the commercials use music?
   b. What is the probability that exactly three of the commercials use music?
   c. What is the probability that at least three of the commercials use music?

80. In October 1986, *Better Homes and Gardens* published the results of a reader survey to which over 30,000 readers had responded. The findings indicated that 34% of the women who responded worked full-time outside the home and 24% worked part-time outside the home.
   a. For a random sample of five women who are *Better Homes and Gardens* readers, what is the probability that three work full-time outside the home?
   b. For a random sample of five women readers, what is the probability that two are part-time workers?
   c. Suppose a random sample of five women (not necessarily *Better Homes and Gardens* readers) was taken. Would your answers to parts (a) and (b) change? Why or why not?

81. Refer to the *Better Homes and Gardens* survey in Exercise 80. In 65% of the two-parent households, the wife does most of the child care; in 1%, the husband does most. For a sample of four respondents from two-parent households, answer the following:
   a. What is the probability that none of the respondents will say the wife does most of the child care?
   b. What is the probability that none of the respondents will say the husband does most of it?
   c. What is the probability that three or more of the respondents will say the wife does most of it?
   d. What is the probability that one of the respondents will say the husband does most of it?

82. A new clothes-washing compound is found to remove excess dirt and stains satisfactorily on 88% of the items washed. Assume that 10 items are to be washed with the new compound.
   a. What is the probability of satisfactory results on all 10 items?
   b. What is the probability at least two items are found with unsatisfactory results?

83. In an audit of a company's billings, an auditor randomly selects five bills. If 3% of all bills contain an error, what is the probability that the auditor will find the following?
   a. exactly one bill in error
   b. at least one bill in error

84. Many companies use a quality-control technique referred to as *acceptance sampling* to monitor incoming shipments of parts, raw materials, and so on. In the electronics industry, it is common to have component parts shipped from suppliers in large lots. Inspection of a sample of *n* components can be viewed as the *n* trials of a binomial experiment. The outcome for each component tested (trial) will be that the component is good or defective. Reynolds Electronics accepts lots from a particular supplier as long as the percent defective in the lot is not greater than 1%. Suppose a random sample of five items from a recent shipment has been tested.

    a. Assume that 1% of the shipment is defective. Compute the probability that no items in the sample are defective.

    b. Assume that 1% of the shipment is defective. Compute the probability that exactly one item in the sample is defective.

    c. What is the probability of observing one or more defective items in the sample if 1% of the shipment is defective?

    d. Would you feel comfortable accepting the shipment if one item was found defective? Why or why not?

CHAPTER 6

CONTINUOUS

PROBABILITY

DISTRIBUTIONS

WHAT YOU WILL LEARN IN THIS CHAPTER

- The use of the continuous uniform probability distribution
- How to compute probabilities for a continuous probability distribution
- The properties of the normal probability distribution
- How to use the standard normal probability distribution to compute probabilities
- How to use the normal probability distribution to approximate binomial probabilities
- The use of the exponential probability distribution

CONTENTS

IQ Scores: Are You Normal?

Intelligence test scores, referred to as intelligence quotient, or IQ scores, are based on characteristics such as verbal skill, abstract reasoning power, numerical ability, and spatial visualization. An IQ score of 100 is considered average. If plotted on a graph with IQ scores on the horizontal axis, the distribution of intelligence test scores approximates a bell-shaped normal probability curve. This distribution shows that the greatest concentration of scores is near 100 and that the frequency of scores decreases gradually and symmetrically as the extremes of intelligence are approached.

Knowing your IQ score gives you an indication of how you rate or compare to other individuals in terms of intelligence. An IQ score above 115 is considered superior. Studies of intellectually gifted children have generally set the lower limit at an IQ score of 140. Approximately 1% of the population has IQ scores of 140 or more. The average IQ score for a high-school graduate is

IQ test scores help teachers determine if students are working up to their abilities.

110. The average IQ score of a college graduate is 120, and the average IQ score of a person with a Ph.D. is 130.

This *Statistics in Practice* is based on "Your Intelligence Quotient," Tom Biracree, in *How You Rate*. New York: Dell Publishing Co., Inc., 1984.

In the previous chapter we discussed discrete random variables and their probability distributions. In this chapter we turn our attention to the study of continuous random variables. Specifically, we discuss three continuous probability distributions: the uniform, the normal, and the exponential probability distributions.

To understand the difference between computing probabilities for discrete and continuous random variables, first recall that for a discrete random variable, we can speak of the probability of the random variable taking on a particular value. For continuous probability distributions, the situation is much different. A continuous random variable may assume any value in an interval on the real line or in a collection of intervals. Since there are an infinite number of values in any interval, it is no longer possible to talk about the probability that the random variable will take on a specific value; instead, we must think in terms of the probability that a continuous random variable will lie within a given interval.

In our discussion of discrete probability distributions, we introduced the concept of a probability function $f(x)$. Recall that this function provided the probability that the random variable x assumed some specific value. In the continuous case, the counterpart of the probability function is the *probability density function,* also denoted by $f(x)$. For a continuous random variable, the probability density function provides the height or value of the function at any particular value of $x;$ it does not directly provide the probability of the random variable taking on some specific value. However, the area under the graph of $f(x)$ corresponding to some interval provides the probability that the continuous random variable will take on a value in that interval. In Section 6.1 we demonstrate these concepts for a continuous random variable that has a uniform probability distribution.

6.1   THE UNIFORM PROBABILITY DISTRIBUTION

The continuous uniform probability distribution is used in situations in which all values of the random variable are equally likely.

EXAMPLE 6.1 ▶ Consider the random variable x that represents the total flight time of an airplane traveling from Chicago to New York. Suppose that the flight time can be any value in the interval from 120 minutes to 140 minutes. Since the random variable x can take on any value in the interval from 120 to 140 minutes, x is a continuous rather than a discrete random variable. Let us assume that sufficient actual flight data are available to conclude that the probability of a flight time within any 1-minute interval is the same as the probability of a flight time within any other 1-minute interval from 120 up to and including 140 minutes. With every 1-minute interval being equally likely, the random variable x is said to have a continuous *uniform probability distribution*. The *probability density function,* which defines the uniform probability distribution for the flight time random variable, is

$$f(x) = \begin{cases} \frac{1}{20} & \text{for } 120 \leq x \leq 140 \\ \\ 0 & \text{elsewhere} \end{cases}$$

A graph of this probability density function is shown in Figure 6.1.   ◀

In general, the continuous uniform probability density function for a random variable x is

---

Continuous Uniform Probability Density Function

$$f(x) = \begin{cases} \dfrac{1}{b-a} & \text{for } a \leq x \leq b \\ \\ 0 & \text{elsewhere} \end{cases}$$ **(6.1)**

FIGURE 6.1   CONTINUOUS UNIFORM PROBABILITY DENSITY FUNCTION FOR FLIGHT TIME

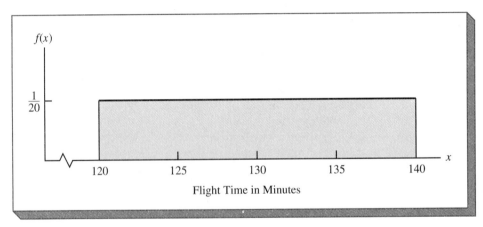

Flight Time in Minutes

In Example 6.1, involving flight time from Chicago to New York, $a = 120$ and $b = 140$.

The graph of the probability density function $f(x)$ provides the height or value of the function at any particular value of x. Note that for a *uniform* probability density function, the height or value of the function is the same for each value of x. For example, in the flight-time example, $f(x) = \frac{1}{20}$ for all values of x between 120 and 140. *In general, the probability density function $f(x)$, unlike the probability function for a discrete random variable, does not represent probability. Rather, it simply provides the height of the function at any particular value of x.* For a continuous random variable, we consider probability only in terms of the likelihood that a random variable has a value within a *specified interval.*

AREA AS A MEASURE OF PROBABILITY

The probability that a continuous random variable will assume a value between a given lower limit c and a given upper limit d is denoted by $P(c \leq x \leq d)$ and is given by the area under the graph of the probability density function between c and d.

EXAMPLE 6.1 ▶
(CONTINUED)

The shaded region in Figure 6.2 shows the area under the graph of the probability density function between 120 and 130 for the uniform random variable corresponding to flight time. To compute the probability the random variable assumes a value in the interval, we must find the shaded area. Since the area is a rectangle, we simply multiply the base $(130 - 120 = 10)$ by the height $(\frac{1}{20})$. Thus, $P(120 \leq x \leq 130) = 10(\frac{1}{20}) = .50$. We conclude that the probability of a flight time between 120 and 130 minutes is .50. ◀

For continuous random variables that have a uniform probability distribution, it is easy to compute the probability that the random variable takes on a value in a certain interval. The area under the probability density function will always be a rectangle, and its area is easy to compute.

Note that $P(120 \leq x \leq 140) = 1$ for the flight-time example. In general, the area under any probability density function is equal to 1. This property holds for all continuous probability distributions and is the analog of the condition that the sum of the probabilities has to equal 1 for a discrete probability function. For a continuous probability density function, we must also require that $f(x) \geq 0$ for all values of x. This is the analog of the requirement that $f(x) \geq 0$ for discrete probability functions.

FIGURE 6.2 **SHADED AREA PROVIDES PROBABILITY OF FLIGHT TIME BETWEEN 120 AND 130 MINUTES**

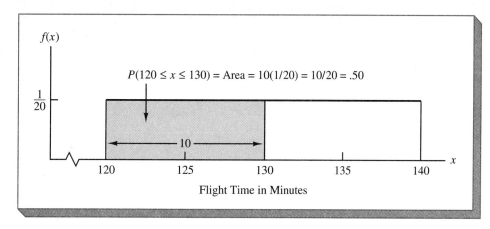

Summarizing, note that when we deal with continuous random variables and their probability distributions, two major differences stand out as compared to the treatment of their discrete counterparts:

1. We no longer talk about the probability of the random variable taking on a particular value. Instead, we talk about the probability of the random variable taking on a value within a given interval.
2. The probability of the random variable taking on a value within a given interval from c to d is defined to be the area under the graph of the probability density function between c and d. *This implies that the probability that a continuous random variable takes on any particular value exactly is zero, since the area under the graph of $f(x)$ at a single point is zero.*

The calculation of the mean and variance for a continuous random variable is analogous to that for a discrete random variable. However, since the computational procedure involves integral calculus, we leave the derivation of the appropriate formulas to more advanced texts.

For the continuous uniform probability distribution introduced in this section, the formulas for the mean and variance are

$$E(x) = \mu = \frac{a + b}{2}$$

$$Var(x) = \sigma^2 = \frac{(b - a)^2}{12}$$

In these formulas, a is the smallest value and b is the largest value that the random variable may take on.

EXAMPLE 6.1
(CONTINUED)

Applying these formulas to the uniform probability distribution for flight time, we obtain

$$E(x) = \mu = \frac{(120 + 140)}{2} = 130$$

$$Var(x) = \frac{(140 - 120)^2}{12} = \frac{400}{12} = 33.33$$

The standard deviation can be found by taking the square root of the variance. Thus $\sigma = 5.77$ minutes.

NOTES AND
COMMENTS

1. Since for any continuous random variable the probability of any single value is zero, we have $P(c \leq x \leq d) = P(c < x < d)$. That is, the probability is the same whether or not the end points of the interval are included.
2. To see more clearly why the height of a probability density function is not a probability, think about a random variable with the following continuous uniform probability distribution:

$$f(x) = \begin{cases} 2 & \text{for } 0 \leq x \leq .5 \\ 0 & \text{elsewhere} \end{cases}$$

The height of the probability density function is 2 for values of x between 0 and .5. But, we know probabilities can never be greater than 1.

3. With the continuous uniform probability distribution, the probability of an interval is proportional to its length.

EXERCISES

METHODS

SELF TEST

1. The random variable x is known to be uniformly distributed between 1.0 and 1.5.
   a. Show the graph of the probability density function.
   b. Find $P(x = 1.25)$.
   c. Find $P(1.0 \leq x \leq 1.25)$.
   d. Find $P(1.20 < x < 1.5)$.

2. The random variable x is known to be uniformly distributed between 10 and 20.
   a. Show the graph of the probability density function.
   b. Find $P(x < 15)$.
   c. Find $P(12 \leq x \leq 18)$.
   d. Find $E(x)$.
   e. Find $Var(x)$.

APPLICATIONS

3. Delta Airlines quotes a flight time of 1 hour, 52 minutes for its flights from Cincinnati to Tampa. Suppose that we believe actual flight times are uniformly distributed between the quoted time and 2 hours, 10 minutes.
   a. Show the graph of the probability density function for flight times.
   b. What is the probability the flight will be no more than 5 minutes late?
   c. What is the probability the flight will be more than 10 minutes late?
   d. What is the expected flight time?

SELF TEST

4. Most computer languages have a function that can be used to generate random numbers. In Microsoft's QuickBASIC, the RND function can be used to generate random numbers between 0 and 1. If we let x denote the random number generated, then x is a continuous random variable with the following probability density function:

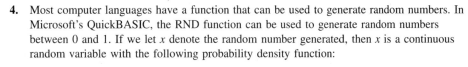

$$f(x) = \begin{cases} 1 & \text{for } 0 \leq x \leq 1 \\ 0 & \text{elsewhere} \end{cases}$$

   a. Graph the probability density function.
   b. What is the probability of generating a random number between .25 and .75?
   c. What is the probability of generating a random number with values less than or equal to .30?
   d. What is the probability of generating a random number with a value greater than .60?

5. The total time to process a loan application is uniformly distributed between 3 and 7 days.
   a. Give a mathematical expression for the probability density function.
   b. What is the probability that the loan application will be processed in fewer than 3 days?
   c. Compute the probability that a loan application will be processed in 5 days or less.
   d. Find the expected processing time and the standard deviation.

6. The label on a bottle of liquid detergent shows the contents to be 12 ounces per bottle. The production operation fills the bottle uniformly according to the following probability density function:

$$f(x) = \begin{cases} 8 & \text{for } 11.975 \leq x \leq 12.100 \\ 0 & \text{elsewhere} \end{cases}$$

   a. What is the probability that a bottle will be filled with between 12 and 12.05 ounces?
   b. What is the probability that a bottle will be filled with 12.02 or more ounces?
   c. Quality control accepts production that is within .02 ounces of the number of ounces shown on the container label. What is the probability that a bottle of this liquid detergent will fail to meet the quality-control standard?

7. Suppose we are interested in bidding on a piece of land and we know there is one other bidder.* The seller has announced that the highest bid in excess of $10,000 will be accepted. Assume that the competitor's bid x is a random variable that is uniformly distributed between $10,000 and $15,000.
   a. Suppose you bid $12,000. What is the probability your bid will be accepted?
   b. Suppose you bid $14,000. What is the probability your bid will be accepted?
   c. What amount should you bid to maximize the probability you get the property?
   d. Suppose you know someone who is willing to pay you $16,000 for the property. Would you consider bidding less than the amount in part (c)? Why or why not?

6.2   THE NORMAL PROBABILITY DISTRIBUTION

Perhaps the most important probability distribution used to describe a continuous random variable is the *normal probability distribution.* The normal probability distribution has been applied in a wide variety of practical applications in which the random variables involved are heights and weights of people, IQ scores, scientific measurements, amounts of rainfall, and so on. In order to use this probability distribution, the random variable must be continuous. However, as we shall see, a continuous normal random variable is often used as an approximation in situations involving discrete random variables. The form, or shape, of the normal probability density function is illustrated by the bell-shaped curve shown in Figure 6.3. The mathematical equation that describes the bell-shaped curve of the normal probability density function is as follows.

Normal Probability Density Function

$$f(x) = \frac{1}{\sqrt{2\pi}\,\sigma} e^{-(x-\mu)^2/2\sigma^2} \tag{6.2}$$

where μ is the mean, σ is the standard deviation, $\pi = 3.14159$ and $e = 2.71828$

The value of $f(x)$ for any choice of x gives the height of the curve (see Figure 6.3).

FIGURE 6.3   BELL-SHAPED CURVE FOR THE NORMAL PROBABILITY DISTRIBUTION

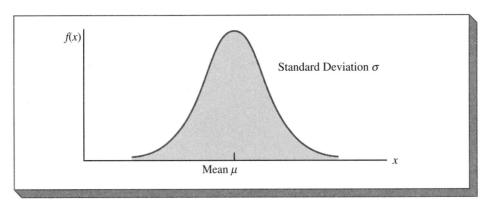

---

*This exercise is based on a problem suggested to us by Professor Roger Myerson of Northwestern University.

Some observations about the characteristics of the normal probability distribution are in order here.

1. There is an entire family of normal probability distributions with each specific normal distribution being differentiated by its mean μ and its standard deviation σ.
2. The highest point on the normal curve occurs at the mean, which is also the median and mode of the distribution.
3. The mean of the distribution can be any numerical value: negative, zero, or positive. Three normal curves with the same standard deviation but three different means (-10, 0, and 20) are shown in Figure 6.4.
4. The normal probability distribution is symmetric, with the tails of the curve extending indefinitely in both directions and theoretically never touching the horizontal axis.
5. The standard deviation determines the width of the probability density function. Larger values of the standard deviation result in wider, flatter curves, showing more dispersion in the random variable. Two normal distributions with the same mean but different values for the standard deviation are shown in Figure 6.5.

FIGURE 6.4   **THREE NORMAL CURVES WITH DIFFERENT MEANS BUT SAME STANDARD DEVIATION**

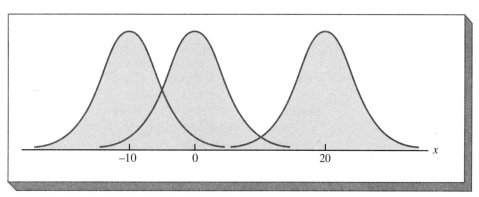

FIGURE 6.5   **TWO NORMAL CURVES WITH SAME MEAN BUT DIFFERENT STANDARD DEVIATIONS**

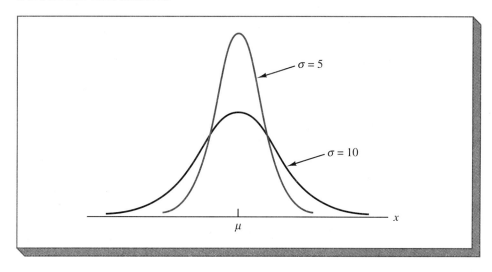

FIGURE 6.6   **AREAS UNDER THE CURVE FOR ANY NORMAL PROBABILITY
DISTRIBUTION**

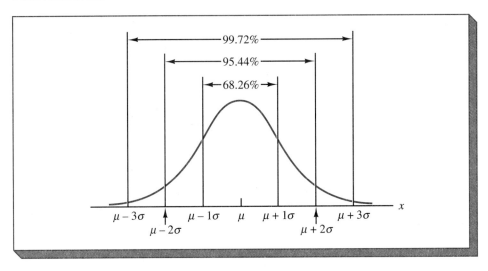

6. Probabilities for the normal random variable are given by areas under the normal curve. For some commonly used intervals, we note the following:
    a. 68.26% of the time, a normal random variable assumes a value within plus or minus one standard deviation of its mean.
    b. 95.44% of the time, a normal random variable assumes a value within plus or minus two standard deviations of its mean.
    c. 99.72% of the time, a normal random variable assumes a value within plus or minus three standard deviations of its mean. Figure 6.6 shows properties (a), (b), and (c) graphically.
7. The total area under the curve for the normal probability distribution is 1. (This is true for all continuous probability distributions.)

THE STANDARD NORMAL PROBABILITY DISTRIBUTION

A random variable that has a normal distribution with a mean of 0 and a standard deviation of 1 is said to have a *standard normal probability distribution*. The letter z is commonly used to designate this particular normal random variable. The graph of the standard normal probability distribution is shown in Figure 6.7. Note that the standard normal probability distribution has the same general appearance as other normal distributions but with the special properties of $\mu_z = 0$ and $\sigma_z = 1$. Because the mean equals 0 and the standard deviation equals 1, the equation for the standard normal probability density function is simpler than equation (6.2).

Standard Normal Probability Density Function

$$f(z) = \frac{1}{\sqrt{2\pi}} e^{-z^2/2} \qquad\qquad (6.3)$$

FIGURE 6.7   THE STANDARD NORMAL PROBABILITY DISTRIBUTION

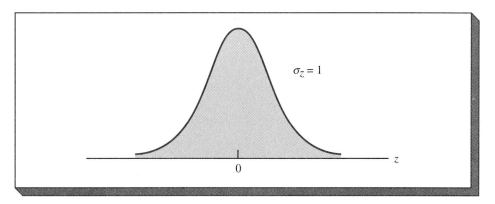

As noted in Section 6.1, with a continuous random variable, probability calculations are always concerned with finding the probability that the random variable assumes any value in an interval between two specific points c and d. The probability that a continuous random variable assumes a value between the two points c and d is the area under the graph of the probability density function between c and d. Because of its wide applicability, areas under the standard normal curve have been computed and are available in tables that can be used in computing the probability values for the standard normal probability distribution. Table 6.1 is such a table. This table is also available as Table 1 of Appendix B and inside the front cover of the text.

Let us show how the table of probabilities for the standard normal probability distribution (Table 6.1) can be used by considering some examples. Later, we will see how the table for the standard normal distribution can be used to compute probabilities for any normal distribution.

EXAMPLE 6.2 ▶ What is the probability that the z value for the standard normal random variable will be between 0.00 and 1.00? That is, what is $P(0.00 \leq z \leq 1.00)$? The shaded region in the following graph shows this area or probability.

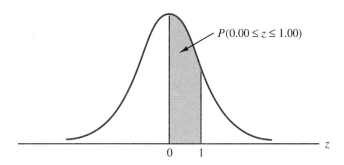

The entries in Table 6.1 give the area under the standard normal curve between the mean, $z = 0$, and a specified positive value of z. In this case, we are interested in the area between $z = 0$ and $z = 1.00$. Thus, we must find the entry in the table corresponding to $z = 1.00$. To do this, we first find 1.0 in the left-hand column of the table and then find .00 in the top row of the table. Then, by looking in the body of the table, we find that the 1.0

TABLE 6.1   PROBABILITIES FOR THE STANDARD NORMAL DISTRIBUTION

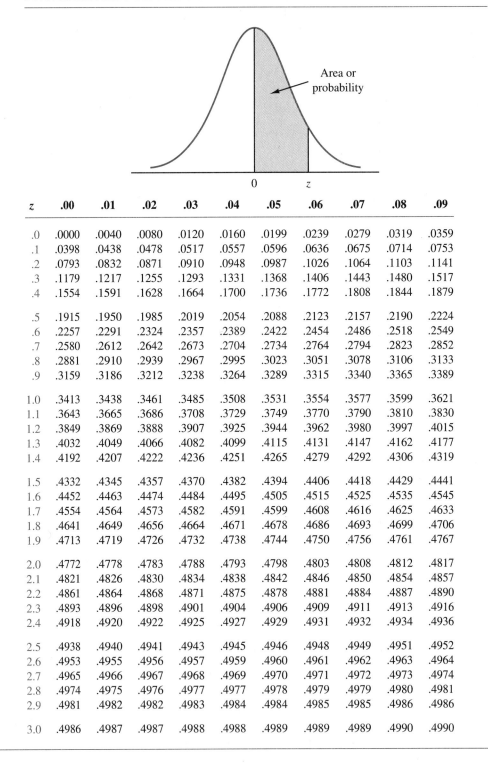

z	.00	.01	.02	.03	.04	.05	.06	.07	.08	.09
.0	.0000	.0040	.0080	.0120	.0160	.0199	.0239	.0279	.0319	.0359
.1	.0398	.0438	.0478	.0517	.0557	.0596	.0636	.0675	.0714	.0753
.2	.0793	.0832	.0871	.0910	.0948	.0987	.1026	.1064	.1103	.1141
.3	.1179	.1217	.1255	.1293	.1331	.1368	.1406	.1443	.1480	.1517
.4	.1554	.1591	.1628	.1664	.1700	.1736	.1772	.1808	.1844	.1879
.5	.1915	.1950	.1985	.2019	.2054	.2088	.2123	.2157	.2190	.2224
.6	.2257	.2291	.2324	.2357	.2389	.2422	.2454	.2486	.2518	.2549
.7	.2580	.2612	.2642	.2673	.2704	.2734	.2764	.2794	.2823	.2852
.8	.2881	.2910	.2939	.2967	.2995	.3023	.3051	.3078	.3106	.3133
.9	.3159	.3186	.3212	.3238	.3264	.3289	.3315	.3340	.3365	.3389
1.0	.3413	.3438	.3461	.3485	.3508	.3531	.3554	.3577	.3599	.3621
1.1	.3643	.3665	.3686	.3708	.3729	.3749	.3770	.3790	.3810	.3830
1.2	.3849	.3869	.3888	.3907	.3925	.3944	.3962	.3980	.3997	.4015
1.3	.4032	.4049	.4066	.4082	.4099	.4115	.4131	.4147	.4162	.4177
1.4	.4192	.4207	.4222	.4236	.4251	.4265	.4279	.4292	.4306	.4319
1.5	.4332	.4345	.4357	.4370	.4382	.4394	.4406	.4418	.4429	.4441
1.6	.4452	.4463	.4474	.4484	.4495	.4505	.4515	.4525	.4535	.4545
1.7	.4554	.4564	.4573	.4582	.4591	.4599	.4608	.4616	.4625	.4633
1.8	.4641	.4649	.4656	.4664	.4671	.4678	.4686	.4693	.4699	.4706
1.9	.4713	.4719	.4726	.4732	.4738	.4744	.4750	.4756	.4761	.4767
2.0	.4772	.4778	.4783	.4788	.4793	.4798	.4803	.4808	.4812	.4817
2.1	.4821	.4826	.4830	.4834	.4838	.4842	.4846	.4850	.4854	.4857
2.2	.4861	.4864	.4868	.4871	.4875	.4878	.4881	.4884	.4887	.4890
2.3	.4893	.4896	.4898	.4901	.4904	.4906	.4909	.4911	.4913	.4916
2.4	.4918	.4920	.4922	.4925	.4927	.4929	.4931	.4932	.4934	.4936
2.5	.4938	.4940	.4941	.4943	.4945	.4946	.4948	.4949	.4951	.4952
2.6	.4953	.4955	.4956	.4957	.4959	.4960	.4961	.4962	.4963	.4964
2.7	.4965	.4966	.4967	.4968	.4969	.4970	.4971	.4972	.4973	.4974
2.8	.4974	.4975	.4976	.4977	.4977	.4978	.4979	.4979	.4980	.4981
2.9	.4981	.4982	.4982	.4983	.4984	.4984	.4985	.4985	.4986	.4986
3.0	.4986	.4987	.4987	.4988	.4988	.4989	.4989	.4989	.4990	.4990

row and the .00 column of the table intersect at the value of .3413. We have found the desired probability; $P(0.00 \leq z \leq 1.00) = .3413$. A portion of Table 6.1 showing these steps is shown below.

z	.00	.01	.02
.			
.			
.			
.9	.3159		
1.0	.3413	.3438	.3461
1.1	.3643		
1.2	.3849		
.	.		
.	.		
.	.	$P(0.00 \leq z \leq 1.00)$	

As another example, use Table 6.1 to show that the area between $z = 0.00$ and $z = 1.25$ is .3944. This area or probability value is found by using the $z = 1.2$ row and the .05 column of the table.

EXAMPLE 6.3 ▶ What is the probability of obtaining a z value between $z = -1.00$ and $z = 1.00$? That is, what is $P(-1.00 \leq z \leq 1.00)$?

First note that we have already used Table 6.1 to show that the probability of a z value between $z = 0.00$ and $z = 1.00$ is .3413. Recall now that the normal probability distribution is *symmetric*. That is, the shape of the curve to the left of the mean is the mirror image of the shape of the curve to the right of the mean. Thus, the probability of a z value between $z = 0.00$ and $z = -1.00$ is the *same* as the probability of a z value between $z = 0.00$ and $z = +1.00$. Hence, the probability of a z value between $z = -1.00$ and $z = +1.00$ is

$$P(-1.00 \leq z \leq 0.00) + P(0.00 \leq z \leq 1.00) = .3413 + .3413 = .6826$$

This area is shown graphically as follows.

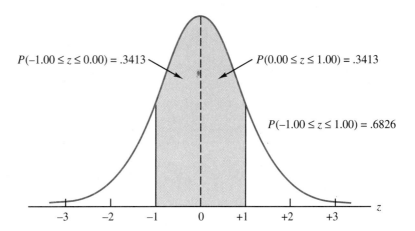

In a manner similar to Example 6.3, we can use the values in Table 6.1 to show that the probability of a z value between -2.00 and $+2.00$ is $.4772 + .4772 = .9544$ and that the probability of a z value between -3.00 and $+3.00$ is $.4986 + .4986 = .9972$. Since we know that the total probability or total area under the curve for any continuous random variable must be one, the probability $.9972$ tells us that the value of z will almost always fall between -3.00 and $+3.00$.

EXAMPLE 6.4 ▶ What is the probability of obtaining a z value of at least 1.58? That is, what is $P(z \geq 1.58)$?

First, we use the $z = 1.5$ row and the .08 column of Table 6.1 to find that $P(0.00 \leq z \leq 1.58) = .4429$. Now, since the normal probability distribution is symmetric and the total area under the curve equals 1, we know that 50% of the area must be above the mean (i.e., $z = 0$) and 50% of the area must be below the mean. Since .4429 is the area between the mean and $z = 1.58$, the area or probability corresponding to $z \geq 1.58$ must be $.5000 - .4429 = .0571$. This probability is shown below.

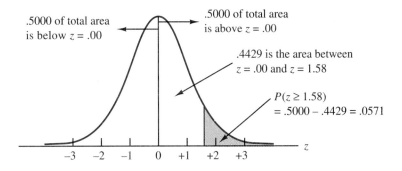

EXAMPLE 6.5 ▶ What is the probability the random variable z assumes a value of $-.50$ or larger? That is, what is $P(z \geq -.50)$?

To make this computation, we note that the probability we are seeking can be written as the sum of two probabilities: $P(z \geq -.50) = P(-.50 \leq z \leq 0.00) + P(z \geq 0.00)$. We have previously seen that $P(z \geq 0.00) = .50$. Also, we know that since the normal distribution is symmetric, $P(-.50 \leq z \leq 0.00) = P(0.00 \leq z \leq .50)$. Referring to Table 6.1, we find that $P(0.00 \leq z \leq .50) = .1915$. Therefore, $P(-.50 \leq z \leq 0.00) = .1915$. Thus, $P(z \geq -.50) = P(-.50 \leq z \leq 0.00) + P(z \geq 0.00) = .1915 + .5000 = .6915$. This area is shown in the graph below.

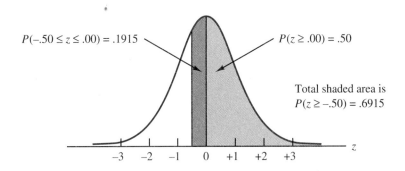

EXAMPLE 6.6 ▶ What is the probability of obtaining a z value between 1.00 and 1.58. That is, what is $P(1.00 \leq z \leq 1.58)$?

From Examples 6.2 and 6.4 we know that there is a .3413 probability of a z value between $z = 0.00$ and $z = 1.00$ and that there is a .4429 probability of a z value between $z = 0.00$ and $z = 1.58$. Thus, there must be a $.4429 - .3413 = .1016$ probability of a z value between $z = 1.00$ and $z = 1.58$; therefore, $P(1.00 \leq z \leq 1.58) = .1016$. This situation is shown graphically in the following figure.

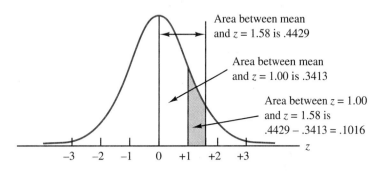

EXAMPLE 6.7 ▶ Find a z value such that the probability of obtaining a larger z value is only .10. This situation is shown graphically as follows:

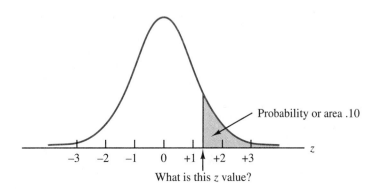

This problem is somewhat different from the examples we have considered thus far in that previously we specified the z value of interest and then found the corresponding probability, or *area*. In this example, we are given the probability, or area, information and asked to find the corresponding z value. This can be found by using the table of areas for the standard normal probability distribution (Table 6.1) a little differently.

Recall that the body of Table 6.1 provides the area under the curve between the mean and a particular z value. Now we are given the information that the area in the upper tail of the curve is .10. Thus, we must determine how much of the area is between the mean and the z value of interest. Since we know .5000 of the area is above the mean, $.5000 - .1000 = .4000$ must be the area under the curve *between* the mean and the desired z value. Scanning the body of the table, we find .3997 as the probability value closest to .4000. The section of the table providing this result is:

z	$\cdots$	.06	.07	.08	.09
.					
.					
.					
1.0				.3599	
1.1				.3810	
1.2		.3962	.3980	.3997	.4015
1.3				.4162	
1.4				.4306	
.					
.					
.					

Reading the z value from the left column and the top row of the table, we find that the corresponding z value is 1.28. Thus, there will be an area of approximately .4000 (actually .3997) between the mean and $z = 1.28$. In terms of the question originally asked, there is approximately a .1000 probability of a z value larger than 1.28. ◀

The preceding examples illustrate that the table of areas for the standard normal probability distribution can be used to find probabilities associated with values of the standard normal random variable z. Two types of questions can be asked. The first type of question specifies a value, or values, for z and asks us to use the table to determine the corresponding areas, or probabilities. The second type of question provides an area, or probability, and asks us to use the table to determine the corresponding z value. Thus, we need to remain flexible in terms of using the standard normal probability table to answer the desired probability question. In most cases, sketching a graph of the standard normal probability distribution and shading the appropriate area helps to visualize the situation and aids in determining the correct answer.

COMPUTING PROBABILITIES FOR ANY NORMAL DISTRIBUTION

The reason that we have been discussing the standard normal distribution so extensively is that probabilities for all normal distributions are computed using the standard normal probability distribution. That is, when we have a normal distribution with any mean μ and any standard deviation σ, we answer probability questions by first converting to the standard normal distribution. Then we can use Table 6.1 and the appropriate z values to find the desired probabilities. The formula used to convert any normal random variable x with mean μ and standard deviation σ to the standard normal random variable z is as follows.

Converting to a Standard Normal Random Variable

$$z = \frac{x - \mu}{\sigma} \tag{6.4}$$

A value of x equal to its mean μ results in $z = (\mu - \mu)/\sigma = 0$. Thus, we see that a value of x equal to its mean μ corresponds to a value of z at its mean 0. Now suppose that x is one standard deviation above its mean; that is, $x = \mu + \sigma$. Applying equation (6.4), we see that the corresponding z value is $z = [(\mu + \sigma) - \mu]/\sigma = \sigma/\sigma = 1$. Thus, a value of x that is one standard deviation above its mean yields $z = 1$. In other words, we can interpret the z value as *the number of standard deviations that the normal random variable x is from its mean μ*.

To see how this conversion enables us to compute probabilities for any normal distribution, suppose we have a normal distribution with $\mu = 10$ and $\sigma = 2$. What is the probability that the random variable x is between 10 and 14? Using equation (6.4), we see that at $x = 10$, $z = (x - \mu)/\sigma = (10 - 10)/2 = 0$ and that at $x = 14$, $z = (14 - 10)/2 = 4/2 = 2$. Thus, the answer to our question about the probability of x being between 10 and 14 is given by the equivalent probability that z is between 0 and 2 for the standard normal distribution. In other words, the probability that we are seeking is the probability that the random variable x is between its mean and two standard deviations above the mean. Using $z = 2.00$ and Table 6.1, we see that the probability is .4772. Hence, the probability that x is between 10 and 14 is .4772.

EXAMPLE 6.8 ▶ IQ scores for a group of sixth graders are normally distributed with a mean of 100 and a standard deviation of 12. What is the probability that a randomly selected student will have an IQ score between 90 and 110?

Letting x be the normally distributed IQ score, we must compute $P(90 \leq x \leq 110)$ based on the information $\mu = 100$ and $\sigma = 12$. Converting the x values to the corresponding z values, we have

$$\text{For } x = 90, \qquad z = \frac{x - \mu}{\sigma} = \frac{90 - 100}{12} = -.83$$

$$\text{For } x = 110, \qquad z = \frac{x - \mu}{\sigma} = \frac{110 - 100}{12} = +.83$$

Using Table 6.1 for $z = +.83$, we find that the probability of z being between zero and $+.83$ is .2967. Also, since $P(-.83 \leq z \leq 0.00) = .2967$, we have $P(-.83 \leq z \leq +.83) = .2967 + .2967 = .5934$. In terms of the IQ scores, we now know the probability that a randomly selected student will have an IQ score between 90 and 110 is .5934. The graphical representation of this probability with the corresponding z values is as follows.

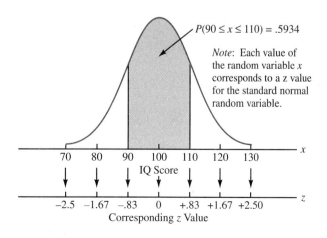

$P(90 \leq x \leq 110) = .5934$

Note: Each value of the random variable x corresponds to a z value for the standard normal random variable.

Using the IQ scores with mean $\mu = 100$ and standard deviation $\sigma = 12$, what is the probability of randomly selecting a student with an IQ score of 120 or more? For $x = 120$, we have

$$z = \frac{x - \mu}{\sigma} = \frac{120 - 100}{12} = 1.67$$

Using Table 6.1, we find an area of .4525 between $z = 0$ and $z = 1.67$. Thus, .4525 is the probability that a student's IQ score is between the mean $\mu = 100$ and the IQ score of 120. This is not the answer to the question seeking the probability that a randomly selected student will have an IQ score of 120 or more. However, since .5000 of the area under a normal curve is above the mean, we see the probability that a randomly selected student will have an IQ score of 120 or more must be $.5000 - .4525 = .0475$. Less than 5% of the students will have an IQ score of 120 or more.

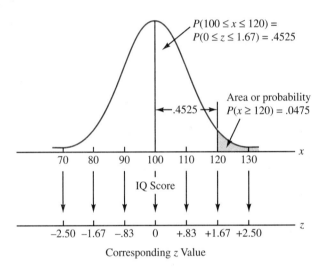

EXAMPLE 6.9 ▶ The Grear Tire Company has just developed a new steel-belted radial tire that will be sold through a national chain of discount stores. From road tests with the tires, it is found that tire mileage is normally distributed with a mean tire mileage of $\mu = 36{,}500$ miles and a standard deviation of $\sigma = 5000$ miles. What is the probability that a tire will last at least 30,000 miles?

The z value corresponding to $x = 30{,}000$ miles is

$$z = \frac{x - \mu}{\sigma} = \frac{30{,}000 - 36{,}500}{5000} = \frac{-6500}{5000} = -1.30$$

Using Table 6.1, we find that $P(-1.30 \leq z \leq 0.00) = .4032$. Thus, the probability a tire will provide at least 30,000 miles of usage is $P(x \geq 30{,}000) = .4032 + .5000 = .9032$. That is, better than 90% of the tires can be expected to wear for at least 30,000 miles. This situation is shown graphically. Again, values for the corresponding standard normal random variable are also shown.

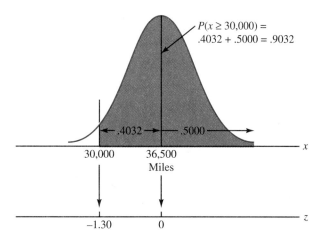

EXAMPLE 6.10 ▶ Test scores for a college midterm examination are normally distributed with a mean of $\mu = 72$ and a standard deviation of $\sigma = 13$. Suppose the professor wishes to assign the grade of A to the 15% of the students obtaining the highest scores on the exam. What is the cutoff score for the A grade?

This is a situation in which the probability is known and the question concerns finding a particular value for the random variable, the exam score. With 15% of the area in the upper tail of the normal distribution, we know that the area between the mean score of $\mu = 72$ and the exam score required to obtain the grade of A must be .5000 − .1500 = .3500. Using the *body* of Table 6.1, we find the probability value closest to .3500 is .3508. Using the left-hand column and the top row of the table, we find that the z value corresponding to .3508 is $z = 1.04$. This tells us that the midterm exam score required to obtain the grade of A must be at least 1.04 standard deviations ($\sigma = 13$) above the mean ($\mu = 72$). Computing the corresponding value of x, we find

$$x = \mu + 1.04\sigma = 72 + 1.04(13) = 85.52$$

as the minimum score a student must obtain to receive the grade of A for the exam. ◀

In summary, the key to answering probability questions about the normal distribution is the conversion of values of the normal random variable x to the corresponding z values and the interpretation of the table of probabilities for the standard normal probability distribution.

EXERCISES

METHODS

8. Using Figure 6.6 as a guide, sketch a normal curve for a random variable x that has a mean $\mu = 100$ and a standard deviation $\sigma = 10$. Label the horizontal axis with values of 70, 80, 90, 100, 110, 120, and 130.

9. A random variable is normally distributed with a mean of $\mu = 50$ and a standard deviation of $\sigma = 5$.

   a. Sketch a normal curve for the random variable. Label the horizontal axis with values of 35, 40, 45, 50, 55, 60, and 65. Figure 6.6 shows that the normal curve almost touches the horizontal line at three standard deviations below and at three standard deviations above the mean (in this case at 35 and 65).

   b. What is the probability that the random variable assumes a value between 45 and 55?

   c. What is the probability that the random variable assumes a value between 40 and 60?

10. Given that z is the standard normal random variable, sketch the standard normal curve. Label the horizontal axis at values of -3, -2, -1, 0, 1, 2, and 3. Then use the table of probabilities for the standard normal distribution to compute the following probabilities:

   a. $P(0 \leq z \leq 1)$ **b.** $P(0 \leq z \leq 1.5)$ **c.** $P(0 < z < 2)$ **d.** $P(0 < z < 2.5)$

11. Given that z is the standard normal random variable, compute the following probabilities:

   a. $P(-1 \leq z \leq 0)$ **b.** $P(-1.5 \leq z \leq 0)$ **c.** $P(-2 < z < 0)$

   d. $P(-2.5 \leq z \leq 0)$ **e.** $P(-3 < z \leq 0)$

12. Given that z is the standard normal random variable, compute the following probabilities:

   a. $P(0 \leq z \leq .83)$ **b.** $P(-1.57 \leq z \leq 0)$ **c.** $P(z > .44)$

   d. $P(z \geq -.23)$ **e.** $P(z < 1.20)$ **f.** $P(z \leq -.71)$

SELF TEST ▷ **13.** Given that z is the standard normal random variable, compute the following probabilities:

   a. $P(-1.98 \leq z \leq .49)$

   b. $P(.52 \leq z \leq 1.22)$

   c. $P(-1.75 \leq z \leq -1.04)$

14. Given that z is the standard normal random variable, find z for each situation.

   a. The area between 0 and z is .4750.

   b. The area between 0 and z is .2291.

   c. The area to the right of z is .1314.

   d. The area to the left of z is .6700.

SELF TEST ▷ **15.** Given that z is the standard normal random variable, find z for each situation.

   a. The area to the left of z is .2119.

   b. The area between $-z$ and z is .9030.

   c. The area between $-z$ and z is .2052.

   d. The area to the left of z is .9948.

   e. The area to the right of z is .6915.

16. Given that z is the standard normal random variable, find z for each situation.

   a. The area to the right of z is .01.

   b. The area to the right of z is .025.

   c. The area to the right of z is .05.

   d. The area to the right of z is .10.

APPLICATIONS

17. The demand for a new product is assumed to be normally distributed with $\mu = 200$ and $\sigma = 40$. Letting x be the number of units demanded, find the following:

   a. $P(180 \leq x \leq 220)$ **b.** $P(x \geq 250)$ **c.** $P(x \leq 100)$ **d.** $P(225 \leq x \leq 250)$

SELF TEST ▷ **18.** The mean cost for employee alcohol rehabilitation programs involving hospitalization is $10,000 (*USA Today*, September 12, 1991). Assume that rehabilitation program cost has a normal probability distribution with a standard deviation of $2200. Answer the following questions:

   a. What is the probability that a rehabilitation program will cost at least $12,000?

   b. What is the probability that a rehabilitation program will cost at least $6000?

   c. What is the cost range for the 10% most expensive rehabilitation programs?

19. The Webster National Bank is reviewing its service charges and interest-paying policies on checking accounts. The bank has found that the average daily balance on personal checking accounts is $550.00, with a standard deviation of $150.00. In addition, the average daily balances have been found to be normally distributed.

   a. What percentage of personal checking account customers carry average daily balances in excess of $800.00?

   b. What percentage of the bank's customers carry average daily balances below $200.00?

   c. What percentage of the bank's customers carry average daily balances between $300.00 and $700.00?

   d. The bank is considering paying interest to customers carrying average daily balances in excess of a certain amount. If the bank does not want to pay interest to more than 5% of its customers, what is the minimum average daily balance it should be willing to pay interest on?

20. Miami University reported admission statistics for 3339 students who were admitted as freshmen for the fall semester of 1991. Of these students, 1590 had taken the Scholastic Aptitude Test (SAT). Assume the SAT verbal test scores were normally distributed with a mean of 530 and a standard deviation of 70.

   a. What percentage of students were admitted with SAT verbal scores between 500 and 600?

   b. What percentage of students were admitted with SAT verbal scores of 600 or more?

   c. What percentage of students were admitted with SAT verbal scores of 480 or less?

21. Mensa is the international high-IQ society. To be a Mensa member, you have to have an IQ of 132 or above (*USA Today*, February 13, 1992). If IQ scores are normally distributed with a mean of 100 and a standard deviation of 15, what percentage of the population qualifies for membership in Mensa?

22. General Hospital's patient account division has compiled data on the age of accounts receivable. The data collected indicate that the age of the accounts follows a normal distribution, with $\mu = 28$ days and $\sigma = 8$ days.

   a. What portion of the accounts are between 20 and 40 days old—that is, $P(20 \leq x \leq 40)$?

   b. The hospital administrator is interested in sending reminder letters to the oldest 15% of accounts. How many days old should an account be before a reminder letter is sent?

   c. The hospital administrator would like to give a discount to those accounts that pay their balance by the 21st day. What percentage of the accounts will receive the discount?

23. The time required to complete a final examination in a particular college course is normally distributed, with a mean of 80 minutes and a standard deviation of 10 minutes. Answer the following questions:

   a. What is the probability of completing the exam in 1 hour or less?

   b. What is the probability a student will complete the exam in more than 60 minutes but less than 75 minutes?

   c. Assume that the class has 60 students and that the examination period is 90 minutes in length. How many students do you expect will be unable to complete the exam in the allotted time?

24. The useful life of a computer terminal at a university computer center is known to be normally distributed, with a mean of 3.25 years and a standard deviation of .5 years.

   a. Historically, 22% of the terminals have had a useful life less than the manufacturer's advertised life. What is the manufacturer's advertised life for the computer terminals?

   b. What is the probability that a computer terminal will have a useful life of at least 3 but less than 4 years?

25. From past experience, the management of a well-known fast-food restaurant estimates that the number of weekly customers at a particular location is normally distributed, with a mean of 5000 and a standard deviation of 800 customers.

   a. What is the probability that on a given week the number of customers will be between 4760 and 5800?

   b. What is the probability of a week with more than 6500 customers?

   c. For 90% of the weeks, the number of customers should exceed what amount?

26. Homes in Chicago, Illinois, in the price range of \$95,000–\$130,000, are on the market an average of 70 days prior to sale (*U.S. News & World Report,* April 6, 1992). Assume that the distribution of days on the market is normal with a standard deviation of 25 days.
   a. What is the probability a house will be on the market 100 days or more?
   b. What is the probability a house sells during the second month it is on the market? That is, $P(31 \leq x \leq 60)$?
   c. How many days are the fastest selling 20% on the market?

6.3 ▽ NORMAL APPROXIMATION OF BINOMIAL PROBABILITIES

In Chapter 5 we introduced the binomial probability distribution as a means for determining the probability of x successes in n trials of a binomial experiment. In cases where the number of trials, n, is large, binomial tables are not usually available and the computations associated with the binomial probability function are not practical. For instance with $n = 120$ and $p = .34$, the probability of 40 successes is given by

$$P(x = 40) = \frac{n!}{x!(n-x)!}p^x(1-p)^{(n-x)} = \frac{120!}{40!\,80!}(.34)^{40}(.66)^{80}$$

Evaluating such an expression is laborious and can lead to rounding errors.

In situations such as this, it is often possible to use the normal probability distribution to obtain good approximations of binomial probabilities. In this section we show how the normal probability distribution can be used for this purpose. The normal approximation provides acceptable accuracy whenever the number of trials, n, and the probability of success on each trial, p, have values such that both $np \geq 5$ and $n(1-p) \geq 5$.

EXAMPLE 6.11 ▶ A particular company has a history of making errors on 10% of its invoices. In a sample of 100 invoices, what is the probability that exactly 12 invoices have an error?

Note that this is a binomial experiment with $n = 100$ trials, $p = .10$, and $x = 12$. Rather than using the binomial probability function directly, we want to show how the normal probability distribution can be used to approximate the desired probability. Checking the requirements for the normal approximation, we find that $np = 100(.10) = 10$ and $n(1-p) = 100(.90) = 90$ are both at least 5. Thus, as previously stated, the normal approximation should provide good results.

Recall from Chapter 5 that the mean of a binomial random variable is $\mu = np$ and standard deviation is $\sigma = \sqrt{np(1-p)}$. For this example, we have $\mu = np = 100(.10) = 10$ and $\sigma = \sqrt{100(.10).90} = 3$. A normal distribution with this mean and standard deviation is shown in Figure 6.8. This is the normal distribution that is used to approximate the probabilities in this situation.

Also recall that with a continuous probability distribution, probabilities are computed as areas under the curve. As a result, the probability of any particular value for a continuous random variable is *zero*. Thus, to approximate the binomial probability of 12 successes, we compute the area under the corresponding normal curve between 11.5 and 12.5. The .5 that we add to and subtract from 12 to enable us to use a continuous distribution to approximate discrete probabilities is called the *continuity correction*. Thus, the interval $11.5 \leq x \leq 12.5$ for the normal random variable is used to approximate the value $x = 12$ for the discrete binomial distribution, and the binomial probability of $f(12)$ is approximated by the normal probability, $P(11.5 \leq x \leq 12.5)$.

FIGURE 6.8   **NORMAL APPROXIMATION OF A BINOMIAL PROBABILITY WITH n = 100 AND p = .10: THE PROBABILITY OF 12 ERRORS IN 100 TRIALS IS APPROXIMATELY .1052**

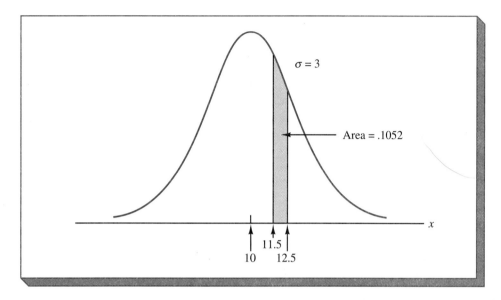

For the normal distribution shown in Figure 6.8, we use the following z values.

$$\text{At } x = 12.5, \qquad z = \frac{x - \mu}{\sigma} = \frac{12.5 - 10}{3} = .83$$

$$\text{At } x = 11.5, \qquad z = \frac{x - \mu}{\sigma} = \frac{11.5 - 10}{3} = .50$$

Consulting Table 6.1, we find the area under the curve between 0 and .83 is .2967. Similarly, the area under the normal curve between 0 and .50 is .1915. Therefore, the area between $z = .50$ and $z = .83$ is $.2967 - .1915 = .1052$. Thus, the normal approximation of exactly 12 errors in the 100 invoices is .1052. As it turns out, the actual binomial probability is .0988, and the error in our approximation is $.1052 - .0988 = .0064$. This is a pretty good approximation.   ◀

EXAMPLE 6.12 ▶ It is believed that 45% of a large population of registered voters favor a particular candidate for the state senate. A public opinion poll uses a randomly selected sample of voters and asks each person polled to indicate his or her preference for the candidates. What is the probability that a weekly poll based on the responses of 200 registered voters will show at least 50% of the voters favoring the candidate? That is, what is the probability at least 100 of the 200 voters will favor the candidate?

First note that this is a binomial probability experiment with $n = 200$ voters and $p = .45$. The binomial probability question asks for the probability of at least 100 successes. With $np = 200(.45) = 90$ and $n(1 - p) = 200(.55) = 110$, we see that the normal probability distribution can be used to approximate the desired binomial probability. The mean for the distribution is $\mu = np = 200(.45) = 90$ and the standard deviation is $\sigma = \sqrt{np(1 - p)} = \sqrt{200(.45)(.55)} = 7.04$. Using the continuity correction, we see that the interval 99.5 to 100.5 under the normal curve with $\mu = 90$ and $\sigma = 7.04$ provides the approximation of the binomial probability of 100 voters favoring the

FIGURE 6.9 **NORMAL APPROXIMATION TO A BINOMIAL PROBABILITY DISTRIBUTION WITH n = 200 AND p = .45: THE APPROXIMATION TO THE BINOMIAL PROBABILITY OF AT LEAST 100 VOTERS IS .0885**

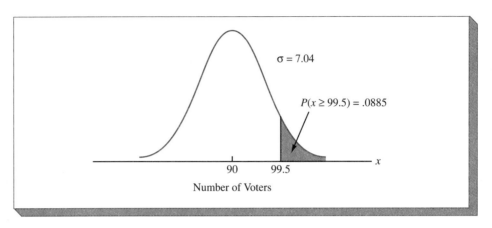

candidate. Since we are asking for the probability of 100 or more voters, we must compute the corresponding normal distribution probability of $x \geq 99.5$. The graph of this normal probability distribution approximation is shown in Figure 6.9.

Converting to the appropriate z value, we find

$$\text{At } x = 99.5, \qquad z = \frac{x - \mu}{\sigma} = \frac{99.5 - 90}{7.04} = 1.35$$

Using Table 6.1, the area between $z = 0$ and $z = 1.35$ is found to be .4115. Thus, we know there must be a $.5000 - .4115 = .0885$ probability of a value of 99.5 or more. We conclude that there is approximately a .0885 probability that the sample of 200 voters will show at least 100 voters favoring the candidate. ◀

EXERCISES

METHODS

27. For a binomial experiment with $n = 20$ and $p = .4$, compute the following probabilities using the normal approximation, and compare the answer with that obtained using the binomial probability tables:
   a. $P(x = 3)$
   b. $P(x = 4)$
   c. $P(x \geq 8)$
   d. $P(x < 6)$

28. For a binomial experiment with $n = 500$ and $p = .07$, compute the following probabilities using the normal approximation:
   a. $P(x > 40)$
   b. $P(20 \leq x \leq 40)$
   c. $P(x > 50)$
   d. $P(x \leq 30)$

APPLICATIONS

29. Thirty percent of the students at a particular university attended Catholic high schools. A random sample of 50 of this university's students has been taken. Use the normal approximation to the binomial probability distribution to answer the following questions:

    a. What is the probability that exactly 10 of the students selected attended Catholic high schools?

    b. What is the probability that 20 or more of the students attended Catholic high schools?

    c. What is the probability that the number of students from Catholic high schools is between 10 and 20, inclusive?

30. To obtain cost savings, a company is considering offering an early retirement incentive for its older management personnel. The consulting firm that designed the early retirement program has found that approximately 22% of the employees qualifying for the program will select early retirement during the first year of eligibility. Assume that the company offers the early retirement program to 50 of its management personnel.

    a. What is the expected number of employees who will select early retirement in the first year?

    b. What is the probability that at least 8 but not more than 12 employees will select early retirement in the first year?

    c. What is the probability that 15 or more employees will select the early retirement option in the first year?

    d. For the program to be judged successful, the company believes that it should entice at least 10 management employees to select early retirement in the first year. What is the probability that the program is successful?

31. Suppose that 54% of a large population of registered voters favor the Democratic candidate for state senator. A public opinion poll uses randomly selected samples of voters and asks each person in the sample his or her preference: the Democratic candidate or the Republican candidate. The weekly poll is based on the response of 100 voters.

    a. What is the expected number of voters who will favor the Democratic candidate?

    b. What is the variance in the number of voters who will favor the Democratic candidate?

    c. What is the probability that 49 or fewer individuals in the sample express support for the Democratic candidate?

32. A Myrtle Beach resort hotel has 120 rooms. In the spring months, hotel room occupancy is approximately 75%. Use the normal approximation to the binomial distribution to answer the following questions:

    a. What is the probability that at least half the rooms are occupied on a given day?

    b. What is the probability that 100 or more rooms are occupied on a given day?

    c. What is the probability that 80 or fewer rooms are occupied on a given day?

33. It is known that 30% of all customers of a major national charge card pay their bills in full before any interest charges are incurred. Use the normal approximation to the binomial distribution to answer the following questions for a group of 150 credit-card holders:

    a. What is the probability that between 40 and 60 customers pay their account charges before any interest charges are incurred? That is, find $P(40 \leq x \leq 60)$.

    b. What is the probability that 30 or fewer customers pay their account charges before any interest charges are incurred?

6.4 ▽ THE EXPONENTIAL PROBABILITY DISTRIBUTION (OPTIONAL)

A continuous probability distribution that is often useful in describing the time it takes to complete a task is the *exponential probability distribution*. The exponential random variable can be used to describe such things as the time between arrivals at a car wash, the time required to load a truck, the distance between major defects in a highway, and so on. The exponential probability density function is as follows.

Exponential Probability Density Function

$$f(x) = \frac{1}{\mu} e^{-x/\mu} \qquad \text{for } x \geq 0,\ \mu > 0 \qquad\qquad \textbf{(6.5)}$$

where μ is the mean of the probability distribution and $e = 2.71828$.

EXAMPLE 6.13 ▶ The time it takes to load a truck at the loading dock for Schips Department Store is described by an exponential probability distribution. The mean, or average, time to load a truck is 15 minutes ($\mu = 15$); thus, the appropriate probability density function is

$$f(x) = \frac{1}{15} e^{-x/15} \qquad \text{for } x \geq 0$$

The graph of this density function is shown in Figure 6.10. Note that when $x = 0$, we have $e^{-0/15} = 1$. So, $f(0) = \frac{1}{15} = .067$ is the intersection of the probability density function with the vertical axis. ◀

COMPUTING PROBABILITIES FOR THE EXPONENTIAL DISTRIBUTION

As with any continuous probability distribution, the area under the curve corresponding to a given interval provides the probability that the random variable takes on a value in that interval.

FIGURE 6.10   EXPONENTIAL PROBABILITY DISTRIBUTION FOR THE SCHIPS LOADING DOCK EXAMPLE

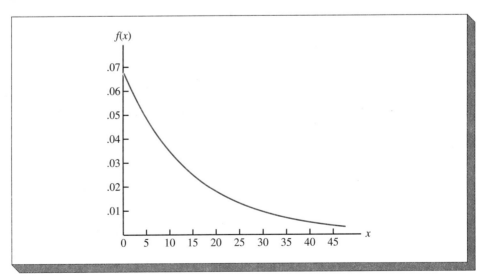

EXAMPLE 6.13
(CONTINUED) ▶ At the Schips loading dock, the probability that it takes 6 minutes or less ($x \leq 6$) to load a truck is given by the area under the curve in Figure 6.10 from $x = 0$ to $x = 6$. Similarly, the probability that a truck is loaded in 18 minutes or less ($x \leq 18$) is the area under the curve from $x = 0$ to $x = 18$. Note also that the probability that it takes between 6 minutes and 18 minutes ($6 \leq x \leq 18$) to load a truck is given by the area from $x = 6$ to $x = 18$. ◀

To compute exponential probabilities such as those described above, we make use of the following formula, which provides the probability of obtaining a value for the exponential random variable of less than or equal to some specific value of x, denoted by d in the following formula.

Computing Exponential Distribution Probabilities

$$P(x \leq d) = 1 - e^{-d/\mu} \qquad \text{(6.6)}$$

EXAMPLE 6.13
(CONTINUED) ▶ For the Schips loading dock, equation (6.6) becomes

$$P(\text{loading time} \leq d) = 1 - e^{-d/15}$$

Hence, the probability that it takes 6 minutes or less ($x \leq 6$) to load a truck is

$$P(x \leq 6) = 1 - e^{-6/15} = .3297$$

Note also the probability that it takes 18 minutes or less ($x \leq 18$) to load a truck is

$$P(x \leq 18) = 1 - e^{-18/15} = .6988$$

Thus, we see that the probability that it takes between 6 minutes and 18 minutes to load a truck is equal to $.6988 - .3297 = .3691$. Probabilities for any other interval can be computed in a similar manner. ◀

▷

EXERCISES

METHODS

34. Answer the following questions concerning the given exponential probability density function

$$f(x) = \frac{1}{8}e^{-x/8} \qquad \text{for } x \geq 0$$

   a. Find $P(x \leq 6)$.   **b.** Find $P(x \leq 4)$.
   c. Find $P(x \geq 6)$.   **d.** Find $P(4 \leq x \leq 6)$.

35. Answer the following questions concerning the given exponential probability distribution

$$f(x) = \frac{1}{4}e^{-x/4} \qquad \text{for } x \geq 0$$

   a. Draw a graph of the probability density function.
   b. Find $P(x \leq 4)$.   **d.** Find $P(2 \leq x \leq 4)$.
   c. Find $P(x \leq 2)$.   **e.** Find $P(1 \leq x \leq 4)$.

SELF TEST **36.** Consider the following exponential probability density function:

$$f(x) = \frac{1}{3}e^{-x/3} \qquad \text{for } x \geq 0$$

    a. Write the formula for $P(x \leq d)$.   **b.** Find $P(x \leq 2)$.
    c. Find $P(x \geq 3)$.   **d.** Find $P(x \leq 5)$.
    e. Find $P(2 \leq x \leq 5)$.

APPLICATIONS

37. There were 34 traffic fatalities in Clermont County, Ohio, during 1987 (The *Cincinnati Enquirer,* December 8, 1988). Assume that, given an average of 34 fatalities per year, an exponential distribution accurately describes the time between fatalities.
    a. What is the probability that the time between fatalities is 1 month or less?
    b. What is the probability that the time between fatalities is 1 week or more?

SELF TEST **38.** The time between arrivals of vehicles at a particular intersection follows an exponential probability distribution with a mean of 12 seconds.
    a. Sketch this exponential probability distribution.
    b. What is the probability that the arrival time between vehicles is 12 seconds or less?
    c. What is the probability that the arrival time between vehicles is 6 seconds or less?
    d. What is the probability that there will be 30 or more seconds between arriving vehicles?

39. The lifetime (hours) of an electronic device is a random variable with the following exponential probability density function:

$$f(x) = \frac{1}{50}e^{-x/50} \qquad \text{for } x \geq 0$$

    a. What is the mean lifetime of the device?
    b. What is the probability the device fails in the first 25 hours of operation?
    c. What is the probability the device operates 100 or more hours before failure?

40. A new automated production process has been averaging two breakdowns per day.
    a. What is the mean time between breakdowns, assuming 8 hours of operation per day?
    b. Show the exponential probability density function that can be used for the time between breakdowns.
    c. What is the probability that the process will run 1 hour or more before another breakdown?
    d. What is the probability that the process can run a full 8-hour shift without a breakdown?

41. The time in minutes for a student using a computer terminal at the computer center of a major university follows an exponential probability distribution with a mean of 36 minutes. Assume a second student arrives at the terminal just as another student is beginning to work on the terminal.
    a. What is the probability that the wait for the second student will be 15 minutes or less?
    b. What is the probability that the wait for the second student will be between 15 and 45 minutes?
    c. What is the probability the second student will have to wait an hour or more?

SUMMARY

This chapter extended the discussion of probability distributions to the case of continuous random variables. The major conceptual difference between discrete and continuous probability distributions is in the method of computing probabilities. With discrete distributions, the probability function $f(x)$ provides the probability that the random variable x assumes various values. With continuous probability distributions, we associate a probability density function, denoted by $f(x)$. The difference is that the probability density function does not provide probability values for a continuous random variable. Probabilities are given by areas under the curve in the graph of the probability density function $f(x)$. Since the area under the curve above a single point is zero, we observe that the probability of any particular value is zero for any continuous random variable.

Three continuous probability distributions, the uniform, normal, and exponential, were presented. Because of its wide range of applicability and because it will be used extensively in the remainder of the text, the normal probability distribution was given greater coverage. It is considered by many to be the most important distribution in probability and statistics. Each normal probability distribution belongs to a family of similar bell-shaped distributions with the specific normal distribution depending on the value of the mean μ and standard deviation σ.

Tables of probabilities are available for a normal probability distribution with a mean of 0 and a standard deviation of 1. This special normal probability distribution is referred to as the standard normal probability distribution. Common notation is to use z to denote the standard normal random variable. The relationship between z and any other normal random variable x with mean μ and standard deviation σ is given by $z = (x - \mu)/\sigma$. The z value indicates the number of standard deviations the normal random variable x is from its mean μ.

Probability questions about a random variable x with a normal distribution can be answered by first converting the random variable x to its corresponding z value and then using the table of areas for the standard normal probability distribution to determine the appropriate probabilities. We saw that two types of probability questions can be asked. An interval of values of the random variable is given and the question is to determine the probability the random variable assumes a value in the interval; alternatively, a probability value is given and the question then is to determine the values of the random variable yielding the given probabilities. Both questions can be answered by using the table of areas for the standard normal probability distribution. In making the probability calculations, it is recommended that a sketch of the appropriate normal curve be made as an aid to visualizing the probability information desired.

Finally, we noted that binomial probabilities can be difficult to compute whenever the number of trials, n, is large. However, if both $np \geq 5$ and $n(1 - p) \geq 5$, the normal probability distribution with $\mu = np$ and $\sigma = \sqrt{np(1-p)}$ provides a good approximation to the binomial probabilities. A continuity correction must be utilized to account for the fact that the discrete binomial probability is being approximated by the continuous normal probability distribution.

GLOSSARY

Continuous uniform probability distribution A continuous probability distribution where the probability that the random variable will assume a value in any interval of equal length is the same for each interval.

Probability density function The function that defines the probability distribution of a continuous random variable.

Normal probability distribution A continuous probability distribution. Its probability density function is bell shaped and determined by the mean μ and standard deviation σ.

Standard normal probability distribution A normal probability distribution with a mean of 0 and a standard deviation of 1.

Continuity correction A value of .5 that is added and subtracted from a value of x when a continuous probability distribution (e.g., the normal) is used to approximate a discrete probability distribution (e.g., the binomial).

Exponential probability distribution A continuous probability distribution that is useful in describing the time, or space, between occurrences of an event.

▼ KEY FORMULAS

Continuous Uniform Probability Density Function

$$f(x) = \begin{cases} \dfrac{1}{b-a} & \text{for } a \le x \le b \\[2mm] 0 & \text{elsewhere} \end{cases} \qquad \text{(6.1)}$$

Normal Probability Density Function

$$f(x) = \frac{1}{\sqrt{2\pi}\,\sigma} e^{-(x-\mu)^2/2\sigma^2} \qquad \text{(6.2)}$$

Standard Normal Probability Density Function

$$f(z) = \frac{1}{\sqrt{2\pi}} e^{-z^2/2} \qquad \text{(6.3)}$$

Converting to the Standard Normal Distribution

$$z = \frac{x-\mu}{\sigma} \qquad \text{(6.4)}$$

Exponential Probability Density Function

$$f(x) = \frac{1}{\mu} e^{-x/\mu} \qquad \text{for } x \ge 0,\ \mu > 0 \qquad \text{(6.5)}$$

Computing Exponential Distribution Probabilities

$$P(x \le d) = 1 - e^{-d/\mu} \qquad \text{(6.6)}$$

▼ REVIEW QUIZ

TRUE/FALSE

1. The continuous uniform probability distribution involves a discrete random variable.
2. For the uniform probability distribution, the height of the probability density function varies but reaches its peak at the mean.
3. The area under the curve to the left of the mean for a uniform probability distribution is .50.
4. As the standard deviation increases, the height of the normal curve increases.
5. The probability the normal random variable assumes a value within one standard deviation of the mean is .50.
6. A standard normal probability distribution has a mean of 0 and a variance of 1.
7. To compute probabilities for a normal random variable, one must first convert to a standard normal probability distribution.
8. A continuity correction is needed whenever normal probabilities are computed.
9. The normal probability distribution may be used to approximate the binomial, provided that $np \ge 5$ and $n(1-p) \ge 5$.
10. The probability of a normal random variable assuming a value within two standard deviations of its mean is approximately .95.

MULTIPLE CHOICE

11. The random variable x is uniformly distributed between 12 and 20.
   a. It has a mean of 16.
   b. It has a variance of $5\frac{1}{3}$.
   c. The probability $x = 15$ is zero.
   d. All the above are true.

For questions 12–14, consider the normally distributed random variable x, which has a mean of 17 and a standard deviation of 3.

12. The probability that x is less than or equal to 14 is
   a. .1587   **c.** .3414
   b. .1765   **d.** .4986

13. If a value of x is randomly selected, the probability that it will be between 20 and 22 is
   a. .1112   **c.** .4525
   b. .3413   **d.** .7938

14. The probability that two randomly selected values of x will both have a value less than 22 is
   a. .2048   **c.** .9073
   b. .4525   **d.** .9525

15. Assume that a normal probability distribution is being used to approximate binomial probabilities for the case of $n = 500$ and $p = .10$. The value of σ that should be used is closest to
   a. 3   **c.** 9
   b. 6   **d.** 40

16. Scores on a reading skills test are normally distributed with a mean of $\mu = 500$ and a standard deviation of 50. An agency hires only people whose scores are in the top 5% of individuals taking the test. This company should consider hiring anyone who achieves at least a score of
   a. 500   **c.** 550
   b. 538   **d.** 583

17. If the average time between fouls in a basketball game is 1 minute, the probability of going 3 minutes with no fouls is closest to
   a. .003   **c.** .050
   b. .010   **d.** .100

SUPPLEMENTARY EXERCISES

42. In an office building, the waiting time for an elevator is found to be uniformly distributed between 0 and 5 minutes.
   a. What is the probability density function $f(x)$ for this uniform distribution?
   b. What is the probability of waiting longer than 3.5 minutes?
   c. What is the probability that the elevator arrives in the first 45 seconds?
   d. What is the probability of a waiting time between 1 and 3 minutes?
   e. What is the expected waiting time?

43. The time required to complete a particular assembly operation is uniformly distributed between 30 and 40 minutes.
   a. What is the mathematical expression for the probability density function?
   b. Compute the probability that the assembly operation will require more than 38 minutes to complete.
   c. If management wants to set a time standard for this operation, what time should be selected such that 70% of the time the operation will be completed within the time specified?
   d. Find the expected value and standard deviation for the assembly time.

44. A particular make of automobile is listed as weighing 4000 pounds. Because of weight differences due to the options ordered with the car, the actual weight varies uniformly between 3900 and 4100 pounds.
   a. What is the mathematical expression for the probability density function?
   b. What is the probability that the car will weigh less than 3950 pounds?

45. Given that z is a standard normal random variable, compute the following probabilities:
   a. $P(-.72 \leq z \leq 0)$ **b.** $P(-.35 \leq z \leq .35)$
   c. $P(.22 \leq z \leq .87)$ **d.** $P(z \leq -1.02)$

46. Given that z is a standard normal random variable, compute the following probabilities:
   a. $P(z \geq -.88)$ **b.** $P(z \geq 1.38)$
   c. $P(-.54 \leq z \leq 2.33)$ **d.** $P(-1.96 \leq z \leq 1.96)$

47. Given that z is a standard normal random variable, find z if it is known that
   a. the area between $-z$ and z is .90
   b. the area to the right of z is .20
   c. the area between -1.66 and z is .25
   d. the area to the left of z is .40
   e. the area between z and 1.80 is .20

48. Motorola used the normal distribution to determine the probability of defects and the number of defects expected in a production process (*APICS—The Performance Advantage,* July 1991). Assume a production process is designed to produce items with a weight of 10 ounces and that the process mean is 10. Calculate the probability of a defect and the expected number of defects for a 1000-unit production run under the following situations.
   a. The process standard deviation is .15 and the process control is set at plus or minus 1 standard deviation. Units with weight less than 9.85 or greater than 10.15 ounces will be classified as defects.
   b. Through process design improvements, the process standard deviation can be reduced to .05. Assume the process control remains the same with weights less than 9.85 or greater than 10.15 ounces being classified as defects.
   c. What is the advantage of reducing process variation and setting process control limits at a greater number of standard deviations from the mean?

49. In 1990, the mean household income for Americans was $37,403 (*Statistical Abstract of the United States,* 1992).
   a. It was noted that 5.2% of the households earned less than $5000. Assuming that household income is normally distributed, what is the standard deviation of household income?
   b. It was also noted that 24.6% of households earned more than $50,000. Does this seem reasonable given the standard deviation computed in part (a)? Explain.
   c. In part (a) we said to assume that household income is normally distributed. Does this assumption appear to be reasonable? Explain.

50. A soup company markets eight varieties of homemade soup throughout the eastern states. The standard-size soup can holds a maximum of 11 ounces, while the label on each can advertises contents of 10¾ ounces. The extra ¼ ounce is to allow for the possibility of the automatic filling machine placing more soup than the company actually wants in a can. Past experience shows that the number of ounces placed in a can is approximately normally distributed, with a mean of 10¾ ounces and a standard deviation of .1 ounce. What is the probability that the machine will attempt to place more than 11 ounces in a can, causing an overflow to occur?

51. The sales of High-Brite Toothpaste are believed to be approximately normally distributed, with a mean of 10,000 tubes per week and a standard deviation of 1500 tubes per week.
   a. What is the probability that more than 12,000 tubes will be sold in any given week?
   b. In order to have a .95 probability that the company will have sufficient stock to cover the weekly demand, how many tubes should be produced?

52. Points scored by the winning team in NCAA college football games are approximately normally distributed, with a mean of 24 and a standard deviation of 6.
 a. What is the probability that a winning team in a football game scores between 20 and 30 points; that is, $P(20 \le x \le 30)$?
 b. How many points does a winning team have to score to be in the highest 20% of scores for college football games?

53. Ward Doering Auto Sales is considering offering a special service contract that will cover the total cost of any service work required on leased vehicles. From past experience, the company manager estimates that yearly service costs are approximately normally distributed, with a mean of $150 and a standard deviation of $25.
 a. If the company offers the service contract to customers for a yearly charge of $200, what is the probability that any one customer's service costs will exceed the contract price of $200?
 b. What is Ward's expected profit per service contract?

54. The attendance at football games at a certain stadium is normally distributed, with a mean of 45,000 and a standard deviation of 3000.
 a. What percentage of the time should attendance be between 44,000 and 48,000?
 b. What is the probability of the attendance exceeding 50,000?
 c. For 80% of the time the attendance should be at least how many?

55. Assume that the test scores from a college admissions test are normally distributed, with a mean of 450 and a standard deviation of 100.
 a. What percentage of the people taking the test score between 400 and 500?
 b. Suppose that someone receives a score of 630. What percentage of the people taking the test score better? What percentage score worse?
 c. If a particular university will not admit anyone scoring below 480, what percentage of the persons taking the test would be acceptable to the university?

56. The Office Products Group of the former Burroughs Corporation manufactures plastic credit cards used in automatic bank teller machines. Any card with a length of less than 3.365 inches is considered defective. One of the dies used in making the credit cards is producing cards with a mean length of 3.367 inches. The lengths are normally distributed with a standard deviation of .001 inch.
 a. What is the probability of obtaining a defective card using this die?
 b. The company does not want to use any die that produces more than 1% defective cards. What should the company do in this instance?
 c. Assuming the standard deviation stays at .001 inch, what is the smallest acceptable mean length for cards manufactured? (*Hint:* For what mean length will no more than 1% of the cards be shorter than 3.365 inches?)

57. A machine fills containers with a particular product. The standard deviation of filling weights is known from past data to be .6 ounces. If only 2% of the containers hold less than 18 ounces, what is the mean filling weight for the machine? That is, what must μ equal? Assume the filling weights have a normal distribution.

58. Consider a multiple-choice examination with 50 questions. Each question has four possible answers. Assume that a student who has done the homework and attended lectures has a .75 probability of answering any question correctly.
 a. A student must answer 43 or more questions correctly to obtain a grade of A. What percentage of the students who have done their homework and attended lectures will obtain a grade of A on this multiple-choice examination?
 b. A student who answers 35–39 questions correctly will receive a grade of C. What percentage of students who have done their homework and attended lectures will obtain a grade of C on this multiple-choice examination?
 c. A student must answer 30 or more questions correctly to pass the examination. What percentage of the students who have done their homework and attended lectures will pass the examination?

d. Assume that a student has not attended class and has not done the homework for the course. Furthermore, assume that the student will simply guess at the answer to each question. What is the probability that this student answers 30 or more questions correctly and passes the examination?

59. The book *100% American* by Daniel Evan Weiss reports that 64% of Americans live in the state where they were born. What is the probability that a random sample of 100 people will find between 60 and 70 people living in the state where they were born?

CHAPTER 7

SAMPLING AND SAMPLING DISTRIBUTIONS

WHAT YOU WILL LEARN IN THIS CHAPTER

- The reasons for sampling
- How to select a simple random sample
- How results from samples can be used to provide estimates of population parameters
- The characteristics and use of the sampling distribution of $\bar{x}$
- What the central limit theorem is and the important role it plays in statistics
- How to calculate the probability of how close a sample mean $\bar{x}$ is to a population mean μ

CONTENTS

232

Charitable Contributions: The People Who Give

Independent Sector, a nonprofit organization located in Washington, D.C., promotes philanthropy and nonprofit groups. As part of a campaign designed to encourage Americans to donate 5% of their income and time to charities, Independent Sector commissioned a survey on private giving in America. The survey, conducted by the Gallup Organization, consisted of a sample of 2775 households throughout the United States.

Households with annual incomes below $10,000 were found to contribute an average of 2.8% of their incomes to charitable causes. Households with annual incomes between $50,000 and $100,000 contributed an average of 1.5%, and households with annual incomes over $100,000 contributed an average of 2.1%. Moreover, households with annual incomes below $30,000 contributed almost half of the total amount given to charity. These results led Brian O'Connell, president of Independent Sector, to conclude that relatively speaking, "People of means cannot be described as particularly caring. For that primary category of humaneness, it is the poor and struggling who generally lead the way."

The survey showed that the average annual contribution for households that donated was $790. Religion, education, and health were the top three categories of giving, with half of all respondents surveyed giving to religious organi-

The United Way collects charitable contributions for numerous organizations.

zations. In addition, the survey found that individuals who volunteer time give an average of 4.7 hours a week.

Independent Sector uses the results of the survey to identify potential donors and plan future advertisements, many of which will be targeted at young professionals. Mr. O'Connell thinks that the survey results may remove confusion on the part of most people about the appropriate level of giving. He believes that when people see these results, they may be more willing to be more generous with their time and money.

This *Statistics in Practice* is based on "Poorer Households Lead in Rate of Charitable Giving," *The Wall Street Journal,* October 19, 1988.

In this chapter we introduce simple random sampling and the process of using a sample mean to provide an estimate of a population mean. In addition, we introduce the important concept of a sampling distribution. It is knowledge of the sampling distribution that enables us to make statements concerning the precision when sample results are used to draw conclusions about a population.

7.1 ▽ INTRODUCTORY SAMPLING CONCEPTS

POPULATIONS AND SAMPLES

As stated in Chapter 1, the primary purpose of statistics is to provide information about a *population* based on information contained in a *sample.* In the *Statistics in Practice* article, a survey of 2775 households conducted by the Gallup Organization was used to learn about charitable giving for the population of all households in the United States. The definitions of a population and a sample are as follows.

> ### Population
> A *population* is the set of all elements of interest for a particular study.

> ### Sample
> A *sample* is a subset of the population selected to represent the whole population.

EXAMPLE 7.1 ▶ Television advertisers paid $1,350,000 per minute for commercials that aired during the 1989 Super Bowl. A total of 49 television commercials were shown during the game. To learn about viewer reaction to the ads, *USA Today* gathered 60 randomly selected individuals and asked them to rate each of the commercials. The 60 individuals formed a sample that was used to represent the population of individuals watching the Super Bowl on television. Results from the sample indicated that American Express and Diet Pepsi provided the two best-liked commercials shown during the 1989 Super Bowl game (*USA Today,* January 23, 1989). ◀

EXAMPLE 7.2 ▶ Sports psychology studies have investigated the mental-training techniques used by the world's greatest athletes (*Peak Performance,* by Charles A. Garfield, 1984). For example, assume a particular study is designed to investigate the training techniques of the world's best professional golfers. The previous year's Professional Golfers Association (PGA) list of the top 100 money winners could be used to identify the population of world's best golfers. However, since the research project requires lengthy interviews and follow-up studies with each golfer studied, the researcher may not have sufficient time or funds to interview every golfer in the population. Consequently, a sample of 10 golfers can be selected from the population of 100 golfers. Data collected from the sample of 10 golfers will be used to make inferences about the mental-training techniques of the population of the world's 100 best golfers. ◀

REASONS FOR SAMPLING

Whenever anyone wants information about a population, two alternatives are available: Collect the information from every item in the population, referred to as a *census,* or collect the information from a subset of the population, referred to as a *sample.* Numerous practical situations indicate that sampling is often the preferred way of collecting the desired information.

Reasons for sampling are largely due to the fact that a sample provides *time* and *cost* savings compared to a census of the entire population. In particular, if the population is large, the time and cost required to conduct a census may be prohibitive. In the case of an extremely large or infinite population, a census of the entire population is impossible. In Example 7.1 we described the *USA Today* study of viewer ratings of television commercials shown during the 1989 Super Bowl. A census of the population of all who watched that Super Bowl on television would have been impractical and essentially impossible.

However, the sample of 60 viewers was a relatively inexpensive way to obtain the viewer ratings information. In addition, the use of this sample enabled *USA Today* to report the ratings for each commercial on Monday, the morning after the Super Bowl. In Example 7.2 a census of the population of 100 top professional golfers would have been possible. However, using a sample of 10 golfers will provide substantial time and cost savings.

The use of sampling is also necessary in situations where the information-gathering process results in damaging or destroying the items being studied. For example, an automobile manufacturer obtains information about car safety by crash-testing a sample of its automobiles. Since the testing damages the automobiles, only a small sample can be used to collect the desired safety information. Finally, and perhaps surprisingly to some, a sample can often result in greater accuracy than a census. This is particularly true when a trained interviewer or a trained scientific technician is needed to collect the information. In this case, a census might necessitate the use of less skilled interviewers or technicians, which could lead to inaccurate or unreliable data being obtained.

NOTES AND COMMENTS

The number of elements in a sample, called the *sample size,* can vary greatly depending on the nature of the study. For example, the *Statistics in Practice* article at the beginning of this chapter reported the results of a survey conducted by the Gallup Organization based upon a sample of 2775 households throughout the United States. Generally, sample sizes of 1000–3000 are typical for the national polls and surveys conducted by organizations such as Gallup and Harris. However, most statistical studies use substantially smaller sample sizes. For example, *The Wall Street Journal* (November 23, 1988) reported information on operating costs and profits for inns located throughout the United States. This study was based on a sample of 72 bed-and-breakfast and full-service inns.

EXERCISES

1. The Federal Bureau of Justice statistics conducted a national study of juveniles currently housed in long-term, state-operated correctional institutions (*Democrat and Chronicle,* September 19, 1988). The results of a sample of juveniles in correctional institutions showed that 1835 of 2621 had been raised in single-parent homes.
    a. Define the population for this study.
    b. Define the sample for this study.
    c. Why was a sample used instead of a census?
    d. Use the sample results to estimate the percent of all juveniles in the population who have been raised in single-parent homes.

2. *The New York Times,* August 14, 1984, reported that a U.S. General Accounting Office (GAO) sample revealed that many college students who were receiving federal aid were not meeting minimum academic standards. The GAO report, based on an analysis of student records at selected colleges and universities, stated that 10% of the students receiving aid had a grade average of F.
    a. Define the population for this study.
    b. Define the sample for this study.
    c. What were the advantages of sampling in this study?

 SELF TEST

3. Researchers at Oklahoma State University conducted a study involving a sample of 163 high school juniors (*The School Counselor,* November 1987). The purpose of the study was to explore the relationships among indices of loneliness, perceptions of school, and grade point

average. The 163 high school students who participated in the study were selected from high schools located in three small cities.

a. Define the sample for this study.

b. The researchers would like to use the results of their study to make conclusions about the population of all high school students in the United States. Do you see any problems with such conclusions? Explain.

4. An official for United Airlines reported that out of 104 United flights arriving at Chicago's O'Hare Airport, only 3 arrived more than 15 minutes late (*The Wall Street Journal,* November 7, 1988). Assuming that the 104 flights are a representative sample of all United Airline flights into O'Hare, answer the following questions:

a. Define the population.

b. Define the sample.

c. Use the sample results to estimate the percentage of all United Airline flights into O'Hare that arrive more than 15 minutes late.

d. The source of statistical information should always be considered when accepting and interpreting reported results. The preceding sample information was provided by United Airlines. Would you feel better about the results if the study of flight arrivals had been conducted by the Federal Aviation Administration? Discuss.

7.2 ▽ SIMPLE RANDOM SAMPLING

There are several methods that can be used to select a sample from a population. One of the most common sampling methods is *simple random sampling*. The definition of a simple random sample and the process of selecting a simple random sample depends on whether the population is *finite* or *infinite* in size.

SAMPLING FROM A FINITE POPULATION

Let us assume that the population of interest is finite and consists of N items. We will also assume that it is possible to obtain a list of all N items in the population. In this situation a simple random sample is defined as follows.

Simple Random Sample (Finite Population)

A *simple random sample* of size n from a finite population of size N is a sample selected such that each possible sample of size n has the same probability of being selected.

EXAMPLE 7.3 ▶ A school district uses five buses (identified as A, B, C, D, and E) to transport elementary students to and from school each day. In selecting a simple random sample of size $n = 2$ buses from the population of size $N = 5$ buses, note that there are 10 different samples of size $n = 2$. These 10 samples consist of buses AB, AC, AD, AE, BC, BD, BE, CD, CE, and DE. If we select a sample in such a way that each of these 10 samples has the same $\frac{1}{10}$ probability of being selected, the sample selected would be a simple random sample. We could find such a sample by writing the two letters corresponding to each of the possible samples on 10 separate, but identical, pieces of paper. Mixing the pieces of paper thoroughly and then randomly selecting one piece of paper would provide a simple random sample of two buses from the finite population of five buses. ◀

Although the method described above enables the selection of a simple random sample, this process becomes cumbersome and impractical as the finite population size increases. Thus, we need a better way to identify a simple random sample from a finite population.

In Example 7.3 the sample of two buses was selected in one random drawing. Another approach to identifying a simple random sample is to select the elements for the sample in a one-at-a-time fashion. At each selection, we make sure that each of the items remaining in the population has the same probability of being selected.

EXAMPLE 7.3 ▶ (CONTINUED)

The one-at-a-time approach to selecting a simple random sample of two school buses can be accomplished as follows:

1. Write the five letters corresponding to each of the buses on five separate and identical pieces of paper.
2. Mix the five pieces of paper thoroughly and select one piece of paper at random; the letter on this piece of paper is the first bus selected for the simple random sample.
3. Mix the remaining four pieces of paper thoroughly and select another piece of paper at random; the letter on this piece of paper is the second bus selected for the simple random sample. ◀

In using the one-at-a-time approach for Example 7.3 we did not replace the first piece of paper after it was selected from the population; this type of sampling is called *sampling without replacement*. If we had replaced the first piece of paper selected prior to choosing the second, we would have been *sampling with replacement*. Although sampling with replacement is a valid way of identifying a simple random sample, sampling without replacement is the sampling procedure used most often. Whenever we refer to simple random sampling, we will make the assumption that the sampling is done without replacement.

Although the one-at-a-time approach is a practical way of selecting a simple random sample, the procedure that we used in Example 7.3 can be improved. In Example 7.4 we describe how random numbers can be used to eliminate the step of using a piece of paper for each item in the population.

EXAMPLE 7.4 ▶

Suppose a university has received 7000 applications for admission. The director of admissions would like to use a sample of 50 applications to obtain information on Scholastic Aptitude Test (SAT) scores of incoming students. How could a simple random sample of 50 applications be selected?

We begin by numbering the 7000 applications in the population from 1 to 7000. Tables of random numbers are available from a variety of handbooks that contain page after page of random numbers.* One such page of random numbers is shown in Table 8 in Appendix B. A portion of this page is also shown in Table 7.1. The digit appearing in any position in the random number table is a random selection of the digits 0, 1, . . . , 9 with each digit having an equal chance of occurring. By selecting four-digit random numbers that range in value from 0001 to 7000, we have a random number corresponding to each of the numbered applications.

To use the random numbers to identify items for the sample, we enter the table at any *arbitrary point* and then select four-digit random numbers by moving systematically down a column or across a row of the table. For example, suppose that we arbitrarily start

---

*For example, *A Million Random Digits with 100,000 Normal Deviates,* by Rand Corporation (New York: The Free Press, 1955). Copyright 1955 and 1983 by Rand Corporation.

TABLE 7.1   RANDOM NUMBERS

63271	59986	71744	51102	15141	80714	58683	93108	13554	79945
88547	09896	95436	79115	08303	01041	20030	63754	08459	28364
55957	57243	83865	09911	19761	66535	40102	26646	60147	15702
46276	87453	44790	67122	45573	84358	21625	16999	13385	22782
55363	07449	34835	15290	76616	67191	12777	21861	68689	03263
69393	92785	49902	58447	42048	30378	87618	26933	40640	16281
13186	29431	88190	04588	38733	81290	89541	70290	40113	08243
17726	28652	56836	78351	47327	18518	92222	55201	27340	10493
36520	64465	05550	30157	82242	29520	69753	72602	23756	54935
81628	36100	39254	56835	37636	02421	98063	89641	64953	99337
84649	48968	75215	75498	49539	74240	03466	49292	36401	45525
63291	11618	12613	75055	43915	26488	41116	64531	56827	30825
70502	53225	03655	05915	37140	57051	48393	91322	25653	06543
06426	24771	59935	49801	11082	66762	94477	02494	88215	27191
20711	55609	29430	70165	45406	78484	31639	52009	18873	96927
41990	70538	77191	25860	55204	73417	83920	69468	74972	38712
72452	36618	76298	26678	89334	33938	95567	29380	75906	91807
37042	40318	57099	10528	09925	89773	41335	96244	29002	46453
53766	52875	15987	46962	67342	77592	57651	95508	80033	69828
90585	58955	53122	10625	84299	53310	67380	84249	25348	04332
32001	96293	37203	64516	51530	37069	40261	61374	05815	06714
62606	64324	46354	72157	67248	20135	49804	09226	64419	29457
10078	28073	85389	50324	14500	15562	64165	06125	71353	77669
91561	46145	24177	15294	10061	98124	75732	00815	83452	97355
13091	98112	53959	79607	52244	63303	10413	63839	74762	50289

with the third column of random numbers in Table 7.1. Since we need only four-digit numbers, we ignore the first digit in the column. Starting at the top and moving downward, the four-digit random numbers and corresponding application numbers are as follows.

	Random Number	Application Number to Include in Sample
	1744	1744
	5436	5436
	3865	3865
	4790	4790
	4835	4835
These numbers are greater than 7000 and cannot be used since applications 9902 and 8190 do not exist.	9902	—
	8190	—
	6836	6836
	.	.
	.	.
	.	.

Continue until 50 different applications are selected.

Since the numbers selected from the table in Example 7.4 are random, this procedure guarantees that each item in the population has the same probability of being included in the sample and that the sample selected will be a simple random sample. In using random numbers for simple sampling, a random number previously used to identify an item for the sample may reappear in the random number table. In selecting the simple random sample *without replacement,* previously used random numbers are ignored because the corresponding item is already in the sample.

SAMPLING FROM AN INFINITE POPULATION

Although many sampling situations involve finite populations, there are other situations in which the population is either infinite or so large that for practical purposes it must be treated as infinite. In sampling from an infinite population, we must give a new definition for a simple random sample. In addition, since the items cannot be listed and numbered, we must use a different process for selecting items for the sample.

Let us consider a situation that can be viewed as requiring a simple random sample from an infinite population. Suppose we want to estimate the average time between placing an order and receiving food for customers arriving at a fast-food restaurant during the 11:30 A.M. to 1:30 P.M. lunch period. If we consider the population as being all customers who arrive in this time period, we see that it would be next to impossible to specify an exact limit on the number of possible customers. In fact, if we view the population as being all customers who could *conceivably* arrive during the lunch period, we can consider the population as being unlimited or infinite. Our task then is to select a simple random sample of *n* customers from this population. With this situation in mind, we now state the definition of a simple random sample from an infinite population.

Simple Random Sample (Infinite Population)

A *simple random sample from an infinite population* is a sample selected such that the following conditions are satisfied:

1. Each item selected comes from the same population.
2. Each item is selected independently.

For the example of selecting a simple random sample of customers at a fast-food restaurant, we find that the first condition is satisfied by any customer who arrives during the 11:30 A.M. to 1:30 P.M. lunch period. The second condition is satisfied by ensuring that the selection of a particular customer does not influence the selection of any other customer.

A well-known fast-food restaurant has implemented a simple random sampling procedure for just such a situation. The sampling procedure is based on the fact that some customers present discount coupons, which provide special prices on sandwiches, drinks, french fries, and so on. Whenever a customer presents a discount coupon, the *next* customer is selected for the sample. Since the customers present discount coupons in a random and independent fashion, the firm is satisfied that the sampling plan satisfies the two conditions for a simple random sample from an infinite population.

In other sampling situations, such as the sampling of parts from a production line, plants for biological study, water for pollution control, and so on, the population can be considered to be infinite in size. In these cases, extra care must be taken to ensure that the sample is representative of the population. Thus, selection patterns such as sampling the water supply only at 8:00 A.M. each day must be avoided. If such precautions can be taken, it is usually reasonable to assume that the properties of a simple random sample have been satisfied.

NOTES AND COMMENTS

1. A population can be classified as being either finite or infinite in size. Finite populations are often defined by lists such as organization membership rosters, enrolled students, mailing lists, credit-card customers, and inventory product numbers. Usually, infinite populations are defined by an ongoing process where a listing of all possible items is impossible, such as all possible parts to be manufactured, all possible customer visits, or all possible bank transactions.

2. The number of different simple random samples of size n that can be selected from a finite population of size N may be determined by using the formula for combinations presented in Chapter 4:

$$\frac{N!}{n!(N-n)!}$$

The $N!$, $n!$, and $(N-n)!$ refer to the factorial computations.

In Example 7.3 we selected a simple random sample of size $n = 2$ from a finite population of size $N = 5$. Using the preceding formula, we find that the number of different simple random samples possible is

$$\frac{5!}{2!(5-2)!} = \frac{(5)(4)(3)(2)(1)}{[(2)(1)][(3)(2)(1)]} = \frac{120}{(2)(6)} = 10$$

With large populations, the number of possible different simple random samples can be quite large.

EXERCISES

METHODS

SELF TEST

5. Consider a finite population with five items labeled A, B, C, D, and E. There are 10 possible simple random samples of size 2 that can be selected.
   a. List the 10 samples beginning with AB, AC, and so on.
   b. Using simple random sampling, what is the probability that each sample of size 2 is selected?
   c. Assume random number 1 corresponds to A, random number 2 corresponds to B, and so on. List the elements in a simple random sample of two items that will be selected using the random digits 8 0 5 7 5 3 2.

6. Assume a finite population has 350 items. Using the last three digits of each set of five-digit random numbers shown below, read across the row, and determine the first four units that will be selected for the simple random sample.

   98601   73022   83448   02147   34229   27553   84147   93289   14209

7. A population consists of four items labeled A, B, C, and D.
   a. How many simple random samples of size $n = 2$ can be selected from this population?
   b. List all possible simple random samples.

8. How many simple random samples are possible when samples of size $n = 3$ are to be selected from a finite population of size $N = 6$?

APPLICATIONS

9. Exercise 1 described a national study of juveniles currently housed in long-term, state-operated correctional institutions. Is the population finite or infinite? Would it be possible to obtain a list of juveniles in these institutions?

10. Exercise 4 described information in *The Wall Street Journal* (November 7, 1988) where United Airlines reported a sample of 104 flights arriving at Chicago's O'Hare Airport. Statistics reported indicated that only 3 of the 104 flights arrived more than 15 minutes late.
    a. If the purpose of the sample is to make inferences about United's flight performance during the preceding month (October 1988), is the population finite or infinite? Explain.
    b. If the purpose of the sample is to make inferences about the ongoing process of all United flights into O'Hare, is the population finite or infinite? Explain.

11. Example 7.2 describes a study of the mental-training techniques of the world's best professional golfers. The population was defined to be the top 100 money winners from the Professional Golfers Association. A simple random sample of 10 golfers will be selected for interviews and follow-up studies. How many different simple random samples of 10 golfers are possible?

12. In Example 7.3, a school district used five buses to transport elementary students to and from school. The five buses were identified as A, B, C, D, and E.
    a. How many samples of size 3 are possible? List the different possible samples.
    b. Using simple random sampling, what is the probability each possible sample will be selected?

13. Consider the following six midwestern states as a population: Iowa, Illinois, Wisconsin, Michigan, Indiana, and Ohio. Assume that a sample of four states will be selected from this population in order to study employment trends in the Midwest.
    a. How many samples of size 4 are possible? List the samples.
    b. How many of the samples contain the state of Illinois?
    c. Using simple random sampling, the possible samples have the same probability of being selected. If this is the case, what is the probability of selecting a sample that contains the state of Illinois?
    d. Repeat part (b) for the state of Indiana.
    e. Using the results from parts (b) and (d), what can you say about the probability that a given state will be included in the simple random sample?

14. Consider a population of five salespersons selling mobile telephone units to a variety of customers. The individuals in the population are identified by the letters A, B, C, D, and E.
    a. Using the 15th row of random numbers in Table 7.1 and the random digits 1, 2, 3, 4, and 5 to correspond to the five salespersons, select a simple random sample of size 2 from the population.
    b. What sample would have been selected if the random numbers in row 20 had been used?

SELF TEST ▷ 15. *Business Week* publishes data on sales, profits, assets, dividends, shares, and earnings per share for America's 1000 most valuable companies (*The 1992 Business Week 1000*). The companies are ranked and then listed in numerical order based on stock-market value. Assume that you wanted to select a simple random sample of 12 companies from the list of 1000 companies. Use column 9 of Table 7.1 beginning with 554. Read down the column, and identify the numbers of the 12 companies that would be selected from the *Business Week 1000*.

16. Based on sales (*The Wall Street Journal,* October 13, 1988), the top 10 athletic footwear manufacturers are

1.	Reebok	6.	L. A. Gear
2.	Nike	7.	Etonic/Tretorn
3.	Converse	8.	New Balance
4.	Avia	9.	ASICS Tiger
5.	Adidas	10.	British Knights

   a. Beginning with the first random digit in Table 7.1 (6) and reading down (8, 5, 4, etc.), use single-digit random numbers from the first column to select a simple random sample of five footwear manufacturers from the population of the top 10 footwear manufacturers.
   b. If the random number 1 corresponds to Reebok, 2 corresponds to Nike, and so on, what single-digit random digit would have to appear in the first column to allow selection of British Knights for the sample?
   c. How many different simple random samples of size 5 can be selected from this population of 10 manufacturers?

17. A student government organization is interested in estimating the proportion of students who favor a mandatory "pass-fail" grading policy for elective courses. A list of names and addresses of the 645 students enrolled during the current quarter is available from the registrar's office. Using row 10 of Table 7.1 and moving across the row from left to right, identify the first 10 students who would be selected by simple random sampling. Note that when every digit in row 10 is used, the three-digit random numbers begin with 816, 283, and 610.

18. The *County and City Data Book,* published by the Bureau of Census, lists information on 3139 counties throughout the United States. Assume that a national study will collect data from 30 randomly selected counties. Use four-digit random numbers from the last column of Table 7.1 to identify the numbers corresponding to the first five counties selected for the sample. Ignore the first digit in the last column and begin with the four-digit random numbers 9945, 8364, 5702, and so on.

19. Assume that we wish to identify a simple random sample of 12 of the 372 doctors located in a particular city. The doctors' names are available from a local medical organization. Use the eighth column of five-digit random numbers in Table 7.1 to identify the 12 doctors for the sample. Ignore the first two random digits in each five-digit grouping of the random numbers. This process begins with random number 108 and proceeds down the column of random numbers.

20. *Business Week,* February 20, 1989, provided detailed information for 640 mutual funds available to investors. Data about the funds included assets, fees, return on investment, price/earnings ratios, and more. Assume that you would like to do a statistical study of the financial characteristics for the population of 640 mutual funds by using a simple random sample of 12 mutual funds. Use the third column of five-digit random numbers in Table 7.1, beginning with 71744. Ignore the 44 and only use the first three digits, 717. Reading down the column of three-digit random numbers, identify the numbers corresponding to the 12 mutual funds to be included in the sample.

21. Schuster's Interior Design, Inc., specializes in a variety of home decorating services for its clients. During the previous year, the firm provided major decorating consultation for 875 homes. Schuster's management was interested in obtaining information about customer satisfaction 6–12 months after the project was complete. To obtain this information, the firm decided to sample 30 of the 875 clients. Using the last three digits in column 10 of Table 7.1, and moving down the column, the random number sequence would be 945, 364, 702, and so on. Use this procedure to identify the first 10 clients that would be included in the sample. Assume that the 875 clients are numbered sequentially in the order in which the decorating projects were conducted.

22. Haskell Public Opinion Poll, Inc., conducts telephone surveys concerning a variety of political and general public interest issues. The households included in the survey are identified by taking a simple random sample from telephone directories in selected metropolitan areas. The telephone directory for a major Midwest area contains 853 pages with 400 lines per page.

    a. Describe a two-stage random selection procedure that could be used to identify a simple random sample of 200 households. The selection process should involve first selecting a page at random (stage 1) and then selecting a line on the sampled page (stage 2). Use the random numbers in Table 7.1 to illustrate this process. Select your own arbitrary starting point in the table.

    b. What would you do if the line selected in part (a) was clearly inappropriate for the study (e.g., the line provided the phone number of a business or restaurant)?

7.3 ▽ SAMPLING DISTRIBUTION OF $\bar{x}$

One of the most common statistical procedures involves using a sample mean $\bar{x}$ to make inferences about an unknown population mean μ. In this case, the sample mean $\bar{x}$ is referred to as a *sample statistic* and the population mean μ is referred to as a *population parameter*. This statistical process is shown in Figure 7.1.

It is important to realize that if we repeat the sampling process shown in Figure 7.1, we can anticipate obtaining a different value for the sample mean $\bar{x}$. As we showed in Section 7.2, several different simple random samples of size n are possible. Since each sample consists of different items from the population, we can expect different samples to provide different values for the sample statistic $\bar{x}$.

FIGURE 7.1   **THE STATISTICAL PROCESS OF USING A SAMPLE MEAN TO MAKE INFERENCES ABOUT A POPULATION MEAN**

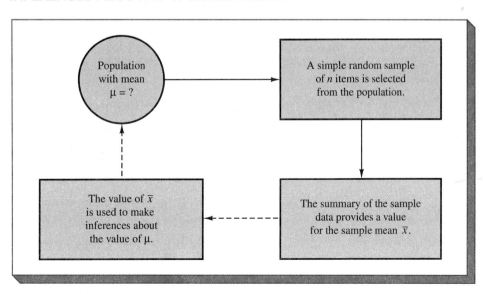

Since each sample has the same probability of being selected, we can associate a known probability with every possible sample and every possible value of $\bar{x}$. As a result, we can identify the probability distribution for the sample mean $\bar{x}$. This probability distribution is called the *sampling distribution of $\bar{x}$*. Because of the importance of the sampling distribution of $\bar{x}$, we restate its definition.

Sampling Distribution of $\bar{x}$

The *sampling distribution of $\bar{x}$* is the probability distribution for all possible values of the sample mean $\bar{x}$.

To illustrate the concept of a sampling distribution, let us reconsider the population of five school buses described in Example 7.3.

EXAMPLE 7.3 ▶
(CONTINUED)

Recall that the population of five school buses consisted of buses A, B, C, D, and E. The number of students riding on each of the buses is shown in Table 7.2. Using the formulas for a population mean and a population standard deviation that were presented in Chapter 3, we can use the data in Table 7.2 to compute the population mean and population standard deviation for the number of students riding the buses as follows:

$$\mu = \frac{\Sigma x_i}{N} = \frac{120}{5} = 24$$

$$\sigma = \sqrt{\frac{\Sigma(x_i - \mu)^2}{N}} = \sqrt{\frac{90}{5}} = \sqrt{18} = 4.24$$

Details of the computations of the values of μ and σ are shown in Table 7.3.

In this situation, the population size is small; thus it is easy to compute the mean and standard deviation for the population. However, to illustrate how a sample mean can be used to estimate a population mean, *let us assume for the moment that μ is unknown* and that we will have to use a simple random sample of two buses to estimate μ. Recall that there are 10 different samples of size 2 that could be selected. Table 7.4 lists these 10 possible samples and their corresponding sample means.

The column labeled "Sample Mean ($\bar{x}$)" in Table 7.4 shows that the value of the sample mean depends on the sample selected. Since we are using simple random sampling, we know that each possible sample—and, therefore, its corresponding sample

---

TABLE 7.2   **NUMBER OF STUDENTS PER BUS FOR THE POPULATION OF FIVE BUSES**

Bus	Number of Students
A	24
B	30
C	21
D	18
E	27

TABLE 7.3   COMPUTATION OF THE POPULATION MEAN AND POPULATION STANDARD DEVIATION FOR EXAMPLE 7.3

Bus	Number of Students (x_i)	$(x_i - \mu)$	$(x_i - \mu)^2$
A	24	$(24 - 24) = 0$	0
B	30	$(30 - 24) = 6$	36
C	21	$(21 - 24) = -3$	9
D	18	$(18 - 24) = -6$	36
E	27	$(27 - 24) = 3$	9
Totals	120		90

$$\mu = \frac{120}{5} = 24 \qquad \sigma = \sqrt{\frac{90}{5}} = \sqrt{18} = 4.24$$

TABLE 7.4   DIFFERENT POSSIBLE SIMPLE RANDOM SAMPLES FOR EXAMPLE 7.3

Buses Selected In Sample	Probability of Sample	Sample Mean $(\bar{x})$
A and B	$1/10$	$\frac{24 + 30}{2} = 27.0$
A and C	$1/10$	$\frac{24 + 21}{2} = 22.5$
A and D	$1/10$	$\frac{24 + 18}{2} = 21.0$
A and E	$1/10$	$\frac{24 + 27}{2} = 25.5$
B and C	$1/10$	$\frac{30 + 21}{2} = 25.5$
B and D	$1/10$	$\frac{30 + 18}{2} = 24.0$
B and E	$1/10$	$\frac{30 + 27}{2} = 28.5$
C and D	$1/10$	$\frac{21 + 18}{2} = 19.5$
C and E	$1/10$	$\frac{21 + 27}{2} = 24.0$
D and E	$1/10$	$\frac{18 + 27}{2} = 22.5$

mean—has the same $1/10$ probability of being selected. Figure 7.2 is a graph showing each possible value of $\bar{x}$ and the corresponding probability of occurrence. For instance, note that a value of $\bar{x} = 24.0$ has a $2/10$ probability of occurring because two samples, BD and CE, provide this value. Thus, we see that Figure 7.2 is simply a probability distribution that shows the probabilities associated with all possible values of the sample mean. This probability distribution is the sampling distribution of $\bar{x}$.

FIGURE 7.2   THE SAMPLING DISTRIBUTION OF $\bar{x}$ FOR EXAMPLE 7.3 BASED
ON SAMPLES OF SIZE 2

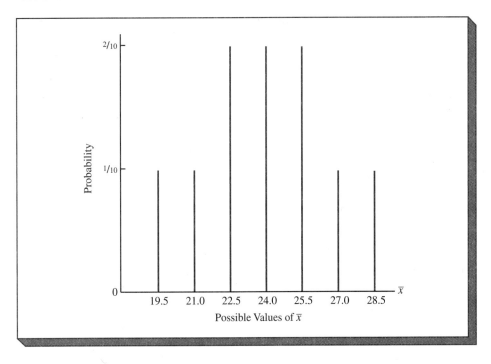

EXAMPLE 7.5 ▶ Consider the following population of five families with the following data indicating the
family size.

Family	A	B	C	D	E
Family Size	2	4	3	5	3

If a simple random sample of size 3 is used to estimate the mean family size for the
population, let us show the sampling distribution of $\bar{x}$.

First, we list all possible samples of size 3. There are 10 such samples consisting of
families ABC, ABD, ABE, ACD, ACE, ADE, BCD, BCE, BDE, and CDE. The sample
mean for each possible sample is shown below.

Sample	Data Values	Sample Mean ($\bar{x}$)
ABC	2, 4, 3	3.00
ABD	2, 4, 5	3.67
ABE	2, 4, 3	3.00
ACD	2, 3, 5	3.33
ACE	2, 3, 3	2.67
ADE	2, 5, 3	3.33
BCD	4, 3, 5	4.00
BCE	4, 3, 3	3.33
BDE	4, 5, 3	4.00
CDE	3, 5, 3	3.67

With a probability of $\frac{1}{10}$ for each sample, the probabilities for each possible value of $\bar{x}$ are shown below. This graph shows the sampling distribution of $\bar{x}$.

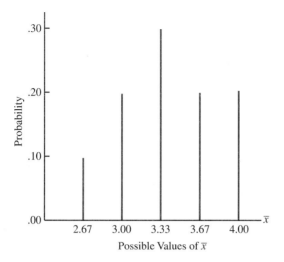

Before continuing the discussion of the sampling distribution of $\bar{x}$, we note that in practice only one sample is actually taken; hence, only one value of $\bar{x}$ is computed and used to make inferences about the population mean μ. The purpose of this section has been to show that there are many possible samples and hence values of $\bar{x}$ that could be obtained. Understanding this provides a much better perspective on the properties and the importance of the sampling distribution of $\bar{x}$. Knowledge of this sampling distribution provides the background to understand the material in Chapter 8 where the focus is on estimating a population mean μ using the information contained in just one sample.

EXPECTED VALUE OF $\bar{x}$

As we have seen, different samples may result in different values for the sample mean. We are often interested in the mean of all possible $\bar{x}$ values that can be generated by the various simple random samples. Let $E(\bar{x})$ denote the expected value of $\bar{x}$, or simply the mean of all possible $\bar{x}$ values. When using simple random sampling from a population with mean μ, the expected value of the sample mean is *equal to μ*.

Expected Value of $\bar{x}$

$$E(\bar{x}) = \mu \tag{7.1}$$

This result states that the expected value of $\bar{x}$—or, stated another way, the mean for all possible values of $\bar{x}$—is the same as the mean of the population from which the samples are taken. Whenever the expected value of a sample statistic is equal to the population parameter, we say the sample statistic is an *unbiased* estimator of the population parameter. Thus, equation (7.1) shows that $\bar{x}$ is an unbiased estimator of the population mean μ.

EXAMPLE 7.3
(CONTINUED) ▶

Recall that in Example 7.2 we computed the population mean for the number of students on the school buses to be $\mu = 24$. Equation (7.1) implies that the mean of the various $\bar{x}$ values must also be 24. Using the 10 values of $\bar{x}$ shown in Table 7.4, we can compute $E(\bar{x})$ as follows:

$$E(\bar{x}) = \frac{27.0 + 22.5 + 21.0 + 25.5 + 25.5 + 24.0 + 28.5 + 19.5 + 24.0 + 22.5}{10}$$

$$= \frac{240.0}{10} = 24$$

Thus, we see that equation (7.1) holds with $E(\bar{x}) = \mu$.　　　　◀

STANDARD DEVIATION OF $\bar{X}$

The standard deviation of $\bar{x}$ measures the dispersion in the possible $\bar{x}$ values. To explore what sampling theory tells us about the standard deviation of $\bar{x}$, we use the following notation:

$$\sigma_{\bar{x}} = \text{standard deviation of the } \bar{x} \text{ values}$$
$$\sigma = \text{standard deviation of the population}$$
$$n = \text{sample size}$$
$$N = \text{population size}$$

The formula for the standard deviation of $\bar{x}$ is as follows.

Standard Deviation of $\bar{x}$

$$\sigma_{\bar{x}} = \sqrt{\frac{N-n}{N-1}}\left(\frac{\sigma}{\sqrt{n}}\right) \tag{7.2}$$

Later, we will see that when only one sample is selected the value of $\sigma_{\bar{x}}$ is helpful in determining how far the sample mean may be from the population mean. Because of the role that $\sigma_{\bar{x}}$ plays in computing possible estimation errors, $\sigma_{\bar{x}}$ is referred to as the *standard error of the mean*. Thus, the standard error of the mean is another name for the standard deviation of $\bar{x}$.

EXAMPLE 7.3
(CONTINUED) ▶

Recall that in Example 7.3 the standard deviation for the population of school buses was computed to be $\sigma = 4.24$ students (see Table 7.3). With a population of size $N = 5$ and a sample of size $n = 2$, equation (7.2) shows that the standard deviation of $\bar{x}$ must be

$$\sigma_{\bar{x}} = \sqrt{\frac{N-n}{N-1}}\left(\frac{\sigma}{\sqrt{n}}\right) = \sqrt{\frac{5-2}{5-1}}\left(\frac{4.24}{\sqrt{2}}\right) = 2.60$$

Table 7.5 shows the computation of the standard deviation of $\bar{x}$ values using the 10 values of $\bar{x}$ generated by the 10 possible samples of two buses. As the computations show, we obtain the same value of $\sigma_{\bar{x}}$ as we found using equation (7.2). However, note that equation (7.2) provides the value of $\sigma_{\bar{x}}$ without having to generate all possible $\bar{x}$ values.　　◀

Consider for a moment the factor $\sqrt{(N-n)/(N-1)}$ that appears in the formula for $\sigma_{\bar{x}}$. This factor is commonly referred to as the *finite population correction factor*. In many

TABLE 7.5	COMPUTATION OF THE STANDARD DEVIATION OF $\bar{x}$ USING ALL POSSIBLE SAMPLE MEANS FOR THE SCHOOL BUS EXAMPLE

Buses Selected in Sample	Sample Mean ($\bar{x}$)	($\bar{x}$ − 24)	($\bar{x}$ − 24)²
A and B	27.0	(27.0 − 24) = 3.0	9.00
A and C	22.5	(22.5 − 24) = −1.5	2.25
A and D	21.0	(21.0 − 24) = −3.0	9.00
A and E	25.5	(25.5 − 24) = 1.5	2.25
B and C	25.5	(25.5 − 24) = 1.5	2.25
B and D	24.0	(24.0 − 24) = 0.0	0.00
B and E	28.5	(28.5 − 24) = 4.5	20.25
C and D	19.5	(19.5 − 24) = −4.5	20.25
C and E	24.0	(24.0 − 24) = 0.0	0.00
D and E	22.5	(22.5 − 24) = −1.5	2.25
Totals	240.0		67.50

$$E(\bar{x}) = \frac{240}{10} = 24 \qquad \sigma_{\bar{x}} = \sqrt{\frac{\Sigma(\bar{x}-24)^2}{10}} = \sqrt{\frac{67.50}{10}} = 2.60$$

practical sampling situations, we find that the population being sampled, although finite, is "large" whereas the sample size is relatively "small." In such cases, the value of $\sqrt{(N - n)/(N - 1)}$ is close to 1. When this occurs, $\sigma_{\bar{x}} = \sigma/\sqrt{n}$ becomes a very good approximation to the standard deviation of $\bar{x}$. We give the following as a general guideline or rule of thumb for computing the standard deviation of $\bar{x}$.

Guideline for Computing the Standard Deviation of $\bar{x}$

Whenever the sample size is less than or equal to 5% of the population size (i.e., $n/N \leq .05$), use

$$\sigma_{\bar{x}} = \frac{\sigma}{\sqrt{n}} \tag{7.3}$$

In the following chapters, we generally assume the population is large relative to the sample size; that is, $n/N \leq .05$. Thus, we use equation (7.3) to compute the standard deviation of $\bar{x}$. If this assumption is not satisfied in a particular application, we use equation (7.2) to compute $\sigma_{\bar{x}}$.

EXAMPLE 7.6 ▶ Assume that a simple random sample of size 49 is to be taken from a large population with mean $\mu = 100$ and standard deviation $\sigma = 21$. We know that repeating the sampling process will generate different sample means $\bar{x}$ due to the different samples selected. What are the mean and standard deviation of the sample means?

Using equation (7.1), we see the mean of the $\bar{x}$ values is

$$E(\bar{x}) = \mu = 100$$

Since the population is large relative to the sample size, equation (7.3) provides the standard deviation of the $\bar{x}$ values, as follows:

$$\sigma_{\bar{x}} = \frac{\sigma}{\sqrt{n}} = \frac{21}{\sqrt{49}} = 3$$

EXERCISES

Note to student: In several of the exercises that follow, the size of the population is not stated. When the population size is *not* provided, you may make the assumption that the population is "large" relative to the sample size; that is, $n/N \leq .05$. In such cases, equation (7.3) may be used to compute the standard deviation of $\bar{x}$.

METHODS

SELF TEST

23. Assume a population of size 5 has the following data: 2, 1, 0, 2, and 3.
    a. If a simple random sample of size 2 is used, show the sampling distribution of $\bar{x}$.
    b. Repeat part (a) if the size of the simple random sample is increased to 3.

24. Simple random samples of size 30 are to be selected from a population of 2000 items. The population standard deviation is $\sigma = 12$.
    a. Use equation (7.2) to compute the standard error of the mean.
    b. Use equation (7.3) to compute the standard error of the mean.
    c. Compare your answers to parts (a) and (b), and comment on why it is acceptable to use equation (7.3) whenever $n/N \leq .05$.
    d. What is the value of the finite population correction factor $\sqrt{(N-n)/(N-1)}$ for this problem?

25. Assume that $\mu = 32$ and standard deviation $\sigma = 5$. Furthermore, assume that the population has 1000 items and that a simple random sample of 30 items is used to obtain information about this population.
    a. What is the expected value of $\bar{x}$?
    b. What is the standard deviation of $\bar{x}$?

26. Assume the population standard deviation is $\sigma = 25$. Compute the standard error of the mean, $\sigma_{\bar{x}}$, for sample sizes of 50, 100, 150, and 200. What can you say about the size of the standard error of the mean as the sample size is increased?

27. Suppose a simple random sample of size 50 is selected from a population with $\sigma = 10$. Find the value of the standard error of the mean in each of the following cases (use the finite population correction factor if appropriate).
    a. The population size is infinite.
    b. The population size is $N = 50,000$.
    c. The population size is $N = 5000$.
    d. The population size is $N = 500$.

APPLICATIONS

28. Using the population of buses as shown in Table 7.2, show the sampling distribution of $\bar{x}$ if a simple random sample of three buses is used to estimate the mean number of students riding a bus.

SELF TEST

29. Four college students are taking the following number of credit hours during the current term.

Student	Albert	Becky	Cindy	David
Number of Credit Hours	15	17	19	17

Treating the four college students as the population, answer the following questions:

a. How many simple random samples of size 2 are possible? List the possible samples.

b. Compute the sample mean for each of the possible simple random samples of size 2.

c. Show a graphical representation of the probability distribution of all possible values of $\bar{x}$.

30. Five sales representatives sell mobile telephone units to private and commercial customers. Assume that the number of units sold for each sales representative is as follows.

Salesperson	Adams (A)	Baker (B)	Collins (C)	Davis (D)	Edwards (E)
Units Sold	14	20	12	8	16

a. Treating the five sales representatives as the population, use the computational procedure of Table 7.3 to compute the population mean μ and the population standard deviation σ.

b. There are 10 possible simple random samples of size 2 that can be selected from the population. Identify each possible sample, and compute its corresponding sample mean $\bar{x}$.

c. Show the sampling distribution of $\bar{x}$.

d. Use the computational procedure of Table 7.5 to compute the mean and standard deviation of the $\bar{x}$ values.

e. Use equations (7.1) and (7.2) to determine the expected value and standard deviation of $\bar{x}$. Compare your results with your answer to part (d).

f. If you wish to compute $E(\bar{x})$ and $\sigma_{\bar{x}}$, do you prefer the approach used in part (d) or in part (e)? Explain.

31. The sizes of the 10 offices on the 12th floor of the new Crosley Tower Bank Building are as follows.

Office	Size (Square Feet)	Office	Size (Square Feet)
1	150	6	300
2	175	7	140
3	180	8	150
4	180	9	150
5	225	10	200

a. Compute the population mean μ and population standard deviation σ for the population of 10 offices.

b. There are many different possible simple random samples of size 3 that can be selected from the population. What are the values of the mean and standard deviation for the $\bar{x}$ values? That is, compute $E(\bar{x})$ and $\sigma_{\bar{x}}$.

c. Compute $E(\bar{x})$ and $\sigma_{\bar{x}}$ if the sample size is increased to four offices.

SELF TEST

32. Weights for males between the ages of 20 and 30 have a mean $\mu = 170$ pounds with a standard deviation of $\sigma = 28$ pounds. If a simple random sample of 40 males in this age group is to be selected and the sample mean weight $\bar{x}$ computed, what are the values of $E(\bar{x})$ and $\sigma_{\bar{x}}$?

33. A statistics class has 80 students. The mean score on the midterm exam was $\mu = 72$ and the standard deviation was $\sigma = 12$. Assume that a simple random sample of 20 students will be selected and the sample mean exam score $\bar{x}$ will be computed. What is the expected value and standard deviation of $\bar{x}$?

7.4 THE CENTRAL LIMIT THEOREM

At this point, we know that simple random samples taken from a population will provide different values for the sample mean $\bar{x}$ due to the fact that different samples consist of different items from the population. The probability distribution showing all possible values of $\bar{x}$ is referred to as the sampling distribution of $\bar{x}$. The final step in identifying the characteristics of the sampling distribution of $\bar{x}$ is to determine the *form* of the sampling distribution of $\bar{x}$. We will consider two cases: one where the population distribution is unknown and one where the population distribution is known to be a normal probability distribution.

For the situation where the population distribution is unknown, we rely on one of the most important theorems in statistics—the *central limit theorem*. A statement of the central limit theorem as it applies to the sampling distribution of $\bar{x}$ is as follows.*

Central Limit Theorem

In selecting simple random samples of size n from a population with mean μ and standard deviation σ, the sampling distribution of $\bar{x}$ approaches a normal probability distribution with mean μ and standard deviation $\sigma/\sqrt{n}$ as the sample size becomes large.

Figure 7.3 shows how the central limit theorem works for three different populations; in each case, the population clearly is not normal. However, note what begins to happen to the sampling distribution of $\bar{x}$ as the sample size is increased. When the samples are of size 2, we see that the sampling distribution of $\bar{x}$ begins to take on an appearance different from the population distribution. For samples of size 5, we see all three sampling distributions beginning to take on a bell-shaped appearance. Finally, for samples of size 30, all three sampling distributions are approximately normally distributed. General statistical practice is to assume that regardless of the population distribution, the sampling distribution of $\bar{x}$ can be approximated by a normal probability distribution whenever the sample size is 30 or more. In effect, the sample size of 30 is the rule of thumb that allows us to assume that the large-sample condition of the central limit theorem has been satisfied. This observation about the sampling distribution of $\bar{x}$ is so important that we restate it.

The sampling distribution of $\bar{x}$ can be approximated by a normal probability distribution whenever the sample size is large. The large-sample-size condition can be assumed for simple random samples of size 30 or more.

---

*The theoretical proof of the central limit theorem requires independent observations or items in the sample. This condition exists for infinite populations or for finite populations where sampling is done with replacement. Although the central limit theorem does not directly address sampling without replacement from finite populations, general statistical practice has been to apply the findings of the central limit theorem in this situation provided the population size is large.

FIGURE 7.3   ILLUSTRATION OF THE CENTRAL LIMIT THEOREM FOR THREE
POPULATIONS

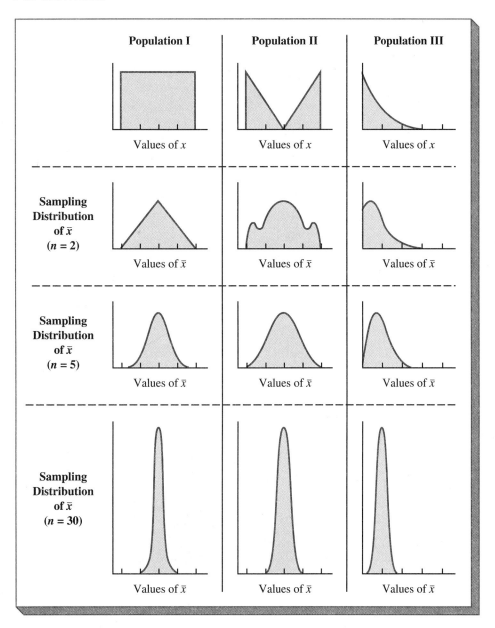

The central limit theorem is the key to identifying the form of the sampling distribution of $\bar{x}$ whenever the population distribution is unknown. However, we may encounter some sampling situations where the population is assumed or believed to have a normal probability distribution. When this condition occurs, it is not necessary to rely on the central limit theorem; in such cases, the following result identifies the form of the sampling distribution of $\bar{x}$.

> Whenever the population being sampled has a normal probability distribution, the sampling distribution of $\bar{x}$ is a normal probability distribution for any sample size.

In summary, whenever we are using a large simple random sample (*rule of thumb: $n \geq 30$*), the central limit theorem enables us to conclude that the sampling distribution of $\bar{x}$ can be approximated by a normal probability distribution. In cases where the simple random sample is small ($n < 30$), the sampling distribution of $\bar{x}$ can be considered to be a normal probability distribution only if the population being sampled has a normal probability distribution.

EXAMPLE 7.6
(CONTINUED)

In Example 7.6 we discussed a situation in which simple random samples of size 49 were to be taken from a population with mean $\mu = 100$ and standard deviation $\sigma = 21$. Show the sampling distribution of $\bar{x}$.

Recall that $E(\bar{x}) = 100$ and $\sigma_{\bar{x}} = 21/\sqrt{49} = 3$. The central limit theorem also tells us that for large samples ($n \geq 30$), the sampling distribution of $\bar{x}$ can be approximated by a normal probability distribution. In such cases, the sampling distribution of $\bar{x}$ is as shown below.

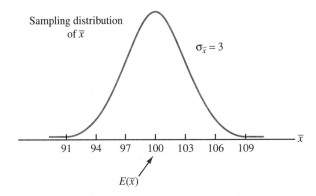

EXAMPLE 7.7

The heights of sixth-grade students in a particular school district are *normally distributed* with a mean of $\mu = 58$ inches and a standard deviation of $\sigma = 3.2$ inches. Show the sampling distribution of $\bar{x}$ if a simple random sample of 16 students is to be used.

We know that

$$E(\bar{x}) = \mu = 58$$

and

$$\sigma_{\bar{x}} = \frac{\sigma}{\sqrt{n}} = \frac{3.2}{\sqrt{16}} = .80$$

Since $n < 30$ and since the population of heights is described as being normally distributed, the sampling distribution of $\bar{x}$ will be a normal probability distribution for any sample size. Thus, the sampling distribution of $\bar{x}$ is as follows:

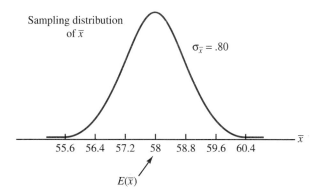

Sampling distribution
of $\bar{x}$

$\sigma_{\bar{x}} = .80$

55.6   56.4   57.2   58   58.8   59.6   60.4

$\bar{x}$

$E(\bar{x})$

EXERCISES

METHODS

34. A sample of size $n = 36$ was selected from a population with a mean of 100 and a standard deviation of 12.
  a. What is the expected value of $\bar{x}$?
  b. What is the standard deviation of $\bar{x}$?
  c. What probability distribution can be used to approximate the sampling distribution of $\bar{x}$?
  d. Sketch a graph of the sampling distribution of $\bar{x}$.

35. A sample of size $n = 9$ was selected from a population that is normally distributed with a mean of 50 and a standard deviation of 15. Sketch the sampling distribution of $\bar{x}$.

SELF TEST **36.** A population has a mean of 200 and a standard deviation of 50. Suppose that a simple random sample of size 100 is selected and the sample mean $\bar{x}$ is used to estimate the population mean.
  a. What is the expected value of $\bar{x}$?
  b. What is the standard deviation of $\bar{x}$?
  c. Show the graph of the sampling distribution of $\bar{x}$.
  d. What does the sampling distribution of $\bar{x}$ show?

37. What important role does the central limit theorem serve whenever $\bar{x}$ is used to estimate μ?

APPLICATIONS

38. The mean and standard deviation for the number of calories in a 12-ounce can of light beer are as follows: $\mu = 105$ and $\sigma = 3$. A random sample of 30 cans will be selected and a laboratory test conducted to determine the exact number of calories present in each of the 30 cans. The sample mean $\bar{x}$ will be computed.
  a. What is the expected value of $\bar{x}$?
  b. What is the standard deviation of $\bar{x}$?
  c. What probability distribution can be used to approximate the sampling distribution of $\bar{x}$?
  d. Sketch a graph of the sampling distribution of $\bar{x}$.

SELF TEST **39.** The length of time of long-distance telephone calls has a mean $\mu = 18$ minutes and a standard deviation $\sigma = 4$ minutes. Sketch the sampling distribution of $\bar{x}$ if a simple random sample of 50 telephone calls is used to estimate the mean length of long-distance telephone calls in the population.

40. A population has a mean of 400 and a standard deviation of 50. The probability distribution of the population is unknown.

a. A research study will use simple random samples of either 10, 20, 30, or 40 items to collect data about the population. In which of these sample-size alternatives will we be able to use a normal probability distribution to describe the sampling distribution of $\bar{x}$? Explain.

b. Show the sampling distribution of $\bar{x}$ for the instances where the normal probability distribution is appropriate.

41. The body length of a certain insect is believed to be *normally distributed* with a mean length of $\mu = 16.5$ mm and a standard deviation of $\sigma = .8$ mm. Describe the sampling distribution of the sample mean body length if 20 insects are to be used in the study. Is it necessary to use the central limit theorem to determine the shape of the sampling distribution? Explain.

42. Assume that the number of points scored in basketball games played by a particular college team is normally distributed with $\mu = 68$ and $\sigma = 5$. Show the sampling distribution of $\bar{x}$ for a sample of 10 games played by this team.

7.5 ▽ COMPUTING PROBABILITIES USING THE SAMPLING DISTRIBUTION OF $\bar{X}$

As the following examples show, we can use the sampling distribution of $\bar{x}$ to compute the probability of selecting a sample that will provide a value of $\bar{x}$ within a specified distance from the population mean μ.

EXAMPLE 7.7 ▶ (CONTINUED)

In Example 7.7 we described the sampling distribution of $\bar{x}$ for simple random samples consisting of the heights of 16 sixth-grade students. Specifically, we showed that the sampling distribution of $\bar{x}$ is a normal probability distribution with $E(\bar{x}) = 58$ and $\sigma_{\bar{x}} = .80$. Suppose that we want to compute the probability of obtaining a simple random sample that provides a sample mean $\bar{x}$ between 57 and 59 inches.

Since the sampling distribution of $\bar{x}$ can be approximated by a normal probability distribution, we can use the standard normal probability distribution to answer this question. First, we need to compute a z value where

$$z = \frac{\bar{x} - \mu}{\sigma_{\bar{x}}} \tag{7.4}$$

This z value is identical to the z value we used for the standard normal probability distribution in Chapter 6, with the exception of the notation $\bar{x}$ and $\sigma_{\bar{x}}$. This notation has been used to indicate that the random variable here is $\bar{x}$ and its standard deviation is denoted by $\sigma_{\bar{x}}$.

The probability of selecting a simple random sample with $\bar{x}$ between 57 and 59 is given by the shaded area under the following normal curve.

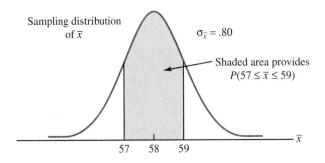

$$\text{At } \bar{x} = 59, \qquad z = \frac{59 - 58}{.80} = 1.25$$

Looking up $z = 1.25$ in the table of areas for the standard normal probability distribution shows an area of .3944 between 58 and 59.

$$\text{At } \bar{x} = 57, \qquad z = \frac{57 - 58}{.80} = -1.25$$

This z value shows that the area between 57 and 58 must also be .3944. Thus, the total probability of selecting a sample with a sample mean $\bar{x}$ between 57 and 59 inches must be

$$.3944 + .3944 = .7888$$

Now, let us see how to compute the probability of obtaining a sample mean $\bar{x}$ of 60 or more for a simple random sample of 16 students. The graph of the sampling distribution of $\bar{x}$ for this situation is as follows.

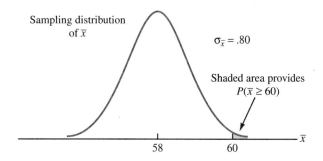

$$\text{At } \bar{x} = 60, \qquad z = \frac{\bar{x} - \mu}{\sigma_{\bar{x}}} = \frac{60 - 58}{.80} = 2.50$$

Using the table of areas for the standard normal probability distribution and $z = 2.50$, we find that the area between $z = 0.00$ and $z = 2.50$ is .4938. Thus, the probability of $z \geq 2.50$ is $.5000 - .4938 = .0062$. Thus, the probability of obtaining a sample mean of $\bar{x} \geq 60$ is .0062.

EXAMPLE 7.8 ▶

Suppose that we have a population of high school students with a mean IQ score of $\mu = 100$ and a standard deviation of $\sigma = 10$. If simple random samples of 36 students are to be taken from this population, $\sigma_{\bar{x}} = \sigma/\sqrt{n} = 10/\sqrt{36} = 1.67$. Let us compute an interval around 100 that includes 95% of all possible sample means that could be obtained.

The table of areas for the standard normal probability distribution shows that 95% of the values in any normal probability distribution fall between $z = -1.96$ and $z = +1.96$. In other words, 95% of all possible $\bar{x}$ values must be within $-1.96\sigma_{\bar{x}}$ and $+1.96\sigma_{\bar{x}}$ of 100. Thus, the range containing 95% of all possible $\bar{x}$ values must be

$$100 - 1.96\sigma_{\bar{x}} = 100 - 1.96(1.67) = 96.7$$

$$100 + 1.96\sigma_{\bar{x}} = 100 + 1.96(1.67) = 103.3$$

That is, there is a .95 probability of selecting a simple random sample having a sample mean IQ score between 96.7 and 103.3. Only 5% of all possible sample means are outside this interval. This situation is as follows:

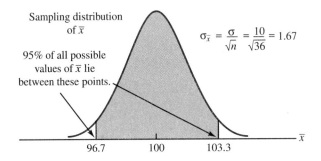

EXERCISES

METHODS

SELF TEST

43. A sample of size $n = 36$ was selected from a population with a mean of 100 and a standard deviation of 12. Compute the probability of finding a sample mean greater than 105.

44. A sample of size $n = 49$ was selected from a population with $\mu = 500$ and $\sigma = 70$. What is the probability that the sample mean will be between 480 and 500?

45. A population has a mean of 100 and a standard deviation of 16. What is the probability that a sample mean will be within 2 of the population mean for each of the following sample sizes?
   a. $n = 50$ **b.** $n = 100$ **c.** $n = 200$ **d.** $n = 400$
   e. What is the advantage of a larger sample size?

APPLICATIONS

SELF TEST

46. Statistics help computer scientists and data-processing specialists understand the operating characteristics of computer systems. Statistical information includes waiting time, running time, central processor time, number of disk accesses, and so on. For a particular class of jobs run on the Amdahl 5880 mainframe computer (*Technical Update,* University of Cincinnati, Spring 1989), the population mean running time is 12.55 minutes per job. The population standard deviation is 4.0 minutes. Assume that a simple random sample of 40 jobs will be used to monitor the running time for the jobs.
   a. Show the graph of the sampling distribution of $\bar{x}$ where $\bar{x}$ is the sample mean running time.
   b. What is the probability that a simple random sample of 40 jobs will provide a sample mean within 1 minute of the population mean?
   c. What is the probability that a simple random sample of 40 jobs will provide a sample mean within 30 seconds of the population mean?

47. Annual surveys of starting salaries for college graduates are conducted by the College Placement Council. The mean annual starting salary for accounting majors is $26,542 (*USA Today,* September 9, 1991). Assume the population of graduates with accounting majors has a mean of $26,542 and a standard deviation of $2000.
   a. What is the probability that a simple random sample of accounting graduates will have a sample mean within $250 of the population mean for each of the following sample sizes: 30, 50, 100, 200, and 400?
   b. What is the advantage of taking a larger sample size when attempting to estimate a population mean?

48. *Money* (February 1989) listed the national mean interest rate for new auto loans at 11.28. This was up from the 10.81 mean interest rate in 1988. Assume that the mean interest rate for the population of all new auto loans is 11.28 and that the population standard deviation is 1.5. Suppose that a simple random sample of 50 loans will be used to monitor interest rates of auto loans.

a. What is the probability that a simple random sample of 50 loans will provide a mean interest rate within .2 of the population mean?

b. What is the probability that a simple random sample of 50 loans will provide a mean interest rate within .1 of the population mean?

49. An electrical component is designed to provide a mean service life of 3000 hours with a standard deviation of 800 hours. A customer purchases a batch of 50 components. What is the probability that the mean life for the sample of 50 components will be at least 2750 hours? At least 3200 hours?

50. The population mean price for a new automobile is $16,012 (*U.S. News & World Report,* September 9, 1991). Assume that the population standard deviation is $4200 and that a sample of 100 new automobile purchases is selected.

a. Show the graph of the sampling distribution for the sample mean price for new automobiles based on the sample of 100.

b. What is the probability that the sample mean for the 100 purchases will be within $1000 of the population mean?

c. Repeat part (b) for $500, $250, and $100.

d. If it were desired to estimate the population mean price to within ±$250 or ±$100, what would you recommend?

51. An automatic machine used to fill cans of soup has the following characteristics: $\mu = 15.9$ ounces and $\sigma = .5$ ounces.

a. Show the graph of the sampling distribution of $\bar{x}$, where $\bar{x}$ is the sample mean for 40 cans selected randomly by a quality-control inspector.

b. What is the probability of finding a sample of 40 cans with a mean $\bar{x}$ greater than 16 ounces?

52. The mean wage rate for workers at General Motors is $14.25 per hour (*Detroit Daily News,* April 14, 1989). Assume that the population standard deviation is $2.00. Answer the following questions if a simple random sample of 50 workers is selected from the population of workers at General Motors.

a. Show the graph of the sampling distribution of $\bar{x}$, where $\bar{x}$ is the sample mean hourly wage rate.

b. What is the probability that the sample mean is at least $13.80 per hour?

c. What is the probability that the sample mean is within $.25 of the population mean of $14.25 per hour?

d. Answer parts (b) and (c) with the sample size increased to 100 workers.

53. A survey reports its results by stating that the standard error of the mean is 20. The population standard deviation was 500.

a. How large was the sample used in this survey?

b. What is the probability that the estimate would be within 25 of the population mean?

54. The grade point average for all juniors at Strausser College has a standard deviation of .50.

a. A random sample of 20 students is to be used to estimate the population mean grade point average. What assumption is necessary to compute the probability of obtaining a sample mean within .2 of the population mean?

b. Provided that this assumption can be made, what is the probability of $\bar{x}$ being within .2 of the population mean?

c. If this assumption cannot be made, what would you recommend doing?

55. A simple random sample of 64 will be used to estimate the mean time required to perform a particular task in a mechanical aptitude test. If the standard deviation in times is $\sigma = 4$ minutes, use the sampling distribution of $\bar{x}$ to compute the probability that the difference between $\bar{x}$ and the population mean μ is

a. 1 minute or less   c. greater than .25 minute

b. .5 minute or less

SUMMARY

In this chapter we introduced the important concepts of simple random sampling and the sampling distribution of $\bar{x}$. We discussed the process of using a sample mean $\bar{x}$ (sample statistic) to estimate a population mean μ (population parameter). Each simple random sample potentially provides a different value for the sample mean $\bar{x}$. The probability distribution for the population of $\bar{x}$ values is called the sampling distribution of $\bar{x}$.

In considering the characteristics of the sampling distribution of $\bar{x}$, we stated that $E(\bar{x}) = \mu$ and that $\sigma_{\bar{x}} = \sigma/\sqrt{n}$, provided $n/N \leq .05$. In addition, the central limit theorem provided the basis for using a normal probability distribution to approximate the sampling distribution of $\bar{x}$. The rule of thumb of $n \geq 30$ provided the large-sample conditions necessary to use the normal probability distribution to approximate the sampling distribution of $\bar{x}$. We also noted that whenever the population being sampled has a normal probability distribution, the sampling distribution of $\bar{x}$ will be a normal probability distribution for any sample size. Knowledge of the sampling distribution of $\bar{x}$ was used to make probability statements about the values of $\bar{x}$ and how close the values of $\bar{x}$ are to the population mean μ.

GLOSSARY

Population The set of all elements of interest for a particular study.

Sample A subset of the population selected to represent the whole population.

Simple random sample (finite population) A sample selected such that each possible sample of size n has the same probability of being selected.

Simple random sample (infinite population) A sample selected such that each item comes from the same population and each item is selected independently.

Sampling without replacement Once an item from the population has been included in the sample, it is removed from further consideration and cannot be selected a second time.

Sampling with replacement As each item is selected for the sample, it is returned to the population. It is possible that a previously selected item may be selected again and therefore appear in the sample more than once.

Sample statistic A numerical measure that is calculated for a sample, such as the sample mean $\bar{x}$.

Population parameter A numerical measure based on all the elements in the population, such as the population mean μ.

Sampling distribution A probability distribution showing all possible values of a sample statistic, such as a sample mean $\bar{x}$.

Standard error of the mean The standard deviation of $\bar{x}$, denoted by $\sigma_{\bar{x}}$.

Finite population correction factor The multiplier term $\sqrt{(N - n)/(N - 1)}$ that is used in the formula for $\sigma_{\bar{x}}$; whenever $n/N \leq .05$, the finite population factor is close to 1 and hence $\sigma_{\bar{x}} = \sigma/\sqrt{n}$.

Central limit theorem A theorem that enables us to use the normal probability distribution to approximate the sampling distribution of $\bar{x}$ whenever the sample size is large. *Rule of thumb:* The central limit theorem applies whenever $n \geq 30$.

Unbiased estimator A property that exists whenever the expected value of a sample statistic is equal to the population parameter. In this chapter, $E(\bar{x}) = \mu$ shows $\bar{x}$ is an unbiased estimator of μ.

KEY FORMULAS

Expected Value of $\bar{x}$

$$E(\bar{x}) = \mu \qquad\qquad (7.1)$$

Standard Deviation of $\bar{x}$

$$\sigma_{\bar{x}} = \sqrt{\frac{N - n}{N - 1}} \left(\frac{\sigma}{\sqrt{n}} \right) \tag{7.2}$$

If the sample size is less than or equal to 5% of the population size, use

$$\sigma_{\bar{x}} = \frac{\sigma}{\sqrt{n}} \tag{7.3}$$

Computation of z Value for the Sampling Distribution $\bar{x}$

$$z = \frac{\bar{x} - \mu}{\sigma_{\bar{x}}} \tag{7.4}$$

REVIEW QUIZ

TRUE/FALSE

1. A simple random sample of size n from a population of size N is a sample selected such that each possible sample of size n has the same probability of being selected.
2. A primary objective of sampling is to choose a sample that is representative of the population being studied.
3. The sample mean and sample standard deviation are not sample statistics.
4. The probability distribution of the sample mean is called the sampling distribution of $\bar{x}$.
5. The standard error of the mean cannot be computed unless the sample size is known.
6. In practice, sampling with replacement is more commonly employed than sampling without replacement.
7. The term $\sqrt{(N - n)/(N - 1)}$ in the formula for the standard deviation of $\bar{x}$ is called the continuity correction factor.
8. The central limit theorem ensures that the sampling distribution of $\bar{x}$ is a normal probability distribution regardless of the sample size.

MULTIPLE CHOICE

9. Consider a population with $\mu = 50$ and $\sigma = 10$. A simple random sample of size $n = 25$ will be used to provide a sample mean $\bar{x}$ to estimate μ. What is the mean value for the sampling distribution of $\bar{x}$?
    a. 2
    b. 10
    c. 50
    d. none of the above

10. Once an item from the population has been included in the sample, it is removed from further consideration and cannot be selected for the sample a second time. This is an example of
    a. nonrandom sampling
    b. probability sampling
    c. sampling without replacement
    d. sampling with replacement

11. Sampling from a large population with $\sigma = 20$ yields a standard error of the mean of 2. What was the sample size used in this situation?
    a. 100
    b. 10
    c. 40
    d. none of the above

12. What condition is required before the central limit theorem justifies approximating the sampling distribution of $\bar{x}$ by a normal probability distribution?
   a. $n/N \geq .05$
   b. $n \geq 30$
   c. $n < 30$
   d. $N \geq 30$

13. Assume a simple random sample of 49 items is taken from a population with $\mu = 16$ and $\sigma = 7$. What is the probability the sample mean $\bar{x}$ will be within 1 of the population mean $\mu = 16$?
   a. .3413
   b. .6826
   c. .1114
   d. cannot be determined from the above information

14. As the sample size increases, variability among the sample means
   a. increases
   b. decreases
   c. remains the same
   d. not enough information given

15. Simple random samples of size 20 are taken from a population that has 200 elements, a mean of 36, and a standard deviation of 8. The distribution of the population is unknown. The mean and standard deviation of $\bar{x}$ are
   a. 10 and 1.79
   b. 36 and 1.79
   c. 36 and 1.70
   d. 36 and 8

16. Which of the following best describes the form of the sampling distribution of the sample mean for the situation in Question 15?
   a. approximately normal because the sample size is small relative to the population size
   b. approximately normal because of the central limit theorem
   c. exactly normal because the population is normally distributed
   d. nothing can be said with the information given

SUPPLEMENTARY EXERCISES

56. Nationwide Supermarkets has 4800 retail stores located in 32 states. At the end of each year, a sample of 35 stores is selected for physical inventories. Results from the inventory samples are used in annual tax reports. Assume that the retail stores are listed sequentially on a computer printout. Begin at the bottom of the second column of the random numbers shown in Table 7.1. Using four-digit random numbers beginning with 8112, read *up* the column to identify the first five stores to be included in the simple random sample.

57. Assume that a simple random sample of size 2 is to be taken from the following list of airlines: American, United, TWA, Delta, Continental, Northwest, and US Air.
   a. How many simple random samples of size 2 are possible? List the possible samples.
   b. What is the probability of each sample being selected?
   c. How many samples include United Airlines?
   d. What is the probability that United Airlines appears in the simple random sample selected?

58. An apartment complex consists of six buildings. The number of units rented in each of the six buildings is as follows: 5, 4, 6, 3, 5, and 4, respectively. Assume that a simple random sample of two buildings will be used to estimate the mean number of rentals per building.
   a. Compute the mean μ and standard deviation σ for the population of six buildings.

    b.   There are 15 possible simple random samples of size 2 that can be selected from this population. Identify each possible sample, and compute its corresponding value of $\bar{x}$.

    c.   Show the sampling distribution of $\bar{x}$.

    d.   Use the computational procedure of Table 7.4 to compute the mean and standard deviation of all possible $\bar{x}$ values.

    e.   Use equations (7.1) and (7.2) and the results of part (a) to determine the expected value and standard deviation of $\bar{x}$. Compare your results with your answer to part (d).

59.   In a population of 4000 employees, a simple random sample of 40 employees is selected in order to estimate the mean age for the population.

    a.   Would you use the finite population correction factor in calculating the standard error of the mean? Explain.

    b.   If the population standard deviation is $\sigma = 8.2$ years, compute the standard error both with and without the finite population correction factor. What is the rationale behind ignoring the finite population correction factor whenever $n/N \leq .05$?

    c.   What is the probability that the sample mean age of the employees will be within 2 years of the population mean age?

60.   *Survey and Analysis of Salary Trends,* 1988, from the research department of the American Federation of Teachers, lists the average annual salary for elementary/secondary school teachers as $28,085. Answer the following questions using this value as the population mean and 3200 as the population standard deviation.

    a.   What is the probability that a simple random sample of 100 teachers will provide a sample mean annual salary within $200 of the population mean?

    b.   What is the probability that a simple random sample of 300 teachers will provide a sample mean annual salary within $200 of the population mean?

    c.   Assuming that we would like a 90% chance of having a sample mean annual salary within $200 of the population mean, how many teachers should be included in the simple random sample?

61.   In a study of the growth rate of a certain plant, a botanist is planning to use a simple random sample of 25 plants for data-collection purposes. After analyzing the data on plant growth rate, the botanist believes that the standard error of the mean is too large. What size simple random sample should the botanist use in order to reduce the standard error to one-half its current value?

62.   The population mean household income in Pittsburgh, Pennsylvania, is $49,000 (*U.S. News & World Report,* April 6, 1992). Assume that the population standard deviation is $12,000. Furthermore, assume that a simple random sample of households in the Pittsburgh area will be selected and the sample mean will be used to estimate the population mean.

    a.   Show the graph of the sampling distribution of the sample mean if a sample of 100 households is used.

    b.   What is the probability that a sample of 100 will provide a sampling error of $1000 or less?

    c.   What is the probability that a sample of 200 will provide a sampling error of $1000 or less?

    d.   What is the probability that a sample of 400 will provide a sampling error of $1000 or less?

    e.   How large of a sample would be required if we wanted a .95 probability of having a sampling error of $1000 or less?

63.   The time it takes a fire department to respond to a request for emergency aid has a mean of $\mu = 14$ minutes with a standard deviation of $\sigma = 4$ minutes. Suppose we randomly sample 50 emergency requests over a 2-month period. Records of aid-request times and arrival times will be used to compute a sample mean response time for the 50 requests.

    a.   Show the sampling distribution of $\bar{x}$.

    b.   What role does the central limit theorem play in identifying this sampling distribution?

    c.   What is the probability that the sample mean will be 15 minutes or less?

d. What is the probability that the sample mean will be within .5 minute of the mean time for the population?

64. The speed of automobiles on a section of I-75 in northern Florida has a mean of $\mu = 67$ miles per hour with a standard deviation of $\sigma = 6$ miles per hour. Answer the following questions if the population can be assumed to have a normal probability distribution and if a sample of 16 automobiles will be selected to compute a sample mean automobile speed.

a. What is the expected value of $\bar{x}$?

b. What is the value of the standard error of the mean?

c. Show the sampling distribution of $\bar{x}$.

d. What is the probability that the value of the sample mean will be 65 miles per hour or more?

e. What is the probability that the value of the sample mean will be between 66 and 68 miles per hour?

65. Consider a population of size $N = 500$ with a mean $\mu = 200$ and a standard deviation $\sigma = 40$. Assume that a simple random sample of size $n = 100$ will be selected from this population.

a. Should the finite population correction factor be used in computing the standard error of the mean?

b. What is the value of the standard error of the mean for this problem?

c. What is the probability of selecting a simple random sample that provides a value of $\bar{x}$ that is within 5 of the population mean μ?

66. In a population of 5000 students, a simple random sample of 50 students is selected to estimate the mean grade point average for the population.

a. Would you use the finite population correction factor in calculating the standard error of the mean? Explain.

b. If the population standard deviation is $\sigma = .4$ year, compute the standard error of the mean, first with and then without the finite population correction factor. What is the rationale for ignoring the finite population correction factor whenever $n/N \leq .05$?

c. What is the probability that the sample grade point average for 50 students will be within .10 of the population mean grade point average?

67. Three firms have inventories that vary in size. Firm A has a population of 2000 items, firm B has a population of 5000 items, and firm C has a population of 10,000 items. The population standard deviation for the cost of the items is $\sigma = 144$. A statistical consultant recommends that each firm take a sample of 50 items from its population to provide statistically valid estimates of the average cost per item. Management of the small firm states that since it has the smallest population, it should be able to obtain the data from a much smaller sample size than required by the larger firms. However, the consultant states that to obtain the same standard error and thus the same precision in the sample results, the size of the sample selected for each firm should be the same and not be dependent on the size of the population.

a. Using the finite population correction factor, compute the standard error for each of the three firms given a sample of size 50.

b. What is the probability that for each firm, the sample mean $\bar{x}$ will be within 25 of the population mean μ?

68. A survey reports its results by stating that the standard error of the mean is 20. The population standard deviation is 500.

a. How large is the sample used in this survey?

b. What is the probability that the sample mean will be within 25 of the population mean?

CHAPTER 8

INFERENCES ABOUT A
POPULATION MEAN

WHAT YOU WILL LEARN IN THIS CHAPTER

- How to construct and interpret an interval estimate of a population mean
- What is meant by the term confidence level
- The concept of sampling error
- How to determine the sample size when estimating a population mean
- What hypotheses are and how they are formulated
- How to use sample results to test hypotheses about a population mean
- How to interpret Type I and Type II errors in hypothesis testing
- What a p-value is and how it is used in hypothesis testing
- The t distribution
- How to use the t distribution to make interval estimates and test hypotheses about a population mean

CONTENTS

Radon Poses Health Hazard

Radon is an odorless gas that is formed in the ground by the natural decay of uranium. As the gas decays, the radioactive particles that are produced travel through the soil and collect in buildings. The problem is that when these radioactive particles are inhaled by humans, the risk of lung cancer is increased. Federal officials have stated that radon is the second-leading cause of lung cancer after smoking. Estimates are that as many as 20,000 deaths can be attributed to this gas each year. The problem is further compounded by the fact that radon is something we cannot see, smell, or feel.

Based on an Environmental Protection Agency (EPA) survey of 11,000 homes in seven states, the federal government has issued a warning that home contamination by radon is significantly more pervasive than previously indicated. In the EPA study, the measurement of radiation used is called a picocurie. Assistant Surgeon General Vernon Houk stated that exposure to 4 picocuries per liter of air is equivalent to having 200–300 chest X rays a year or smoking half a pack of cigarettes a day.

Based on the sample data, the study found North Dakota, with an average home rating of 7 picocuries, had the highest level of radon. Arizona, with an average home rating of 1.6 picocuries, had the lowest level of radon. Although there is no federal standard for a "safe" level of radon, the EPA has stated that if a home has a radon level of 4 or more picocuries, the residence should be monitored. If the level of radon persists, the homeowner should take steps to vent the contaminated area.

Easy-to-use test kits can identify the level of radon in your home.

Based on the results of the states surveyed in 1988 and the results from 10 other states surveyed in 1987, the EPA estimates that more than 3 million homes in the 17 states are contaminated at a potentially health-threatening level. The statistical evidence led government officials to conclude that the potential for radon contamination is so widespread that the only people who do not need to test their homes are those living in apartments above the second floor.

This *Statistics in Practice* is based on "Federal Health Officials Urge Nationwide Tests For Radon," *Charlotte Observer,* September 13, 1988.

In this chapter we continue the discussion of how a sample mean $\bar{x}$ can be used to make inferences about a population mean μ. First we consider the statistical process known as *estimation,* where the value of $\bar{x}$ is used to estimate the value of μ. This was the process used by the EPA in *Statistics in Practice*; the EPA's sample results provided an estimate of the mean radon level for homes in each of the states in the survey.

We also discuss the statistical process known as *hypothesis testing*. In this process, we hypothesize that the population mean has a specific numerical value; then we use the value of the sample mean $\bar{x}$ to *test the hypothesis* about the value of the population mean.

We begin by presenting inferences about a population mean based on a simple random sample that contains at least 30 items. We refer to this situation as the *large-sample* case.

In Sections 8.2 and 8.8 we discuss inferences about a population mean for the *small-sample* case, where the sample consists of less than 30 items. As you will see, the methodology for the small-sample case requires the use of a new probability distribution known as the *t* distribution.

8.1 ▽ INTERVAL ESTIMATION OF A POPULATION MEAN: LARGE-SAMPLE CASE

The central limit theorem, introduced in Chapter 7, allows us to conclude that the sampling distribution of $\bar{x}$ can be approximated by a normal probability distribution whenever the sample size is large. Recall that the large-sample condition is assumed to be satisfied whenever a simple random sample of size 30 or more is used. To illustrate an *interval estimate* of a population mean for the large-sample case, let us consider the following example.

EXAMPLE 8.1 ▶ The Statewide Insurance Company provides a variety of life, health, disability, and business insurance policies for customers located throughout the United States. As part of an annual review of life insurance policies, a simple random sample of 36 Statewide policyholders is selected. The corresponding life insurance policies are reviewed in terms of the amount of the coverage, the cash value of the policy, the disability options, and so on. For the current policy review study, the project manager has requested information on the ages of the life insurance policyholders. Table 8.1 shows the age data collected from the simple random sample of 36 policyholders. Let us use these data to develop an interval estimate of the mean age of the population of life insurance policyholders covered by Statewide.

The interval estimation procedure that we will develop is based on the assumption that the value of the population standard deviation is *known*. For the Statewide study, previous studies on policyholder ages permit us to use a known population standard deviation of $\sigma = 7.2$ years. Later in this section we will show how to develop an interval estimate when the value of the population standard deviation is unknown.

TABLE 8.1   **AGES OF LIFE INSURANCE POLICYHOLDERS FROM A SIMPLE RANDOM SAMPLE OF 36 STATEWIDE POLICYHOLDERS**

Policyholder	Age	Policyholder	Age	Policyholder	Age
1	32	13	39	25	23
2	50	14	46	26	36
3	40	15	45	27	42
4	24	16	39	28	34
5	33	17	38	29	39
6	44	18	45	30	34
7	45	19	27	31	35
8	48	20	43	32	42
9	44	21	54	33	53
10	47	22	36	34	28
11	31	23	34	35	49
12	36	24	48	36	39

Let x_1 indicate the age of the first policyholder in the sample, x_2 the age of the second policyholder, and so on. The sample mean $\bar{x}$ provides a *point estimate* of the population mean μ. Using the data in Table 8.1, we obtain

$$\bar{x} = \frac{\Sigma x_i}{n} = \frac{1422}{36} = 39.5$$

Thus, the point estimate of the mean age of the population of Statewide life insurance policyholders is 39.5 years. ◀

In discussing the practical value of the sampling distribution of $\bar{x}$ in Chapter 7, we indicated that we cannot expect the value of a sample mean $\bar{x}$ to *exactly* equal the value of the population mean μ. Thus, anytime a sample mean is used to provide a point estimate of a population mean, someone may ask, How good is the estimate? The "how good" question is a way of asking about the error involved when the value of $\bar{x}$ is used as a point estimate of the population mean μ. In general, we refer to the absolute value of the difference between an unbiased point estimator and the population parameter as the *sampling error*. For the case of a sample mean estimating a population mean, the sampling error is

$$\text{Sampling Error} = |\bar{x} - \mu| \tag{8.1}$$

Note that even after a sample is selected and the sample mean is computed, we will not be able to use equation (8.1) to find the value of the sampling error because the population mean μ is unknown. However, we shall see that the sampling distribution of $\bar{x}$ developed in Chapter 7 can be used to make probability statements about the sampling error.

From the central limit theorem, we know that whenever the sample size is large ($n \geq 30$), the sampling distribution of $\bar{x}$ can be approximated by a normal probability distribution. For the Statewide Insurance study, where $\sigma = 7.2$ years and $n = 36$, this theorem enables us to conclude that the sampling distribution of $\bar{x}$ is approximately normal with mean μ and standard deviation $\sigma_{\bar{x}} = \sigma/\sqrt{n} = 7.2/\sqrt{36} = 1.2$ years.* This sampling distribution is shown in Figure 8.1.

Although the population mean μ is unknown, the sampling distribution in Figure 8.1 shows how the $\bar{x}$ values are distributed around μ. In effect, this distribution is providing us with information about the possible differences between $\bar{x}$ and μ and, as a result, about the possible sampling error.

PROBABILITY STATEMENTS ABOUT THE SAMPLING ERROR

Since the sampling distribution of $\bar{x}$ can be approximated by a normal probability distribution, we can use the standard normal probability distribution table to make probability statements about the sampling error. For example, we find that 95% of the values of a normally distributed random variable lie within ±1.96 standard deviations of the mean. Hence, for the sampling distribution of $\bar{x}$ shown in Figure 8.1, 95% of all $\bar{x}$ values are within ±1.96$\sigma_{\bar{x}}$ of the mean μ. Since $1.96\sigma_{\bar{x}} = 1.96(1.2) = 2.35$, we can state that 95% of all sample means lie within ±2.35 years of the population mean μ. The location of all the sample means that provide a sampling error of 2.35 years or less is shown in Figure 8.2. Note, however, that the sample mean could fall in one of the two tails of the sampling distribution, which would result in a sampling error greater than 2.35

---

*In this chapter we will be assuming that $n/N \leq .05$. Thus, the finite population correction factor is not needed in the computation of $\sigma_{\bar{x}}$, and we use $\sigma_{\bar{x}} = \sigma/\sqrt{n}$.

FIGURE 8.1   **SAMPLING DISTRIBUTION OF THE SAMPLE MEAN AGE FROM SIMPLE RANDOM SAMPLES OF 36 STATEWIDE POLICYHOLDERS**

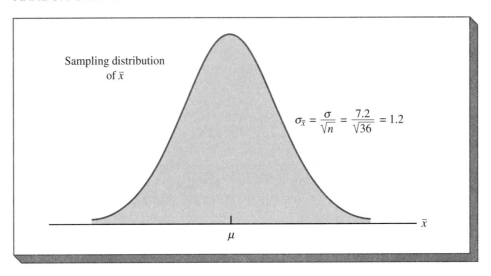

FIGURE 8.2   **SAMPLING DISTRIBUTION OF $\bar{x}$ SHOWING THE LOCATION OF SAMPLE MEANS THAT PROVIDE A SAMPLING ERROR OF 2.35 YEARS OR LESS**

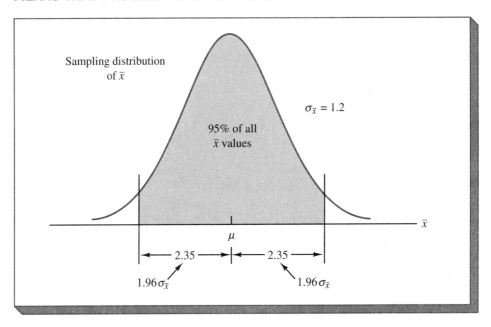

years. However, we see from Figure 8.2 that the probability of this occurring is only $1 - .95 = .05$. Thus, knowledge of the sampling distribution of $\bar{x}$ enables us to make the following probability statement about the sampling error whenever a simple random sample of 36 Statewide policyholders is used to estimate the mean age of the population:

There is a .95 probability that the sample mean will provide a sampling error of 2.35 years or less.

The above probability statement about the sampling error is a statement of the *precision* of the estimate. If the manager is not satisfied with this degree of precision, a larger sample size will be necessary. We shall discuss a procedure for determining the sample size necessary to obtain a desired precision in Section 8.3.

Note that in the above analysis, the .95 probability used in the statement about the sampling error was arbitrary. Although a .95 probability is frequently used in making such statements, other probability values can be selected. Probabilities of .90 and .99 are commonly used alternatives. Let us consider what would have happened if a probability of .99 had been selected. Figure 8.3 shows the location of 99% of the sample means for the Statewide Insurance problem. From the standard normal probability distribution table, we find that 99% of the $\bar{x}$ values lie within ±2.575 standard deviations of the population mean μ. Since $2.575\sigma_{\bar{x}} = 2.575(1.2) = 3.09$, we can make the following statement about the sampling error whenever a simple random sample of 36 Statewide policyholders is used to estimate the mean age of the population:

> There is a .99 probability that the sample mean will provide a sampling error of 3.09 years or less.

A similar calculation with a .90 probability shows that there is a .90 probability that the sample mean will provide a sampling error of $1.645\sigma_{\bar{x}} = 1.645(1.2) = 1.97$ years or less.

These results show that there are various probability statements that can be made about the sampling error. They also show that there is a trade-off between the probability specified and the stated limit on the sampling error. In particular, note that the higher probability statements possess larger values for the sampling error.

Let us now generalize the procedure we are using to make probability statements about the sampling error whenever a sample mean is used to estimate a population mean. We will use the Greek letter α (alpha) to indicate the probability that a sampling error is *larger* than the sampling error mentioned in the precision statement. Refer to Figure 8.4. We see that $\alpha/2$ will be the area or probability in each tail of the distribution, and $1 - \alpha$

FIGURE 8.3 **SAMPLING DISTRIBUTION OF $\bar{x}$ SHOWING THE LOCATION OF 99% OF THE $\bar{x}$ VALUES**

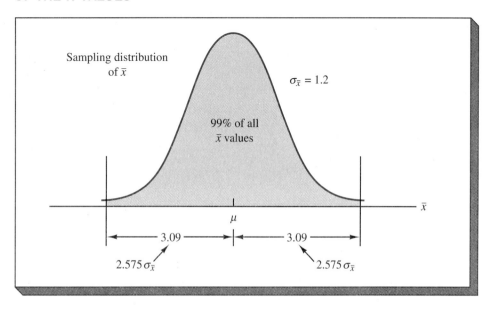

FIGURE 8.4   **AREAS OF A SAMPLING DISTRIBUTION OF $\bar{x}$ USED TO MAKE PROBABILITY STATEMENTS ABOUT THE SAMPLING ERROR**

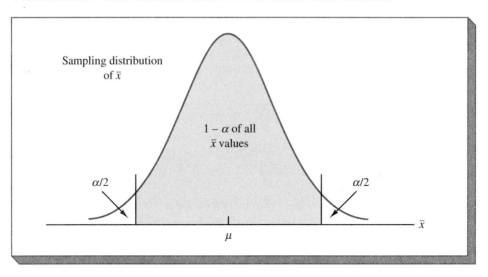

will be the area or probability that a sample mean will provide a sampling error *less than or equal to* the sampling error used in the precision statement.

Refer to the Statewide Insurance study. The statement that there is a .95 probability that the value of a sample mean will provide a sampling error of 2.35 years or less is based on $\alpha = .05$ and $1 - \alpha = .95$. The area in each tail of the sampling distribution is $\alpha/2 = .025$ (see Figure 8.2).

Using z to denote the value from a standard normal probability distribution, we will place a subscript on z to denote the *area in the upper tail* of the probability distribution. Thus, $z_{.025}$ will correspond to the z value with .025 of the area in the upper tail of the probability distribution. As can be found in the standard normal probability distribution table, $z_{.025} = 1.96$. If we wanted a .99 probability statement, $\alpha = .01$ in this case, we would be interested in an area of $\alpha/2 = .005$ in the upper tail of the distribution and hence, $z_{.005} = 2.575$.

In general, $z_{\alpha/2}$ denotes the value corresponding to an area of $\alpha/2$ in the upper tail of the standard normal probability distribution, and $\sigma_{\bar{x}}$ denotes the standard deviation of the sampling distribution of $\bar{x}$ (also called the standard error of the mean). We now have the following general procedure for making a probability statement about the sampling error whenever $\bar{x}$ is used to estimate μ.

> **Probability Statement About the Sampling Error**
>
> There is a $1 - \alpha$ probability that the value of a sample mean will provide a sampling error of $z_{\alpha/2}\sigma_{\bar{x}}$ or less.

CALCULATING AN INTERVAL ESTIMATE

We have the ability to make probability statements about the sampling error. We can now combine the point estimate with the probability information about the sampling error to obtain an interval estimate of the population mean. The rationale for the interval-estimation procedure is as follows: We have already stated that there is a $1 - \alpha$

probability that the value of a sample mean will provide a sampling error of $z_{\alpha/2}\sigma_{\bar{x}}$ or less. This means that there is a $1 - \alpha$ probability that the sample mean *will not miss* the population mean *by more than* $z_{\alpha/2}\sigma_{\bar{x}}$. Thus, if we form an interval by subtracting $z_{\alpha/2}\sigma_{\bar{x}}$ from the sample mean $\bar{x}$ and then adding $z_{\alpha/2}\sigma_{\bar{x}}$ to the sample mean $\bar{x}$, we would have a $1 - \alpha$ probability of obtaining an interval that *includes* the population mean μ. This condition is stated as follows:

> There is a $1 - \alpha$ probability that the interval formed by $\bar{x} \pm z_{\alpha/2}\sigma_{\bar{x}}$ will contain the population mean μ.

EXAMPLE 8.1
(CONTINUED) ▶

Let us return to the Statewide Insurance study. We previously stated that there is a .95 probability that the value of a sample mean will provide a sampling error of 2.35 years or less. Looking at the sampling distribution of $\bar{x}$ as shown in Figure 8.5, let us consider possible values of the sample mean $\bar{x}$ that could be obtained from three different simple random samples, each containing 36 policyholders. Remember, in each case we will form an interval estimate of the population mean by subtracting 2.35 from $\bar{x}$ and adding 2.35 to $\bar{x}$.

Consider what happens if the first sample mean turns out to have the value shown in Figure 8.5 as $\bar{x}_1$. Note that in this case, the interval formed by subtracting 2.35 from $\bar{x}_1$ and adding 2.35 to $\bar{x}_1$ includes the population mean μ. Now consider what happens if the sample mean turns out to have the value shown in Figure 8.5 as $\bar{x}_2$. Although this sample mean is different from the first sample mean, we see that the interval based on

FIGURE 8.5   **INTERVALS FORMED FROM SELECTED SAMPLE MEANS AT LOCATIONS $\bar{X}_1, \bar{X}_2$, AND $\bar{X}_3$**

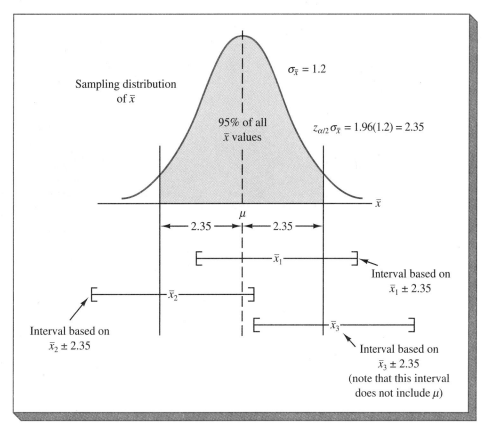

$\bar{x}_2$ also includes the population mean μ. However, the interval based on the third sample mean, denoted by $\bar{x}_3$, does not include the population mean. The reason for this is that the sample mean $\bar{x}_3$ lies in a tail of the probability distribution and is thus more than 2.35 years from μ. Thus, subtracting and adding 2.35 to $\bar{x}_3$ forms an interval that does not include μ.

Now think of repeating the sampling process many times, each time computing the value of the sample mean and then forming an interval from $\bar{x} - 2.35$ to $\bar{x} + 2.35$. Any sample mean $\bar{x}$ that falls between the vertical lines in Figure 8.5 will provide an interval that includes the population mean μ. Since 95% of the sample means are in the shaded region, 95% of all intervals that could be formed will include μ. As a result, we say that we are 95% confident that an interval constructed from $\bar{x} - 2.35$ to $\bar{x} + 2.35$ will include the population mean. We refer to this interval as a *confidence interval*. Moreover, since 95% of the sample means lead to a confidence interval including μ, we say the interval is established at the 95% *confidence level*. The value .95 is referred to as the *confidence coefficient*.

Recall that the sample mean age for the sample of 36 Statewide life insurance policyholders is $\bar{x} = 39.5$. A 95% confidence interval estimate of the mean age for the population of Statewide life insurance policyholders is 39.5 ± 2.35, or 37.15 years to 41.85 years. Thus, at the 95% confidence level, Statewide can conclude that the mean age for the population of life insurance policyholders is between 37.15 and 41.85 years. ◀

Let us now state the general procedure for computing an interval estimate of a population mean. As previously noted, there is a $1 - \alpha$ probability that the interval formed by $\bar{x} \pm z_{\alpha/2}\sigma_{\bar{x}}$ includes the population mean μ. Using the fact that $\sigma_{\bar{x}} = \sigma/\sqrt{n}$, the general procedure for calculating the interval estimate of population mean using a large sample as follows.

Interval Estimate of a Population Mean
Large-Sample Case ($n \geq 30$)

$$\bar{x} \pm z_{\alpha/2}\frac{\sigma}{\sqrt{n}} \tag{8.2}$$

where $1 - \alpha$ is the confidence coefficient and $z_{\alpha/2}$ is the z value providing an area of $\alpha/2$ in the upper tail of the standard normal probability distribution.

If the population standard deviation σ is unknown, the sample standard deviation s may be substituted for σ in the above expression.

The values of $z_{\alpha/2}$ for the most commonly used confidence levels are shown in Table 8.2.

TABLE 8.2   VALUES OF $z_{\alpha/2}$ FOR THE MOST COMMONLY USED CONFIDENCE LEVELS

Confidence Level	$z_{\alpha/2}$
90%	1.645
95%	1.96
99%	2.575

A difficulty in using equation (8.2) is that in most sampling situations the value of the population standard deviation σ is unknown. In the large-sample case ($n \geq 30$), we simply use the value of the sample standard deviations s as the point estimate of the population standard deviation σ and obtain the confidence interval, $\bar{x} \pm z_{\alpha/2} \, s/\sqrt{n}$. In the following example, we provide an illustration as to how a confidence interval is developed when $n \geq 30$ and σ is unknown.

EXAMPLE 8.2 ▶

The *1993 Information Please Almanac* lists a variety of information on colleges and universities in the United States. A sample of 40 colleges and universities from this population was used to estimate the mean annual room-and-board cost associated with attending college. The sample of 40 provided a sample mean of $\bar{x} = \$4275$ per year and a sample standard deviation of $s = \$1264$. Develop a 90% confidence interval estimate of the population mean annual room-and-board cost.

At 90% confidence, Table 8.2 shows that $z_{.05} = 1.645$. Using the sample standard deviation $s = 1264$ to estimate σ, equation (8.2) can be used to provide the confidence interval

$$4275 \pm 1.645\left(\frac{1264}{\sqrt{40}}\right)$$

$$4275 \pm 329$$

Thus, we can be 90% confident that the interval \$3946 to \$4604 contains the mean annual room-and-board cost for the population of colleges and universities.   ◀

NOTES AND COMMENTS

1. In using equation (8.2) to develop an interval estimate of the population mean, we specify the desired confidence coefficient ($1 - \alpha$) before selecting the sample. Thus, prior to selecting the sample, we conclude that there is a $1 - \alpha$ probability that the confidence interval we eventually compute will contain the population mean μ. However, once the sample is taken and the confidence interval is computed, the resulting interval *may or may not* contain μ. If $1 - \alpha$ is reasonably large, we can be confident that the resulting interval contains μ because we know that if we use this procedure in the long term, $100(1 - \alpha)$ percent of all possible intervals developed in this manner will contain μ.

2. Note that the sample size n appears in the denominator of the interval-estimation expression (8.2). Thus, if a particular sample size provides too wide an interval to be of practical use, we may want to consider increasing the sample size. With n in the denominator, a larger sample size will reduce width of the confidence interval and provide a more precise estimate of μ. The procedure for determining the size of a simple random sample required to obtain a desired precision is discussed in Section 8.3.

3. The justification for equation (8.2) comes from the fact that the central limit theorem enables us to *approximate* the sampling distribution of $\bar{x}$ by a normal probability distribution whenever the sample size is large. In addition, in most applications the sample standard deviation s is used to *approximate* the population standard deviation σ. Thus, statisticians often refer to equation (8.2) as an *approximate* confidence interval for a population mean.

EXERCISES

METHODS

1. A simple random sample of 40 items resulted in a sample mean of 25. The population standard deviation is $\sigma = 5$.
   a. What is the standard error of the mean?
   b. At a 95% probability, what can be said about the size of the sampling error?

SELF TEST

2. A simple random sample of 50 items resulted in a sample mean of 32 and a sample standard deviation of 6.
   a. Compute the 90% confidence interval for the population mean.
   b. Compute the 95% confidence interval for the population mean.
   c. Compute the 99% confidence interval for the population mean.

3. A sample of 60 items resulted in a sample mean of 80 and a sample standard deviation of 15.
   a. Compute the 95% confidence interval for the population mean.
   b. Assume that the same sample mean and the sample standard deviation were obtained from a sample of 120 items. Provide a 95% confidence interval for the population mean.
   c. What is the effect of a larger sample size on the interval estimate of a population mean?

4. The 95% confidence interval for a population mean is reported to be 122 to 130. If the sample mean was 126 and the sample standard deviation was 16.07, what sample size was used in this study?

APPLICATIONS

SELF TEST

5. In an effort to estimate the mean amount spent per customer for dinner meals at a major Atlanta restaurant, the manager collected data for a sample of 49 customers over a 3-week period.
   a. Assume a population standard deviation of $2.50. What is the standard error of the mean?
   b. With a .95 probability, what statement can be made about the sampling error?
   c. If the sample mean is $12.60, what is the 95% confidence interval for the population mean?

6. Data on the automotive industry and automobile purchasing characteristics were presented in *Financial World,* April 14, 1992. The mean car payment per month was reported as $310. Assume that this result was based on a sample of 250 car payment records and that the sample standard deviation was $100. Compute the 95% confidence interval for the mean monthly car payments.

7. The results of an annual survey of 1404 mutual funds were presented in *Forbes,* September 2, 1991. A sample of 75 funds showed a mean return over the previous 12 months of 10.2%. The sample standard deviation was 3%. Compute the 95% confidence interval for the mean return for the population of mutual funds.

8. The mean annual income of U.S. factory workers is $24,000 (*Barron's,* April 10, 1989). Assume that this estimate was based on a sample of 250 U.S. factory workers and that the population standard deviation is $\sigma = 5000.
   a. Compute the 90% confidence interval for the population mean.
   b. Compute the 95% confidence interval for the population mean.
   c. Compute the 99% confidence interval for the population mean.
   d. Discuss what happens to the width of the interval estimate as the confidence level is increased. Why does this seem reasonable?

9. A production filling operation has a historical standard deviation of 5.5 ounces. A quality-control inspector periodically selects 36 containers at random and uses the sample

mean filling weight to estimate the population mean filling weight for the production process.

a. What is the standard error of the mean, $\sigma_{\bar{x}}$?

b. With .75, .90, and .99 probabilities, what statements can be made about the sampling error? What happens to the statement about the sampling error when the probability is increased? Why does this happen?

c. What is the 99% confidence interval for the population mean filling weight for the process if the sample mean is 48.6 ounces?

10. A survey of readers of *Money* magazine found that the sample mean age of men was 47 years and the sample mean age of women was 44 years (*Money Extra,* Fall 1988). Altogether, 454 people were included in the reader poll—340 men and 114 women. Assume that the population standard deviation of age for both men and women is 8 years.

a. Compute the 95% confidence interval for the mean age of the population of men who read *Money* magazine.

b. Compute the 95% confidence interval for the mean age of the population of women who read *Money* magazine.

c. Compare the widths of the two interval estimates from parts (a) and (b). Did the estimate of the mean age of men or the mean age of women have the better precision? Why?

11. E. Lynn and Associates is an energy research firm that provides estimates of monthly heating costs for new homes based on style of house, square footage, insulation, and so on. The firm's service is used by both builders and potential buyers of new homes who wish advance information on heating costs. For winter months, the standard deviation in the home heating bills for residential homes in a certain area is $100. Assume that a sample of 36 homes in a particular subdivision will be used to estimate the mean monthly heating bill for the population all homes in this type of subdivision.

a. What is the standard error of the mean, $\sigma_{\bar{x}}$?

b. Show the sampling distribution for the sample mean heating bill.

c. At an 80% probability, what can be said about the sampling error? Show this probability on the graph of the sampling distribution in part (b).

d. What is the 98% confidence interval for the population mean monthly heating bill if the sample mean is $196.50?

12. *Consumer Research,* April 1989, reports information on the time required for caffeine from products such as coffee and soft drinks to leave the body after consumption. Assume that the 95% confidence interval of the population mean time for adults is 5.6 hours to 6.4 hours.

a. What is the point estimate of the mean time for caffeine to leave the body of adults after consumption?

b. If the population standard deviation is 2 hours, how large a sample was used to provide the interval estimate?

13. Data were collected on the golf-ball driving distances by professional golfers in a recent tournament. Using data on the first drives of 30 randomly selected golfers, it was found that the sample mean distance was 250 yards and the sample standard deviation was 10 yards. Compute the 95% confidence interval for the mean driving distance for the population.

14. Researchers at the University of Illinois used a sample of 70 students from five high schools in a large metropolitan area to learn about the performance of students on the American College Test (ACT) (*Journal of College Student Development,* February 1989). Of the students in the sample, 47 went on to college and 23 did not. The ACT scores for the sample of students who went on to college had a mean $\bar{x} = 25.23$ and a standard deviation $s = 3.21$.

a. Compute the 95% confidence interval for the population mean ACT score for students going on to college.

    b. Compute the 98% confidence interval for the population mean ACT score for students going on to college.

8.2   INTERVAL ESTIMATION OF A POPULATION MEAN: SMALL-SAMPLE CASE

For a large sample ($n \geq 30$), the central limit theorem enables us to approximate the sampling distribution of $\bar{x}$ by a normal probability distribution. This normal distribution approximation is what makes the use of $z_{\alpha/2}$ appropriate in equation (8.2). However, what happens to the interval-estimation procedure when the sample size is small ($n < 30$), and we cannot use the central limit theorem to justify the use of a normal probability distribution to approximate the sampling distribution of $\bar{x}$?

In the small-sample case, the sampling distribution of $\bar{x}$ depends on the distribution of the population. *If the population has a normal probability distribution,* the methodology that follows can be used to develop a confidence interval for a population mean. However, if the assumption of a normal probability distribution for the population is not appropriate, we recommend increasing the sample size to $n \geq 30$ and relying on the large-sample interval-estimation procedure given by equation (8.2).

If the population has a normal probability distribution, the sampling distribution of $\bar{x}$ will be normal regardless of the sample size. In this case, if the population standard deviation σ is *known,* equation (8.2) can still be used to compute an interval estimate of a population mean even with a small sample. However, if the population standard deviation σ is *unknown,* which is the usual situation, and the sample standard deviation s is used to estimate σ, the resulting confidence interval $\bar{x} \pm z_{\alpha/2}s/\sqrt{n}$ is only appropriate provided $n \geq 30$; in the small-sample case within $n < 30$, the appropriate confidence interval is based on a probability distribution known as the t *distribution.*

The t distribution is actually a family of similar probability distributions, with a specific t distribution depending on a parameter known as the *degrees of freedom.* That is, there is a unique t distribution with 1 degree of freedom, with 2 degrees of freedom, with 3 degrees of freedom, and so on. As the number of degrees of freedom increases, the difference between the t distribution and the standard normal probability distribution becomes smaller and smaller. Figure 8.6 shows t distributions with 10 and 20 degrees of freedom and their relationship to the standard normal probability distribution. Note that a t distribution with more degrees of freedom has less dispersion and more closely resembles the standard normal distribution. Note also that the mean of the distribution is zero.

We will use a subscript for t to indicate the area in the upper tail of the t distribution. For example, just as we used $z_{.025}$ to indicate the z value providing a .025 area in the upper tail of a standard normal probability distribution, we will use $t_{.025}$ to indicate a .025 area in the upper tail of the t distribution. In general, we will use the notation $t_{\alpha/2}$ to represent a t value with an area of $\alpha/2$ in the upper tail of the t distribution. See Figure 8.7.

A table for the t distribution is provided in Table 8.3. This table is also shown inside the front cover of the text. Note, for example, that for a t distribution with 10 degrees of freedom, $t_{.025} = 2.228$. Similarly, for a t distribution with 20 degrees of freedom, $t_{.025} = 2.086$. As the degrees of freedom continue to increase, $t_{.025}$ approaches $z_{.025}$.

Now that we have an idea of what the t distribution is, let us show how it is used to develop an interval estimate of a population mean. Assume that the population has a normal probability distribution and that the sample standard deviation s is used as a point estimate of the population standard deviation σ. The interval-estimation procedure shown at the top of page 281 is applicable.

FIGURE 8.6   **COMPARISON OF THE STANDARD NORMAL DISTRIBUTION WITH t DISTRIBUTIONS HAVING 10 AND 20 DEGREES OF FREEDOM**

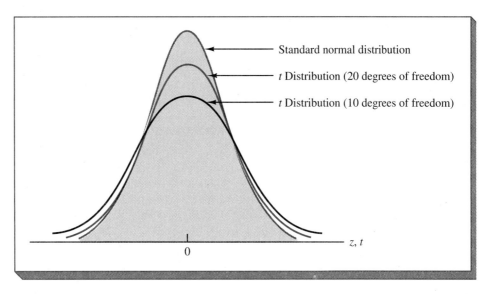

FIGURE 8.7   **t DISTRIBUTION WITH $\alpha/2$ AREA OR PROBABILITY IN THE UPPER TAIL**

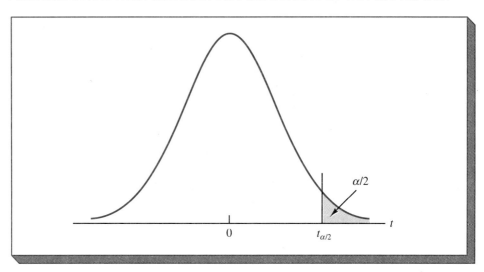

TABLE 8.3

t DISTRIBUTION TABLE FOR AREAS IN THE UPPER TAIL. EXAMPLE: WITH 10 DEGREES OF FREEDOM $t_{.025} = 2.228$

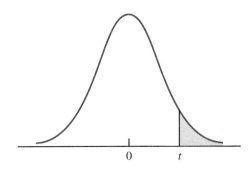

Degrees of Freedom	Upper-Tail Area (Shaded)				
	.10	.05	.025	.01	.005
1	3.078	6.314	12.706	31.821	63.657
2	1.886	2.920	4.303	6.965	9.925
3	1.638	2.353	3.182	4.541	5.841
4	1.533	2.132	2.776	3.747	4.604
5	1.476	2.015	2.571	3.365	4.032
6	1.440	1.943	2.447	3.143	3.707
7	1.415	1.895	2.365	2.998	3.499
8	1.397	1.860	2.306	2.896	3.355
9	1.383	1.833	2.262	2.821	3.250
10	1.372	1.812	2.228	2.764	3.169
11	1.363	1.796	2.201	2.718	3.106
12	1.356	1.782	2.179	2.681	3.055
13	1.350	1.771	2.160	2.650	3.012
14	1.345	1.761	2.145	2.624	2.977
15	1.341	1.753	2.131	2.602	2.947
16	1.337	1.746	2.120	2.583	2.921
17	1.333	1.740	2.110	2.567	2.898
18	1.330	1.734	2.101	2.552	2.878
19	1.328	1.729	2.093	2.539	2.861
20	1.325	1.725	2.086	2.528	2.845
21	1.323	1.721	2.080	2.518	2.831
22	1.321	1.717	2.074	2.508	2.819
23	1.319	1.714	2.069	2.500	2.807
24	1.318	1.711	2.064	2.492	2.797
25	1.316	1.708	2.060	2.485	2.787
26	1.315	1.706	2.056	2.479	2.779
27	1.314	1.703	2.052	2.473	2.771
28	1.313	1.701	2.048	2.467	2.763
29	1.311	1.699	2.045	2.462	2.756
30	1.310	1.697	2.042	2.457	2.750
40	1.303	1.684	2.021	2.423	2.704
60	1.296	1.671	2.000	2.390	2.660
120	1.289	1.658	1.980	2.358	2.617
∞	1.282	1.645	1.960	2.326	2.576

> **Interval Estimate of a Population Mean Small-Sample Case ($n < 30$)**
>
> $$\bar{x} \pm t_{\alpha/2}\frac{s}{\sqrt{n}} \tag{8.3}$$
>
> where $1 - \alpha$ is the confidence coefficient, $t_{\alpha/2}$ is the t value providing an area of $\alpha/2$ in the upper tail of a t distribution with $n - 1$ *degrees of freedom*, and s is the sample standard deviation. It is assumed that the population has a normal probability distribution.*

The reason the number of degrees of freedom associated with the t value in equation (8.3) is $n - 1$ has to do with the use of s as an estimate of the population standard deviation σ. The expression for the sample standard deviation is

$$s = \sqrt{\frac{\Sigma(x_i - \bar{x})^2}{n - 1}}$$

Degrees of freedom here refers to the number of independent pieces of information that go into the computation of $\Sigma(x_i - \bar{x})^2$. The pieces of information involved in computing $\Sigma(x_i - \bar{x})^2$ are $x_1 - \bar{x}, x_2 - \bar{x}, \ldots, x_n - \bar{x}$. In Section 3.2 we indicated that $\Sigma(x_i - \bar{x}) = 0$ for any data set. Thus, only $n - 1$ of the $x_i - \bar{x}$ values are independent; that is, if we know $n - 1$ of the values, the remaining value can be determined exactly by using the condition that the $x_i - \bar{x}$ values must sum to 0. Thus, $n - 1$ is the number of degrees of freedom associated with $\Sigma(x_i - \bar{x})^2$ and hence the t distribution used in equation (8.3).

EXAMPLE 8.3 ▶ The director of manufacturing for Scheer Industries is interested in a computer-assisted training program that can be used to train the firm's maintenance employees for machine-repair operations. It is anticipated that the computer-assisted training will reduce training time and costs. To evaluate the training method, the director of manufacturing has requested an estimate of the mean training time required with the computer-assisted training program.

Suppose that management has agreed to train 15 employees with the new approach. The data on training days required for each employee in the sample are shown in Table 8.4. The sample mean and sample standard deviation for these data are as follows:

$$\bar{x} = \frac{\Sigma x_i}{n} = \frac{808}{15} = 53.87 \text{ days}$$

$$s = \sqrt{\frac{\Sigma(x_i - \bar{x})^2}{n - 1}} = \sqrt{\frac{651.73}{14}} = 6.82 \text{ days}$$

The point estimate of the mean training time for the population of employees is $\bar{x} = 53.87$ days. We can obtain information about the precision of this estimate by developing an interval estimate of the population mean. Since the population standard deviation is unknown, we will use the sample standard deviation $s = 6.82$ days as the

---

*The population standard deviation σ is usually unknown and is estimated by the sample standard deviation s. However, if σ is known and if the population has a normal probability distribution, the small-sample case would use z rather than t; $\bar{x} \pm z_{\alpha/2}(\sigma/\sqrt{n})$ could then be used to develop the interval estimate of the population mean.

TABLE 8.4	TRAINING TIME IN DAYS FOR THE COMPUTER-ASSISTED TRAINING PROGRAM AT SCHEER INDUSTRIES

Employee	Time	Employee	Time	Employee	Time
1	52	6	59	11	54
2	44	7	50	12	58
3	55	8	54	13	60
4	44	9	62	14	62
5	45	10	46	15	63

point estimate of σ. With the small-sample size, $n = 15$, we will use equation (8.3) to develop the interval estimate of the population mean with a 95% confidence level. Assuming that the population of training times has a normal probability distribution, the t distribution with $n - 1 = 14$ degrees of freedom is the appropriate probability distribution. We see from Table 8.3 that with 14 degrees of freedom, $t_{\alpha/2} = t_{.025} = 2.145$. Using equation (8.3), we have

$$\bar{x} \pm t_{.025} \frac{s}{\sqrt{n}}$$

$$53.87 \pm 2.145 \left(\frac{6.82}{\sqrt{15}} \right)$$

$$53.87 \pm 3.78$$

Thus, the 95% confidence interval estimate of the population mean training time is 50.09 days to 57.65 days. ◀

The approach which uses the t distribution to develop an interval estimate of μ is applicable whenever the population standard deviation is unknown and the population has a normal probability distribution. However, statistical research has shown that equation (8.3) is applicable even if the population being sampled is not quite normal. That is, confidence intervals based on the t distribution give good results when the population distribution does not differ extensively from a normal probability distribution. The fact that the t distribution can give satisfactory results even when the population distribution is not normal is referred to as the *robustness* property of the t distribution.

COMPUTER-GENERATED CONFIDENCE INTERVALS

Computer software packages are available for computing confidence intervals for population means. The Minitab output shown in Figure 8.8 illustrates how Minitab can be used to develop the confidence intervals for two of the examples discussed previously in this chapter: the mean annual college room-and-board cost and the mean training time for Scheer Industries.

Minitab has two interval-estimation commands, ZINTERVAL and TINTERVAL. In Figure 8.8, we see that the user selects the Minitab RETRIEVE command to obtain the previously stored data on annual room-and-board costs for a sample of 40 colleges and universities. The STD C1 command is used to compute the sample standard deviation for

FIGURE 8.8 MINITAB CONFIDENCE INTERVALS FOR THE MEAN ANNUAL COLLEGE ROOM-AND-BOARD COST AND THE MEAN TRAINING TIME FOR SCHEER INDUSTRIES

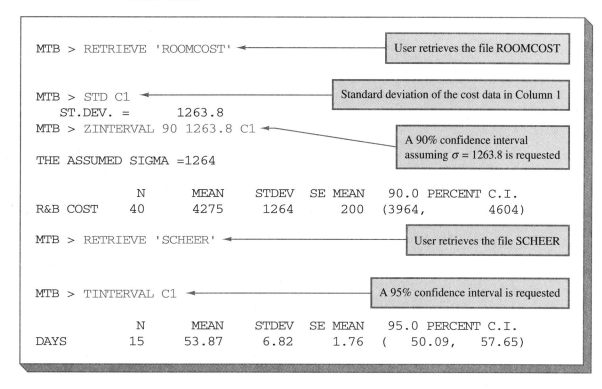

the annual room-and-board cost, which is shown to be $1263.80. The C1 notes that the data are stored in Column 1 of the Minitab worksheet. Since the sample size of 40 is considered large, the user continues the Minitab session by selecting the ZINTERVAL procedure to develop the confidence interval. The specific command ZINTERVAL 90 1263.8 C1 indicates that a 90% confidence interval is desired with the population standard deviation estimated by the sample standard deviation of 1263.8. The printout shown in Figure 8.8 is self-explanatory. After providing the sample size, the sample mean of 4275, the sample standard deviation of 1264, and the standard error of the mean of 200, the printout shows the 90% confidence interval for the population mean annual room-and-board cost is $3946 to $4604.

Referring to Figure 8.8, we see that the user continues by retrieving the previously stored Scheer Industries training-time data that were presented in Table 8.4. With the small sample size of $n = 15$, the user selects the TINTERVAL command to indicate that the t distribution is to be used to develop the confidence interval. Whenever the ZINTERVAL or TINTERVAL command does not specify a confidence level, Minitab automatically provides a 95% confidence interval. Thus, since the Scheer Industries example specified a 95% confidence interval, the user does not have to include 95 in the TINTERVAL command. Finally, since the TINTERVAL command does not specify a value for the standard deviation, the sample standard deviation is automatically computed from the data. Thus, the printout as shown in Figure 8.8 indicates that the 95% confidence interval estimate of the population mean training time is 50.09 days to 57.65 days.

NOTES AND
COMMENTS

We would like to point out that the t distribution is not restricted to the small-sample situation. Actually, the t distribution is applicable whenever the population is normal or near normal and whenever the sample standard deviation is used to estimate the population standard deviation. If these conditions exist, the t distribution can be used for any sample size. However, equation (8.2) shows that with a large sample ($n \geq 30$), interval estimation of a population mean can be based on the standard normal probability distribution and the value $z_{\alpha/2}$. Thus, with equation (8.2) available for the large-sample case, we generally do not consider the use of the t distribution until we encounter a small-sample case.

EXERCISES

METHODS

15. For a t distribution with 12 degrees of freedom, find the area, or probability, that lies in each of the following regions.
    a. to the left of 1.782
    b. to the right of -1.356
    c. to the right of 2.681
    d. to the left of -1.782
    e. between -2.179 and $+2.179$
    f. between -1.356 and $+1.782$

16. Find the t value(s) for each of the following:
    a. upper-tail area of .05 with 18 degrees of freedom
    b. lower-tail area of .10 with 22 degrees of freedom
    c. upper-tail area of .01 with 5 degrees of freedom
    d. 90% of the area is between these two t values with 14 degrees of freedom
    e. 95% of the area is between these two t values with 28 degrees of freedom

SELF TEST

17. The following data have been collected from a sample of eight items:

    10, 8, 12, 15, 13, 11, 6, 5.

    a. What is the point estimate of the population mean?
    b. What is the point estimate of the population standard deviation?
    c. What is the 95% confidence interval for the population mean?

18. A simple random sample of 20 items resulted in a sample mean of 17.25 and a sample standard deviation of 3.3.
    a. Compute the 90% confidence interval for the population mean.
    b. Compute the 95% confidence interval for the population mean.
    c. Compute the 99% confidence interval for the population mean.

APPLICATIONS

SELF TEST

19. In testing a new production method, 18 employees were randomly selected and asked to try the new method. The sample mean production rate for the 18 employees was 80 parts per hour, and the sample standard deviation was 10 parts per hour. Provide 90% and 95% confidence intervals for the mean production rate for the new method assuming the population has a normal probability distribution.

20. The Money & Investing section of the *The Wall Street Journal* contains a summary of the daily performances of stocks. The P/E ratio for each stock is determined by dividing the price of a share of stock by the earnings per share. A sample of 10 stocks taken from *The Wall Street Journal,* May 19, 1992, provided the following data on P/E ratios:

    5, 7, 9, 10, 14, 23, 20, 15, 3, 26.

   a. What is the point estimate of the mean P/E ratio for the population of stocks.

   b. What is the point estimate of the standard deviation of the P/E ratios for the population of stocks.

   c. For a .95 confidence coefficient, what is the interval estimate of the mean P/E ratio for the population of stocks.

   d. Comment on the precision of the results.

21. The following data are family sizes from a simple random sample of households in a new test market area.

Household	Family Size	Household	Family Size
1	4	7	3
2	3	8	2
3	2	9	3
4	2	10	6
5	4	11	3
6	5	12	2

Provide a 95% confidence interval for the mean family size for the population.

22. The American Association of Advertising Agencies records data on nonprogramming minutes per half hour of prime-time television programming (*U.S. News & World Report,* April 13, 1992). Representative data for a sample of prime-time programs on major networks at 8:30 P.M. are shown below. Provide a point estimate and the 95% confidence interval for the mean number of minutes of nonprogramming on half-hour prime-time television shows at 8:30 P.M.

$$6.0 \quad 6.6 \quad 5.8 \quad 7.0 \quad 6.3 \quad 6.2 \quad 7.2 \quad 5.7 \quad 6.4 \quad 7.0$$
$$6.5 \quad 6.2 \quad 6.0 \quad 6.5 \quad 7.2 \quad 7.3 \quad 7.6 \quad 6.8 \quad 6.0 \quad 6.2$$

23. Hertz announced that it would impose a surcharge of $15–56 per day on residents of Queens, Brooklyn, and the Bronx who rent cars in the New York metropolitan area (*The New York Times,* January 3, 1992). Hertz, the nation's largest car-rental company, was imposing the charge to recoup losses it had suffered under a New York state law that makes a car-rental agency liable for damage caused by a renter. Typical daily rental costs without the surcharge are shown in the following sample of 12 rentals:

$$65, 58, 40, 48, 52, 60, 75, 38, 50, 51, 59, 40.$$

   a. What is the point estimate of the mean daily car rental cost in the New York area?

   b. Compute the 95% confidence interval for the mean daily car-rental cost in the New York area?

24. Sales personnel for Skillings Distributors are required to submit weekly reports listing the customer contacts made during the week. A sample of 61 weekly contact reports showed a mean of 22.4 customer contacts per week and a sample standard deviation of five contacts.

   a. Compute the 95% confidence interval for the mean number of weekly customer contacts for the population of sales personnel.

   b. Assume that the population of weekly customer contacts is normally distributed. Use the *t* distribution with 60 degrees of freedom to compute the 95% confidence interval for the mean number of weekly customer contacts.

   c. Compare your answers for parts (a) and (b). Comment on why in the large-sample case it is permissible to base interval estimates on the procedure used in part (a) even though the *t* distribution may also be applicable.

25. Researchers at the University of Georgia used a sample of college students to study psychological issues associated with attending college (*Journal of College Student Development,* January 1989). One variable of interest was how the students progressed in terms of working on autonomy. The autonomy score was lowest in the fall quarter, better in the winter quarter, and highest in the spring quarter. A sample of 16 students for the fall quarter resulted in a sample mean of $\bar{x} = 51$ and sample standard deviation of $s = 10.18$.
   a. Compute the 95% confidence interval for the population mean autonomy score for the fall quarter.
   b. What assumption about the population was necessary to compute the interval estimate in part (a)?
   c. Assume that it was desirable to estimate the population mean autonomy score with a sampling error of 3 points or less. Do the statistical data provide this desired level of precision? What action, if any, would you recommend be taken?

26. The duration (in minutes) for a sample of 20 flight-reservation telephone calls is shown below.

| 2.1 | 4.8 | 5.5 | 10.4 | 3.3 | 3.5 | 4.8 | 5.8 | 5.3 | 5.5 |
| 2.8 | 3.6 | 5.9 | 6.6 | 7.8 | 10.5 | 7.5 | 6.0 | 4.5 | 4.8 |

   a. What is the point estimate of the population mean time for flight-reservation phone calls?
   b. Assuming that the population has a normal distribution, compute the 95% confidence interval for the population mean time.

8.3   DETERMINING THE SAMPLE SIZE

In Section 8.1 we made the following probability statement about the sampling error whenever a sample mean was used to provide an estimate of a population mean:

> There is a $1 - \alpha$ probability that the value of the sample mean will provide a sampling error of $z_{\alpha/2}\sigma_{\bar{x}}$ or less.

Since $\sigma_{\bar{x}} = \sigma/\sqrt{n}$, we can rewrite this statement as follows:

> There is a $1 - \alpha$ probability that the value of the sample mean will provide a sampling error of $z_{\alpha/2}(\sigma/\sqrt{n})$ or less.

From this statement we see that the values of $z_{\alpha/2}$, σ, and the sample size n combine to determine the sampling error mentioned in the precision statement. Once we select a confidence coefficient or probability of $1 - \alpha$, $z_{\alpha/2}$ can be determined. Given values for $z_{\alpha/2}$ and σ, we can determine the sample size n needed to achieve any sampling error. The formula used to compute the required sample size n is developed as follows.

Let E = the maximum value of the sampling error mentioned in the statement about the desired precision. We have

$$E = z_{\alpha/2}\frac{\sigma}{\sqrt{n}} \tag{8.4}$$

Using equation (8.4) to solve for $\sqrt{n}$, we have

$$\sqrt{n} = \frac{z_{\alpha/2}\sigma}{E}$$

Squaring both sides of this equation, we obtain the following equation for the sample size.

Sample Size for an Interval Estimate of a Population Mean

$$n = \frac{(z_{\alpha/2})^2\sigma^2}{E^2} \tag{8.5}$$

This sample size will provide a precision statement with a $1 - \alpha$ probability that the sampling error will be *E or less*.

In equation (8.5) the value E is the maximum sampling error that the user is willing to accept, and the value of $z_{\alpha/2}$ follows directly from the confidence level to be used in developing the interval estimate. Although user preference must be considered, 95% confidence is the most frequently chosen value ($z_{.025} = 1.96$).

Finally, use of equation (8.5) requires a value for the population standard deviation σ. In most cases, σ will be unknown. However, to be able to use equation (8.5), we must have a preliminary or *planning value* for σ. In practice, one of the following procedures can be used:

1. Use the sample standard deviation from a previous sample of the same or similar units.
2. Conduct a pilot study in order to select a preliminary sample of units. The sample standard deviation from the preliminary sample can be used as the planning value for σ.
3. Use judgment or a "best guess" for the value of σ. For example, we might begin by estimating the largest and smallest data values in the population. The difference between the largest and smallest values provides an estimate of the range for the data. Finally, the range divided by 4 is often suggested as a rough approximation of the standard deviation and thus, an acceptable planning value for σ.

EXAMPLE 8.3 ▶
(CONTINUED)

Let us return to the Scheer Industries study to see how equation (8.5) can be used to determine the sample size for the study. Previously we showed that with a 95% level of confidence, a sample of 15 Scheer employees resulted in an interval estimate of the population mean training-time of 53.87 ± 3.78 days. Assume that after viewing these results, Scheer's director of manufacturing is not satisfied with this degree of precision. Furthermore, suppose that the director makes the following statement about the desired precision: "I would like a .95 probability that the value of the sample mean will provide a sampling error of 2 days or less." We can see that the director is specifying a maximum sampling error of $E = 2$ days. In addition, the .95 probability indicates that a 95% confidence level is to be used; thus, $z_{\alpha/2} = z_{.025} = 1.96$. Lastly, we need a planning value for σ to use equation (8.5) in order to determine the sample size. Do we have a planning value for σ in the Scheer Industries study? Although σ is unknown, let us take advantage of the data provided for the 15 employees in Table 8.4. We can view these data as being a pilot study, with the sample standard deviation $s = 6.82$ days providing the planning value for σ. Thus, using equation (8.5), we have

$$n = \frac{(z_{\alpha/2})^2 \sigma^2}{E^2} = \frac{(1.96)^2 (6.82)^2}{2^2} = 44.67$$

In cases where the computed n is a fraction, we round up to the next integer value; thus, the recommended sample size is 45 employees. Since Scheer already has data for 15 employees, an additional $45 - 15 = 30$ employees should be tested if the director wishes to obtain the desired precision of ±2 days at a 95% confidence level.

Finally, note that $z_{.025}$ was used to determine the sample size even though the original computations for 15 employees employed the t distribution. The reason for the use of $z_{.025}$ is that since the sample size is yet to be determined, we are anticipating that n will be larger than 30, making $z_{.025}$ the appropriate value. In addition, if n is yet to be determined, we do not know the $(n - 1)$ degrees of freedom necessary to use the t distribution. Thus, the use of equation (8.5) to determine the sample size will always be based on a z value rather than a t value.

EXERCISES

METHODS

27. How large a sample should be selected to be 95% confident that the sampling error is 5 or less? Assume that the population standard deviation is 25.

SELF TEST

28. The range for a set of data is estimated to be 36.
  a. What is the planning value for the population standard deviation?
  b. How large a sample should be taken to be 95% confident that the sampling error is 3 or less?
  c. How large a sample should be taken to be 95% confident that the sampling error is 2 or less?

APPLICATIONS

SELF TEST

29. What sample size would have been recommended for the Scheer Industries study if the director of manufacturing had specified a .95 probability for a sampling error of 1.5 days or less? How large a sample would have been necessary if the precision statement had specified a .90 probability for a sampling error of 2 days or less?

30. In Example 8.1 the Statewide Insurance Company used a simple random sample of 36 policyholders to estimate the mean age of the population of policyholders. The resulting precision statement was reported to have a .95 probability that the value of the sample mean provided a sampling error of 2.35 years or less. This statement was based on a population standard deviation of 7.2 years.
  a. How large a simple random sample would have been necessary to reduce the sampling error to 2 years or less? To 1.5 years or less? To 1 year or less?
  b. Would you recommend that Statewide attempt to estimate the mean age of the policyholders with $E = 1$ year? Explain.

31. Starting annual salaries for college graduates are believed to have a standard deviation of approximately $2000. Assume that a 95% confidence interval estimate of the mean annual starting salary is desired. How large a sample size should be taken if the size of the maximum sampling error in the precision statement is
  a. $500 **b.** $200 **c.** $100

32. The mean number of days a house is on the market prior to selling was reported for 100 different cities (*U.S. News & World Report,* April 6, 1992). In a particular city the standard deviation of the number of days a house is on the market prior to selling is 20. How many house sales records would have to be collected to estimate the population mean with a maximum sampling error of 2 days? Use a 95% level of confidence.

33. Exercise 8, which provided the mean annual income for U.S. factory workers (*Barron's,* April 10, 1989), assumed the population standard deviation of annual income is $\sigma = \$5000$. If we wanted to estimate the mean annual income for the population of U.S. factory workers with a maximum sampling error of $500, what sample size should be used? Assume 95% confidence.

34. From Exercise 20, the sample standard deviation of P/E ratios for stocks is $s = 7.8$ (*The Wall Street Journal,* May 19, 1992). Assume that we are interested in estimating the population mean P/E ratio for all stocks. How many stocks should be included in the sample if we would like a .95 probability that the sampling error is 2 or less?

35. A gasoline service station shows a standard deviation of $6.25 for the charges made by the credit-card customers. Assume that the station's management would like to estimate the population mean gasoline bill for its credit-card customers to within $1.00. For a 95% confidence level, how large a sample would be necessary?

36. A national survey research firm has past data that indicate that the interview time for a consumer opinion study has a standard deviation of 6 minutes.
    a. How large a sample should be taken if the firm desires a .98 probability of estimating the mean interview time to within 2 minutes or less?
    b. Assume that the simple random sample you recommended in part (a) is taken and that the mean interview time for the sample is 32 minutes. What is the 98% confidence interval estimate for the mean interview time for the population of interviews?

8.4 ▽ HYPOTHESIS TESTS ABOUT A POPULATION MEAN

In Sections 8.1 and 8.2 we showed how a sample could be used to develop an interval estimate of a population mean. In the following sections we continue the discussion of statistical inference by showing how *hypothesis testing* can be used to determine whether or not a statement about the value of a population mean should be rejected.

In hypothesis testing we begin by making a tentative assumption about a population parameter. This tentative assumption is called the *null hypothesis* and is denoted by H_0. We then define another hypothesis, called the *alternative hypothesis,* which is the opposite of what is stated in the null hypothesis. This alternative hypothesis is denoted by H_a. The hypothesis-testing procedure involves using data from a sample to test the two competing statements indicated by H_0 and H_a.

The situation encountered in hypothesis testing is similar to the one encountered in a criminal trial. In a criminal trial, the assumption is that the defendant is innocent. Thus, the null hypothesis is one of innocence. The opposite of the null hypothesis is the alternative hypothesis—that the defendant is guilty. Thus, the hypotheses for a criminal trial would be written

$$H_0: \text{The defendant is innocent}$$

$$H_a: \textit{The defendant is guilty}$$

To test these competing statements, or hypotheses, a trial is held. The testimony and evidence obtained during the trial provide the sample information. If the sample information is not inconsistent with the assumption of innocence, the null hypothesis that the defendant is innocent cannot be rejected. However, if the sample information is inconsistent with the assumption of innocence, the null hypothesis will be rejected. In this case, action will be taken based on the alternative hypothesis that the defendant is guilty.

In the above illustration, we have shown that in hypothesis testing we develop two competing statements, or hypotheses, about a population. One statement is called the null hypothesis H_0, and the other statement is called the alternative hypothesis H_a. In some applications, it may not be obvious how the null and alternative hypotheses should be formulated. Care must be taken to be sure that the hypotheses are structured appropriately and that the hypothesis-testing conclusion provides the information that the researcher or decision maker desires.

Guidelines for establishing the null and alternative hypotheses about a population mean will be given for three types of situations that frequently employ hypothesis-testing procedures. A discussion of each of these situations follows.

TESTING RESEARCH HYPOTHESES

A particular model automobile currently has an average fuel consumption of 24 miles per gallon. A product-research group has developed a new carburetor specifically designed to increase the miles-per-gallon rating. To evaluate the new carburetors, several carburetors will be manufactured, installed in these automobiles, and subjected to research-controlled driving tests. Note that the product-research group is looking for evidence to conclude that the new design *increases* the mean miles-per-gallon rating. In this case, the research hypothesis is that the new carburetor will provide a mean miles-per-gallon rating exceeding 24; that is, $\mu > 24$. As a general guideline, a research hypothesis such as this should be formulated as the *alternative hypothesis*. Thus, the appropriate null and alternative hypotheses for the study are as follows:

$$H_0: \ \mu \leq 24$$

$$H_a: \ \mu > 24$$

If the sample results indicate that H_0 cannot be rejected, we will not be able to conclude that the new carburetor is better. Perhaps more research and subsequent testing should be conducted. However, if the sample results indicate H_0 can be rejected, the inference can be made that $H_a: \ \mu > 24$ is true. With this conclusion, the researcher has the statistical support necessary to conclude that the new carburetor increases the mean number of miles per gallon.

In research studies such as these, the research hypothesis should be expressed as the alternative hypothesis; in this way, rejection of H_0 will provide the researcher with the conclusion being sought. Thus, we see that action is taken only if the null hypothesis is rejected.

TESTING THE VALIDITY OF A CLAIM

As an illustration of testing the validity of claims that companies make about their products, consider the situation of a manufacturer of soft drinks who states that 2-liter containers of its products have an average of at least 67.6 fluid ounces. A sample of 2-liter containers will be selected, and the contents will be measured to test the manufacturer's claim. In this type of hypothesis-testing situation, we generally follow the rationale suggested by the criminal trial analogy. That is, the manufacturer's claim should be assumed true (innocent) unless the sample evidence proves otherwise (guilty). Using this approach for the soft-drink example, the null and alternative hypotheses would be stated as follows:

$$H_0: \ \mu \geq 67.6$$

$$H_a: \ \mu < 67.6$$

If the sample results indicate H_0 cannot be rejected, the manufacturer's claim cannot be challenged. Note that this does not mean that we have proven that $\mu \geq 67.6$, but simply that the results observed are not inconsistent with the manufacturer's claim. However, if the sample results indicate H_0 can be rejected, the inference will be made that H_a: $\mu < 67.6$ is true. With this conclusion, statistical evidence indicates that the manufacturer's claim is incorrect and that the soft-drink containers are being filled with a mean less than the claimed 67.6 ounces; in this case, appropriate action against the manufacturer is taken. Note that, as was true for research hypotheses, action is only taken when H_0 is rejected.

In any situation that involves testing the validity of a product claim, the null hypothesis is generally formulated based on the assumption that the claim is true. The alternative hypothesis is then formulated so that rejection of H_0 will provide the statistical evidence that the stated assumption is incorrect. Action to correct the claim should be taken whenever H_0 is rejected.

TESTING IN DECISION-MAKING SITUATIONS

Testing research hypotheses or testing the validity of a claim involves taking action only if H_0 is rejected. In many instances, however, action must be taken if H_0 cannot be rejected or H_0 can be rejected. In general, this type of situation occurs when a decision maker must choose between two courses of action, one associated with the null hypothesis and another associated with the alternative hypothesis. For example, on the basis of a sample of parts from a shipment that has just been received, a quality-control inspector must decide whether to accept the entire shipment or to return the shipment to the supplier because it does not meet specifications. Assume that specifications for a particular part indicate a mean length of 2 inches per part is required. If the average length of the parts is greater or less than the 2-inch standard, the parts will cause quality problems in the assembly operation. In this case, the null and alternative hypotheses would be formulated as follows:

$$H_0:\ \mu = 2$$

$$H_a:\ \mu \neq 2$$

If the sample results indicate H_0 cannot be rejected, the quality-control inspector will have no reason to doubt that the shipment meets specifications, and the shipment will be accepted. However, if the sample results indicate that H_0 should be rejected, the conclusion can be made that the parts do not meet specifications. In this case, the quality-control inspector has sufficient evidence to return the shipment to the supplier. Thus, we see that for decision making situations, action is taken when H_0 cannot be rejected and when H_0 can be rejected.

A SUMMARY OF FORMS FOR NULL AND ALTERNATIVE HYPOTHESES

Let μ_0 denote the specific numerical value of the population mean being considered in the null and alternative hypotheses. In general, a hypothesis test concerning the values of a population mean μ must take one of the following three forms:

$$H_0:\ \mu \geq \mu_0 \qquad H_0:\ \mu \leq \mu_0 \qquad H_0:\ \mu = \mu_0$$

$$H_a:\ \mu < \mu_0 \qquad H_a:\ \mu > \mu_0 \qquad H_a:\ \mu \neq \mu_0$$

In many situations, the choice of H_0 and H_a is not obvious; in such cases, judgment on the part of the user is needed to select the proper form of H_0 and H_a. However, as the above forms show, the equality part of the expression (either $\geq$, $\leq$, or $=$) *always* appears in the null hypothesis. In selecting the proper form of H_0 and H_a, keep in mind that the alternative hypothesis is what the sampling study is attempting to establish. Thus, asking whether the user is looking for evidence to support $\mu < \mu_0$, $\mu > \mu_0$, or $\mu \neq \mu_0$ will help determine H_a. The following exercises are designed to provide practice in choosing the proper form for a hypothesis test.

EXERCISES

APPLICATIONS

37. The manager of the Danvers-Hilton Resort Hotel has stated that the mean guest bill for a weekend is $400 or less. A member of the hotel's accounting staff has noticed that the total charges for guest bills have been increasing in recent months. The accountant will use a sample of weekend guest bills to test the manager's claim.

   a. Which form of the hypotheses should be used to test the manager's claim? Explain.

$$H_0:\ \mu \geq 400 \qquad H_0:\ \mu \leq 400 \qquad H_0:\ \mu = 400$$
$$H_a:\ \mu < 400 \qquad H_a:\ \mu > 400 \qquad H_a:\ \mu \neq 400$$

   b. What conclusion is appropriate when H_0 cannot be rejected?
   c. What conclusion is appropriate when H_0 can be rejected?

SELF TEST

38. The manager of an automobile dealership is considering a new bonus plan that is designed to increase sales volume. Currently, the mean sales volume is 14 automobiles per month. The manager would like to conduct a research study to see if there is evidence that the new bonus plan increases sales volume. To collect data on the plan, a sample of sales personnel will be allowed to sell under the new bonus plan for a 1-month period.

   a. Develop the null and alternative hypotheses that are most appropriate for this research situation.
   b. Comment on the conclusion when H_0 cannot be rejected.
   c. Comment on the conclusion when H_0 can be rejected.

39. A production-line operation is designed to fill cartons of laundry detergent with a mean weight of 32 ounces. A sample of cartons is periodically selected and weighed to determine if underfilling or overfilling exists. If the sample data generate a conclusion of underfilling or overfilling, the production line will be shut down and adjusted to obtain proper filling.

   a. Formulate the null and alternative hypotheses that will help in deciding whether or not to shut down and adjust the production line.
   b. Comment on the conclusion and the decision when H_0 cannot be rejected.
   c. Comment on the conclusion and the decision when H_0 can be rejected.

40. Because of high production-changeover time and costs, a director of manufacturing must convince management that a proposed manufacturing method reduces costs before the new method can be implemented. The current production method operates with a mean cost of $220 per hour. A research study will be conducted with the cost of the new method measured over a sample production period.

   a. Develop the null and alternative hypotheses that are most appropriate for this study.
   b. Comment on the conclusion when H_0 cannot be rejected.
   c. Comment on the conclusion when H_0 can be rejected.

8.5 TYPE I AND TYPE II ERRORS

The null and alternative hypotheses are competing statements about the state of nature. Either the null hypothesis H_0 is true or the alternative hypothesis H_a is true, but not both. Ideally, the hypothesis-testing procedure should lead to the acceptance of H_0 when H_0 is the state of nature and the rejection of H_0 when H_a is the state of nature. Unfortunately, these results are not always possible. Since hypothesis tests are based on sample information, we must allow for the possibility of errors. Table 8.5 illustrates the two kinds of errors that can be made in hypothesis testing.

The first row of Table 8.5 shows what can happen if we make the conclusion to accept H_0. If H_0 is the state of nature, this conclusion is correct. However, if H_a is the state of nature, we have made a *Type II error;* that is, we have accepted H_0 when it is false.

The second row of Table 8.5 shows what can happen if we make the conclusion to reject H_0. If the state of nature is H_0, we have made a *Type I error;* that is, we rejected H_0 when it is true. However, if H_a is the state of nature, then rejecting H_0 is correct.

Although we cannot eliminate the possibility of errors in hypothesis testing, we can consider the probability of their occurrence. Using common statistical notation, we denote the probabilities of making the two errors as follows:

$$\alpha = \text{the probability of making a Type I error}$$

$$\beta = \text{the probability of making a Type II error}$$

For example, recall the hypothesis-testing illustration discussed in Section 8.4 in which an automobile product-research group had developed a new carburetor designed to increase the miles-per-gallon rating of a particular automobile. With the current model obtaining an average of 24 miles per gallon, the hypothesis test was formulated as follows:

$$H_0: \ \mu \leq 24$$

$$H_a: \ \mu > 24$$

The alternative hypothesis, $H_a: \ \mu > 24$, indicates that the researchers are looking for sample evidence that will permit the rejection of H_0, thus supporting H_a and the conclusion that the mean miles per gallon is greater than 24.

TABLE 8.5   **TYPES OF ERRORS IN HYPOTHESIS TESTING**

		State of Nature	
		H_0 *True*	H_a *True*
Conclusion	*Accept H_0*	Correct Conclusion	Type II Error
	Reject H_0	Type I Error	Correct Conclusion

In this application, the Type I error of rejecting H_0 when it is true corresponds to the researchers claiming that the new carburetor improves the miles per gallon rating ($\mu > 24$) when in fact the new carburetor is not any better than the current carburetor. On the other hand, the Type II error of accepting H_0 when it is false corresponds to the researchers accepting the position that the new carburetor is not any better than the current carburetor ($\mu \leq 24$) when in fact the new carburetor provides an improved miles-per-gallon performance.

In practice, the person conducting the hypothesis test specifies the maximum allowable probability of making a Type I error, called the *level of significance* for the test. Common choices for the level of significance are .05 and .01. Referring to the second row of Table 8.5, note that the conclusion to *reject H_0* indicates that either a Type I error or a correct conclusion has been made. Thus, if the probability of making a Type I error is controlled for by selecting a low value for the level of significance, we have a high degree of confidence that the conclusion to reject H_0 is correct. In such cases, we have statistical support to conclude that H_0 is false and H_a is true. Any action suggested by the alternative hypothesis H_a is appropriate.

Although most applications of hypothesis testing control the probability of making a Type I error, they do not always control for the probability of making a Type II error. Thus, whenever we decide to accept H_0, we cannot determine how confident we can be with the decision to accept H_0. Because of the uncertainty associated with making a Type II error, statisticians often recommend that we use the statement *do not reject H_0* instead of *accept H_0*. Using the statement *do not reject H_0* carries the recommendation to withhold both judgment and action. In effect, by never directly accepting H_0, the statistician avoids the risk of making a Type II error. Whenever the probability of making a Type II error has not been determined and controlled, we will not make the conclusion to accept H_0. In such cases, only two conclusions are possible: *do not reject H_0* or *reject H_0*.

Just because it is not common to control for the Type II error in hypothesis testing, this does not mean that it is not possible to do so. In fact, in Sections 8.9 and 8.10, we will illustrate procedures for determining and controlling for the probability of making a Type II error. If proper controls have been established for this error, it can be appropriate to take action based on the decision to accept H_0.

NOTES AND COMMENTS

Many applications of hypothesis testing have a decision-making goal. The *reject H_0* provides the statistical support to conclude that H_a is true and take whatever action is appropriate. The statement *do not reject H_0*, although inconclusive, often forces management to behave as if H_0 is true. In this case, management needs to be aware of the fact that such behavior may result in a Type II error.

EXERCISES

APPLICATIONS

SELF TEST

41. The average American buys 6.08 books per year (*Louis Rukeyser's Business Almanac,* 1988). A University of Iowa researcher believes that adults in Des Moines purchase books at an annual rate higher than the national average. The following null and alternative hypotheses have been formulated by the researcher:

$$H_0: \ \mu \leq 6.08$$

$$H_a: \ \mu > 6.08$$

    a. What is the Type I error in this situation? What are the consequences of making this error?

    b. What is the Type II error in this situation? What are the consequences of making this error?

42. The label on a 3-quart container of orange juice claims that the orange juice contains an average of 1 gram of fat or less. Answer the following questions for a hypothesis test that could be used to test this claim:

    a. Write the appropriate null and alternative hypotheses.

    b. What is the Type I error in this situation? What are the consequences of making this error?

    c. What is the Type II error in this situation? What are the consequences of making this error?

43. Carpetland salespersons have been selling an average of $8000 of carpeting per week. Steve Contois, the firm's vice president, has proposed a compensation plan with new selling incentives. Steve hopes that the results of a trial selling period will enable him to conclude that the compensation plan increases the average sales per salesperson.

    a. Write the appropriate null and alternative hypotheses.

    b. What is the Type I error in this situation? What are the consequences of making this error?

    c. What is the Type II error in this situation? What are the consequences of making this error?

44. Suppose that a new production method will be implemented if a hypothesis test supports the conclusion that the new method reduces the mean operating cost per hour.

    a. State the appropriate null and alternative hypotheses if the mean cost for the current production method is $220 per hour.

    b. What is the Type I error in this situation? What are the consequences of making this error?

    c. What is the Type II error in this situation? What are the consequences of making this error?

8.6 ▽ ONE-TAILED HYPOTHESIS TESTS ABOUT A POPULATION MEAN: LARGE-SAMPLE CASE

EXAMPLE 8.4 ▷ The Federal Trade Commission (FTC) periodically conducts studies designed to test the claims manufacturers make about their products. For example, the label on a large can of Hilltop Coffee states that the can contains at least 3 pounds of coffee. Suppose that we wish to test this claim using hypothesis testing.

    The first step is to develop the null and the alternative hypotheses. We begin by tentatively assuming that the manufacturer's claim is correct. Note that if the population of coffee cans has a mean weight of 3 or more pounds per can, Hilltop's claim about its product is correct. However, if the population of coffee cans has a mean weight less than 3 pounds per can, Hilltop's claim is invalid.

    With μ denoting the mean weight for the population, the null and the alternative hypotheses are formulated as follows:

$$H_0: \mu \geq 3$$

$$H_a: \mu < 3$$

If the sample data indicate that H_0 cannot be rejected, the statistical evidence does not support the conclusion that a label violation has occurred. Thus, no action would be taken

FIGURE 8.9   **SAMPLING DISTRIBUTION OF $\bar{x}$ FOR THE HILLTOP COFFEE STUDY WHEN THE NULL HYPOTHESIS IS TRUE ($\mu = 3$)**

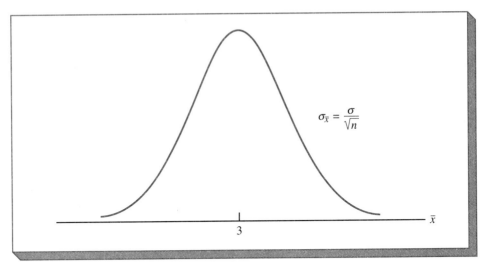

against Hilltop. However, if sample data indicate that H_0 can be rejected, we will conclude that the alternative hypothesis, H_a: $\mu < 3$, is true. In this case, an FTC claim of underfilling and a charge of a label violation would be appropriate.

Suppose that a random sample of 36 cans of coffee is selected. Note that if the mean filling weight for the sample of 36 cans is less than 3 pounds, the sample results will begin to cast doubt on the null hypothesis H_0: $\mu \geq 3$. But how much less than 3 must $\bar{x}$ be before we would be willing to risk making a Type I error and falsely accuse the company of a label violation?

To answer this question, let us tentatively assume that the null hypothesis is true with $\mu = 3$. From our study of sampling distributions in Chapter 7, we know that whenever the sample size is large ($n \geq 30$), the sampling distribution of $\bar{x}$ can be approximated by a normal probability distribution. Figure 8.9 shows the sampling distribution of $\bar{x}$ when the null hypothesis is true at $\mu = 3$.

The value of $z = (\bar{x} - 3)/\sigma_{\bar{x}}$ gives the number of standard deviations $\bar{x}$ is from $\mu = 3$. For hypothesis tests about a population mean, z will be the *test statistic* used to determine whether $\bar{x}$ deviates enough from $\mu = 3$ to justify rejecting the null hypothesis. Note that a value of $z = -1$ means that $\bar{x}$ is one standard deviation below $\mu = 3$, a value of $z = -2$ means that $\bar{x}$ is two standard deviations below $\mu = 3$, and so on. Obtaining a value of $z < -3$ is very unlikely if the null hypothesis is true. The key question is, How small must the test statistic z be before we have enough evidence to reject the null hypothesis?

Figure 8.10 shows that the probability of observing a value of $\bar{x}$ more than 1.645 standard deviations below the mean of $\mu = 3$ is .05. Thus, if we were to reject the null hypothesis whenever the value of the test statistic $z = (\bar{x} - 3)/\sigma_{\bar{x}}$ is less than -1.645, the probability of making a Type I error would be .05. If the FTC considered .05 to be an acceptable level for the probability of making a Type I error, then we would reject the null hypothesis whenever the test statistic indicated that the sample mean was more than 1.645 standard deviations below $\mu = 3$. Thus, we would reject H_0 if $z < -1.645$.

The methodology of hypothesis testing requires that we specify the maximum allowable probability of a Type I error. As noted in the previous section, this maximum probability is called the level of significance for the test; it is denoted by α, and it

FIGURE 8.10   **THE PROBABILITY $\bar{x}$ IS MORE THAN 1.645 STANDARD DEVIATIONS BELOW THE MEAN OF $\mu = 3$**

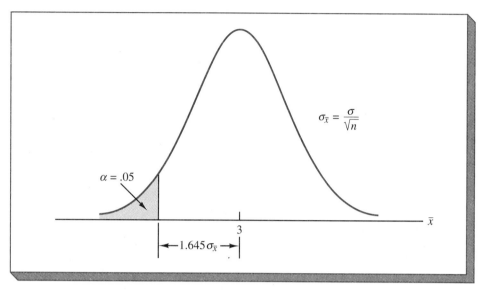

represents the probability of making a Type I error when the null hypothesis is true as an equality. Specifying the level of significance is a job for the manager. If the cost of making a Type I error is high, then a small value should be chosen for the level of significance. If the cost is not too great, a larger value may be appropriate.

In the Hilltop Coffee study, the director of the weight-testing program has made the following statement: "If the company is meeting its weight specifications exactly ($\mu = 3$), I would like a 99% chance of not taking any action against the company. While I do not want to accuse the company wrongly of underfilling its product, I am willing to live with a 1% chance of making this error."

From the director's statement, the maximum probability of a Type I error is .01. Thus, the level of significance for the hypothesis test is $\alpha = .01$. Figure 8.11 shows both the sampling distributions of $\bar{x}$ and $z = (\bar{x} - \mu)/\sigma_{\bar{x}}$ for the Hilltop Coffee study. Note that when the null hypothesis is true at $\mu = 3$, the probability is .01 that $\bar{x}$ is more than 2.33 standard deviations below the mean of 3. Therefore, we establish the following rejection rule:

$$\text{Reject } H_0 \text{ if } z = \frac{\bar{x} - \mu}{\sigma_{\bar{x}}} < -2.33$$

If the value of $\bar{x}$ is such that the test statistic z is in the rejection region, then we reject H_0 and conclude that H_a is true. On the other hand, if the value of $\bar{x}$ is such that the test statistic z is not in the rejection region, then we cannot reject H_0. Note that the rejection region shown in Figure 8.11 is in only one tail of the sampling distribution. Whenever this occurs, we say the test is a *one-tailed* hypothesis test.

Suppose that a sample of 36 cans provides a mean of $\bar{x} = 2.92$ pounds and that it is known from previous studies that the population standard deviation is $\sigma = .18$. With $\sigma_{\bar{x}} = \sigma/\sqrt{n}$, the value of the test statistic is given by

$$z = \frac{\bar{x} - 3}{\sigma/\sqrt{n}} = \frac{2.92 - 3}{.18/\sqrt{36}} = -2.67$$

FIGURE 8.11   **HILLTOP COFFEE REJECTION RULE HAS A LEVEL OF SIGNIFICANCE OF $\alpha = .01$**

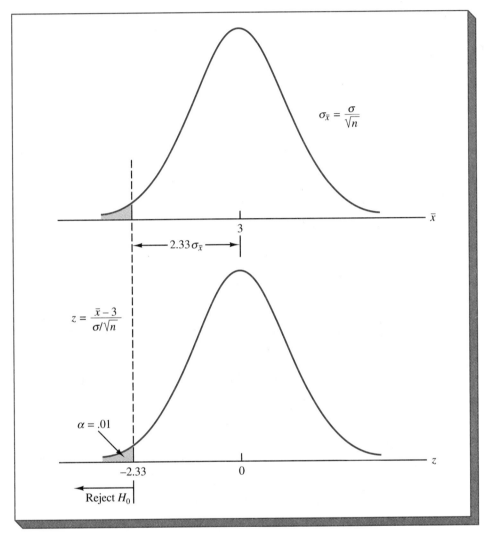

Figure 8.12 shows that this value of the test statistic is in the rejection region; thus, we can conclude that $\mu < 3$ at a .01 level of significance. Consequently, the director should take action against Hilltop Coffee for underfilling its product.

Suppose, instead, that the sample of 36 cans had provided a sample mean of $\bar{x} = 2.97$. In this case, the value of the test statistic would be:

$$z = \frac{\bar{x} - 3}{\sigma/\sqrt{n}} = \frac{2.97 - 3}{.18/\sqrt{36}} = -1.00$$

Since $z = -1.00$ is greater than -2.33, the value of the test statistic is not in the rejection region (see Figure 8.13). Since we cannot reject the null hypothesis, no action should be taken against Hilltop Coffee.

The value of z that establishes the boundary of the rejection region is called the *critical value*. In establishing the critical value, we tentatively assume the null hypothesis is true.

FIGURE 8.12   **VALUE OF THE TEST STATISTIC FOR X̄ = 2.92 IS IN THE REJECTION REGION**

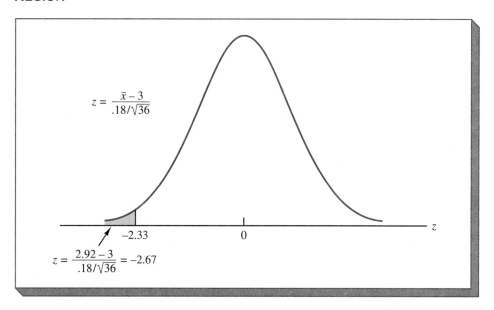

FIGURE 8.13   **VALUE OF THE TEST STATISTIC FOR X̄ = 2.97 IS NOT IN THE REJECTION REGION**

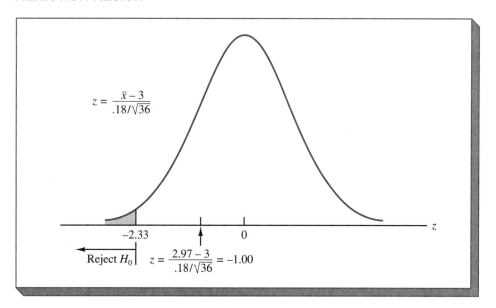

Note that for Hilltop Coffee, the null hypothesis is true whenever $\mu \geq 3$; however, we considered only the case when $\mu = 3$. What about the case when $\mu > 3$? If $\mu > 3$, the probability of making a Type I error will be less than it is when $\mu = 3$; that is, in this case, it is even less likely that we will find a value of the test statistic that is in the rejection region. Since the objective of the hypothesis-testing procedure is to limit the maximum probability of making a Type I error, the critical value for the test is established assuming that $\mu = 3$.   ◀

SUMMARY: ONE-TAILED TESTS ABOUT A POPULATION MEAN

Let us generalize the hypothesis-testing procedure for one-tailed tests about a population mean. We consider here only the large-sample case ($n \geq 30$) where the central limit theorem permits us to assume a normal sampling distribution for $\bar{x}$. In cases where σ is unknown, we substitute the sample standard deviation s for σ in computing the test statistic. The general form of a lower-tailed test, where μ_0 is a stated value for the population mean, is shown below.

Large-Sample ($n \geq 30$) Hypothesis Test About a Population Mean for a One-Tailed Test of the Form

$$H_0: \ \mu \geq \mu_0$$

$$H_a: \ \mu < \mu_0$$

Test Statistic:

$$z = \frac{\bar{x} - \mu_0}{\sigma/\sqrt{n}}$$

If σ is unknown, substitute s for σ in computing z.

Rejection Rule at a Level of Significance of α

Reject H_0 if $z < -z_\alpha$ **(8.6)**

A second form of the one-tailed test rejects the null hypothesis when the test statistic is in the upper tail of the sampling distribution. This one-tailed test and rejection rule are summarized below (see Figure 8.14). Again, we are considering the large-sample case; when σ is unknown, we substitute s for σ in computing z.

Large-Sample ($n \geq 30$) Hypothesis Test About a Population Mean for a One-Tailed Test of the Form

$$H_0: \ \mu \leq \mu_0$$

$$H_a: \ \mu > \mu_0$$

Test Statistic:

$$z = \frac{\bar{x} - \mu_0}{\sigma/\sqrt{n}}$$

If σ is unknown, substitute s for σ in computing z.

Rejection Rule at a Level of Significance of α

Reject H_0 if $z > z_\alpha$ **(8.7)**

FIGURE 8.14   REJECTION REGION FOR AN UPPER-TAILED HYPOTHESIS TEST ABOUT A POPULATION MEAN

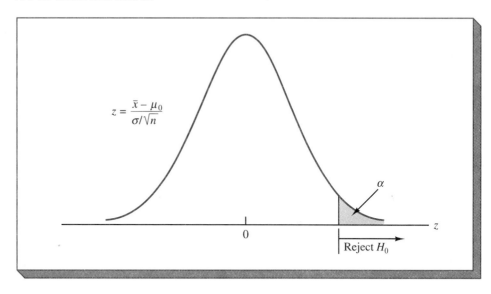

$$z = \frac{\bar{x} - \mu_0}{\sigma/\sqrt{n}}$$

α

0

Reject H_0

z

THE USE OF P-VALUES

Another approach used to make the decision whether or not to reject H_0 is based on what is called a *p-value*. We will show how the *p*-value for a sample can be computed from the value of the test statistic *z*.

Assuming that the null hypothesis is true, the *p*-value is the probability of obtaining a sample result that is at least as unlikely as what is observed. In the Hilltop Coffee study, the rejection region is in the lower tail; therefore the *p*-value is the probability of observing a sample mean less than or equal to what is observed.

EXAMPLE 8.4
(CONTINUED)

Let us compute the *p*-value associated with the sample mean $\bar{x} = 2.92$ in the Hilltop Coffee study. The *p*-value in this case is the probability of obtaining a value for the sample mean that is less than or equal to the observed value of $\bar{x} = 2.92$, given the hypothesized value for the population mean of $\mu = 3$. Previously we showed that the test statistic $z = -2.67$ corresponded to $\bar{x} = 2.92$. Thus, as shown in Figure 8.15, the *p*-value is the area in the tail of the standard normal probability distribution for $z = -2.67$. Using the standard normal probability distribution table, we find that the area between the mean and $z = -2.67$ is .4962. Thus, there is a $.5000 - .4962 = .0038$ probability of obtaining a sample mean that is less than or equal to the observed $\bar{x} = 2.92$. The *p*-value is therefore .0038. This *p*-value shows us that there is a very small probability of obtaining a sample mean of $\bar{x} = 2.92$ or less when sampling from a population with $\mu = 3$. ◀

The *p*-value can be used to make the decision for a hypothesis test by noting that if the *p-value is less than the level of significance* α, the value of the test statistic must be in the *rejection region*. Similarly, if the *p*-value is *greater than or equal to* α, the value of the test statistic is not in the rejection region. For the Hilltop Coffee study, the fact that the *p*-value of .0038 is less than the level of significance, $\alpha = .01$, indicates that the null hypothesis should be rejected. Given the stated level of significance α for any hypothesis

FIGURE 8.15   *p*-VALUE FOR THE HILLTOP COFFEE STUDY WHEN $\bar{x}$ = 2.92
AND Z = −2.67

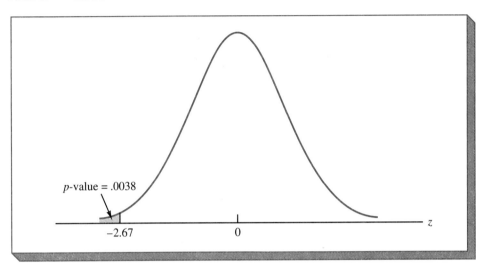

test, the decision of whether or not to reject H_0 can be made in terms of the *p*-value as follows.

---

p-Value Criterion for Hypothesis Testing

Reject H_0 if *p*-value $< \alpha$

The *p*-value and the corresponding test statistic will always provide the same hypothesis-testing conclusion at a given level of significance α. When the rejection region is in the lower tail of the sampling distribution, the *p*-value is in the lower tail; that is, the area under the curve less than or equal to the test statistic. When the rejection region is in the upper tail of the sampling distribution, the *p*-value is in the upper tail; that is, the area under the curve greater than or equal to the test statistic. A small *p*-value thus indicates a sample result that is unusual given the assumption that H_0 is true; therefore, *p*-values lead to rejection of H_0. On the other hand, a relatively large *p*-value indicates that the null hypothesis cannot be rejected.

THE STEPS OF HYPOTHESIS TESTING

In conducting the Hilltop Coffee hypothesis test, we carried out the steps that are required for any hypothesis-testing procedure. A summary of the steps that can be applied to any hypothesis test are as follows.

Steps of Hypothesis Testing

1. Determine the null and alternative hypotheses that are appropriate for the application.

—Continued on the next page

—Continued from previous page

2. Select the test statistic that will be used to decide whether or not to reject the null hypothesis.
3. Specify the level of significance α for the test.
4. Use the level of significance to develop the rejection rule that indicates the values of the test statistic that will lead to the rejection of H_0.
5. Collect the sample data, and compute the value of the test statistic.
6. Compare the value of the test statistic to the critical value(s) specified in the rejection rule to determine whether or not H_0 should be rejected.
7. If desired, compute the *p*-value for the test; reject H_0 if the *p*-value $< \alpha$.

**EXAMPLE 8.4
(CONTINUED)** ▶

For Hilltop Coffee, the null and alternative hypotheses (step 1) were as follows:

$$H_0: \mu \geq 3$$

$$H_a: \mu < 3$$

With the large sample ($n \geq 30$) and the population standard deviation given as $\sigma = .18$, the test statistic (step 2) was

$$z = \frac{\bar{x} - \mu}{\sigma/\sqrt{n}}$$

The level of significance (step 3) was given as $\alpha = .01$. Using $\alpha = .01$, the corresponding rejection rule for the test statistic (step 4) was reject H_0 if $z < -2.33$.

With a simple random sample of $n = 36$ cans of coffee and a sample mean of $\bar{x} = 2.92$ pounds, the value of the test statistic (step 5) was computed to be

$$z = \frac{2.92 - 3.00}{.18/\sqrt{36}} = -2.67$$

A comparison of $z = -2.67$ to the critical value specified in the rejection rule (step 6) showed that H_0 should be rejected. Finally, the *p*-value (step 7) associated with $z = -2.67$ was shown to be .0038. Since $.0038 < \alpha$, H_0 should be rejected. ◀

**NOTES AND
COMMENTS** ▽

The *p*-value is often called the *observed level of significance* for the test. It is a measure of how unlikely the sample results are assuming the null hypothesis is true. The smaller the *p*-value, the less likely the sample results. Most statistical software packages print the *p*-value associated with a hypothesis test.

▷

EXERCISES

METHODS

45. Consider the following hypothesis test:

$$H_0: \mu \geq 10$$

$$H_a: \mu < 10$$

A sample of 50 provides a sample mean of 9.46 and sample standard deviation of 2.
a.   Using $\alpha = .05$, what is the critical value for z? What is the rejection rule?
b.   Compute the value of the test statistic z. What is your conclusion?

SELF TEST ▷   **46.**   Consider the following hypothesis test:

$$H_0: \mu \leq 15$$

$$H_a: \mu > 15$$

A sample of 40 provides a sample mean of 16.5 and sample standard deviation of 7.
a.   Using $\alpha = .02$, what is the critical value for z, and what is the rejection rule?
b.   Compute the value of the test statistic z.
c.   What is the p-value?
d.   What is your conclusion?

47.   Consider the following hypothesis test:

$$H_0: \mu \geq 25$$

$$H_a: \mu < 25$$

A sample of 100 is used and the population standard deviation is 12. Use $\alpha = .05$. Provide the value of the test statistic z and your conclusion for each of the following sample results:
a. $\bar{x} = 22.0$ b. $\bar{x} = 24.0$ c. $\bar{x} = 23.5$ d. $\bar{x} = 22.8$

48.   Consider the following hypothesis test:

$$H_0: \mu \leq 5$$

$$H_a: \mu > 5$$

Assume the test statistics are as shown below. Compute the corresponding p-values and make the appropriate conclusions based on $\alpha = .05$.
a. $z = 1.82$ b. $z = .45$ c. $z = 1.50$
d. $z = 3.30$ e. $z = -1.00$

APPLICATIONS

SELF TEST ▷   **49.**   According to *Business Week* (February 6, 1989), the mean cost of a heart-bypass operation is $26,100, and approximately 230,000 operations are performed annually. A sample of 36 bypass operations in a particular city showed a mean cost of $\bar{x} = \$25,000$ and $s = \$2400$.
a.   Develop appropriate hypotheses to test whether or not the mean bypass operation cost is less than $26,100 in this city.
b.   Use $\alpha = .05$. What is your conclusion?
c.   What is the p-value for this test?

50.   In 1990, *The Motor Vehicle Manufacturers Association* of Detroit, Michigan, reported statistics on the average number of years passenger cars were being used. In 1980, the population mean was reported to be 6.5 years. Assume that the 1990 data were obtained from a sample of 100 passenger cars and showed a sample mean of 7.8 years and a sample standard deviation of 2.2 years.
a.   Formulate the null and alternative hypotheses if the researcher is looking for evidence to show that individuals are driving cars longer in 1990.
b.   Do the data support the conclusion that individuals are driving cars longer? Use a .01 level of significance.
c.   What implications does this have for vehicle manufacturers?

51.   The average annual income in the United States is $19,780 (*St. Petersburg Times*, December 1989). A sample of 150 individuals in Japan resulted in a sample mean income of $21,040 in U.S. dollars. Assuming a sample standard deviation of $6000, do these data support the conclusion that the mean annual income in Japan is greater than the mean annual income in the United States? Use a .05 level of significance. What is the p-value? What is your conclusion?

52. Fightmaster and Associates Real Estate, Inc., advertises that the mean selling time of a residential home is 40 days or less after it is listed with the company. A sample of 50 recently sold residential homes shows a sample mean selling time of 45 days and a sample standard deviation of 20 days. Using a .02 level of significance, test the validity of the company's claim.

53. Fowle Marketing Research, Inc., bases charges to a client on the assumption that telephone surveys can be completed with a mean time of 15 minutes or less. If the mean time required to conduct the survey exceeds 15 minutes, a premium rate is charged the client. Suppose that a sample of 35 surveys shows a sample mean of 17 minutes and a sample standard deviation of 4 minutes. Is the premium rate justified? Test at the $\alpha = .01$ level of significance.

54. New tires manufactured by a company in Findlay, Ohio, are designed to provide a mean of at least 28,000 miles. Tests with 30 tires show a sample mean of 27,500 miles with a sample standard deviation of 1000 miles. Using a .05 level of significance, test whether or not there is sufficient evidence to reject the claim of a mean of at least 28,000 miles. What is the p-value?

55. A company currently pays its production employees a mean wage of $15.00 per hour. The company is planning to build a new factory, and several locations are being considered. The availability of labor at a rate less than $15.00 per hour is a major factor in the location decision. For one location, a sample of 40 workers showed a current mean hourly wage of $\bar{x} = \$14.00$ and a sample standard deviation of $s = \$2.40$.
   a. Using a .10 level of significance, do the sample data indicate that the location has a mean wage rate significantly below the $15.00 per hour rate?
   b. What is the p-value?

56. A new diet program claims that participants will lose on average at least 8 pounds during the first week of the program. A random sample of 40 people participating in the program showed a sample mean weight loss of 7 pounds. The sample standard deviation was 3.2 pounds.
   a. What is the rejection rule with $\alpha = .05$?
   b. What is your conclusion about the claim made by the diet program?
   c. What is the p-value?

8.7 TWO-TAILED HYPOTHESIS TESTS ABOUT A POPULATION MEAN: LARGE-SAMPLE CASE

Two-tailed hypothesis tests differ from one-tailed tests in that a rejection region is placed in both the lower and the upper tails of the sampling distribution. Let us use the following example to show how and why two-tailed tests are conducted.

EXAMPLE 8.5 ▶ The United States Golf Association (USGA) has established rules that manufacturers of golf equipment must meet for their products to be acceptable for use in USGA events. One of the rules regarding the manufacture of golf balls states that "A brand of golf ball, when tested on apparatus approved by the USGA on the outdoor range at the USGA Headquarters ... shall not cover an average distance in carry and roll exceeding 280 yards. ..." Suppose that Superflight, Inc., has recently developed a high-technology manufacturing method that can produce golf balls with an average distance in carry and roll of 280 yards.

Superflight realizes, however, that if the new manufacturing process goes out of adjustment, the process may produce balls with an average distance of less than 280 yards or with an average distance of greater than 280 yards. In the former case, Superflight may experience a downturn in sales as a result of marketing an inferior product, and in the latter case, Superflight may have its golf balls rejected by the USGA. As a result, management at Superflight has instituted a quality-control program to carefully monitor the new manufacturing process.

As part of the quality-control program, an inspector periodically selects a sample of golf balls from the production line and subjects them to tests that are equivalent to those performed by the USGA. Assuming that the manufacturing process is functioning correctly, we establish the following null and alternative hypotheses:

$$H_0: \; \mu = 280$$

$$H_a: \; \mu \neq 280$$

As usual, we make the tentative assumption that the null hypothesis is true. A rejection region must be established for the test statistic z. We want to reject the claim that $\mu = 280$ when the z value indicates that the sample mean $\bar{x}$ is significantly less than 280 yards or when the z value indicates that the sample mean is significantly greater than 280 yards. Thus, H_0 should be rejected for values of the test statistic in either the lower tail or the upper tail of the sampling distribution. As a result, the test will be referred to as a *two-tailed* hypothesis test.

Following the hypothesis-testing procedure developed in the previous sections, we first specify a level of significance by determining a maximum allowable probability of making a Type I error. Suppose we choose $\alpha = .05$ as the level of significance. This means that there will be a .05 probability of concluding that the mean distance is not 280 yards when in fact it is. The test statistic is

$$z = \frac{\bar{x} - \mu}{\sigma/\sqrt{n}}$$

Figure 8.16 shows the sampling distribution of z with the two-tailed rejection region for $\alpha = .05$. With two-tailed hypothesis tests, we will always determine the rejection region by placing an area or probability of $\alpha/2$ in each tail of the distribution. The values of z that provide an area of .025 in each tail can be found from the standard normal probability distribution table. We see in Figure 8.16 that $-z_{.025} = -1.96$ identifies an area of .025 in the lower tail and $z_{.025} = +1.96$ identifies an area of .025 in the upper tail. The corresponding rejection rule is as follows.

Reject H_0 if $z < -1.96$ or if $z > 1.96$

Suppose that a sample of 36 golf balls provides a sample mean distance of $\bar{x} = 278.5$ yards and a sample standard deviation of $s = 12$ yards. Using the value of μ from the null hypothesis and the sample standard deviation of $s = 12$ as an estimate of the population standard deviation σ, the value of the test statistic is

$$z = \frac{\bar{x} - \mu}{\sigma/\sqrt{n}} = \frac{278.5 - 280}{12/\sqrt{36}} = -.75$$

FIGURE 8.16   **REJECTION REGION (SHADED AREA) FOR THE TWO-TAILED HYPOTHESIS TEST FOR SUPERFLIGHT, INC.**

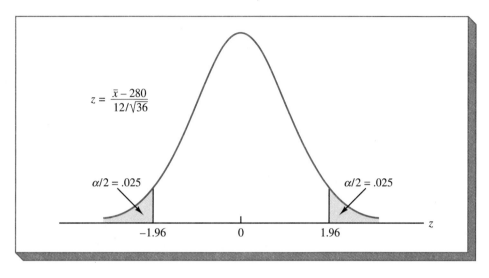

According to the rejection rule, H_0 cannot be rejected. The sample results indicate that the quality-control manager has no reason to doubt the assumption that the manufacturing process is producing golf balls with a mean distance of 280 yards.   ◀

SUMMARY: TWO-TAILED TESTS ABOUT A POPULATION MEAN

Let μ_0 represent the value of the mean as claimed in the null hypothesis. The general form of the two-tailed hypothesis test about a population mean is as follows.

Large-Sample ($n \geq 30$) Hypothesis Test About a Population Mean Two-Tailed Test of the Form

$$H_0\colon \mu = \mu_0$$

$$H_a\colon \mu \neq \mu_0$$

Test Statistic:

$$z = \frac{\bar{x} - \mu_0}{\sigma/\sqrt{n}}$$

If σ is unknown, substitute s for σ in computing z.

Rejection Rule at a Level of Significance of α

Reject H_0 if $z < -z_{\alpha/2}$ or if $z > z_{\alpha/2}$ **(8.8)**

p-VALUES FOR TWO-TAILED TESTS

Assuming that the null hypothesis is true, the p-value is the probability of obtaining a sample result that is at least as unlikely as what is observed. A small p-value indicates the sample result is unusual given the assumption that H_0 is true. Thus, as with the one-tailed hypothesis tests, a small p-value leads to the rejection of H_0.

EXAMPLE 8.5
(CONTINUED)

Let us compute the p-value for the Superflight golf ball study. The sample mean of $\bar{x} = 278.5$ has a corresponding z value of $-.75$. The table for the standard normal probability distribution shows that the area between the mean and $z = -.75$ is .2734. Thus, the area in the lower tail less than or equal to $\bar{x} = 278.5$ and $z = -.75$ is $.5000 - .2734 = .2266$. Looking at Figure 8.16, we see that the lower-tailed portion of the rejection region has an area or probability of $\alpha/2 = .05/2 = .025$. Thus, with $.2266 > .025$, the test statistic does not fall in the rejection region, and the null hypothesis cannot be rejected.

One question remains: What value should we report as the p-value for the two-tailed test? At first glance, you may be inclined to say the p-value is .2266. If this is your choice, you will have to remember two different rules: one for the one-tailed test, which is to reject H_0 if the p-value $< \alpha$, and another for the two-tailed test, which is to reject H_0 if the p-value $< \alpha/2$. Alternatively, suppose we define the p-value for a two-tailed test as *double* the area found in the tail of the distribution? Thus, for the Superflight study, we would define the p-value to be $2(.2266) = .4532$. The advantage of this definition of the p-value for a two-tailed test is that the p-value can be compared directly to the level of significance α. Thus, with $.4532 > .05$, we see that the null hypothesis cannot be rejected. By remembering that the p-value for a two-tailed test is simply double the area found in the tail of the distribution, the previous rule to reject H_0 if the p-value $< \alpha$ will be true for all hypothesis tests.

COMPUTER SOFTWARE AND HYPOTHESIS TESTING

Computer software packages are helpful in performing the computations required for a hypothesis test. The Minitab printout for the Superflight golf ball hypothesis test is shown in Figure 8.17. The data for the yards driven by each of the 36 golf balls has previously been entered into a Minitab worksheet. Since the sample size is large, the user selects the command ZTEST to indicate a z test statistic is to be used in the hypothesis test

FIGURE 8.17   MINITAB OUTPUT FOR THE SUPERFLIGHT GOLF BALL HYPOTHESIS TEST

```
MTB > ZTEST 280 12 C1

TEST OF MU = 280.000 VS MU N.E. 280.000
THE ASSUMED SIGMA = 12.0

            N       MEAN     STDEV    SE MEAN        Z     P VALUE
C1         36    278.500    12.000     2.000     -0.75       0.45
```

computation. The full command ZTEST 280 12 C1 indicates the test is to be conducted for the null hypothesis H_0: $\mu = 280$ with the population standard deviation assumed to be 12. The C1 indicates that the data for the 36 golf balls are stored in Column 1 of the Minitab worksheet.

The output in Figure 8.17 is easily interpreted by the user. The number of items in the data set, the sample mean, the sample standard sample deviation, and the standard error of the mean are provided. The z value of -0.75 and the associated p-value of 0.45 show that at the .05 level of significance, the null hypothesis $\mu = 280$ cannot be rejected.

THE RELATIONSHIP BETWEEN INTERVAL ESTIMATION AND HYPOTHESIS TESTING

In the case of interval estimation, the population mean μ was unknown. Once the sample was selected and the sample mean $\bar{x}$ computed, we developed an interval around the value of $\bar{x}$ that had a good chance of including the value of the parameter μ. The interval estimate computed was referred to as a confidence interval with $1 - \alpha$ defined as the confidence coefficient. In the large-sample case, the interval estimate of a population mean was given by

$$\bar{x} \pm z_{\alpha/2}\frac{\sigma}{\sqrt{n}} \tag{8.9}$$

Conducting a hypothesis test requires us first to make an assumption about the value of a population parameter. In the case of the population mean, the two-tailed hypothesis test has the form

$$H_0: \mu = \mu_0$$

$$H_a: \mu \neq \mu_0$$

where μ_0 is the hypothesized value for the population mean. Using the rejection rule provided by equation (8.8), we see that the region over which we do not reject H_0 includes all values of the sample mean $\bar{x}$ that are within $-z_{\alpha/2}$ and $+z_{\alpha/2}$ standard errors of μ_0. Thus, the following expression provides the do-not-reject region for the sample mean $\bar{x}$ in a two-tailed hypothesis test with a level of significance of α:

$$\mu_0 \pm z_{\alpha/2}\frac{\sigma}{\sqrt{n}} \tag{8.10}$$

A close look at equations (8.9) and (8.10) will provide insight into the relationship between the interval estimation and hypothesis-testing approaches to statistical inference. Note in particular that both procedures require the computation of the values $z_{\alpha/2}$ and $\sigma/\sqrt{n}$. Focusing on α, we see that a confidence coefficient of $(1 - \alpha)$ for interval estimation corresponds to a level of significance of α in hypothesis testing. For example, a 95% confidence level for interval estimation corresponds to a .05 level of significance for hypothesis testing. Furthermore, equations (8.9) and (8.10) show that since $z_{\alpha/2}(\sigma/\sqrt{n})$ is the plus or minus value for both expressions, if $\bar{x}$ falls in the do-not-reject region defined by equation (8.10), the hypothesized value μ_0 will be in the confidence interval defined by equation (8.9). Conversely, if the hypothesized value μ_0 falls in the confidence interval defined by equation (8.9), the sample mean $\bar{x}$ will be in the do-not-reject region for the hypothesis H_0: $\mu = \mu_0$. These observations lead to the following procedure for using confidence interval results to draw hypothesis-testing conclusions.

A Confidence Interval Approach to Testing a Hypothesis of the Form

$$H_0: \ \mu = \mu_0$$

$$H_a: \ \mu \neq \mu_0$$

1. Select a simple random sample from the population, and use the value of the sample mean $\bar{x}$ to develop the confidence interval for the population mean μ.

$$\bar{x} \pm z_{\alpha/2} \frac{\sigma}{\sqrt{n}}$$

2. If the confidence interval contains the hypothesized value μ_0, do not reject H_0. Otherwise, reject H_0.

EXAMPLE 8.5
(CONTINUED) ▶

The Superflight golf ball study resulted in the following two-tailed test:

$$H_0: \ \mu = 280$$

$$H_a: \ \mu \neq 280$$

To test this hypothesis with a level of significance of $\alpha = .05$, we sampled 36 golf balls and found a sample mean distance of $\bar{x} = 278.5$ yards and a sample standard deviation of $s = 12$ yards. Using these results with $z_{.025} = 1.96$, the 95% confidence interval estimate of the population mean distance becomes

$$\bar{x} \pm z_{.025} \frac{\sigma}{\sqrt{n}}$$

$$278.5 \pm 1.96 \frac{12}{\sqrt{36}}$$

$$278.5 \pm 3.92$$

or

$$274.58 \text{ to } 282.42$$

This finding enables the quality-control manager to conclude with 95% confidence that the mean distance for the population of golf balls is between 274.58 and 282.42 yards. Since the hypothesized value for the population mean, $\mu_0 = 280$, is in this interval, the hypothesis-testing conclusion is that the null hypothesis, $H_0: \mu = 280$, cannot be rejected. ◀

Note that this discussion and example have been devoted to two-tailed hypothesis tests about a population mean. However, the same confidence interval and hypothesis-testing relationship exists for other population parameters as well. In addition, the relationship can be extended to make one-tailed tests about population parameters. However, this requires the development of one-sided confidence intervals.

NOTES AND
COMMENTS

1. The *p*-value is called the observed level of significance; it depends only on the sample outcome. However, it is necessary to know whether the hypothesis test being investigated is one-tailed or two-tailed. Given the value of $\bar{x}$ in a sample, the *p*-value for a two-tailed test will always be *twice* the area in the tail of the sampling distribution at the value of $\bar{x}$. Computer software packages, such as Minitab, automatically take this into consideration when computing the *p*-value.

2. The interval-estimation approach to hypothesis testing helps to highlight the role of the sample size. From equation (8.9) it can be seen that larger sample sizes *n* lead to more narrow confidence intervals. Thus, for a given level of significance α, a larger sample is less likely to lead to an interval containing μ_0 when the null hypothesis is false. That is, the larger sample size will provide a higher probability of rejecting H_0 when H_0 is false.

EXERCISES

METHODS

57. Consider the following hypothesis test:

$$H_0: \mu = 10$$

$$H_a: \mu \neq 10$$

A sample of 36 provides a sample mean of 11 and sample standard deviation of 2.5.
a. Using $\alpha = .05$, what is the rejection rule?
b. Compute the value of the test statistic *z*. What is your conclusion?

SELF TEST

58. Consider the following hypothesis test:

$$H_0: \mu = 15$$

$$H_a: \mu \neq 15$$

A sample of 50 provides a sample mean of 14.2 and sample standard deviation of 5.
a. Using $\alpha = .02$, what is the rejection rule?
b. Compute the value of the test statistic *z*.
c. What is the *p*-value?
d. What is your conclusion?

59. Consider the following hypothesis test:

$$H_0: \mu = 25$$

$$H_a: \mu \neq 25$$

A sample of 80 is used, and the population standard deviation is 10. Use $\alpha = .05$. Compute the value of the test statistic *z* and specify your conclusion for each of the following sample results:
a. $\bar{x} = 22.0$ b. $\bar{x} = 27.0$ c. $\bar{x} = 23.5$ d. $\bar{x} = 28.0$

60. Consider the following hypothesis test:

$$H_0: \mu = 5$$

$$H_a: \mu \neq 5$$

Assume the test statistics are as shown below. Compute the corresponding p-values and specify your conclusions based on $\alpha = .05$.

a. $z = 1.80$ **b.** $z = -.45$ **c.** $z = 2.05$
d. $z = -3.50$ **e.** $z = -1.00$

SELF TEST ▷

APPLICATIONS

61. The U.S. Bureau of the Census reported mean hourly earnings in the wholesale trade industry of \$9.70 per hour. A sample of 49 wholesale trade workers in a particular city showed a sample mean hourly wage of $\bar{x} = \$9.30$ with a standard deviation of $s = \$1.05$. Use H_0: $\mu = 9.70$ and H_a: $\mu \neq 9.70$.

a. Test to see if wage rates in the city differ significantly from the reported \$9.70. Use $\alpha = .05$.

b. What is the p-value for this test?

62. A study of the operation of a city-owned parking garage shows a historical mean parking time of 220 minutes per car. The garage area has recently been remodeled, and the parking charges have been increased. The city manager would like to know whether these changes have had any effect on the mean parking time. Test the hypotheses H_0: $\mu = 220$ and H_a: $\mu \neq 220$ at a .05 level of significance.

a. What is your conclusion if a sample of 50 cars resulted in $\bar{x} = 208$ and $s = 80$?

b. What is the p-value?

63. A production line operates with a filling weight standard of 16 ounces per container. Overfilling or underfilling is a serious problem, and the production line should be shut down if either occurs. From past data, σ is known to be .8 ounce. A quality-control inspector samples 30 items every 2 hours and at that time makes the decision of whether or not to shut the line down for adjustment.

a. With a .05 level of significance, what is the rejection rule for the hypothesis-testing procedure?

b. If a sample mean of $\bar{x} = 16.32$ ounces occurs, what action would you recommend?

c. If a sample mean of $\bar{x} = 15.82$ ounces occurs, what action would you recommend?

d. What is the p-value for parts (b) and (c)?

64. An automobile assembly-line operation has a scheduled mean completion time of 2.2 minutes. Because of the effect of completion time on both earlier and later assembly operations, it is important to maintain the 2.2-minute standard. A random sample of 45 times shows a sample mean completion time of 2.39 minutes, with a sample standard deviation of .20 minute. Use a .02 level of significance and test whether or not the operation is meeting its 2.2-minute standard.

65. Historically, evening long-distance phone calls from a particular city have averaged 15.20 minutes per call. In a random sample of 35 calls, the sample mean time was 14.30 minutes per call, with a sample standard deviation of 5 minutes. At the .05 level of significance, test whether or not there has been a change in the mean duration of long distance phone calls. What is the p-value?

66. The mean salary for full professors at public universities is \$45,300 (from the *Statistical Abstract of the United States*, 1988). A sample of 36 full professors of chemistry showed $\bar{x} = \$52,000$ and $s = \$5000$. Suppose H_0: $\mu = 45,300$ and H_a: $\mu \neq 45,300$.

a. Develop a 95% confidence interval for the mean salary of full professors of chemistry using the sample data.

b. Use the confidence interval to conduct the hypothesis test. What is your conclusion?

67. At Western University, the historical mean scholastic examination score of entering students has been 900, with a standard deviation of 180. Each year, a sample of applications is taken to see whether the examination scores are at the same level as in previous years. The null hypothesis tested is H_0: $\mu = 900$. A sample of 200 students in this year's class shows a sample mean score of 935. Use a .05 level of significance.

    a. Use a confidence interval approach to test this hypothesis.
    b. Use a test statistic to test this hypothesis.
    c. What is the *p*-value for this test?

68. An industry pays an average wage rate of $9.00 per hour. A sample of 36 workers from one company showed a mean wage of $\bar{x}$ = $8.50 and a sample standard deviation of *s* = $.60.

    a. A one-sided confidence interval uses the sample results to establish either an upper limit or a lower limit for the value of the population parameter. For this exercise, establish an upper 95% confidence limit for the hourly wage rate paid by the company. The form of this one-sided confidence interval requires that we be 95% confident that the population mean is this value or less. What is the 95% confidence statement for this one-sided confidence interval?

    b. Use the one-sided confidence interval result to test the hypothesis H_0: $\mu \geq 9$. What is your conclusion? Explain.

8.8 ▽ HYPOTHESIS TESTS ABOUT A POPULATION MEAN: SMALL-SAMPLE CASE

The methods of hypothesis testing that we have discussed thus far require sample sizes of at least 30. The reason for this is that in the large-sample case ($n \geq 30$), the central limit theorem enables us to approximate the sampling distribution of $\bar{x}$ with a normal probability distribution. In this case, the test statistic

$$z = \frac{\bar{x} - \mu_0}{\sigma/\sqrt{n}} \tag{8.11}$$

is a standard normal random variable. The sample standard deviation *s* can be used in equation (8.11) when the population standard deviation σ is unknown.

    Assume that the sample size is small ($n < 30$) and that the sample standard deviation *s* is used to estimate the population standard deviation σ.* If the population has a normal probability distribution, the *t* distribution can be used to make inferences about the value of the population mean. In using the *t* distribution for hypothesis tests about a population mean, the test statistic is

$$t = \frac{\bar{x} - \mu_0}{s/\sqrt{n}} \tag{8.12}$$

This test statistic has a *t* distribution with $n - 1$ degrees of freedom. Note the similarities of the test statistics defined by equations (8.11) and (8.12). It should not be surprising that the small-sample procedure for hypothesis tests about a population mean is very similar to the large-sample procedures presented in Sections 8.6 and 8.7.

EXAMPLE 8.6 ▷ A company that produces products for home gardening has developed a new plant food designed to increase the growing height of plants. The new plant food was tested on a sample of 12 plants that are known to have a population mean growing height of 18 inches. Test results showed a sample mean height of 19.4 inches and a sample standard deviation of 3 inches. Assume that the heights of the plants are normally distributed.

---

*Suppose $n < 30$. If the population has a normal probability distribution and the population standard deviation σ is *known*, equation (8.11) can also be used as the test statistic for the small-sample hypothesis test.

At the a .10 level of significance, is there reason to believe that the new plant food increases plant height? The null and alternative hypotheses are as follows:

$$H_0: \ \mu \leq 18$$

$$H_a: \ \mu > 18$$

The new plant food will be judged to increase plant height if H_0 can be rejected.

The rejection region is located in the upper tail of the sampling distribution. With $n - 1 = 12 - 1 = 11$ degrees of freedom, Table 8.3 shows that $t_{.10} = 1.363$. Thus, the rejection rule is

$$\text{Reject } H_0 \text{ if } t > 1.363$$

Using equation (8.12) with $\bar{x} = 19.4$ and $s = 3$, we have the following value for the test statistic:

$$t = \frac{\bar{x} - \mu_0}{s/\sqrt{n}} = \frac{19.4 - 18}{3/\sqrt{12}} = 1.62$$

Since 1.62 is greater than 1.363, the null hypothesis is rejected. At the .10 level of significance, it can be concluded that the mean plant height exceeds 18 inches when the new plant food is used. Figure 8.18 shows that the value of the test statistic is in the rejection region. ◀

p-VALUES AND THE t DISTRIBUTION

As discussed in Sections 8.6 and 8.7, the p-value can be interpreted as the observed level of significance for the hypothesis test. The usual rule applies: If p-value $< \alpha$, the null hypothesis can be rejected. Unfortunately, the format of the t distribution table provided in most statistics textbooks does not have sufficient detail to determine the exact p-value

FIGURE 8.18   **VALUE OF THE TEST STATISTIC (t = 1.62) FOR THE PLANT FOOD HYPOTHESIS TEST**

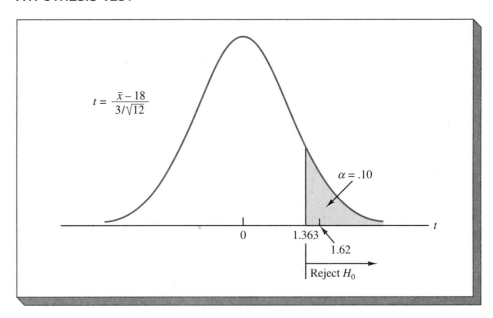

for the test. However, we can still use the t distribution table to identify a range for the p-value. For example, the t distribution used in the plant food hypothesis test has 11 degrees of freedom. Referring to Table 8.3, we see that row 11 provides the following information about a t distribution with 11 degrees of freedom.

Upper Tail Area	.10	.05	.025	.01	.005
t Value	1.363	1.796	2.201	2.718	3.106

The computed t value for the hypothesis test was $t = 1.62$; thus, the p-value is the area in the upper tail corresponding to $t = 1.62$. From the information above, we see 1.62 is between 1.363 and 1.796. Thus, although we cannot determine the exact p-value associated with $t = 1.62$, we do know that the p-value must be less than .10 but greater than .05. Since the p-value must be less than $\alpha = .10$, the null hypothesis is rejected.

An advantage of computer software packages is that the exact p-value can be computed for the t distribution test. The Minitab printout for the plant food hypothesis test is shown in Figure 8.19. The TTEST command is selected so that the t distribution will be used in the computations. The null hypothesis value of 18 is specified in the TTEST command. The Minitab subcommand of ALTERNATIVE = 1 indicates that the one-tailed test has the alternative hypothesis H_a: $\mu > 18$. The subcommand ALTERNATIVE = -1 would have been used for the alternative hypothesis H_a: $\mu < 18$. The printout shown in Figure 8.19 provides summary statistics, the value of the test statistic $t = 1.62$, and the p-value = 0.067. Thus, since the p-value (0.067) is less than the level of significance (.10), the null hypothesis can be rejected.

A TWO-TAILED TEST

As an example of a two-tailed hypothesis test about a population mean using a small sample, consider the following production problem.

EXAMPLE 8.7 ▶ A production process is designed to fill containers with a mean filling weight of $\mu = 16$ ounces. An undesirable condition exists if the process is underfilling containers and the consumer is not receiving the amount of product indicated on the container label. In addition, an equally undesirable condition exists if the process is overfilling containers; in this case, the firm is losing money since the process is placing more product in the container than is required. Quality-assurance personnel periodically select a simple random sample of eight containers and test the following two-tailed hypotheses:

FIGURE 8.19   MINITAB OUTPUT FOR THE PLANT FOOD HYPOTHESIS TEST

```
MTB > TTEST 18 C1;
SUBC> ALTERNATIVE = 1.


TEST OF MU = 18.000 VS MU G.T. 18.000

          N       MEAN      STDEV    SE MEAN        T    P VALUE
C1       12     19.400      3.000      0.866     1.62      0.067
```

$$H_0: \; \mu = 16$$

$$H_a: \; \mu \neq 16$$

If H_0 is rejected, the production manager will request that the production process be stopped and that the mechanism for regulating filling weights be readjusted to provide a mean filling weight of 16 ounces. If the sample provides the eight data values of 16.02, 16.22, 15.82, 15.92, 16.22, 16.32, 16.12, and 15.92 ounces, what action should be taken at a .05 level of significance? Assume that the population of filling weights is normally distributed.

Since the data have not been summarized, we must first compute the sample mean and sample standard deviation. Doing so provides the following results:

$$\bar{x} = \frac{\Sigma x_i}{n} = \frac{128.56}{8} = 16.07 \text{ ounces}$$

and

$$s = \sqrt{\frac{\Sigma(x_i - \bar{x})^2}{n - 1}} = \sqrt{\frac{.22}{7}} = .18 \text{ ounces}$$

With a two-tailed test and a level of significance of $\alpha = .05$, $-t_{.025}$ and $t_{.025}$ determine the rejection region for the test. Using the table for the t distribution, we find that with $n - 1 = 8 - 1 = 7$ degrees of freedom, $-t_{.025} = -2.365$ and $t_{.025} = +2.365$. Thus, the rejection rule is written

$$\text{Reject } H_0 \text{ if } t < -2.365 \text{ or if } t > 2.365$$

Using $\bar{x} = 16.07$ and $s = .18$, we have

$$t = \frac{\bar{x} - \mu_0}{s/\sqrt{n}} = \frac{16.07 - 16.00}{.18/\sqrt{8}} = 1.10$$

Since $t = 1.10$ is not in the rejection region, the null hypothesis $\mu = 16$ ounces cannot be rejected. There is not enough evidence to stop the production process.

Using Table 2 of Appendix B and the row for 7 degrees of freedom, we see that the computed t value of 1.10 has an upper tail area of *more than* .10. Although the format of the t distribution table prevents us from being more specific, we can at least conclude that the two-tailed p-value is greater than 2(.10) = .20. Since this is greater than the .05 level of significance, we see that the p-value leads to the same conclusion; that is, do not reject H_0. The computer solution for this hypothesis test shows $t = 1.10$ and the exact p-value = .31.

EXERCISES

METHODS

69. Consider the following hypothesis test:

$$H_0: \; \mu \leq 10$$

$$H_a: \; \mu > 10$$

A sample of 16 provides a sample mean of 11 and sample standard deviation of 3.
a. Using $\alpha = .05$, what is the rejection rule?
b. Compute the value of the test statistic t. What is your conclusion?

SELF TEST

70. Consider the following hypothesis test:

$$H_0: \mu = 20$$

$$H_a: \mu \neq 20$$

Data from a sample of six items are as follows:

18, 20, 16, 19, 17, 18.

a. Compute the sample mean.
b. Compute the sample standard deviation.
c. Using $\alpha = .05$, what is the rejection rule?
d. Compute the value of the test statistic t.
e. What is your conclusion?

71. Consider the following hypothesis test:

$$H_0: \mu \geq 15$$

$$H_a: \mu < 15$$

A sample of 22 is used and the sample standard deviation is 8. Use $\alpha = .05$. Compute the value of the test statistic t and state your conclusion for each of the following sample results:

a. $\bar{x} = 13.0$ b. $\bar{x} = 11.5$ c. $\bar{x} = 15.0$ d. $\bar{x} = 19.0$

72. Consider the following hypothesis test:

$$H_0: \mu \leq 50$$

$$H_a: \mu > 50$$

Assume a sample of 16 items provides the test statistics as shown below. What can you say about the p-values in each case? What are your conclusions based on $\alpha = .05$?

a. $t = 2.602$
b. $t = 1.341$
c. $t = 1.960$
d. $t = 1.055$
e. $t = 3.261$

APPLICATIONS

SELF TEST

73. City Homes bought 101 badly deteriorated Baltimore rowhouses and turned them into housing for the poor. The mean monthly rental for these houses was $200 (*Financial World,* November 29, 1988). Nine hundred homes in another city were built and turned over to a foundation to rent. A sample of 10 of these homes showed the following monthly rental rates:

$220, $190, $250, $230, $185, $210, $240, $260, $200, $195.

a. Using $\alpha = .05$, test to see if the mean monthly rental rate for the population of 900 homes exceeds the $200 mean monthly rental rate in Baltimore. What is your conclusion?
b. What is the p-value for this test?

74. The average hourly wage in the United States is $10.05 (*The Tampa Tribune,* December 15, 1991). Assume that a sample of 25 individuals in Phoenix, Arizona, showed a sample mean wage of $10.83 per hour with a sample standard deviation of $3.25 per hour. Test the hypotheses $H_0: \mu = 10.05$ and $H_a: \mu \neq 10.05$ to see if the population mean in Phoenix differs from the mean throughout the United States. Using a .05 level of significance, what is your conclusion?

75. It is estimated that, on the average, a housewife with a husband and two children works 55 hours or less per week on household-related activities. Shown below are the hours worked during a week for a sample of eight housewives:

$$58, 52, 64, 63, 59, 62, 62, 55.$$

a. Using $\alpha = .05$, test the hypotheses H_0: $\mu \leq 55$ and H_a: $\mu > 55$. What is your conclusion about the mean number of hours worked per week?

b. What can you say about the p-value?

76. A study of a drug designed to reduce blood pressure used a sample of 25 men between the ages of 45 and 55. With μ indicating the mean change in blood pressure for the population of men receiving the drug, the hypotheses in the study were written H_0: $\mu \geq 0$ and H_a: $\mu < 0$. Rejection of H_0 shows that the mean change is negative, indicating that the drug is effective in lowering blood pressure.

a. At a .05 level of significance, what conclusion should be drawn if $\bar{x} = -10$ and $s = 15$?

b. What can you say about the p-value?

77. Last year the number of lunches served at an elementary-school cafeteria was normally distributed with a mean of 300 lunches per day. At the beginning of the current year, the price of a lunch was raised by 25¢. A sample of 6 days during the months of September, October, and November provided the following number of children being served lunches: 290, 275, 310, 260, 270, and 275. Do these data indicate that the mean number of lunches severed per day has dropped compared to last year? Test the hypothesis H_0: $\mu \geq 300$ against the alternative hypothesis H_a: $\mu < 300$ at a .05 level of significance.

8.9 ▽ CALCULATING THE PROBABILITY OF TYPE II ERRORS (OPTIONAL)

In this section we show how to calculate the probability of making a Type II error for a hypothesis test concerning a population mean. We will illustrate the procedure using the following example.

EXAMPLE 8.8 ▶ A quality-control manager must decide to accept a shipment of batteries from a supplier or to return the shipment due to poor quality. Assume that design specifications require batteries from the supplier to have a mean useful life of at least 120 hours. To evaluate the quality of an incoming shipment, a sample of 36 batteries will be selected and tested. On the basis of the sample, a decision must be made to accept the shipment or to reject it return it to the supplier. Let μ denote the mean hours of useful life for batteries in the shipment. The null and alternative hypotheses about the population mean are as follows:

$$H_0: \ \mu \geq 120$$

$$H_a: \ \mu < 120$$

If H_0 is rejected, the decision will be made to return the shipment to the supplier because the mean hours of useful life is less than the specified 120 hours. If H_0 is not rejected, the decision will be made to accept the shipment.

Suppose that a level of significance of $\alpha = .05$ is used to conduct the hypothesis test. The test statistic is

$$z = \frac{\bar{x} - \mu}{\sigma/\sqrt{n}} = \frac{\bar{x} - 120}{\sigma/\sqrt{n}}$$

With $z_{.05} = 1.645$, the rejection rule for the lower-tailed test becomes

$$\text{Reject } H_0 \text{ if } z < -1.645$$

Suppose that a sample of 36 batteries will be selected and that it is known from previous testing that the standard deviation for the population is $\sigma = 12$ hours. The preceding rejection rule indicates that we will reject H_0 if

$$z = \frac{\bar{x} - 120}{12/\sqrt{36}} < -1.645$$

Solving for $\bar{x}$ in the preceding expression indicates that we will reject H_0 whenever

$$\bar{x} < 120 - 1.645\left(\frac{12}{\sqrt{36}}\right) = 116.71$$

Rejecting H_0 when $\bar{x} < 116.71$ means that we will make the decision to accept the shipment whenever

$$\bar{x} \geq 116.71$$

Based on this information, we are now ready to compute probabilities associated with making a Type II error. First, recall that we make a Type II error whenever the true mean useful life is less than 120 hours and we make the decision to accept H_0: $\mu \geq 120$. Thus, to compute the probability of making a Type II error, we must select a value of μ less than 120 hours. For example, suppose that the shipment is considered to be of poor quality if the batteries have a mean life of $\mu = 112$ hours. If $\mu = 112$ is really true, what is the probability of accepting H_0: $\mu \geq 120$, and hence committing a Type II error? To find this probability, note that it is the probability that the sample mean $\bar{x}$ is greater than or equal to 116.71 when $\mu = 112$.

Figure 8.20 shows the sampling distribution of $\bar{x}$ when the mean is $\mu = 112$. The shaded area in the upper tail shows the probability of obtaining $\bar{x} \geq 116.71$. Using Figure 8.20, we see that

FIGURE 8.20   **PROBABILITY OF A TYPE II ERROR WHEN $\mu = 112$**

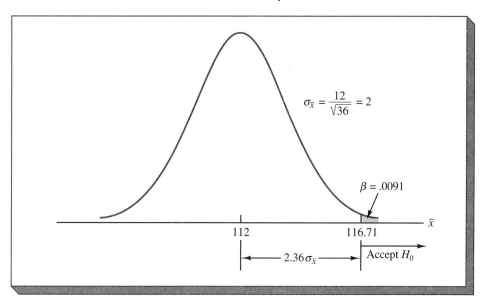

$$z = \frac{\bar{x} - \mu}{\sigma/\sqrt{n}} = \frac{116.71 - 112}{12/\sqrt{36}} = 2.36$$

The standard normal probability distribution table shows that with $z = 2.36$, the area in the upper tail is $.5000 - .4909 = .0091$; this value is the probability of accepting H_0 when $\mu = 112$. In other words, $.0091$ is the probability of making a Type II error when $\mu = 112$. Denoting the probability of making a Type II error as β, we see that when $\mu = 112$, $\beta = .0091$. Therefore, we can conclude that if the mean of the population is 112 hours, the probability of making a Type II error is only $.0091$.

We can repeat these calculations for other values of μ less than 120. Doing this will show that there is a different probability of making a Type II error for each value of μ less than 120. For example, suppose that we consider the case where the shipment of batteries has a mean useful life of $\mu = 115$ hours. Since we will accept H_0 whenever $\bar{x} \geq 116.71$, the z value for $\mu = 115$ is given by

$$z = \frac{\bar{x} - \mu}{\sigma/\sqrt{n}} = \frac{116.71 - 115}{12/\sqrt{36}} = .86$$

From the standard normal probability distribution table, we find that the area in the upper tail of the standard normal probability distribution for $z = .86$ is $.5000 - .3051 = .1949$. Thus, the probability of making a Type II error is $\beta = .1949$ when the true mean is $\mu = 115$.

In Table 8.6 we show the probability of making a Type II error for a variety of values of μ less than 120. Note that as μ increases toward 120, the probability of making a Type II error increases toward an upper bound of .95. However, as μ decreases to values farther below 120, the probability of making a Type II error becomes smaller and smaller. This is what we should expect. When the true population mean μ is close to the null hypothesis value of $\mu = 120$, there is a high probability that we will accept H_0 and make a Type II error. However, when the true population mean μ is far below the null hypothesis value of $\mu = 120$, there is a low probability that we will accept H_0 and make a Type II error. ◀

The probability of correctly rejecting H_0 when it is false is called the *power* of the test. For any particular value of μ, the power is $1 - \beta$. That is, the probability of correctly

---

TABLE 8.6	**PROBABILITIES OF MAKING A TYPE II ERROR FOR THE LOT-ACCEPTANCE HYPOTHESIS TEST**

Value of μ	$z = \dfrac{116.71 - \mu}{12/\sqrt{36}}$	**Probability of Making a Type II Error (β)**	**Power $(1 - \beta)$**
112	2.36	.0091	.9909
114	1.36	.0869	.9131
115	.86	.1949	.8051
116.71	.00	.5000	.5000
117	−.15	.5596	.4404
118	−.65	.7422	.2578
119.999	−1.645	.9500	.0500

rejecting the null hypothesis is 1 minus the probability of making a Type II error. Values of power are also shown in Table 8.6. Using these values, the power associated with each value of μ is shown graphically in Figure 8.21. Such a graph is called a *power curve*. Note that the power curve extends over the values of μ for which the null hypothesis is false. The height of the power curve at any value of μ provides the probability of correctly rejecting H_0 when H_0 is false.*

In summary, the following step-by-step procedure can be used to compute the probability of making a Type II error in hypothesis tests about a population mean.

1. Formulate the null and alternative hypotheses.
2. Use the level of significance α to establish a rejection rule based on the test statistic.
3. Using the rejection rule, solve for the value of the sample mean that identifies the rejection region for the test.
4. Use the results from step 3 to state the values of the sample mean that lead to the acceptance of H_0; this defines the acceptance region for the test.
5. Using the sampling distribution of $\bar{x}$ for any value of μ from the alternative hypothesis, and the acceptance region from step 4, compute the probability that the sample mean will be in the acceptance region. This probability is the probability of making a Type II error at the chosen value of μ.
6. Repeat step 5 for other values of μ from the alternative hypothesis.

---

*Another graph, called the *operating characteristic curve,* is sometimes used to provide information about the probability of making a Type II error. The operating characteristic curve shows the probability of accepting H_0 and thus provides β for the values of μ where the null hypothesis is false. The probabilities of making Type II errors can be read directly from this graph.

FIGURE 8.21   **POWER CURVE FOR THE LOT-ACCEPTANCE HYPOTHESIS TEST**

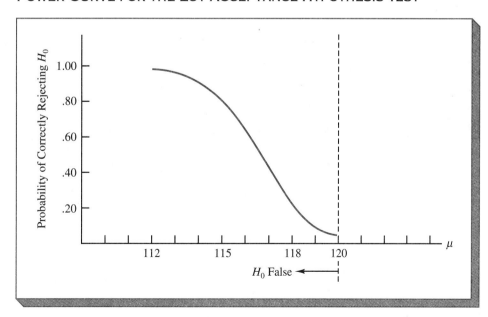

EXERCISES

METHODS

SELF TEST

78. Consider the following hypothesis test:

$$H_0: \mu \geq 10$$

$$H_a: \mu < 10$$

The sample size is 120, and the population standard deviation is 5. Use $\alpha = .05$.

a. If the actual population mean is 9, what is the probability that the sample mean leads to the conclusion do not reject H_0?

b. What type of error would be made if the actual population mean is 9 and we conclude that $H_0: \mu \geq 10$ is true?

c. What is the probability of making a Type II error if the actual population mean is 8?

79. Consider the following hypothesis test:

$$H_0: \mu = 20$$

$$H_a: \mu \neq 20$$

A sample of 200 items will be taken, and the population standard deviation is 10. Use $\alpha = .05$. Compute the probability of making a Type II error if the population mean is as shown below:

a. $\mu = 18.0$ **b.** $\mu = 22.5$ **c.** $\mu = 21.0$

APPLICATIONS

80. Fowle Marketing Research, Inc., bases charges to a client on the assumption that the mean time to complete telephone surveys is 15 minutes or less. If more time is required, a premium rate is charged to the client. For a sample of 35 surveys, a standard deviation of 4 minutes, and a level of significance of .01, the sample mean will be used to test the null hypothesis $H_0: \mu \leq 15$.

a. What is your interpretation of the Type II error for this problem? What is its impact on the firm?

b. What is the probability of making a Type II error when the actual mean time is $\mu = 17$ minutes?

c. What is the probability of making a Type II error when the actual mean time is $\mu = 18$ minutes?

d. Sketch the general shape of the power curve for this test.

SELF TEST

81. A consumer-research group is interested in testing an automobile manufacturer's claim that a new economy model will provide a mean fuel consumption of at least 25 miles per gallon of gasoline. That is, $H_0: \mu \geq 25$.

a. Using a .02 level of significance and a sample of 30 cars, what is the rejection rule based on the value of $\bar{x}$ for the test to determine if the manufacturer's claim should be rejected? Assume that σ is 3 miles per gallon.

b. What is the probability of committing a Type II error if the actual mileage is 23 miles per gallon?

c. What is the probability of committing a Type II error if the actual mileage is 24 miles per gallon?

d. What is the probability of committing a Type II error if the actual mileage is 25.5 miles per gallon?

82. *Young Adult* magazine states the following hypotheses about the mean age of its subscribers:

$$H_0: \mu = 28$$

$$H_a: \mu \neq 28$$

a. What would a Type II error correspond to in this situation?
b. Suppose that the population standard deviation is $\sigma = 6$ years and the sample size is 100. With $\alpha = .05$, what is the probability of accepting H_0 for μ equal to 26, 27, 29, and 30?
c. What is the power at $\mu = 26$? What does this tell you?

83. A production line operation is tested for filling-weight accuracy with the following hypotheses.

Hypothesis	Conclusion and Action
$H_0: \mu = 16$	Process operating correctly; keep running
$H_a: \mu \neq 16$	Process operating incorrectly; stop and adjust machine

Suppose that the sample size is 30 and the population standard deviation is $\sigma = .8$. Use $\alpha = .05$.
a. What would a Type II error correspond to in this situation?
b. What is the probability of making a Type II error when the machine is overfilling by .5 ounce?
c. What is the power of the statistical test when the machine is overfilling by .5 ounce?
d. Show the power curve for this hypothesis test. What information does it contain for the production manager?

84. Refer to Exercise 80. Repeat parts (b) and (c) if the firm selects a sample of 50 surveys. What observation can you make about how increasing the sample size affects the probability of making a Type II error?

85. Sparr Investments, Inc., specializes in tax-deferred investment opportunities for its clients. Recently, Sparr has offered a payroll deduction investment program for the employees of a particular company. Sparr estimates that the employees are currently averaging $100 or less per month in tax deferred investments. A sample of 40 employees will be used to test Sparr's hypothesis about the current level of investment activity among the population of employees. Assume that the employee monthly tax-deferred investment amounts have a standard deviation of $75 and that a .05 level of significance will be used in the hypothesis test.
a. What is the Type II error in this situation?
b. What is the probability of the Type II error if the population mean employee monthly investment is $120?
c. What is the probability of the Type II error if the population mean employee monthly investment is $130?
d. Repeat parts (b) and (c) if a sample size of 80 employees is used.

8.10 DETERMINING THE SAMPLE SIZE FOR A HYPOTHESIS TEST ABOUT A POPULATION MEAN (OPTIONAL)

Assume that a hypothesis test is to be conducted about the value of a population mean. The user's specified level of significance sets the maximum probability of making a Type I error for the test. By controlling the sample size, the user can also control the probability of making a Type II error. Let us show how a sample size can be determined for the following one-tailed test about a population mean.

$$H_0: \ \mu \geq \mu_0$$

$$H_a: \ \mu < \mu_0$$

where μ_0 is the hypothesized value for the population mean.

The upper part of Figure 8.22 shows the sampling distribution of $\bar{x}$ when H_0 is true and $\mu = \mu_0$. Note that the user's specified level of significance α determines the rejection region for the test. Let c denote the critical value such that $\bar{x} < c$ determines the rejection region for the test. Using the upper part of Figure 8.22 with z_α indicating the z value corresponding to an area of α in the tail of the standard normal probability distribution, we compute c as follows:

$$c = \mu_0 - z_\alpha \frac{\sigma}{\sqrt{n}} \tag{8.13}$$

Now consider the sampling distribution shown in the lower part of Figure 8.22. Specifically, we have selected a value of the population mean, denoted by μ_a, which corresponds to the case when H_0 is false and H_a is true with $\mu_a < \mu_0$. Let us assume that the user specifies the probability of a Type II error that can be tolerated if the true

FIGURE 8.22   DETERMINING THE SAMPLE SIZE FOR SPECIFIED LEVELS OF THE TYPE I (α) AND TYPE II (β) ERRORS

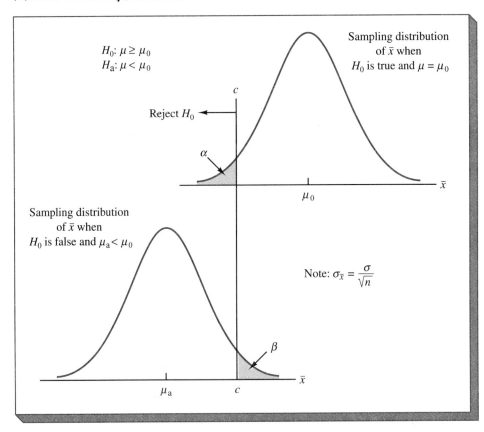

population mean is μ_a. This probability is shown as β in Figure 8.22. Using the lower part of Figure 8.22 with z_β indicating the z value corresponding to an area of β in the tail of the standard normal probability distribution, we can compute the critical value c as follows:

$$c = \mu_a + z_\beta \frac{\sigma}{\sqrt{n}} \tag{8.14}$$

Since equations (8.13) and (8.14) provide two expressions for c, we know they must be equal; thus

$$\mu_0 - z_\alpha \frac{\sigma}{\sqrt{n}} = \mu_a + z_\beta \frac{\sigma}{\sqrt{n}}$$

To determine the expression that will provide the desired sample size, we first solve for $\sqrt{n}$ as follows:

$$\mu_0 - \mu_a = z_\alpha \frac{\sigma}{\sqrt{n}} + z_\beta \frac{\sigma}{\sqrt{n}}$$

$$\mu_0 - \mu_a = \frac{(z_\alpha + z_\beta)\sigma}{\sqrt{n}}$$

and

$$\sqrt{n} = \frac{(z_\alpha + z_\beta)\sigma}{(\mu_0 - \mu_a)}$$

Squaring both sides of the above expression provides the following sample-size formula for a one-tailed hypothesis test about a population mean.

Sample Size for a One-Tailed Hypothesis Test About a Population Mean

$$n = \frac{(z_\alpha + z_\beta)^2 \sigma^2}{(\mu_0 - \mu_a)^2} \tag{8.15}$$

where

z_α = z value providing an area of α in the tail of a standard normal distribution
z_β = z value providing an area of β in the tail of a standard normal distribution
σ = the population standard deviation
μ_0 = the value of the population mean in the null hypothesis
μ_a = the value of the population mean used for the Type II error
Note: For a two-tailed hypothesis test, use equation (8.15) with $z_{\alpha/2}$ replacing z_α.

Although the logic of equation (8.15) was developed for the hypothesis test shown in Figure 8.22, it holds for any one-tailed test about a population mean. Note that in a two-tailed hypothesis test about a population mean, $z_{\alpha/2}$ is used instead of z_α.

**EXAMPLE 8.8
(CONTINUED)**

The design specification for the shipment of batteries indicated a mean useful life of at least 120 hours for the batteries. Shipments were rejected if the hypothesis H_0: $\mu \geq 120$

was rejected. Let us assume that the quality-control manager makes the following statements about the allowable probabilities for the Type I and Type II errors:

Type I error statement: If the mean life of the batteries in the shipment is $\mu = 120$, I want the maximum probability of making a Type I error to be $\alpha = .05$.

Type II error statement: If the mean life of the batteries in the shipment is 5 hours under the specification (i.e., $\mu = 115$), I am willing to risk a $\beta = .10$ probability of accepting the shipment.

These statements are based on the judgment of the manager. Someone else might specify different restrictions on the probabilities. However, statements about the allowable probabilities of both errors must be made before the sample size can be determined.

Thus, in our example, $\alpha = .05$ and $\beta = .10$. Using the standard normal probability distribution, we have $z_{.05} = 1.645$ and $z_{.10} = 1.28$. From the statements about the error probabilities, we note that $\mu_0 = 120$ and $\mu_a = 115$. Finally, recall that population standard deviation was assumed known at a value of $\sigma = 12$. Using equation (8.15), the recommended sample size is

$$n = \frac{(1.645 + 1.28)^2(12)^2}{(120 - 115)^2} = 49.3$$

Rounding up, the recommended sample size is 50.

Since both the Type I and Type II error probabilities have been controlled at allowable levels with $n = 50$, the quality-control manager is now justified in using the accept H_0 and reject H_0 statements for the hypothesis test.   ◀

Before closing this section, let us make some comments about the relationship among α, β, and the sample size n.

1. Once two of the three values are known, the other value can be computed.
2. For a given level of significance α, increasing the sample size will reduce β.
3. For a given sample size, decreasing α will increase β, whereas increasing α will decrease β.

The third observation is something to keep in mind when the probability of a Type II error is not being controlled. It suggests that one should not choose unnecessarily small values for the level of significance α. For a given sample size, choosing a smaller level of significance α will increase the probability of making a Type II error. Inexperienced users of hypothesis testing often think that smaller values of α are always better; note, however, that this is true if we are concerned *only* about the Type I error.

EXERCISES

METHODS

86. Consider the following hypothesis test:

SELF TEST ▷

$$H_0: \mu \geq 10$$

$$H_a: \mu < 10$$

The sample size is 120, and the population standard deviation is 5. Use $\alpha = .05$. If the actual population mean is 9, the probability of making a Type II error is .2912. Suppose that the researcher would like to reduce the probability of making a Type II error to .10 when the actual population mean is 9. What sample size is recommended?

87. Consider the following hypothesis test:

$$H_0: \ \mu = 20$$

$$H_a: \ \mu \neq 20$$

The population standard deviation is 10. Use $\alpha = .05$. How large a sample should be taken if the researcher is willing to accept a .05 probability of making a Type II error when the actual population mean is 22?

APPLICATIONS

88. Suppose that the project director for the Hilltop Coffee study (Example 8.4) had asked for a .10 probability of claiming that Hilltop was not in violation when it really was underfilling by 1 ounce ($\mu = 2.9375$ pounds). What sample size would have been recommended?

SELF TEST

89. A special industrial battery must have a life of at least 400 hours. A hypothesis test is to be conducted with $\alpha = .02$. If the batteries from a particular production run have a population mean life of 385 hours, the production manager would like a sampling procedure that provides a .10 probability of erroneously concluding that the batch is acceptable. What sample size is recommended for the hypothesis test? Use 30 hours as an estimate of the population standard deviation.

90. For the *Young Adult* magazine study, $H_0: \ \mu = 28$ years was tested at a .05 level of significance. If the manager conducting the test is willing to accept a .15 probability of making a Type II error when the population mean age is 29, what sample size should be selected? Assume $\sigma = 6$.

91. An automobile mileage study tested the following hypotheses.

Hypothesis	Conclusion
$H_0: \ \mu \geq 25$ mpg	Manufacturer's claim supported
$H_a: \ \mu < 25$ mpg	Manufacturer's claim rejected; average mileage per gallon is less than stated

For $\sigma = 3$ and a .02 level of significance, what sample size would be recommended if the researcher would like an 80% chance of detecting that μ is less than 25 miles per gallon when μ is 24?

SUMMARY

In this chapter we presented methods for developing a confidence interval for a population mean. The purpose of developing a confidence interval is to provide the user with an understanding of the precision when a sample mean $\bar{x}$ is used to estimate a population mean. A wide confidence interval indicates poor precision in that a large sampling error may be present; in such cases, the sample size can be increased to reduce the width of the interval and improve the precision of the estimate.

Figure 8.23 summarizes the interval-estimation procedures for a population mean and provides a practical guide for computing the interval estimate. The figure shows that the expression used to

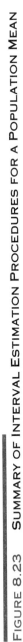

FIGURE 8.23   SUMMARY OF INTERVAL ESTIMATION PROCEDURES FOR A POPULATION MEAN

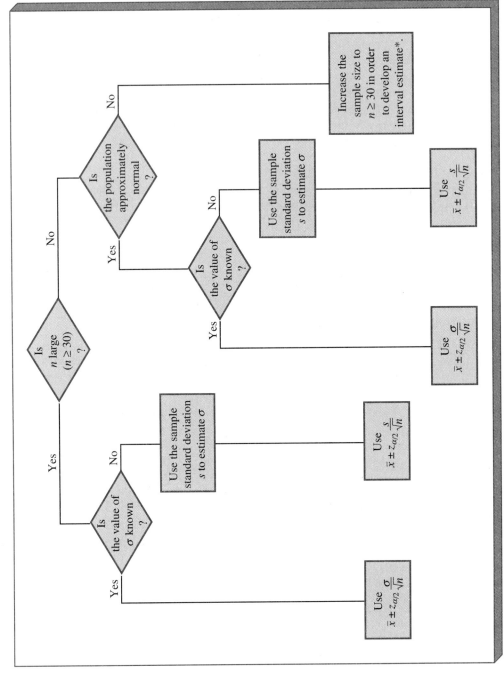

In some cases, methods from nonparametric statistics may be used to develop confidence intervals for location parameters of a population. However, these procedures are beyond the scope of this text, so increasing the sample size to $n \geq 30$ is recommended here.

compute an interval estimate depends on whether the sample size is large ($n \geq 30$) or small ($n < 30$), whether the population standard deviation is known, and in some cases, whether or not the population has a normal or approximately normal probability distribution. If the sample size is large, no assumption is required about the distribution of the population and $z_{\alpha/2}$ is used in the computation of the interval estimate. If the sample size is small, the population must have a normal or approximately normal probability distribution to compute an interval estimate of μ. In such cases, $z_{\alpha/2}$ is used in the computation of the interval estimate when σ is known, and $t_{\alpha/2}$ is used when σ is estimated by the sample standard deviation s. Finally, if the sample size is small and the assumption of a normally distributed population is inappropriate, we recommend increasing the sample size to $n \geq 30$.

We also discussed methods for performing hypothesis tests about a population mean. Hypothesis testing is a statistical procedure that uses sample data to determine whether or not a statement about the value of a population parameter should be rejected. The hypotheses, which come from a variety of sources, must be two competing statements: a null hypothesis H_0 and an alternative hypothesis H_a. In some applications, it is not obvious how the null and alternative hypotheses should be formulated. We suggested guidelines for developing hypotheses based on three types of situations.

1. *Testing research hypotheses:* The research hypothesis should be formulated as the alternative hypothesis. Whenever the sample data contradict the null hypothesis, the research hypothesis is supported.
2. *Testing the validity of a claim:* Generally, this situation corresponds to the "innocent until proven guilty" analogy. The claim made is chosen as the null hypothesis; the challenge to the claim is chosen as the alternative hypothesis. Action against the claim will be taken if the null hypothesis is rejected.
3. *Testing in decision-making situations:* This occurs when a decision maker must choose between two courses of action, one associated with the null hypothesis and one associated with the alternative hypothesis. In these situations, it is often suggested that the hypotheses be formulated such that the Type I error is the more serious error.

Figure 8.24 summarizes the test statistics used in hypothesis tests about a population mean and provides a practical guide for selecting the hypothesis-testing procedure. The figure shows that the test statistic depends on whether the sample size is large ($n \geq 30$) or small ($n < 30$), whether the population standard deviation is known, and in some cases, whether or not the population has a normal or approximately normal probability distribution. If the sample size is large, the z test statistic is used to conduct the hypothesis test. If the sample size is small, the population must have a normal or approximately normal distribution to conduct a hypothesis test about the value of μ. In such cases, the z test statistic is used if σ is known, and the t test statistic is used if σ is estimated by the sample standard deviation s. Finally, note that if the sample size is small and the assumption of a normally distributed population is inappropriate, we recommend increasing the sample size of $n \geq 30$.

The rejection rule for all the hypothesis-testing procedures involves comparing the value of the test statistic with a critical value. For lower-tail tests, the null hypothesis is rejected if the value of the test statistic is less than the critical value. For upper-tail tests, the null hypothesis is rejected if the value of the test statistic is greater than the critical value. For two-tailed tests, the null hypothesis is rejected for values of the test statistic in either tail of the sampling distribution.

We also saw that p-values could be used for hypothesis testing. The p-value yields the probability, when the null hypothesis is true, of obtaining a sample result at least as unlikely as what is observed. When p-values are used to conduct a hypothesis test, the rejection rule calls for rejecting the null hypothesis whenever the p-value is less than α. The p-value is often called the observed level of significance because the null hypothesis will be rejected if the p-value is less than α.

Extensions of the hypothesis-testing procedure to control the probability of making a Type II error were also presented. In Section 8.9 we showed how to compute the probability of making a Type II error. In Section 8.10 we showed how to determine a sample size that would enable us to control the probability of making both a Type I and a Type II error.

FIGURE 8.24 SUMMARY OF THE TEST STATISTICS TO BE USED FOR HYPOTHESIS TESTS ABOUT A POPULATION MEAN

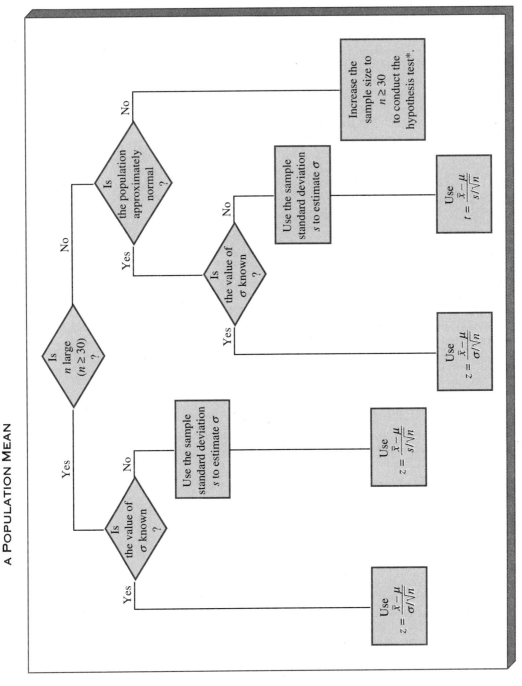

In some cases, methods of nonparametric statistics may be used. Nonparametric methods are discussed in Chapter 17.

▽
GLOSSARY

Point estimate A single numerical value used to estimate a population parameter.

Interval estimate An estimate of a population parameter that provides an interval or range for the value of the parameter.

Sampling error The absolute value of the difference between the value of an unbiased point estimator, such as the sample mean $\bar{x}$, and the value of the population parameter it estimates, such as the population mean μ; in this case, the sampling error is $|\bar{x} - \mu|$.

Precision A statement about the size of the sampling error.

Confidence level The confidence associated with an interval estimate. For example, if an interval-estimation procedure provides intervals such that 95% of the intervals developed will include the population parameter, an interval estimate is said to be constructed at the 95% confidence level; note that .95 is referred to as the *confidence coefficient.*

***t* Distribution** A family of probability distributions that can be used to develop interval estimates of a population mean whenever the population standard deviation is unknown and the population has a normal or near-normal probability distribution.

Degrees of freedom A parameter of the *t* distribution. When the *t* distribution is used in the computation of an interval estimate of a population mean, the appropriate *t* distribution has $n - 1$ degrees of freedom, where n is the sample size.

Null hypothesis The hypothesis tentatively assumed true in the hypothesis-testing procedure.

Alternative hypothesis The hypothesis opposite to the null hypothesis.

Type I error The error of rejecting H_0 when it is true.

Type II error The error of accepting H_0 when it is false.

Critical value A value that is compared with the test statistic to determine whether or not H_0 should be rejected.

Level of significance The maximum probability of making a Type I error.

One-tailed test A hypothesis test in which rejection of the null hypothesis occurs for values of the test statistic in one tail of the sampling distribution.

Two-tailed test A hypothesis test in which rejection of the null hypothesis occurs for values of the test statistic in either tail of the sampling distribution.

***p*-Value** The probability, when the null hypothesis is true, of obtaining a sample result that is at least as unlikely as what is observed. It is often called the observed level of significance.

Power The probability of correctly rejecting H_0 when it is false.

Power curve A graph showing the probability of rejecting H_0 when it is false.

▽
KEY FORMULAS

Sampling Error When Estimating μ

$$|\bar{x} - \mu| \tag{8.1}$$

Interval Estimate of a Population Mean (Large-Sample Case)

$$\bar{x} \pm z_{\alpha/2}\frac{\sigma}{\sqrt{n}} \tag{8.2}$$

Interval Estimate of a Population Mean (Small-Sample Case)

$$\bar{x} \pm t_{\alpha/2}\frac{s}{\sqrt{n}} \tag{8.3}$$

Sample Size for an Interval Estimate of a Population Mean

$$n = \frac{(z_{\alpha/2})^2 \sigma^2}{E^2}$$ (8.5)

Test Statistic About a Population Mean (Large-Sample Case)

$$z = \frac{\bar{x} - \mu_0}{\sigma/\sqrt{n}}$$ (8.11)

When σ is unknown, substitute s for σ.

Test Statistic About a Population Mean (Small-Sample Case with s Used to Estimate σ)

$$t = \frac{\bar{x} - \mu_0}{s/\sqrt{n}}$$ (8.12)

Sample Size for a One-Tailed Hypothesis Test About a Population Mean

$$n = \frac{(z_\alpha + z_\beta)^2 \sigma^2}{(\mu_0 - \mu_a)^2}$$ (8.15)

In a two-tailed test, replace z_α with $z_{\alpha/2}$.

REVIEW QUIZ

TRUE/FALSE

1. Whenever the central limit theorem is applicable, the sampling error will be zero when using $\bar{x}$ to estimate μ.
2. Interval estimates provide information on how close the value of the sample mean $\bar{x}$ is to the population mean μ.
3. A value of $\bar{x}$ does not contain information about the size of the sampling error.
4. An interval estimate contains information about the size of the sampling error.
5. To determine the appropriate sample size, the user must specify a maximum allowable sampling error.
6. In hypothesis testing, we do not assume that the null hypothesis is true until after the sample results have been analyzed.
7. In hypothesis testing, the z value can be interpreted as the number of standard errors the sample mean is from the hypothesized value of μ.
8. The Type I error is the error of accepting H_0 when it is false.
9. If the level of significance is made smaller, the rejection region becomes larger.
10. Decision rules based on critical z values and p-values will always provide the same hypothesis-testing conclusions.
11. In order to use the t distribution, we must be willing to assume that the population has a normal probability distribution.
12. When we use the t distribution to form a 95% confidence interval, the interval will be smaller than if a z value from a standard normal probability distribution is used.
13. The t distribution can never be used if the sample size is 30 or more.

MULTIPLE CHOICE

Use the following information for Questions 14–16. A random sample of 81 automobile tires has a mean tread life of 36,000 miles. It is known that the standard deviation of tread life of tires is $\sigma = 4500$ miles.

14. A 95% confidence interval for the population mean is
   a. 35,500 to 36,500
   b. 35,177.5 to 36,822.5
   c. 35,020 to 36,980
   d. none of the above

15. If the sample mean of 36,000 had been from a random sample size 50, the 95% confidence interval would have been
   a. the same
   b. a wider interval
   c. a narrower interval
   d. none of the above

16. A 90% confidence interval for the population mean is
   a. 35,000 to 37,000
   b. 35,177.5 to 36,822.5
   c. 35,020 to 36,980
   d. none of the above

17. The useful life of a certain type of light bulb is known to have a standard deviation of $\sigma = 40$ hours. How large a sample should be taken if it is desired to have a sampling error of 10 hours or less at a 95% level of confidence?
   a. 62
   b. 44
   c. 37
   d. 8

18. If a hypothesis test leads to the rejection of the null hypothesis
   a. a Type I error is always committed
   b. a Type II error is always committed
   c. a Type I error may have been committed
   d. a Type II error may have been committed

Use the following information for Questions 19–21. The ABC Electronics Company claims that the batteries it produces have a useful life of at least 100 hours. It is known that the standard deviation is 20 hours. A test is undertaken to check the validity of this claim.

19. With the level of significance set at .05, the critical value or values for the test based on a sample of 49 batteries is
   a. $z = -1.645$
   b. $z = -1.435$
   c. $z = -1.96$ and $+1.96$
   d. $z = +1.645$

20. If the random sample of 49 batteries resulted in an average life of 96 hours, can the manufacturer's claim be rejected at the .05 level of significance?
   a. Yes, the null hypothesis can be rejected.
   b. No, do not reject the null hypothesis.
   c. Not enough information is given to answer this question.

21. What is the p-value associated with the sample mean of 96 hours?
   a. .042
   b. .081
   c. .419
   d. .96

22. If the level of significance of a hypothesis test is increased from .01 to .05, the probability of a Type II error
   a. will also be increased from .01 to .05
   b. will not be changed
   c. will be decreased
   d. Not enough information is given to answer this question.

SUPPLEMENTARY EXERCISES

92. In the United States, 99% of the households have at least one television set, and 98% of Americans watch some television every day (*In Health,* January 1992). Part of the study indicated a mean of 2.25 television sets per household. Assume that the mean number of sets per households was based on a sample of 300 households and that the sample standard deviation was 1.2 television sets. Provide a 95% confidence interval estimate of the population mean number of television sets per household.

93. The North Carolina Savings and Loan Association would like to develop an estimate of the mean size of home improvement loans granted by its member institutions. A sample of 100 loans granted by member institutions resulted in a sample mean of $3400 and a sample standard deviation of $650. With these data, develop a 98% confidence interval for the population mean dollar amount of home improvement loans.

94. A sample of 1033 recreational fishermen was used in a study reported in the *Journal of Marketing Research,* November 1987. Data on temperature were collected to study the relationship between temperature and a fisherman's decision to go fishing. The sample mean temperature was 55.6° and the sample standard deviation was 7.37°. Construct a 95% confidence interval for the population mean temperature for the recreational fishermen considered in this study.

95. Dailey Paints, Inc., implemented a long-term test study designed to check the wear resistance of its major brand of paint. The test consisted of painting eight houses in various parts of the United States and observing the number of months until signs of peeling were observed. The following data were obtained from a normal population:

House	1	2	3	4	5	6	7	8
Months Until Signs of Peeling	60	51	64	45	48	62	54	56

   a. What is a point estimate of the mean number of months until signs of peeling are observed?

   b. Develop a 95% confidence interval to estimate the population mean number of months until signs of peeling are observed.

   c. Develop a 99% confidence interval for the population mean.

96. What is the mean annual salary/bonus paid to the chief executive officers of the largest firms in the United States? A random sample of seven firms provided the annual salary/bonus data (*Business Week,* May 1, 1989) shown below.

Firm	Annual Salary/ Bonus ($000s)
Bank of Boston	1200
Citicorp	1798
DuPont	1611
Abbott Laboratories	1920
Teledyne	860
Emerson Electric	1681
Conagra	1312

   a. What is the point estimate of the population mean annual salary/bonus for chief executives?

b. What is the point estimate of the population standard deviation for annual salaries?

c. What is the 95% confidence interval estimate of the population mean annual salary/bonus for chief executives?

97. The Atlantic Fishing and Tackle Company has developed a new synthetic fishing line. To estimate the breaking strength of this line (pounds), testers subjected six lengths of line to breakage testing. The following data were obtained from a normal population.

Line	1	2	3	4	5	6
Breaking Strength (pounds)	18	24	19	21	20	18

Develop a 95% confidence interval for the mean breaking strength of the new line.

98. Sample assembly times for a particular manufactured part were 8, 10, 10, 12, 15, and 17 minutes. If the mean of the sample is used to estimate the mean of the population of assembly times, provide a point estimate and a 90% confidence interval for the population mean. Assume that the population has a normal probability distribution.

99. A utility company finds that a sample of 100 delinquent accounts yields an average amount owed of $131.44, with a sample standard deviation of $16.19. Develop a 90% confidence interval for the population mean amount owed.

100. In Exercise 99 the utility company sampled 100 delinquent accounts to estimate the mean amount owed by these accounts. The sample standard deviation was $16.19. How large a sample should be taken if the company wants to be 90% confident that the estimate of the population mean will have a sampling error of $1 or less?

101. Consider the Atlantic Fishing and Tackle Company problem presented in Exercise 97. How large a sample would be necessary to estimate the mean breaking strength of the new line with a .99 probability of a sampling error of 1 pound or less?

102. Mileage tests are conducted for a particular model of automobile. If the desired precision is stated such that there is a .98 probability of a sampling error of 1 mile per gallon or less, how many automobiles should be used in the test? Assume that preliminary mileage tests indicate the standard deviation to be 2.6 miles per gallon.

103. In developing patient appointment schedules, a medical center desires an estimate of the mean time that a staff member spends with each patient. How large a sample should be taken if the precision of the estimate is to be ±2 minutes at a 95% level of confidence? How large a sample should be taken for a 99% level of confidence? Use a planning value for the population standard deviation of 8 minutes.

104. Exercise 96 provided annual salary/bonus data for chief executives of firms in the United States (*Business Week,* May 1, 1989). The sample standard deviation was $375, with data provided in thousands of dollars. How many chief executives should be in the sample if we would like to estimate the population mean annual salary/bonus with a margin of error of $100,000? Note that the margin of error should be stated as $100 since the data were provided in thousands of dollars. Use a 95% confidence level.

105. The manager of the Keeton Department Store has assumed that the mean annual income of the store's credit-card customers is at least $38,000 per year. A sample of 58 credit-card customers shows a sample mean of $37,200 and a sample standard deviation of $3000. At the .05 level of significance should this assumption be rejected? What is the *p*-value?

106. The chamber of commerce of a Florida Gulf Coast community advertises area residential property available at a mean cost of $75,000 or less per lot. Using a .05 level of significance, test the validity of this claim. Suppose a sample of 32 properties provided a sample mean of $76,000 per lot and a sample standard deviation of $2500. What is the *p*-value?

107. A bath soap manufacturing process is designed to produce a mean of 120 bars of soap per batch. Quantities over or under this standard are undesirable. A sample of 10 batches shows the following numbers of bars of soap:

108, 118, 120, 122, 119, 113, 124, 122, 120, 123.

Using a .05 level of significance, test to see if the sample results indicate that the manufacturing process is functioning properly.

108. The monthly rent for a two-bedroom apartment in a particular city is reported to average $350. Suppose we would like to test the hypothesis H_0: $\mu = 350$ versus the hypothesis H_a: $\mu \neq 350$. A sample of 36 two-bedroom apartments is selected. The sample mean turns out to be $\bar{x} = \$362$, with a sample standard deviation of $s = \$40$.
  a. Conduct this hypothesis test with a .05 level of significance.
  b. Compute the p-value.
  c. Use the sample results to construct a 95% confidence interval for the population mean. What hypothesis-testing conclusion would you draw based on the confidence interval result?

109. Stout Electric Company operates a fleet of trucks that provide electrical service to the construction industry. Monthly mean maintenance costs have been $275 per truck. A random sample of 40 trucks shows a sample mean maintenance cost of $282.50 per month, with a sample standard deviation of $30. Management would like a test to determine whether or not the mean monthly maintenance cost has increased.
  a. Using a .05 level of significance, what is the rejection rule for this test?
  b. What is your conclusion based on the sample mean of $282.50?
  c. What is the p-value associated with this sample result? What is your conclusion based on the p-value?

110. In making bids on building projects, Sonneborn Builders, Inc., assumes construction workers are idle no more than 15% of the time. For a normal 8-hour shift, this means that the mean idle time per worker should be 72 minutes or less per day. A sample of 30 construction workers provided a mean idle time of 80 minutes per day. The sample standard deviation was 20 minutes. Suppose a hypothesis test is to be designed to test the validity of the company's assumption.
  a. What is the p-value associated with the sample result?
  b. Using a .05 level of significance and the p-value, test the hypothesis H_0: $\mu \leq 72$. What is your conclusion?

111. Refer to Exercise 110.
  a. What is the probability of making a Type II error when the population mean idle time is 80 minutes?
  b. What is the probability of making a Type II error when the population mean idle time is 75 minutes?
  c. What is the probability of making a Type II error when the population mean idle time is 70?
  d. Sketch the power curve for this problem.

112. A federal funding program is available to low-income neighborhoods. To qualify for the funding a neighborhood must have a mean household income of less than $9000 per year. Neighborhoods with mean annual household incomes of $9000 or more do not qualify. Funding decisions are based on a sample of residents in the neighborhood. A hypothesis test with a .02 level of significance is conducted. If the funding guidelines call for a maximum probability of .05 of not funding a neighborhood with a mean annual household income of $8500, what sample size should be used in the funding decision study? Use $\sigma = \$2000$ as a planning value.

113. The bath soap production process uses the hypothesis H_0: $\mu = 120$ and H_a: $\mu \neq 120$ to test whether or not the production process is meeting the standard output of 120 bars per batch. Use a .05 level of significance for the test and a planning value of 5 for the standard deviation.

a. If the mean output drops to 117 bars per batch, the firm would like a 98% chance of concluding that the standard production output is not being met. How large a sample should be selected?

b. Using your sample size from part (a), what is the probability of concluding that the process is operating satisfactorily for each of the following actual mean outputs: 117, 118, 119, 121, 122, and 123 bars per batch? That is, what is the probability of a Type II error in each case?

COMPUTER CASE: METROPOLITAN RESEARCH, INC.

Metropolitan Research, Inc., is a consumer research organization that conducts surveys designed to evaluate a wide variety of products and services available to consumers. In one particular study, Metropolitan was interested in learning about consumer satisfaction with the performance of automobiles produced by a major Detroit manufacturer. A questionnaire sent to owners of one of the manufacturer's full-sized cars revealed several complaints about faulty transmissions. To learn more about the transmission failures, Metropolitan used a sample of actual transmission repairs provided by a transmission repair firm located in the Detroit area. The following data show the actual number of miles that 50 vehicles had been driven at the time of transmission failure. The data are available in the data set AUTO.

85,092	32,609	59,465	77,437	32,534	64,090	32,464	59,902
39,323	89,641	94,219	116,803	92,857	63,436	65,605	85,861
64,342	61,978	67,998	59,817	101,769	95,774	121,352	69,568
74,276	66,998	40,001	72,069	25,066	77,098	69,922	35,662
74,425	67,202	118,444	53,500	79,294	64,544	86,813	116,269
37,831	89,341	73,341	85,288	138,114	53,402	85,586	82,256
77,539	88,798						

REPORT

1. Use appropriate descriptive statistics to summarize the transmission failure data.
2. Develop a 95% confidence interval for the mean number of miles driven until transmission failure for the population of automobiles that have experienced transmission failure. Provide an interpretation of the interval estimate.
3. Discuss the implication of your statistical finding in terms of the feeling that some owners of the automobiles have experienced early transmission failures.
4. How many repair records should be sampled if the research firm would like the population mean number of miles driven until transmission failure to be estimated to within ±5000 miles at 95% confidence?
5. What other information would you like to gather to more fully evaluate the transmission failure problem?

COMPUTER CASE: QUALITY ASSOCIATES, INC.

Quality Associates, Inc., is a consulting firm that advises its clients about sampling and statistical procedures that can be used to control their clients' manufacturing processes. In one particular application, a client supplied Quality Associates with a sample of 800 observations taken during a time in which this client's process was operating satisfactorily. The sample standard deviation for these data was .21; thus, the population standard deviation was assumed to be .21. Quality Associates then suggested that random samples of size 30 be taken periodically to monitor the process in an ongoing manner. By analyzing the new samples, the client could quickly learn whether the process was operating satisfactorily. When the process was not operating satisfactorily, corrective action could be taken to eliminate the problem. The design specification indicated the process's mean should be 12. The hypotheses test suggested by Quality Associates was as follows:

$$H_0: \ \mu = 12$$
$$H_a: \ \mu \neq 12$$

 QUALITY

Corrective action will be taken any time H_0 is rejected.

The following samples were collected at hourly intervals during the first day of operation of the new statistical process-control procedure. These data are available in the data set named QUALITY.

Sample 1	Sample 2	Sample 3	Sample 4	Sample 1	Sample 2	Sample 3	Sample 4
11.55	11.62	11.91	12.02	11.93	12.00	12.01	12.35
11.62	11.69	11.36	12.02	11.85	11.92	12.06	12.09
11.52	11.59	11.75	12.05	11.76	11.83	11.76	11.77
11.75	11.82	11.95	12.18	12.16	12.23	11.82	12.20
11.90	11.97	12.14	12.11	11.77	11.84	12.12	11.79
11.64	11.71	11.72	12.07	12.00	12.07	11.60	12.30
11.80	11.87	11.61	12.05	12.04	12.11	11.95	12.27
12.03	12.10	11.85	11.64	11.98	12.05	11.96	12.29
11.94	12.01	12.16	12.39	12.30	12.37	12.22	12.47
11.92	11.99	11.91	11.65	12.18	12.25	11.75	12.03
12.13	12.20	12.12	12.11	11.97	12.04	11.96	12.17
12.09	12.16	11.61	11.90	12.17	12.24	11.95	11.94
11.93	12.00	12.21	12.22	11.85	11.92	11.89	11.97
12.21	12.28	11.56	11.88	12.30	12.37	11.88	12.23
12.32	12.39	11.95	12.03	12.15	12.22	11.93	12.25

REPORT

1. Conduct the hypothesis test for each sample at the .01 level of significance and determine what action, if any, should be taken. Provide the test statistic and p-value for each test.
2. Consider the standard deviation for each of the four samples. Does the assumption of .21 for the population standard deviation appear reasonable?

CHAPTER 9

INFERENCES ABOUT A POPULATION PROPORTION

WHAT YOU WILL LEARN IN THIS CHAPTER

- How to use a sample proportion to make inferences about a population proportion
- The characteristics of the sampling distribution of the sample proportion
- How to construct and interpret an interval estimate of a population proportion
- How to conduct hypothesis tests about a population proportion

CONTENTS

Cost Is the Major Reason for Not Going to College

The Council for the Advancement and Support of Education used a Gallup survey of 1001 people between the ages of 13 and 21 in order to learn about their attitudes toward college. The results showed that 48% of the respondents believe that the major reason more students do not go to college is that colleges are too expensive. Although 59% of the high school juniors and seniors in the survey indicated that they had savings available for their college education, 41% did not have savings for college. Nonetheless, 92% of the juniors and seniors plan to go to college or attend some type of vocational or trade school.

An interesting result of the study was that 53% of the people surveyed who were between the ages of 13 and 15 agreed with the statement that "the higher the tuition costs of a college, the better the quality of education a student will receive." Although 41% of the 16- or 17-year-old respondents also agreed with this statement, only 27% of current college students and college graduates felt that way. In addition, 70% of those surveyed believed that public colleges provide just as good an education as private colleges, and 60% felt that 2-year schools are as good as 4-year schools.

A proud family on graduation day.

When asked which factors were "extremely important" in selecting a college, 67% selected availability of particular courses and curricula, whereas 44% cited the academic reputation of the college. In addition, 45% said that the expense of attending a particular college was crucial in the selection decision. In contrast, only 20% of those surveyed said that the social life and athletic reputation of the college were "extremely important."

This *Statistics in Practice* is based on "Cost Seen as Key Reason for Not Going to College," The Cincinnati Enquirer, October 10, 1988.

In Chapters 7 and 8 we showed how a sample mean can be used to develop interval estimates and conduct hypothesis tests about a population mean. In some applications, however, we find that the population proportion is of more interest than the population mean. The *Statistics in Practice* article describes a survey that was conducted to estimate the *proportion* of people between the ages of 13 and 21 who support statements involving a variety of college issues. Several proportions were reported in the article, including the proportion who believe that the major reason for not going to college is that college is too expensive, the proportion who believe that higher tuition costs are associated with a better quality of education, and so on. Other situations include studies designed to estimate the proportion of voters who prefer a particular political candidate, the proportion of adults who have a particular disease, the proportion of college students who are in-state residents, and so on. In situations such as these, we use the sample proportion $\bar{p}$ to make statistical inferences about the population proportion p. This statistical process is depicted in Figure 9.1.

EXAMPLE 9.1 ▶ A sample of 200 married couples selected from throughout the United States showed that for 84 of the couples, both the husband and the wife held full-time jobs. Use the sample

FIGURE 9.1   **THE STATISTICAL PROCESS OF USING A SAMPLE PROPORTION TO MAKE INFERENCES ABOUT A POPULATION PROPORTION**

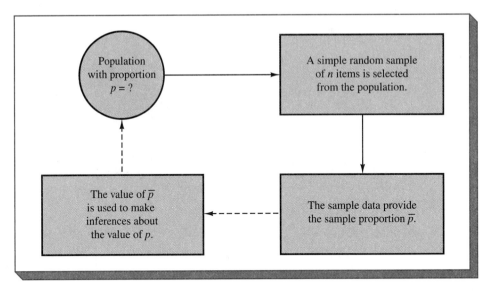

results to estimate the proportion of all married couples in the United States for which both the husband and wife hold full-time jobs. Let

x = the number of couples in the sample for which
both the husband and the wife hold full-time jobs
n = the number of couples in the sample

The sample proportion is computed as follows:

$$\bar{p} = \frac{x}{n} \tag{9.1}$$

Using the given data, we have

$$\bar{p} = \frac{x}{n} = \frac{84}{200} = .42$$

The sample proportion $\bar{p}$ is referred to as the point estimator of p; the value of the point estimator is called the point estimate. In this example, the point estimate of p is $\bar{p} = .42$. Thus, we would estimate that for 42% of the married couples in the United States, both the husband and wife hold full-time jobs.   ◀

NOTES AND
COMMENTS

The notation $\bar{p}$ is used for the sample proportion because it is analogous to the sample mean $\bar{x}$. In Example 9.1 the sample proportion $\bar{p}$ represents the proportion of couples in a sample of 200 in which both the husband and the wife hold full-time jobs. If we define the random variable

$$x_i = \begin{cases} 1 & \text{if for the } i\text{th couple in the sample both the husband} \\ & \text{and the wife hold full-time jobs} \\ 0 & \text{otherwise} \end{cases}$$

then the sample proportion $\bar{p} = \Sigma x_i/n$ is also a sample mean.

9.1 ▽ THE SAMPLING DISTRIBUTION OF $\bar{p}$

If we repeat the sampling process shown in Figure 9.1 several times, we can expect the different random samples to provide different values for the sample statistic $\bar{p}$. This situation would be observed in Example 9.1 if we selected a new random sample of 200 married couples. Perhaps this time we would find 96 couples for which both the husband and the wife hold full-time jobs; in this case, $\bar{p} = 96/200 = .48$.

Because different simple random samples can provide different values for the sample proportion, whenever we make inferences about a population proportion p, we are interested in the probability distribution of all possible values of the sample proportion $\bar{p}$—that is, the *sampling distribution of $\bar{p}$*. If we can identify the properties of the sampling distribution of $\bar{p}$, we can use this distribution to make inferences about a population proportion p in *exactly* the same way that we used the sampling distribution of $\bar{x}$ to make inferences about the population mean μ. Let us begin by defining the properties of the sampling distribution of $\bar{p}$.

EXPECTED VALUE OF $\bar{p}$

The *expected value of $\bar{p}$* denoted $E(\bar{p})$, is the mean value of all possible $\bar{p}$ values that can be observed when simple random samples are selected from a population.

Expected Value of $\bar{p}$

$$E(\bar{p}) = p \tag{9.2}$$

This shows that since the expected value of $\bar{p}$ is equal to the population proportion p, $\bar{p}$ is an unbiased estimator of p.

STANDARD DEVIATION OF $\bar{p}$

The *standard deviation* of $\bar{p}$, denoted $\sigma_{\bar{p}}$, is a measure of dispersion in the possible $\bar{p}$ values. When simple random samples of size n are selected from a population with proportion p, the expression for the standard deviation of $\bar{p}$ is as follows.*

Standard Deviation of $\bar{p}$

$$\sigma_{\bar{p}} = \sqrt{\frac{p(1-p)}{n}} \tag{9.3}$$

The standard deviation of $\bar{p}$ is also called the *standard error of the proportion*.

---

*In this chapter we assume that the population is *large* relative to the sample size, with $n/N \leq .05$. Since the finite population correction factor, $\sqrt{(N-n)/(N-1)}$ is approximately 1 in this case, it is not needed in the formula for the standard deviation or standard error. As a result, equation (9.3) may be used to compute $\sigma_{\bar{p}}$. If $n/N > .05$, however, the finite population correction factor should appear in equation (9.3) with $\sigma_{\bar{p}} = \sqrt{(N-n)/(N-1)} \sqrt{p(1-p)/n}$.

FORM OF THE SAMPLING DISTRIBUTION OF $\bar{p}$

In Chapter 7 we stated that the central limit theorem enables us to conclude that the sampling distribution of $\bar{x}$ can be approximated by a normal probability distribution whenever the sample size is large. Applying the central limit theorem as it relates to the sample proportion $\bar{p}$, we have the following statement about the form of the sampling distribution of $\bar{p}$.

> If a simple random sample of size n is selected from a population with proportion p, the sampling distribution of $\bar{p}$ can be approximated by a normal probability distribution with mean p and standard deviation $\sqrt{p(1-p)/n}$ if the sample size is large. The sample size can be considered large whenever the following two conditions are satisfied:
>
> $$np \geq 5$$
>
> $$n(1-p) \geq 5$$

In the following examples, we will show how the sampling distribution of $\bar{p}$ can be used to make probability statements about the value of the sample proportion $\bar{p}$ based upon the results of a simple random sample.

EXAMPLE 9.2 ▶ Consider the population of fourth-grade students in the Cleveland, Ohio, public school system. Assume that the proportion of students in this population who wear eyeglasses is $p = .25$. Suppose that a random sample of 50 students is selected and $\bar{p}$, the proportion of students in the sample who wear eyeglasses, is computed.

From equation (9.2), we know the mean of the sampling distribution is

$$E(\bar{p}) = p = .25$$

From equation (9.3), we know that the standard deviation of $\bar{p}$ is

$$\sigma_{\bar{p}} = \sqrt{\frac{p(1-p)}{n}} = \sqrt{\frac{.25(1-.25)}{50}} = .0612$$

Finally, since $n = 50$, we have $np = 50(.25) = 12.5$ and $n(1-p) = 50(1-.25) = 37.5$; thus, the large-sample conditions are satisfied and thus the sampling distribution of $\bar{p}$ can be approximated by a normal probability distribution. The sampling distribution of $\bar{p}$ for this example is shown in Figure 9.2.   ◀

EXAMPLE 9.2
(CONTINUED) ▶ Let us use the sampling distribution of $\bar{p}$ in Figure 9.2 to compute the probability that the sample proportion of students who wear eyeglasses will be between .20 and .30. The sampling distribution of $\bar{p}$ and the area showing the probability of $.20 \leq \bar{p} \leq .30$ is shown in Figure 9.3.

Since the sampling distribution of $\bar{p}$ can be approximated by a normal probability distribution, we can use the areas under the curve of the standard normal probability distribution to answer the probability question. In this case, the z value is given

$$z = \frac{\bar{p} - p}{\sigma_{\bar{p}}} \tag{9.4}$$

FIGURE 9.2   **SAMPLING DISTRIBUTION OF $\bar{p}$ FOR THE SAMPLE PROPORTION OF STUDENTS WEARING EYEGLASSES**

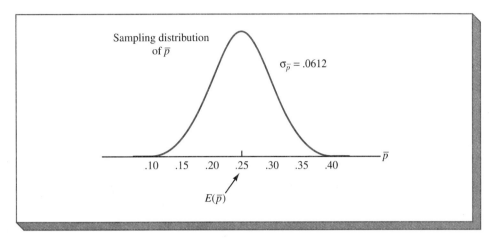

FIGURE 9.3   **SAMPLING DISTRIBUTION OF $\bar{p}$ SHOWING $P(.20 \leq \bar{p} \leq .30)$ FOR EXAMPLE 9.2**

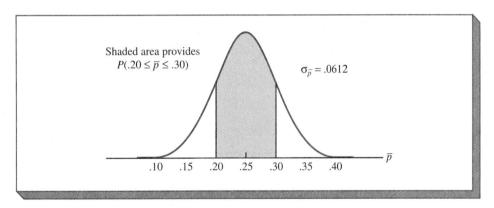

At $\bar{p} = .30$,

$$z = \frac{\bar{p} - p}{\sigma_{\bar{p}}} = \frac{.30 - .25}{.0612} = +.82$$

Looking up $z = .82$ in the table of areas for the standard normal probability distribution, we find an area between .25 and .30 of .2939. At $\bar{p} = .20$,

$$z = \frac{\bar{p} - p}{\sigma_{\bar{p}}} = \frac{.20 - .25}{.0612} = -.82$$

Thus, the area between .20 and .25 is also .2939. Therefore, the probability that the sample proportion $\bar{p}$ is between .20 and .30 is .2939 + .2939 = .5878.   ◀

EXAMPLE 9.3 ▶   According to the Census Bureau (*The Wall Street Journal,* October 20, 1988), 28% of all households are alternative family households; that is, households where people live alone or with unrelated people. If a simple random sample of 100 households is selected, what is the probability that the proportion of alternative family households in the sample will be .37 or more?

Let $\bar{p}$ denote the proportion of alternative family households in the sample. The sampling distribution of $\bar{p}$ corresponding to $p = .28$ and $n = 100$ is shown in Figure 9.4. At $\bar{p} = .37$,

$$z = \frac{\bar{p} - p}{\sigma_{\bar{p}}} = \frac{.37 - .28}{.0449} = 2.00$$

Using the table of areas for the standard normal probability distribution, the area between .28 and .37 is .4772. Thus, the probability the sample proportion $\bar{p}$ will be .37 or more is $.5000 - .4772 = .0228$.   ◀

EXAMPLE 9.4 ▶   At a large university, 40% of all undergraduate women students are members of sororities; thus, $p = .40$. If simple random samples of size 80 were selected from this population, we would expect the different samples to consist of different students and

FIGURE 9.4   **SAMPLING DISTRIBUTION OF $\bar{p}$ FOR THE PROPORTION OF ALTERNATIVE FAMILY HOUSEHOLDS IN A SAMPLE OF 100 HOUSEHOLDS**

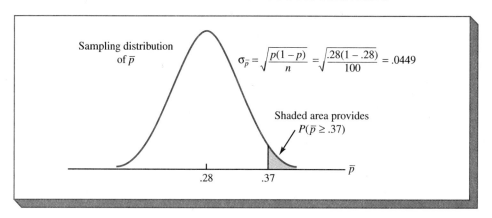

FIGURE 9.5   **SAMPLING DISTRIBUTION OF $\bar{p}$ FOR THE PROPORTION OF WOMEN STUDENTS BELONGING TO A SORORITY**

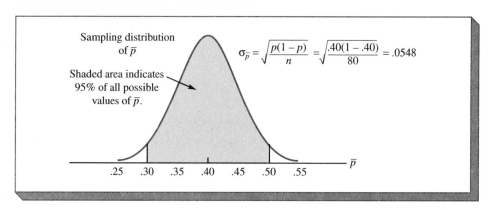

therefore provide a variety of values for the sample proportion $\bar{p}$. Let us compute a range of values for $\bar{p}$ that will include 95% of all possible sample proportions that can be observed.

The sampling distribution of $\bar{p}$, corresponding to $p = .40$ and $n = 80$, is shown in Figure 9.5. Using the standard normal probability distribution, we know that 95% of the area under the curve is between $z = -1.96$ and $z = 1.96$. Thus, the range containing 95% of all possible $\bar{p}$ values is

$$p - 1.96\sigma_{\bar{p}} = .40 - 1.96\sigma_{\bar{p}} = .40 - 1.96(.0548) = .2926$$

$$p + 1.96\sigma_{\bar{p}} = .40 + 1.96\sigma_{\bar{p}} = .40 + 1.96(.0548) = .5074$$

Thus, there is a .95 probability of selecting a simple random sample of 80 women students and obtaining a sample proportion between .2926 and .5074.   ◀

NOTES AND COMMENTS

To understand the rationale behind the large sample requirements that $np \geq 5$ and $n(1 - p) \geq 5$, first note that the population proportion p is equivalent to the probability of success associated with the binomial probability distribution. In fact, the exact sampling distribution of $\bar{p}$ can be determined by using the binomial probability distribution. However, as we saw in Chapter 6, whenever n is large, the normal probability distribution can be used to approximate the binomial distribution. Recall that the normal approximation of the binomial distribution is appropriate if $np \geq 5$ and $n(1 - p) \geq 5$; thus, these requirements also provide guidelines for when the normal approximation is appropriate for the sampling distribution of $\bar{p}$.

EXERCISES

METHODS

1. A simple random sample of size 100 is selected from a population with $p = .40$.
    a. What is the expected value of $\bar{p}$?
    b. What is the standard deviation of $\bar{p}$?
    c. Show the sampling distribution of $\bar{p}$.

SELF TEST

2. A population proportion is .40. A simple random sample of size 200 will be taken with the sample proportion $\bar{p}$ used to estimate the population proportion.
    a. What is the probability that the sample proportion will be within .03 of the population proportion?
    b. What is the probability that the sample proportion will be within .05 of the population proportion?

3. Assume that the population proportion is .55. Compute the standard error of the proportion, $\sigma_{\bar{p}}$, for sample sizes of 100, 200, 500, and 1000. What can you say about the size of the standard error of the proportion as the sample size is increased?

4. The population proportion is .30. What is the probability that a sample proportion will be within ±.04 of the population proportion for each of the following sample sizes?
    a. $n = 100$
    b. $n = 200$
    c. $n = 500$
    d. $n = 1000$
    e. What is the advantage of a larger sample size?

APPLICATIONS

SELF TEST ▷

5. The president of Doerman Distributors, Inc., believes that 30% of the firm's orders come from new or first-time customers. A simple random sample of 100 orders will be used to estimate the proportion of new or first-time customers. The results of the sample will be used to verify the president's claim of $p = .30$.
   a. Assume that the president is correct and $p = .30$. What is the sampling distribution of $\bar{p}$ for this study?
   b. What is the probability that the sample proportion $\bar{p}$ will be between .20 and .40?
   c. What is the probability that the sample proportion will be within .05 of the population proportion $p = .30$?

6. Consider the following population of 25-year-old males, where Yes indicates that the individual has a life insurance policy and No indicates that the individual does not have such a policy.

Individual	1	2	3	4	5	6
Response	Yes	No	No	Yes	No	Yes

   a. Selecting simple random samples of size 4 provides a total of 15 possible samples. List the 15 samples.
   b. Compute the proportion of Yes responses for each sample, and show a graph of the sampling distribution of $\bar{p}$.

7. A research finding published in the *Journal of Clinical Gastroenterology* indicates that one-third of all adults experience gastrointestinal problems after consuming as little as 10 grams of sorbitol, a common artificial sweetener (*Journal of Clinical Gastroenterology, 1988*). A sample of 50 individuals will be taken to verify this result.
   a. Assuming $p = .33$, describe the sampling distribution of $\bar{p}$.
   b. Assuming $p = .33$ and the sample size is increased to $n = 100$, describe the sampling distribution of $\bar{p}$.

8. A 1993 *Newsweek* poll conducted by The Gallup Organization used a national sample of adults to obtain information on the American public views of the job being done by President Bill Clinton (*Newsweek*, July 12, 1993). Assume the poll used a sample of 750 adults and reported a margin of error of ±4 percentage points.
   a. For the question, do you approve or disapprove of the way Bill Clinton is handling his job? assume that 40% of the population would approve; that is, $p = .40$. What is the probability the sample proportion $\bar{p}$ from the sample of 750 adults would be within 4 percentage points of this population proportion?
   b. What is the probability the sample proportion from the sample of 750 adults would be within 2 percentage points of the population proportion?
   c. Why would the Newsweek poll quote a 4% margin of error rather than a 2% margin of error?

9. Louis Harris & Associates, Inc., conducted a survey of 1253 adults to learn how individuals feel about the position of the United States in the global economy (*Business Week*, April 6, 1992). One question asked how concerned the individual was about U.S. industry becoming less competitive in the global economy. Assume that for the entire population, 55% of the adults are very concerned about U.S. industry becoming less competitive. Let $\bar{p}$ be the sample proportion of the polled adults who are very concerned on this issue.
   a. Show the sampling distribution of $\bar{p}$ if the population proportion is $p = .55$.
   b. What is the probability that the Harris poll sample proportion will be within .02 of the population proportion?
   c. What is the probability that the Harris poll sample proportion will be within .03 of the population proportion?

    d. Comment on why the Harris poll stated, "Results should be accurate to within 3 percentage points."

10. *American Association of Individual Investors* (April 1992) provided results from a survey on a range of investment-related topics. One question was, Do you favor relaxed SEC rules for financial reporting by foreign corporations to allow more foreign stocks to be traded in the United States? Assume that the population proportion that answered yes is .42 and that the sample size is 300.

    a. What is the probability that the sample proportion will be within ±.03 of the population proportion?

    b. What is the probability that the sample proportion will be .45 or greater?

    c. What is the probability that the sample proportion will show 50% or more of the population favoring relaxed SEC rules for foreign corporations?

11. A particular county in West Virginia has a 9% unemployment rate. A monthly survey of 800 individuals is conducted by a state agency to monitor the unemployment rate of the county.

    a. Assume that $p = .09$. What is the sampling distribution of $\bar{p}$ when a sample of size 800 is used?

    b. What is the probability that a sample proportion $\bar{p}$ of at least .08 will be observed?

12. Surveys of subscribers of *The Wall Street Journal* and *Investor's Daily* in 1988 provided reader profile information for these two business publications (*Investor's Daily,* March 23, 1989). The profile information included median reader age, percentage of college graduates, percentage in top management, mean income, and mean net worth. Based on data provided by these surveys, assume that 80% of the population of all subscribers of these publications are college graduates. Answer the following questions about the results of the simple random sample of subscribers. The statistic of interest is the sample proportion of subscribers found to be college graduates.

    a. Show the sampling distribution of $\bar{p}$ if a simple random sample of 400 subscribers is used. Assume $p = .80$.

    b. With a sample size of 400, what is the probability that the margin of error will be 3% or less? That is, what is the probability the sample proportion of college graduates among the 400 subscribers will be within .03 of the population proportion $p = .80$?

    c. Answer part (b) when the size of the simple random sample is increased to 750 subscribers.

13. A doctor believes that 80% of all patients having a particular disease will be fully recovered within 3 days after receiving a new drug.

    a. A simple random sample of 20 medical records will be used to develop an estimate of the proportion of patients who were fully recovered within 3 days after receiving the drug. If a data analyst suggests using the normal probability distribution to approximate the sampling distribution of $\bar{p}$, what would you say? Explain.

    b. If the sample of patient records is increased to 60, what is the probability that the sample proportion will be within .10 of the population proportion? (Assume that the population proportion is .80.)

14. Assume that 15% of the items produced in an assembly-line operation are defective, but that the firm's production manager is not aware of this situation. Assume further that 50 parts are tested by the quality assurance department to determine the quality of the assembly operation. Let $\bar{p}$ be the sample proportion defective found by the quality assurance test.

    a. Show the sampling distribution for $\bar{p}$.

    b. What is the probability that the sample proportion will be within ±.03 of the population proportion defective?

    c. If the test shows $\bar{p} = .10$ or more, the assembly-line operation will be shut down to check for the cause of the defects. What is the probability that the sample of 50 parts will lead to the conclusion that the assembly line should be shut down?

15. Baskin Robbins offers frozen yogurt at 900 of its 2500 outlets (*Advertising Age,* April 10, 1989).

   a. What is the population proportion of Baskin Robbins outlets offering frozen yogurt?

   b. Assume that a simple random sample of 40 Baskin Robbins outlets will be used and that the same proportion $\bar{p}$ will be used to estimate the population proportion of Baskin Robbins outlets offering frozen yogurt. What is the probability that the sample proportion will be within .05 of the population proportion?

   c. What is the probability that the sample proportion will be within .05 of the population proportion if the sample size is increased to 120 outlets?

9.2 ▽ INTERVAL ESTIMATION OF A POPULATION PROPORTION

In Chapter 8 we used the value of the sample mean $\bar{x}$ to develop an interval estimate of the population mean μ. The procedure that we used was based on knowledge of the properties of the sampling distribution of $\bar{x}$. In this section we show how the value of the population proportion $\bar{p}$ can be used to develop an interval estimate of the population proportion p. Since the properties of the sampling distribution of $\bar{p}$ are known, we can use the logic we used to develop an interval estimate of μ to develop an interval estimate of p.

For the case of estimating the population proportion p using the sample proportion $\bar{p}$, the sampling error is defined as the absolute value of the difference between $\bar{p}$ and p; that is,

$$\text{Sampling Error} = |\bar{p} - p| \tag{9.5}$$

The probability statements that we can make about the sampling error for the proportion take the following form.

> There is a $1 - \alpha$ probability that the value of the sample proportion will provide a sampling error of $z_{\alpha/2}\sigma_{\bar{p}}$ or less.

The rationale for the preceding statement is the same as we used when the value of a sample mean was used as an estimate of a population mean. Specifically, since we know that the sampling distribution of $\bar{p}$ can be approximated by a normal probability distribution, we can use the value of $z_{\alpha/2}$ and the value of the standard error of the proportion, $\sigma_{\bar{p}}$, to make the probability statement about the sampling error.

Once we see that the probability statement concerning the sampling error is based on $z_{\alpha/2}\sigma_{\bar{p}}$, we can subtract this value from $\bar{p}$ and add it to $\bar{p}$ to obtain an interval estimate of the population proportion. Such an interval estimate is given by

$$\bar{p} \pm z_{\alpha/2}\sigma_{\bar{p}} \tag{9.6}$$

where $1 - \alpha$ is the confidence coefficient. Since $\sigma_{\bar{p}} = \sqrt{p(1 - p)/n}$, we can rewrite equation (9.6) as follows:

$$\bar{p} \pm z_{\alpha/2}\sqrt{\frac{p(1 - p)}{n}} \tag{9.7}$$

However, note that in using equation (9.7) to develop an interval estimate of a population proportion p, the value of p would have to be *known*. Since the value of p is what we are trying to estimate, and is thus *unknown*, we substitute the sample proportion

$\bar{p}$ for p. The general expression for a confidence interval estimate of a population proportion is as follows.*

Interval Estimation of a Population Proportion

$$\bar{p} \pm z_{\alpha/2} \sqrt{\frac{\bar{p}(1 - \bar{p})}{n}} \tag{9.8}$$

where $1 - \alpha$ is the confidence coefficient and $z_{\alpha/2}$ is the z value providing an area of $\alpha/2$ in the upper tail of the standard normal probability distribution.

EXAMPLE 9.5 ▶ Based on a sample of 250 credit-card holders, a department store found that 185 card holders incurred a monthly interest charge on an unpaid balance. Let us develop 95% and 90% confidence interval estimates of the population of credit-card holders who incur a monthly interest charge.

The point estimate of the population proportion p is $\bar{p} = 185/250 = .74$. For a 95% confidence interval, we have $\alpha = .05$ and $z_{\alpha/2} = z_{.025} = 1.96$. Using equation (9.8) provides

$$\bar{p} \pm z_{.025} \sqrt{\frac{\bar{p}(1 - \bar{p})}{n}} = .74 \pm 1.96 \sqrt{\frac{.74(1 - .74)}{250}} = .74 \pm .0544$$

Thus, the 95% confidence interval estimate of the population proportion is .6856 to .7944.

At 90% confidence, $z_{\alpha/2} = z_{.05} = 1.645$. Using equation (9.8) provides

$$.74 \pm 1.645 \sqrt{\frac{.74(1 - .74)}{250}} = .74 \pm .0456$$

Thus, the 90% confidence interval estimate of the population proportion p is .6944 to .7856. ◀

EXAMPLE 9.6 ▶ A survey conducted by the U.S. Department of Labor found that 48 out of 500 heads of households were unemployed. Develop a 98% confidence interval estimate of the proportion of unemployed heads of households in the population.

The point estimate of the population proportion is $\bar{p} = 48/500 = .096$. At a 98% confidence level, $z_{.01} = 2.33$. Using equation (9.8) gives the confidence interval

$$.096 \pm 2.33 \sqrt{\frac{.096(1 - .096)}{500}} = .096 \pm .0307$$

Thus, the 98% confidence interval estimate of the population proportion of unemployed heads of households is .0653 to .1267. ◀

DETERMINING THE SIZE OF THE SAMPLE

Let us consider the question of how large the sample size should be to obtain an estimate of a population proportion at a specified level of precision. The rationale for the

---

*An unbiased estimate of the standard error of the proportion is given by $\sqrt{\bar{p}(1 - \bar{p})/(n - 1)}$. The bias introduced by using n in the denominator does not cause any difficulty because large samples are usually used to estimate a population proportion; in such cases, the numerical difference between the results using n and $n - 1$ is negligible.

sample-size determination in developing an interval estimate of p is very similar to the rationale used in Section 8.3 to determine the sample size for developing an interval estimate of a population mean μ.

Earlier in this section we provided the following probability statement about the sampling error.

> There is a $1 - \alpha$ probability that the value of the sample proportion will provide a sampling error of $z_{\alpha/2}\sigma_{\bar{p}}$ or less.

This is a statement concerning the precision of the interval estimate; smaller values of $z_{\alpha/2}\sigma_{\bar{p}}$ provide more precision. With $\sigma_{\bar{p}} = \sqrt{p(1 - p)/n}$, the size of the sampling error in the above statement is based on the values of $z_{\alpha/2}$, the population proportion p, and the sample size n. For a given confidence coefficient $1 - \alpha$, $z_{\alpha/2}$ can be determined. Then, since the value of the population proportion is fixed, the size of the sampling error mentioned in the precision statement depends upon the sample size n. Larger sample sizes provide better precision.

Let E = the maximum value of the sampling error mentioned in the statement about the desired precision. Thus, we have

$$E = z_{\alpha/2}\sqrt{\frac{p(1 - p)}{n}} \tag{9.9}$$

Solving equation (9.9) for n provides the following formula for the sample size.

Sample Size for Interval Estimation of a Population Proportion

$$n = \frac{z_{\alpha/2}^2 p(1 - p)}{E^2} \tag{9.10}$$

Equation (9.10) shows that in order to determine the sample size, we must have a preliminary idea of the value of the population proportion p. Past data, a preliminary sample, or an educated guess are suggested ways for obtaining the necessary planning value for p. However, in some cases it may be difficult or impossible to specify the planning value. To handle these situations, note that the numerator of equation (9.10) shows that the sample size is proportional to the quantity $p(1 - p)$. In Table 9.1 we show some possible values for this quantity. To be on the conservative side, we need to consider the largest possible value for $p(1 - p)$, since this will provide the largest sample size. As the values in Table 9.1 suggest, if an appropriate planning value for p cannot be obtained, use $p = .50$. This planning value will provide the largest possible recommended sample size. If the population proportion is different than the .50 planning value, the estimate of the sample proportion will have more precision than requested. However, in using $p = .50$ to determine n, we have guaranteed that the desired precision will be obtained.

EXAMPLE 9.7 ▶   A medical experiment is being conducted to determine the recovery rate of patients given a new drug. In particular, the researcher would like to estimate the proportion of patients who fully recover within 2 weeks from when they begin taking the drug. The desired precision of the estimate is expressed as follows: At a 98% level of confidence, the

TABLE 9.1   POSSIBLE VALUES FOR THE QUANTITY $p(1 - p)$

Value of p	Value of $p(1 - p)$
.10	$(.10)(.90) = .09$
.30	$(.30)(.70) = .21$
.40	$(.40)(.60) = .24$
.50	$(.50)(.50) = .25 \leftarrow$ Largest value
.60	$(.60)(.40) = .24$
.70	$(.70)(.30) = .21$
.90	$(.90)(.10) = .09$

sampling error should be .04 or less. For planning purposes, the researcher anticipates that .80 of the patients receiving the new drug will fully recover within the 2-week period. How large should the sample size be in this study?

Using equation (9.10) with $z = 2.33$ for a 98% confidence level, a planning value of $p = .80$, and $E = .04$, we have

$$n = \frac{z_{\alpha/2}^2 p(1 - p)}{E^2} = \frac{(2.33)^2(.80)(1 - .80)}{(.04)^2} = 542.89$$

Rounding up, we obtain a sample size of $n = 543$.   ◀

EXAMPLE 9.8 ▶   A national survey of registered voters is being conducted to determine the proportion of voters who favor a particular candidate. Assume that the desired confidence level is 95% and that the desired maximum sampling error is .02.

1.   How large a sample is needed if it is believed that approximately 35% of the population currently support the candidate?
2.   How large a sample is needed if no information is available on the proportion of voters currently supporting the candidate?

In part (1) the planning value for p is .35. With $z_{.025} = 1.96$ and $E = .02$, the necessary sample size is

$$n = \frac{z_{\alpha/2}^2 p(1 - p)}{E^2} = \frac{(1.96)^2(.35)(1 - .35)}{(.02)^2} = 2184.91$$

Rounding up, we obtain a sample size of $n = 2185$.

In part (2) no planning value is available for p. Using the conservative approach discussed earlier, we base the sample size on a planning value of $p = .50$. Doing so provides

$$n = \frac{z_{\alpha/2}^2 p(1 - p)}{E^2} = \frac{(1.96)^2(.50)(1 - .50)}{(.02)^2} = 2401$$

Note that this recommended sample size of 2401 voters is larger than the sample size of 2185 voters recommended in part (1). This larger size should have been anticipated due to the fact that $p = .50$ is a conservative planning value and guarantees that the sample size will be large enough to satisfy the precision requirement regardless of the actual value of p.   ◀

NOTES AND
COMMENTS

The desired maximum sampling error for estimating a population proportion is almost always .10 or less. In national public opinion polls conducted by organizations such as Gallup and Harris, a .03 or .04 maximum sampling error is generally reported. The use of these values (E in equation 9.10) will generally provide a sample size that is large enough to satisfy the central limit theorem requirements of $np \geq 5$ and $n(1 - p) \geq 5$.

EXERCISES

METHODS

SELF TEST

16. A simple random sample of 400 items provides 100 Yes responses.
   a. What is the point estimate of the proportion of the population who would provide Yes responses?
   b. What is the standard error of the proportion?
   c. Compute the 95% confidence interval for the population proportion.

17. A simple random sample of 800 units generates a sample proportion $\bar{p} = .70$.
   a. Provide a 90% confidence interval for the population proportion.
   b. Provide a 95% confidence interval for the population proportion.

18. In a sample design, the planning value for the population proportion p is given as .35. How large a sample should be taken to be 95% confident that the sample proportion is within .05 of the population proportion?

19. How large a sample should be taken to be 95% confident that the sampling error for the estimation of a population proportion is .03 or less? Assume past data are not available for developing a planning value for p.

APPLICATIONS

20. Researchers who studied prescriptions from 2000 doctors found that most doctors failed to adjust dosages for elderly people or for weight (*Archives of Internal Medicine*, 1987). Assuming that 1600 of the 2000 doctors studied failed to adjust dosages, develop a 95% confidence interval estimate of the population proportion.

SELF TEST

21. A Louis Harris survey of 400 senior executives found that 248 of the executives stated that the U.S. legal system significantly hampers the ability of U.S. companies to compete with Japanese and European companies (*Business Week*, April 13, 1992).
   a. What is the point estimate of the population proportion of executives who believe the legal system hampers the ability to compete?
   b. What is the 90% confidence interval for the population proportion?

22. What is the public opinion toward the U.S. Supreme Court ruling on abortion that stated that women may end a pregnancy during the first 3 months? A survey of 1227 adults in 12 southern states found that only 429 adults favored the Supreme Court ruling (*Atlanta Constitution*, April 13, 1989). What is the 95% confidence interval for the proportion of adults in the 12 southern states favoring the Supreme Court ruling? What is your interpretation of this interval estimate?

23. A *USA Today* poll (January 11, 1990) reported that 79% of Americans say that, if they had evidence, they would turn in a relative who killed someone. Experts were not surprised by the poll results, saying the poll reflects the "socially desirable response" rather than real-life action. The telephone poll of 305 adults was conducted by the Gordon S. Black Corporation. What is the 95% confidence interval for the population proportion?

24. Medical researchers at Cornell University studied the effect of tight neckties on the flow of blood to the head and the possible decrease in the brain's ability to respond to visual information (*Medical Self Care,* July–August 1988). Results of a sample of businessmen found that 67% wear their ties too tight. Assuming a sample size of 250 businessmen, what is the 98% confidence interval estimate of the proportion of the population of businessmen who wear their ties too tight?

25. In an election campaign, a campaign manager requests that a sample of voters be polled to determine public support for the candidate. In a sample of 120 voters, 64 expressed plans to support the candidate.
    a. What is the point estimate of the proportion of voters in the population who will support the candidate?
    b. Develop and interpret the 95% confidence interval for the proportion of voters in the population who will support the candidate.
    c. Given the result from part (b), is the campaign manager justified in feeling confident that the candidate has the support of at least 50% of the voters? Explain.
    d. How many voters should be sampled if we want to estimate the population proportion with a maximum sampling error of 5%? Continue to use the 95% confidence level.

26. It is estimated that 29% of all minimum-wage workers are teenagers (*Boston Globe,* April 12, 1989). Using a 95% confidence level, provide an interval estimate of the proportion of the population of minimum-wage workers who are teenagers if the estimate is based on
    a. a simple random sample of 200
    b. a simple random sample of 600
    c. a simple random sample of 1000
    d. In general, what happens to the interval estimate of a population proportion as the sample size is increased?

27. The Tourism Institute for the State of Florida plans to sample visitors at major beaches throughout the state in order to estimate the proportion of beach visitors who are not residents of Florida. Preliminary estimates are that 55% of the beach visitors are not residents of Florida.
    a. How large a sample should be taken to estimate the proportion of out-of-state visitors to within 3% of the actual value? Use a 95% confidence level.
    b. How large a sample should be taken if the maximum sampling error is increased to 6%?

28. Where do people place their investment dollars? A sample conducted by *The Wall Street Journal* showed that 47% of all investors own some real estate (*The Wall Street Journal,* December 2, 1988).
    a. Compute a 95% confidence interval for the proportion of the population of investors who own some real estate. Assume that a sample size of 250 investors was used in the preceding study.
    b. How large a sample of investors should be used if a 95% confidence level that the maximum sampling error is 5% is desired?

29. A firm provides national survey and interview services designed to estimate the proportion of the population who have certain beliefs or preferences. Typical questions seek to find the proportion favoring gun control, abortion, a particular political candidate, and so on. Assume that all interval estimates of population proportions are conducted at the 95% confidence level. How large a sample size would you recommend if the firm desired the sampling error to be
    a. 3% or less?   b. 2% or less?   c. 1% or less?

30. A Gallup poll asked the following question: "Would you favor or oppose federal legislation banning the manufacture, sale and possession of semiautomatic assault guns,

such as the AK-47?'' (*The Wall Street Journal,* April 7, 1989). If it is believed approximately 75% of the population would favor such legislation, how many individuals should be sampled for each of the following maximum sampling errors? Use a 95% confidence level.

   a. 10%

   b. 7.5%

   c. 5%

   d. 3%

   e. In general, what happens to the sample size as the size of the maximum sampling of error desired decreases?

31. The National Automobile Dealers Association collects data on sales numbers, prices, and usage of both new and used automobiles. One statistic of interest is the percentage of automobiles that are still on the road after 10 years (*U.S. News & World Report,* September 9, 1991).

   a. How large a sample should be taken if we want to be 95% confident that the sample percentage is within 2.5% of the actual percentage of automobiles that are still on the road after 10 years? Use $p = .25$ as a planning value for the population proportion.

   b. Using your sample size in part (a), assume that 357 of the automobiles sampled in 1991 were still on the road after 10 years. Provide the point estimate and a 95% confidence interval for the population proportion.

   c. In 1980, only 21% of automobiles were still on the road after 10 years. What conclusion can you make after viewing the confidence interval results for 1991 in part (b)?

32. A sample of 200 people were asked to identify their major source of news information; 110 stated that their major source was television news coverage.

   a. Construct a 95% confidence interval for the proportion of the people in the population who consider television news their major source.

   b. How large a sample would be necessary to estimate the population proportion with a maximum sampling error of .05 at a 95% confidence level?

33. A survey is to be taken to estimate the proportion of high school graduates in a particular school district who plan to attend college. How large a sample of students should be selected if the survey is to provide a 95% confidence of reporting a sample proportion that is within .025 of the population proportion? Use $p = .35$ as a planning value for the proportion of high school students who plan to attend college.

9.3   HYPOTHESIS TESTS ABOUT A POPULATION PROPORTION

Hypothesis tests about a population proportion are based on the sampling distribution of $\bar{p}$. The mechanics of conducting a test follow the procedure used in Chapter 8 to make hypothesis tests about a population mean μ. The only difference is that in this section we use the sample proportion $\bar{p}$ and the sampling distribution of $\bar{p}$ in the analysis.

We begin by formulating a null hypothesis and an alternative hypothesis about the value of the population proportion. We then consider the sampling distribution of $\bar{p}$ under the assumption that the null hypothesis is true. Based on whether the hypotheses are one-tailed or two-tailed and the level of significance, a critical value(s) is selected for the test statistic z. Then, using the value of the sample proportion $\bar{p}$, we compute a value for z; comparing this value to the critical value enables us to determine whether or not the null hypothesis should be rejected.

EXAMPLE 9.9 ▶ A newspaper article contains the statement that, nationwide, 60% of all college seniors have a job prior to graduation. The director of a college placement office at a large university is interested in testing this claim for her university. The hypotheses about the proportion of the students having a job prior to graduation are as follows:

$$H_0: p = .60$$

$$H_a: p \neq .60$$

If a random sample of 75 recent graduates showed that 40 had a job prior to graduation, what conclusion can be drawn? Use a .05 level of significance.

Under the assumption that the null hypothesis is true, we have $np = 75(.60) = 45$ and $n(1 - p) = 75(.40) = 30$. Since both of these values exceed 5, the sampling distribution of $\bar{p}$ can be approximated by a normal probability distribution. Assuming $p = .60$, the standard error of the proportion is

$$\sigma_{\bar{p}} = \sqrt{\frac{p(1 - p)}{n}} = \sqrt{\frac{.60(1 - .60)}{75}} = .0566$$

Thus

$$z = \frac{\bar{p} - p}{\sigma_{\bar{p}}} \qquad (9.11)$$

is a standard normal random variable that can be used to determine the number of standard errors an observed value of a sample proportion $\bar{p}$ is from the hypothesized value p. Using a .05 level of significance, the rejection regions for the two-tailed hypothesis test are shown in Figure 9.6. The rejection rule is

Reject H_0 if $z < -1.96$ or $z > +1.96$

With the sample proportion $\bar{p} = 40/75 = .5333$ and a hypothesized value for the population proportion $p = .60$, using equation (9.11) we have the following value for z:

$$z = \frac{\bar{p} - p}{\sigma_{\bar{p}}} = \frac{.5333 - .60}{.0566} = -1.18$$

This value of z is not in the rejection region of Figure 9.6. Thus, the hypothesis that 60% of the graduating seniors have a job prior to graduation cannot be rejected.

Using the table of areas for the standard normal probability distribution, the area in the tail of the distribution at $z = 1.18$ is $.5000 - .3810 = .1190$. For a two-tailed test, we

FIGURE 9.6   REJECTION REGION FOR A TWO-TAILED TEST WITH $\alpha = .05$

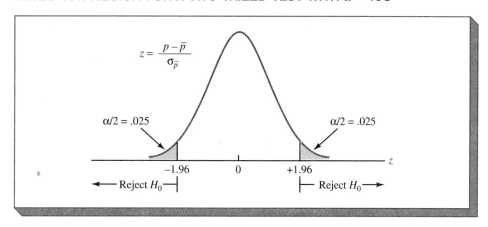

have a p-value of $2 \times .1190 = .2380$. With $.2380 > \alpha$, we see that the p-value criterion also indicates that the null hypothesis cannot be rejected. The interpretation of the p-value for tests about a population proportion is the same as we used for tests about a population mean. ◀

EXAMPLE 9.10 ▶ The manager of an Italian restaurant is considering opening a carryout food service. However, the manager is concerned that not all individuals placing orders by phone actually pick up the order. If 90% or less of the phone orders will be picked up, the restaurant will not have a profitable carryout operation. However, if it can be concluded that more than 90% of the phone orders will be picked up, the carryout operation will be a worthwhile addition for the restaurant. The hypothesis test of interest is

$$H_0: \ p \leq .90$$

$$H_a: \ p > .90$$

This one-tailed test indicates that the restaurant should implement the carryout operation if H_0 can be rejected. During a 2-week test period, 234 orders in a sample of 250 phone-in orders were picked up. Using a .05 level of significance, what conclusion should be drawn?

With the one-tailed test and $\alpha = .05$, rejection of the null hypothesis will occur in the upper tail of the distribution with a critical value of $z = 1.645$. Thus, if the data show $z > 1.645$, the null hypothesis can be rejected.

Given the hypothesized value of $p = .90$ and the sample proportion $\bar{p} = 234/250 = .936$, the value of z is as follows:

$$z = \frac{\bar{p} - p}{\sigma_{\bar{p}}} = \frac{\bar{p} - p}{\sqrt{p(1 - p)/n}} = \frac{.936 - .90}{\sqrt{.90(1 - .90)/250}} = 1.90$$

Since this value is greater than 1.645, the null hypothesis is rejected. The manager of the restaurant should be safe in concluding that the proportion of carryout orders picked up for the population exceeds .90. Thus, it is recommended the restaurant begin the carryout food service. With $z = 1.90$, the corresponding p-value is $.5000 - .4713 = .0287$. Since the p-value is less than $\alpha = .05$, the p-value criterion indicates H_0 should be rejected. ◀

EXAMPLE 9.11 ▶ During a water shortage in Florida, restaurants were asked not to serve water with meals unless requested to do so by the customers. In the initial 3-month period, 45% of the customers served at a particular restaurant requested water with their meal. Recently, the restaurant placed a card at each table describing the water-shortage problem and pointing out that considering the drinking water, the ice, and the water to wash the glass, it requires 24 ounces of water for every 8 ounces of water served. The restaurant would like to use a statistical test to determine if placing these cards at each table significantly decreases the proportion of customers requesting water with their meal. If a sample of 150 customers showed 53 customers ordering water, what is your conclusion? Use a .02 level of significance.

The hypothesis test is as follows:

$$H_0: \ p \geq .45$$

$$H_a: \ p < .45$$

Rejection of H_0 will occur in the lower tail of the distribution. With an area of $\alpha = .02$ in the lower tail, the table of areas for the standard normal distribution shows the critical z value to be -2.05. Thus, the rejection rule is to reject H_0 if $z < -2.05$.

When we use the hypothesized value, $p = .45$, and the sample proportion $\bar{p} = 53/150 = .3533$, the value of z is

$$z = \frac{\bar{p} - p}{\sigma_{\bar{p}}} = \frac{\bar{p} - p}{\sqrt{p(1-p)/n}} = \frac{.3533 - .45}{\sqrt{.45(1-.45)/150}} = -2.38$$

Since $z = -2.38$ is in the rejection region, the null hypothesis H_0: $p \geq .45$ is rejected. The restaurant can conclude that the cards have helped reduce the proportion of customers who order water with their meal.

NOTES AND COMMENTS

We have not shown a procedure for small-sample hypothesis tests involving population proportions. In the small-sample case, the sampling distribution of $\bar{p}$ follows the binomial distribution, and thus the normal approximation is not applicable. More advanced texts show how hypothesis tests are conducted for this situation. However, in practice, small-sample tests are rarely conducted for a population proportion.

EXERCISES

METHODS

34. Consider the following hypothesis test:

$$H_0: p \leq .50$$

$$H_a: p > .50$$

A sample of 200 provided a sample proportion $\bar{p} = .57$.
a. Using $\alpha = .05$, what is the rejection rule?
b. Compute the value of the test statistic z. What is your conclusion?

SELF TEST   **35.** Consider the following hypothesis test:

$$H_0: p = .20$$

$$H_a: p \neq .20$$

A sample of 400 provided a sample proportion of $\bar{p} = .175$.
a. Using $\alpha = .05$, what is the rejection rule?
b. Compute the value of the test statistic z.
c. What is the p-value?
d. What is your conclusion?

36. Consider the following hypothesis test:

$$H_0: p \geq .75$$

$$H_a: p < .75$$

A sample of 300 is selected. Use $\alpha = .05$. Provide the value of the test statistic z, the p-value, and your conclusion for each of the following sample results:
a. $\bar{p} = .68$ b. $\bar{p} = .72$ c. $\bar{p} = .70$ d. $\bar{p} = .77$

APPLICATIONS

37. The Honolulu Board of Water Supply suggested that a water-by-request rule be adopted at restaurants on the island of Oahu to conserve water. A restaurant owner stated that 30% of the patrons never touch their water (*The Honolulu Advertiser,* December 28, 1991). Test H_0: $p = .30$ versus H_a: $p \neq .30$. Assume a sample of 480 patrons at restaurants showed that 128 patrons never touched their water. Test the owner's claim at a .05 level of significance. What is the p-value and what is your conclusion?

SELF TEST ▷ 38. The American Association of Individual Investors conducted a survey to learn about member responses to the October 1987 stock market crash. (*AAII Journal,* May 1988). Results indicated that 11.3% of the AAII members owned stock on margin. A sample of 200 individual clients with a large brokerage firm showed that 29 owned stock on margin. Let p be the proportion of all the brokerage firm's clients owning stock on margin. Use H_0: $p \leq .113$ and H_a: $p > .113$ in a hypothesis test. Does the sample information indicate that the proportion of the brokerage firm's clients owning stock on margin is greater than .113? Use $\alpha = .05$.

39. The director of a college placement office claims that at least 50% of graduating seniors have made employment commitments 1 month prior to graduation. At a .05 level of significance, what is your conclusion if a sample of 100 seniors shows that 42 made employment commitments 1 month prior to graduation? Should the director's claim be rejected? What is the p-value?

40. A magazine claims that 25% of its readers are college students. A random sample of 200 readers is taken. It is found that 42 of these readers are college students. Use a .10 level of significance to test H_0: $p = .25$ and H_a: $p \neq .25$. What is the p-value?

41. A new television series must prove that it has more than 25% of the viewing audience after its initial 13-week run in order to be judged successful. Assume that in a sample of 400 households, 112 were watching the series.
    a. At a .10 level of significance, can the series be judged successful based on the sample information?
    b. What is the p-value for the sample results? What is your hypothesis-testing conclusion?

42. An accountant believes that the company's cash-flow problems are a direct result of the slow collection of accounts receivable. The accountant claims that at least 70% of the current accounts receivable are over 2 months old. A sample of 120 accounts receivable shows 78 over 2 months old. Test the accountant's claim at the .05 level of significance.

43. In 1987, it was reported that 26.8% of adults in California smoked (*Newsweek,* April 6, 1992). In 1988, California passed Proposition 99, a popular initiative designed to help discourage smoking. Researchers conducted a study in 1991 to learn about California's smoking rate and to identify the impact of Proposition 99 on the smoking population. Assume that a sample of 1000 adults showed 222 smokers. At a .01 level of significance, can it be concluded that the proportion of Californians smoking in 1991 is less than the proportion of Californians smoking in 1987? What is your conclusion?

44. The Gordon S. Black Corporation surveyed 439 working parents with a total of 736 children younger than 12; 73% of those surveyed said that one parent would stay at home if money were not a factor (*Democrat and Chronicle,* November 28, 1988). Is there sufficient evidence to conclude that p, the population proportion who would stay at home, is greater than .70? Use $\alpha = .02$.

45. A fast-food restaurant plans to initiate a special offer that will enable customers to purchase specially designed drink glasses featuring well-known cartoon characters. If more than 15% of the customers will purchase the glasses, the special offer will be initiated. A preliminary test has been set up at several locations, and 88 of 500 customers have purchased the glasses. Should the special glass offer be made? Conduct a hypothesis test that will support your decision. Use a .01 level of significance. What is your recommendation?

46. It has been claimed that 5% of the students at one college make blood donations during a given year. If, in a random sample, 10 of 250 students have given blood during the past year, test $H_0: p = .05$ versus $H_a: p \neq .05$. Use a .05 level of significance in reaching your conclusion. What is the p-value?

SUMMARY

In this chapter we discussed procedures for making statistical inferences about a population proportion. After describing the properties of the sampling distribution of $\bar{p}$, we showed how to develop an interval estimate and conduct a hypothesis test for a population proportion. In the large-sample case where both $np \geq 5$ and $n(1 - p) \geq 5$, the central limit theorem enables us to use the normal probability distribution to approximate the sampling distribution of $\bar{p}$.

The procedure for developing a confidence interval estimate of a population proportion uses the same logic as the procedure for developing a confidence interval estimate of a population mean. The essential difference is that the sampling distribution of $\bar{p}$ is used as the basis for developing the confidence interval instead of the sampling distribution of $\bar{x}$.

A procedure was presented for determining the sample size that will meet a desired precision when estimating a population proportion. This procedure requires a planning value for the population proportion p. If p is unknown, using a value of $p = .50$ provides a sample size that will satisfy the precision requirements regardless of the actual value of the population proportion.

The procedure for testing hypotheses about a population proportion follows the logic used for testing hypotheses about a population mean. Examples of both one-tailed and two-tailed hypothesis tests about a population proportion were presented.

GLOSSARY

Sampling distribution of $\bar{p}$ The probability distribution showing all possible values of the sample proportion $\bar{p}$.

Standard error of the proportion The standard deviation of all possible values of the sample proportion $\bar{p}$.

KEY FORMULAS

Sample Proportion

$$\bar{p} = \frac{x}{n} \tag{9.1}$$

Expected Value of $\bar{p}$

$$E(\bar{p}) = p \tag{9.2}$$

Standard Deviation of $\bar{p}$

$$\sigma_{\bar{p}} = \sqrt{\frac{p(1 - p)}{n}} \tag{9.3}$$

Standard Normal Random Variable

$$z = \frac{p - \bar{p}}{\sigma_{\bar{p}}} \tag{9.4}$$

Interval Estimation of a Population Proportion

$$\bar{p} \pm z_{\alpha/2} \sqrt{\frac{\bar{p}(1 - \bar{p})}{n}} \qquad (9.8)$$

Sample Size for Interval Estimation of a Population Proportion

$$n = \frac{z_{\alpha/2}^2 p(1 - p)}{E^2} \qquad (9.10)$$

REVIEW QUIZ

TRUE/FALSE

1. The sample proportion $\bar{p}$ is not a point estimator of the population proportion p.
2. The central limit theorem cannot be applied to the sampling distribution of $\bar{p}$.
3. The standard error of the proportion, $\sigma_{\bar{p}}$, depends on the value of the population proportion p.
4. In computing confidence intervals for a population proportion, the sample proportion $\bar{p}$ can be used to obtain an estimate of the standard error of the proportion, $\sigma_{\bar{p}}$.
5. In determining the sample size, the larger the planning value for p, the larger the sample size.
6. In conducting hypothesis tests about a population proportion, the value of p specified in the null hypothesis is used to compute the standard error of the proportion, $\sigma_{\bar{p}}$.

MULTIPLE CHOICE

Use the following information for Questions 7–9. A random sample of 300 voters showed 47% in favor of a certain ballot proposal.

7. Which of the following best describes the form of the sampling distribution of the sample proportion?
   a. When standardized, it is exactly the standard normal probability distribution.
   b. When standardized, it is the t distribution.
   c. It is approximately normal because $n \geq 30$.
   d. It is approximately normal because $n\bar{p} \geq 5$ and $n(1 - \bar{p}) \geq 5$.
8. The estimate of σ_p is
   a. .0164
   b. .0255
   c. .0288
   d. .0350
9. A 90% confidence interval estimate for the population proportion of voters favoring the proposal is
   a. .38 to .56
   b. .4412 to .4988
   c. .4136 to .5264
   d. none of the above
10. In choosing a sample size for a public-opinion survey, what hypothesized value of the population proportion will lead to the largest sample size when the confidence level and the maximum sampling error are specified?
    a. $p = .1$
    b. $p = .5$
    c. $p = .99$
    d. The confidence level must be known before an answer can be given.

SUPPLEMENTARY EXERCISES

47. Consider a population of six people, four of whom are college graduates.
   a. Identify the 15 different possible samples of four people that can be selected from this population.
   b. Compute the sample proportion of college graduates for each sample.
   c. Show the sampling distribution of $\bar{p}$ by showing the probability distribution of the values of $\bar{p}$ found in part (b).

48. Assume that the proportion of persons having a college degree is $p = .35$.
   a. Explain how the sampling distribution of $\bar{p}$ results from random samples of size 80 being used to estimate the proportion of individuals having a college degree.
   b. Show the sampling distribution for $\bar{p}$ in this case.
   c. If the sample size is increased to 200, what happens to the sampling distribution of $\bar{p}$? Compare the standard error of the proportion for the $n = 80$ and $n = 200$ alternatives.

49. CinemaScore is a movie research firm that uses a sample of film audiences to learn why members of the audience chose to attend a particular motion picture (*USA Today,* April 11, 1989). Possible responses include the subject matter, the cast, and other reasons. Assume that for a particular new motion picture, 65% of the audience population chose the movie due to subject matter. In addition, assume that a simple random sample will be used to estimate the proportion of the population who chose the movie due to subject matter.
   a. What is the probability that a simple random sample of 100 people will provide a sample proportion within .04 of the population proportion?
   b. What is the probability that a simple random sample of 200 people will provide a sample proportion within .04 of the population proportion?
   c. Assume that we would like to select a simple random sample that will provide a 90% chance of obtaining a sample proportion within .04 of the population proportion. How many people should be included in the simple random sample?

50. A market research firm conducts telephone surveys with a 40% historical response rate. What is the probability that in a new sample of 400 telephone numbers, at least 150 individuals will cooperate and respond to the questions? In other words, what is the probability the sample proportion will be at least $150/400 = .375$?

51. A production run is not acceptable for shipment if a sample of 100 items contains 5% or more defective items. If a production run has a population proportion defective of $p = .10$, what is the probability that $\bar{p}$ will be at least .05?

52. The proportion of individuals insured by the All-Driver Automobile Insurance Company who have received at least one traffic ticket during a 5-year period is .15.
   a. Show the sampling distribution of $\bar{p}$ if a random sample of 150 insured individuals is used to estimate the proportion having received at least one ticket.
   b. What is the probability that the sample proportion will be within .03 of the population proportion?

53. Historical records show that .50 of all orders placed at Big Burger fast-food restaurants include a soft drink. With a simple random sample of 40 orders, what is the probability that between .45 and .55 of the sampled orders will include a soft drink?

54. A Boston University survey of 1600 employees in the Northeast found that 11% of the respondents were part of the "traditional" work force; that is, married males with wives at home full time (Boston University: "Balancing Job and Homelife Study," 1987, as reported in *The Wall Street Journal,* November 1, 1988). Assuming that this is a representative sample of all employees in the Northeast, develop a 95% confidence interval estimate of the population proportion.

55. The New Orleans Beverage Company has been experiencing problems with the automatic machine that places labels on bottles. The company desires an estimate of the percentage of bottles that have improperly applied labels. A simple random sample of 400 bottles resulted in 18 bottles with improperly applied labels. Using these data, develop a 90% confidence interval for the population proportion of bottles with improperly applied labels.

56. H. G. Forester and Company is a distributor of lumber supplies throughout the southwest United States. Management at H. G. Forester would like to check a shipment of over 1 million pine boards to determine if excessive warpage exists for the boards. A sample of 50 boards resulted in the identification of 7 boards with excessive warpage. With these data, develop a 95% confidence interval for the proportion of boards defective in the whole shipment.

57. A University of Michigan study (January 1992) reported that drug use among college students had continued its decade-long decline. However, the survey showed alcohol consumption remained steady. Among the 1400 college students surveyed, 602 indicated they had five or more drinks within the 2 weeks prior to the survey. Compute a 95% confidence interval for the proportion of all college students who have had five or more drinks within the 2-week period.

58. Towers Perrin, a compensation consultant, asked 500 U.S. companies if they encouraged quality by giving top performers recognition and/or rewards such as cash and stock (*Business Week,* December 1991). Results showed that 56% of the companies used recognition to encourage quality, while 26% of the companies used cash and stock rewards. Develop 95% confidence intervals for the population in terms of both the proportion that uses recognition and the proportion that uses cash and stock rewards to encourage quality.

59. A Time/CNN telephone poll of 1400 American adults asked "Where would you rather go in your spare time?" (*Time,* April 6, 1992). The top response by 504 adults was a shopping mall.
  a. What is the point estimate of the proportion of adults who would prefer going to a shopping mall in their spare time?
  b. At 95% confidence, what is the maximum sampling error associated with this estimate?

60. A well-known bank credit-card firm is interested in estimating the proportion of credit-card holders that carry a nonzero balance at the end of the month and incur an interest charge. Assume that the desired precision for the proportion estimate is 3% at a 98% confidence level.
  a. How large a sample should be recommended if it is anticipated that roughly 70% of the firm's cardholders carry a nonzero balance at the end of the month?
  b. How large a sample would be recommended if no planning value for the population proportion can be specified?

61. *Newsweek* (April 6, 1992) reported data on the percentage of adults who smoke in the United States. Assume that the study designed to collect the data for this report had a preliminary estimate that 30% of the population smoke.
  a. How large a sample should be taken to estimate the current proportion of smokers in the population to within 2% at 95% confidence?
  b. Assume that the study used your sample-size recommendation in part (a) and found 555 smokers. What is the point estimate of the proportion of smokers in the population, and what is the 95% confidence interval?

62. A survey was used to learn about how people view the quality of education provided by public school systems (*Fortune,* March, 1989). Assume that a 90% confidence level, the study was designed to have a 4% margin of error when the population proportion is approximately .60.
  a. How many individuals should be included in this sample?

b. Using your sample size, what is the 90% confidence interval for the population proportion rating public schools "fair/poor" if 313 people in the sample respond "fair/poor"?

c. Using your sample size, what is the 90% confidence interval for the population proportion responding "public schools have deteriorated in the last 10 years" if 260 respond Yes to this statement?

d. Do the interval estimates in parts (b) and (c) provide the desired 4% margin of error? Explain.

63. Sixty percent of Americans believe that business profits are distributed unfairly (*General Social Surveys,* National Opinion Research Center, University of Chicago). Suppose a sample of 40 midwesterners showed that 27 believe business profits are distributed unfairly.

a. Do these results justify making the inference that a larger proportion of midwesterners believe that business profits are distributed unfairly? Use $\alpha = .05$.

b. What is the p-value?

64. In the past, the Dumont Clothing Store has recorded 72% charge purchases and 28% cash purchases. A sample of 200 recent purchases shows that 160 were charge purchases. Does this suggest a change in the paying practices of the Dumont customers? Test with $\alpha = .05$. What is the p-value?

65. The manager of K-Mark Supermarkets assumes that at least 30% of the Saturday customers purchase the price-reduced special advertised in the Friday newspaper. Use $\alpha = .05$ and test the validity of the manager's assumption if a sample of 250 customers shows that 60 purchased the advertised special.

66. The Gallup Organization conducted a survey of 1350 people for the National Occupational Information Coordinating Committee, a panel that Congress created to make better use of job information (*The Arizona Republic,* January 12, 1990). A research question related to the study was, Do individuals hold jobs that they planned to hold or do they hold jobs for such reasons as chance or lack of choice? Let p indicate the population proportion of individuals who hold jobs that they planned to hold.

a. If the hypotheses are stated H_0: $p \geq .50$ and H_a: $p < .50$, discuss the research hypothesis H_a in terms of what the researcher is investigating.

b. The Gallup poll reported 41% of the respondents hold jobs that they planned to hold. What is your conclusion at a .01 level of significance? Discuss.

67. A well-known doctor hypothesized that 75% of women wear shoes that are too small. A 1991 study of 356 women by the American Orthopedic Foot and Ankle Society found 313 women wore shoes that were at least a size too small (*New York Times,* March 10, 1991). Test H_0: $p = .70$ and H_a: $p \neq .70$ at $\alpha = .01$. What is your conclusion?

68. The filling machine for a production operation must be adjusted if more than 8% of all items being produced are underfilled. A random sample of 80 items from the day's production contained 9 underfilled items. Does the sample evidence indicate that the filling machine should be adjusted? Use $\alpha = .02$. What is the p-value?

69. A radio station in a major resort area announced that at least 90% of the hotels and motels would be full for the Memorial Day weekend. The station went on to advise listeners to make reservations in advance if they planned to be in the resort over the weekend. On Saturday night a sample of 58 hotels and motels showed 49 with a no-vacancy sign and 9 with vacancies. What is your reaction to the radio station's claim based on the sample evidence? Use $\alpha = .05$ in making this statistical test. What is the p-value for the sample results?

70. It is assumed that at least 90% of juvenile first-time criminals are given probation upon their admission of guilt. Test this hypothesis at a .02 level of significance, if a sample of 92 juvenile criminal convictions shows 78 juveniles receiving probation. What is the p-value?

CHAPTER 10

INFERENCES ABOUT
MEANS AND
PROPORTIONS WITH
TWO POPULATIONS

WHAT YOU WILL LEARN IN THIS CHAPTER

- How to construct interval estimates and conduct hypothesis tests about the difference between the means of two populations
- When and how to use the t distribution to make inferences about the difference between the means of two populations
- The difference between independent and matched samples
- How to compute a pooled variance estimate
- How to construct interval estimates and conduct hypothesis tests about the difference between the proportions of two populations

CONTENTS

Nicotine Chewing Gum Helps Stop Smoking

A group of physicians from Copenhagen, Denmark, reported the results of a study on the effectiveness of nicotine chewing gum in helping people stop smoking. In a sample of 173 smokers, 60 were classified as having a high dependence on nicotine and 113 were classified as having a medium or low dependence. Of the 60 subjects classified as highly dependent on nicotine, 27 were given gum containing 4 milligrams of nicotine and the remaining 33 were given gum containing 2 milligrams of nicotine. For the 113 smokers with medium or low dependence, 60 were given gum containing 2 milligrams of nicotine, and 53 were given a placebo gum.

In carrying out the experiment, a double-blind study was used. That is, neither the subjects nor the individuals who monitored their performance knew what type of gum each smoker had been given. At the end of 6 weeks, 1 year, and 2 years of treatment, each smoker was chemically tested to determine whether he or she had abstained from smoking. The results obtained are summarized in the two tables below. The entries in each table show the percentage of the group that was abstinent after each of the 3 follow-up periods.

With the exception of the 2-mg nicotine gum versus the placebo gum at 1 year, each of the differences in outcomes can be shown to be

Even with good intentions, many have difficulty quitting the smoking habit.

statistically significant at the 5% level. Thus, the study indicates that the effectiveness of nicotine gum is not due to random effects and that it is related to the amount of nicotine in the gum. The researchers concluded that, "The use of nicotine gum in appropriate doses should be helpful to persons who are attempting to stop smoking."

Type of Gum	Sample Size	Percentage of Highly Dependent Smokers Abstinent After		
		6 Weeks	1 Year	2 Years
4-mg gum	27	81.5%	44.4%	33.3%
2-mg gum	33	54.5%	12.1%	6.1%

Type of Gum	Sample Size	Percentage of Medium- or Low-Dependence Smokers Abstinent After		
		6 Weeks	1 Year	2 Years
2-mg gum	60	73.3%	38.3%	28.3%
Placebo gum	53	41.5%	22.6%	9.4%

This *Statistics in Practice* is based on "Effect of Nicotine Chewing Gum in Combination with Group Counseling on the Cessation of Smoking," *New England Journal of Medicine,* January 1988.

In the preceding three chapters we showed how confidence intervals and hypothesis tests can be developed for population means and population proportions. However, the statistical procedures we have discussed thus far have considered only single-population situations. In this chapter we consider cases where *two populations* are involved. Specifically, we will be selecting random samples and performing statistical analyses that will enable us to draw conclusions about the difference between the means and/or the proportions for two populations. An example of this type of situation appeared in *Statistics in Practice,* where differences between the proportion of smokers who remain abstinent were reported for smokers given either nicotine gum (population 1) or a placebo (population 2). We begin the study of two-population situations by showing how to estimate the difference between the means of two populations.

10.1 ESTIMATION OF THE DIFFERENCE BETWEEN THE MEANS OF TWO POPULATIONS: INDEPENDENT SAMPLES

In many practical situations, we are faced with two separate populations where the difference between the means of the two populations is of prime importance. We know from Chapters 7 and 8 that we can take a simple random sample from a single population and use the sample mean $\bar{x}$ to estimate the population mean. In the two-population case, we will select two separate and independent simple random samples, one from population 1 and another from population 2. Let

μ_1 = mean of population 1
μ_2 = mean of population 2
$\bar{x}_1$ = sample mean for the simple random sample from population 1
$\bar{x}_2$ = sample mean for the simple random sample from population 2

The difference between the two population means is $\mu_1 - \mu_2$. The point estimator of $\mu_1 - \mu_2$ is as follows.

**Point Estimator of the Difference Between
the Means of Two Populations**

$$\bar{x}_1 - \bar{x}_2 \tag{10.1}$$

Thus, we see that the point estimator of the difference between two population means is the difference between the sample means of the two independent simple random samples. This situation is depicted in Figure 10.1.

SAMPLING DISTRIBUTION OF $\bar{x}_1 - \bar{x}_2$

In the study of the difference between the means of two populations, $\bar{x}_1 - \bar{x}_2$ is the point estimator of interest. This point estimator, just like the point estimators discussed previously, has its own sampling distribution. If we can identify the sampling distribution of $\bar{x}_1 - \bar{x}_2$, we can use it to develop an interval estimate of the difference between the two population means in much the same way that we used the sampling distribution of $\bar{x}$ to develop an interval estimate of a single population mean μ. The properties of the sampling distribution of $\bar{x}_1 - \bar{x}_2$ are as follows.

Sampling Distribution of $\bar{x}_1 - \bar{x}_2$

Expected Value: $E(\bar{x}_1 - \bar{x}_2) = \mu_1 - \mu_2$ **(10.2)**

Standard Deviation: $\sigma_{\bar{x}_1 - \bar{x}_2} = \sqrt{\dfrac{\sigma_1^2}{n_1} + \dfrac{\sigma_2^2}{n_2}}$ **(10.3)**

where

σ_1 = standard deviation of population 1
σ_2 = standard deviation of population 2
n_1 = sample size for the simple random sample
    from population 1
n_2 = sample size for the simple random sample   from population 2

Distribution form: If the sample sizes are both *large* ($n_1 \geq 30$ and $n_2 \geq 30$), the sampling distribution of $\bar{x}_1 - \bar{x}_2$ can be approximated by a normal probability distribution.

FIGURE 10.1   **STATISTICAL PROCESS OF USING THE DIFFERENCE BETWEEN SAMPLE MEANS TO ESTIMATE THE DIFFERENCE BETWEEN POPULATION MEANS**

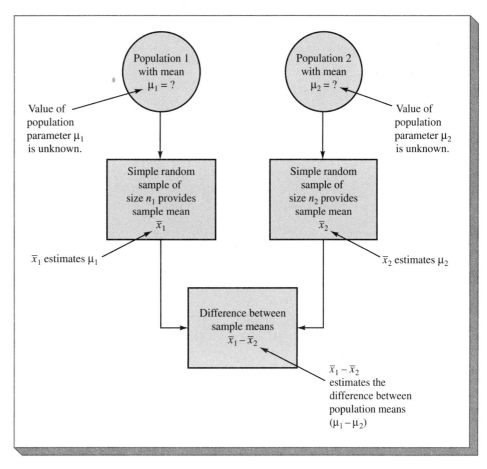

Figure 10.2 shows the sampling distribution of $\bar{x}_1 - \bar{x}_2$ and its relationship to the individual sampling distributions of $\bar{x}_1$ and $\bar{x}_2$.

Let us use the sampling distribution of $\bar{x}_1 - \bar{x}_2$ to develop an interval estimate of the difference between the means of two populations. We shall consider two cases, one where the sample sizes are large ($n_1 \geq 30$ and $n_2 \geq 30$) and the other where one or both sample sizes are small ($n_1 < 30$ and/or $n_2 < 30$).

LARGE-SAMPLE CASE

In the large-sample case, we know that the sampling distribution of $\bar{x}_1 - \bar{x}_2$ can be approximated by a normal probability distribution. With this approximation, we can use the following expression to develop an interval estimate of the difference between the means of the two populations.

FIGURE 10.2   SAMPLING DISTRIBUTION OF $\bar{x}_1 - \bar{x}_2$ AND ITS RELATIONSHIP TO THE INDIVIDUAL SAMPLING DISTRIBUTIONS OF $\bar{x}_1$ AND $\bar{x}_2$

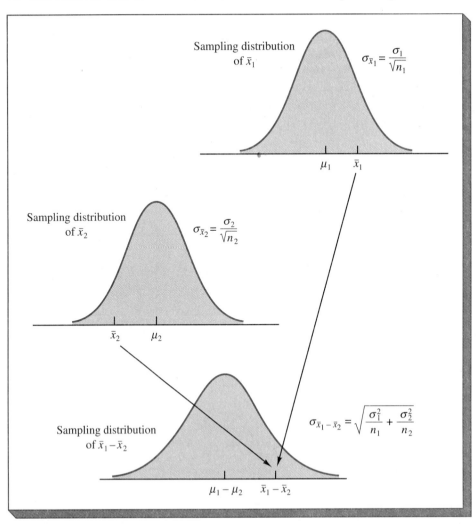

> **Interval Estimate of the Difference Between the Means of Two Populations**
> **(Large-Sample Case with $n_1 \geq 30$ and $n_2 \geq 30$)**
>
> $$\bar{x}_1 - \bar{x}_2 \pm z_{\alpha/2}\sigma_{\bar{x}_1-\bar{x}_2} \qquad (10.4)$$
>
> where $1 - \alpha$ is the confidence coefficient.

Since the population standard deviations σ_1 and σ_2 are unknown, we cannot use equation (10.3) to calculate $\sigma_{\bar{x}_1-\bar{x}_2}$. However, we can use the sample standard deviations as estimates of the population standard deviations and estimate $\sigma_{\bar{x}_1-\bar{x}_2}$ as follows.

> **Point Estimator of $\sigma_{\bar{x}_1-\bar{x}_2}$**
>
> $$s_{\bar{x}_1-\bar{x}_2} = \sqrt{\frac{s_1^2}{n_1} + \frac{s_2^2}{n_2}} \qquad (10.5)$$

With large sample sizes, $s_{\bar{x}_1-\bar{x}_2}$ can be accepted as a good estimate of $\sigma_{\bar{x}_1-\bar{x}_2}$.

EXAMPLE 10.1 ▶ The Educational Testing Service conducted a study to investigate possible differences between the scores of males and females on the Scholastic Aptitude Test (SAT) (*Journal of Educational Measurement,* Spring 1987). A random sample of 562 females and 852 males provided a sample mean SAT verbal score of $\bar{x}_1 = 547$ for the females and $\bar{x}_2 = 525$ for the males. The sample standard deviations were $s_1 = 83$ for the females and $s_2 = 78$ for the males. Using a 95% confidence level, estimate the difference between the mean SAT verbal scores for the population of females and the population of males.

The point estimate of the difference between mean scores is $\bar{x}_1 - \bar{x}_2 = 547 - 525 = 22$ points. Thus, we estimate that females score an average of 22 points higher on the verbal test than the males. Using $s_1 = 83$ and $s_2 = 78$ to estimate σ_1 and σ_2, equation (10.5) can be used to estimate $\sigma_{\bar{x}_1-\bar{x}_2}$.

$$s_{\bar{x}_1-\bar{x}_2} = \sqrt{\frac{s_1^2}{n_1} + \frac{s_2^2}{n_2}} = \sqrt{\frac{(83)^2}{562} + \frac{(78)^2}{852}} = 4.4$$

Thus, the 95% confidence interval estimate of the difference between mean verbal scores can be computed using expression (10.4):

$$\bar{x}_1 - \bar{x}_2 \pm z_{\alpha/2}\sigma_{\bar{x}_1-\bar{x}_2} = 547 - 525 \pm 1.96(4.404)$$

$$= 22 \pm 8.6$$

Thus, the 95% confidence interval estimate of 22 ± 8.6, or approximately 22 ± 9, tells us that the population mean verbal score for females is from 13 to 31 points higher than the population mean verbal score for males. ◀

EXAMPLE 10.2 ▶ A firm in Atlanta, Georgia, has department stores located in the inner city and stores located in suburban shopping centers. In a study designed to learn about the different characteristics of the inner-city and suburban customer populations, a sample of 60 inner-city customers and a sample of 80 suburban customers were taken. The data obtained on customer ages is summarized below.

Store Type	Sample Size	Sample Mean Age	Sample Standard Deviation
Inner city	60	$\bar{x}_1 = 40$ years	$s_1 = 9$ years
Suburban	80	$\bar{x}_2 = 35$ years	$s_2 = 10$ years

Using 90% and 99% confidence intervals, estimate the difference between the mean ages of the two populations of customers.

Using equation (10.5), we first develop an estimate of $\sigma_{\bar{x}_1 - \bar{x}_2}$.

$$s_{\bar{x}_1 - \bar{x}_2} = \sqrt{\frac{s_1^2}{n_1} + \frac{s_2^2}{n_2}} = \sqrt{\frac{(9)^2}{60} + \frac{(10)^2}{80}} = 1.61$$

At 90%, $z_{\alpha/2} = z_{.05} = 1.645$. Using expression (10.4), we have

$$40 - 35 \pm 1.645 \sqrt{\frac{(9)^2}{60} + \frac{(10)^2}{80}} = 5 \pm 1.645(1.61)$$

$$= 5 \pm 2.65$$

Thus, the 90% confidence estimate of the difference between the mean ages of the two populations is 2.35 to 7.65 years with the inner-city stores having the older customers.

At 99%, $z_{\alpha/2} = z_{.005} = 2.575$. Using expression (10.4), we have

$$40 - 35 \pm 2.575 \sqrt{\frac{(9)^2}{60} + \frac{(10)^2}{80}} = 5 \pm 2.575(1.61)$$

$$= 5 \pm 4.15$$

Hence, the 99% confidence interval for the difference between the mean ages of the two populations is .85 to 9.15 years. As expected, the interval estimate with a 99% confidence level results in a wider interval than the interval estimate with a 90% confidence level. ◄

INTERVAL ESTIMATION: SMALL-SAMPLE CASE

Let us now consider the interval-estimation procedure for the difference between the means of two populations whenever one or both sample sizes are less than 30—that is, $n_1 < 30$ and/or $n_2 < 30$. This will be referred to as the small-sample case.

In Chapter 8 we presented a procedure for interval estimation of the mean for a single population whenever a small sample was used. Recall that the procedure required the assumption that the population had a normal probability distribution. With the sample standard deviation s used as an estimate of the population standard deviation σ, the t distribution was used to develop an interval estimate of the population mean.

To develop interval estimates for the two-population small-sample case, we will make two assumptions about the two populations and the samples selected from the two populations:

1. Both populations have normal probability distributions.
2. The variances of the populations are equal ($\sigma_1^2 = \sigma_2^2 = \sigma^2$).

Given these assumptions, the sampling distribution of $\bar{x}_1 - \bar{x}_2$ is normally distributed regardless of the sample sizes involved. The expected value of $\bar{x}_1 - \bar{x}_2$ is $\mu_1 - \mu_2$. Because of the equal variances assumption, equation (10.3) can be written

$$\sigma_{\bar{x}_1 - \bar{x}_2} = \sqrt{\frac{\sigma^2}{n_1} + \frac{\sigma^2}{n_2}} = \sqrt{\sigma^2 \left(\frac{1}{n_1} + \frac{1}{n_2} \right)} \tag{10.6}$$

The sampling distribution of $\bar{x}_1 - \bar{x}_2$ is shown in Figure 10.3.

If the variance of the populations is known, expression (10.4) can be used to develop the interval estimate of the difference between the means of the two populations. However, in most cases, σ^2 is unknown; thus, the two sample variances s_1^2 and s_2^2 must be used to develop the estimate of σ^2 in equation (10.6). Since equation (10.6) is based on the assumption that $\sigma_1^2 = \sigma_2^2 = \sigma^2$, we do not need separate estimates of σ_1^2 and σ_2^2. In fact, we can combine the data from both samples to provide the best single estimate of σ^2. The process of combining the results of two independent random samples to provide one estimate of σ^2 is referred to as *pooling*. The *pooled estimator* of σ^2, denoted by s^2, is a weighted average of the two sample variances s_1^2 and s_2^2. The formula for the pooled estimator of σ^2 is as follows.

Pooled Estimator of σ^2

$$s^2 = \frac{(n_1 - 1)s_1^2 + (n_2 - 1)s_2^2}{n_1 + n_2 - 2} \tag{10.7}$$

With the pooled estimator s^2 and equation (10.6), we can compute the following estimator of the standard deviation of $\bar{x}_1 - \bar{x}_2$.

FIGURE 10.3   SAMPLING DISTRIBUTION OF $\bar{x}_1 - \bar{x}_2$ WHEN THE POPULATIONS HAVE NORMAL DISTRIBUTIONS WITH EQUAL VARIANCES

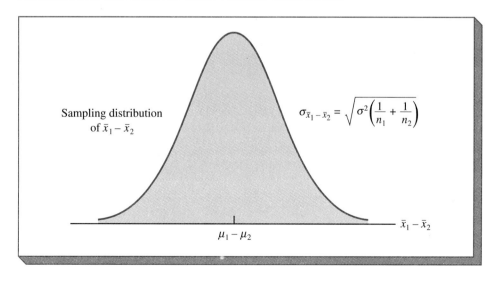

Point Estimator of $\sigma_{\bar{x}_1 - \bar{x}_2}$ When $\sigma_1^2 = \sigma_2^2$

$$s_{\bar{x}_1 - \bar{x}_2} = \sqrt{s^2 \left(\frac{1}{n_1} + \frac{1}{n_2} \right)} \qquad (10.8)$$

The t distribution can now be used to compute an interval estimate of the difference between the means of the two populations. Since there are $n_1 - 1$ degrees of freedom associated with the random sample from population 1 and $n_2 - 1$ degrees of freedom associated with the random sample from population 2, the t distribution will have $n_1 + n_2 - 2$ degrees of freedom. The interval-estimation procedure is as follows.

Interval Estimate of the Difference Between the Means of Two Populations (Small-Sample Case with $n_1 < 30$ and/or $n_2 < 30$)

$$\bar{x}_1 - \bar{x}_2 \pm t_{\alpha/2} s_{\bar{x}_1 - \bar{x}_2} \qquad (10.9)$$

where the t value is based on a t distribution with $n_1 + n_2 - 2$ degrees of freedom and where $1 - \alpha$ is the confidence coefficient.

EXAMPLE 10.3 ▶ An urban-planning group is interested in estimating the difference between the mean household incomes for two neighborhoods in a large metropolitan area. Independent random samples of households in the neighborhoods provided the following results.

	Neighborhood 1	Neighborhood 2
Sample size	n_1 = 8 households	n_2 = 12 households
Sample mean	$\bar{x}_1$ = \$15,700	$\bar{x}_2$ = \$13,500
Sample standard deviation	s_1 = \$700	s_2 = \$850

The point estimate of the difference between the mean household incomes for the two neighborhoods is

$$\bar{x}_1 - \bar{x}_2 = \$15,700 - \$13,500 = \$2200$$

Making the assumptions that the incomes are normally distributed in both neighborhoods and that the population variances are equal, equation (10.9) can be used to develop a 95% confidence interval for the difference between the mean incomes for the two neighborhoods. First, we use equation (10.7) to compute the pooled estimate of σ^2 as follows:

$$s^2 = \frac{(n_1 - 1)s_1^2 + (n_2 - 1)s_2^2}{n_1 + n_2 - 2}$$

$$= \frac{(8 - 1)(700)^2 + (12 - 1)(850)^2}{8 + 12 - 2}$$

$$= \frac{7(490,000) + 11(722,500)}{18} = 632,083$$

With $n_1 + n_2 - 2 = 18$, we use the t distribution with 18 degrees of freedom to find $t_{.025} = 2.101$. (See Table 2 in Appendix B.) Thus, using expression (10.9), we obtain the 95% confidence interval of

$$\bar{x}_1 - \bar{x}_2 \pm t_{.025}\sqrt{s^2\left(\frac{1}{n_1} + \frac{1}{n_2}\right)} = 2200 \pm (2.101)\sqrt{(632,083)\left(\frac{1}{8} + \frac{1}{12}\right)}$$

$$= 2200 \pm (2.101)(362.88)$$

$$= 2200 \pm 762$$

Subtracting and adding 762 to the estimate of 2200 provides the 95% confidence interval of $1438 to $2962.   ◀

EXAMPLE 10.4 ▶ A sociology class project involves the study of dating practices on a major college campus. One aspect of the study compared the frequency of dating by freshman women and freshman men. A sample of 15 freshman women showed a mean of 8.2 dates per month with a standard deviation of 2.5. A sample of 10 freshman men showed a mean of 6.2 dates per month with a standard deviation of 2.2. What is the 90% confidence interval estimate of the difference between the mean number of dates per month for freshman women and freshman men?

Making the assumptions that the populations are normally distributed with equal variances, the pooled estimate of the variance is

$$s^2 = \frac{(n_1 - 1)s_1^2 + (n_2 - 1)s_2^2}{n_1 + n_2 - 2}$$

$$= \frac{(14)(2.5)^2 + 9(2.2)^2}{15 + 10 - 2} = 5.7$$

At 90% confidence with $n_1 + n_2 - 2 = 15 + 10 - 2 = 23$ degrees of freedom, the t distribution table shows $t_{.05} = 1.714$. Thus, we have

$$\bar{x}_1 - \bar{x}_2 \pm t_{.05}\sqrt{s^2\left(\frac{1}{n_1} + \frac{1}{n_2}\right)} = 8.2 - 6.2 \pm (1.714)\sqrt{5.7\left(\frac{1}{15} + \frac{1}{10}\right)}$$

$$= 2.0 \pm 1.714(.975)$$

$$= 2.0 \pm 1.67$$

Thus, the 90% confidence interval shows that the population of freshman women students average from .33 to 3.67 more dates per month than do the population of freshman men.   ◀

NOTES AND COMMENTS

1. In developing a confidence interval estimate of the difference between two population means, we can refer to either of the populations as population 1. For example, assume that two independent samples were selected from the population of service stations in Miami and in Boston. The purpose of the samples is to estimate the difference between the mean price for unleaded gasoline in the two cities. Assume that the Miami sample resulted in a mean of $1.04 per gallon and the Boston sample resulted in a mean of $1.09 per gallon. If we let μ_1 denote the mean price in Miami and μ_2 denote the mean

—Continues on next page

—Continued from previous page

price in Boston, then the point estimate of the difference between the population means would be $1.04 − $1.09 = −$.05. Alternatively, we could have defined μ_1 as the mean price in Boston and μ_2 as the mean price in Miami. In this case, the point estimate of the difference between the means of the two populations, $1.09 − $1.04 = $.05, would be positive. Many experimenters prefer to use the second approach to avoid the use of a negative sign in presenting the results.

2. The use of the t distribution in the small-sample procedure is based on the assumptions that both populations have a normal probability distribution and that $\sigma_1^2 = \sigma_2^2$. Fortunately, this procedure is a *robust* statistical procedure, meaning that it is relatively insensitive to these assumptions. For instance, if $\sigma_1^2 \neq \sigma_2^2$, the procedure provides acceptable results if n_1 and n_2 are approximately equal.

EXERCISES

METHODS

SELF TEST

1. Consider the results shown below for independent random samples taken from two populations.

Sample 1	Sample 2
$n_1 = 50$	$n_2 = 35$
$\bar{x}_1 = 13.6$	$\bar{x}_2 = 11.6$
$s_1 = 2.2$	$s_2 = 3.0$

a. What is the point estimate of the difference between the two population means?
b. Provide a 90% confidence interval for the difference between the two population means.
c. Provide a 95% confidence interval for the difference between the two population means.

2. Consider the following results for independent random samples taken from two populations.

Sample 1	Sample 2
$n_1 = 10$	$n_2 = 8$
$\bar{x}_1 = 22.5$	$\bar{x}_2 = 20.1$
$s_1 = 2.5$	$s_2 = 2.0$

a. What is the point estimate of the difference between the two population means?
b. What is the pooled estimate of the population variance?
c. Develop a 95% confidence interval for the difference between the two population means.

3. Consider the following data for independent random samples taken from two populations.

Sample 1	10	12	9	7	7	9
Sample 2	8	8	6	7	4	9

a. Compute the two sample means.
b. Compute the two sample standard deviations.
c. What is the point estimate of the difference between the two population means?
d. What is the pooled estimate of the population variance?
e. Develop a 95% confidence interval for the difference between the two population means.

APPLICATIONS

4. A survey of salaries for professors of statistics was conducted at the University of Iowa (*Amstat News,* December 1989). The survey compared annual salaries for faculty at the ranks of associate professor and assistant professor. A sample of 41 professors who have held the rank of associate professor for 1–2 years showed a sample mean of $40,800. A sample of 71 professors who have held the rank of assistant professor for 1–2 years showed a sample mean of $36,000. Assume that the standard deviations for associate professors and assistant professors were $3600 and $1400, respectively.
   a. What is the point estimate of the difference between the mean annual salaries for the population of associate professors and the population of assistant professors?
   b. What is the 95% confidence interval for the difference between the two population means?

5. A college admissions board is interested in estimating the difference between the mean grade point averages of students from two high schools. Independent simple random samples of students at the two high schools provided the following results.

Mt. Washington	Country Day
$n_1 = 46$	$n_2 = 33$
$\bar{x}_1 = 3.02$	$\bar{x}_2 = 2.72$
$s_1 = .38$	$s_2 = .45$

   a. What is the point estimate of the difference between the means of the two populations?
   b. Develop a 90% confidence interval for the difference between the two population means.
   c. Develop a 95% confidence interval for the difference between the two population means.

6. *Working Woman* (January 1989) reported on a survey of salary for women in a variety of occupations. The mean entry-level salary for women accountants in large public accounting firms was $23,750. The mean entry-level salary for women accountants in large corporations was $21,000. Assume that the following sample sizes and sample standard deviations were available.

Public Accounting	Corporations
$n_1 = 220$	$n_2 = 250$
$s_1 = \$1200$	$s_2 = \$1000$

Provide a 95% confidence interval for the difference between the mean entry-level salaries for women in large public accounting firms and in large corporations.

7. The Butler County Bank and Trust Company is interested in estimating the difference between the mean credit-card balances at two of its branch banks. Independent random samples of credit-card customers generated the following results.

Branch 1	Branch 2
$n_1 = 32$	$n_2 = 36$
$\bar{x}_1 = \$500$	$\bar{x}_2 = \$375$
$s_1 = \$150$	$s_2 = \$130$

a. Develop a point estimate of the difference between the mean balances at the two branches.
b. Develop a 99% confidence interval for the difference between the mean balances.

SELF TEST ▷ 8. *Traveler* magazine (August 1993) reported on the hours of refresher training time airline crews were receiving in order to improve both emergency and security operations for airline passengers. Delta Airlines was reported as having an average of 10 hours of refresher training while USAir was reported as having an average of 8 hours. Assume that these data were based upon samples of training times for 10 Delta crews and 12 USAir crews. Assume sample standard deviations of 1.2 hours and 1.0 hours respectively.
   a. Develop a point estimate of the difference between the population mean refresher training times for Delta and USAir.
   b. Develop a 95% confidence interval for the difference between the population mean training times.
   c. What assumptions were made to compute the interval estimate in part (b)?

9. Production quantities for two assembly-line workers are shown below. Each data value indicates the amount produced during a randomly selected 1-hour period.

Worker 1	20	18	21	22	20
Worker 2	22	18	20	23	24

   a. Develop a point estimate of the difference between the population mean hourly production rates of the two workers. Which worker appears to have the higher population mean production rate?
   b. Develop a 90% confidence interval for the difference between the population mean production rates of the two workers. Does the confidence interval provide support for the conclusion that the worker having the higher sample mean production rate is the worker with the overall higher production rate? Explain.

10. A sample of 15 graduates from Eastern University showed that the mean time until they received their first job promotion was 5.2 years, with a sample standard deviation of 1.4 years. A sample of 12 graduates from Midwestern University showed a sample mean of 2.7 years, with a standard deviation of 1.1 years.
   a. What is the pooled estimate of the population variance?
   b. Develop a 95% confidence interval for the difference between the mean time until the first job promotion for the populations of Eastern and Midwestern University graduates.

10.2 HYPOTHESIS TESTS ABOUT THE DIFFERENCE BETWEEN THE MEANS OF TWO POPULATIONS: INDEPENDENT SAMPLES

In this section we present procedures that can be used to test hypotheses about the difference between the means of two populations. The methodology is again divided into large-sample ($n_1 \geq 30$, $n_2 \geq 30$) and small-sample ($n_1 < 30$ and/or $n_2 < 30$) cases. Note that the procedures assume that independent simple random-samples have been selected from each of the populations.

LARGE-SAMPLE CASE

Hypothesis tests about the difference between the means of two populations use the same logic as the hypothesis tests in Chapters 8 and 9, except that the tests are based on the sample statistic $\bar{x}_1 - \bar{x}_2$. The null and alternative hypotheses for a two-tailed test is written as follows:

$$H_0: \mu_1 - \mu_2 = 0$$

$$H_a: \mu_1 - \mu_2 \neq 0$$

As we did for the hypothesis-testing procedures presented in Chapters 8 and 9, we will make the tentative assumption that H_0 is true. Using the difference between the sample means as the point estimator of the difference between the population means, we consider the sampling distribution of $\bar{x}_1 - \bar{x}_2$ when H_0 is true. In the large-sample case, $n_1 \geq 30$ and $n_2 \geq 30$, the sampling distribution of $\bar{x}_1 - \bar{x}_2$ can be approximated by a normal probability distribution (see Figure 10.4), and thus the following test statistic can be used:

$$z = \frac{(\bar{x}_1 - \bar{x}_2) - (\mu_1 - \mu_2)}{\sqrt{\sigma_1^2/n_1 + \sigma_2^2/n_2}} \tag{10.10}$$

Since σ_1^2 and σ_2^2 are generally unknown, we will use s_1^2 and s_2^2 as their estimates in order to compute the above test statistic.

The value of z given by equation (10.10) can be interpreted as the number of standard errors $\bar{x}_1 - \bar{x}_2$ is from the value of $\mu_1 - \mu_2$ specified in H_0. For $\alpha = .05$ and thus

---

FIGURE 10.4   **SAMPLING DISTRIBUTION OF $\bar{x}_1 - \bar{x}_2$ WITH H_0: $\mu_1 - \mu_2 = 0$**

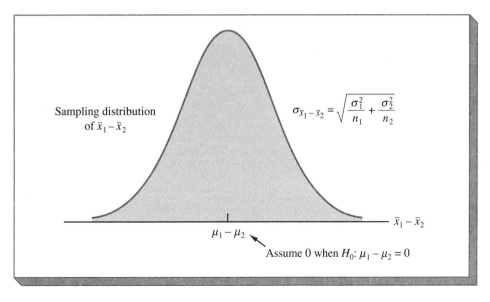

Sampling distribution of $\bar{x}_1 - \bar{x}_2$

$$\sigma_{\bar{x}_1 - \bar{x}_2} = \sqrt{\frac{\sigma_1^2}{n_1} + \frac{\sigma_2^2}{n_2}}$$

$\bar{x}_1 - \bar{x}_2$

$\mu_1 - \mu_2$

Assume 0 when $H_0: \mu_1 - \mu_2 = 0$

$z_{\alpha/2} = z_{.025} = 1.96$, the rejection region for the two-tailed hypothesis test is shown in Figure 10.5. The rejection rule is as follows:

$$\text{Reject } H_0 \text{ if } z < -1.96 \text{ or if } z > +1.96$$

EXAMPLE 10.5▶   A medical research study was conducted to determine whether there is a difference between the effectiveness of two pain-relief medicines used for headaches. Over a 6-month period, a sample of individuals used one of the medicines and a second sample of individuals used the other medicine. Data collected during the study showed the time required to receive pain relief. Letting

$$\mu_1 = \text{mean pain-relief time for medicine 1}$$
$$\mu_2 = \text{mean pain-relief time for medicine 2}$$

the hypothesis test is expressed as follows:

$$H_0: \mu_1 - \mu_2 = 0$$

$$H_a: \mu_1 - \mu_2 \neq 0$$

Note that these hypotheses are equivalent to $H_0: \mu_1 = \mu_2$ and $H_a: \mu_1 \neq \mu_2$. In either case, if the null hypothesis is rejected, the test will have shown that the two medicines differ in terms of pain-relief speed.

Using the following results, conduct the test and draw a conclusion comparing the two medicines. Use $\alpha = .05$.

	Individuals Using Medicine 1	Individuals Using Medicine 2
Sample size	$n_1 = 248$	$n_2 = 225$
Sample mean	$\bar{x}_1 = 24.8$ minutes	$\bar{x}_2 = 26.1$ minutes
Sample standard deviation	$s_1 = 3.3$ minutes	$s_2 = 4.2$ minutes

With $\alpha = .05$ and a two-tailed test, the critical z values are -1.96 and $+1.96$. The rejection rule can be stated as follows:

$$\text{Reject } H_0 \text{ if } z < -1.96 \text{ or } z > +1.96$$

Using equation (10.10), the value of z is

$$z = \frac{(\bar{x}_1 - \bar{x}_2) - (\mu_1 - \mu_2)}{\sqrt{\sigma_1^2/n_1 + \sigma_2^2/n_2}} = \frac{(24.8 - 26.1) - 0}{\sqrt{(3.3)^2/248 + (4.2)^2/225}}$$

$$= \frac{-1.3}{.35} = -3.71$$

With this value of z, we reject H_0 and conclude that there is a significant difference between the mean pain-relief times for the two medicines.

Using the p-value approach, the table of areas in the standard normal probability distribution only goes up to $z = 3.09$. With this value of z, the area in the tail of the distribution is .001. Thus, $z = -3.71$ has an area less than .001 in the tail of the distribution.

FIGURE 10.5   REJECTION REGION FOR THE TWO-TAILED HYPOTHESIS TEST
WITH $\alpha = .05$

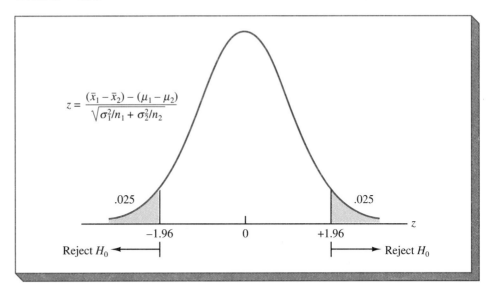

$$z = \frac{(\bar{x}_1 - \bar{x}_2) - (\mu_1 - \mu_2)}{\sqrt{\sigma_1^2/n_1 + \sigma_2^2/n_2}}$$

.025   .025

−1.96   0   +1.96   z

Reject H_0 ◄———┤   ├———► Reject H_0

For this two-tailed hypothesis test, the p-value would be less than $.001 \times 2 = .002$. Thus, since the p-value is less than $\alpha = .05$, we would reject H_0. ◄

EXAMPLE 10.6 ▶   It has been suggested that college students learn more and obtain higher grades in small classes (40 students or less) than in large classes (150 students or more). To test this claim, a university assigned a professor to teach a small class and a large class of the same course. At the end of the course, students from the two classes were given the same final exam. Final grade differences for the two classes would provide a basis for testing the difference between the small-class and large-class situations.

Letting μ_1 denote the mean exam score for the population of students taking a small class and μ_2 denote the mean exam score for the population of students taking a large class, the one-tailed hypothesis test is as follows:

$$H_0: \mu_1 - \mu_2 \leq 0$$

$$H_a: \mu_1 - \mu_2 > 0$$

Rejecting H_0 will lead to the conclusion that the mean exam score is higher in the small class.

Viewing the students currently taking the courses as independent samples from the populations of students in small and large classes, the following data were obtained.

	Individuals Taking Small Class	Individuals Taking Large Class
Sample size	$n_1 = 35$	$n_2 = 170$
Sample mean	$\bar{x}_1 = 74.2$	$\bar{x}_2 = 71.7$
Sample standard deviation	$s_1 = 14$	$s_2 = 13$

Using $\alpha = .05$, test the hypothesis and draw a conclusion about the mean exam scores for the small and large classes.

For the one-tailed test, the critical z value is 1.645. The corresponding rejection rule becomes

$$\text{Reject } H_0 \text{ if } z > +1.645$$

The point estimate of $\mu_1 - \mu_2$ is given by $\bar{x}_1 - \bar{x}_2 = 74.2 - 71.7 = 2.5$; thus, the small class group has a sample mean that is 2.5 points higher than the sample mean for the large class. Using equation (10.10), the value of z becomes

$$z = \frac{(\bar{x}_1 - \bar{x}_2) - (\mu_1 - \mu_2)}{\sqrt{\sigma_1^2/n_1 + \sigma_2^2/n_2}} = \frac{(2.5) - 0}{\sqrt{(14)^2/35 + (13)^2/170}}$$

$$= \frac{2.5}{2.57} = .97$$

Since $z < 1.645$, we can not reject H_0. As a result, we can not conclude that the population mean exam score for small classes is greater than the population mean exam score for large classes. With $z = .97$, the p-value for the test is $.5000 - .3340 = .1660$.

The analysis of the data indicates H_0 can not be rejected. However, we must note that these results were based on an experiment with only one professor. While the evidence is not sufficient to conclude that grade performance improves in smaller classes, further testing may be warranted before arriving at a final conclusion. ◀

SMALL-SAMPLE CASE

Let us now consider hypothesis tests about the difference between the means of two populations for the small-sample case; that is, where $n_1 < 30$ and/or $n_2 < 30$. The procedure we will use is based on the t distribution with $n_1 + n_2 - 2$ degrees of freedom. As we did in Section 10.1, we assume that both populations have normal probability distributions and that the variances of the populations are equal.

EXAMPLE 10.7▶ A new computer software package has been developed to help systems analysts reduce the time required to design, develop, and implement an information system. To evaluate the benefits of the new software package, a random sample of 24 systems analysts was selected. Each analyst was given specifications for a hypothetical information system, and 12 of the analysts were instructed to produce the information system using current technology. The other 12 analysts were first trained in the use of a new software package and then instructed to use it to produce the information system.

In this study, there are two populations: a population of systems analysts using the current technology and a population of systems analysts using the new software package. In terms of the time required to complete the information systems design project, the population means are as follows:

$$\mu_1 = \text{the mean project-completion time for analysts using the current technology}$$
$$\mu_2 = \text{the mean project-completion time for analysts using the new software package}$$

The researcher in charge of the software-evaluation project hopes to show that the new software package will provide a smaller mean project-completion time. Thus, the researcher is looking for evidence to conclude that μ_2 is less than μ_1; in this case, the difference between the two population means $\mu_1 - \mu_2$ will be greater than zero. In formulating the hypotheses, the research hypothesis $\mu_1 - \mu_2 > 0$ is stated as the alternative hypothesis. Thus, we have

$$H_0: \mu_1 - \mu_2 \leq 0$$

$$H_a: \mu_1 - \mu_2 > 0$$

In tentatively assuming H_0 to be true, we are taking the position that using the new software package takes the same time or perhaps even longer than the current technology. The researcher is looking for evidence to reject H_0 and conclude that the new software package possesses a smaller mean completion time.

Let us assume that the 24 analysts completed the study with the results shown below:

	Current Technology	New Software Package
Sample size	$n_1 = 12$	$n_2 = 12$
Sample mean	$\bar{x}_1 = 325$ hours	$\bar{x}_2 = 288$ hours
Sample standard deviation	$s_1 = 40$ hours	$s_2 = 44$ hours

Under the assumption that the variances of the populations are equal, equation (10.7) is used to compute the pooled estimate of σ^2.

$$s^2 = \frac{(n_1 - 1)s_1^2 + (n_2 - 1)s_2^2}{n_1 + n_2 - 2} = \frac{11(40)^2 + 11(44)^2}{12 + 12 - 2} = 1768$$

The test statistic for the small-sample case is

$$t = \frac{(\bar{x}_1 - \bar{x}_2) - (\mu_1 - \mu_2)}{\sqrt{s^2\left(\dfrac{1}{n_1} + \dfrac{1}{n_2}\right)}} \tag{10.11}$$

In the case of two independent simple random samples of sizes n_1 and n_2, the t distribution will have $n_1 + n_2 - 2$ degrees of freedom. For $\alpha = .05$, the t distribution table shows that with $12 + 12 - 2 = 22$ degrees of freedom, $t_{.05} = 1.717$. Thus, the rejection region for the one-tailed test is as follows:

Reject H_0 if $t > 1.717$

The sample data and equation (10.11) provide the following value for the test statistic:

$$t = \frac{(325 - 288) - 0}{\sqrt{1768\left(\dfrac{1}{12} + \dfrac{1}{12}\right)}} = 2.16$$

Since $t = 2.16 > 1.717$, we can reject H_0 at the .05 level of significance. Thus, the researcher can conclude that the new software package provides a lower mean completion time.

COMPUTER SOLUTION

Minitab and other computer software packages provide the capability for testing hypotheses concerning the difference between the means of two populations.

EXAMPLE 10.7
(CONTINUED)

The Minitab output in Figure 10.6 shows the hypothesis-testing results for the software-evaluation project. The completion times for the 12 systems analysts who used current technology to produce the information system have been entered into column C1 of the Minitab worksheet, while the completion times for the 12 systems analysts who used the new software package have been entered into column C2. The command TWOSAMPLE C1 C2 requests a two-independent-sample test for a difference between the means of the data in columns C1 and C2. The Minitab subcommand ALTERNATIVE = 1 indicates a one-tailed test with H_a: $\mu_1 - \mu_2 > 0$, and the subcommand POOLED indicates that the statistical analysis is to be based on the pooled estimate of σ^2. The first part of the output shows the mean completion time and the standard deviation of completion time for the two samples. The row TTEST provides the hypothesis-testing results. With $t = 2.16$ and a p-value $= 0.021$, the null hypothesis can be rejected at the .05 level of significance; thus, the conclusion is that the new software package provides a lower mean completion time.

Note that the computer solution also provides the 95% confidence interval for the difference between the two population means. Although the hypothesis test enabled us to conclude that the new software is better, the 95% confidence interval shows that on average the improvement that can be expected with the new software may be as little as 1 hour or as high as 73 hours. The wide confidence interval suggests that further study may be desirable to obtain a more precise estimate of how much improvement can be anticipated with the new software package.

FIGURE 10.6   MINITAB OUTPUT FOR THE HYPOTHESIS TEST ON THE SOFTWARE-EVALUATION PROJECT

```
MTB > TWOSAMPLE C1 C2;
SUBC> ALTERNATIVE = 1;
SUBC> POOLED.


TWOSAMPLE T FOR CURRENT VS NEW
            N       MEAN      STDEV     SE MEAN
CURRENT    12      325.0      40.0        12
NEW        12      288.0      44.0        13

95 PCT CI FOR MU CURRENT - MU NEW: (1, 73)

TTEST MU CURRENT = MU NEW (VS GT): T= 2.16   P=0.021   DF=   22

POOLED STDEV =        42.1
```

Notes and Comments

1. In some hypothesis tests, the concern is not simply testing whether or not the two population means are equal, but instead testing if the two population means differ by some amount. For example, in Example 10.6, suppose that because of the increased cost of small classes, the administration had decided to use large classes unless it could be concluded that the mean score for small classes (μ_1) is at least 5 points higher than the mean score for large classes (μ_2). In this case, the relevant hypotheses would be

$$H_0: \mu_1 - \mu_2 \leq 5$$

$$H_a: \mu_1 - \mu_2 > 5$$

and the value of z would be

$$z = \frac{(\bar{x}_1 - \bar{x}_2) - 5}{\sqrt{\sigma_1^2/n_1 + \sigma_2^2/n_2}}$$

In general, if we let D_0 denote the hypothesized difference between the two population means, z can be written as follows:

$$z = \frac{(\bar{x}_1 - \bar{x}_2) - D_0}{\sqrt{\sigma_1^2/n_1 + \sigma_2^2/n_2}}$$

2. In Section 8.4 we showed that confidence interval estimation was equivalent to a two-tailed hypothesis test about a population mean. The same conclusion is valid for conducting two-tailed hypothesis tests for two-population situations. For example, if we let D_0 denote the hypothesized difference between the means, the hypotheses for a two-tailed test can be written

$$H_0: \mu_1 - \mu_2 = D_0$$

$$H_a: \mu_1 - \mu_2 \neq D_0$$

To use a confidence interval approach to test these hypotheses, we first use equation (10.4) to develop an interval estimate of the difference between the two population means. If the value of D_0 is contained within this confidence interval, we cannot reject H_0; if the value of D_0 is *not* contained within the confidence interval, we can reject H_0.

3. In the previous section we stated that using the t distribution for inferences about the means of two populations is fairly insensitive to the assumptions of normal populations and equal variances. However, if a user feels strongly that these assumptions are not appropriate for a particular application, one of the following actions should be taken:

   - Consider the nonparametric Wilcoxon rank sum test presented in Chapter 17.
   - If the populations are approximately normal but the variances may not be equal ($\sigma_1^2 \neq \sigma_2^2$), use equation (10.5) to estimate $\sigma_{\bar{x}_1-\bar{x}_2}$. The t distribution can still be used with the degrees of freedom given by

$$df = \frac{(1/n_1 + s_2^2/s_1^2 n_2)^2}{[1/n_1^2(n_1 - 1)] + [s_2^2/s_1^2 n_2^2(n_2 - 1)]}$$

   - Increase both sample sizes to the large-sample case with $n_1 \geq 30$ and $n_2 \geq 30$.

EXERCISES

METHODS

SELF TEST

11. Consider the following hypothesis test:

$$H_0: \mu_1 - \mu_2 \leq 0$$

$$H_a: \mu_1 - \mu_2 > 0$$

The following results are for independent random samples taken from two populations.

Sample 1	Sample 2
$n_1 = 40$	$n_2 = 50$
$\bar{x}_1 = 25.2$	$\bar{x}_2 = 22.8$
$s_1 = 5.2$	$s_2 = 6.0$

    a. Using $\alpha = .05$, what is your hypothesis-testing conclusion?
    b. What is the p-value?

12. Consider the following hypothesis test:

$$H_0: \mu_1 - \mu_2 = 0$$

$$H_a: \mu_1 - \mu_2 \neq 0$$

The following results are for independent random samples taken from two populations.

Sample 1	Sample 2
$n_1 = 80$	$n_2 = 70$
$\bar{x}_1 = 104$	$\bar{x}_2 = 106$
$s_1 = 8.4$	$s_2 = 7.6$

    a. Using $\alpha = .05$, what is your hypothesis-testing conclusion?
    b. What is the p-value?

13. Consider the following hypothesis test:

$$H_0: \mu_1 - \mu_2 = 0$$

$$H_a: \mu_1 - \mu_2 \neq 0$$

The results shown below are for independent random samples taken from two populations. Using $\alpha = .05$, what is your hypothesis-testing conclusion?

Sample 1	Sample 2
$n_1 = 8$	$n_2 = 7$
$\bar{x}_1 = 1.4$	$\bar{x}_2 = 1.0$
$s_1 = 0.4$	$s_2 = 0.6$

APPLICATIONS

14. Do starting-salary differentials exist for male and female graduates? Miami University in Oxford, Ohio, reported that a sample of 30 males received an average starting salary of $29,300 while a sample of 36 females received an average starting salary of $28,800 (*Miami University Class Profile,* 1990). If the sample standard deviations were $2000 and $1800, respectively, is there any statistical support for the conclusion that a differential exists between the mean starting salaries for the two populations of males and females? Test H_0: $\mu - \mu_2 \leq 0$ and H_a: $\mu_1 - \mu_2 > 0$ at a .05 level of significance. What is the *p*-value, and what is your conclusion?

SELF TEST

15. A firm is studying the delivery times for two raw material suppliers. The firm is basically satisfied with supplier A and is prepared to stay with this supplier provided that the population mean delivery time is the same as or less than that of supplier B. However, if the firm finds that the population mean delivery time from supplier B is less than that of supplier A, it will begin making raw material purchases from supplier B.
   a. What are the null and alternative hypotheses for this situation?
   b. Assume that independent samples show the following delivery time characteristics for the two suppliers.

Supplier A	Supplier B
$n_1 = 50$	$n_2 = 30$
$\bar{x}_1 = 14$ days	$\bar{x}_2 = 12.5$ days
$s_1 = 3$ days	$s_2 = 2$ days

Using $\alpha = .05$, what is your conclusion for the hypotheses from part (a)? What action do you recommend in terms of supplier selection?

16. In a wage discrimination case involving male and female employees, independent samples of male and female employees with 5 years or more experience provided the following hourly wage results.

Male Employees	Female Employees
$n_1 = 44$	$n_2 = 32$
$\bar{x}_1 = \$9.25$	$\bar{x}_2 = \$8.70$
$s_1 = \$1.00$	$s_2 = \$.80$

The null hypothesis is stated such that male employees have a population mean hourly wage less than or equal to that of the female employees. Rejection of H_0 leads to the conclusion that male employees have a population mean hourly wage exceeding the female employees' population mean hourly wage. Test the hypothesis with $\alpha = .01$. Does wage discrimination appear to exist in this case?

17. A production line is designed on the assumption that the difference between mean assembly times for two operations is 5 minutes. Independent tests for the two assembly operations show the following results.

Operation A	Operation B
$n_1 = 100$	$n_2 = 50$
$\bar{x}_1 = 14.8$ minutes	$\bar{x}_2 = 10.4$ minutes
$s_1 = .8$ minute	$s_2 = .6$ minute

Using $\alpha = .02$, test the hypothesis that the difference between the population mean assembly times is $\mu_1 - \mu_2 = 5$ minutes.

18. Starting-salary data for college graduates is reported by the College Placement Council (*USA Today*, April 6, 1992). Annual salaries in thousands of dollars for a sample of accounting majors and a sample of finance majors are shown below.

Accounting		Finance	
28.8	28.1	26.3	29.0
25.3	24.7	23.6	27.4
26.2	25.2	25.0	23.5
27.9	29.2	23.0	26.9
27.0	29.7	27.9	26.2
26.2	29.3	24.5	24.0

a. Use a .05 level of significance to test the hypothesis that there is no difference between the mean annual starting salaries for the populations of accounting majors and finance majors. What is your conclusion?
b. Provide the point estimate and the 95% confidence interval for the difference between the population mean starting salaries for the two majors.

10.3 INFERENCES ABOUT THE DIFFERENCE BETWEEN THE MEANS OF TWO POPULATIONS: MATCHED SAMPLES

A manufacturing company has two methods available for performing a production task. The company would like to identify the method with the smallest mean completion time per unit. Let μ_1 denote the mean completion time for method 1 and μ_2 denote the mean completion time for method 2. With no preliminary indication of the preferred production method, we begin by tentatively assuming that the two methods have the same mean completion time. Thus, the null hypothesis becomes H_0: $\mu_1 - \mu_2 = 0$. If this hypothesis is rejected, we can conclude that a difference between the mean completion times exists. In this case, the method providing the smaller mean completion time would be recommended. The null and alternative hypotheses are written as follows:

$$H_0: \ \mu_1 - \mu_2 = 0$$

$$H_a: \ \mu_1 - \mu_2 \neq 0$$

In designing the sampling procedure that will be used to collect completion time data and test the above hypotheses, we consider two alternative designs. One design is based on *independent samples,* and the other design is based on *matched samples.* The designs are described as follows:

1. *Independent-sample design:* A simple random sample of workers is selected, and each worker uses method 1. A second independent simple random sample of workers is selected and each worker uses method 2. The test of the difference between means is based on the procedures of Section 10.2.

2. *Matched-sample design:* One simple random sample of workers is selected with each worker first using one method and then using the other method. The order of the two methods is assigned randomly to the workers, with some workers performing method 1 first and others performing method 2 first. Each worker provides a pair of data values, one value for method 1 and another value for method 2.

Our interest in the matched-sample design is that since both production methods are tested under similar conditions (i.e., same workers), this design often leads to a smaller sampling error than the independent-sample design. The primary reason for this is that in a matched-sample design, variation between workers is eliminated as a source of the sampling error.

EXAMPLE 10.8 ▷ A matched-sample design was used for two production methods. A sample of six workers was taken; each worker performed the task for both production methods. The data on completion times for the six workers are as follows.

Worker	Completion Time, Method 1 (Minutes)	Completion Time, Method 2 (Minutes)	Difference in Completion Times (d_i)
1	6.0	5.4	.6
2	5.0	5.2	−.2
3	7.0	6.5	.5
4	6.2	5.9	.3
5	6.0	6.0	.0
6	6.4	5.8	.6

Note that each worker provides a pair of data values, one for each production method. Also note that the last column contains the difference (d_i) in completion times for each worker in the sample. For example, $d_1 = .6$ shows that worker 1 required .6 minute more for method 1, $d_2 = −.2$ shows that worker 2 required .2 minute more for method 2, and so on. Thus, we see that a negative d_i value simply indicates that the worker required more time for method 2.

The key to analyzing a matched-sample design is to use the *difference data* only. Doing so converts the situation to a single sample containing six values. Then the procedures introduced in Chapter 8 can be used to make an inference about the mean for the population of all possible differences.

Let μ_d = the mean of the *difference* values for the population of workers. With this notation, the null and alternative hypotheses are rewritten as follows.

Hypothesis	Conclusion
H_0: $\mu_d = 0$	Unable to conclude that a difference exists between the mean completion times for the two methods
H_a: $\mu_d \neq 0$	A difference exists between the mean completion times

The sample mean and sample standard deviation for the six difference values are as follows:

$$\bar{d} = \frac{\Sigma d_i}{n} = \frac{[.6 + (-.2) + .5 + .3 + .0 + .6]}{6} = \frac{1.8}{6} = .3$$

$$s_d = \sqrt{\frac{\Sigma(d_i - \bar{d})^2}{n - 1}}$$

$$= \sqrt{\frac{(.3)^2 + (-.5)^2 + (.2)^2 + (0)^2 + (-.3)^2 + (.3)^2}{6 - 1}}$$

$$= \sqrt{\frac{.56}{5}} = .335$$

In Chapter 8 we stated that if the population has a normal probability distribution and s is used to estimate σ, the t distribution with $n - 1$ degrees of freedom can be used to test a hypothesis about the population mean. The test statistic t was given by

$$t = \frac{\bar{x} - \mu}{s/\sqrt{n}}$$

Assuming the population of difference values (d_i) has a normal probability distribution and using the sample standard deviation of differences, s_d, to estimate the population standard deviation of differences, this same formula can be used to test the hypothesis H_0: $\mu_d = 0$. To indicate that we are using the d_i values for a matched-sample design, we write the test statistic t as follows:

$$t = \frac{\bar{d} - \mu_d}{s_d/\sqrt{n}} \tag{10.12}$$

With a sample of six workers, we have $n - 1 = 5$ degrees of freedom. Using $\alpha = .05$, we find that $t_{.025} = 2.571$. The rejection rule is to reject H_0 if $t < -2.571$ or $t > +2.571$. Using the sample results, we have

$$t = \frac{\bar{d} - \mu_d}{s_d/\sqrt{n}} = \frac{.3 - 0}{.335/\sqrt{6}} = 2.19$$

Since t is between -2.571 and 2.571, H_0 cannot be rejected. The sample data do not provide sufficient evidence to reject the assumption of no difference between the mean completion times of the two methods.

Using the sample results, we could also develop an interval estimate of the difference between the means of the two populations using the methodology of interval estimation for one population, as introduced in Chapter 8. Using this approach, we have

$$\bar{d} \pm t_{.025} \frac{s_d}{\sqrt{n}} \tag{10.13}$$

or

$$.3 \pm 2.571 \frac{.335}{\sqrt{6}} = .3 \pm .35$$

Thus, the 95% confidence interval estimate of the difference between the means of the two production methods is −.05 to .65 minute. ◀

NOTES AND COMMENTS

1. In Example 10.8 workers performed the production task using first one method and then the other method. This is an example of a matched-sample design, where each sampled item (worker) provides a pair of data values. Although this is often the procedure used in the matched-sample analysis, it is possible to use different but "similar" items to provide the pair of data values. In this sense, a worker at one location could be matched with a similar worker at another location (similarity based on age, education, sex, experience, etc.). The pairs of workers would provide the data that could be used in the matched-sample analysis.

2. Since a matched-sample procedure for inferences about two population means generally provides better precision than the independent-sample approach, it is the recommended design. However, in some applications the matching cannot be achieved, or perhaps the time and cost associated with matching is excessive. In these cases, the independent-sample design should be used.

3. Example 10.8 used a sample size of six workers. As such, the small-sample case existed, and the t distribution was used in both the test of hypothesis and interval-estimation computations. If the sample size is large ($n \geq 30$), the use of the t distribution is unnecessary; in such cases, statistical inferences can be based on the z values of the standard normal probability distribution.

EXERCISES

METHODS

19. Consider the following hypothesis test:

$$H_0: \mu_d \leq 0$$

$$H_a: \mu_d > 0$$

The following data are from matched samples taken from two populations.

	Population	
Element	1	2
1	21	20
2	28	26
3	18	18
4	20	20
5	26	24

    a. Compute the difference value for each element.
    b. Compute $\bar{d}$.
    c. Compute the standard deviation s_d.
    d. Test the hypothesis using $\alpha = .05$. What is your conclusion?

20. The following data are from matched samples taken from two populations.

	Population	
Element	1	2
1	11	8
2	7	8
3	9	6
4	12	7
5	13	10
6	15	15
7	15	14

    a. Compute the difference value for each element.
    b. Compute $\bar{d}$.
    c. Compute the standard deviation s_d.
    d. What is the point estimate of the difference between the two population means?
    e. Provide a 95% confidence interval for the difference between the two population means.

APPLICATIONS

SELF TEST

21. A market research firm used a sample of individuals to rate the purchase potential for a particular product before and after the individuals saw a new television commercial about the product. The purchase-potential ratings were based on a 0 to 10 scale, with higher values indicating a higher purchase potential. The null hypothesis stated that the population mean rating "after" would be less than or equal to the population mean rating "before." Rejection of this hypothesis would provide the conclusion that the commercial improved the mean purchase-potential rating. Use $\alpha = .05$ and the following data to test the hypothesis and comment on the value of the commercial.

| | Purchase Rating | | | Purchase Rating | |
Individual	After	Before	Individual	After	Before
1	6	5	5	3	5
2	6	4	6	9	8
3	7	7	7	7	5
4	4	3	8	6	6

22. Earnings per share for a sample of 10 corporations (*Business Week,* 1989 Bonus Issue) are shown below for the years 1987 and 1988.

Corporation	1987	1988
IBM	8.72	9.27
Sears Roebuck	4.35	2.72
Chevron	3.65	5.17
Walt Disney	2.85	3.80
American Express	1.20	2.31
McDonalds	2.89	3.43
Anheuser-Busch	2.04	2.45
Kellogg	3.20	3.90
J. C. Penney	4.11	6.02
Motorola	2.39	3.43

a. Based on the preceding data, is it appropriate to conclude that the population mean earnings per share was significantly higher in 1988? Test at a .05 level of significance.
b. Provide a 95% confidence interval for the increase in population mean earnings per share for the 1-year period.

23. Transportation costs from the airport to the downtown area depend on the method of transportation. One-way costs for taxi and shuttle bus transportation for a sample of 10 major cities are shown below (*USA Today,* February 13, 1992). Provide a 95% confidence interval for the population mean cost increase associated with taxi transportation.

City	Taxi	Shuttle Bus	City	Taxi	Shuttle Bus
Atlanta	15	7	Minneapolis	16.5	7.5
Chicago	22	12.5	New Orleans	18	7
Denver	11	5	New York (LaGuardia)	16	8.5
Houston	15	4.5	Philadelphia	20	8
Los Angeles	26	11	Washington, D.C.	10	5

24. A survey was made of Book-of-the-Month-Club members to see if members spend more time watching television than they do reading (*The Cincinnati Enquirer,* November 21, 1991). Assume a small sample of respondents in this survey provided the weekly hours of

watching television and weekly hours of reading as shown below. Using a .05 level of significance, can it be concluded that even Book-of-the-Month-Club members spend more time per week, on average, watching television than reading?

Respondent	Television	Reading	Respondent	Television	Reading
1	10	6	9	4	7
2	14	16	10	8	8
3	16	8	11	16	5
4	18	10	12	5	10
5	15	10	13	8	3
6	14	8	14	19	10
7	10	14	15	11	6
8	12	14			

25. A manufacturer produces both a deluxe and a standard model automatic sander designed for home use. Selling prices obtained from a sample of retail outlets are as follows.

Retail Outlet	Model Price		Retail Outlet	Model Price	
	Deluxe	Standard		Deluxe	Standard
1	39	27	5	40	30
2	39	28	6	39	34
3	45	35	7	35	29
4	38	30			

a. The manufacturer's suggested retail prices for the two models show a $10 differential in prices. Using a .05 level of significance, test that the mean difference between the population mean prices of the two models is $10.

b. What is the 95% confidence interval for the difference between the population mean prices for the two models?

26. A company attempts to evaluate the potential for a new bonus plan by selecting a random sample of five salespersons to use the bonus plan for a trial period. The weekly sales volumes before and after implementing the bonus plan are shown below.

Salesperson	Weekly Sales	
	Before	After
1	15	18
2	12	14
3	18	19
4	15	18
5	16	18

a. Use $\alpha = .05$ and test to see if the bonus plan will result in an increase in the population mean weekly sales.

b. Provide a 90% confidence interval for the population mean increase in weekly sales that can be expected if a new bonus plan is implemented.

27. Word-processing systems are often justified on the basis of improved efficiencies for a secretarial staff. Shown below are typing rates in words per minute for seven secretaries who previously used electronic typewriters and who are now using computer-based word processors. Test at the .05 level of significance to see if there has been an increase in the population mean typing rate due to the word-processing system.

Secretary	Electronic Typewriter	Word Processor	Secretary	Electronic Typewriter	Word Processor
1	72	75	5	52	55
2	68	66	6	55	57
3	55	60	7	64	64
4	58	64			

10.4 INFERENCES ABOUT THE DIFFERENCE BETWEEN THE PROPORTIONS OF TWO POPULATIONS

We now consider the case where two populations are involved and we are interested in making inferences about the difference between the proportions of the two populations based on two *independent* simple random samples. Let

p_1 = proportion for population 1
p_2 = proportion for population 2
$\bar{p}_1$ = sample proportion for the simple random sample from population 1
$\bar{p}_2$ = sample proportion for the simple random sample from population 2

The difference between the two population proportions is given by $p_1 - p_2$. The point estimator of $p_1 - p_2$ is as follows.

Point Estimator of the Difference Between the Proportions of Two Populations

$$\bar{p}_1 - \bar{p}_2 \qquad \textbf{(10.14)}$$

Thus, the point estimator of the difference between the proportions of two populations is the difference between the proportions of the two independent simple random samples.

SAMPLING DISTRIBUTION OF $\bar{p}_1 - \bar{p}_2$

In the study of the difference between two population proportions, $\bar{p}_1 - \bar{p}_2$ is the point estimator of interest. As we have seen in several previous cases, the sampling distribution of the point estimator is a key factor in developing interval estimates and in testing hypotheses about the parameter of interest. The properties of the sampling distribution of $\bar{p}_1 - \bar{p}_2$ are as follows.

Sampling Distribution of $\bar{p}_1 - \bar{p}_2$

Expected Value: $E(\bar{p} - \bar{p}_2) = p_1 - p_2$ **(10.15)**

Standard Deviation: $\sigma_{\bar{p}_1 - \bar{p}_2} = \sqrt{\dfrac{p_1(1 - p_1)}{n_1} + \dfrac{p_2(1 - p_2)}{n_2}}$ **(10.16)**

where

n_1 = sample size for the simple random sample from population 1
n_2 = sample size for the simple random sample from population 2

Distribution form: If the sample sizes are large [$n_1 p_1$, $n_1(1 - p_1)$, $n_2 p_2$, and $n_2(1 - p_2)$ are all greater than or equal to 5], the sampling distribution of $\bar{p}_1 - \bar{p}_2$ can be approximated by a normal probability distribution.

Figure 10.7 shows the sampling distribution of $\bar{p}_1 - \bar{p}_2$. Note that the formula for the standard deviation of $\bar{p}_1 - \bar{p}_2$, given by equation (10.16), requires that we know the values for the population proportions p_1 and p_2. Since p_1 and p_2 are unknown, we cannot use equation (10.16) to compute $\sigma_{\bar{p}_1 - \bar{p}_2}$. However, using $\bar{p}_1$ as the point estimator of p_1 and $\bar{p}_2$ as the point estimator of p_2, the point estimator of $\sigma_{\bar{p}_1 - \bar{p}_2}$ is as follows.

Point Estimator of $\sigma_{\bar{p}_1 - \bar{p}_2}$

$s_{\bar{p}_1 - \bar{p}_2} = \sqrt{\dfrac{\bar{p}_1(1 - \bar{p}_1)}{n_1} + \dfrac{\bar{p}_2(1 - \bar{p}_2)}{n_2}}$ **(10.17)**

FIGURE 10.7   SAMPLING DISTRIBUTION OF $\bar{p}_1 - \bar{p}_2$

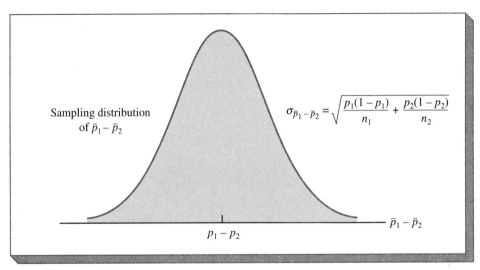

Sampling distribution of $\bar{p}_1 - \bar{p}_2$

$\sigma_{\bar{p}_1 - \bar{p}_2} = \sqrt{\dfrac{p_1(1 - p_1)}{n_1} + \dfrac{p_2(1 - p_2)}{n_2}}$

$\bar{p}_1 - \bar{p}_2$

$p_1 - p_2$

INTERVAL ESTIMATION OF $p_1 - p_2$

With the sampling distribution of $\bar{p}_1 - \bar{p}_2$ known and with $s_{\bar{p}_1 - \bar{p}_2}$ providing an estimate of $\sigma_{\bar{p}_1 - \bar{p}_2}$, the following expression can be used to develop an interval estimate of the difference between the proportions of two populations.

Interval Estimate of the Difference Between the Proportions of Two Populations [Large-Sample Case with $n_1 p_1$, $n_1(1 - p_1)$, $n_2 p_2$, and $n_2(1 - p_2) \geq 5$]

$$\bar{p}_1 - \bar{p}_2 \pm z_{\alpha/2} s_{\bar{p}_1 - \bar{p}_2} \qquad (10.18)$$

where $1 - \alpha$ is the confidence coefficient.

EXAMPLE 10.9 ▶ A firm that specializes in preparing tax returns for its clients is interested in comparing the quality of work at two of its regional offices. A random sample of tax returns prepared at each office is selected and verified for accuracy. Of concern to the firm is the proportion of erroneous returns prepared at each office. In particular, the firm would like to estimate the difference between the proportions of erroneous returns prepared at the two offices.

Sample results show that of 250 returns prepared at office 1, 35 were in error and that of 300 returns prepared at office 2, 27 were in error. Develop 95% and 90% confidence interval estimates for the difference between the two population proportions.

The sample proportions at the two offices are as follows:

$$\bar{p}_1 = \frac{35}{250} = .14$$

$$\bar{p}_2 = \frac{27}{300} = .09$$

The point estimate of the difference between the proportion of erroneous tax returns for the two populations is $\bar{p}_1 - \bar{p}_2 = .14 - .09 = .05$. Specifically, we are led to believe that office 1 possesses a 5% greater error rate than office 2. Substituting into expression (10.18) yields the 95% confidence interval for the difference between the two population proportions:

$$\bar{p}_1 - \bar{p}_2 \pm z_{.025} \sqrt{\frac{\bar{p}_1(1 - \bar{p}_1)}{n_1} + \frac{\bar{p}_2(1 - \bar{p}_2)}{n_2}}$$

$$.14 - .09 \pm 1.96 \sqrt{\frac{.14(1 - .14)}{250} + \frac{.09(1 - .09)}{300}}$$

$$.05 \pm 1.96(.0275) = .05 \pm .0539$$

Thus, the 95% confidence interval estimate of the difference in error rates at the two offices is −.0039 to .1039.

The 90% confidence interval uses $z_{.05} = 1.645$. Thus, replacing the 1.96 with 1.645 in expression (10.18) provides the 90% confidence interval estimate. This substitution will result in a confidence interval of $.05 \pm 1.645(.0275)$, or $.05 \pm .0452$. Thus, the 90% confidence interval for the difference between the population error rates at the two offices is .0048 to .0952. ◀

EXAMPLE 10.10 ▶ In a study of household television-viewing habits, individuals in sampled households were asked to participate by keeping a daily diary of television viewing. Letters requesting participation in the study were sent to two groups. The first group received letters that included $1.00 as a token of appreciation for participating in the study. The second group received only a letter requesting participation in the study. Of 300 letters sent to the first group, 141 households participated in the study. Of 500 letters sent to the second group, 150 households participated in the study. Provide a 95% confidence interval estimate for the difference in participation rate that exists if $1.00 is included with the request for participation.

Using the sample results, we have

$$\bar{p}_1 = \frac{141}{300} = .47$$

$$\bar{p}_2 = \frac{150}{500} = .30$$

Using expression (10.18) with $z_{.025} = 1.96$, the 95% confidence interval is

$$.47 - .30 \pm 1.96\sqrt{\frac{.47(1 - .47)}{300} + \frac{.30(1 - .30)}{500}} = .17 \pm 1.96(.0354)$$

$$= .17 \pm .0694$$

Thus, the 95% confidence interval shows that including $1.00 with the letter increases the participation rate to between .1006 and .2394, or approximately from 10% to 24%. ◀

HYPOTHESIS TESTS ABOUT $p_1 - p_2$

Let us now consider statistical inferences involving hypothesis tests about the difference between the proportions of two populations. As with other hypothesis-testing situations, we use the standardized normal random variable z to compare the value of the sample statistic with the hypothesized value for the population parameter. For hypothesis tests concerning the proportions of two populations, z is as follows:

$$z = \frac{(\bar{p}_1 - \bar{p}_2) - (p_1 - p_2)}{\sigma_{\bar{p}_1 - \bar{p}_2}} \qquad (10.19)$$

We may be tempted to use equation (10.17) to compute $s_{\bar{p}_1 - \bar{p}_2}$ and use this value as an estimate of $\sigma_{\bar{p}_1 - \bar{p}_2}$. However, whenever the null hypothesis for a hypothesis test about the difference between two population proportions involves 0 (that is, H_0: $p_1 - p_2 = 0$, H_0: $p_1 - p_2 \geq 0$, or H_0: $p_1 - p_2 \leq 0$), the hypothesis-testing procedure requires us to assume H_0 is true with $p_1 - p_2 = 0$. Whenever we make this assumption, there is no need to use the individual values of $\bar{p}_1$ and $\bar{p}_2$ to estimate $\sigma_{\bar{p}_1 - \bar{p}_2}$. Using $p_1 = p_2 = p$, equation (10.16) can be rewritten

$$\sigma_{\bar{p}_1 - \bar{p}_2} = \sqrt{\frac{p(1 - p)}{n_1} + \frac{p(1 - p)}{n_2}} = \sqrt{p(1 - p)\left(\frac{1}{n_1} + \frac{1}{n_2}\right)} \qquad (10.20)$$

With equation (10.20), we see that we need an estimate of p in order to estimate $\sigma_{\bar{p}_1 - \bar{p}_2}$. The pooled estimator of p is

$$\bar{p} = \frac{n_1\bar{p}_1 + n_2\bar{p}_2}{n_1 + n_2} \qquad (10.21)$$

In effect, equation (10.21) provides a combined, or pooled estimate of the population proportion p. The following expression for $s_{\bar{p}_1 - \bar{p}_2}$ can then be used to estimate $\sigma_{\bar{p}_1 - \bar{p}_2}$ in equation (10.19).

Point Estimator of $\sigma_{\bar{p}_1 - \bar{p}_2}$ When $p_1 = p_2$

$$s_{\bar{p}_1 - \bar{p}_2} = \sqrt{\bar{p}(1 - \bar{p})\left(\frac{1}{n_1} + \frac{1}{n_2}\right)} \qquad (10.22)$$

EXAMPLE 10.11 ▶ A sample of driving records over a 2-year period show that 16 of 400 adult drivers had received traffic citations, whereas 24 of 300 teenage drivers had received traffic citations. Test the hypothesis that there is no difference between the traffic citation rate for adult and teenage drivers.

Letting p_1 = the population proportion of adult drivers with citations and p_2 = the population proportion of teenage drivers with citations, we want to test the hypotheses

$$H_0: p_1 - p_2 = 0$$

$$H_a: p_1 - p_2 \neq 0$$

With $\alpha = .05$, we reject H_0 if $z < -1.96$ or $z > 1.96$.

Sample results show the following values for the sample proportions:

$$\text{Adults:} \qquad \bar{p}_1 = \frac{16}{400} = .04$$

$$\text{Teenagers:} \quad \bar{p}_2 = \frac{24}{300} = .08$$

Using equation (10.21), we have

$$\bar{p} = \frac{n_1 \bar{p}_1 + n_2 \bar{p}_2}{n_1 + n_2} = \frac{400(.04) + 300(.08)}{400 + 300} = .0571$$

as the pooled estimate of p. The value of z is then computed as follows:

$$z = \frac{(\bar{p}_1 - \bar{p}_2) - (p_1 - p_2)}{\sqrt{\bar{p}(1 - \bar{p})\left(\frac{1}{n_1} + \frac{1}{n_2}\right)}} = \frac{(.04 - .08) - 0}{\sqrt{.0571(1 - .0571)\left(\frac{1}{400} + \frac{1}{300}\right)}}$$

$$= \frac{-.04}{.0177} = -2.26$$

With this value for z, we reject H_0 and conclude that there is a significant difference between the traffic-citation proportions for adult and teenage drivers. ◀

EXAMPLE 10.12 ▶ In the validation of examination questions used in a physics course at a large university, an instructor would like to compare the proportion of A students and the proportion of F students answering the questions correctly. A particular examination question is judged to be a good discriminator if the proportion of A students (p_1) answering the question correctly is greater than the proportion of F students (p_2) answering the question correctly. That is, rejecting H_0 in the following hypothesis test will support the conclusion that the examination question is a good discriminator between A and F students.

$$H_0: \; p_1 - p_2 \leq 0$$

$$H_a: \; p_1 - p_2 > 0$$

Using $\alpha = .05$, H_0 will be rejected if $z > 1.645$.

At the end of the term, sample results showed that 85 of the 110 students who received an A in the course had answered the examination question correctly, whereas 50 of the 95 students who received an F in the course answered the examination question correctly. The computations required to conduct this hypothesis test are as follows:

$$\text{A Students:} \quad \bar{p}_1 = \frac{85}{110} = .7727$$

$$\text{F Students:} \quad \bar{p}_2 = \frac{50}{95} = .5263$$

$$\text{Overall:} \quad \bar{p} = \frac{110(.7727) + 95(.5263)}{110 + 95} = .6585$$

$$z = \frac{(\bar{p}_1 - \bar{p}_2) - (p_1 - p_2)}{\sqrt{\bar{p}(1 - \bar{p})\left(\dfrac{1}{n_1} + \dfrac{1}{n_2}\right)}} = \frac{(.7727 - .5263) - 0}{\sqrt{.6585(1 - .6585)\left(\dfrac{1}{110} + \dfrac{1}{95}\right)}}$$

$$= \frac{.2464}{.0664} = 3.71$$

Thus since $z = 3.71 > 1.645$ we reject H_0 and conclude that A students do significantly better than F students on this examination question.

Alternatively, we can test these hypotheses using the p-value. To compute the p-value for this one-sided hypothesis test, we must determine the probability of getting a value for z that is greater than 3.71. Since this probability is approximately 0 and clearly less than $\alpha = .05$, the null hypothesis can be rejected.

EXERCISES

METHODS

28. Consider the following results for independent random samples taken from two populations.

Sample 1	Sample 2
$n_1 = 400$	$n_2 = 300$
$\bar{p}_1 = .48$	$\bar{p}_2 = .36$

    a. What is the point estimate of the difference between the two population proportions?

    b. Develop a 90% confidence interval for the difference between the two population proportions.

    c. Develop a 95% confidence interval for the difference between the two population proportions.

SELF TEST **29.** Consider the following hypothesis test:

$$H_0: p_1 - p_2 \geq 0$$

$$H_a: p_1 - p_2 < 0$$

The following results are for independent random samples taken from two populations.

Sample 1	Sample 2
$n_1 = 200$	$n_2 = 300$
$\bar{p}_1 = .22$	$\bar{p}_2 = .16$

    a. Using $\alpha = .05$, what is your hypothesis-testing conclusion?
    b. What is the p-value?

APPLICATIONS

    30. *Business Week*/Harris polls (*Business Week,* April 6, 1992) compared the views adults had about their children's future in 1989 with the views adults held in 1992. In a 1989 poll, 59% of the adults sampled felt their children would have a better life than they had. In a 1992 poll, 34% of the adults sampled felt their children would have a better life. Assume that 1250 adults were used in both polls. Provide a 95% confidence interval estimate of the difference between the population proportions in 1989 and 1992. What is your interpretation of the interval estimate and the difference shown?

SELF TEST   **31.** During the primary elections of 1992, a particular presidential candidate had the preelection voter support in Wisconsin and Illinois shown below. Compute a 95% confidence interval for the difference between the proportion of voters favoring the candidate in the two states.

State	Voters Surveyed	Voters Favoring the Candidate
Wisconsin	500	270
Illinois	360	162

    32. Leo J. Shapiro & Associates, a Chicago market research firm, surveys consumers on a variety of issues (*The Wall Street Journal,* October 17, 1988). Every year in late summer and early fall, the firm surveys consumers to learn about spending plans for the forthcoming holiday season. In 1988, 36% of the consumers surveyed indicated they planned to "spend less" during the 1988 season. In 1987, 28% of the consumers surveyed indicated they planned to "spend less" during the 1987 season. Assume that 400 consumers were surveyed each year.
    a. Compare the results for the two years. Use $\alpha = .05$ to see if there has been a significant increase in the population proportion of consumers planning to spend less during the 1988 season. What is your conclusion? What is the p-value?
    b. Assume you are a retailer in October 1988. Comment on the value of the 2-year study and what the results might mean to your business.
    c. Develop an interval estimate of the 1-year increase in the population proportion of consumers who indicate they will spend less during the 1988 season. Use a 95% confidence level.

33. Two loan officers at the North Ridge National Bank show the following data for defaults on loans that they have approved (the data are based on samples of loans granted over the past 5 years).

Loan Officer	Loans Reviewed in the Sample	Defaulted Loans
A	60	9
B	80	6

Using $\alpha = .05$, test the hypothesis that the population default rates are the same for the two loan officers.

34. A Media General/Associated Press Poll (*USA Today,* April 27, 1989) reported that 16% of men and 5% of women would want to be president of the United States. Assume 500 men and 500 women participated in the poll. Test H_0: $p_1 - p_2 = 0$ versus H_a: $p_1 - p_2 \neq 0$. Use $\alpha = .05$. What is your conclusion about the population proportions of men and women wanting to be president of the United States?

35. A survey firm conducts door-to-door surveys on a variety of issues. Some individuals cooperate with the interviewer and complete the interview questionnaire, while others do not. The sample data are shown below.

Respondents	Sample Size	Number Cooperating
Men	200	110
Women	300	210

a. Using $\alpha = .05$, test the hypothesis that the response rate is the same for both populations of men and women.
b. Compute the 95% confidence interval for the difference between the population proportions of men and women who cooperate with the survey.

36. In a test of the quality of two television commercials, each commercial was shown in a separate test area six times over a 1-week period. The following week a telephone survey was conducted to identify individuals who had seen the commercials. The individuals who had seen the commercials were asked to state the primary message in the commercial. The following results were recorded.

Commercial	Number Who Saw Commercial	Number Who Recalled Primary Message
A	150	63
B	200	60

a. Using $\alpha = .05$, test the hypothesis that there is no difference in the population recall proportions for the two commercials.

b. Compute a 95% confidence interval for the difference between the population recall proportions for the two populations.

37. *The Los Angeles Times* polled Californians to learn whether they felt the state government was on the right track in terms of economic and social service programs (*Business Week,* December 30, 1991). In May of 1991, 520 of 1679 respondents felt the state was on the right track. In December of 1991, 293 of 1629 respondents felt the state was on the right track.

    a. Considering the populations in May and December as two populations, was there a shift in the proportion who felt the state was on the right track with its economic and social service programs? Test H_0: $p_1 - p_2 = 0$ and H_a: $p_1 - p_2 \neq 0$. Use a .01 level of significance.

    b. Provide a 95% confidence interval for any difference between the population proportion who felt the state was on the right track. What is your interpretation of this interval estimate?

SUMMARY

In this chapter we discussed procedures for developing interval estimates and conducting hypothesis tests involving two populations. First, we showed how to make inferences about the differences between the means of two populations when independent simple random samples are selected. We considered both the large- and small-sample cases. The *z* values from the standard normal probability distribution are used for inferences about the difference between two population means when the sample sizes are large. The *t* distribution is used for the inferences when the populations are assumed to be normally distributed with equal variances. This permits inferences in the small-sample case.

Inferences about the difference between the means of two populations were then discussed for the matched-sample design. In the matched-sample design, each element provides a pair of data values, one from each population. The difference between the paired data values is then used in the statistical analysis. The matched-sample design is generally preferred over the independent-sample design because the matched-sample procedure often reduces the size of the sampling error and consequently improves the precision of the estimate.

Finally, interval estimation and hypothesis testing about the difference between two population proportions were discussed. Statistical procedures for analyzing the difference between proportions for two populations are similar to the procedures for analyzing the difference between means for two populations.

GLOSSARY

Independent samples Samples selected from two (or more) populations where the elements making up one sample are chosen independently of the elements making up the other sample(s).

Matched samples Samples where each data value in one sample is matched with a corresponding data value in the other sample. In many cases, one sample provides two data values, one for population 1 and another for population 2.

Pooled estimator of σ^2 An estimate of the variance of a population based on the combination of two (or more) sample results. The pooled variance estimate is appropriate whenever the variances of two (or more) populations are assumed equal.

▽

KEY FORMULAS

Expected Value of $\bar{x}_1 - \bar{x}_2$

$$E(\bar{x}_1 - \bar{x}_2) = \mu_1 - \mu_2 \tag{10.2}$$

Standard Deviation of $\bar{x}_1 - \bar{x}_2$

$$\sigma_{\bar{x}_1 - \bar{x}_2} = \sqrt{\frac{\sigma_1^2}{n_1} + \frac{\sigma_2^2}{n_2}} \tag{10.3}$$

Interval Estimate of the Difference Between the Means of Two Populations (Large-Sample Case with $n_1 \geq 30$ and $n_2 \geq 30$)

$$\bar{x}_1 - \bar{x}_2 \pm z_{\alpha/2}\sigma_{\bar{x}_1 - \bar{x}_2} \tag{10.4}$$

Point Estimator of $\sigma_{\bar{x}_1 - \bar{x}_2}$

$$s_{\bar{x}_1 - \bar{x}_2} = \sqrt{\frac{s_1^2}{n_1} + \frac{s_2^2}{n_2}} \tag{10.5}$$

Standard Deviation of $\bar{x}_1 - \bar{x}_2$ When $\sigma_1^2 = \sigma_2^2 = \sigma^2$

$$\sigma_{\bar{x}_1 - \bar{x}_2} = \sqrt{\frac{\sigma^2}{n_1} + \frac{\sigma^2}{n_2}} = \sqrt{\sigma^2\left(\frac{1}{n_1} + \frac{1}{n_2}\right)} \tag{10.6}$$

Pooled Estimator of σ^2

$$s^2 = \frac{(n_1 - 1)s_1^2 + (n_2 - 1)s_2^2}{n_1 + n_2 - 2} \tag{10.7}$$

Point Estimator of $\sigma_{\bar{x}_1 - \bar{x}_2}$ When $\sigma_1^2 = \sigma_2^2$

$$s_{\bar{x}_1 - \bar{x}_2} = \sqrt{s^2\left(\frac{1}{n_1} + \frac{1}{n_2}\right)} \tag{10.8}$$

Interval Estimate of the Difference Between the Means of Two Populations (Small-Sample Case with $n_1 < 30$ and/or $n_2 < 30$)

$$\bar{x}_1 - \bar{x}_2 \pm t_{\alpha/2}s_{\bar{x}_1 - \bar{x}_2} \tag{10.9}$$

Test Statistic for Hypothesis Tests About the Difference Between the Means of Two Populations

$$z = \frac{(\bar{x}_1 - \bar{x}_2) - (\mu_1 - \mu_2)}{\sqrt{\sigma_1^2/n_1 + \sigma_2^2/n_2}} \tag{10.10}$$

Test Statistic for Matched Samples

$$t = \frac{\bar{d} - \mu_d}{s_d/\sqrt{n}} \tag{10.12}$$

Expected Value of $\bar{p}_1 - \bar{p}_2$

$$E(\bar{p}_1 - \bar{p}_2) = p_1 - p_2 \tag{10.15}$$

Standard Deviation of $\bar{p}_1 - \bar{p}_2$

$$\sigma_{\bar{p}_1 - \bar{p}_2} = \sqrt{\frac{p_1(1 - p_1)}{n_1} + \frac{p_2(1 - p_2)}{n_2}} \tag{10.16}$$

Point Estimator of $\sigma_{\bar{p}_1 - \bar{p}_2}$

$$s_{\bar{p}_1 - \bar{p}_2} = \sqrt{\frac{\bar{p}_1(1 - \bar{p}_1)}{n_1} + \frac{\bar{p}_2(1 - \bar{p}_2)}{n_2}} \tag{10.17}$$

Interval Estimate of the Difference Between the Proportions of Two Populations
Large-Sample Case with $n_1 p_1$, $n_1(1 - p_1)$, $n_2 p_2$, and $n_2(1 - p_2) \geq 5$

$$\bar{p}_1 - \bar{p}_2 \pm z_{\alpha/2} s_{\bar{p}_1 - \bar{p}_2} \tag{10.18}$$

Test Statistic for Hypothesis Tests About the Difference
Between the Proportions of Two Populations

$$z = \frac{(\bar{p}_1 - \bar{p}_2) - (p_1 - p_2)}{\sigma_{\bar{p}_1 - \bar{p}_2}} \tag{10.19}$$

Pooled Estimator of the Population Proportion

$$\bar{p} = \frac{n_1 \bar{p}_1 + n_2 \bar{p}_2}{n_1 + n_2} \tag{10.21}$$

Point Estimator of $\sigma_{\bar{p}_1 - \bar{p}_2}$ When $p_1 = p_2$

$$s_{\bar{p}_1 - \bar{p}_2} = \sqrt{\bar{p}(1 - \bar{p})\left(\frac{1}{n_1} + \frac{1}{n_2}\right)} \tag{10.22}$$

REVIEW QUIZ

TRUE/FALSE

1. A point estimator of the difference between two population means is the corresponding difference between the two sample means.
2. The central limit theorem cannot be applied to the sampling distribution of $\bar{x}_1 - \bar{x}_2$.
3. The pooled estimate of σ^2 is a weighted average of the two sample variances.
4. The only assumption necessary in using the t distribution for interval estimation of the difference between population means is that the variances of the two populations are equal.
5. The normal probability distribution cannot be used for large-sample hypothesis tests concerning the difference between population means.
6. If sampling n_1 items from one population and n_2 items from a second population, the large-sample case is applicable as long as $n_1 + n_2 \geq 60$.
7. When the matched-sample approach is used for inferences about the difference between population means, each item in the sample provides two data values.
8. The advantage of the matched-sample design is that it allows the experimenter to control some factors to obtain a better measure of others.
9. The sampling distribution of $\bar{p}_1 - \bar{p}_2$ can be approximated by a normal probability distribution, provided that one of the sample sizes is large.
10. When conducting the hypothesis test H_0: $p_1 - p_2 = 0$, a pooled estimate of the population proportion is computed from the two sample proportions.

MULTIPLE CHOICE

11. Independent simple random samples are obtained from two normal populations with equal variances to construct a confidence interval estimate for the difference between the population means. If the first sample contains 16 items and the second sample contains 21 items, the correct form to use for the sampling distribution is
    a. normal distribution
    b. t distribution with 15 degrees of freedom
    c. t distribution with 37 degrees of freedom
    d. t distribution with 35 degrees of freedom

12. The t distribution can be used in the estimation of $\mu_1 - \mu_2$
    a. if σ_1^2 and σ_2^2 are equal
    b. if either $n_1 \geq 30$ or $n_2 \geq 30$
    c. always
    d. only when both populations are normal

Use the following for Questions 13–16. A testing company is checking to see whether there is any significant difference in the coverage of two different brands of paint for a hardware store chain. The results are summarized below.

	Amazon Paint	Coverup Paint
Mean coverage (in square feet)	305	295
Standard deviation	20	25
Sample size	31	41

13. A point estimate of the difference between the population mean is
    a. −5   b. 0
    c. 5   d. 10

14. A point estimate of the standard deviation of the difference between the sample means is
    a. −5   b. 5.3
    c. 28.1   d. 32

15. The form of the sampling distribution of the difference between the sample means is the
    a. normal distribution, approximately
    b. t distribution with 30 degrees of freedom
    c. t distribution with 35 degrees of freedom
    d. t distribution with 40 degrees of freedom

16. If a two-tailed test is used with a .05 level of significance, the critical z values are
    a. −1.96 and +1.96
    b. −1.645 and +1.645
    c. −10.4 and +10.4
    d. none of the above

SUPPLEMENTARY EXERCISES

38. Starting annual salaries for individuals with master's and bachelor's degrees were collected in two independent random samples. Use the data shown below to develop a 90% confidence interval estimate of the increase in starting salary that can be expected upon completion of the master's degree.

	Master's Degree	Bachelor's Degree
	$n_1 = 60$	$n_2 = 80$
	$\overline{x}_1 = \$23{,}000$	$\overline{x}_2 = \$21{,}000$
	$s_1 = \$2500$	$s_2 = \$2000$

39. Safegate Foods, Inc., is redesigning the checkout lanes in its supermarkets throughout the country. Two designs have been suggested. Tests on customer checkout times have been collected at stores where the two new systems have been installed. The sample data are as follows.

	System A	System B
	$n_1 = 120$	$n_2 = 100$
	$\overline{x}_1 = 4.1$ minutes	$\overline{x}_2 = 3.3$ minutes
	$s_1 = 2.2$ minutes	$s_2 = 1.5$ minutes

Test at the .05 level of significance to determine whether there is a difference between the population mean checkout times for the two systems. Which system is preferred?

40. Samples of final examination scores for two statistics classes with different instructors provided the following results.

	Instructor A	Instructor B
	$n_1 = 12$	$n_2 = 15$
	$\overline{x}_1 = 72$	$\overline{x}_2 = 78$
	$s_1 = 8$	$s_2 = 10$

With $\alpha = .05$, test whether these data are sufficient to conclude that the population mean grades differ for the two instructors.

41. In a study of job attitudes and job satisfaction, a sample of 50 men and 50 women were asked to rate their overall job satisfaction on a 1 to 10 scale. A high rating indicates a higher degree of job satisfaction. Using the sample results shown below, does there appear to be a significant difference between the levels of job satisfaction for the populations of men and women? Use $\alpha = .05$.

	Men	Women
	$\overline{x}_1 = 7.2$	$\overline{x}_2 = 6.4$
	$s_1 = 1.7$	$s_2 = 1.4$

42. Figure Perfect, Inc., is a women's figure salon that specializes in weight-reduction programs. Weights for a sample of clients before and after a 6-week introductory program are as follows:

	Weight	
Client	Before	After
1	140	132
2	160	158
3	210	195
4	148	152
5	190	180
6	170	164

Using $\alpha = .05$, test to determine if the introductory program provides a statistically significant weight loss.

43. A cable television firm is considering submitting bids for rights to operate in two regions of the state of Florida. Surveys of the two regions provided the following data on customer acceptance of the cable television service.

Region I	**Region II**
$n_1 = 500$	$n_2 = 800$
Number indicating	Number indicating
an intent to purchase $= 175$	an intent to purchase $= 360$

Develop a 99% confidence interval for the difference between population proportions of customer acceptance in the two regions.

44. The chapter opening *Statistics in Practice* reported that a group of physicians from Denmark conducted a yearlong study on the effectiveness of nicotine chewing gum in helping people stop smoking (*New England Journal of Medicine,* 1988). The 113 people who participated in the study were all smokers. Of these, 60 were given chewing gum with 2 milligrams of nicotine, and 53 were given a placebo chewing gum with no nicotine content. No one in the study knew which type of gum he or she had been given. All were told to use the gum and refrain from smoking.
  a. Define the null and alternative hypotheses that would be appropriate if the researchers hoped to show that the group given nicotine chewing gum had a higher population proportion of nonsmokers 1 year after the study began.
  b. Results showed that 23 of the smokers given nicotine chewing gum had remained nonsmokers for the 1-year period while 12 of the smokers given the placebo had remained nonsmokers during the same period. Do these results support the conclusion that nicotine gum can help stop smoking? Test using $\alpha = .05$. What is the *p*-value?

45. A large automobile-insurance company selected samples of single and married male policyholders and recorded the number who had made an insurance claim over the previous 3-year period.

Single Policyholders	**Married Policyholders**
$n_1 = 400$	$n_2 = 900$
Number making claims $= 76$	Number making claims $= 90$

a. Using $\alpha = .05$, test to determine if the claim rates differ between the populations of single and married male policyholders.

b. Provide a 95% confidence interval for the difference between the proportions for the two populations.

46. Medical tests were conducted to learn about drug-resistant cases of tuberculosis (*The New York Times,* January 24, 1992). Of 142 cases tested in New Jersey, 9 were found to be drug-resistant. Of 268 cases tested in Texas, 5 were found to be drug-resistant. Do these data suggest a statistically significant difference between the proportion of drug-resistant cases in the two states? Test H_0: $p_1 - p_2 = 0$ at the .02 level of significance. What is the *p*-value, and what is your conclusion?

COMPUTER CASE: PAR, INC.

Par, Inc., is a major manufacturer of golf equipment. Management believes that Par's market share could be increased with the introduction of a cut-resistant, longer-lasting golf ball. As a result, the research group at Par has been investigating a new golf ball coating that is designed to resist cuts and provide a more durable ball. The tests of golf balls with the new coating have been very promising.

A concern raised by one of the researchers has to do with the effect that the new coating will have on driving distances. Par would like the new cut-resistant ball to still offer good driving distances when compared to their current-model golf ball. To compare the driving distances for the two balls, 40 balls of both the new and current models were subjected to distance tests. The testing was performed with a mechanical hitting machine so that if a difference exists between the mean distances for the two models, it can be attributed to a difference in the design. The results of the tests, with distances measured to the nearest yard, are shown below. These data are available on the data disk in the file named GOLF.

GOLF

Model		Model		Model		Model	
Current	New	Current	New	Current	New	Current	New
264	277	270	272	263	274	281	283
261	269	287	259	264	266	274	250
267	263	289	264	284	262	273	253
272	266	280	280	263	271	263	260
258	262	272	274	260	260	275	270
283	251	275	281	283	281	267	263
258	262	265	276	255	250	279	261
266	289	260	269	272	263	274	255
259	286	278	268	266	278	276	263
270	264	275	262	268	264	262	279

REPORT

1. Formulate and present the rationale for a hypothesis test that Par could use to compare the driving distances of the current and new golf ball products.
2. Analyze the data to provide the hypothesis-testing conclusion. What is the *p*-value for your test? What is your recommendation for Par, Inc.?
3. Provide descriptive statistical summaries of the data for each model.
4. What are the 95% confidence intervals for the population mean of each model, and what is the 95% confidence interval for the difference between the means of the two populations?
5. Do you feel there is a need for larger sample sizes and more testing with the golf balls? Discuss.

CHAPTER 11

INFERENCES ABOUT POPULATION VARIANCES

GAO Audits Pollution Control

The U.S. General Accounting Office (GAO) is an independent, nonpolitical audit organization in the legislative branch of the federal government. GAO evaluators determine the effectiveness of existing or proposed federal programs. To carry out their duties, evaluators must be proficient in records review, legislative research, and statistical analysis.

In one case, GAO evaluators studied a Department of Interior program established to help clean up the nation's rivers and lakes. As part of this program, federal grants were made to small cities throughout the United States. Congress asked the GAO to determine how effectively these programs were operating. To do this, the GAO examined records and visited the sites of several waste treatment plants.

One objective of the GAO audit was to ensure that the effluent (treated sewage) at the plants met certain standards. Among other things, the audits reviewed sample data on the oxygen content, the pH level, and the amount of suspended solids in the effluent. A requirement of the program was that a variety of tests be taken daily at each plant with the collected data periodically sent to the state engineering department. The GAO's investigation of the data was concerned with determining whether various characteristics of the effluent were within acceptable limits.

For example, the mean or average pH level of the effluent was looked at carefully. In addition, the variance in the reported pH levels was also reviewed. The following hypothesis test was conducted concerning the variance in pH level for the population of effluent:

$$H_0: \ \sigma^2 = \sigma_0^2$$

$$H_a: \ \sigma^2 \neq \sigma_0^2$$

GAO evaluators were concerned with effluent standards maintained by waste treatment plants.

In this test, σ_0^2 was the population variance in pH level expected at a properly functioning plant. In one particular plant, the null hypothesis was rejected. Further analysis showed that this plant had a variance in pH level that was significantly less than normal.

The auditors visited the plant to examine the measuring equipment and to discuss their statistical findings with the plant manager. The auditors found that the measuring equipment was not being used because the operator did not know how to operate it. Instead, the operator had been told by an engineer what an acceptable pH level was and had simply recorded similar values without actually conducting the test. The unusually low variance in this plant's data had resulted in rejecting H_0. The GAO suspected that similar problems might exist at other plants, and they recommended an operator training program to improve the data-collection aspect of the pollution-control program.

This *Statistics in Practice* was provided by Mr. Dale Ledman of the U.S. General Accounting Office.

In the previous four chapters we discussed methods of statistical inference involving means and proportions. In this chapter we continue the discussion of statistical inference by considering methods for making inferences about population variances and standard deviations. As an example of where variance is important, consider the process of filling containers with a liquid detergent. The filling mechanism is adjusted so that the mean filling weight is 16 ounces per container. Although a mean of 16 ounces is desired, the variance of the filling weights is also critical. That is, even with the filling mechanism properly adjusted for the mean of 16 ounces, we cannot expect every container to have exactly 16 ounces. By selecting a sample of containers, we can compute a sample variance for the filling weights. This value serves as an estimate of the variance for the population of containers being filled by the production process. If the sample variance is modest, the production process is continued. However, if the sample variance is excessive, overfilling and underfilling can create problems even though the mean filling weight is correct (16 ounces). In this case, the filling mechanism must be readjusted to reduce the filling variance for the containers. Quality control applications such as this are an area in which inferences about population variances are often necessary.

In the following section we consider methods for making inferences about the variance of a single population. Later we discuss procedures for making inferences about the variances of two populations. Since the standard deviation is the square root of the variance, the methods introduced in this chapter can also be used to make inferences about population standard deviations.

11.1 INFERENCES ABOUT A POPULATION VARIANCE

Consider a population with an unknown variance σ^2. The sample variance s^2 will be used to make inferences about the population variance σ^2. In such cases we refer to s^2 as the *point estimator* of the population variance σ^2; the numerical value obtained for s^2 is called the *point estimate* of σ^2. The formula for computing a sample variance is restated below.

Sample Variance

$$s^2 = \frac{\Sigma(x_i - \bar{x})^2}{n - 1}$$ (11.1)

where

$\bar{x}$ = sample mean
n = sample size

The process of using s^2 to make an inference about σ^2 is shown in Figure 11.1.

THE SAMPLING DISTRIBUTION OF $(n - 1)s^2/\sigma^2$

In previous chapters we showed that knowledge of a sampling distribution is essential for computing interval estimates and conducting hypothesis tests about a population mean or a population proportion. Thus, it should not be surprising to learn that in order to make inferences about a population variance, we again work with a sampling distribution. It can be shown that for normally distributed populations, the sampling distribution of $(n - 1)s^2/\sigma^2$ is a *chi-square distribution*. Since tables of probabilities are available for the chi-square distribution, it is relatively easy to use the chi-square distribution to make interval estimates and test hypotheses about the value of a population variance.

FIGURE 11.1   THE STATISTICAL PROCESS OF USING A SAMPLE VARIANCE TO MAKE
INFERENCES ABOUT A POPULATION VARIANCE

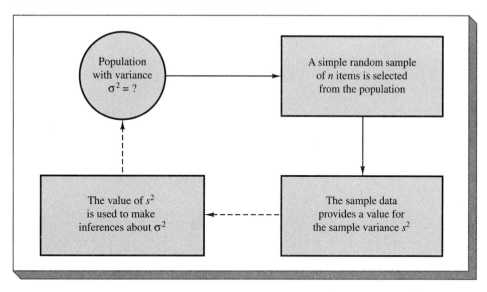

The sampling distribution of $(n - 1)s^2/\sigma^2$ is described as follows.

Sampling Distribution of $(n - 1)s^2/\sigma^2$

Whenever a simple random sample of size n is selected from a *normally distributed population,*

$$\frac{(n - 1)s^2}{\sigma^2} \qquad\qquad (11.2)$$

has a *chi-square distribution* with $n - 1$ degrees of freedom, where s^2 is the sample variance and σ^2 is the population variance.

Typical graphs of the sampling distributions of $(n - 1)s^2/\sigma^2$ are shown in Figure 11.2.

THE CHI-SQUARE DISTRIBUTION

Like the t distribution, the chi-square distribution is a family of similar probability distributions. Each specific chi-square distribution depends on its *degrees of freedom* (abbreviated df) parameter. That is, there is a chi-square distribution with 1 degree of freedom, a chi-square distribution with 2 degrees of freedom, and so on. Figure 11.2 shows the graphs of the chi-square distributions for 2, 5, and 10 degrees of freedom. In using a sample of size n to make inferences about a population variance, the appropriate chi-square distribution has $n - 1$ degrees of freedom.

To see how tables of chi-square values can be used to make inferences about a population variance, consider a situation where the sample size is 20. The degrees of

FIGURE 11.2   EXAMPLES OF THE SAMPLING DISTRIBUTION OF $(n - 1)s^2/\sigma^2$ WITH 2, 5, AND 10 DEGREES OF FREEDOM

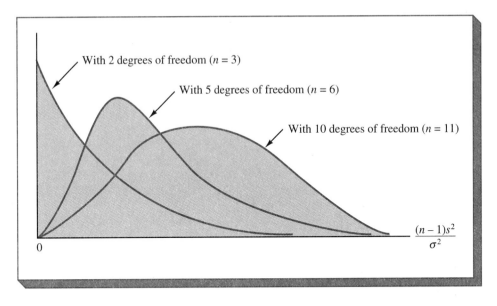

FIGURE 11.3   A CHI-SQUARE DISTRIBUTION WITH 19 DEGREES OF FREEDOM

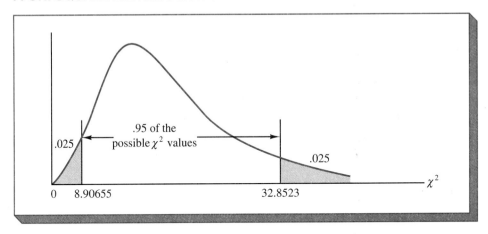

freedom for this case are $n - 1 = 20 - 1 = 19$. A graph of the chi-square distribution with 19 degrees of freedom is shown in Figure 11.3. Using the symbol χ^2 to refer to the chi-square value, note that the distribution shows that 95% of the possible χ^2 values are between 8.90655 and 32.8523. That is, when a random sample is selected from a normal population, there is a .95 probability that the chi-square value [i.e., $\chi^2 = (n - 1)s^2/\sigma^2$] will be between 8.90655 and 32.8523.

Table 11.1 contains a table of values for selected chi-square distributions. We use a subscript on χ^2 to denote the area or probability under the curve to the *right* of the stated χ^2 value. For example, the area under the curve to the right of $\chi^2_{.025}$ is .025. Thus, $\chi^2_{.025}$ corresponds to a chi-square value in the upper tail of the distribution. Similarly, $\chi^2_{.975}$ (97.5% of the chi-square values are to the right of $\chi^2_{.975}$) corresponds to a chi-square value in the lower tail of the distribution.

TABLE 11.1   CHI-SQUARE DISTRIBUTION TABLE

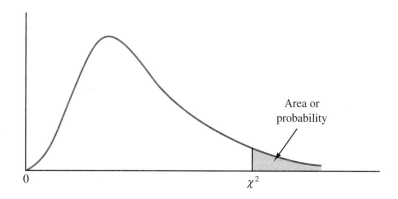

Degrees of Freedom	Area in Upper Tail					
	.99	.975	.95	.05	.025	.01
1	$157,088 \times 10^{-9}$	$982,069 \times 10^{-9}$	$393,214 \times 10^{-8}$	3.84146	5.02389	6.63490
2	.0201007	.0506356	.102587	5.99147	7.37776	9.21034
3	.114832	.215795	.351846	7.81473	9.34840	11.3449
4	.297110	.484419	.710721	9.48773	11.1433	13.2767
5	.554300	.831211	1.145476	11.0705	12.8325	15.0863
6	.872085	1.237347	1.63539	12.5916	14.4494	16.8119
7	1.239043	1.68987	2.16735	14.0671	16.0128	18.4753
8	1.646482	2.17973	2.73264	15.5073	17.5346	20.0902
9	2.087912	2.70039	3.32511	16.9190	19.0228	21.6660
10	2.55821	3.24697	3.94030	18.3070	20.4831	23.2093
11	3.05347	3.81575	4.57481	19.6751	21.9200	24.7250
12	3.57056	4.40379	5.22603	21.0261	23.3367	26.2170
13	4.10691	5.00874	5.89186	22.3621	24.7356	27.6883
14	4.66043	5.62872	6.57063	23.6848	26.1190	29.1413
15	5.22935	6.26214	7.26094	24.9958	27.4884	30.5779
16	5.81221	6.90766	7.96164	26.2962	28.8454	31.9999
17	6.40776	7.56418	8.67176	27.5871	30.1910	33.4087
18	7.01491	8.23075	9.39046	28.8693	31.5264	34.8053
19	7.63273	8.90655	10.1170	30.1435	32.8523	36.1908
20	8.26040	9.59083	10.8508	31.4104	34.1696	37.5662
21	8.89720	10.28293	11.5913	32.6705	35.4789	38.9321
22	9.54249	10.9823	12.3380	33.9244	36.7807	40.2894
23	10.19567	11.6885	13.0905	35.1725	38.0757	41.6384
24	10.8564	12.4011	13.8484	36.4151	39.3641	42.9798
25	11.5240	13.1197	14.6114	37.6525	40.6465	44.3141
26	12.1981	13.8439	15.3791	38.8852	41.9232	45.6417
27	12.8786	14.5733	16.1513	40.1133	43.1944	46.9630
28	13.5648	15.3079	16.9279	41.3372	44.4607	48.2782
29	14.2565	16.0471	17.7083	42.5569	45.7222	49.5879
30	14.9535	16.7908	18.4926	43.7729	46.9792	50.8922
40	22.1643	24.4331	26.5093	55.7585	59.3417	63.6907
50	29.7067	32.3574	34.7642	67.5048	71.4202	76.1539
60	37.4848	40.4817	43.1879	79.0819	83.2976	88.3794
70	45.4418	48.7576	51.7393	90.5312	95.0231	100.425
80	53.5400	57.1532	60.3915	101.879	106.629	112.329
90	61.7541	65.6466	69.1260	113.145	118.136	124.116
100	70.0648	74.2219	77.9295	124.342	129.561	135.807

Additional values of chi-square can be found in Table 3 of Appendix B.

Now let us use the chi-square distribution information in Table 11.1 to show how the values of χ^2 in Figure 11.3 were obtained. Since the chi-square distribution in Figure 11.3 is based on 19 degrees of freedom, we see that $\chi^2_{.975} = 8.90655$ and $\chi^2_{.025} = 32.8523$ are found in row 19 and the .975 and .025 columns of Table 11.1. Thus, there is a .95 probability that a chi-square value will be between $\chi^2_{.975}$ and $\chi^2_{.025}$. This statement holds for all chi-square distributions. However, the numerical values of $\chi^2_{.975}$ and $\chi^2_{.025}$ change depending on the number of degrees of freedom. If we had wanted an interval containing 90% of the chi-square values, we could have used the interval from $\chi^2_{.95}$ to $\chi^2_{.05}$. With 19 degrees of freedom, we see from Table 11.1 that this interval is from 10.1170 to 30.1435.

EXAMPLE 11.1 ▶ Consider a chi-square distribution with 15 degrees of freedom. Find the value of χ^2 that provides an area of .01 in the upper tail of the distribution.

Referring to Table 11.1, we find that with 15 degrees of freedom $\chi^2_{.01} = 30.5779$ provides an area of .01 in the upper tail.   ◀

EXAMPLE 11.2 ▶ Consider a chi-square distribution with 9 degrees of freedom. Find the value of χ^2 that provides an area of .05 in the lower tail of the distribution.

Referring to Table 11.1, we find that with 9 degrees of freedom $\chi^2_{.95} = 3.32511$ provides an area of .05 in the lower tail of the distribution.   ◀

EXAMPLE 11.3 ▶ Consider a chi-square distribution with 50 degrees of freedom. Find values of χ^2 that provide an area of .01 in each tail of the distribution.

Referring to Table 11.1, we find that with 50 degrees of freedom $\chi^2_{.99} = 29.7067$ provides an area of .01 in the lower tail and $\chi^2_{.01} = 76.1539$ provides an area of .01 in the upper tail of the distribution.   ◀

INTERVAL ESTIMATION OF σ^2

Based on our previous discussion of the chi-square distribution, we can conclude that there is a $1 - \alpha$ probability that χ^2 will be between $\chi^2_{(1-\alpha/2)}$ and $\chi^2_{\alpha/2}$. That is, there is a $1 - \alpha$ probability $\chi^2_{(1-\alpha/2)} \leq \chi^2 \leq \chi^2_{\alpha/2}$. From formula (11.2) we know the quantity $(n - 1)s^2/\sigma^2$ follows the chi-square distribution. Therefore, there must be a $1 - \alpha$ probability that

$$\chi^2_{(1-\alpha/2)} \leq \frac{(n - 1)s^2}{\sigma^2} \leq \chi^2_{\alpha/2} \tag{11.3}$$

Working with the right-hand side of inequality (11.3), we have

$$\frac{(n - 1)s^2}{\sigma^2} \leq \chi^2_{\alpha/2} \tag{11.4}$$

Multiplying inequality (11.4) by σ^2 provides the following inequality:

$$(n - 1)s^2 \leq \sigma^2 \chi^2_{\alpha/2}$$

Dividing both sides of this inequality by $\chi^2_{\alpha/2}$, we obtain inequality (11.5):

$$\frac{(n - 1)s^2}{\chi^2_{\alpha/2}} \leq \sigma^2 \tag{11.5}$$

With n, s^2, and $\chi^2_{\alpha/2}$ known, we can use inequality (11.5) to compute a lower limit for the value of the population variance σ^2. Using a similar approach with the left-hand inequality in expression (11.3), we obtain an upper limit for the value of σ^2 expressed in terms of n, s^2, and $\chi^2_{(1-\alpha/2)}$.

Based on these results, inequality (11.6) can be used to find a confidence interval estimate of a population variance. By altering the values of α in inequality (11.6), we can obtain the desired level of confidence. For example, $\chi^2_{.025}$ and $\chi^2_{.975}$ will provide a 95% confidence interval, $\chi^2_{.05}$ and $\chi^2_{.95}$ will provide a 90% confidence interval, and $\chi^2_{.005}$ and $\chi^2_{.995}$ will provide a 99% confidence interval.

Interval Estimation for a Population Variance

Whenever a simple random sample of size n is selected from a *normally distributed population*,

$$\frac{(n-1)s^2}{\chi^2_{\alpha/2}} \leq \sigma^2 \leq \frac{(n-1)s^2}{\chi^2_{(1-\alpha/2)}} \tag{11.6}$$

where the values of χ^2 are based on the chi-square distribution with $n - 1$ degrees of freedom and where $(1 - \alpha)$ is the confidence coefficient.

EXAMPLE 11.4 ▶ A production process is designed to fill 16-ounce containers with liquid detergent. Suppose that the filling weights are normally distributed, and that the production manager is concerned about the variance. A sample of 20 containers provides a sample variance of $s^2 = .0025$. Develop a 95% confidence interval for the population variance of the filling weights.

With $n - 1 = 19$ degrees of freedom, we have $\chi^2_{.025} = 32.8523$ and $\chi^2_{.975} = 8.90655$. Using inequality (11.6), the 95% confidence interval becomes

$$\frac{(20-1)(.0025)}{32.8523} \leq \sigma^2 \leq \frac{(20-1)(.0025)}{8.90655}$$

or

$$.001446 \leq \sigma^2 \leq .005333$$

By taking the square root of these terms, we can also find the 95% confidence interval for the population standard deviation σ:

$$.038 \leq \sigma \leq .073$$

Thus, we are 95% confident the population variance is between .001446 and .005333 and the population standard deviation is between .038 and .073. ◀

EXAMPLE 11.5 ▶ Twenty-eight children were given a language test with the scores recorded on a scale of 0 to 100. The test scores are normally distributed, and the variance in the test scores serves as a measure of the homogeneity in language skills for the children. The sample standard deviation of the test scores was found to be $s = 12$. Provide a 90% confidence interval estimate of the variance and standard deviation of the test scores for the population of children.

With $n - 1 = 27$ degrees of freedom, Table 11.1 shows that $\chi^2_{.95} = 16.1513$ and $\chi^2_{.05} = 40.1133$. Using the sample variance of $s^2 = (12)^2 = 144$ in inequality (11.6), we have

$$\frac{(28-1)(144)}{40.1133} \leq \sigma^2 \leq \frac{(28-1)(144)}{16.1513}$$

or

$$96.93 \leq \sigma^2 \leq 240.72$$

This shows that a 90% confidence interval estimate of the population variance is 96.93 to 240.72. Taking the square root of the above values provides a 90% confidence interval estimate of the population standard deviation (9.85 to 15.52).   ◀

HYPOTHESIS TESTS ABOUT σ^2

Hypothesis tests about the value of a population variance are based on the value of $(n-1)s^2/\sigma^2$ and the chi-square distribution with $n-1$ degrees of freedom. The value of σ^2 specified in the null hypothesis is used in the denominator of $(n-1)s^2/\sigma^2$. The rejection rule is based on the value of χ^2 in a manner similar to the use of z and t values in previous hypothesis-testing applications.

EXAMPLE 11.6 ▶   The St. Louis Metro Bus Company has recently made a concerted effort to improve reliability by encouraging its drivers to maintain consistent schedules. As a standard policy, the company expects arrival times at a bus stop to have low variability. Specifically, the company desires an arrival time standard deviation of 2 minutes or less, which indicates a variance of 4 or less. A simple random sample of 10 arrival times shows a sample variance of 5. Using a .05 level of significance, should the company reject the hypothesis that the arrival time variance for the population is less than or equal to the allowable variance of 4?

The hypotheses for the study are

$$H_0\colon \ \sigma^2 \leq 4$$

$$H_a\colon \ \sigma^2 > 4$$

Rejection of H_0 will imply the company is not meeting the variance guideline.

Assuming arrival times are normally distributed, we can use the χ^2 distribution for the hypothesis test. With $\alpha = .05$, the one-tailed test decision rule is based on the upper-tail chi-square value of $\chi^2_{.05}$. With $n-1 = 9$, Table 11.1 shows $\chi^2_{.05} = 16.9190$. Therefore, if $(n-1)s^2/\sigma^2$ is greater than 16.9190, we will reject H_0. The chi-square distribution with the appropriate rejection region is shown in Figure 11.4.

The value of χ^2 is computed as follows:

$$\chi^2 = \frac{(n-1)s^2}{\sigma^2} = \frac{(10-1)(5)}{4} = 11.25$$

Comparing 11.25 to the $\chi^2_{.05} = 16.9190$ value, we see that the null hypothesis cannot be rejected. Thus, the sample of 10 bus arrivals provides insufficient evidence to conclude that the bus arrival time variance is not meeting the company standard.   ◀

EXAMPLE 11.7 ▶   Historically, test scores for individuals applying for driver's licenses have been normally distributed with $\sigma^2 = 100$. Although a new examination has been designed, motor vehicle administrators believe that it is desirable for the variance in the test scores to remain at the historical level. A simple random sample of 30 individuals is given the new version of the driver's examination; the sample variance is 64. Assume test scores continue to be

FIGURE 11.4   REJECTION REGION FOR THE ST. LOUIS METRO BUS COMPANY STUDY

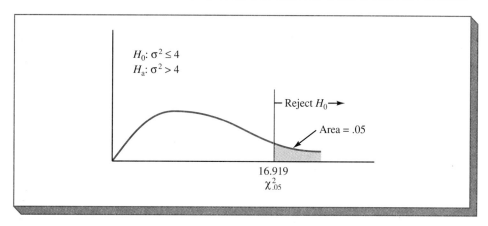

normally distributed. Is there reason to believe the variance of the test scores has changed? Use $\alpha = .05$.

The hypothesis test is

$$H_0: \sigma^2 = 100$$

$$H_a: \sigma^2 \neq 100$$

The two-tailed test with $\alpha = .05$ requires the use of $\chi^2_{.975}$ and $\chi^2_{.025}$. With $n - 1 = 29$ degrees of freedom, Table 11.1 shows $\chi^2_{.975} = 16.0471$ and $\chi^2_{.025} = 45.7222$. Thus, H_0 should be rejected if the value of χ^2 is less than 16.0471 or greater than 45.7222.

With $\sigma^2 = 100$, the value of χ^2 is

$$\chi^2 = \frac{(n-1)s^2}{\sigma^2} = \frac{29(64)}{100} = 18.56$$

With this value of χ^2, we cannot reject H_0.

EXERCISES

METHODS

1. Find the following chi-square distribution values from Table 3 of Appendix B.
    a. $\chi^2_{.05}$ with df $= 12$ b. $\chi^2_{.025}$ with df $= 15$
    c $\chi^2_{.975}$ with df $= 20$ d. $\chi^2_{.01}$ with df $= 10$
    e. $\chi^2_{.95}$ with df $= 18$

SELF TEST

2. A sample of 20 items from a normally distributed population provides a sample standard deviation of 5.
    a. Compute a 90% confidence interval for the population variance.
    b. Compute a 95% confidence interval for the population variance.
    c. Compute a 95% confidence interval for the population standard deviation.

3. A sample of 16 items from a normally distributed population provides a sample standard deviation of 8. Test the following hypotheses using $\alpha = .05$. What is your conclusion?

$$H_0: \sigma^2 \leq 50$$

$$H_a: \sigma^2 > 50$$

APPLICATIONS

4. The variance in dosage weights is critical in the pharmaceutical industry. For a specific drug, with weights measured in grams, a sample of 18 units provided a sample variance of $s^2 = .36$.

    a. Construct a 90% confidence interval for the population variance for the weights of this drug.

    b. Construct a 90% confidence interval for the population standard deviation.

5. A sample of 12 cans of soup produced by Carle Foods provides the weights, measured in ounces, shown below. Provide 95% confidence intervals for the variance and the standard deviation of the population.

12.2	11.9	12.0	12.2	11.7
11.6	11.9	12.0	12.1	12.3
11.8	11.9			

6. A study of workers' attitudes about their jobs was conducted for the airline industry (*Industrial Relations*, Winter 1988). A sample of airline employees provided a sample mean age of 40 years and a sample standard deviation of 9.5 years. For purposes of illustration, assume the population is normally distributed, and compute the 95% confidence interval estimate of the population standard deviation for the following sample sizes:

    a. $n = 30$ **b.** $n = 51$ **c.** $n = 101$

    d. What happens to the interval estimate of the population standard deviation as the sample size increases?

7. In the St. Louis Metro Bus Company study, the sample of 10 bus arrivals showed a sample variance of $s^2 = 4.8$.

    a. Provide a 95% confidence interval for the population variance of arrival times.

    b. Assume that a sample variance of $s^2 = 4.8$ had been obtained from a sample of 25 bus arrivals. Provide a 95% confidence interval for the population variance of arrival times.

    c. What effect does a larger sample size have on the interval estimate of a population variance? Does this seem reasonable?

8. *Barron's* (March 30, 1992) reported the percent changes in major investment markets around the world. Based on market performances during 1991, which ranged from +42.8% in Hong Kong to −16.7% in Finland, the variance in the percent returns over a 3-month period was 48. A sample of 24 markets during the first 3 months of 1992 showed a sample variance in the percent returns to be 67.6. Assume the percent returns are normally distributed.

    a. Test at a .05 level of significance to determine if the variance in percent returns for the investment markets appears to have increased in the first 3 months of 1992. What is your conclusion?

    b. Compute a 95% confidence interval for the variance in percent returns over a 3-month period based on the 1992 sample.

    c. Compute a 95% confidence interval for the standard deviation in the percent returns.

SELF TEST ▷   **9.** A certain part must be machined to very close tolerances, or it is not acceptable to customers. Production specifications call for a maximum variance in the lengths of the parts of .0004. Suppose that the sample variance for 30 parts turns out to be $s^2 = .0005$. Using $\alpha = .05$, test to see if the population variance specification is being violated. Assume the population is normally distributed.

10. City Trucking, Inc., claims consistent delivery times for its routine customer deliveries. A sample of 22 truck deliveries shows a sample variance of 1.5. Test to determine if H_0: $\sigma^2 \leq 1$ can be rejected. Use $\alpha = .10$, and assume the population is normally distributed.

11. The variance in the filling amounts of cups of soft drink from an automatic drink machine is an important consideration to the owner of the soft-drink service. If the variance is too large, overfilling and underfilling of cups will cause customer dissatisfaction with the service. An acceptable variance in filling amounts is $\sigma^2 \leq .25$ where filling amounts are measured in ounces. In a test of filling amounts for a particular machine, a sample of 18 cups showed a sample variance of .40. Assume the population is normally distributed.

   a. Do the sample results indicate that the filling mechanism on the machine should be adjusted due to a large variance in filling amounts? Use a .05 level of significance.

   b. Provide a 90% confidence interval for the variance in the filling amounts for this machine.

12. From a sample of 9 days over the past 6 months, a dentist has seen the following number of patients: 22, 25, 20, 18, 15, 22, 24, 19, and 26. Assuming that the number of patients seen per day is normally distributed, would an analysis of these sample data reject the hypothesis that the variance in the number of patients seen per day is equal to 10? Use a .10 level of significance. What is your conclusion?

11.2 ▽ INFERENCES ABOUT THE VARIANCES OF TWO POPULATIONS

In some statistical applications, it is desirable to compare the variances of two populations. For instance, we might want to compare the variability in product quality resulting from two different production processes, the variability in assembly times for two assembly methods, or the variability in temperatures for two heating devices.

THE SAMPLING DISTRIBUTION OF s_1^2/s_2^2

In making comparisons about the variances of two normally distributed populations, we use data collected from two independent simple random samples, one from population 1 and another from population 2. The sample variances s_1^2 and s_2^2 serve as the estimates of the corresponding population variances σ_1^2 and σ_2^2. The statistic of interest is the ratio of the two sample variances, s_1^2/s_2^2. If the two populations involved both have normal probability distributions and equal variances, the sampling distribution of s_1^2/s_2^2 is an F *distribution*. Tables of areas or probabilities are available for this distribution.

The sampling distribution of s_1^2/s_2^2 is described as follows.

Sampling Distribution of s_1^2/s_2^2 When $\sigma_1^2 = \sigma_2^2$

Whenever independent simple random samples of sizes n_1 and n_2 are selected from normally distributed populations with equal variances, the ratio

$$F = \frac{s_1^2}{s_2^2} \tag{11.7}$$

has an F distribution with $n_1 - 1$ degrees of freedom for the numerator and $n_2 - 1$ degrees of freedom for the denominator, where s_1^2 is the sample variance for the simple random sample of n_1 items from population 1 and s_2^2 is the sample variance for the simple random sample of n_2 items from population 2.

A graph of the sampling distribution of s_1^2/s_2^2 with 20 degrees of freedom associated with s_1^2 and 20 degrees of freedom associated with s_2^2 is shown in Figure 11.5. This distribution, which is an F distribution, is the sampling distribution of s_1^2/s_2^2 if two independent simple random samples of size $n_1 = 21$ and $n_2 = 21$ are taken from two normally distributed populations with equal variances.

FIGURE 11.5   F DISTRIBUTION WITH 20 NUMERATOR AND 20 DENOMINATOR DEGREES OF FREEDOM

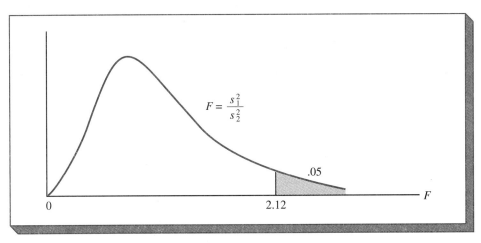

THE F DISTRIBUTION

Each specific F distribution depends on the number of degrees of freedom associated with the numerator and the number of degrees of freedom associated with the denominator. We use F_α to denote the F value that provides an area or probability of α in the upper tail of the F distribution. For instance, $F_{.05}$ provides an area in the upper tail of .05. Similarly, $F_{(1-\alpha)}$ provides an area of α in the lower tail of the F distribution. Thus, $F_{.95}$ results in an area of .05 in the lower tail of the F distribution. Table 11.2 contains $F_{.05}$ values for various numerator and denominator degrees of freedom. A more complete listing for the F distribution is provided in Table 4 of Appendix B.

EXAMPLE 11.8▶ Suppose that a simple random sample of size 21 is selected from population 1 and a random sample of size 21 is selected from population 2. Assume both populations have normal distributions. Find $F_{.05}$ for the F distribution.

Referring to Table 11.2, we see that with 20 degrees of freedom in the numerator and 20 degrees of freedom in the denominator, $F_{.05} = 2.12$. This value is shown in Figure 11.5. ◀

Table 4 in Appendix B provides F values with areas of .05, .025, and .01 in the upper tail of the distribution. However, F values that provide these same areas in the lower tail of the F distribution can be easily computed from the inverse relationship shown below.

Inverse Relationship for the F Distribution

$$F_{(1-\alpha)} = \frac{1}{F_\alpha} \qquad (11.8)$$

where $F_{(1-\alpha)}$ is from an F distribution with ν_1 degrees of freedom in the numerator and ν_2 degrees of freedom in the denominator and F_α is from an F distribution with ν_2 degrees of freedom in the *numerator* and ν_1 degrees of freedom in the *denominator*.

TABLE 11.2   VALUES OF $F_{.05}$ FOR THE F DISTRIBUTION

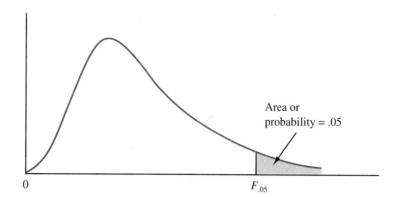

Denominator Degrees of Freedom	Numerator Degrees of Freedom							
	1	2	3	4	5	6	7	8
1	161.4	199.5	215.7	224.6	230.2	234.0	236.8	238.9
2	18.51	19.00	19.16	19.25	19.30	19.33	19.35	19.37
3	10.13	9.55	9.28	9.12	9.01	8.94	8.89	8.85
4	7.71	6.94	6.59	6.39	6.26	6.16	6.09	6.04
5	6.61	5.79	5.41	5.19	5.05	4.95	4.88	4.82
6	5.99	5.14	4.76	4.53	4.39	4.28	4.21	4.15
7	5.59	4.74	4.35	4.12	3.97	3.87	3.79	3.73
8	5.32	4.46	4.07	3.84	3.69	3.58	3.50	3.44
9	5.12	4.26	3.86	3.63	3.48	3.37	3.29	3.23
10	4.96	4.10	3.71	3.48	3.33	3.22	3.14	3.07
11	4.84	3.98	3.59	3.36	3.20	3.09	3.01	2.95
12	4.75	3.89	3.49	3.26	3.11	3.00	2.91	2.85
13	4.67	3.81	3.41	3.18	3.03	2.92	2.83	2.77
14	4.60	3.74	3.34	3.11	2.96	2.85	2.76	2.70
15	4.54	3.68	3.29	3.06	2.90	2.79	2.71	2.64
16	4.49	3.63	3.24	3.01	2.85	2.74	2.66	2.59
17	4.45	3.59	3.20	2.96	2.81	2.70	2.61	2.55
18	4.41	3.55	3.16	2.93	2.77	2.66	2.58	2.51
19	4.38	3.52	3.13	2.90	2.74	2.63	2.54	2.48
20	4.35	3.49	3.10	2.87	2.71	2.60	2.51	2.45
21	4.32	3.47	3.07	2.84	2.68	2.57	2.49	2.42
22	4.30	3.44	3.05	2.82	2.66	2.55	2.46	2.40
23	4.28	3.42	3.03	2.80	2.64	2.53	2.44	2.37
24	4.26	3.40	3.01	2.78	2.62	2.51	2.42	2.36
25	4.24	3.39	2.99	2.76	2.60	2.49	2.40	2.34
26	4.23	3.37	2.98	2.74	2.59	2.47	2.39	2.32
27	4.21	3.35	2.96	2.73	2.57	2.46	2.37	2.31
28	4.20	3.34	2.95	2.71	2.56	2.45	2.36	2.29
29	4.18	3.33	2.93	2.70	2.55	2.43	2.35	2.28
30	4.17	3.32	2.92	2.69	2.53	2.42	2.33	2.27
40	4.08	3.23	2.84	2.61	2.45	2.34	2.25	2.18
60	4.00	3.15	2.76	2.53	2.37	2.25	2.17	2.10
120	3.92	3.07	2.68	2.45	2.29	2.17	2.09	2.02
∞	3.84	3.00	2.60	2.37	2.21	2.10	2.01	1.94

Table continues on next page

Entries in the table give $F_{.05}$ values, where .05 is the area or probability in the upper tail of the F distribution. For example, with 12 numerator and 15 denominator degrees of freedom, $F_{.05} = 2.48$. Additional values for the F distribution can be found in Table 4 of Appendix B.

TABLE 11.2   VALUES OF $F_{.05}$ FOR THE F DISTRIBUTION (CONTINUED)

Denominator Degrees of Freedom	Numerator Degrees of Freedom							
	9	10	12	15	20	24	30	40
1	240.5	241.9	243.9	245.9	248.0	249.1	250.1	251.1
2	19.38	19.40	19.41	19.43	19.45	19.45	19.46	19.47
3	8.81	8.79	8.74	8.70	8.66	8.64	8.62	8.59
4	6.00	5.96	5.91	5.86	5.80	5.77	5.75	5.72
5	4.77	4.74	4.68	4.62	4.56	4.53	4.50	4.46
6	4.10	4.06	4.00	3.94	3.87	3.84	3.81	3.77
7	3.68	3.64	3.57	3.51	3.44	3.41	3.38	3.34
8	3.39	3.35	3.28	3.22	3.15	3.12	3.08	3.04
9	3.18	3.14	3.07	3.01	2.94	2.90	2.86	2.83
10	3.02	2.98	2.91	2.85	2.77	2.74	2.70	2.66
11	2.90	2.85	2.79	2.72	2.65	2.61	2.57	2.53
12	2.80	2.75	2.69	2.62	2.54	2.51	2.47	2.43
13	2.71	2.67	2.60	2.53	2.46	2.42	2.38	2.34
14	2.65	2.60	2.53	2.46	2.39	2.35	2.31	2.27
15	2.59	2.54	2.48	2.40	2.33	2.29	2.25	2.20
16	2.54	2.49	2.42	2.35	2.28	2.24	2.19	2.15
17	2.49	2.45	2.38	2.31	2.23	2.19	2.15	2.10
18	2.46	2.41	2.34	2.27	2.19	2.15	2.11	2.06
19	2.42	2.38	2.31	2.23	2.16	2.11	2.07	2.03
20	2.39	2.35	2.28	2.20	2.12	2.08	2.04	1.99
21	2.37	2.32	2.25	2.18	2.10	2.05	2.01	1.96
22	2.34	2.30	2.23	2.15	2.07	2.03	1.98	1.94
23	2.32	2.27	2.20	2.13	2.05	2.01	1.96	1.91
24	2.30	2.25	2.18	2.11	2.03	1.98	1.94	1.89
25	2.28	2.24	2.16	2.09	2.01	1.96	1.92	1.87
26	2.27	2.22	2.15	2.07	1.99	1.95	1.90	1.85
27	2.25	2.20	2.13	2.06	1.97	1.93	1.88	1.84
28	2.24	2.19	2.12	2.04	1.96	1.91	1.87	1.82
29	2.22	2.18	2.10	2.03	1.94	1.90	1.85	1.81
30	2.21	2.16	2.09	2.01	1.93	1.89	1.84	1.79
40	2.12	2.08	2.00	1.92	1.84	1.79	1.74	1.69
60	2.04	1.99	1.92	1.84	1.75	1.70	1.65	1.59
120	1.96	1.91	1.83	1.75	1.66	1.61	1.55	1.50
∞	1.88	1.83	1.75	1.67	1.57	1.52	1.46	1.39

EXAMPLE 11.9 ▶ Consider a case in which a simple random sample of size 25 has been taken from population 1 and an independent simple random sample of size 10 has been taken from population 2. Assume both populations are normally distributed. Find $F_{.05}$ and $F_{.95}$.

Refer to Table 11.2 in the column corresponding to 24 degrees of freedom and the row corresponding to 9 degrees of freedom. We find $F_{.05} = 2.90$.

To find $F_{.95}$, we must employ the inverse relationship. From equation (11.8) we have

$$F_{.95} = \frac{1}{F_{.05}}$$

where the numerator and denominator degrees of freedom for $F_{.05}$ are reversed. Using Table 11.2, we find that for 9 numerator degrees of freedom and 24 denominator degrees of freedom, $F_{.05} = 2.30$. Therefore

$$F_{.95} = \frac{1}{F_{.05}} = \frac{1}{2.30} = .435$$

Here $F_{.95}$ has 24 numerator and 9 denominator degrees of freedom, respectively, since a sample of size 25 was taken from population 1 and a sample of size 10 was taken from population 2. ◀

HYPOTHESIS TESTS ABOUT σ_1^2 AND σ_2^2

Let us now see how the F distribution can be used for hypothesis tests concerning the variances of two normally distributed populations. Hypothesis tests about the variances of two populations are based on the value of s_1^2/s_2^2. The rejection rule is based on the F value in a manner similar to the way in which z, t, and χ^2 values have been used in previous hypothesis-testing applications.

EXAMPLE 11.10 ▶ Dullus County Schools is renewing its school bus service contract for the coming year and must select one of the two bus companies, the Milbank Company or the Gulf Park Company. We will use the variance of pickup and delivery time as a primary measure of the quality of the bus service. Low variance values will indicate the more consistent and higher-quality service. If the population variances for the two services are the same, Dullus School administrators will select the company offering the better financial terms. However, if sample data on bus pickup and delivery times for the two companies indicate that a significant difference exists between the variances, the administrators may want to give special consideration to the company with the better or lower-variance service. The appropriate hypotheses and their associated conclusions and actions are as follows.

Hypothesis	Conclusion and Action
H_0: $\sigma_1^2 = \sigma_2^2$	No significant difference in quality of service; base service selection decision on financial terms
H_a: $\sigma_1^2 \neq \sigma_2^2$	Unequal quality of service; give special consideration to the low-variance service

Assume that the probability distributions of arrival times are normally distributed. A simple random sample of 25 arrival times is available for the Milbank service (population 1), and an independent simple random sample of 16 arrival times is available for the Gulf Park service (population 2). The F distribution with $n_1 - 1 = 24$ degrees of freedom and $n_2 - 1 = 15$ degrees of freedom is shown in Figure 11.6. Note that for $\alpha = .10$, the two-tailed rejection regions are indicated by the critical values at $F_{.95}$ and $F_{.05}$.

Using Table 11.2, we find that with 24 degrees of freedom in the numerator and 15 degrees of freedom in the denominator, $F_{.05} = 2.29$. While the $F_{.95}$ value is not available in this table, it can be found from the inverse relationship of equation (11.8). With 15 and 24 numerator and denominator degrees of freedom, respectively, Table 11.2 shows that $F_{.05} = 2.11$. Substituting into equation (11.8), we find that $F_{.95}$ for 24 numerator and 15 denominator degrees of freedom is

$$F_{.95} = \frac{1}{2.11} = .474$$

FIGURE 11.6   **F DISTRIBUTION WITH 24 NUMERATOR AND 15 DENOMINATOR DEGREES OF FREEDOM**

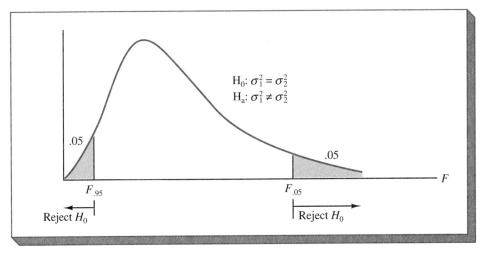

Thus, the rejection rule for the hypothesis test is

$$\text{Reject } H_0 \text{ if } F < .474 \text{ or if } F > 2.29$$

Suppose that the samples of pickup and delivery times show a sample variance of $s_1^2 = 48$ for Milbank and a sample variance of $s_2^2 = 20$ for Gulf Park. The corresponding F value is

$$F = \frac{s_1^2}{s_2^2} = \frac{48}{20} = 2.4$$

Since $F = 2.40 > 2.29$, we reject H_0. The bus services appear to differ in terms of pickup and delivery-time variances. The recommendation is that the Dullus school administrators give special consideration to the better, or lower-variance, service offered by the Gulf Park Company.

One-tailed tests involving two population variances are also possible. Again the F distribution is used, with the one-tailed rejection region enabling us to conclude whether one population variance is significantly greater or significantly less than the other. Only upper-tail F values are needed. For any one-tailed test, we set up the null hypothesis so that the rejection region is in the upper tail. This can be accomplished by a judicious choice of which population is labeled population 1. That is, the sample with the largest variance is treated as if it was selected from population 1 and the hypothesis test is stated in the following form:

$$H_0: \ \sigma_1^2 \le \sigma_2^2$$

$$H_a: \ \sigma_1^2 > \sigma_2^2$$

In this form, rejection of H_0 will only occur when the value of s_1^2/s_2^2 is in the upper tail of the F distribution. Thus, the population with the variance that involves "greater than" in the alternative hypothesis should be labeled population 1.

EXAMPLE 11.11 ▶ Both men and women were selected to participate in a study designed to measure attitudes about current political issues. The objective of the study is to learn whether women demonstrate a greater variation in attitude about political affairs than men. A significance level of $\alpha = .05$ is desired.

In formulating the hypotheses to be tested, we note that for women to demonstrate a significantly greater variance, women must show the larger variance in the alternative hypothesis. Thus, the hypothesis test should be stated as

$$H_0: \sigma^2_{\text{women}} \leq \sigma^2_{\text{men}}$$

$$H_a: \sigma^2_{\text{women}} > \sigma^2_{\text{men}}$$

In conducting the F test, women will be population 1 and men will be population 2.

The attitude scores for a simple random sample of 41 women provided a sample variance of 120, and the attitude scores for an independent simple random sample of 31 men provided a sample variance of 80. Thus, the corresponding F distribution has 40 numerator and 30 denominator degrees of freedom. Table 11.2 shows that with these degrees of freedom, $F_{.05} = 1.79$. Computing the F value from the sample results, we have

$$F = \frac{s_1^2}{s_2^2} = \frac{120}{80} = 1.5$$

Since 1.5 is less than $F_{.05} = 1.79$, H_0 cannot be rejected. Thus, the sample results do not support the position that women have significantly greater variation in attitudes about current political issues. ◀

EXERCISES

METHODS

13. Find the following F distribution values from Table 4 of Appendix B.
    a. $F_{.05}$ with 10 numerator degrees of freedom and 10 denominator degrees of freedom
    b. $F_{.025}$ with 20 numerator degrees of freedom and 15 denominator degrees of freedom
    c. $F_{.01}$ with 8 numerator degrees of freedom and 12 denominator degrees of freedom
    d. $F_{.975}$ with 10 numerator degrees of freedom and 20 denominator degrees of freedom

14. A sample of 16 from population 1 has a sample variance $s_1^2 = 5.8$, and a sample of 20 from population 2 has a sample variance $s_2^2 = 2.4$. Assume normally distributed populations, and test the following hypotheses at the .05 level of significance:

$$H_0: \sigma_1^2 \leq \sigma_2^2$$

$$H_a: \sigma_1^2 > \sigma_2^2$$

What is your conclusion?

SELF TEST ▶ 15. Consider the following hypothesis test:

$$H_0: \sigma_1^2 = \sigma_2^2$$

$$H_a: \sigma_1^2 \neq \sigma_2^2$$

What is your conclusion if $n_1 = 25$, $s_1^2 = 4.0$, $n_2 = 21$, and $s_2^2 = 8.2$? Assume normally distributed populations and use $\alpha = .05$.

APPLICATIONS

16. The average price of a new automobile in 1991 was $16,700, an increase of 4.3% over the 1990 average price of $16,012 (*U.S. News & World Report,* September 9, 1991). Assume that samples of 121 new 1991 automobiles showed a sample standard deviation in price of $4200 and 121 new 1990 automobiles showed a sample standard deviation in price of $3850. Using a .05 level of significance, can it be concluded that the variance in prices of new automobiles also increased in 1991?

SELF TEST

17. Most individuals are aware of the fact that the average annual repair cost for an automobile depends on the age of the automobile. For example, the average annual repair cost (*Consumer Reports 1992 Buyers Guide*) for automobiles 4 years old ($400) is almost twice as large as the average annual repair cost for automobiles 2 years old ($220). A researcher is interested in studying the variance of the annual repair costs to see if the variance in the repair costs also increases with the age of the automobile. A sample of 25 automobiles 4 years old showed a sample standard deviation for annual repair costs of $170 while a sample of 25 automobiles 2 years old showed a sample standard deviation for annual repair costs of $100.

a. State the null and alternative hypotheses for the research hypothesis that the variance in annual repair costs is larger for the older automobiles.

b. Using a .01 level of significance, what is your conclusion? Discuss the reasonableness of your findings.

18. The Educational Testing Service has conducted studies designed to identify differences between the scores of males and females on the Scholastic Aptitude Test (*Journal of Educational Measurement,* Spring 1987). For a sample of females, the standard deviation of test scores was 83 on the verbal portion of the SAT. For a sample of males, the standard deviation was 78 on the same test. Assume that standard deviations were based on random samples of 121 females and 121 males. Do the data indicate there are differences between the variances of females and males on the verbal portion of the SAT? Use $\alpha = .05$.

19. Independent random samples of parts manufactured by two suppliers provided the following results.

Supplier	Sample Size	Sample Variance of Part Sizes
Durham Electric	41	$s_1^2 = 3.8$
Raleigh Electronics	31	$s_2^2 = 2.0$

The firm making the supplier-selection decision is prepared to use the Durham supplier unless the test results show that the Raleigh supplier provides a significantly lower variance in part sizes. Use $\alpha = .05$, and conduct the statistical test that will help the firm select a supplier. Which supplier do you recommend?

20. The following sample data have been collected from two independent random samples.

Population	Sample Size	Sample Mean	Sample Variance
A	$n_A = 25$	$\bar{x}_A = 40$	$s_A^2 = 5$
B	$n_B = 21$	$\bar{x}_B = 50$	$s_B^2 = 11$

Test for the equality of the variances of population A and population B. Use $\alpha = 10$. What is your conclusion?

21. Two secretaries are each given eight typing assignments of equal difficulty. The sample standard deviations of the completion times were 3.8 minutes and 5.2 minutes, respectively. Do the data suggest that there is a difference in the variability of completion times for the two secretaries? Test the hypothesis at a .10 level of significance.

22. A research hypothesis is that the variance of stopping distances of automobiles on wet pavement is substantially greater than the variance of stopping distances of automobiles on dry pavement. In the research study, 16 automobiles traveling at the same speed are tested with respect to stopping distances on wet pavement and then tested with respect to stopping distances on dry pavement. On wet pavement, the standard deviation of stopping distances was 32 feet. On dry pavement, the standard deviation was 16 feet.
   a. At a .05 level of significance, do the sample data justify the conclusion that the variance in stopping distances on wet pavement is greater than the variance in stopping distances on dry pavement?
   b. What are the implications of your statistical conclusions in terms of driving safety recommendations?

SUMMARY

In this chapter we presented statistical procedures that can be used to make inferences about population variances. We introduced two new probability distributions: the chi-square distribution and the F distribution. The chi-square distribution can be used as the basis for interval estimation and hypothesis tests concerning the variance of a normal population. We showed that for simple random samples of size n selected from a normal population, the sampling distribution of $(n-1)s^2/\sigma^2$ is a chi-square distribution with $n-1$ degrees of freedom.

We discussed the use of the F distribution in making hypothesis tests concerning the variances of two normally distributed populations. In particular, we showed that with independent simple random samples of sizes n_1 and n_2 selected from two normal populations with equal variances $\sigma_1^2 = \sigma_2^2$, the sampling distribution s_1^2/s_2^2 is an F distribution with $n_1 - 1$ degrees of freedom for the numerator and $n_2 - 1$ degrees of freedom for the denominator.

KEY FORMULAS

Interval Estimation For a Population Variance

$$\frac{(n-1)s^2}{\chi^2_{\alpha/2}} \leq \sigma^2 \leq \frac{(n-1)s^2}{\chi^2_{(1-\alpha/2)}} \qquad (11.6)$$

Sampling Distribution of s_1^2/s_2^2 When $\sigma_1^2 = \sigma_2^2$

$$F = \frac{s_1^2}{s_2^2} \qquad (11.7)$$

▽
REVIEW QUIZ

TRUE/FALSE

1. The sample variance should not be used as a point estimator of the population variance.
2. The sampling distribution of $(n - 1)s^2/\sigma^2$ is a normal probability distribution.
3. The random variable for the chi-square probability distribution may not assume negative values.
4. The chi-square probability distribution is symmetric.
5. In order to determine the appropriate number of degrees of freedom for a chi-square distribution, one must know the sample size.
6. When applying the chi-square distribution, we must be able to assume the population being sampled follows a normal probability distribution.
7. For the F distribution, the number of degrees of freedom in the numerator must be greater than or equal to the number of degrees of freedom in the denominator.
8. An F test can be used for a hypothesis test concerning the equality of two population variances.

MULTIPLE CHOICE

9. The sampling distribution of the quantity $(n - 1)s^2/\sigma^2$ is the
   a. chi-square distribution
   b. normal distribution
   c. F distribution
   d. t distribution
10. The sampling distribution of the ratio of independent sample variances from two normally distributed populations with equal variances is the
    a. chi-square distribution
    b. normal distribution
    c. F distribution
    d. t distribution
11. When a sample variance of 25 is obtained from a sample of 10 items, the 90% confidence interval for a population variance is
    a. 12.3 to 57.1
    b. 13.3 to 67.7
    c. 14.1 to 46.25
    d. 15.3 to 53.98
12. Compared to a 95% confidence interval, a 99% confidence interval would be
    a. a narrower interval
    b. a wider interval
    c. the same width because you use the same data
    d. either narrower or wider

Use the following information for Questions 13–15. These sample results were obtained for independent random samples from two normally distributed populations with equal variances.

	Sample 1	Sample 2
Sample size	10	16
Sample variance	25	20

13. Using a .05 level of significance, one critical value needed to use the F test for the equality of the population variances is

      a. 2.59

      b. 3.01

      c. 3.12

      d. 3.89

14. The value of the test statistic F is

      a. .13

      b. .5

      c. .8

      d. 1.25

15. Which conclusion would be reached for these data?

      a. There is a statistically significant difference between the variances of the two populations.

      b. There is no statistically significant difference between the variances of the two populations.

      c. Insufficient data—can't tell in this case.

SUPPLEMENTARY EXERCISES

23. Because of staffing decisions, management of the Gibson-Marimont Hotel is interested in the variability for the number of rooms occupied per day during a particular season of the year. A sample of 20 days of operation shows a sample mean of 290 rooms occupied per day and a sample standard deviation of 30 rooms.

    a. What is the point estimate of the population variance?

    b. Provide a 90% confidence interval for the population variance.

    c. Provide a 90% confidence interval for the population standard deviation.

24. Initial public offerings (IPOs) of stocks are on average underpriced (*Financial Analysts Journal,* December 1987). The standard deviation measures the dispersion or variation in the underpricing-overpricing indicators. A sample of 13 Canadian IPOs that were subsequently traded on the Toronto Stock Exchange had a standard deviation of 14.95. Develop a 95% confidence interval for the population standard deviation for the underpricing-overpricing indicator.

25. Historical delivery times for Buffalo Trucking, Inc., have had a mean of 3 hours and a standard deviation of .5 hour. A sample of 22 deliveries over the past month provides a sample mean of 3.1 hours and a sample standard deviation of .75 hour.

    a. Use a hypothesis test to determine if the sample results lead to rejection of the historical delivery variance of H_0: $\sigma^2 = (.5)^2 = .25$. Use $\alpha = .05$.

    b. Compute the 95% confidence intervals for the population variance and the population standard deviation.

26. Part variability is critical in the manufacturing of ball bearings. Large variances in the size of the ball bearings cause bearing failure and rapid wearout. Production standards call for a maximum variance of .0001 when the bearing sizes are measured in inches. A sample of 15 bearings shows a sample standard deviation of .014 inch.

    a. Using $\alpha = .10$, determine if the sample indicates that the maximum variance is being exceeded.

    b. Compute the 90% confidence interval estimate for the variance of the ball bearings in the population.

27. The filling variance for boxes of cereal is designed to be .02 ounce or less. A sample of 41 boxes of cereal shows a sample standard deviation of .16 ounce. Using $\alpha = .05$, determine if the variance in the cereal box fillings is exceeding the standard.

28. A sample standard deviation for the number of passengers taking a particular airline flight is 8. A 95% confidence interval estimate for the standard deviation is 5.86 passengers to 12.62 passengers.
   a. Was a sample size of 10 or 15 used in the above statistical analysis?
   b. If the sample standard deviation of $s = 8$ had been based on a sample of 25 flights, what change would you expect in the confidence interval for the population standard deviation? Compute a 95% confidence interval for σ if a sample of size 25 had been used.

29. A firm gives a mechanical aptitude test to all job applicants. A sample of 20 male applicants shows a sample variance of 80 for the test scores. A sample of 16 female applicants shows a sample variance of 220. Using $\alpha = .05$, determine if the test score variances differ for male and female job applicants. If a difference in variances exists, which group has the higher variance in mechanical aptitude?

30. The grade point averages of 352 students who completed a college course in financial accounting has a standard deviation of .940 (*The Accounting Review,* January 1988). The grade point averages of 73 students who dropped out of the same course has a standard deviation of .797. Do these data indicate that there is a difference between the variances of grade point averages for students who complete the financial accounting course and students who drop out? Use a .05 level of significance. *Note:* $F_{.025}$ with 351 and 72 degrees of freedom is approximately 1.45.

31. The accounting department analyzes the variance of the weekly unit costs reported by two production departments. A sample of 16 cost reports for each of the departments shows cost variances of 2.3 and 5.4, respectively. Is this sample sufficient to conclude that the two production departments differ in terms of unit cost variances? Use $\alpha = .10$.

32. Two new assembly methods are tested with the variances in assembly times shown below. Using $\alpha = .10$, test for equality of the two population variances.

Method	Sample Size	Sample Variance
A	31	$s_1^2 = 25$
B	25	$s_2^2 = 12$

COMPUTER CASE: AIR FORCE TRAINING PROGRAM

An Air Force introductory course in electronics uses a personalized system of instruction whereby each student views a video-taped lecture and then is given a programmed instruction text. The students work independently with the text until they have completed the training and passed a test. Of concern is the varying pace at which the students complete this portion of their training program. Some students are able to cover the programmed instruction text relatively quickly, while other students work much longer with the text and require additional time to complete the course. The faster students wait until the slower students complete the introductory course before the entire group proceeds together with other aspects of their training.

A proposed method of instruction involves the use of computer-assisted instruction. Under this method, each student will view the same video-taped lecture and then be assigned to a computer terminal for further instruction. Each student will work independently with the computer guiding the student through the self-training portion of the course.

To compare the proposed and current methods of instruction, an entering class of 122 students was randomly assigned to one of the two methods. One group of 61 students used the current programmed-text method, and the other group of 61 students used the proposed computer-assisted method. The time in hours was recorded for each student in the study. The following data are provided on the data disk in the file named TRAINING.

Course-Completion Times (hours) for Current Programmed-Text Method

76	76	77	74	76	74	74	77	72	78	73
78	75	80	79	72	69	79	72	70	70	81
76	78	72	82	72	73	71	70	77	78	73
79	82	65	77	79	73	76	81	69	75	75
77	79	76	78	76	76	73	77	84	74	74
69	79	66	70	74	72					

 TRAINING

Course-Completion Times (hours) for Proposed Computer-Assisted Method

74	75	77	78	74	80	73	73	78	76	76
74	77	69	76	75	72	75	72	76	72	77
73	77	69	77	75	76	74	77	75	78	72
77	78	78	76	75	76	76	75	76	80	77
76	75	73	77	77	77	79	75	75	72	82
76	76	74	72	78	71					

REPORT

1. Use appropriate descriptive statistics to summarize the training-time data for each method. What similarities and/or differences do you observe from the sample data?
2. Use the methods of Chapter 10 to comment on any difference between the population means for the two methods. Discuss your findings.
3. Compute the standard deviation and variance for each training method. Conduct a hypothesis test about the equality of population variances for the two training methods. Discuss your findings.
4. What conclusion can you reach about any differences between the two methods? What is your recommendation? Explain.
5. Can you suggest other data or testing that might be desirable before making a final decision on the training program to be used in the future?

CHAPTER 12

ANALYSIS OF

VARIANCE AND

EXPERIMENTAL DESIGN

WHAT YOU WILL LEARN IN THIS CHAPTER

- How to use analysis of variance to test if the means of three or more populations are equal
- What an analysis of variance table is and how it is constructed
- How to conduct statistical comparisons between pairs of population means
- How to collect and analyze data for completely randomized, randomized block, and factorial experimental designs

CONTENTS

High Self-Esteem: A Key to Academic Achievement

Self-esteem has been the subject of a large amount of research, which has revealed the importance of this characteristic in the school setting. Students with high self-esteem are more receptive to the educational process and respond more positively to the teacher, assignments, and school in general.

Robert H. Phillips, a researcher at Fordham University, reports on a study designed to investigate ways in which teacher interaction with elementary school students can improve student self-esteem. A sample of 30 elementary school students was selected from 10 low-income areas of New York City. Each student was assigned to one of three groups. The first group, the experimental group, was placed in an environment in which the students were given positive reinforcement by the teacher for any positive statements the students made about themselves. For example, if a student offered a legitimate positive self-statement, the teacher was instructed to respond with statements such as, "I'm proud of you too," "Yes, you did do well," "You are doing beautifully," and so on. The second group, the control group, was placed in the same physical environment, but the teacher made no comment when students offered positive statements about themselves. The third group, referred to as the inventory group, received no special instructions and operated under normal conditions.

All students in the research project were given self-esteem tests, at the beginning and at the end of the 7-week study. A measure of how each student's self-esteem score changed during the study was recorded. The statistical technique known as analysis of variance was used to see whether the three groups differed in terms of their change in self-esteem scores. The statistical conclusion was that there was a significant difference among the mean self-esteem scores

A teacher's individual attention and positive statements help students improve self-esteem and academic achievement.

for the three groups. The experimental group with its reinforcing teacher statements showed a significantly greater increase in self-esteem scores.

Because self-esteem is important in children's academic experiences, this research suggests that teacher responses designed to enhance student self-esteem should be part of regular educational programs. Future research is planned to gather additional evidence concerning the degree to which improvements in self-esteem scores lead to corresponding improvements in academic achievement.

This *Statistics in Practice* is based on "Increasing Positive Self-Referent Statements to Improve Self-Esteem in Low-Income Elementary School Children" by Robert H. Phillips, *The Journal of School Psychology* (Summer 1984).

In Chapter 10 we discussed how to test whether or not the means of two populations are equal. Recall that the test required the selection of an independent simple random sample from each of the two populations. The statistical approach that we introduce in this chapter, referred to as *analysis of variance* (ANOVA), can be thought of as an extension of the Chapter 10 methods to the case of more than two populations. For example, in the *Statistics in Practice,* analysis of variance was used to test whether the mean self-esteem scores for three groups of students were equal.

Experiments are undertaken to discover something not yet known or to examine the validity of a hypothesis. In an experimental study, the variables of interest are first identified. Then, one or more factors in the study are controlled so that data may be obtained about how the factors influence the variables. These data are then analyzed in order to reach a conclusion. The process of planning an experiment so that appropriate data can be collected and analyzed using statistical methods is referred to as the *statistical design of experiments.*

For example, a pharmaceutical firm might be interested in conducting an experiment designed to determine the effect of three different drugs, referred to as A, B, and C, on the cholesterol level of individuals. Note that this experiment involves one variable, the cholesterol level of individuals, and one factor, the type of drug administered. Suppose that the objective of the experimenter is to determine whether the mean change in cholesterol level for individuals is the same for all three drugs. To obtain data on the effect of the three drugs, a sample of individuals would be selected. In the simplest type of experimental design, one-third of the individuals would be given drug A, one-third would be given drug B, and one-third would be given drug C. Data on cholesterol level would then be collected for each group. Statistical analysis of the data would help determine if the mean change in cholesterol level is the same for all three drugs.

Every statistical problem involving an experiment such as the one just described consists of three steps: (1) designing the experiment; (2) collecting the data; and (3) analyzing the data. These three steps are closely related, since how the data are collected and analyzed depends on the type of design selected. In this chapter we discuss three different experimental designs: completely randomized designs, randomized block designs, and factorial designs.

12.1 AN INTRODUCTION TO ANALYSIS OF VARIANCE

In order to introduce the concept of analysis of variance, let us consider the situation described in the following example.

EXAMPLE 12.1 ▶ National Computer Products, Inc. (NCP) manufactures printers and fax machines at plants located in Charlotte, Houston, and San Diego. To measure how much employees at these plants know about total quality management, a random sample of six employees was selected from each plant and given a quality-awareness examination. The examination scores obtained for these 18 employees are shown in Table 12.1. The corresponding sample means, sample variances, and sample standard deviations for each group are provided. Management would like to use these data to test the hypothesis that the mean examination score is the same at each plant.

We will define population 1 as all employees at the Charlotte plant, population 2 as all employees at the Houston plant, and population 3 as all employees at the San Diego plant.

TABLE 12.1   EXAMINATION SCORES FOR 18 NCP EMPLOYEES

Observation	Plant 1 Charlotte	Plant 2 Houston	Plant 3 San Diego
1	85	71	59
2	75	75	64
3	82	73	62
4	76	74	69
5	71	69	75
6	85	82	67
Sample mean	79	74	66
Sample variance	34	20	32
Sample standard deviation	5.83	4.47	5.66

Let

$$\mu_1 = \text{mean examination score for population 1}$$
$$\mu_2 = \text{mean examination score for population 2}$$
$$\mu_3 = \text{mean examination score for population 3}$$

Although we will never know the actual values of μ_1, μ_2, and μ_3, we would like to use the sample results to test the following hypotheses:

$$H_0:\ \mu_1 = \mu_2 = \mu_3$$

$$H_a:\ \text{Not all population means are equal} \qquad \blacktriangleleft$$

As we will demonstrate shortly, analysis of variance is a statistical procedure for testing hypotheses such as those presented in Example 12.1.

ASSUMPTIONS FOR ANALYSIS OF VARIANCE

To test the above hypotheses using analysis of variance, the following three assumptions must be made:

1. *Each population is normally distributed.* Implication: In Example 12.1 the examination scores (response variable) must be normally distributed at each plant.
2. *The variance, denoted σ^2, is the same for each population.* Implication: In Example 12.1 the variance of examination scores must be the same at each plant.
3. *The observations must be independent.* Implication: In Example 12.1 the examination score for each employee must be independent of the examination score for any other employee.

A CONCEPTUAL OVERVIEW

Suppose that the assumptions for analysis of variance are satisfied in Example 12.1. If the null hypothesis is true ($\mu_1 = \mu_2 = \mu_3 = \mu$), each sample observation would have been drawn from the same normal probability distribution with mean μ and variance σ^2. To provide a visual perspective of the situation, consider the Minitab dotplots for the NCP data shown in Figure 12.1. Assuming that the three populations are normally distributed with variance σ^2, does it appear that the observations in each sample have been drawn from populations with the same mean? Although this is purely a subjective observation, you might agree that the employees at the Charlotte plant appear to have a higher mean examination score, while the employees at the San Diego plant have a lower mean examination score.

If the means for the three populations are equal, we would expect the three sample means to be close together. In fact, the closer the three sample means are to one another, the more evidence we have supporting the conclusion that the population means are equal. Alternatively, the more the sample means differ, the more evidence we have that the population means are not equal.

If the null hypothesis, H_0: $\mu_1 = \mu_2 = \mu_3$, is true, we can use the variability among the sample means to develop an estimate of the population variance σ^2. And, if the assumptions for analysis of variance are satisfied, each sample will have come from the same normal probability distribution with mean μ and variance σ^2. Recall from Chapter 7 that the sampling distribution of the sample mean $\bar{x}$ for a simple random sample of size n from a normal population will be normally distributed with mean μ and standard deviation $\sigma_{\bar{x}} = \sigma/\sqrt{n}$. Figure 12.2 illustrates such a sampling distribution; note that the variance is $\sigma_{\bar{x}}^2 = \sigma^2/n$.

If the null hypothesis is true, we can think of each of the three sample means, $\bar{x}_1 = 79$, $\bar{x}_2 = 74$, and $\bar{x}_3 = 66$ from Table 12.1 as values drawn at random from the sampling distribution shown in Figure 12.2. In this case, the mean and variance of the three $\bar{x}$ values can be used to estimate the mean and variance of the sampling distribution.

EXAMPLE 12.1
(CONTINUED)

The best estimate of the mean of the sampling distribution of $\bar{x}$ is the mean or average of the three sample means. That is, $(79 + 74 + 66)/3 = 73$. We refer to this estimate as the *overall sample mean*. To estimate the variance of the sampling distribution of $\bar{x}$, denoted $s_{\bar{x}}^2$ we compute the variance using the three sample means:

$$s_{\bar{x}}^2 = \frac{(79 - 73)^2 + (74 - 73)^2 + (66 - 73)^2}{3 - 1} = \frac{86}{2} = 43$$

Since $\sigma_{\bar{x}}^2 = \sigma^2/n$, solving for σ^2 gives

$$\sigma^2 = n\sigma_{\bar{x}}^2$$

FIGURE 12.1   MINITAB DOTPLOT FOR NCP EXAMINATION SCORES

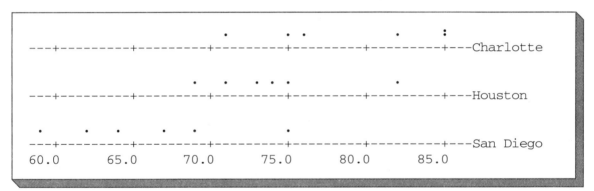

FIGURE 12.2   SAMPLING DISTRIBUTION OF $\bar{X}$ GIVEN H_0 IS TRUE

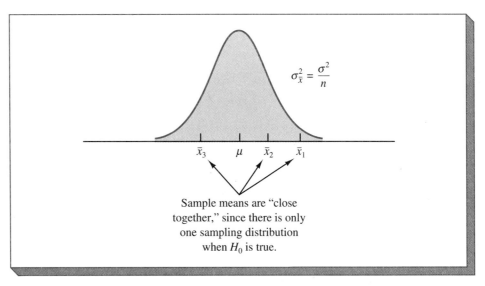

$$\sigma_{\bar{x}}^2 = \frac{\sigma^2}{n}$$

$\bar{x}_3 \quad \mu \quad \bar{x}_2 \quad \bar{x}_1$

Sample means are "close together," since there is only one sampling distribution when H_0 is true.

Hence,

$$\text{Estimate of } \sigma^2 = n(\text{Estimate of } \sigma_{\bar{x}}^2) = ns_{\bar{x}}^2 = 6(43) = 258$$

The result, $ns_{\bar{x}}^2 = 258$, is referred to as the *between-samples* estimate of σ^2.   ◄

The between-samples estimate of σ^2 is based on the assumption that the null hypothesis is true. In this case, each sample comes from the same population, and there is only one sampling distribution of $\bar{x}$. To illustrate what happens when H_0 is false, suppose that the population means *all differ*. Note that since the three samples are from normal populations with different means, there will be different sampling distributions. Figure 12.3 shows that in this case, the sample means are not as close together as they were when H_0 was true. Thus, $s_{\bar{x}}^2$ will be larger, causing the between-samples estimate of σ^2 to be larger. In general, when the population means are not equal, the between-samples estimate will overestimate the population variance σ^2.

The variation within each of the samples can also have an effect on the conclusion we reach in analysis of variance. When a simple random sample is selected from each population, each of the sample variances provides an unbiased estimate of the population variance σ^2. Thus, we can combine or pool the individual estimates of σ^2 into one overall estimate. The estimate of σ^2 obtained in this fashion is called the *pooled* or *within-samples* estimate of σ^2. Because each sample variance provides an estimate of σ^2 based only on the variation within each sample, the within-samples estimate of σ^2 is not affected by whether or not the population means are equal. When the sample sizes are equal, the within-samples estimate of σ^2 can be obtained by computing the average of the individual sample variances.

EXAMPLE 12.1 ►
(CONTINUED)

Using the sample variances from Table 12.1, we obtain

$$\text{Within-Samples Estimate of } \sigma^2 = \frac{34 + 20 + 32}{3} = \frac{86}{3} = 28.67$$

FIGURE 12.3   SAMPLING DISTRIBUTION FOR $\bar{x}$ GIVEN H_0 IS FALSE

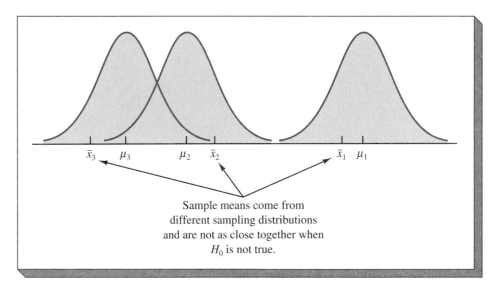

Sample means come from
different sampling distributions
and are not as close together when
H_0 is not true.

Note that the between-samples estimate of σ^2 (258) is much larger than the within-samples estimate of σ^2 (28.67). In fact, the ratio of these two estimates is

$$258/28.67 = 9.00$$ ◄

The between-samples approach provides a good estimate of σ^2 only if the null hypothesis is true; if the null hypothesis is false, the between-samples approach *overestimates* σ^2. The within-samples approach provides a good estimate of σ^2 in either case. Thus, if the null hypothesis is true, both estimates will be similar and their ratio will be close to 1. If the null hypothesis is false, the between-samples estimate will be larger than the within-samples estimate, and their ratio will be large. In the next section we will show how large this ratio must be to reject H_0: $\mu_1 = \mu_2 = \mu_3$.

In summary then, the logic behind ANOVA is based on the development of two independent estimates of the population variance σ^2. One estimate of σ^2 is based on the variability among the sample means themselves, and the other estimate of σ^2 is based on the variability of the data within each sample. By comparing these two estimates of σ^2, we will be able to determine whether or not we should reject H_0: $\mu_1 = \mu_2 = \mu_3$. Since the methodology uses a comparison of variances, it is referred to as analysis of variance.

NOTES AND COMMENTS

1. In Chapter 10 we presented statistical methods for testing the hypothesis that the means of two populations are equal. Although in the introduction to this chapter we stated that the analysis of variance is a statistical technique for testing the hypothesis that the means of three or more populations are equal, ANOVA can also be used to test the hypothesis that the means of two populations are equal. In practice, however, analysis of variance is usually thought of as a technique for testing for the equality of three or more population means.

—Continues on next page

—Continued from previous page

2. In Chapter 10 we discussed how to test for the equality of two population means whenever one or both sample sizes are less than 30. As part of that discussion, we illustrated the process of combining the results of two independent simple random samples to provide one estimate of σ^2; this process was referred to as pooling, and the resulting sample variance we obtained was referred to as the pooled estimator of σ^2. In analysis of variance, the within-samples estimate of σ^2 is simply the generalization of this concept to the case of more than two samples; this is why we also referred to the within-samples estimator as the pooled estimator of σ^2.

12.2 ANALYSIS OF VARIANCE: TESTING FOR THE EQUALITY OF k POPULATION MEANS

In general, analysis of variance can be used to test for the equality of k population means. The general form of the hypotheses tested is

$$H_0: \mu_1 = \mu_2 = \cdots = \mu_k$$

$$H_a: \text{ Not all the means are equal}$$

where

$$\mu_j = \text{mean of the } j\text{th population}$$

We assume that a simple random sample of size n_j has been selected from each of the k populations. Let

$$x_{ij} = \text{the } i\text{th observation in the } j\text{th sample}$$
$$n_j = \text{the number of observations in the } j\text{th sample}$$
$$\bar{x}_j = \text{the mean of the } j\text{th sample}$$
$$s_j^2 = \text{the variance of the } j\text{th sample}$$
$$s_j = \text{the standard deviation of the } j\text{th sample}$$

The formulas for the jth sample mean and variance are as follows:

$$\bar{x}_j = \frac{\sum_{i=1}^{n_j} x_{ij}}{n_j} \tag{12.1}$$

$$s_j^2 = \frac{\sum_{i=1}^{n_j} (x_{ij} - \bar{x}_j)^2}{n_j - 1} \tag{12.2}$$

The overall sample mean, denoted $\bar{\bar{x}}$, is the sum of all the observations divided by the total number of observations. That is,

$$\bar{\bar{x}} = \frac{\sum_{j=1}^{k} \sum_{i=1}^{n_j} x_{ij}}{n_T} \tag{12.3}$$

where

$$n_T = n_1 + n_2 + \cdots + n_k \tag{12.4}$$

If the size of each sample is n, $n_T = kn$; in this case, expression (12.3) reduces to

$$\bar{\bar{x}} = \frac{\sum_{i=1}^{k}\sum_{i=1}^{n_j} x_{ij}}{kn} = \frac{\sum_{j=1}^{k}\sum_{i=1}^{n_j} x_{ij}/n}{k} = \frac{\sum_{j=1}^{k} \bar{x}_j}{k} \tag{12.5}$$

In other words, whenever the sample sizes are the same, the overall sample mean is just the average of the k sample means.

EXAMPLE 12.1 ▶ (CONTINUED)

Since each sample consists of $n = 6$ observations, the overall sample mean can be computed using equation (12.5). For the data shown in Table 12.1 we obtained the following result:

$$\bar{\bar{x}} = \frac{79 + 74 + 66}{3} = 73$$

Thus, if the null hypothesis is true, the overall sample mean of 73 is the best estimate of the population mean μ.   ◀

BETWEEN-SAMPLES ESTIMATE OF POPULATION VARIANCE

In the previous section we introduced the concept of a between-samples estimate of σ^2. This estimate of σ^2 is called the *mean square between* and is denoted MSB. The general formula for computing MSB is as follows:

$$\text{MSB} = \frac{\sum_{j=1}^{k} n_j(\bar{x}_j - \bar{\bar{x}})^2}{k - 1} \tag{12.6}$$

The numerator in equation (12.6) is called the *sum of squares between* and is denoted SSB. The denominator, $k - 1$, represents the degrees of freedom associated with SSB. Thus, the mean square between can be computed as follows.

Mean Square Between

$$\text{MSB} = \frac{\text{SSB}}{k - 1} \tag{12.7}$$

where

$$\text{SSB} = \sum_{j=1}^{k} n_j(\bar{x}_j - \bar{\bar{x}})^2 \tag{12.8}$$

If H_0 is true, MSB provides an unbiased estimate of σ^2. However, if the means of the k populations are not equal, MSB is not an unbiased estimate of σ^2; in fact, in this case, MSB should overestimate σ^2.

EXAMPLE 12.1 ▶
(CONTINUED)

For the NCP data shown in Table 12.1, we obtain the following results:

$$\text{SSB} = \sum_{j=1}^{k} n_j(\bar{x}_j - \bar{\bar{x}})^2 = 6(79 - 73)^2 + 6(74 - 73)^2 + 6(66 - 73)^2 = 516$$

$$\text{MSB} = \frac{\text{SSB}}{k - 1} = \frac{516}{2} = 258$$

◀

WITHIN-SAMPLES ESTIMATE OF POPULATION VARIANCE

The second estimate of σ^2 is based on the variation of the sample observations within each sample. This estimate of σ^2 is called the *mean square within* and is denoted MSW. The general formula for computing MSW is as follows:

$$\text{MSW} = \frac{\displaystyle\sum_{j=1}^{k} (n_j - 1)s_j^2}{n_T - k} \tag{12.9}$$

The numerator in equation (12.9) is called the *sum of squares within* and is denoted SSW. The denominator of MSW is referred to as the degrees of freedom associated with SSW. Thus, the formula for MSW can also be stated as follows.

Mean Square Within

$$\text{MSW} = \frac{\text{SSW}}{n_T - k} \tag{12.10}$$

where

$$\text{SSW} = \sum_{j=1}^{k} (n_j - 1)s_j^2 \tag{12.11}$$

Note that MSW is based on the variation within each of the samples; it is not influenced by whether or not the null hypothesis is true. Thus, MSW always provides an unbiased estimate of σ^2.

EXAMPLE 12.1 ▶
(CONTINUED)

For the NCP data in Table 12.1, we obtain the following results:

$$\text{SSW} = \sum_{j=1}^{k} (n_j - 1)s_j^2 = (6 - 1)34 + (6 - 1)20 + (6 - 1)32 = 430$$

$$\text{MSW} = \frac{\text{SSW}}{n_T - k} = \frac{430}{18 - 3} = \frac{430}{15} = 28.67$$

◀

COMPARING THE VARIANCE ESTIMATES: THE F TEST

Let us assume for the moment that the null hypothesis is true. In this case, MSB and MSW provide two independent, unbiased estimates of σ^2. Recall from Chapter 11 that for normal populations, the sampling distribution of the ratio of two independent estimates of σ^2 follows an F distribution. Thus, if the null hypothesis is true and the ANOVA assumptions are valid, the sampling distribution of MSB/MSW is an F distribution with numerator degrees of freedom equal to $k - 1$ and denominator degrees of freedom equal to $n_T - k$.

If the means of the k populations are not equal, the value of MSB/MSW will be inflated because MSB overestimates σ^2. Hence, we will reject H_0 if the resulting value of MSB/MSW appears to be too large to have been selected at random from an F distribution with degrees of freedom $k - 1$ in the numerator and $n_T - k$ in the denominator. The value of MSB/MSW that will cause us to reject H_0 depends on α, the level of significance. Once α is selected, a critical value can be determined. Figure 12.4 shows the sampling distribution of MSB/MSW and the rejection region associated with a level of significance equal to α where F_α denotes the critical value.

EXAMPLE 12.1 ▶
(CONTINUED)

Suppose the manager responsible for making the decision at National Computer Products was willing to accept a probability of a Type I error of $\alpha = .05$. From Table 4 of Appendix B, we can determine the critical F value by locating the value corresponding to numerator degrees of freedom equal to $k - 1 = 3 - 1 = 2$ and denominator degrees of freedom equal to $n_T - k = 18 - 3 = 15$. Thus, we obtain the value $F_{.05} = 3.68$. Note that this tells us that if we were to select a value at random from an F distribution with 2 numerator and 15 denominator degrees of freedom, only 5% of the time would we observe a value greater than 3.68. Moreover, the theory behind the analysis of variance tells us that if the null hypothesis is true, the ratio of MSB/MSW would be a value from this F distribution. Hence, the appropriate rejection rule for the NCP data is written

Reject H_0 if MSB/MSW > 3.68

Recall that MSB = 258 and MSW = 28.67. Since MSB/MSW = 258/28.67 = 9.00 is greater than the critical value, $F_{.05} = 3.68$, there is sufficient evidence to reject the null hypothesis that the population mean examination scores at the three NCP plants are equal. ◀

FIGURE 12.4   SAMPLING DISTRIBUTION OF MSB/MSW; THE CRITICAL VALUE
FOR REJECTING THE NULL HYPOTHESIS OF EQUALITY OF MEANS IS F_α

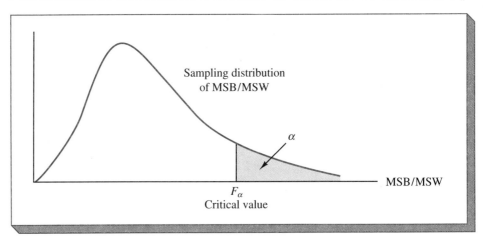

THE ANOVA TABLE

The results of the preceding calculations can be conveniently displayed in a table referred to as the *analysis of variance table.* Table 12.2 shows the analysis of variance table for Example 12.1. The sum of squares associated with the source of variation referred to as "Total" is called the total sum of squares (SST). Note that the results for the NCP problem suggest that SST = SSB + SSW, and that the degrees of freedom associated with the total sum of squares is the sum of the degrees of freedom associated with the between-samples estimate of σ^2 and the within-samples estimate of σ^2.

We should point out that SST divided by its degrees of freedom $n_T - 1$, is nothing more than the overall sample variance that would be obtained if we treated the entire set of 18 observations as one data set. Using the entire data set as one sample, the formula for computing the total sum of squares, SST, is as follows:

$$SST = \sum_{j=1}^{k} \sum_{i=1}^{n_j} (x_{ij} - \overline{\overline{x}})^2 \tag{12.12}$$

It can be shown that the results we observed for the analysis of variance table for Example 12.1 also apply to other problems. That is,

$$SST = SSB + SSW \tag{12.13}$$

In other words, SST can be partitioned into two sums of squares: the sum of squares between and the sum of squares within. Note also that the degrees of freedom corresponding to SST, $n_T - 1$, can be partitioned into the degrees of freedom corresponding to SSB, $k - 1$, and the degrees of freedom corresponding to SSW, $n_T - k$. The analysis of variance can be viewed as the process of *partitioning* the total sum of squares and degrees of freedom into their corresponding sources: between and within. Dividing the sum of squares by the appropriate degrees of freedom provides the variance estimates and the F value used to test the hypothesis of equal population means.

COMPUTER RESULTS FOR ANALYSIS OF VARIANCE

Because of the widespread availability of statistical computer packages, analysis of variance computations with large sample sizes and/or a large number of populations can be easily handled. In Figure 12.5 we show output for the NCP problem obtained from the Minitab computer package. The first part of the computer output contains the familiar ANOVA table format. Comparing Figure 12.5 with Table 12.2, we see that the same information is available, although some of the headings are a little different. The heading SOURCE is used for the source of variation column, FACTOR identifies the between-samples row, and ERROR identifies the within-samples row. The sum of squares and

---

TABLE 12.2   **ANALYSIS OF VARIANCE TABLE FOR THE NCP DATA**

Source of Variation	Sum of Squares	Degrees of Freedom	Mean Square	F
Between	516	2	258.00	9.00
Within	430	15	28.67	
Total	946	17		

FIGURE 12.5   MINITAB OUTPUT FOR THE NCP ANALYSIS OF VARIANCE

```
ANALYSIS OF VARIANCE
SOURCE   DF       SS      MS         F        p
FACTOR    2    516.0   258.0      9.00     0.003
ERROR    15    430.0    28.7
TOTAL    17    946.0
                              INDIVIDUAL 95 PCT CI'S FOR MEAN
                              BASED ON POOLED STDEV
  LEVEL   N    MEAN   STDEV   ---+---------+---------+---------+---
PLANT 1   6  79.000   5.831                        (------*------)
PLANT 2   6  74.000   4.472                (------*-----)
PLANT 3   6  66.000   5.657   (-----*------)
                              ---+---------+---------+---------+---
POOLED STDEV =      5.354    63.0      70.0      77.0      84.0
```

degrees of freedom columns are also interchanged, and a *p*-value is provided for the *F* test.

Note that below the ANOVA table, the computer output contains the respective sample sizes, the sample means, and the standard deviations. In addition, Minitab provides a figure that shows individual 95% confidence interval estimates of each population mean. In developing these confidence interval estimates, Minitab uses MSW as the estimate of σ^2. Thus, the square root of MSW provides the best estimate of the population standard deviation σ. This estimate of σ is referred to on the computer output as the POOLED STDEV; it is equal to 5.354. To illustrate how these interval estimates are developed, we will compute a 95% confidence interval estimate of the population mean for the Charlotte plant, identified as PLANT 1 in the computer output.

From our study of interval estimation in Chapter 8, we know that the general form of an interval estimate of a population mean is

$$\bar{x} \pm t_{\alpha/2}\frac{s}{\sqrt{n}} \tag{12.14}$$

where *s* is the estimate of the population standard deviation σ. Since in the analysis of variance the best estimate of σ is provided by the square root of MSW or the POOLED STDEV, we use a value of 5.354 for *s* in expression (12.14). The degrees of freedom for the *t* value is 15, the degrees of freedom associated with the within-samples estimate of σ^2. Thus, with $t_{.025} = 2.131$, we obtain

$$79 \pm 2.131 \frac{5.354}{\sqrt{6}} = 79 \pm 4.66$$

From this calculation, we see that the figure shown on the Minitab output for plant 1 depicts an interval that goes from 74.34 to 83.66. Since the sample sizes are all equal for Example 12.1, the confidence intervals for plants 2 and 3 are also constructed by adding and subtracting 4.66 from each sample mean. Thus, in the figure provided by Minitab, we see that the width of each confidence interval is the same.

NOTES AND
COMMENTS

1. The overall sample mean can also be computed as a weighted average of the k sample means.

$$\bar{\bar{x}} = \frac{n_1\bar{x}_1 + n_2\bar{x}_2 + \cdots + n_k\bar{x}_k}{n_T}$$

In problems where the sample means have been provided, this formula is simpler than equation (12.3) for computing the overall mean.

2. If each sample consists of n observations, equation (12.6) can be written as

$$\text{MSB} = \frac{n\sum_{j=1}^{k}(\bar{x}_j - \bar{\bar{x}})^2}{k-1} = n\left[\frac{\sum_{j=1}^{k}(\bar{x}_j - \bar{\bar{x}})^2}{k-1}\right] = ns_{\bar{x}}^2$$

Note that this is the same result that we presented in Section 12.1 when we introduced the concept of the between-samples estimate of σ^2. Equation (12.6) is simply a generalization of this result to the unequal sample-size case.

3. If each sample has n observations, $n_T = kn$; thus, $n_T - k = k(n-1)$, and equation (12.9) can be rewritten as

$$\text{MSW} = \frac{\sum_{j=1}^{k}(n-1)s_j^2}{k(n-1)} = \frac{(n-1)\sum_{j=1}^{k}s_j^2}{k(n-1)} = \frac{\sum_{j=1}^{k}s_j^2}{k}$$

In other words, if the sample sizes are the same, the within-samples estimate of σ^2 is just the average of the k sample variances. Note that this is the result we used in Section 12.1 when we introduced the concept of the within-samples estimate of σ^2.

EXERCISES

METHODS

SELF TEST

1. Samples of five observations were selected from each of three populations. The data obtained are shown below.

Observation	Sample 1	Sample 2	Sample 3
1	32	44	33
2	30	43	36
3	30	44	35
4	26	46	36
5	32	48	40
Sample mean	30	45	36
Sample variance	6.00	4.00	6.50

a. Develop the dotplots for these data. Based on your subjective evaluation of the dotplots, does it appear that the observations in each sample have been drawn from the same population?

b. Compute the between-samples estimate of σ^2.

c. Compute the within-samples estimate of σ^2.

d. At the $\alpha = .05$ level of significance, can we reject the null hypothesis that the means of the three populations are equal?

e. Set up the ANOVA table for this problem.

2. Four observations were selected from each of three populations. The data obtained are shown below.

Observation	Sample 1	Sample 2	Sample 3
1	165	174	169
2	149	164	154
3	156	180	161
4	142	158	148
Sample mean	153	169	158
Sample variance	96.67	97.33	82.00

a. Compute the between-samples estimate of σ^2.

b. Compute the within-samples estimate of σ^2.

c. At the $\alpha = .05$ level of significance, can we reject the null hypothesis that the three population means are equal? Explain.

d. Set up the ANOVA table for this problem.

3. Samples were selected from three populations. The data obtained are shown below.

	Sample 1	Sample 2	Sample 3
	93	77	88
	98	87	75
	107	84	73
	102	95	84
		85	75
		82	
Sample mean	100	85	79
Sample variance	35.33	35.60	43.50

a. Compute the between-samples estimate of σ^2.

b. Compute the within-samples estimate of σ^2.

c. At the $\alpha = .05$ level of significance, can we reject the null hypothesis that the three population means are equal? Explain.

d. Set up the ANOVA table for this problem.

4. A random sample of 16 observations was selected from each of four populations. A portion of the ANOVA table is shown below.

Source of Variation	Sum of Squares	Degrees of Freedom	Mean Square	F
Between			400	
Within				
Total	1500			

a. Complete the missing entries in the ANOVA table.
b. At the $\alpha = .05$ level of significance, can we reject the null hypothesis that the means of the four populations are equal?

5. Random samples of 25 observations were selected from each of three populations. For these data, SSB = 120 and SSW = 216.
  a. Set up the ANOVA table for this problem.
  b. At the $\alpha = .05$ level of significance, what is the critical F value?
  c. At the $\alpha = .05$ level of significance, can we reject the null hypothesis that the three population means are equal?

APPLICATIONS

SELF TEST

6. To test if the mean time needed to mix a batch of material is the same for machines produced by three manufacturers, the Jacobs Chemical Company obtained the following data on the time (in minutes) needed to mix the material. Use these data to test if the population mean times needed to mix a batch of material differ for the three manufacturers. Use $\alpha = .05$.

	Manufacturer		
	1	2	3
	20	28	20
	26	26	19
	24	31	23
	22	27	22
Sample mean	23	28	21
Sample variance	6.67	4.67	3.33

7. A 100-question science test was given to 13-year-old children from several different countries (*Newsweek*, February 17, 1992); a portion of the data showing the number of questions that were answered correctly for three samples of six students is shown below.

	South Korea	Soviet Union	United States
	81	71	63
	71	78	61
	85	62	69
	70	71	70
	74	68	75
	87	76	64
Sample mean	78	71	67
Sample variance	53.6	32.8	27.6

At the $\alpha = .05$ level of significance, test if the population mean test scores differ.

 INFO

8. Managers at all levels of an organization need to have the information necessary to perform their respective tasks. A study investigated the effect the source has on the dissemination of the information (*Journal of Management Information Systems*, Fall 1988). In this particular study, the sources of information were a superior, a peer, and a subordinate. In each case, a measure of dissemination was obtained with higher values indicating greater dissemination of information. Using $\alpha = .05$ and the data shown below, test whether the source of information significantly affects dissemination. What is your conclusion, and what does this suggest about the use and dissemination of information?

	Superior	Peer	Subordinate
	8	6	6
	5	6	5
	4	7	7
	6	5	4
	6	3	3
	7	4	5
	5	7	7
	5	6	5
Sample mean	5.75	5.5	5.25
Sample variance	1.64	2.00	1.93

9. A study investigated the perception of corporate ethical values among individuals specializing in marketing (*Journal of Marketing Research*, July 1989). Suppose that the data shown below were obtained in a similar study (higher scores indicate higher ethical values). Using $\alpha = .05$, test to see if there are significant differences in perception for the three groups of specialists.

	Marketing Managers	Marketing Research	Advertising
	6	5	6
	5	5	7
	4	4	6
	5	4	5
	6	5	6
	4	4	6
Sample mean	5	4.5	6
Sample variance	.8	.3	.4

 MACHINES

10. To test for any significant difference in the number of hours between breakdowns for four machines, the following data were obtained.

	Machine			
	1	2	3	4
	6.4	8.7	11.1	9.9
	7.8	7.4	10.3	12.8
	5.3	9.4	9.7	12.1
	7.4	10.1	10.3	10.8
	8.4	9.2	9.2	11.3
	7.3	9.8	8.8	11.5
Sample mean	7.1	9.1	9.9	11.4
Sample variance	1.21	.93	.70	1.02

At the $\alpha = .05$ level of significance, is there any difference in the population mean times among the four machines?

12.3 ▼ MULTIPLE COMPARISON PROCEDURES

When we use analysis of variance to test if the means of k populations are equal, we must keep in mind that rejection of the null hypothesis only allows us to conclude that the population means are *not all equal*. In such cases, we can determine where the differences among means occur by conducting statistical comparisons between pairs of population means.

FISHER'S LSD

Fisher's least significant difference (LSD) procedure is one of the oldest and perhaps most widely used methods for making pairwise comparisons of population means. In Example 12.1 there were three populations. The hypotheses were written as follows:

$$H_0:\ \mu_1 = \mu_2 = \mu_3$$

$$H_a:\ \text{Not all population means are equal}$$

Using analysis of variance, we rejected H_0 and concluded that the population mean examination scores are not the same at each plant. In this case, the follow-up question is, We believe that the population means differ, but where do the differences occur? That is, do the means of populations 1 and 2 differ? Or do those of populations 1 and 3? Or those of populations 2 and 3?

For example, let us test to see if there is a significant difference between the means of population 1 (Charlotte) and population 2 (Houston). Although Table 12.1 shows that the sample mean is 79 for the Charlotte plant and 74 for the Houston plant, is this sample information sufficient to justify the conclusion that there is a difference between the population mean examination scores for these two plants? In statistical terms, we state the following hypotheses:

$$H_0:\ \mu_1 = \mu_2$$

$$H_a:\ \mu_1 \neq \mu_2$$

In Chapter 10 we presented a statistical procedure for testing the hypothesis that the means of two populations are equal. With a slight modification in how we estimate the population variance, Fisher's LSD procedure is based on the t test statistic presented for the two-population case. The test statistic for Fisher's LSD procedure is

$$t = \frac{\bar{x}_1 - \bar{x}_2}{\sqrt{\text{MSW}\left(\frac{1}{n_1} + \frac{1}{n_2}\right)}} \tag{12.15}$$

We reject H_0 if $t < -t_{\alpha/2}$ or $t > t_{\alpha/2}$.

EXAMPLE 12.1
(CONTINUED)

Table 12.2 shows that the value of MSW is 28.67. This is the estimate of σ^2 and is based on 15 degrees of freedom. At the .05 level of significance, the t distribution table shows that with 15 degrees of freedom, $t_{.025} = 2.131$. Thus, if the value of t found using equation (12.15) is less than -2.131 or greater than 2.131, we reject H_0. For the NCP data, we obtain the following t value:

$$t = \frac{79 - 74}{\sqrt{28.67\left(\frac{1}{6} + \frac{1}{6}\right)}} = 1.62$$

Since $t = 1.62$, we do not have sufficient statistical evidence to reject the null hypothesis; thus, we cannot conclude that the population mean score at the Charlotte plant is different from the population mean score at the Houston plant. ◀

Many practitioners find it easier to determine how large a difference between the sample means must exist to reject H_0. Thus, solving for $\bar{x}_1 - \bar{x}_2$ in equation (12.15) we see that if

$$\bar{x}_1 - \bar{x}_2 > t_{\alpha/2}\sqrt{\text{MSW}\left(\frac{1}{n_1} + \frac{1}{n_2}\right)}$$

or

$$\bar{x}_1 - \bar{x}_2 < -t_{\alpha/2}\sqrt{\text{MSW}\left(\frac{1}{n_1} + \frac{1}{n_2}\right)}$$

we reject H_0. If we define the least significant difference (LSD) as

$$\text{LSD} = t_{\alpha/2}\sqrt{\text{MSW}\left(\frac{1}{n_1} + \frac{1}{n_2}\right)} \tag{12.16}$$

we will reject H_0 if

$$\bar{x}_1 - \bar{x}_2 > \text{LSD}$$

or

$$\bar{x}_1 - \bar{x}_2 < -\text{LSD}$$

If we consider only the magnitude or absolute value of the difference, we can write the rejection rule for Fisher's LSD test as

$$\text{Reject } H_0 \text{ if } |\bar{x}_1 - \bar{x}_2| > \text{LSD}$$

EXAMPLE 12.1
(CONTINUED)

The value of LSD for the NCP data is

$$\text{LSD} = 2.131 \sqrt{28.67\left(\frac{1}{6} + \frac{1}{6}\right)} = 6.59$$

Note that when the sample sizes are equal, only one value for LSD is computed. In such cases, we can simply compare the magnitude of the difference between any two means with the value of LSD. For example, the difference between the sample means for population 1 and population 3 is $79 - 66 = 13$. Since this difference is greater than 6.59, we can reject the null hypothesis that the population mean examination score for the Charlotte plant is equal to the population mean examination score for the San Diego plant. Similarly, since the difference between the sample means for populations 2 and 3 is $74 - 66 = 8 > 6.59$, we also can reject the hypothesis that the population mean examination score for the Houston plant is equal to the population mean examination score at the San Diego plant. In effect, our conclusion is that the Charlotte and Houston plants both differ from the San Diego plant. ◄

Fisher's LSD procedure can also be used to develop a confidence interval estimate of the difference between two population means. For example, a confidence interval estimate of the difference between the means of populations 1 and 2 is given by the following expression:

$$\bar{x}_1 - \bar{x}_2 \pm t_{\alpha/2} \sqrt{\text{MSW}\left(\frac{1}{n_1} + \frac{1}{n_2}\right)} \tag{12.17}$$

Using equation (12.16) we can write this interval as

$$\bar{x}_1 - \bar{x}_2 \pm \text{LSD} \tag{12.18}$$

If the confidence interval in expression (12.18) includes the value of 0, we cannot reject the hypothesis that the two population means are equal. However, if the confidence interval does not include the value 0, we conclude that there is a difference between the population means.

EXAMPLE 12.1
(CONTINUED)

Recall that LSD = 6.59 (corresponding to $t_{.025} = 2.131$). Thus, a 95% confidence interval estimate of the difference between the means of populations 1 and 2 is $79 - 74 \pm 6.59 = 5 \pm 6.59 = -1.59$ to 11.59; since this interval includes 0, we cannot reject the hypothesis that the two population means are equal. ◄

TYPE I ERROR RATES

We began the discussion of Fisher's LSD procedure with the premise that in analysis of variance we found statistical evidence to reject the null hypothesis of equal population means. In such cases, we showed how Fisher's LSD procedure can be used to determine where the differences occur. Technically, this is referred to as a *protected* or *restricted* LSD test since it is only employed if we find a significant F value using analysis of variance. To see why this is important in multiple comparison tests, we need to explain the difference between a *comparisonwise* Type I error rate and an *experimentwise* Type I error rate.

Consider the following three hypotheses tests.

Test 1	Test 2	Test 3
H_0: $\mu_1 = \mu_2$	H_0: $\mu_1 = \mu_3$	H_0: $\mu_2 = \mu_3$
H_a: $\mu_1 \neq \mu_2$	H_a: $\mu_1 \neq \mu_3$	H_a: $\mu_2 \neq \mu_3$

Assume that these three hypotheses show all possible pairwise comparisons for the problem. Suppose that we use Fisher's LSD procedure to test each of these three pairwise comparisons. If we carry out each test using a level of significance of $\alpha = .05$ and are not able to reject the null hypothesis for any test, it would seem reasonable to conclude that the three population means must be equal. Before jumping to any conclusions, however, let us consider what can happen if we follow this approach.

To begin with, let us consider using Fisher's LSD procedure for test 1. If the null hypothesis is true ($\mu_1 = \mu_2$), the probability that we will make a Type I error is $\alpha = .05$; hence, the probability that we will not make a Type I error is $1 - .05 = .95$. Now, suppose that we follow this same procedure for test 2; the probability that we will make a Type I error for this test is also $\alpha = .05$, and the probability that we will not make a Type I error for test 2 is also .95. Clearly, when performing a single statistical test, the probability of making a Type I error is $\alpha = .05$. In discussing *multiple comparison procedures,* we refer to $\alpha = .05$ as the *comparisonwise Type I error rate.* In essence, comparisonwise Type I error rates indicate the level of significance associated with a single statistical test.

Let us now consider a slightly different question. What is the probability that in using this sequential approach to hypothesis testing, we will commit a Type I error on at least one of the first two tests? To answer this question, note that the probability that we will not make a Type I error for tests 1 and 2 is $(.95)(.95) = .9025$.* Since the sum of the probabilities of making zero, one, or two Type I errors is 1, the probability of making at least one Type I error is $1 - .9025 = .0975$. Thus, when we use Fisher's LSD procedure to sequentially test two sets of hypotheses, the Type I error rate associated with this approach is not .05, but actually .0975; we refer to this error rate as the *experimentwise Type I error rate.*

Continuing with the analysis, suppose we also use Fisher's LSD procedure for test 3. The comparisonwise error rate still remains at $\alpha = .05$; however, the probability that we will commit a Type I error on at least one of the three tests has now increased to $1 - (.95)(.95)(.95) = 1 - .8574 = .1426$. Thus, if we sequentially apply Fisher's LSD to all pairwise comparisons, the *overall* or *experimentwise Type I error rate* associated with this sequential approach is .1426. To avoid confusion, we will denote the experimentwise Type I error rate as α_{EW}.

To write a general expression for the experimentwise Type I error rate, let C denote the number of possible pairwise comparisons. For a problem with k populations, the value of C is the number of combinations of k populations taken 2 at a time; that is

---

*This assumes that the two tests are independent, and hence the joint probability of the two events can be obtained by simply multiplying the individual probabilities. In fact, the two tests are not independent since MSW is used in each test; hence, the error involved is even greater than that shown.

$$C = \text{Number of Pairwise Comparisons} = \binom{k}{2} = \frac{k!}{(k-2)!2!} = \frac{k(k-1)}{2} \qquad \textbf{(12.19)}$$

For example, if $k = 5$, there are $[5(5-1)]/2 = 10$ possible pairwise comparisons. In general then, for a problem involving C pairwise comparisons, the probability of making at least one Type I error is $1 - (1 - \alpha)^C$. That is,

$$\alpha_{EW} = \text{Experimentwise Type I Error Rate} = 1 - (1 - \alpha)^C \qquad \textbf{(12.20)}$$

For example, when there are five populations and we want to test all possible pairwise comparisons using Fisher's LSD with a comparisonwise error rate of $\alpha = .05$, the experimentwise Type I error rate is $1 - (1 - .05)^{10} = .40$. With this large of an experimentwise Type I error rate, many practitioners look to alternatives that provide better control over the experimentwise error rate.

BONFERRONI ADJUSTMENT

A problem with Fisher's LSD procedure is that the experimentwise Type I error rate is really dependent on the comparisonwise error rate α and the number of pairwise comparisons. Instead of specifying a comparisonwise error rate then, suppose we specify the value of α_{EW} that is desired and then find the value of α that will provide this value.

In the preceding discussion, we stated that the probability of making at least one Type I error for problems involving k possible pairwise comparisons is

$$\alpha_{EW} = \text{Experimentwise Type I Error Rate} = 1 - (1 - \alpha)^C$$

The Italian mathematician Bonferroni proved that

$$1 - (1 - \alpha)^C \leq C\alpha \qquad \textbf{(12.21)}$$

for any value of C whenever α is between 0 and 1. Thus, since $\alpha_{EW} = 1 - (1 - \alpha)^C$, the probability of making at least one Type I error whenever C pairwise comparisons are tested is less than or equal to $C\alpha$. If we want the maximum probability of making a Type I error for the overall experiment to be α_{EW}, we simply use a comparisonwise Type I error rate equal to α_{EW}/C.

EXAMPLE 12.1 ▶
(CONTINUED)

Suppose that we want to use Fisher's LSD procedure to test all three NCP pairwise comparisons. Furthermore, suppose that we want the maximum experimentwise Type I error rate to be .05. If we set the comparisonwise error rate to be $\alpha = .05/3 = .017$, we will ensure that the overall or experimentwise error rate will be less than or equal to .05. Thus, when using Fisher's LSD procedure with the Bonferroni adjustment, the t value is $t_{.017/2} = t_{.0085}$. Since we do not have tables to determine the exact t value corresponding to this level of significance, we note that this value is approximately $t_{.01} = 2.602$. Recall that the value of LSD is

$$\text{LSD} = 2.131 \sqrt{28.67\left(\frac{1}{6} + \frac{1}{6}\right)} = 6.59$$

If we use the t value of 2.602 instead of 2.131, we obtain a value of LSD that we will refer to as the Bonferroni significant difference, denoted BSD:

$$\text{BSD} = 2.602 \sqrt{28.67\left(\frac{1}{6} + \frac{1}{6}\right)} = 8.04$$

Thus, to reject the hypothesis that the two means are equal using Fisher's LSD with the Bonferroni adjustment, we must observe a difference between the sample means of more than 8.04 points. Note that this is a larger difference between the two sample means than would be required when using Fisher's LSD procedure.

The difference between the sample means at the Charlotte and Houston plants is $79 - 74 = 5$; since this difference does not exceed BSD = 8.04, we cannot reject the null hypothesis that the two population means are equal. However, since the difference between the Charlotte and San Diego plants is $79 - 66 = 13 > 8.04$, we can reject the hypothesis that the population mean examination score for the Charlotte plant is the same as the population mean examination score for the San Diego plant. Finally, since the difference between the mean at Houston and the mean at San Diego is $74 - 66 = 8$, we cannot conclude that the population means for these two plants are significantly different.

◀

If a problem consisted of five populations and hence 10 possible pairwise comparisons, the Bonferroni adjustment would suggest a comparisonwise Type I error rate of $.05/10 = .005$. For a fixed sample size, any decrease in the probability of making a Type I error will result in an increase in the probability of making a Type II error, which corresponds to accepting the hypothesis that the two means of the populations are equal when in fact they are not equal. As a result, many practitioners feel uncomfortable in performing individual tests with a very low comparisonwise Type I error rate since it carries an increased risk of making a Type II error. In the following discussion we introduce a procedure that was developed to help in this regard.

TUKEY'S PROCEDURE

Tukey's procedure allows an experimenter to perform tests of all possible pairwise comparisons and still maintain an overall experimentwise Type I error rate such as $\alpha_{EW} = .05$. The basis for the test is a probability distribution referred to as the "studentized range" distribution. Let $\bar{x}_{max}$ denote the largest sample mean and $\bar{x}_{min}$ denote the smallest sample mean. In situations where the size of each sample is the same size (n) and the population variances are equal, the sampling distribution of

$$q = \frac{\bar{x}_{max} - \bar{x}_{min}}{\sqrt{\dfrac{MSW}{n}}} \tag{12.22}$$

follows a studentized range distribution. Table 11 of Appendix B presents critical values of the studentized range distribution for both $\alpha = .05$ and $\alpha = .01$. A portion of this table for $\alpha_{EW} = .05$ is shown in Table 12.3. To show how the above result can be used to perform tests of all pairwise comparisons, we will illustrate the procedure for Example 12.1.

EXAMPLE 12.1 ▶
(CONTINUED)

We have three samples, each of size $n = 6$. Suppose we want to perform tests on all possible pairwise comparisons and ensure an overall experimentwise error rate of $\alpha_{EW} = .05$. The columns in Table 12.3 correspond to the number of populations we are sampling from and the rows correspond to the degrees of freedom associated with the best estimate of the population variance σ^2. Thus, we select column 3, since there are $k = 3$ populations. We must also select row 15, since the best estimate of σ^2 is MSW and this estimate has 15 degrees of freedom. Note that the critical value is $q = 3.67$.

TABLE 12.3   CRITICAL VALUES OF THE STUDENTIZED RANGE DISTRIBUTION FOR $\alpha_{EW} = .05$

Degrees of Freedom	Number of Populations								
	2	3	4	5	6	7	8	9	10
1	18.0	27.0	32.8	37.1	40.4	43.1	45.4	47.4	49.1
2	6.08	8.33	9.80	10.9	11.7	12.4	13.0	13.5	14.0
3	4.50	5.91	6.82	7.50	8.04	8.48	8.85	9.18	9.46
4	3.93	5.04	5.76	6.29	6.71	7.05	7.35	7.60	7.83
5	3.64	4.60	5.22	5.67	6.03	6.33	6.58	6.80	6.99
6	3.46	4.34	4.90	5.30	5.63	5.90	6.12	6.32	6.49
7	3.34	4.16	4.68	5.06	5.36	5.61	5.82	6.00	6.16
8	3.26	4.04	4.53	4.89	5.17	5.40	5.60	5.77	5.92
9	3.20	3.95	4.41	4.76	5.02	5.24	5.43	5.59	5.74
10	3.15	3.88	4.33	4.65	4.91	5.12	5.30	5.46	5.60
11	3.11	3.82	4.26	4.57	4.82	5.03	5.20	5.35	5.49
12	3.08	3.77	4.20	4.51	4.75	4.95	5.12	5.27	5.39
13	3.06	3.73	4.15	4.45	4.69	4.88	5.05	5.19	5.32
14	3.03	3.70	4.11	4.41	4.64	4.83	4.99	5.13	5.25
15	3.01	3.67	4.08	4.37	4.59	4.78	4.94	5.08	5.20
16	3.00	3.65	4.05	4.33	4.56	4.74	4.90	5.03	5.15
17	2.98	3.63	4.02	4.30	4.52	4.70	4.86	4.99	5.11
18	2.97	3.61	4.00	4.28	4.49	4.67	4.82	4.96	5.07
19	2.96	3.59	3.98	4.25	4.47	4.65	4.79	4.92	5.04
20	2.95	3.58	3.96	4.23	4.45	4.62	4.77	4.90	5.01
24	2.92	3.53	3.90	4.17	4.37	4.54	4.68	4.81	4.92
30	2.89	3.49	3.85	4.10	4.30	4.46	4.60	4.72	4.82
40	2.86	3.44	3.79	4.04	4.23	4.39	4.52	4.63	4.73
60	2.83	3.40	3.74	3.98	4.16	4.31	4.44	4.55	4.65
120	2.80	3.36	3.68	3.92	4.10	4.24	4.36	4.47	4.56
∞	2.77	3.31	3.63	3.86	4.03	4.17	4.29	4.39	4.47

Using $q = 3.67$, $n = 6$, and MSW $= 28.67$, we can use equation (12.22) to determine how large the difference between any two sample means has to be to reject the null hypothesis that the corresponding population means are equal. We refer to this value as *Tukey's significant difference,* denoted TSD.

$$TSD = \text{Tukey's Significant Difference} = q\sqrt{\frac{MSW}{n}} \qquad (12.23)$$

For the NCP data

$$TSD = 3.67\sqrt{\frac{28.67}{6}} = 8.02$$

Thus, if the absolute value of the difference between any two sample means exceeds 8.02, there is sufficient evidence to conclude that the corresponding population means are not equal. Note that TSD is greater than LSD and slightly less than the value of BSD;

however, Tukey's procedure is similar to using the Bonferroni adjustment in that a larger difference between two sample means must be observed to conclude that the population means are not equal.

The difference between the sample means at the Charlotte and Houston plants is $79 - 74 = 5$; thus, we cannot reject the hypothesis that the two population means are equal since 5 does not exceed TSD = 8.02. However, since the difference between the Charlotte and San Diego plants is $79 - 66 = 13 > 8.02$, we can reject the hypothesis that the population mean examination score for the Charlotte plant is the same as the population mean examination score for the San Diego plant. Finally, since the difference between the mean at Houston and the mean at San Diego is $74 - 66 = 8$, we cannot conclude that the population means for these two plants are significantly different. ◀

We can also use the value of TSD to compute confidence intervals for each pair of differences. We refer to the resulting set of interval estimates as the *simultaneous confidence interval estimates*. The general expression for computing these interval estimates is as follows:

$$\bar{x}_i - \bar{x}_j \pm q\sqrt{\frac{\text{MSW}}{n}} \tag{12.24}$$

EXAMPLE 12.1 ▶
(CONTINUED)

The value of TSD is 8.02, regardless of which two populations we are considering. Thus, the set of simultaneous confidence intervals is

Charlotte–Houston

$(79 - 74) \pm 8.02 = -3.02$ to 13.02

Charlotte–San Diego

$(79 - 66) \pm 8.02 = 4.98$ to 21.02

Houston–San Diego

$(74 - 66) \pm 8.02 = -.02$ to 16.02

Since 0 is contained in the Charlotte–Houston interval and the Houston–San Diego interval, there is not sufficient statistical evidence to conclude that the population means of these plants differ. However, since the Charlotte–San Diego interval does not include the value of 0, we have statistical evidence to conclude that the population means of these two plants are not equal. The key point about using Tukey's procedure to form these confidence intervals is that we are 95% confident that these intervals hold true for all the intervals formed. ◀

RECOMMENDATIONS FOR PERFORMING MULTIPLE COMPARISONS

We have introduced three procedures for performing multiple comparisons: Fisher's LSD, Bonferroni adjustment, and Tukey's procedure. In using Fisher's LSD procedure for Example 12.1, we concluded that there is a significant difference between the population means for the Charlotte and San Diego plants as well as for the Houston and San Diego plants. When using Fisher's LSD with the Bonferroni adjustment or Tukey's procedure, we concluded that the only difference is between the Charlotte and San Diego plants. Thus, different conclusions can be drawn depending on which procedure is used.

There is considerable controversy in the statistical community as to which procedure is "best." The truth of the matter is that no one procedure is best for all types of problems. However, if the number of pairwise comparisons that you want to make is small, we recommend using the Bonferroni adjustment. If the number of pairwise comparisons is very large, then Tukey's procedure provides a better alternative. For example, suppose you only want to test two or three pairwise comparisons, each of which has been identified prior to looking at the data; in this case, we recommend use of the Bonferroni adjustment. Note however, that this virtually rules out the use of the Bonferroni adjustment as a method for testing all possible pairwise comparisons. If the latter is your objective, we recommend that Tukey's procedure be used.

In situations where you view the outcome of your analysis as nothing more than a way to suggest what really needs to be investigated in future studies, Fisher's LSD procedure is preferred. In such cases, Fisher's LSD procedure allows us to base our rejection of the hypothesis that the two population means are equal on smaller observed differences than are required using the other two approaches. Thus, there is less likelihood we might discard something that may at a later time show a much more significant difference. Situations like this are often referred to as hypothesis generators.

NOTES AND COMMENTS

Tukey's procedure is an *unprotected* or *unrestrictive* testing approach. That is, it is not necessary to find a significant difference when using analysis of variance before applying Tukey's procedure. Thus, Tukey's procedure provides an alternative to analysis of variance for testing if the means of k populations are equal. However, to use Tukey's procedure we still have to estimate the population variance using MSW.

EXERCISES

METHODS

SELF TEST

11. In Exercise 1, five observations were selected from each of three populations. For these data, $\bar{x}_1 = 30$, $\bar{x}_2 = 45$, $\bar{x}_3 = 36$, and MSW = 5.5. At the $\alpha = .05$ level of significance, the null hypothesis of equal population means was rejected. In answering the following questions, use $\alpha = .05$.
  a. Using Fisher's LSD procedure, test to see if there is a significant difference between the means of populations 1 and 2, populations 1 and 3, and populations 2 and 3.
  b. Use Fisher's LSD procedure to develop a 95% confidence interval estimate of the difference between the means of populations 1 and 2.
  c. Using the Bonferroni adjustment, which means appear to be different?
  d. Using Tukey's procedure, which means appear to be different?
  e. Use Tukey's procedure to compute 95% confidence intervals for each pair of differences.

12. Four observations were selected from each of three populations. The data obtained is shown below.

	Sample 1	Sample 2	Sample 3
	63	82	69
	47	72	54
	54	88	61
	40	66	48
Sample mean	51	77	58
Sample variance	96.67	97.34	81.99

In answering the following questions, use $\alpha = .05$.
a.   Using analysis of variance, test if there is a significant difference among the means of the three populations.
b.   Using Fisher's LSD procedure, which means appear to be different?
c.   Using the Bonferroni adjustment, which means appear to be different?
d.   Using Tukey's procedure, which means appear to be different?

APPLICATIONS

SELF TEST

13.   Refer to Exercise 6. For these data $\bar{x}_1 = 23$, $\bar{x}_2 = 28$, $\bar{x}_3 = 21$, and MSW = 4.89. At the $\alpha = .05$ level of significance, use Fisher's LSD procedure to test for the equality of the means for manufacturers 1 and 3. What conclusion can you make after carrying out this test?

SELF TEST

14.   Refer to Exercise 13. Use Tukey's procedure to develop a 95% confidence interval estimate of the difference between the means of population 1 and population 2.

15.   In Exercise 7 scores on a 100-question science test were provided for 13-year-old children from South Korea, the Soviet Union, and the United States; the sample means were 78, 71, and 67, respectively, and MSW = 38.0. Use Tukey's procedure to test for the equality of the means for South Korea and the United States. What conclusion can you make after carrying out the test? Use $\alpha_{EW} = .05$

16.   Refer to Exercise 15. Use the Bonferroni adjustment to test for the equality of the means for South Korea and the United States.

17.   Refer to Exercise 15. Use Tukey's procedure to develop a 95% confidence interval for the mean difference between South Korea and the United States.

18.   Refer to Exercise 9. At the $\alpha = .05$ level of significance, we can conclude that there are differences in the mean perceptions for specialists in marketing ($\bar{x}_1 = 5$), marketing research ($\bar{x}_2 = 4.5$), and advertising ($\bar{x}_3 = 6$); for these data MSW = .5. Use the procedures in this section to determine where the differences occur. Use $\alpha = .05$.

12.4   AN INTRODUCTION TO EXPERIMENTAL DESIGN

Statistical studies can be classified as being either experimental or observational. In an *experimental study,* variables of interest are identified. Then, one or more factors in the study are controlled so that data may be obtained about how the factors influence the variables. In *observational* or *nonexperimental* studies, no attempt is made to control the influences of factors on the variable or variables of interest. A sample survey (see Chapter 18) is perhaps the most common type of observational study.

Example 12.1, which we used to introduce analysis of variance, is an illustration of an observational statistical study. To measure how much NCP employees knew about total quality management, a random sample of six employees was selected from each of NCP's three plants and given a quality-awareness examination. The examination scores for these employees were then analyzed using analysis of variance to test the hypothesis that the population mean examination scores were equal for the three plants.

As an example of an experimental statistical study, let us consider the problem facing Chemitech, Inc.

EXAMPLE 12.2

Chemitech has developed a new filtration system for municipal water supplies. The components for the new filtration system will be purchased from several different suppliers, and Chemitech will assemble the components at their plant in Columbia, South

Carolina. The industrial engineering group has been given the responsibility of determining the best assembly method for the new filtration system. After considering a variety of possible approaches, the group has narrowed the alternatives to three possibilities: method A, method B, and method C. Each of these methods differs in terms of the sequence of steps used to assemble the product. Management at Chemitech would like to determine which assembly method results in the greatest number of filtration systems produced per week. ◀

In the Chemitech experiment, the assembly method is referred to as a *factor*. Since there are three assembly methods, or levels, corresponding to this factor, we say that there are three *treatments* associated with this experiment: one treatment corresponds to method A, another to method B, and the third to method C. In general, a factor is just a variable that the experimenter has selected for investigation, and a treatment corresponds to a level of a factor. The Chemitech problem is an example of a *single-factor experiment* involving a qualitative factor (method of assembly). Other experiments may consist of multiple factors; some may be qualitative and some may be quantitative.

The three assembly methods or treatments define the three populations of interest for the Chemitech experiment. One population corresponds to all Chemitech employees who use assembly method A, another corresponds to those who use method B, and the third to those who use method C. Note that for each population the random variable of interest (the response variable) is the number of filtration systems assembled per week, and the primary statistical objective for the experiment is to determine whether the mean number of units produced per week is the same for all three populations. In experimental design terminology, the random variable of interest is referred to as the *dependent variable,* the *response variable,* or simply the *response.*

Suppose that a random sample of three employees is selected from all assembly workers at the Chemitech production facility. In experimental design terminology, the three randomly selected workers are referred to as the *experimental units.* The experimental design that we will use for the Chemitech problem is referred to as a *completely randomized design.* This type of design requires that each of the three assembly methods or treatments be randomly assigned to one of the experimental units or workers. For example, method A might be assigned to the second worker, method B to the first worker, and method C to the third worker. The concept of *randomization,* as illustrated here, is an important principle of all experimental designs.

Note that the experiment as described above would only result in one measurement or number of units assembled for each treatment. In other words, we have a sample size of 1 corresponding to each treatment. Thus, to obtain additional data for each assembly method, we must repeat or replicate the basic experimental process. For example, suppose that instead of selecting just three workers at random we had selected 15 workers and then randomly assigned each of the three treatments to five of the workers. Since each method of assembly is assigned five workers, we say that five replicates have been obtained. The process of *replication* is another important principle of experimental design. Figure 12.6 shows the completely randomized design for the Chemitech experiment.

DATA COLLECTION

Once we are satisfied with the experimental design, we proceed by collecting and analyzing the data. In this case, the employees would be instructed in how to perform the assembly method that they have been assigned to and then would begin assembling the new filtration systems using that method. Suppose that this has been done and the number

FIGURE 12.6   COMPLETELY RANDOMIZED DESIGN FOR THE CHEMITECH PROBLEM

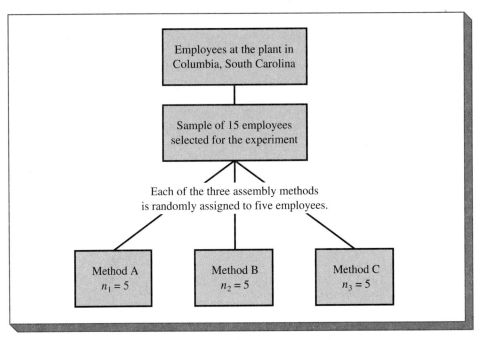

TABLE 12.4   NUMBER OF UNITS PRODUCED FOR 15 WORKERS

| | **Method** | | |
Observation	A	B	C
1	58	58	48
2	64	69	57
3	55	71	59
4	66	64	47
5	67	68	49
Sample mean	62	66	52
Sample variance	27.5	26.5	31.0
Sample standard deviation	5.24	5.15	5.57

of units assembled by each employee during 1 week are as shown in Table 12.4. The sample mean number of units produced for the three assembly methods are 62 for method A, 66 for method B, and 52 for method C. From these data it appears that method B provides the highest production rate.

The real issue is whether the three sample means observed are different enough for us to conclude that the means of the populations corresponding to the three methods of assembly are different. To write this question in statistical terms, we introduce the following notation:

μ_1 = mean number of units produced per week for method A
μ_2 = mean number of units produced per week for method B
μ_3 = mean number of units produced per week for method C

Although we will never know the actual values of μ_1, μ_2, and μ_3, what we want to do is use the sample means to test the following hypotheses:

$$H_0: \ \mu_1 = \mu_2 = \mu_3$$

$$H_a: \ \text{Not all population means are equal}$$

The problem that we face in analyzing data from a completely randomized experimental design is the same problem we faced when we first introduced analysis of variance as a method for testing whether the means of more than two populations are equal. In the next section we will show how analysis of variance is applied in problem situations such as this.

NOTES AND COMMENTS

1. Randomization in experimental design is the analog of probability sampling in an observational experiment.
2. In many medical experiments, potential bias is eliminated by using a double-blind study. In these studies, neither the physician applying the treatment nor the subject receiving the treatment know which treatment is being applied. Many other types of experiments could benefit from this type of study.

12.5 COMPLETELY RANDOMIZED DESIGNS

The hypotheses that we want to test when analyzing the data from a completely randomized design are exactly the same as the general form of the hypotheses we presented in Section 12.2.

$$H_0: \ \mu_1 = \mu_2 = \cdots = \mu_k$$

$$H_a: \ \text{Not all population means are equal}$$

where

$$\mu_j = \text{mean of the } j\text{th population}$$

We will proceed based on the assumptions stated in Section 12.1 for analysis of variance. Thus, to test for the equality of means in situations where the data have been collected using a completely randomized experimental design, we can use analysis of variance as introduced in Sections 12.1 and 12.2. Recall that analysis of variance requires the calculation of two independent estimates of the population variance σ^2.

BETWEEN-TREATMENTS ESTIMATE OF POPULATION VARIANCE

In the context of experimental design, the between-samples estimate of σ^2 is referred to as the *mean square due to treatments* and is denoted MSTR. This is the same as what we called mean square between, MSB, in Section 12.2. It is also referred to as the *mean square between treatments.* The formula for computing MSTR is as follows:

$$\text{MSTR} = \frac{\sum_{j=1}^{k} n_j(\bar{x}_j - \bar{\bar{x}})^2}{k - 1} \tag{12.25}$$

The numerator in equation (12.25) is called the *sum of squares between* or *sum of squares due to treatments* and is denoted by SSTR. The denominator $k - 1$ represents the degrees of freedom associated with SSTR.

EXAMPLE 12.2
(CONTINUED)

For the Chemitech data shown in Table 12.4, we obtain the following results:

$$\bar{\bar{x}} = \frac{\sum_{j=1}^{k}\sum_{i=1}^{n_j} x_{ij}}{n_T} = \frac{900}{15} = 60$$

$$\text{SSTR} = \sum_{j=1}^{k} n_j(\bar{x}_j - \bar{\bar{x}})^2 = 5(62 - 60)^2 + 5(66 - 60)^2 + 5(52 - 60)^2 = 520$$

$$\text{MSTR} = \frac{\text{SSTR}}{k - 1} = \frac{520}{3 - 1} = 260$$

◀

WITHIN-TREATMENTS ESTIMATE OF POPULATION VARIANCE

The second estimate of σ^2 is based on the variation of the sample observations within each sample or treatment. In our discussion of analysis of variance, we referred to this estimate of σ^2 as the within-samples estimate of population variance. In the context of experimental design, this estimate is referred to as the *mean square due to error* and is denoted MSE. This is the same as what we called mean square within, MSW, in Section 12.2. It is also referred to as the *mean square within treatments.* The formula for computing MSE is as follows:

$$\text{MSE} = \frac{\sum_{j=1}^{k} (n_j - 1)s_j^2}{n_T - k} \tag{12.26}$$

The numerator in equation (12.26) is given the name *sum of squares within* or *sum of squares due to error* and is denoted SSE. The denominator of MSE is referred to as the degrees of freedom associated with the within-treatment variance estimate.

EXAMPLE 12.2
(CONTINUED)

For the Chemitech data shown in Table 12.4, we obtain the following results:

$$\text{SSE} = \sum_{j=1}^{k} (n_j - 1)s_j^2 = 4(27.5) + 4(26.5) + 4(31) = 340$$

$$\text{MSE} = \frac{\text{SSE}}{n_T - k} = \frac{340}{15 - 3} = 28.33$$

◀

COMPARING THE VARIANCE ESTIMATES: THE F TEST

If the null hypothesis is true and the ANOVA assumptions are valid, the sampling distribution of MSTR/MSE is an F distribution with numerator degrees of freedom equal to $k - 1$ and denominator degrees of freedom equal to $n_T - k$. Recall also that if the means of the k populations are not equal, the value of MSTR/MSE will be inflated because MSTR overestimates σ^2. Hence, we will reject H_0 if the resulting value of MSTR/MSE appears to be too large to have been selected at random from an F distribution with degrees of freedom $k - 1$ in the numerator and $n_T - k$ in the denominator.

EXAMPLE 12.2
(CONTINUED)

The value of F = MSTR/MSE = 260/28.33 = 9.18. The critical F value is based upon 2 numerator and 12 denominator degrees of freedom. For a .05 level of significance, Table 4 of Appendix B shows a value of $F_{.05} = 3.89$. Since the observed value of F is greater than the critical value, we reject the null hypothesis and conclude that not all the population means are equal. ◀

THE ANOVA TABLE FOR COMPLETELY RANDOMIZED DESIGNS

Using the terminology we have introduced for the completely randomized experimental design, we can now write the result that shows how the total sum of squares, SST, is partitioned:

$$SST = SSTR + SSE \tag{12.27}$$

Note that this result also holds true for the degrees of freedom associated with each of these sums of squares; that is, the total degrees of freedom is the sum of the degrees of freedom associated with SSTR and SSE. The general form of the ANOVA table for a completely randomized design is shown in Table 12.5; Table 12.6 shows the corresponding ANOVA table for the Chemitech problem.

PAIRWISE COMPARISONS

We can use Tukey's procedure to test all possible pairwise comparisons for the Chemitech problem. At the 5% level of significance, Table 12.3 can be used to provide the critical value of the studentized range distribution with $k = 3$ populations and 12 degrees of

TABLE 12.5   **ANOVA TABLE FOR A COMPLETELY RANDOMIZED DESIGN**

Source of Variation	Sum of Squares	Degrees of Freedom	Mean Square	F
Treatments	SSTR	$k - 1$	$MSTR = \dfrac{SSTR}{k - 1}$	$\dfrac{MSTR}{MSE}$
Error	SSE	$n_T - k$	$MSE = \dfrac{SSE}{n_T - k}$	
Total	SST	$n_T - 1$		

TABLE 12.6   **ANOVA TABLE FOR THE CHEMITECH PROBLEM**

Source of Variation	Sum of Squares	Degrees of Freedom	Mean Square	F
Treatments	520	2	260.00	9.18
Error	340	12	28.33	
Total	860	14		

freedom; the value obtained is $q = 3.77$. Using MSE $= 28.33$ in place of MSW in expression (12.23), we obtain Tukey's significant difference.

$$TSD = q\sqrt{\frac{MSE}{n}} = 3.77\sqrt{\frac{28.33}{5}} = 8.97$$

Thus, if the magnitude of the difference between any two sample means exceeds 8.97, we can reject the hypothesis that the corresponding population means are equal. For the Chemitech data in Table 12.4, we obtain the following results.

Sample Differences	Significant?
Method A − Method B = 62 − 66 = −4	No
Method A − Method C = 62 − 52 = 10	Yes
Method B − Method C = 66 − 52 = 14	Yes

Thus, the difference in the population means is attributable to the difference between the means for method A and method C and the difference between the means for method B and method C. Thus, methods A and B are preferred to method C. However, more testing should be done to compare method A with method B. Based on the current study, there is not sufficient evidence to conclude that these two methods differ.

NOTES AND COMMENTS

> The computational aspect of the analysis of variance procedure is devoted primarily to computing the appropriate sums of squares. When a hand calculator is used to compute the sum of squares, some computational help can be obtained by using alternate forms of the sums-of-squares formulas. In Appendix 12.1 we provide a step-by-step procedure that illustrates the use of these alternate formulas.

EXERCISES

METHODS

SELF TEST   **19.** The following data are from a completely randomized design.

	Treatment		
Observation	A	B	C
1	162	142	126
2	142	156	122
3	165	124	138
4	145	142	140
5	148	136	150
6	174	152	128
Sample mean	156	142	134
Sample variance	164.4	131.2	110.4

a. Compute the sum of squares between treatments.
b. Compute the mean square between treatments.
c. Compute the sum of squares due to error.
d. Compute the mean square due to error.
e. At the $\alpha = .05$ level of significance, test if the means for the three treatments are equal.

20. Refer to Exercise 19.
 a. Set up the ANOVA table.
 b. At the $\alpha = .05$ level of significance, use Tukey's procedure to test all possible comparisons. What conclusion can you make after carrying out this procedure?

21. In a completely randomized experimental design, seven experimental units were used for each of the five levels of the factor. Complete the ANOVA table shown.

Source of Variation	Sum of Squares	Degrees of Freedom	Mean Square	F
Treatments	300			
Error				
Total	460			

22. Refer to Exercise 21.
 a. What hypotheses are implied in this problem?
 b. At the $\alpha = .05$ level of significance, can we reject the null hypothesis in part (a)? Explain.

23. In an experiment designed to test the output levels of three different treatments, the following results were obtained: SST = 400, SSTR = 150, $n_T = 19$. Set up the ANOVA table, and test for any significant difference between the mean output levels of the three treatments. Use $\alpha = .05$.

24. In a completely randomized experimental design, 12 experimental units were used for the first treatment, 15 experimental units for the second treatment, and 20 experimental units for the third treatment. Complete the analysis of variance table shown at the top of page 468. Using a .05 level of significance, is there a significant difference between the treatments?

Source of Variation	Sum of Squares	Degrees of Freedom	Mean Square	F
Treatments	1200			
Error				
Total	1800			

25. Develop the analysis of variance computations for the experimental results shown below. Using $\alpha = .05$, is there a significant difference between the treatment means?

	Treatment		
	A	B	C
	136	107	92
	120	114	82
	113	125	85
	107	104	101
	131	107	89
	114	109	117
	129	97	110
	102	114	120
		104	98
		89	106
Sample mean	119	107	100
Sample variance	146.86	96.44	173.78

APPLICATIONS

26. Three different methods for assembling a product were proposed by an industrial engineer. To investigate the number of units assembled correctly using each method, 30 employees were randomly selected and randomly assigned to the three proposed methods such that 10 workers were associated with each method. The number of units assembled correctly was recorded, and the analysis of variance procedure was applied to the resulting data set. The following results were obtained: SST = 10,800, SSTR = 4560.
    a. Set up the ANOVA table for this problem.
    b. Using $\alpha = .05$, test for any significant difference in the means for the three assembly methods.

27. In an experiment designed to test the breaking strength of four types of cables, the following results were obtained: SST = 85.05, SSTR = 61.64, $n_T = 24$. Set up the ANOVA table, and test for any significant difference in the mean breaking strength of the four cables. Use $\alpha = .05$.

28. To study the effect of temperature upon yield in a chemical process, five batches were produced under each of three temperature levels. The results are given below. Construct an analysis of variance table. Using a .05 level of significance, test to see if the temperature level appears to have an effect on the mean yield of the process.

	Temperature		
	50°C	60°C	70°C
	34	30	23
	24	31	28
	36	34	28
	39	23	30
	32	27	31
Sample mean	33	29	28
Sample variance	32	17.5	9.5

JUDGMENT

29. Auditors must make judgments concerning various aspects of an audit based on their own direct experience, indirect experience, or a combination of the two. In one study, auditors were asked to make judgments about the frequency of errors to be found in an audit (*Journal of Accounting Research,* Autumn 1988). The judgments by the auditors were then compared to the actual results. Suppose the data shown below were obtained from a similar study; lower scores indicate better judgments. Using $\alpha = .05$, test to see if the basis for the judgment affects the quality of the judgment. What is your conclusion?

	Direct	**Indirect**	**Combination**
	17.0	16.6	25.2
	18.5	22.2	24.0
	15.8	20.5	21.5
	18.2	18.3	26.8
	20.2	24.2	27.5
	16.0	19.8	25.8
	13.3	21.2	24.2
Sample mean	17.0	20.4	25.0
Sample variance	5.01	6.26	4.01

PAINT

30. Four different paints are advertised as having the same drying time. To check the manufacturer's claims, five paint samples were tested for each make of paint. The time in minutes until the paint was dry enough for a second coat to be applied was recorded. The following data were obtained.

	Paint 1	**Paint 2**	**Paint 3**	**Paint 4**
	128	144	133	150
	137	133	143	142
	135	142	137	135
	124	146	136	140
	141	130	131	153
Sample mean	133	139	136	144
Sample variance	47.5	50	21	54.5

At the $\alpha = .05$ level of significance, test to see if the mean drying time is the same for each type of paint.

31. Three top-of-the-line intermediate-sized automobiles manufactured in the United States have been test-driven and compared on a variety of criteria by a well-known automotive magazine. In the area of gasoline mileage performance, five automobiles of each brand were each test-driven 500 miles; the miles per gallon data obtained are shown in the table below.

	Automobile		
	A	B	C
	19	19	24
	21	20	26
	20	22	23
	19	21	25
	21	23	27
Sample mean	20	21	25
Sample variance	1	2.5	2.5

Use the analysis of variance procedure with $\alpha = .05$ to determine if there is a significant difference in the mean miles per gallon for the three types of automobiles.

32. Refer to Exercise 29. Use Tukey's procedure to test all possible comparisons. What conclusion can you make after carrying out this procedure? Use $\alpha = .05$.

33. Refer to Exercise 31. Use Tukey's procedure to test all possible comparisons. What conclusion can you make after carrying out this procedure? Use $\alpha = .05$.

12.6 RANDOMIZED BLOCK DESIGNS

Thus far we have considered the completely randomized experimental design. Recall that to test for a difference among treatment means, we computed an F value using the ratio

$$F = \frac{\text{MSTR}}{\text{MSE}} \tag{12.28}$$

A problem can arise whenever differences due to extraneous factors (ones not considered in the experiment) cause the MSE term in this ratio to become large. In such cases, the F value in equation (12.28) can become small, signaling no difference among treatment means when in fact such a difference exists.

In this section we present an experimental design referred to as a *randomized block design*. The purpose of this design is to control some of the extraneous sources of variation by removing such variation from the MSE term. This design tends to provide a better estimate of the true error variance and leads to a more powerful hypothesis test in terms of the ability to detect differences among treatment means. To illustrate, let us consider a stress study for air traffic controllers.

EXAMPLE 12.3▶ A study directed at measuring the fatigue and stress on air traffic controllers has resulted in proposals for modification and redesign of the controller's work station. After consideration of several designs for the work station, three specific alternatives have been selected as having the best potential for reducing controller stress. The key question is, To what extent do the three alternatives differ in terms of their effect on controller stress? To answer this question, we need to design an experiment that will provide measurements of air traffic controller stress under each alternative.

In a completely randomized design, a random sample of controllers would be assigned to each work station alternative. However, it is believed that controllers differ substantially in terms of their ability to handle stressful situations. What is high stress to one controller might be only moderate or even low stress to another. Thus, when considering the within-group source of variation (MSE), we must realize that this variation includes both random error and error due to individual controller differences. In fact, for this study, management expected controller variability to be a major contributor to the MSE term.

One way to separate the effect of the individual differences is to use a randomized block design. This design will identify the variability stemming from individual controller differences and remove it from the MSE term. The randomized block design calls for a single sample of controllers. Each controller in the sample is tested using each of the three work station alternatives. In experimental design terminology, the work station is the *factor of interest,* and the controllers are referred to as the *blocks.* The three treatments or populations associated with the work station factor correspond to the three work station alternatives. For simplicity, we will refer to the work station alternatives as system A, system B, and system C.

The *randomized* aspect of the randomized block design refers to the fact that the order in which the treatments (systems) are assigned to the controllers is chosen randomly. If every controller were to test the three systems in the same order, any observed difference in systems might be due to the order of the test rather than to true differences in the systems.

To provide the necessary data, the three types of work stations were installed at the Cleveland Control Center in Oberlin, Ohio. Six controllers were selected at random and assigned to operate each of the systems. A follow-up interview and a medical examination of each controller participating in the study provided a measure of the stress for each controller on each system. The data are shown in Table 12.7.

A summary of the stress data collected is shown in Table 12.8. In this table we have included column totals (treatments) and row totals (blocks) as well as some sample means that will be helpful in making the sum of squares computations for the ANOVA

TABLE 12.7   **A RANDOMIZED BLOCK DESIGN FOR THE AIR TRAFFIC CONTROLLER STRESS TEST**

		Treatments		
		System A	*System B*	*System C*
	Controller 1	15	15	18
	Controller 2	14	14	14
Blocks	*Controller 3*	10	11	15
	Controller 4	13	12	17
	Controller 5	16	13	16
	Controller 6	13	13	13

TABLE 12.8   SUMMARY OF STRESS DATA FOR THE AIR TRAFFIC CONTROLLER STRESS TEST

		Treatments		Row or Block Totals	Block Means
	System A	*System B*	*System C*		
Controller 1	15	15	18	48	$\bar{x}_{1.} = 48/3 = 16.0$
Controller 2	14	14	14	42	$\bar{x}_{2.} = 42/3 = 14.0$
Blocks *Controller 3*	10	11	15	36	$\bar{x}_{3.} = 36/3 = 12.0$
Controller 4	13	12	17	42	$\bar{x}_{4.} = 42/3 = 14.0$
Controller 5	16	13	16	45	$\bar{x}_{5.} = 45/3 = 15.0$
Controller 6	13	13	13	39	$\bar{x}_{6.} = 39/3 = 13.0$
Column or Treatment Totals	81	78	93	252	$\bar{\bar{x}} = \dfrac{252}{18} = 14.0$
Treatment Means	$\bar{x}_{.1} = \dfrac{81}{6}$ $= 13.5$	$\bar{x}_{.2} = \dfrac{78}{6}$ $= 13.0$	$\bar{x}_{.3} = \dfrac{93}{6}$ $= 15.5$		

procedure. Since lower stress values are viewed as better, the sample data available would seem to favor system B with its mean stress rating of 13. However, the usual question remains: Do the sample results justify the conclusion that the mean stress levels for the three systems differ? That is, are the differences statistically significant? An analysis of variance computation similar to the one performed for the completely randomized design can be used to answer this statistical question.   ◀

THE ANALYSIS OF VARIANCE FOR A RANDOMIZED BLOCK DESIGN

The analysis of variance for a randomized block design requires us to partition the sum of squares total (SST) into three groups: sum of squares due to treatments, sum of squares due to blocks, and sum of squares due to error. The formula for this partitioning is as follows:

$$\text{SST} = \text{SSTR} + \text{SSBL} + \text{SSE} \tag{12.29}$$

This sum of squares partition is summarized in the ANOVA table for a randomized block design as shown in Table 12.9. The notation used in this table is as follows:

$$k = \text{the number of treatments}$$
$$b = \text{the number of blocks}$$
$$n_T = \text{the total sample size } (n_T = kb)$$

Note that the ANOVA table in Table 12.9 also shows how the $n_T - 1$ total degrees of freedom are partitioned such that $k - 1$ go to treatments, $b - 1$ go to blocks, and $(k - 1)(b - 1)$ go to the error term. The mean square column shows the sum of squares divided by the degrees of freedom, and $F = \text{MSTR}/\text{MSE}$ is the F ratio used to test for a

TABLE 12.9   **ANOVA TABLE FOR A RANDOMIZED BLOCK DESIGN WITH k TREATMENTS AND b BLOCKS**

Source of Variation	Sum of Squares	Degrees of Freedom	Mean Square	F
Treatments	SSTR	$k - 1$	$\text{MSTR} = \dfrac{\text{SSTR}}{k-1}$	$\dfrac{\text{MSTR}}{\text{MSE}}$
Blocks	SSBL	$b - 1$	$\text{MSBL} = \dfrac{\text{SSBL}}{b-1}$	
Error	SSE	$(k-1)(b-1)$	$\text{MSE} = \dfrac{\text{SSE}}{(k-1)(b-1)}$	
Total	SST	$n_T - 1$		

significant difference among the treatment means. In general, *blocking* refers to the process of using the same or similar experimental units for all treatments. The purpose of blocking is to remove a source of variation from the error term and hence provide a more powerful test for a difference in population or treatment means.

COMPUTATIONS AND CONCLUSIONS

To compute the F statistic needed to test for a difference among treatment means using a randomized block design, we need to compute MSTR and MSE. To calculate these two mean squares, we must first compute SSTR and SSE; in doing so, we will also compute SSBL and SST. To simplify the presentation, we will perform the calculations using four steps. In addition to k, b, and n_T as previously defined, the following notation is used:

$$x_{ij} = \text{value of the observation under treatment } j \text{ in block } i$$
$$\bar{x}_{\cdot j} = \text{sample mean of the } j\text{th treatment}$$
$$\bar{x}_{i\cdot} = \text{sample mean for the } i\text{th block}$$
$$\bar{\bar{x}} = \text{overall sample mean}$$

Step 1   Compute the total sum of squares (SST):

$$\text{SST} = \sum_{i=1}^{b} \sum_{j=1}^{k} (x_{ij} - \bar{\bar{x}})^2 \tag{12.30}$$

Step 2   Compute the sum of squares due to treatments (SSTR):

$$\text{SSTR} = b \sum_{j=1}^{k} (\bar{x}_{\cdot j} - \bar{\bar{x}})^2 \tag{12.31}$$

Step 3   Compute the sum of squares due to blocks (SSBL):

$$\text{SSBL} = k \sum_{i=1}^{b} (\bar{x}_{i\cdot} - \bar{\bar{x}})^2 \tag{12.32}$$

Step 4   Compute the sum of squares due to error (SSE):

$$\text{SSE} = \text{SST} - \text{SSTR} - \text{SSBL} \tag{12.33}$$

EXAMPLE 12.3 ▶
(CONTINUED)

For the air traffic controller data in Table 12.8, these steps lead to the following sum of squares:

Step 1 $SST = (15 - 14)^2 + (15 - 14)^2 + (18 - 14)^2 + \cdots + (13 - 14)^2 = 70$

Step 2 $SSTR = 6[(13.5 - 14)^2 + (13.0 - 14)^2 + (15.5 - 14)^2] = 21$

Step 3 $SSBL = 3[(16 - 14)^2 + (14 - 14)^2 + (12 - 14)^2 + (14 - 14)^2 + (15 - 14)^2 + (13 - 14)^2] = 30$

Step 4 $SSE = 70 - 21 - 30 = 19$

These sums of squares divided by their degrees of freedom provide the corresponding mean square values shown in Table 12.10. The F ratio used to test for differences between treatment means is $MSTR/MSE = 10.5/1.9 = 5.53$. Checking the F values in Table 4 of Appendix B, we find that the critical F value at $\alpha = .05$ (2 numerator and 10 denominator degrees of freedom) is 4.10. With $F = 5.53$, we reject the null hypothesis H_0: $\mu_1 = \mu_2 = \mu_3$ and conclude that the work station designs differ in terms of the mean stress effects on air traffic controllers. ◀

Before leaving this section, let us make some general comments about the randomized block design. The blocking as described in this section is referred to as a *complete* block design; the word "complete" indicates that each block is subjected to all k treatments. That is, all controllers (blocks) were tested using all three systems (treatments). Experimental designs employing blocking where some, but not all, treatments are applied to each block are referred to as *incomplete* block designs. A discussion of incomplete block designs is beyond the scope of this text.

In addition, note that in the air traffic controller stress test, each controller in the study was required to use all three systems. While this guarantees a complete block design, in some cases blocking is carried out with "similar" experimental units in each block. For example, assume that in a pretest of air traffic controllers, the population of controllers was divided into groups ranging from extremely high stress individuals to extremely low stress individuals. The blocking could still have been accomplished by having three controllers from each of the stress classifications participate in the study. Each block would then be formed from three controllers in the same stress class. The randomized aspect of the block design would be conducted by randomly assigning the three controllers in each block to the three systems.

Finally, note that the ANOVA table shown in Table 12.10 provides an F value to test for treatment effects but *not* for blocks. The reason for this is that the experiment was designed to test a single factor—work station design. The blocking based on individual stress differences was conducted to remove this variation from the MSE term. However, the study was not designed to test specifically for individual differences in stress.

TABLE 12.10   **ANOVA TABLE FOR THE AIR TRAFFIC CONTROLLER STRESS TEST**

Source of Variation	Sum of Squares	Degrees of Freedom	Mean Square	F
Treatments	21	2	10.5	$10.5/1.9 = 5.53$
Blocks	30	5	6.0	
Error	19	10	1.9	
Total	70	17		

Some analysts compute $F = MSB/MSE$ and use this statistic to test for significance of the blocks. Then they use the result as a guide to whether this type of blocking would be desirable in future experiments. However, if individual stress difference is to be a factor in the study, a different experimental design should be used. A test of significance on blocks should not be performed to attempt to draw such a conclusion about a second factor.

NOTES AND COMMENTS

1. The matched-samples t test introduced in Chapter 10 is an example of a randomized block design with two treatments.
2. Alternate formulas for SST, SSTR, and SSBL can be developed that can ease the computational burden when using hand calculation. In Appendix 12.2 we have included a step-by-step procedure that illustrates the use of these alternate formulas.

EXERCISES

METHODS

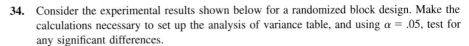

SELF TEST

34. Consider the experimental results shown below for a randomized block design. Make the calculations necessary to set up the analysis of variance table, and using $\alpha = .05$, test for any significant differences.

		Treatments		
		A	B	C
	1	10	9	8
	2	12	6	5
Blocks	3	18	15	14
	4	20	18	18
	5	8	7	8

35. The following data were obtained for a randomized block design involving five treatments and three blocks: SST = 430, SSTR = 310, SSBL = 85. Set up the ANOVA table and test for any significant differences. Use $\alpha = .05$.

36. An experiment has been conducted for four treatments using eight blocks. Complete the following analysis of variance table.

Source of Variation	Sum of Squares	Degrees of Freedom	Mean Square	F
Treatments	900			
Blocks	400			
Error				
Total	1800			

Using $\alpha = .05$, test for any significant differences.

APPLICATIONS

37. An automobile dealer conducted a test to determine if the time needed to complete a minor engine tuneup depends on whether a computerized engine analyzer or an electronic analyzer is used. Because tuneup time varies among compact, intermediate, and full-sized cars, the three types of cars were used as blocks in the experiment. The data obtained are shown below.

		Analyzer	
		Computerized	*Electronic*
Car	*Compact*	50	42
	Intermediate	55	44
	Full-size	63	46

Using $\alpha = .05$, test for any significant differences.

38. Five different auditing procedures were compared with respect to total audit time. To control for possible variation due to the person conducting the audit, four accountants were selected randomly and treated as blocks in the experiment. The following values were obtained using the ANOVA procedure: SST = 100, SSTR = 45, SSBL = 36. Using $\alpha = .05$, test to see if there is any significant difference in total audit time stemming from the auditing procedure used.

39. An important factor in selecting software for word-processing and data base management systems is the time required to learn how to use a particular system. To evaluate three file management systems, a firm designed a test involving five different word-processing operators. Since operator variability was believed to be a significant factor, each of the five operators was trained on each of the three file management systems. The data obtained are shown below.

		System		
		A	*B*	*C*
Operator	*1*	16	16	24
	2	19	17	22
	3	14	13	19
	4	13	12	18
	5	18	17	22

Using $\alpha = .05$, test to see if there is any difference in mean training times for the three systems.

12.7 FACTORIAL EXPERIMENTS

The experimental designs we have considered thus far enable statistical conclusions to be drawn about one factor. However, in some experiments we want to draw conclusions about more than one factor. *Factorial experiments* and their corresponding ANOVA computations are valuable designs when simultaneous conclusions are required about two or more factors. The term "factorial" is used because the experimental conditions include all possible combinations of the factors involved. For example, if there are *a* levels of factor A and *b* levels of factor B, the experiment will involve collecting data on *ab* treatment combinations. In this section we will show the analysis of a two-factor factorial experiment. This basic approach can be extended to experiments involving more than two factors. As an illustration of a two-factor factorial experiment, we will consider the following study. The Graduate Management Admissions Test (GMAT) is a standardized test used by graduate schools of business to evaluate an applicant's ability to pursue a graduate program in that field. Scores on the GMAT range from 200 to 800, with higher scores implying higher aptitude.

EXAMPLE 12.4 ▶ In an attempt to improve the performance of students on the GMAT exam, a university is considering offering the following three GMAT preparation programs:

1. A 3-hour review session covering the types of questions generally asked on the GMAT.
2. A 1-day program covering relevant exam material, along with the taking and grading of a sample exam.
3. An intensive 10-week course involving the identification of each student's weaknesses and the setting up of individualized programs for improvement.

Thus, one factor in this study is the GMAT preparation program, which has three treatments: 3-hour review, 1-day program, and 10-week course. Before adopting one of the preparation programs, further study will be conducted to determine how the proposed programs affect GMAT scores.

In addition, it was noted that the GMAT is usually taken by students from three colleges: the College of Business, the College of Engineering, and the College of Arts and Sciences. Thus, also of interest in the experiment is whether or not a student's undergraduate college affects the GMAT score. This second factor, undergraduate college, also has three treatments: business, engineering, and arts and sciences. The factorial design for this experiment with three treatments corresponding to factor A, the preparation program, and three treatments corresponding to factor B, the undergraduate college, will have total of $3 \times 3 = 9$ treatment combinations. These treatment combinations or experimental conditions are summarized in Table 12.11.

Assume that a sample of two students will be selected corresponding to each of the nine treatment combinations shown in Table 12.11: two business students will take the 3-hour review, two will take the 1-day program, and two will take the 10-week course. In addition, two engineering students and two arts and sciences students will take each of the three preparation programs. In experimental design terminology, the sample size of 2 for

---

TABLE 12.11   **TREATMENT COMBINATIONS FOR THE TWO-FACTOR GMAT EXPERIMENT**

		Factor B: College		
		Business	*Engineering*	*Arts and Sciences*
Factor A:	*3-hour review*	1	2	3
Preparation	*1-day program*	4	5	6
Program	*10-week course*	7	8	9

each treatment combination indicates that we have two replications. Additional replications and an increased sample size could easily be made, but we elected not to do so to minimize the computational aspects for this illustration.

This experimental design requires that six students who plan to attend graduate school be randomly selected from *each* of the three undergraduate colleges. Then two students from each college should be assigned randomly to each preparation program, resulting in a total of 18 students being used in the study.

Let us assume that the students have been randomly selected, have participated in the preparation program, and have taken the GMAT. The scores obtained are shown in Table 12.12.

The analysis of variance computations using the data in Table 12.12 will provide answers to the following questions:

- *Main effect (factor A):* Do the preparation programs differ in terms of effect on GMAT scores?
- *Main effect (factor B):* Do the undergraduate colleges differ in terms of student ability to perform on the GMAT?
- *Interaction effect (factors A and B):* Do students in some colleges do better on one type of preparation program while others do better on a different type of preparation program?

The term *interaction* refers to a new effect that we can now study because we have used a factorial experiment. If the interaction effect has a significant impact on the GMAT scores, it will mean that the effect of the type of preparation program depends on the undergraduate college. ◀

THE ANALYSIS OF VARIANCE FOR A FACTORIAL EXPERIMENT

The analysis of variance for a two-factor factorial experiment is similar to the completely randomized experiment and the randomized block experiment in that we once again partition the sum of squares and the degrees of freedom into their respective sources. The formula for partitioning the sum of squares for the two-factor factorial experiments is as follows:

$$SST = SSA + SSB + SSAB + SSE \qquad (12.34)$$

TABLE 12.12   **GMAT SCORES FOR THE TWO-FACTOR GMAT EXPERIMENT**

		Factor B: College		
		Business	*Engineering*	*Arts and Sciences*
	3-hour review	500	540	480
		580	460	400
Factor A:	*1-day program*	460	560	420
Preparation		540	620	480
Program	*10-week course*	560	600	480
		600	580	410

The partitioning of the sum of squares and degrees of freedom is summarized in Table 12.13. The following notation is used:

$$a = \text{number of levels of factor A}$$
$$b = \text{number of levels of factor B}$$
$$r = \text{number of replications}$$
$$n_T = \text{total number of observations; } n_T = abr$$

COMPUTATIONS AND CONCLUSIONS

To compute the F statistics needed to test for the significance of factor A, factor B, and interaction, we need to compute MSA, MSB, MSAB, and MSE. To calculate these four mean squares, we must first compute SSA, SSB, SSAB, and SSE; in doing so, we will also compute SST. In addition to a, b, r, and n_T as previously defined, the following notation is used:

$x_{ijk} = $ observation corresponding to the kth replicate taken from treatment i of factor A and treatment j of factor B

$\bar{x}_{i\cdot} = $ sample mean for the observations in treatment i (factor A)

$\bar{x}_{\cdot j} = $ sample mean for the observations in treatment j (factor B)

$\bar{x}_{ij} = $ sample mean for the observations corresponding to the combination of treatment i (factor A) and treatment j (factor B)

$\bar{\bar{x}} = $ overall sample mean of all n_T observations

We will perform the computations using the following 5-step procedure.

Step 1 Compute the total sum of squares:

$$\text{SST} = \sum_{i=1}^{a} \sum_{j=1}^{b} \sum_{k=1}^{r} (x_{ijk} - \bar{\bar{x}})^2 \tag{12.35}$$

Step 2 Compute the sum of squares for factor A:

$$\text{SSA} = br \sum_{i=1}^{a} (\bar{x}_{i\cdot} - \bar{\bar{x}})^2 \tag{12.36}$$

TABLE 12.13   **ANOVA TABLE FOR A TWO-FACTOR FACTORIAL EXPERIMENT WITH r REPLICATIONS**

Source of Variation	Sum of Squares	Degrees of Freedom	Mean Square	F
Factor A	SSA	$a - 1$	$\text{MSA} = \dfrac{\text{SSA}}{a - 1}$	$\dfrac{\text{MSA}}{\text{MSE}}$
Factor B	SSB	$b - 1$	$\text{MSB} = \dfrac{\text{SSB}}{b - 1}$	$\dfrac{\text{MSB}}{\text{MSE}}$
Interaction	SSAB	$(a - 1)(b - 1)$	$\text{MSAB} = \dfrac{\text{SSAB}}{(a - 1)(b - 1)}$	$\dfrac{\text{MSAB}}{\text{MSE}}$
Error	SSE	$ab(r - 1)$	$\text{MSE} = \dfrac{\text{SSE}}{ab(r - 1)}$	
Total	SST	$n_T - 1$		

Step 3   Compute the sum of squares for factor B:

$$SSB = ar \sum_{j=1}^{b} (\bar{x}_{\cdot j} - \bar{\bar{x}})^2 \tag{12.37}$$

Step 4   Compute the sum of squares for interaction:

$$SSAB = r \sum_{i=1}^{a} \sum_{j=1}^{b} (\bar{x}_{ij} - \bar{x}_{i\cdot} - \bar{x}_{\cdot j} + \bar{\bar{x}})^2 \tag{12.38}$$

Step 5   Compute the sum of squares due to error:

$$SSE = SST - SSA - SSB - SSAB \tag{12.39}$$

EXAMPLE 12.4
(CONTINUED)

Table 12.14 shows the data collected in the experiment, along with the various sums that will help us with the sum of squares computations. Using equations (12.35) to (12.39) we have the following sum of squares for the GMAT two-factor factorial experiment:

Step 1   SST = $(500 - 515)^2 + (580 - 515)^2 + (540 - 515)^2 + \cdots +$
$(410 - 515)^2 = 82,450$

Step 2   SSA = $(3)(2)[(493.33 - 515)^2 + (513.33 - 515)^2 + (538.33 - 515)^2] = 6100$

Step 3   SSB = $(3)(2)[(540 - 515)^2 + (560 - 515)^2 + (445 - 515)^2] = 45,300$

Step 4   SSAB = $2[(540 - 493.33 - 540 + 515)^2 + (500 - 493.33 - 560 +$
$515)^2 + \cdots + (445 - 538.33 - 445 + 515)^2] = 11,200$

Step 5   SSE = $82,450 - 6100 - 45,300 - 11,200 = 19,850$

These sums of squares divided by their corresponding degrees of freedom, as shown in Table 12.15, provide the appropriate mean square values for testing the two main effects (preparation program and undergraduate college) and the interaction effect. The F ratio used to test for differences among preparation programs is 1.38. The critical F value at $\alpha = .05$ (with 2 numerator and 9 denominator degrees of freedom) is 4.26. With $F = 1.38$, we cannot reject the null hypothesis and must conclude that there is not a significant difference in the three preparation programs. However, for the undergraduate college effect, $F = 10.27$ exceeds the critical F value of 4.26. Thus, the analysis of variance results allow us to conclude that there is a difference in GMAT test scores among the three undergraduate colleges; that is, the three undergraduate colleges do not provide the same preparation for performance on the GMAT. Finally, the interaction F value of $F = 1.27$ (critical F value = 3.63 at $\alpha = .05$) means that we cannot identify a significant interaction effect. Thus, there is no reason to believe that the three preparation programs differ in their ability to prepare students from the different colleges for the GMAT.

Undergraduate college was found to be a significant factor. Checking the calculations in Table 12.14, we see that the sample means are as follows: business students $\bar{x}_{\cdot 1} = 540$, engineering students $\bar{x}_{\cdot 2} = 560$, and arts and sciences students $\bar{x}_{\cdot 3} = 445$. Tests on individual treatment means can be conducted; yet after reviewing the three sample means, we would anticipate no difference in preparation for business and engineering graduates. However, the arts and sciences students appear to be significantly less prepared for the GMAT than students in the other colleges. Perhaps this observation will lead the university to consider other options for assisting these students in preparing for the Graduate Management Admission Test.

TABLE 12.14 GMAT SUMMARY DATA FOR THE TWO-FACTOR EXPERIMENT

Treatment combination totals	Factor B: College			Row Totals	Factor A Means
	Business	Engineering	Arts and Sciences		
Factor A: Preparation Program — 3-hour review	500 580 1080 $\bar{x}_{11} = \frac{1080}{2} = 540$	540 460 1000 $\bar{x}_{12} = \frac{1000}{2} = 500$	480 400 880 $\bar{x}_{13} = \frac{880}{2} = 440$	2960	$\bar{x}_{1\cdot} = \frac{2960}{6} = 493.33$
1-day program	460 540 1000 $\bar{x}_{21} = \frac{1000}{2} = 500$	560 620 1180 $\bar{x}_{22} = \frac{1180}{2} = 590$	420 480 900 $\bar{x}_{23} = \frac{900}{2} = 450$	3080	$\bar{x}_{2\cdot} = \frac{3080}{6} = 513.33$
10-week course	560 600 1160 $\bar{x}_{31} = \frac{1160}{2} = 580$	600 580 1180 $\bar{x}_{32} = \frac{1180}{2} = 590$	480 410 890 $\bar{x}_{33} = \frac{890}{2} = 445$	3230	$\bar{x}_{3\cdot} = \frac{3230}{6} = 538.33$
Column Totals	3240	3360	2670	9270	**Overall Total**
Factor B Means	$\bar{x}_{\cdot1} = \frac{3240}{6} = 540$	$\bar{x}_{\cdot2} = \frac{3360}{6} = 560$	$\bar{x}_{\cdot3} = \frac{2670}{6} = 445$		$\bar{\bar{x}} = \frac{9270}{18} = 515$

TABLE 12.15   **ANOVA TABLE FOR THE TWO-FACTOR GMAT STUDY**

Source of Variation	Sum of Squares	Degrees of Freedom	Mean Square	F
Factor A	6,100	2	3,050	3,050/2,206 = 1.38
Factor B	45,300	2	22,650	22,650/2,206 = 10.27
Interaction	11,200	4	2,800	2,800/2,206 = 1.27
Error	19,850	9	2,206	
Total	82,450	17		

FIGURE 12.7   **COMPUTER OUTPUT FOR THE GMAT TWO-FACTOR DESIGN**

```
ANALYSIS OF VARIANCE   GMAT

SOURCE        DF        SS        MS
FACTOR A       2       6100      3050
FACTOR B       2      45300     22650
INTERACTION    4      11200      2800
ERROR          9      19850      2206
TOTAL         17      82450
```

Because of the computational effort involved in any modest to large-size factorial experiment, the computer usually plays an important role in making and summarizing the analysis of variance computations. The Minitab analysis of variance output for the GMAT two-factor factorial experiment is shown in Figure 12.7.

NOTES AND COMMENTS

Alternate formulas for SST, SSA, SSB, SSAB, and SSE can be developed that can ease the computational burden when using hand calculation. In Appendix 12.3 we have included a step-by-step procedure that illustrates the use of these alternate formulas.

EXERCISES

METHODS

SELF TEST

40. A factorial experiment involving three levels of factor A and two levels of factor B resulted in the following data.

	Factor B		
	Level 1	*Level 2*	*Level 3*
Factor A *Level 1*	135 165	90 66	75 93
Level 2	125 95	127 105	120 136

Test for any significant main effects and any interaction. Use $\alpha = .05$.

41. The calculations for a factorial experiment involving four levels of factor A, three levels of factor B, and three replications resulted in the following data: SST = 280, SSA = 26, SSB = 23, SSAB = 175. Set up the ANOVA table and test for any significant main effects and any interaction effect. Use $\alpha = .05$.

APPLICATIONS

42. A mail-order catalog firm designed a factorial experiment to test the effect of the size of a magazine advertisement and the advertisement design on the number of catalog requests received (1000's). Three advertising designs and two different-size advertisements were considered. The data obtained are shown below. Test for any significant effects due to type of design, size of advertisement, or interaction. Use $\alpha = .05$.

		Size of Advertisement	
		Small	*Large*
	A	8 12	12 8
Design	B	22 14	26 30
	C	10 18	18 14

43. An amusement park has been studying methods for decreasing the waiting time (minutes) on rides by loading and unloading riders more efficiently. Two alternative loading/unloading methods have been proposed. To account for potential differences due to the type of ride and the possible interaction between the method of loading and unloading and the type of ride, a factorial experiment was designed. Using the data shown below, test for any significant effect due to the loading and unloading method, the type of ride, and interaction. Use $\alpha = .05$.

	Type of Ride		
	Roller Coaster	*Screaming Demon*	*Log Flume*
Method 1	41 43	52 44	50 46
Method 2	49 51	50 46	48 44

44. Jack's Restaurant is considering a new specialty sandwich. To determine the effect of sandwich price and sandwich size on sales, the new sandwich was test-marketed in selected company restaurants. The data in terms of the number of sandwiches sold per day, are shown below. Test for any significant differences due to price, size, and interaction. Use $\alpha = .05$.

			Price	
		$1.49	*$1.79*	*$1.99*
Size	¼ *pound*	955	845	820
		985	860	845
	⅓ *pound*	945	910	860
		875	905	935

SUMMARY

In this chapter we have shown how analysis of variance can be used to test for differences among means of several populations or treatments. In addition, we introduced the single-factor completely randomized, the randomized block, and the two-factor factorial experimental designs. The completely randomized design and the randomized block designs are used to draw conclusions about differences in the means of a single factor. The primary purpose of blocking in the randomized block design is to remove extraneous sources of variation from the error term. This blocking provides a better estimate of the error variance and a better test to determine whether the population or treatment means of the factor differ significantly.

We showed that the basis for the statistical tests used in analysis of variance and experimental design is the development of independent estimates of the population variance σ^2. In the single-factor case, one estimator is based on the variation between the treatments; this estimator provides an unbiased estimate of σ^2 only if the means $\mu_1, \mu_2, \ldots, \mu_k$ are all equal. A second estimator of σ^2 is based on the variation of the observations within each sample; this estimator will always provide an unbiased estimate of σ^2. By computing the ratio of these two estimators using the F distribution, we developed a rejection rule for determining whether or not to reject the null hypothesis that the population or treatment means are equal. In all the experimental designs considered, the partitioning of the sum of squares and degrees of freedom into their various sources enabled us to compute the appropriate values for making the analysis of variance calculations and tests. We also showed how Fisher's LSD procedure, the Bonferroni adjustment, and Tukey's procedure can be used to perform pairwise comparisons to determine which means are different.

GLOSSARY

Analysis of variance (ANOVA) A statistical procedure for determining whether the means of two or more populations are equal.

ANOVA table A table used to summarize the analysis of variance computations and results. It contains columns showing the source of variation, the degrees of freedom, the sum of squares, the mean squares, and the F values.

Partitioning The process of allocating the total sum of squares and degrees of freedom into the various components.

Multiple comparison procedures Statistical procedures used to conduct statistical comparisons between pairs of population or treatment means.

Factor Another word for the variable of interest in an experiment.

Treatment Different levels of a factor.

Single-factor experiment An experiment involving only one factor with k populations or treatments.

Experimental units The objects of interest in the experiment.

Completely randomized design An experimental design where the treatments are randomly assigned to the experimental units.

Replication The number of times each experimental condition is repeated in an experiment. It is the sample size associated with each treatment combination.

Mean square The sum of squares divided by its corresponding degrees of freedom.

Blocking The process of using the same or similar experimental units for all treatments. The purpose of blocking is to remove a source of variation from the error term and hence provide a more powerful test for a difference among population or treatment means.

Randomized block design An experimental design employing blocking.

Factorial experiments An experimental design that permits statistical conclusions about two or more factors. All levels of each factor are considered with all levels of the other factors to specify the experimental conditions for the experiment.

Interaction The effect produced when the levels of one factor interact with the levels of another factor.

KEY FORMULAS

Analysis of Variance

jth Sample Mean

$$\bar{x}_j = \frac{\sum_{i=1}^{n_j} x_{ij}}{n_j} \tag{12.1}$$

jth Sample Variance

$$s_j^2 = \frac{\sum_{i=1}^{n_j} (x_{ij} - \bar{x}_j)^2}{n_j - 1} \tag{12.2}$$

Overall Sample Mean

$$\bar{\bar{x}} = \frac{\sum_{j=1}^{k} \sum_{i=1}^{n_j} x_{ij}}{n_T} \tag{12.3}$$

$$n_T = n_1 + n_2 + \cdots + n_k \tag{12.4}$$

Mean Square Between

$$\text{MSB} = \frac{\text{SSB}}{k-1} \tag{12.7}$$

Sum of Squares Between

$$\text{SSB} = \sum_{j=1}^{k} n_j (\bar{x}_j - \bar{\bar{x}})^2 \tag{12.8}$$

Mean Square Within

$$MSW = \frac{SSW}{n_T - k} \tag{12.10}$$

Sum of Squares Within

$$SSW = \sum_{j=1}^{k} (n_j - 1)s_j^2 \tag{12.11}$$

Total Sum of Squares

$$SST = \sum_{j=1}^{k} \sum_{i=1}^{n_j} (x_{ij} - \bar{\bar{x}})^2 \tag{12.12}$$

Multiple Comparison Procedures

Test Statistic for Fisher's LSD

$$t = \frac{\bar{x}_1 - \bar{x}_2}{\sqrt{MSW\left(\dfrac{1}{n_1} + \dfrac{1}{n_2}\right)}} \tag{12.15}$$

Fisher's LSD

$$LSD = t_{\alpha/2}\sqrt{MSW\left(\frac{1}{n_1} + \frac{1}{n_2}\right)} \tag{12.16}$$

Test Statistic for Tukey's Test

$$q = \frac{\bar{x}_{max} - \bar{x}_{min}}{\sqrt{\dfrac{MSW}{n}}} \tag{12.22}$$

Tukey's Significant Difference

$$TSD = q\sqrt{\frac{MSW}{n}} \tag{12.23}$$

Completely Randomized Designs

Mean Square Due to Treatments

$$MSTR = \frac{\sum_{j=1}^{k} n_j(\bar{x}_j - \bar{\bar{x}})^2}{k - 1} \tag{12.25}$$

Mean Square Due to Error

$$MSE = \frac{\sum_{j=1}^{k} (n_j - 1)s_j^2}{n_T - k} \tag{12.26}$$

The F Value

$$F = \frac{\text{MSTR}}{\text{MSE}}$$ **(12.28)**

Randomized Block Designs

Total Sum of Squares

$$\text{SST} = \sum_{i=1}^{b} \sum_{j=1}^{k} (x_{ij} - \overline{\overline{x}})^2$$ **(12.30)**

Sum of Squares Due to Treatments

$$\text{SSTR} = b \sum_{j=1}^{k} (\overline{x}_{\cdot j} - \overline{\overline{x}})^2$$ **(12.31)**

Sum of Squares Due to Blocks

$$\text{SSBL} = k \sum_{i=1}^{b} (\overline{x}_{i\cdot} - \overline{\overline{x}})^2$$ **(12.32)**

Sum of Squares Due to Error

$$\text{SSE} = \text{SST} - \text{SSTR} - \text{SSBL}$$ **(12.33)**

Factorial Experiments

Total Sum of Squares

$$\text{SST} = \sum_{i=1}^{a} \sum_{j=1}^{b} \sum_{k=1}^{r} (x_{ijk} - \overline{\overline{x}})^2$$ **(12.35)**

Sum of Squares Due to Factor A

$$\text{SSA} = br \sum_{i=1}^{a} (\overline{x}_{i\cdot} - \overline{\overline{x}})^2$$ **(12.36)**

Sum of Squares Due to Factor B

$$\text{SSB} = ar \sum_{j=1}^{b} (\overline{x}_{\cdot j} - \overline{\overline{x}})^2$$ **(12.37)**

Sum of Squares Due to Interaction

$$\text{SSAB} = r \sum_{i=1}^{a} \sum_{j=1}^{b} (\overline{x}_{ij} - \overline{x}_{i\cdot} - \overline{x}_{\cdot j} + \overline{\overline{x}})^2$$ **(12.38)**

Sum of Squares Due to Error

$$\text{SSE} = \text{SST} - \text{SSA} - \text{SSB} - \text{SSAB}$$ **(12.39)**

▽ **REVIEW QUIZ**

TRUE/FALSE

1. In a completely randomized design, the treatments must be randomly assigned to the experimental units.
2. The analysis of variance can be used to test the null hypothesis that k population means are equal.
3. The only assumption required for the analysis of variance is that the response variable is normally distributed.
4. In computing the between-treatments estimate of the population variance, we cannot assume that the null hypothesis is true.
5. The degrees of freedom associated with the sum of squares between treatments is the same as the number of treatments.
6. The within-treatments estimate of the population variance is called the mean square due to error.
7. In a completely randomized design, the sum of squares due to error has a number of degrees of freedom equal to the total sample size minus the number of treatments minus 1.
8. In a completely randomized design, the degrees of freedom associated with the total sum of squares is the sum of the degrees of freedom for SSTR and the degrees of freedom for SSE.
9. Whenever a randomized block design is used, the F test should not be used.
10. If the null hypothesis that k population means are equal is rejected, we can then use a chi-square test to determine which individual means differ.
11. In a two-factor factorial experiment, the total sum of squares is partitioned into two components: main effect factor A and main effect factor B.
12. In a factorial experiment, the number of levels of factor A must be equal to the number of levels of factor B.

MULTIPLE CHOICE

Use the following information to answer Questions 13–17. A statistics professor wishes to know the effect of class format on student learning, as measured by improvement on examination scores from the beginning to the end of the semester. The five class formats to be studied reflect different emphases on homework problems and computer exercises. Sixty students are randomly selected for this study; 12 students are randomly assigned to each class format.

13. In this example, the term *factor* is illustrated by
    a. the change in exam scores
    b. the class formats
    c. the different emphases on homework problems and computer exercises
    d. the 12 students in the sample from each class format
    e. the 60 students in the random sample
14. The term *treatment* is illustrated by
    a. the change in exam scores
    b. the class formats
    c. the different amounts of homework and computer work in the different class formats
    d. the 12 students assigned to each class format
    e. the 60 students in the random sample
15. The term *replication* is illustrated by
    a. the change in exam scores
    b. the class formats
    c. the different amounts of homework and computer work in the different class formats
    d. the 12 students assigned to each class format
    e. the 60 students in the random sample

16. The term *response* is illustrated by
  a. the change in exam scores
  b. the class formats
  c. the different amounts of homework and computer work in the different class formats
  d. the 12 students in the sample from each class format
  e. the 60 students in the random sample
17. This example best reflects which experimental design?
  a. randomized block design   **b.** completely randomized design
  c. factorial experiments   **d.** individual treatment means
18. Which of the following is a required condition for using the analysis of variance?
  a. The data are obtained from independently selected samples.
  b. The populations are all normally distributed.
  c. The populations have the same variance.
  d. All of the above are necessary conditions.
19. A completely randomized experimental design resulted in data from three sample groups, each comprising four observations. The degrees of freedom for the critical value of F is
  a. 11   **b.** 2, 9
  c. 2, 11   **d.** 3, 11
20. A factorial experiment that involved four levels of factor A, five levels of factor B, and three replications would have error degrees of freedom equal to
  a. 60   **b.** 24
  c. 40   **d.** 59
21. To compute the F statistics needed to test the significance of factor A, factor B, and interaction in a two-factorial experiment, we need to compute
  a. MSA and MSB   **b.** MSA, MSB, and MSAB
  c. MSA, MSB, and MSE   **d.** MSA, MSB, MSAB, and MSE
  e. MSAB and MSE

SUPPLEMENTARY EXERCISES

45. In your own words, explain what the analysis of variance is used for.

46. What has to be true for the mean square between, MSB, to provide a good estimate of σ^2? Explain.

47. Why do we assume that the populations sampled all have the same variance when we use the analysis of variance?

48. Explain why the mean square between, MSB, and the mean square within, MSW, provide two independent estimates of σ^2.

49. Explain why the mean square between, MSB, provides an inflated estimate of σ^2 when the population means are not the same.

50. A simple random sample of the asking price ($1000's) of four houses currently for sale in each of two residential areas resulted in the following data.

	Area 1	Area 2
	92	90
	89	102
	98	96
	105	88
Sample mean	96	94
Sample variance	50	40

a. Use the procedure developed in Chapter 10 to test if the mean asking price is the same in both areas. Use $\alpha = .05$.

b. Use the analysis of variance to test if the mean asking price is the same. Compare your analysis with part (a). Use $\alpha = .05$.

51. Suppose that in Exercise 50 data were collected for another residential area. The asking prices for the simple random sample from the third area are as follows: $81,000, $86,000, $75,000, and $90,000; the sample mean and sample variance for these data are 83 and 42, respectively. Is the mean asking price for all three areas the same? Use $\alpha = .05$.

52. An analysis of the number of units sold by 10 salespersons in each of four sales territories resulted in the following data.

Category	Sales Territory			
	1	2	3	4
Number of salespersons	10	10	10	10
Mean number sold ($\bar{x}$)	130	120	132	114
Sample variance (s^2)	72	64	69	67

Test at the $\alpha = .05$ level to see if there is any significant difference among the mean number of units sold in the four sales territories.

53. Suppose that in Exercise 52 the number of salespersons in each territory was as follows: $n_1 = 10$, $n_2 = 12$, $n_3 = 10$, and $n_4 = 15$. Using the same data for $\bar{x}$ and s^2 as given in Exercise 52, test at the $\alpha = .05$ level to see if there is any significant difference among the population mean number of units sold in the four sales territories.

 SERVICE

54. Executives rated service quality in each of several American industries (*Journal of Accountancy,* February 1992). Assume a sample of ratings are as shown below for the airline, retail, hotel, and automotive industries; higher scores indicate a higher service quality rating. At the $\alpha = .05$ level of significance, test for a significant difference among the population mean quality ratings for the four industries. What is your conclusion?

	Airlines	Retail	Hotel	Automotive
	59	63	70	49
	56	49	68	55
	47	60	62	48
	46	54	69	49
	55	56	59	50
	54	55		
	48			
Sample mean	52.14	56.17	65.6	50.20
Sample variance	25.81	23.77	23.20	7.70

 ASSEMBLY

55. Three different assembly methods have been proposed for a new product. A completely randomized experimental design was used to determine which assembly method results in the greatest number of parts produced per hour, and 30 workers were randomly selected and assigned to use one of the proposed methods. The number of units produced by each worker is given at the top of page 491.

	Method		
	A	B	C
	97	93	99
	73	100	94
	93	93	87
	100	55	66
	73	77	59
	91	91	75
	100	85	84
	86	73	72
	92	90	88
	95	83	86
Sample mean	90	84	81
Sample variance	98.00	168.44	159.78

Use these data, and test to see if the mean number of parts produced with each method is the same. Use $\alpha = .05$.

56. In a completely randomized experimental design, three brands of paper towels were tested for their ability to absorb water. Equal-sized towels were used, with four sections of towels tested per brand. The absorbency rating data are given. Using a .05 level of significance, does there appear to be a difference in the ability of the different brands to absorb water?

	Brand		
	X	Y	Z
	91	99	83
	100	96	88
	88	94	89
	89	99	76
Sample mean	92	97	84
Sample variance	30	6	35.33

57. To test to see if there is any significant difference in the mean number of units produced per week by each of three production methods, a completely randomized experimental design was used to obtain the data shown below. At the $\alpha = .05$ level of significance, is there any difference among the means for the three methods?

	Method 1	Method 2	Method 3
	58	52	48
	64	63	57
	55	65	59
	66	58	47
	67	62	49
Sample mean	62	60	52
Sample variance	27.5	26.5	31

 DowJones

58. Shown below are the percent changes in the Dow Jones Industrial Averages for stock market performance in each of the four years of six presidential terms. (*The Beacon Street Financial*, February 1992). Does there appear to be any significant effect due to the year of the presidential term on stock market performance? Use $\alpha = .05$.

	First Year	Second Year	Third Year	Fourth Year
	10.9	−18.9	15.2	4.3
	−15.2	4.8	6.1	14.6
	−16.7	−27.6	38.3	17.9
	−17.3	−3.1	4.2	14.9
	−9.2	19.6	20.3	−3.7
	27.7	22.6	2.3	11.9
	27.0	−4.3	20.3	8.8
Sample mean	1.03	−0.99	15.24	9.81
Sample variance	416.93	343.04	159.31	55.43

59. Hargreaves Automotive Parts, Inc., would like to compare the mileage for four different types of brake linings. Thirty linings of each type were produced and placed on a fleet of rental cars. The number of miles that each brake lining lasted until it no longer met the required federal safety standard was recorded, and an average value was computed for each type of lining. The following data were obtained.

Type	Sample Size	Sample Mean	Standard Deviation
A	30	32,000	1450
B	30	27,500	1525
C	30	34,200	1650
D	30	30,300	1400

Test if the corresponding population means are equal. Use $\alpha = .05$.

60. A manufacturer of batteries for electronic toys and calculators is considering three new battery designs. An attempt was made to determine if the mean lifetime in hours is the same for each of the three designs, and the following data were collected.

	Design A	Design B	Design C
	78	112	115
	98	99	101
	88	101	100
	96	116	120
Sample mean	90	107	109
Sample variance	82.67	68.67	100.67

Test to see if the population means are equal. Use $\alpha = .05$.

61. A study was conducted to investigate the browsing activity by shoppers (*Journal of the Academy of Marketing Science,* Winter 1989). Each shopper was initially classified as nonbrowser, light browser, or heavy browser. For each shopper in the study, a measure was obtained to determine how comfortable the shopper was in a store. Higher scores indicated greater comfort. Suppose the following data are from a related study.

	Nonbrowser	Light Browser	Heavy Browser
	4	5	5
	5	6	7
	6	5	5
	3	4	7
	3	7	4
	4	4	6
	5	6	5
	4	5	7
Sample mean	4.25	5.25	5.75
Sample variance	1.07	1.07	1.36

a. Using $\alpha = .05$, test for differences among the mean comfort levels for the three types of browsers.

b. Use Tukey's procedure to compare the mean comfort levels of nonbrowsers and light browsers. Use $\alpha = .05$. What is your conclusion?

62. A research firm tests the miles-per-gallon characteristics of three brands of gasoline. Because of different gasoline performance characteristics in different brands of automobiles, five brands of automobiles are selected and treated as blocks in the experiment. That is, each brand of automobile is tested with each type of gasoline. The results of the experiment (in miles per gallon) are shown below.

		Gasoline Brands		
		I	II	III
	A	18	21	20
	B	24	26	27
Blocks: Automobiles	C	30	29	34
	D	22	25	24
	E	20	23	24

Using $\alpha = .05$, is there a significant difference among the mean miles-per-gallon characteristics of the three brands of gasoline?

63. Analyze the experimental data provided in Exercise 62 using the analysis of variance for completely randomized designs. Compare your findings with those obtained in Exercise 62. What is the advantage of attempting to remove the block effect?

64. Three different road-repair compounds were tested at four different highway locations. At each location, three sections of the road were repaired, with each section using one of the three compounds. Data were then collected on the number of days of traffic usage before additional repair was required. These data are shown below. Using $\alpha = .01$, test to see if there are significant differences among the compounds.

| | | \multicolumn{4}{c}{Location} |
		1	2	3	4
	A	99	73	85	103
Compound	B	82	72	85	97
	C	81	79	82	86

65. A factorial experiment was designed to test for any significant differences in the mean time needed to perform English-to-foreign language translations using two computerized language translators. Since the type of language translated was also considered a significant factor, translations were made using both systems for three different languages: Spanish, French, and German. Use the following data for translation time shown in hours.

| | \multicolumn{3}{c}{**Language**} |
	Spanish	*French*	*German*
System 1	8	10	12
	12	14	16
System 2	6	14	16
	10	16	22

Test for any significant differences due to language translator, type of language, and interaction. Use $\alpha = .05$.

66. A manufacturing company designed a factorial experiment to determine if the number of defective parts produced by two machines differed and if the number of defective parts produced also depended on whether raw material needed by each machine was loaded manually or using an automatic feed system. The following data show the number of defective parts produced. Using $\alpha = .05$, test for any significant effect due to machine, loading system, and interaction.

| | \multicolumn{2}{c}{**Loading System**} |
	Manual	*Automatic*
Machine 1	30	30
	34	26
Machine 2	20	24
	22	28

COMPUTER CASE: WENTWORTH MEDICAL CENTER

As part of a long-term study of individuals 65 years of age or older, sociologists and physicians at the Wentworth Medical Center in upstate New York conducted a study to investigate the relationship between geographic location and depression. A sample of 60 individuals, all in reasonably good health, was selected; 20 individuals were residents of Florida, 20 were residents of North Carolina, and 20 were residents of New York. Each of the individuals sampled was given a standardized test to measure depression. The data collected are shown below; higher test scores

MEDICAL1

indicate higher levels of depression. These data are available on the data disk in the file MEDICAL1.

A second part of the study considered the relationship between geographic location and depression for individuals 65 years of age or older who had a chronic health condition such as arthritis, hypertension, or heart ailment. A sample of 60 individuals with such conditions was identified. Again 20 were residents of Florida, 20 were residents of New York, and 20 were residents of North Carolina. The levels of depression recorded for this study are also shown below. These data are available on the data disk in the file MEDICAL2.

MEDICAL2

Data from MEDICAL1			Data from MEDICAL2		
Florida	New York	North Carolina	Florida	New York	North Carolina
3	8	10	13	14	10
7	11	7	12	9	12
7	9	3	17	15	15
3	7	5	17	12	18
8	8	11	20	16	12
8	7	8	21	24	14
8	8	4	16	18	17
5	4	3	14	14	8
5	13	7	13	15	14
2	10	8	17	17	16
6	6	8	12	20	18
2	8	7	9	11	17
6	12	3	12	23	19
6	8	9	15	19	15
9	6	8	16	17	13
7	8	12	15	14	14
5	5	6	13	9	11
4	7	3	10	14	12
7	7	8	11	13	13
3	8	11	17	11	11

REPORT

1. Use descriptive statistics to summarize the data in the two studies. What are your preliminary observations about the depression scores?
2. Use analysis of variance on both data sets. State the hypothesis being tested in each case. What are your conclusions?
3. Use inferences about individual treatment means where appropriate. What are your conclusions?
4. Discuss extensions of this study or other analyses that you feel might be helpful.

APPENDIX 12.1

COMPUTATIONAL PROCEDURE FOR A COMPLETELY RANDOMIZED DESIGN

The following step-by-step procedure is designed to ease the burden in computing the appropriate sums of squares for completely randomized designs. The formulas shown below can be applied to both balanced and unbalanced designs.

NOTATION

$$x_{ij} = \text{value of the } i\text{th observation under treatment } j$$
$$T_j = \text{sum of all observations for treatment } j$$
$$T = \text{sum of all observations}$$
$$n_j = \text{sample size for the } j\text{th treatment}$$
$$n_T = \text{total sample size for the experiment}$$
$$k = \text{number of treatments}$$

PROCEDURE

Step 1 Compute the total sum of squares:

$$\text{SST} = \sum_{j=1}^{k} \sum_{i=1}^{n_j} x_{ij}^2 - \frac{T^2}{n_T}$$

Step 2 Compute the sum of squares due to treatments:

$$\text{SSTR} = \sum_{j=1}^{k} \frac{T_j^2}{n_j} - \frac{T^2}{n_T}$$

Step 3 Compute the sum of squares due to error:

$$\text{SSE} = \text{SST} - \text{SSTR}$$

EXAMPLE

Using this computational procedure with the Chemitech data in Table 12.4, we obtain the following results:

Step 1   SST = 54,860 − 810,000/15 = 860

Step 2   SSTR = $(310)^2/5 + (330)^2/5 + (260)^2/5$ − 810,000/15 = 520

Step 3   SSE = SST − SSTR = 860 − 520 = 340

APPENDIX 12.2

COMPUTATIONAL PROCEDURE FOR A RANDOMIZED BLOCK DESIGN

The following step-by-step procedure is designed to help in computing the appropriate sums of squares for randomized block designs.

NOTATION

$$x_{ij} = \text{value of the observation for treatment } j \text{ in block } i$$
$$T_{i\cdot} = \text{the total of all observations in block } i$$
$$T_{\cdot j} = \text{the total of all observations in treatment } j$$
$$T = \text{the total of all observations}$$
$$k = \text{the number of treatments}$$
$$b = \text{the number of blocks}$$
$$n_T = \text{the total sample size } (n_T = kb)$$

PROCEDURE

Step 1   Compute the total sum of squares:

$$\text{SST} = \sum_{i=1}^{b} \sum_{j=1}^{k} x_{ij}^2 - \frac{T^2}{n_T}$$

Step 2   Compute the sum of squares due to treatments:

$$\text{SSTR} = \frac{\displaystyle\sum_{j=1}^{k} T_{\cdot j}^2}{b} - \frac{T^2}{n_T}$$

Step 3   Compute the sum of squares due to blocks:

$$\text{SSBL} = \frac{\displaystyle\sum_{i=1}^{b} T_{i\cdot}^2}{k} - \frac{T^2}{n_T}$$

Step 4   Compute the sum of squares due to error:

$$\text{SSE} = \text{SST} - \text{SSTR} - \text{SSBL}$$

EXAMPLE

For the air traffic controller data in Table 12.8, these steps lead to the sum of squares:

Step 1 $\text{SST} = 3598 - \dfrac{(252)^2}{18} = 70$

Step 2 $\text{SSTR} = \dfrac{(81)^2 + (78)^2 + (93)^2}{6} - \dfrac{(252)^2}{18} = 21$

Step 3 $\text{SSBL} = \dfrac{(48)^2 + (42)^2 + \cdots + (39)^2}{3} - \dfrac{(252)^2}{18} = 30$

Step 4 $\text{SSE} = 70 - 21 - 30 = 19$

APPENDIX 12.3

COMPUTATIONAL PROCEDURE FOR A TWO-FACTOR FACTORIAL DESIGN

NOTATION

x_{ijk} = observation corresponding to the kth replicate taken from treatment i of factor A and treatment j of factor B

$T_{i\cdot}$ = total of all observations in treatment i (factor A)

$T_{\cdot j}$ = total of all observations in treatment j (factor B)

T_{ij} = total of all observations in the combination of treatment i (factor A) and treatment j (factor B)

T = total of all observations

a = number of levels of factor A
b = number of levels of factor B
r = number of replications
n_T = total number of observations; $n_T = abr$

PROCEDURE

Step 1   Compute the total sum of squares:

$$\text{SST} = \sum_{i=1}^{a} \sum_{j=1}^{b} \sum_{k=1}^{r} x_{ijk}^2 - \frac{T^2}{n_T}$$

Step 2   Compute the sum of squares due to factor A:

$$\text{SSA} = \frac{\sum_{i=1}^{a} T_{i\cdot}^2}{br} - \frac{T^2}{n_T}$$

Step 3   Compute the sum of squares due to factor B:

$$\text{SSB} = \frac{\sum_{j=1}^{b} T_{\cdot j}^2}{ar} - \frac{T^2}{n_T}$$

Step 4   Compute the sum of squares due to interaction:

$$\text{SSAB} = \frac{\sum_{i=1}^{a} \sum_{j=1}^{b} T_{ij}^2}{r} - \frac{T^2}{n_T} - \text{SSA} - \text{SSB}$$

Step 5   Compute the sum of squares due to error:

$$\text{SSE} = \text{SST} - \text{SSA} - \text{SSB} - \text{SSAB}$$

EXAMPLE

For the GMAT data in Table 12.14, these steps lead to the following sum of squares:

Step 1 $\text{SST} = 4{,}856{,}500 - \dfrac{(9270)^2}{18} = 82{,}450$

Step 2 $\text{SSA} = \dfrac{(2960)^2 + (3080)^2 + (3230)^2}{6} - \dfrac{(9270)^2}{18} = 6100$

Step 3 $\text{SSB} = \dfrac{(3240)^2 + (3360)^2 + (2670)^2}{6} - \dfrac{(9270)^2}{18} = 45{,}300$

Step 4 $\text{SSAB} = \dfrac{(1080)^2 + (1000)^2 + \cdots + (890)^2}{2} - \dfrac{(9270)^2}{18} - 6100 - 45{,}300 = 11{,}200$

Step 5 $\text{SSE} = 82{,}450 - 6100 - 45{,}300 - 11{,}200 = 19{,}850$

LINEAR REGRESSION

AND CORRELATION

WHAT YOU WILL LEARN IN THIS CHAPTER

- How to use the least squares method to develop an equation that estimates the relationship between two variables
- How to compute and interpret the coefficient of determination
- How to use the t and F distributions to test for significant relationships between variables
- How to use a regression equation for estimation and prediction
- How to test model assumptions using residual analysis
- How to use residual analysis to identify outliers and influential observations
- How to compute and interpret the sample correlation coefficient

CONTENTS

Relating the Change in Film Speed to the Film's Age

Polaroid markets a wide variety of instant photographic systems, cameras, components, and films for professional, industrial, scientific, and medical uses. Other businesses include magnetics, sunglasses, industrial polarizers, chemicals, custom coating, and holography.

Sensitometry is a field of special interest to Polaroid; it measures the sensitivity of photographic materials and provides information on many characteristics of film, such as its useful exposure range. Within Polaroid's central sensitometry laboratory, scientists systematically sample and analyze instant films that have been stored at temperature and humidity levels approximating those the films will be subjected to once in the hands of the consumers. To investigate the relationship between film speed and the age of a Polaroid extended range color professional print film, Polaroid's sensitometry lab selected film samples ranging in age from 1 to 13 months after manufacture. The data showed that film speed decreases with age, and that a straight-line or linear relationship could be used to approximate the relationship between change in film speed and age of the film.

Using regression analysis, Polaroid was able to develop the following equation, which relates the change in the film speed to the film's age:

$$\hat{y} = -19.8 - 7.6x$$

where

$\hat{y}$ = change in film speed
x = film age in months

This equation shows that the average decrease in film speed is 7.6 units per month. The information provided by this analysis, when coupled with consumer purchase and use patterns, enables Polaroid to make manufacturing adjustments that produce films with the performance levels their customers require.

*This *Statistics in Practice* was provided by Mr. Lawrence Friedman, Manager, Photographic Quality.

Regression analysis is a statistical procedure that can be used to develop a mathematical equation showing how variables are related. In regression terminology, the variable that is being predicted by the mathematical equation is called the *dependent* variable. The variable or variables being used to predict the value of the dependent variable are called the *independent* variables.

In this chapter we consider the simplest type of regression: situations involving one independent and one dependent variable for which the relationship between the variables is approximated by a straight line. This is called *simple linear regression.* Regression analysis involving two or more independent variables, called multiple regression analysis, is covered in Chapters 14 and 15.

Another topic discussed in this chapter is correlation. In correlation analysis, we are not concerned with identifying a mathematical equation relating an independent and dependent variable; we are concerned only with determining the extent to which the variables are linearly related. Correlation analysis is a procedure for making this determination and, if such a relationship exists, for providing a measure of the relative strength of the relationship.

We caution the reader before beginning this chapter that neither regression nor correlation analysis can be interpreted as establishing cause-and-effect relationships. Regression and correlation analyses can indicate only how or to what extent variables are associated with each other. Any conclusions about a cause-and-effect relationship must be based on the *judgment of the analyst.*

13.1 ▽ THE LEAST SQUARES METHOD

In this section we show how the least squares method can be used to develop a linear equation relating two variables. As an illustration, let us consider the problem currently being faced by the management of Armand's Pizza Parlors, a chain of Italian-food restaurants located in a five-state area. One of the most successful types of locations for Armand's has been near a college campus.

EXAMPLE 13.1 ▶ Prior to opening a new restaurant, Armand's management requires an estimate of annual sales revenues. Such an estimate is used in planning the appropriate restaurant capacity, making initial staffing decisions, and deciding whether the potential revenues justify the cost of the operation.

Suppose management believes that the size of the student population on the nearby campus is related to the annual sales revenues. On an intuitive basis, management believes restaurants located near larger campuses generate more revenue than those near small campuses. To evaluate the relationship between student population x and annual sales y, Armand's collected data from a sample of 10 of its restaurants located near college campuses. These data are summarized in Table 13.1. For the ith observation, x_i represents the value of the independent variable and y_i represents the value of the dependent variable. We see that restaurant 1, with $x_1 = 2$ and $y_1 = 58$, is located near a campus with 2000 students and has annual sales of \$58,000; restaurant 2 is located near a campus with 6000 students and has annual sales of \$105,000; and so on.

Figure 13.1 shows graphically the data presented in Table 13.1. The size of the student population is shown on the horizontal axis, with annual sales on the vertical axis. A graph such as this is known as a *scatter diagram*. Values for the independent variable are shown on the horizontal axis, and the corresponding values for the dependent variable are shown on the vertical axis. The scatter diagram provides an overview of the data and enables us to draw preliminary conclusions about a possible relationship between the variables.

What preliminary conclusions can we draw from Figure 13.1? It appears that low sales volumes are associated with small student populations and higher sales volumes are associated with larger student populations. It also appears that the relationship between

TABLE 13.1   STUDENT POPULATION AND ANNUAL SALES FOR 10 ARMAND'S PIZZA PARLORS

Restaurant i	Student Population (1000s) x_i	Annual Sales (\$1000s) y_i
1	2	58
2	6	105
3	8	88
4	8	118
5	12	117
6	16	137
7	20	157
8	20	169
9	22	149
10	26	202

FIGURE 13.1   SCATTER DIAGRAM OF ANNUAL SALES VERSUS STUDENT POPULATION FOR ARMAND'S PIZZA PARLORS

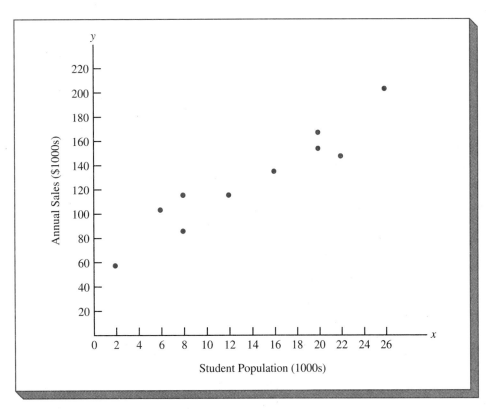

the two variables can be approximated by a straight line. In Figure 13.2 we have drawn a straight line through the data that appears to provide a good linear approximation of the relationship between the variables. The equation for this line is $y = 50 + 5.5x$. The y-intercept, the point at which the line intersects the y-axis, is 50, and the slope, the amount of change in y per unit change in x, is 5.5.

Clearly, there are many different straight lines that we could have drawn in Figure 13.2 to represent the relationship between x and y. The question is, Which of the straight lines that could be drawn best represents the relationship? ◀

The *least squares method* is a procedure that is used to find the straight line that provides the best approximation for the relationship between the independent and dependent variables. We refer to the equation of the line developed using the least squares method as the *estimated regression line,* or the *estimated regression equation.*

Estimated Regression Equation

$$\hat{y} = b_0 + b_1 x \tag{13.1}$$

where

b_0 = y-intercept of the line
b_1 = slope of the line
$\hat{y}$ = estimated value of the dependent variable

FIGURE 13.2   STRAIGHT-LINE APPROXIMATION FOR ARMAND'S PIZZA PARLORS

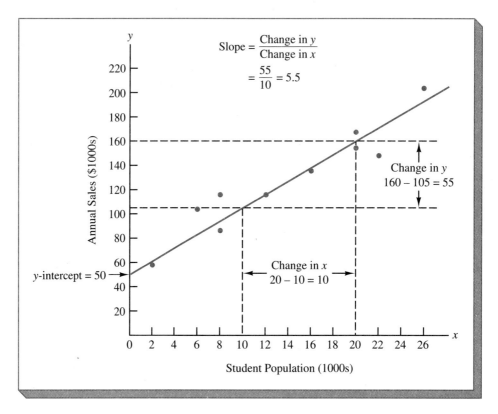

For any particular value of the independent variable x_i, the corresponding value on the estimated regression line is denoted by $\hat{y}_i = b_0 + b_1 x_i$.

Application of the least squares method provides the values of b_0 and b_1 that make the sum of the squared deviations between the observed values of the dependent variable y_i and the estimated values of the dependent variable $\hat{y}_i$ a minimum. The criterion for the least squares method is given by expression (13.2).

Least Squares Criterion

$$\text{Min } \Sigma(y_i - \hat{y}_i)^2 \tag{13.2}$$

where

y_i = observed value of the dependent variable for the ith observation

$\hat{y}_i$ = estimated value of the dependent variable for the ith observation

---

Using differential calculus, it can be shown (see the appendix to this chapter) that the values of b_0 and b_1 that minimize expression (13.2) can be found using equations (13.3) and (13.4).

Slope and *y*-Intercept for the Estimated Regression Equation

$$b_1 = \frac{\Sigma(x_i - \bar{x})(y_i - \bar{y})}{\Sigma(x_i - \bar{x})^2} = \frac{\Sigma x_i y_i - (\Sigma x_i \Sigma y_i)/n}{\Sigma x_i^2 - (\Sigma x_i)^2/n} \qquad \text{(13.3)}$$

$$b_0 = \bar{y} - b_1 \bar{x} \qquad \text{(13.4)}$$

where

x_i = value of the independent variable for the *i*th observation
y_i = value of the dependent variable for the *i*th observation
$\bar{x}$ = mean value for the independent variable
$\bar{y}$ = mean value for the dependent variable
n = total number of observations

The second form of equation (13.3) is normally used for computing b_1 with a calculator because it avoids the tedious calculations involving the computation of each $(x_i - \bar{x})$ and $(y_i - \bar{y})$. However, to avoid rounding errors, it is best to carry as many significant digits as possible in the calculation; we recommend carrying at least four significant digits.

EXAMPLE 13.1
(CONTINUED)

Some of the calculations necessary to develop the least squares estimated regression equation for Armand's Pizza Parlors are shown in Table 13.2. In this example, there are 10 restaurants, or observations; hence $n = 10$. Using equations (13.3) and (13.4), we can now compute the slope and intercept of the estimated regression equation for Armand's Pizza Parlors. The calculation of the slope (b_1) proceeds as follows:

$$b_1 = \frac{\Sigma x_i y_i - (\Sigma x_i \Sigma y_i)/n}{\Sigma x_i^2 - (\Sigma x_i)^2/n}$$

$$= \frac{21{,}040 - (140)(1300)/10}{2528 - (140)^2/10}$$

$$= \frac{2840}{568}$$

$$= 5$$

The calculation of the *y*-intercept (b_0) is as follows:

$$\bar{x} = \frac{\Sigma x_i}{n} = \frac{140}{10} = 14$$

$$\bar{y} = \frac{\Sigma y_i}{n} = \frac{1300}{10} = 130$$

$$b_0 = \bar{y} - b_1 \bar{x}$$

$$= 130 - 5(14)$$

$$= 60$$

Thus, the estimated regression equation found by using the method of least squares is

$$\hat{y} = 60 + 5x$$

TABLE 13.2   CALCULATIONS FOR THE LEAST SQUARES ESTIMATED REGRESSION EQUATION FOR ARMAND'S PIZZA PARLORS

Restaurant i	x_i	y_i	$x_i y_i$	x_i^2
1	2	58	116	4
2	6	105	630	36
3	8	88	704	64
4	8	118	944	64
5	12	117	1,404	144
6	16	137	2,192	256
7	20	157	3,140	400
8	20	169	3,380	400
9	22	149	3,278	484
10	26	202	5,252	676
Totals	140	1,300	21,040	2,528
	Σx_i	Σy_i	$\Sigma x_i y_i$	Σx_i^2

We show the graph of this equation in Figure 13.3.

The slope of the estimated regression equation ($b_1 = 5$) is positive, implying that as student population increases, annual sales increase. In fact, we can conclude (since sales are measured in \$1000s and student population in 1000s) that an increase in the student population of 1000 is associated with an increase of \$5000 in expected annual sales; that is, sales are expected to increase by \$5.00 per student.

If we believe that the least squares estimated regression equation adequately describes the relationship between x and y, then it would seem reasonable to use the estimated regression equation to predict the value of y for a given value of x. For example, if we wanted to predict annual sales for a restaurant to be located near a campus with 16,000 students, we would compute

$$\hat{y} = 60 + 5(16)$$

$$= 140$$

Hence, we would predict sales of \$140,000 per year. ◄

In the following sections we will discuss methods for assessing the appropriateness of using the estimated regression equation for estimation and prediction.

NOTES AND COMMENTS

The least squares method provides an estimated regression equation that minimizes the sum of squared deviations between the observed values of the dependent variable y_i and the estimated values of the dependent variable $\hat{y}_i$. This is the least squares criterion for choosing the equation that provides the best fit. If some other criterion were used, such as minimizing the sum of the absolute deviations between y_i and $\hat{y}_i$, a different equation would be obtained. In practice, the least squares method is the most widely used.

FIGURE 13.3 GRAPH OF THE ESTIMATED REGRESSION EQUATION FOR ARMAND'S PIZZA PARLORS: $\hat{y} = 60 + 5x$

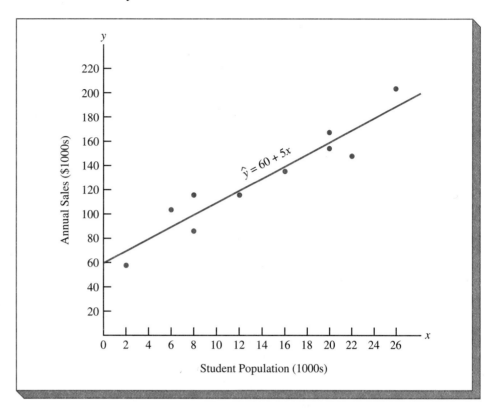

EXERCISES

METHODS

SELF TEST

1. Given are five observations for two variables, x and y.

x_i	1	2	3	4	5
y_i	3	7	5	11	14

a. Develop a scatter diagram for these data.
b. What does the scatter diagram developed in part (a) indicate about the relationship between the two variables?
c. Try to approximate the relationship between x and y by drawing a straight line through the data.
d. Develop the estimated regression equation by computing the values of b_0 and b_1 using equations (13.3) and (13.4).
e. Use the estimated regression equation to predict the value of y when $x = 4$.

2. Given are five observations for two variables, x and y.

x_i	2	3	5	1	8
y_i	25	25	20	30	16

   a. Develop a scatter diagram for these data.

   b. What does the scatter diagram developed in part (a) indicate about the relationship between the two variables?

   c. Try to approximate the relationship between x and y by drawing a straight line through the data.

   d. Develop the estimated regression equation by computing the values of b_0 and b_1 using equations (13.3) and (13.4).

   e. Use the estimated regression equation to predict the value of y when $x = 6$.

3. Given below are five observations collected in a regression study on two variables.

x_i	2	4	5	7	8
y_i	2	3	2	6	4

   a. Develop a scatter diagram for these data.

   b. Develop the estimated regression equation for these data.

   c. Use the estimated regression equation to predict the value of y when $x = 4$.

APPLICATIONS

SELF TEST

4. The following data were collected on the height (inches) and weight (pounds) of women swimmers.

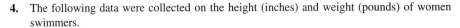

Height	68	64	62	65	66
Weight	132	108	102	115	128

   a. Develop a scatter diagram for these data with height as the independent variable.

   b. What does the scatter diagram developed in part (a) indicate about the relationship between the two variables?

   c. Try to approximate the relationship between height and weight by drawing a straight line through the data.

   d. Develop the estimated regression equation by computing the values of b_0 and b_1 using equations (13.3) and (13.4).

   e. If a swimmer's height is 63 inches, what would you estimate her weight to be?

5. The data shown below were collected on the monthly starting salaries and the grade point averages (GPA) for undergraduate students who had obtained a degree in political science.

GPA	Monthly Salary ($)	GPA	Monthly Salary ($)
2.6	1400	3.2	1600
3.4	1700	3.5	2000
3.6	2100	2.9	1700

   a. Develop a scatter diagram for these data with GPA as the independent variable.

   b. What does the scatter diagram developed in part (a) indicate about the relationship between the two variables?

   c. Draw a straight line through the data to approximate a linear relationship between GPA and salary.

   d. Use the least squares method to develop the estimated regression equation.

   e. Predict the monthly starting salary for a student with a 3.0 GPA and for a student with a 3.5 GPA.

6. Performance data for a Century Coronado 21 with a 310-hp MerCruiser V-8 gasoline inboard engine was reported in *Boating*, September 1991. Data on how the boat speed in miles per hour (mph) affected fuel consumption in gallons per hour (gph) are shown below.

Speed (mph)	6.1	10.7	20.9	27.5	31.5
Fuel Consumption (gph)	2.3	4.8	7.5	9.2	12.4

   a. Develop a scatter diagram for these data.
   b. What does the scatter diagram developed in part (a) indicate about the relationship between these two variables?
   c. Develop the estimated regression equation showing how fuel consumption is related to boat speed.
   d. What is the estimated fuel consumption if the boat speed is 25 mph?

7. Eddie's Restaurants collected the data shown below on the relationship between advertising and sales (both in $1000s) using a sample of seven restaurants. Develop a scatter diagram for these data with advertising expenditures as the independent variable and sales as the dependent variable. Does there appear to be a linear relationship?

Advertising Expenditures	Sales	Advertising Expenditures	Sales
1.0	19.0	10.0	52.0
2.0	32.0	14.0	53.0
4.0	44.0	20.0	54.0
6.0	40.0		

8. *Consumer Reports* uses a survey to collect data on the annual cost of repairs for over 300 makes and models of automobiles (*Consumer Reports 1992 Buying Guide*). The following data show the average annual repair cost ($) and the age of the automobile (years).

Age	1	2	3	4	5
Repair	135	175	320	300	450

   a. Develop the estimated regression equation showing how annual repair cost is related to the age of an automobile.
   b. What is the estimated annual repair cost for a 3-year-old automobile?

9. A sales manager collected the following data concerning annual sales and years of experience.

Salesperson	Years of Experience	Annual Sales ($1000s)	Salesperson	Years of Experience	Annual Sales ($1000s)
1	1	80	6	8	111
2	3	97	7	10	119
3	4	92	8	10	123
4	4	102	9	11	117
5	6	103	10	13	136

a. Develop a scatter diagram for these data with years of experience as the independent variable.

b. Develop the estimated regression equation that can be used to predict annual sales given the years of experience.

c. Use the estimated regression equation to predict annual sales for a salesperson with 9 years of experience.

10. Tyler Realty collected the following data regarding the selling price of new homes and the size of the homes measured in terms of square footage of living space.

Square Footage	Selling Price
2500	$124,000
2400	$108,000
1800	$ 92,000
3000	$146,000
2300	$110,000

a. Develop a scatter diagram for these data with square footage on the horizontal axis.

b. Try to approximate the relationship between square footage and selling price by drawing a straight line through the data.

c. Does there appear to be a linear relationship?

d. Develop an estimated regression equation using the least squares method.

e. Predict the selling price for a home with 2700 square feet.

11. The following data show the number of golf courses and the number of paid rounds of golf (in millions) for the Myrtle Beach, South Carolina, area over an 8-year period (*Myrtle Beach Magazine*, October 1991).

Number of Golf Courses	Number of Paid Rounds of Golf	Number of Golf Courses	Number of Paid Rounds of Golf
26	1.0	33	1.4
30	1.1	35	1.6
31	1.2	38	1.8
32	1.3	43	2.0

a. Develop a scatter diagram for these data.

b. Does there appear to be a linear relationship?

c. Develop the estimated regression equation relating the number of paid rounds of golf to the number of golf courses.

d. Suppose that next year 57 golf courses will be available. What is an estimate of the number of paid rounds of golf? Discuss any potential problem associated with making this estimate.

12. The following data show the percentage of women working in each company (*x*) and the percentage of management jobs held by women in that company (*y*); the data shown represent companies in retailing and trade (*Louis Rukeyser's Business Almanac*).

Company	x_i	y_i
Federated Department Stores	72	61
Kroger	47	16
Marriott	51	32
McDonald's	57	46
Sears, Roebuck	55	36

a. Develop a scatter diagram for these data.
b. What does the scatter diagram developed in part (a) indicate about the relationship between x and y?
c. Develop the estimated regression equation for these data.
d. Predict the percentage of management jobs held by women in a company that has 60% women employees.
e. Use the estimated regression equation to predict the percentage of management jobs held by women in a company where 55% of the jobs are held by women. How does this predicted value compare to the 36% value observed for Sears, Roebuck, a company in which 55% of the employees are women?

13. A university medical center has developed a test designed to measure a patient's stress level. The test is designed so that higher scores on the test correspond to higher levels of stress. As part of a research study, the blood pressure (low reading) of patients who took the test was recorded. The following data were obtained.

Stress Test Score	Blood Pressure	Stress Test Score	Blood Pressure
53	70	73	78
94	91	82	85
64	78	90	84

a. Develop a scatter diagram for these data with stress test score on the horizontal axis. Does a linear relationship between the two variables appear to be appropriate?
b. Develop the estimated regression equation for these data.
c. Estimate an individual's blood pressure if he or she scored 85 on the stress test.

14. The annual growth rate and cash flow for nine pharmaceutical companies are shown below (*Forbes,* May 1989).

Firm	Annual Growth Rate (%)	Cash Flow ($1,000,000s)
Pfizer	12	986
Bristol Meyers	14	957
Merck	16	1412
American Home Products	11	1074
Abbott Laboratories	18	1023
Eli Lilly	11	965
SmithKline Beckman	3	450
Upjohn	10	456
Warner-Lambert	7	437

    a. Develop a scatter diagram for these data, plotting cash flow on the vertical axis. Does it appear that the two variables are related?

    b. Use the least squares method to fit a straight line to the data.

    c. Estimate the cash flow for a company that has a 10% growth rate. How does the predicted value compare with the observed results for Upjohn?

13.2 ▽ THE COEFFICIENT OF DETERMINATION

In Section 13.1 we developed the estimated regression equation $\hat{y} = 60 + 5x$ to approximate the linear relationship between student population x and annual sales y for Armand's Pizza Parlors. A question that might occur is, How good is the fit? In this section we show that the *coefficient of determination* provides a measure of goodness of fit of the estimated regression equation to the data.

Recall that the least squares method is a technique for finding the values of b_0 and b_1 that minimize the sum of squared deviations between the observed values of the dependent variable y_i and the predicted values of the dependent variable $\hat{y}_i$. The difference between y_i and $\hat{y}_i$ represents the error in using $\hat{y}_i$ to estimate y_i; the difference for the ith observation is $y_i - \hat{y}_i$. This difference is referred to as the ith *residual*. Thus, the sum of squares minimized by the least squares method is referred to as the sum of squares due to error, or the residual sum of squares. We use SSE to represent this quantity.

Sum of Squares Due to Error

$$\text{SSE} = \Sigma(y_i - \hat{y}_i)^2 \qquad \qquad \textbf{(13.5)}$$

EXAMPLE 13.1
(CONTINUED) ▶ Table 13.3 shows the calculations required to compute SSE for Armand's Pizza Parlors. The entry SSE = 1530 is a measure of the error involved in using the estimated regression equation $\hat{y} = 60 + 5x$ to predict the y_i values.

TABLE 13.3   CALCULATION OF SSE FOR ARMAND'S PIZZA PARLORS

Restaurant i	x_i = Student Population (1000s)	y_i = Annual Sales ($1000s)	$\hat{y}_i = 60 + 5x_i$	$y_i - \hat{y}_i$	$(y_i - \hat{y}_i)^2$
1	2	58	70	−12	144
2	6	105	90	15	225
3	8	88	100	−12	144
4	8	118	100	18	324
5	12	117	120	−3	9
6	16	137	140	−3	9
7	20	157	160	−3	9
8	20	169	160	9	81
9	22	149	170	−21	441
10	26	202	190	12	144
					SSE = 1530

Now suppose that we were asked to develop an estimate of annual sales without using the size of the student population. We could not use the estimated regression equation and would have to use the value of the sample mean, $\bar{y} = 130$, as the best estimate of annual sales. In Table 13.4 we show the errors that would result from using $\bar{y}$ to estimate annual sales at the 10 Armand's Pizza Parlors. The corresponding sum of squares about the mean, denoted by SST, is as follows:

Total Sum of Squares

$$SST = \Sigma(y_i - \bar{y})^2 \qquad (13.6)$$

For Armand's Pizza Parlors, SST = 15,730. ◀

In Figure 13.4 we show the least squares regression line $\hat{y} = 60 + 5x$ and the line corresponding to $\bar{y} = 130$. Note that, in general, the points cluster more closely around the estimated regression line than they do about the line $\bar{y} = 130$. For example, for the 10th restaurant we see that the error is much larger when $\bar{y} = 130$ is used as an estimate of y_{10} than when $\hat{y}_{10} = 190$ is used. We can think of SST as a measure of how well the observations cluster about the $\bar{y}$ line and SSE as a measure of how well the observations cluster about the $\hat{y}$ line.

To measure how much the predicted values $\hat{y}$ on the estimated regression line deviate from $\bar{y}$, another sum of squares is computed. This sum of squares is called the *sum of squares due to regression* and is denoted by SSR. The sum of squares due to regression can be written as follows.

Sum of Squares Due to Regression

$$SSR = \Sigma(\hat{y}_i - \bar{y})^2 \qquad (13.7)$$

TABLE 13.4   COMPUTATION OF THE TOTAL SUM OF SQUARES FOR ARMAND'S PIZZA PARLORS

Restaurant i	x_i = Student Population (1000s)	y_i = Annual Sales ($1000s)	$y_i - \bar{y}$	$(y_i - \bar{y})^2$
1	2	58	−72	5,184
2	6	105	−25	625
3	8	88	−42	1,764
4	8	118	−12	144
5	12	117	−13	169
6	16	137	7	49
7	20	157	27	729
8	20	169	39	1,521
9	22	149	19	361
10	26	202	72	5,184
			SST =	15,730

FIGURE 13.4   **DEVIATIONS ABOUT THE ESTIMATED REGRESSION LINE AND THE LINE**
$y = \bar{y}$ **FOR ARMAND'S PIZZA PARLORS**

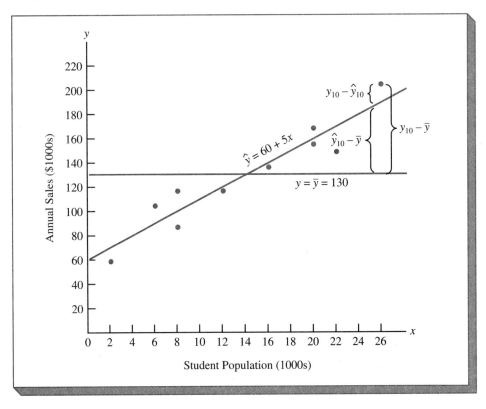

From the preceding discussion, we should expect that SSE, SST, and SSR are related. Indeed, they are. The relationship among SSE, SST, and SSR is shown below.

Relationship Among SST, SSR, and SSE

$$SST = SSR + SSE \tag{13.8}$$

where

$$SST = \text{total sum of squares}$$
$$SSR = \text{sum of squares due to regression}$$
$$SSE = \text{sum of squares due to error}$$

EXAMPLE 13.1
(CONTINUED)

Using equation (13.8), we can conclude that the sum of squares due to regression for Armand's Pizza Parlors is

$$SSR = SST - SSE = 15{,}730 - 1530 = 14{,}200$$

Now let us see how these sums of squares can be used to provide a measure of the goodness of fit for the regression relationship. We would have the best possible fit if every observation happened to lie on the least squares line; the line would pass through each point, and we would have SSE = 0. Hence, for a perfect fit, SSR must equal SST, and thus the ratio SSR/SST = 1. On the other hand, a poorer fit to the observed data results in a larger SSE. Since SST = SSR + SSE, however, the largest SSE (and hence worst fit) occurs when SSR = 0. In this case, the estimated regression equation does not help predict y. Thus, the worst possible fit yields the ratio SSR/SST = 0.

If we were to use the ratio SSR/SST to evaluate the goodness of fit for the regression relationship, we would have a measure that could take on values between 0 and 1, with values closer to 1 implying a better fit. The fraction SSR/SST is called the *coefficient of determination* and is denoted r^2.

Coefficient of Determination

$$r^2 = \frac{\text{SSR}}{\text{SST}}$$

(13.9)

EXAMPLE 13.1
(CONTINUED)

The value of the coefficient of determination for Armand's Pizza Parlors is

$$r^2 = \frac{\text{SSR}}{\text{SST}} = \frac{14,200}{15,730} = .903$$

◀

To better interpret r^2, we can think of SST as the measure of how good $\bar{y}$ is as a predictor of annual sales volume. After developing the estimated regression equation, we compute SSE as the measure of the goodness of $\hat{y}$ as a predictor of annual sales volume. Thus, SSR (the difference between SST and SSE) really measures the portion of SST that is explained by the estimated regression equation. We can think of r^2 as

$$r^2 = \frac{\text{Sum of Squares Explained by Regression}}{\text{Total Sum of Squares}}$$

When it is expressed as a percentage, r^2 can be interpreted as the percentage of the total sum of squares (SST) that can be explained using the estimated regression equation.

EXAMPLE 13.1
(CONTINUED)

For Armand's Pizza Parlors, we conclude that the estimated regression equation has accounted for 90.3% of the total sum of squares. We should be very pleased with such a good fit. ◀

COMPUTATIONAL EFFICIENCIES

When using a calculator to compute the value of the coefficient of determination, computational efficiencies can be realized by computing SSR directly using the following formula:

$$\text{SSR} = \frac{[\Sigma x_i y_i - (\Sigma x_i \Sigma y_i)/n]^2}{\Sigma x_i^2 - (\Sigma x_i)^2/n}$$

(13.10)

In addition, we need not compute SST using the expression $\Sigma(y_i - \bar{y})^2$; this expression can be algebraically expanded to provide

$$\text{SST} = \Sigma y_i^2 - \frac{(\Sigma y_i)^2}{n}$$

(13.11)

TABLE 13.5   **CALCULATIONS USED IN COMPUTING SSR AND SST FOR ARMAND'S PIZZA PARLORS**

Restaurant i	x_i	y_i	$x_i y_i$	x_i^2	y_i^2
1	2	58	116	4	3,364
2	6	105	630	36	11,025
3	8	88	704	64	7,744
4	8	118	944	64	13,924
5	12	117	1,404	144	13,689
6	16	137	2,192	256	18,769
7	20	157	3,140	400	24,649
8	20	169	3,380	400	28,561
9	22	149	3,278	484	22,201
10	26	202	5,252	676	40,804
Totals	140	1,300	21,040	2,528	184,730
	Σx_i	Σy_i	$\Sigma x_i y_i$	Σx_i^2	Σy_i^2

EXAMPLE 13.1 ▶
(**CONTINUED**)

For Armand's Pizza Parlors, part of the calculations needed to compute SSR and SST using the above formulas are shown in Table 13.5. Using the values in this table along with equations (13.10) and (13.11), we can compute SSR and SST as follows:

$$SSR = \frac{[21,040 - (140)(1300)/10]^2}{2528 - (140)^2/10}$$

$$= \frac{8,065,600}{568}$$

$$= 14,200$$

$$SST = 184,730 - \frac{(1300)^2}{10}$$

$$= 15,730$$

Note that since these are equivalent formulas, we get the same values for SSR, SST, and r^2 that we obtained previously:

$$r^2 = \frac{SSR}{SST} = \frac{14,200}{15,730} = .903$$ ◀

NOTES AND COMMENTS

1. In developing the least squares estimated regression equation and computing the coefficient of determination, no probabilistic assumptions and no statistical inferences have been made. Larger values of r^2 simply imply that the least squares line provides a better fit to the data; that is, the observations are more closely grouped about the least squares line. But, using only r^2, no conclusion can be made regarding whether the relationship between x and y is statistically significant. Such a conclusion must be based on considerations that involve the sample size and the properties of the appropriate sampling distributions of the least squares estimators.

—Continues on next page

—*Continued from previous page*

2. As a practical matter, for typical data found in the social sciences, values of r^2 as low as .25 are often considered useful. For data in the physical and medical sciences, r^2 values of .60 or greater are often found; in fact, in some cases, r^2 values greater than .90 can be found.

EXERCISES

METHODS

SELF TEST

15. The data from Exercise 1 are shown below.

x_i	1	2	3	4	5
y_i	3	7	5	11	14

The estimated regression equation for these data is $\hat{y} = .20 + 2.60x$.
a. Compute SSE, SST, and SSR using equations (13.5), (13.6), and (13.8).
b. Compute the coefficient of determination r^2. Comment on the goodness of fit.
c. Recompute SSR and SST using equations (13.10) and (13.11). Do you get the same results as in part (a)?

16. The data from Exercise 2 are shown below.

x_i	2	3	5	1	8
y_i	25	25	20	30	16

The estimated regression equation for these data is $\hat{y} = 30.33 - 1.88x$.
a. Compute SSE, SST, and SSR.
b. Compute the coefficient of determination r^2. Comment on the goodness of fit.

17. The data from Exercise 3 are shown below.

x_i	2	4	5	7	8
y_i	2	3	2	6	4

The estimated regression equation for these data is $\hat{y} = .75 + .51x$. What percentage of the total sum of squares can be accounted for by the estimated regression equation?

APPLICATIONS

SELF TEST

18. In Exercise 5 data were collected regarding the monthly salaries y and the grade point averages x for undergraduate students who had obtained a degree in political science. The data from Exercise 5 are shown below. The estimated regression equation for these data is $\hat{y} = -109.46 + 581.08x$.

GPA	Monthly Salary ($)	GPA	Monthly Salary ($)
2.6	1400	3.2	1600
3.4	1700	3.5	2000
3.6	2100	2.9	1700

    a. Compute SSE, SST, and SSR.

    b. Compute the coefficient of determination r^2. Comment on the goodness of fit.

19. The data from Exercise 7 (both in $1000s) are shown below.

Advertising Expenditures (x_i)	Sales (y_i)	Advertising Expenditures (x_i)	Sales (y_i)
1.0	19.0	10.0	52.0
2.0	32.0	14.0	53.0
4.0	44.0	20.0	54.0
6.0	40.0		

The estimated regression equation for these data is $\hat{y} = 29.4 + 1.55x$. What percentage of the total sum of squares can be accounted for by the estimated regression equation? Comment on the goodness of fit.

20. A medical laboratory at Duke University estimates the amount of protein in liver samples through the use of regression analysis. A spectrometer emitting light shines through a substance containing the sample, and the amount of light absorbed is used to estimate the amount of protein in the sample. A new estimated regression equation is developed daily because of differing amounts of dye in the solution. On one day six samples with known protein concentrations gave the following absorbence readings.

Absorbence Reading (x_i)	Milligrams of Protein (y_i)	Absorbence Reading (x_i)	Milligrams of Protein (y_i)
.509	0	1.400	80
.756	20	1.570	100
1.020	40	1.790	127

    a. Use these data to develop an estimated regression equation relating the light absorbence reading to milligrams of protein present in the sample.

    b. Compute r^2. Would you feel comfortable using the estimated regression equation developed in part (a) to estimate the amount of protein in a sample?

    c. In a sample just received, the light absorbence reading was .941. Estimate the amount of protein in the sample.

21. A list of the best-selling cars for 1987 whose sales in units varied between 175,000 and 300,000 (rounded to the nearest thousand) is shown in the following table (*The World Almanac*, 1989). The 1988 suggested retail price (in thousands of dollars, rounded to the nearest hundred dollars) is also shown.

Model	1988 Price ($1000s)	Number Sold ($1000s)
Hyundai	5.4	264
Oldsmobile Ciera	11.4	245
Nissan Sentra	6.4	236
Ford Tempo	9.1	219
Chevrolet Corsica/Beretta	10.0	214
	—Table continues on next page	

Model	1988 Price ($1000s)	Number Sold ($1000s)
Pontiac Grand Am	10.3	211
Toyota Camry	11.2	187
Chevrolet Caprice	12.5	177

a. Use these data to develop an estimated regression equation that could be used to predict the number sold given the price.

b. Compute r^2. Would you feel comfortable using the estimated regression equation developed in part (a) to estimate the number sold given the price? Explain.

13.3 ▽ THE REGRESSION MODEL AND ITS ASSUMPTIONS

An important concept that must be understood before we consider testing for significance in regression analysis involves the distinction between a *deterministic model* and a *probabilistic model*. In a deterministic model, the relationship between the dependent variable y and the independent variable x is such that if we specify the value of the independent variable, the value of the dependent variable can be determined *exactly*.

EXAMPLE 13.2 ▶ To illustrate a relationship between two variables that is deterministic, suppose that a major oil company leases a service station under a contractual agreement of $500 per month plus 10% of the gross sales. The relationship between the dealer's monthly payment y and the gross sales value x can be expressed as

$$y = 500 + .10x$$

With this relationship, a June gross sales of $6000 would provide a monthly payment of $y = 500 + .10(6000) = \$1100$, and a July gross sales of $7200 would provide a monthly payment of $y = 500 + .10(7200) = \$1220$. This type of relationship is deterministic: Once the gross sales value x is specified, the monthly payment y is determined exactly. Figure 13.5 shows graphically the relationship between gross sales and monthly payment. ◀

EXAMPLE 13.1 ▶
(CONTINUED)

To illustrate a relationship between two variables that is probabilistic rather than deterministic, recall the Armand's Pizza Parlors situation; the data for this problem are presented in Table 13.1. Note that restaurants 3 and 4 are both located near college campuses having 8000 students. Restaurant 3 shows annual sales of $88,000; however, restaurant 4 shows annual sales of $118,000. Thus, we see that the relationship between y and x cannot be deterministic, since different values of y are observed for the same value of x. Note that this is also the case for restaurants 7 and 8, where a campus size of 20,000 students generates annual sales of $157,000 for restaurant 7 and $169,000 for restaurant 8. Since the value of y cannot be determined exactly from the value of x, we say that the model relating x and y is *probabilistic*.

Next, let us reconsider Figure 13.1, the scatter diagram for the Armand's Pizza Parlors data. We concluded that the relationship between student population x and annual sales y could be approximated by a straight line. As a result, we used the least squares method to develop the following estimated regression equation:

$$\hat{y} = 60 + 5x$$

FIGURE 13.5   ILLUSTRATION OF A DETERMINISTIC RELATIONSHIP

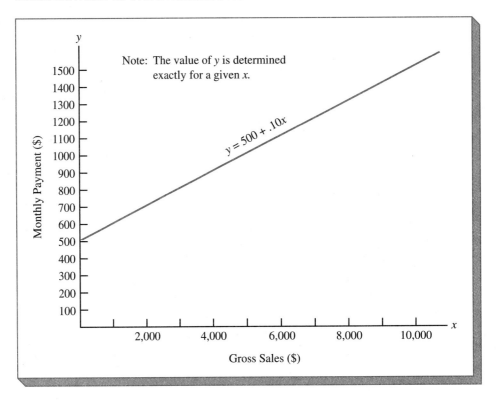

However, when we graphed this estimated regression equation in Figure 13.3, we saw that the relationship described by this equation was not perfect; that is, none of the observations fell exactly on the regression line.

Since we are unable to guarantee a single value of y for each value of x, the underlying relationship for Armand's Pizza Parlors can be explained only with a probabilistic model. Based on the observation that the relationship between student population and annual sales can be approximated by a straight line, we now make the assumption that the following probabilistic model—referred to as the *regression model*—is a realistic representation of the true relationship between the two variables.

Regression Model

$$y = \beta_0 + \beta_1 x + \epsilon \qquad\qquad (13.12)$$

where

$\beta_0 = y$-intercept of the line given by $\beta_0 + \beta_1 x$

$\beta_1 = $ the slope of the line given by $\beta_0 + \beta_1 x$

$\epsilon = $ the error or deviation of the actual y value from the line given by $\beta_0 + \beta_1 x$

Using equation (13.12) as a model of the relationship between x and y, we are saying that we believe the two variables are related in such a fashion that the line given by $\beta_0 + \beta_1 x$ provides a good approximation of the y value at each x. However, to identify the exact value of y, we must also consider the error term ϵ (the Greek letter epsilon), which is the measure of how far the actual y value is above or below the line $\beta_0 + \beta_1 x$. In the regression model, the independent variable x is treated as being known; the model is used to predict y given knowledge of x. We refer to β_0 (the y-intercept) and β_1 (the slope) as the *parameters* of the model.

The following assumptions are made about the error term ϵ in the regression model $y = \beta_0 + \beta_1 x + \epsilon$.

Assumptions About the Error Term ϵ in the Regression Model

$$y = \beta_0 + \beta_1 x + \epsilon$$

1. The error term ϵ is a random variable with a mean or expected value of 0; that is, $E(\epsilon) = 0$.
   Implication: Since β_0 and B_1 are constants, $E(\beta_0) = \beta_0$ and $E(\beta_1) = \beta_1$; thus for a given value of x, the expected value of y is

   $$E(y) = \beta_0 + \beta_1 x \qquad (13.13)$$

   Equation (13.13) is referred to as the *regression equation.*
2. The variance of ϵ, denoted by σ^2, is the same for all values of x.
   Implication: The variance of y equals σ^2 and is the same for all values of x.
3. The values of ϵ are independent.
   Implication: The value of ϵ for a particular value of x is not related to the value of ϵ for any other value of x; thus, the value of y for a particular value of x is not related to the value of y for any other value of x.
4. The error term ϵ is a normally distributed random variable.
   Implication: Since y is a linear function of ϵ, y is also a normally distributed random variable.

Figure 13.6 illustrates the model assumptions and their implications; note that in this graphical interpretation, the value of $E(y)$ depends upon the specific value of x considered. However, regardless of the x value, the probability distribution of ϵ and hence the probability distribution of y are normally distributed, each with the same variance. The specific value of the error ϵ at any particular point depends on whether the actual value of y is greater than or less than $E(y)$.

At this point, we must keep in mind that we are also making an assumption or hypothesis about the relationship between x and y. That is, we have assumed that a straight line represented by $\beta_0 + \beta_1 x$ is the basis for the relationship between the variables. We must not lose sight of the fact that some other model, for instance $\beta_0 + \beta_1 x^2$, may turn out to be a better model for the underlying relationship. After using the sample data to estimate the parameters of the regression model (β_0 and β_1), we will want to conduct further analysis to determine whether the model assumed or hypothesized appears to be valid.

FIGURE 13.6   ASSUMPTIONS FOR THE REGRESSION MODEL

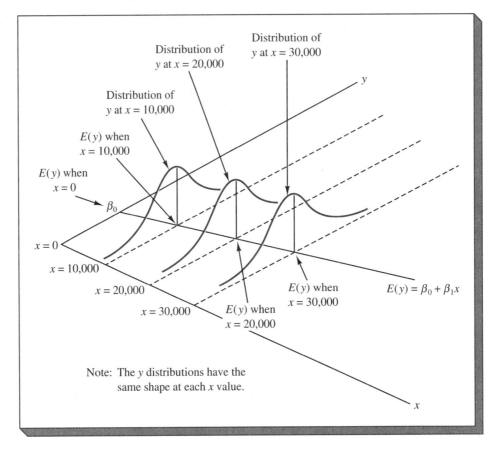

THE RELATIONSHIP BETWEEN THE REGRESSION EQUATION AND THE ESTIMATED REGRESSION EQUATION

Recall from Chapter 8 that when data were available for just one variable, the objective was to use a sample statistic (e.g., the sample mean) to make inferences about the corresponding population parameter (e.g., population mean). When we discussed the least squares method in Section 13.1, we presented formulas for computing the y-intercept (b_0) and the slope (b_1) of the estimated regression equation. The value of b_0 is a sample statistic that provides an estimate of the β_0 parameter in the regression model, and the value of b_1 is a sample statistic that provides an estimate of the β_1 parameter. Since the regression equation is $E(y) = \beta_0 + \beta_1 x$, the best estimate of the regression equation is provided by the estimated regression equation $\hat{y} = b_0 + b_1 x$; thus, we say that $\hat{y}$ provides the best estimate of $E(y)$. Figure 13.7 summarizes these concepts.

13.4 ▽ TESTING FOR SIGNIFICANCE

In Section 13.2 we saw how the coefficient of determination (r^2) could be used as a measure of the goodness of fit of the estimated regression line. Larger values of r^2 indicated a better fit. However, the value of r^2 does not allow us to conclude whether a

FIGURE 13.7   ESTIMATING THE POPULATION REGRESSION EQUATION USING SAMPLE DATA

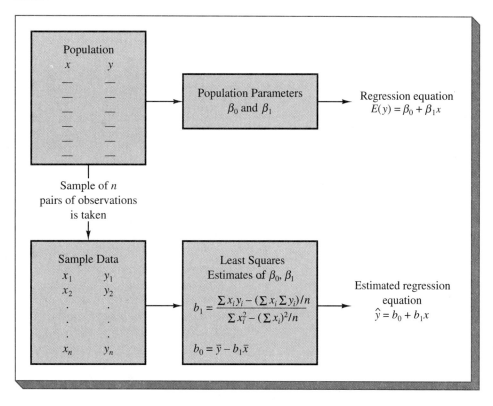

regression relationship is statistically significant. To draw conclusions concerning statistical significance, we must, among other things, take the sample size into consideration. In this section we show how to conduct significance tests that will allow us to draw conclusions about the existence of a regression relationship.

AN ESTIMATE OF σ^2

As stated in Section 13.3, σ^2 is the variance of the error term ϵ in the regression model $y = \beta_0 + \beta_1 x + \epsilon$. In the following discussion, we show how to obtain an estimate of σ^2 using the sum of squares due to error, SSE. First, recall that SSE is a measure of the variability of the actual observations about the estimated regression equation. With $\hat{y}_i = b_0 + b_1 x_i$, SSE can be written:

$$SSE = \Sigma(y_i - \hat{y}_i)^2 = \Sigma(y_i - b_0 - b_1 x_i)^2$$

Every sum of squares has associated with it a number called its degrees of freedom. The degrees of freedom indicates how many independent pieces of information involving the n independent values $y_1, y_2, \ldots, y_n$ are used to compute the sum of squares. Statisticians have shown that SSE has $n - 2$ degrees of freedom since two parameters (β_0 and β_1) have to be estimated in order to compute SSE.

Mean square is a number computed by dividing a sum of squares by its degrees of freedom. Thus, the mean square due to error, also referred to as mean square error, is computed by dividing SSE by its degrees of freedom, $n - 2$. Statisticians have shown that the mean square error, denoted MSE, provides an unbiased estimate of σ^2. Since MSE is an estimate of σ^2, the notation s^2 is also used.

Mean Square Error (Estimate of σ^2)

$$s^2 = \text{MSE} = \frac{\text{SSE}}{n - 2} \qquad (13.14)$$

EXAMPLE 13.1
(CONTINUED)

From the data in Table 13.3, we see that SSE = 1530; thus, for Armand's Pizza Parlors we have

$$s^2 = \text{MSE} = \frac{1530}{8} = 191.25$$

An unbiased estimate of σ^2 is therefore equal to $s^2 = 191.25$. To estimate σ, we take the square root of s^2. The resulting value s is referred to as the *standard error of the estimate*.

Standard Error of the Estimate

$$s = \sqrt{\text{MSE}} = \sqrt{\frac{\Sigma(y_i - \hat{y}_i)^2}{n - 2}} \qquad (13.15)$$

For Armand's Pizza Parlors, $s = \sqrt{\text{MSE}} = \sqrt{191.25} = 13.829$. In the discussion that follows, we will use this estimate of σ in tests for the significance of the regression equation.

t TEST

Recall that the regression equation is assumed to be $E(y) = \beta_0 + \beta_1 x$. If there really exists a relationship of this form between x and y in which the value of x influences the value of y, β_1 could not equal 0. Thus, to test for a significant relationship between the two variables, we use the following hypotheses:

$$H_0: \beta_1 = 0$$

$$H_a: \beta_1 \neq 0$$

Before using the t test for these hypotheses, we need to consider the properties of b_1, the least squares estimator of β_1.

First, let us consider what would have happened if we had used a different random sample for the same regression study. For example, suppose that in the Armand's Pizza Parlors example we had used the sales records of a different sample of 10 restaurants. A regression analysis of this new sample might result in an estimated regression equation similar to our previous estimated regression equation $\hat{y} = 60 + 5x$. However, it is doubtful that we would obtain exactly the same equation (with an intercept of exactly 60 and a slope of exactly 5). Indeed, b_0 and b_1, the least squares estimators, are sample statistics that have their own sampling distributions. The properties of the sampling distribution of b_1 are shown at the top of page 525.

Sampling Distribution of b_1

Expected Value

$$E(b_1) = \beta_1$$

Standard Deviation

$$\sigma_{b_1} = \frac{\sigma}{\sqrt{\Sigma x_i^2 - (\Sigma x_i)^2/n}} \qquad (13.16)$$

Distribution Form

Normal

Note that the expected value of b_1 is equal to β_1, so b_1 is an unbiased estimator of β_1.

Since we do not know the value of σ, we develop an estimate of σ_{b_1}, denoted s_{b_1}, by estimating σ with s in equation (13.16). Thus, we obtain the following estimate of σ_{b_1}.

Estimated Standard Deviation of b_1

$$s_{b_1} = \frac{s}{\sqrt{\Sigma x_i^2 - (\Sigma x_i)^2/n}} \qquad (13.17)$$

EXAMPLE 13.1 **(CONTINUED)**

For Armand's Pizza Parlors, $s = 13.829$. Thus,

$$s_{b_1} = \frac{13.829}{\sqrt{2528 - (140)^2/10}} = .5803$$

The t test regarding β_1 is based on the fact that the test statistic

$$t = \frac{b_1 - \beta_1}{s_{b_1}}$$

follows a t distribution with $n - 2$ degrees of freedom. If the null hypothesis is true, then $\beta_1 = 0$ and $t = b_1/s_{b_1}$. Using b_1/s_{b_1} as the test statistic, the rejection rule to test H_0: $\beta_1 = 0$ versus H_a: $\beta_1 \neq 0$ is as follows:

$$\text{Reject } H_0 \text{ if } \frac{b_1}{s_{b_1}} < -t_{\alpha/2} \text{ or if } \frac{b_1}{s_{b_1}} > t_{\alpha/2}$$

For Armand's Pizza Parlors, we have $b_1 = 5$ and $s_{b_1} = .5803$. Thus, we have $b_1/s_{b_1} = 5/.5803 = 8.62$. From Table 2 of Appendix B we find that the t value corresponding to $\alpha = .01$ and 8 degrees of freedom is $t_{.005} = 3.355$. Since $b_1/s_{b_1} = 8.62 > 3.355$, we reject H_0 and conclude at the .01 level of significance that β_1 is not equal to zero. ◀

F TEST

The t test has been used to test the null hypothesis H_0: $\beta_1 = 0$. An F test also exists for testing this null hypothesis. In regression models with only one independent variable, the

t test and the F test yield the same conclusion; that is, if the t test results in the rejection of H_0, the F test will also lead to the rejection of H_0. But with more than one independent variable, only the F test can be used to test for a significant relationship between a dependent variable and a set of independent variables. Here we introduce the F test and show that it leads to the same conclusion as the t test.

The hypotheses we will be testing are the same as before:

$$H_0: \beta_1 = 0$$

$$H_a: \beta_1 \neq 0$$

The logic behind the use of the F test for determining whether the relationship between x and y is statistically significant is based on our being able to develop two independent estimates of σ^2. We have just seen that MSE provides an estimate of σ^2. If the null hypothesis $H_0: \beta_1 = 0$ is true, the *mean square due to regression* or *mean square regression,* denoted MSR, provides another *independent* estimate of σ^2.

To compute MSR, recall that for any sum of squares, the mean square is the sum of squares divided by its degrees of freedom. Thus,

$$MSR = \frac{SSR}{\text{Regression Degrees of Freedom}}$$

Since the number of degrees of freedom for SSR is equal to the number of independent variables, we can write

$$MSR = \frac{SSR}{\text{Number of Independent Variables}} \tag{13.18}$$

In this chapter we only consider models involving one independent variable; in this case, MSR = SSR/1 = SSR. Thus, for Armand's Pizza Parlors, MSR = SSR = 14,200.

If the null hypothesis $(H_0: \beta_1 = 0)$ is true, MSR and MSE are two independent estimates of σ^2. In this case, the sampling distribution of MSR/MSE follows an F distribution with numerator degrees of freedom equal to 1 and denominator degrees of freedom equal to $n - 2$. The test concerning the significance of the regression relationship is based on the following F statistic:

$$F = \frac{MSR}{MSE} \tag{13.19}$$

Given any sample size, the numerator of the F statistic will increase as more of the variability in y is explained by the regression model and decrease as less is explained. Similarly, the denominator will increase if there is more variability about the estimated regression line and decrease if there is less variability. Thus, one would intuitively expect large values of $F = MSR/MSE$ to cast doubt on the null hypothesis and lead us to the conclusion that $\beta_1 \neq 0$. Indeed, this is correct; large values of F lead to rejection of H_0 and the conclusion that the relationship between x and y is statistically significant.

EXAMPLE 13.1 ▶
(CONTINUED)

Let us now conduct the F test for Armand's Pizza Parlors. Assume that the level of significance is $\alpha = .01$. From Table 4 of Appendix B, we can determine the critical F value by locating the value corresponding to numerator degrees of freedom equal to 1 (the number of independent variables) and denominator degrees of freedom equal to $n - 2 = 10 - 2 = 8$. Thus, we obtain $F = 11.26$. Hence, the appropriate rejection rule for Armand's Pizza Parlors is written

Reject H_0 if MSR/MSE > 11.26

Since MSR/MSE = 14,200/191.25 = 74.25 is greater than the critical value $F_{.01} = 11.26$, we can reject H_0 and conclude that there is a statistically significant relationship at the .01 level of significance between annual sales and the size of the student population. ◀

We see that for Armand's Pizza Parlors, the F test leads to the same conclusion as the t test. In fact, as we stated earlier, in simple linear regression these two tests are equivalent. It is interesting to note, however, that the t test required both a lower- and upper-tailed rejection region, each corresponding to a level of significance of $\alpha/2$. However, when using the F test, the rejection region only appears in the upper tail of the F distribution.

A CAUTION REGARDING STATISTICAL SIGNIFICANCE

It is important to note here that rejection of H_0 does not permit us to conclude that the relationship between x and y is *linear*. However, it is valid to conclude that x and y are related and that a linear relationship explains a significant amount of the variability in y over the range of x values observed in the sample. To illustrate this qualification, we call your attention to Figure 13.8, where an F test (on $\beta_1 = 0$) yielded the conclusion that x and y were related. The figure shows that the actual relationship is nonlinear. In the graph we see that the linear approximation is very good for the values of x used in developing the least squares line, but it is very bad for larger values of x.

Given a significant relationship, we should feel confident in using the regression equation for predictions corresponding to x values within the range of the x values for the sample. For Armand's Pizza Parlors, this corresponds to values of x between 2 and 26. But unless there are reasons to believe the model is valid beyond this range, predictions outside the range of the independent variable should be made with caution. For Armand's Pizza Parlors, since the regression relationship has been found significant at the .01 level,

FIGURE 13.8   EXAMPLE OF A LINEAR APPROXIMATION OF A NONLINEAR RELATIONSHIP

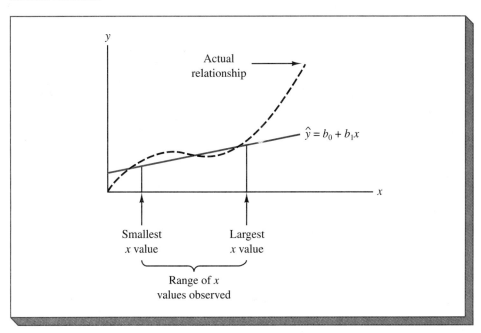

we should feel confident using it to predict sales whenever the student population is between 2000 and 26,000.

1. The assumptions made about the error term (Section 13.3) are what permit the tests of statistical significance in this section. The properties of the sampling distribution of b_1 and the subsequent F and t tests directly follow from these assumptions.

2. Do not confuse statistical significance with practical significance. With very large sample sizes, it is possible to obtain statistically significant results for values of b_1 that are relatively small; in such cases, one must exercise care in concluding that the relationship has practical significance.

3. The reason that the F test and the t test yield the same result *for simple linear regression* is that $F = t^2$. The critical value for the F test is the square of the critical value for the t test, and the test statistic for the F test is the square of the test statistic for the t test.

EXERCISES

METHODS

 SELF TEST

22. The data from Exercise 1 are shown below.

x_i	1	2	3	4	5
y_i	3	7	5	11	14

a. Compute the mean square error using equation (13.14).
b. Compute the standard error of the estimate using equation (13.15).
c. Compute the standard deviation of b_1 using equation (13.17).
d. Use the t test to test the following hypotheses ($\alpha = .05$):

$$H_0: \beta_1 = 0$$

$$H_a: \beta_1 \neq 0$$

e. Use the F test to test the hypotheses in part (d) at the $\alpha = .05$ level of significance.

23. The data from Exercise 2 are shown below.

x_i	2	3	5	1	8
y_i	25	25	20	30	16

a. Compute the mean square error using equation (13.14).
b. Compute the standard error of the estimate using equation (13.15).
c. Compute the standard deviation of b_1 using equation (13.17).
d. Use the t test to test the following hypotheses ($\alpha = .05$):

$$H_0: \beta_1 = 0$$

$$H_a: \beta_1 \neq 0$$

e. Use the F test to test the hypotheses in part (d) at the $\alpha = .05$ level of significance.

24. The data from Exercise 3 are shown.

x_i	2	4	5	7	8
y_i	2	3	2	6	4

Test whether x and y are related at the $\alpha = .05$ level of significance.

APPLICATIONS

SELF TEST   **25.** The data from Exercise 5 are shown below.

GPA	Monthly Salary ($)	GPA	Monthly Salary ($)
2.6	1400	3.2	1600
3.4	1700	3.5	2000
3.6	2100	2.9	1700

a. Use the t test to test the following hypotheses ($\alpha = .05$):

$$H_0: \beta_1 = 0$$

$$H_a: \beta_1 \neq 0$$

b. Use the F test to test the hypotheses in part (a) at the $\alpha = .05$ level of significance.

26. Refer to Exercise 10, where an estimated regression line relating square footage to selling prices of new homes was developed. Test whether selling price and square footage are related at the $\alpha = .01$ level of significance.

27. Refer to Exercise 12, where an estimated regression line relating the percentage of management jobs held by women and the percentage of women employed was developed. Test whether these two variables are related at the $\alpha = .05$ level of significance.

28. Refer to Exercise 9, where an estimated regression line relating years of experience and annual sales was developed. At the $\alpha = .05$ level of significance, determine whether annual sales and years of experience are related.

29. Refer to Exercise 20, where an estimated regression line relating light absorbence readings and milligrams of protein present in a liver sample was developed. Test whether the absorbence readings and amount of protein present are related at the $\alpha = .01$ level of significance.

13.5 ESTIMATION AND PREDICTION

In Example 13.1 we concluded that annual sales y for Armand's Pizza Parlors and the size of the student population x are related. Moreover, the estimated regression equation $\hat{y} = 60 + 5x$ provides an approximation of the relationship between x and y. Now we can begin to use the estimated regression line to develop interval estimates of annual sales for a given student population.

There are two types of interval estimates to consider. The first is an interval estimate of the mean value of y for a particular value of x. We refer to this type of interval estimate as a *confidence interval estimate*. For instance, we might want a confidence interval estimate of the *expected* annual sales for all restaurants located near a campus with a student population of 10,000. In this case, the expected annual sales represents the average of the annual sales for all restaurants located near a college campus with 10,000 students.

The second type of interval estimate that we will consider is appropriate in situations where we want to predict an individual value of *y* corresponding to a given value of *x*. We refer to this type of interval estimate as a *prediction interval estimate*. For instance, we might be interested in developing a prediction interval estimate for the annual sales for one specific restaurant located near Talbot College, a school with 10,000 students. In this case, our interest is in predicting the annual sales for one specific restaurant, as opposed to predicting the average sales for all restaurants located near campuses with 10,000 students.

CONFIDENCE INTERVAL ESTIMATE OF THE MEAN VALUE OF *y*

Suppose we wanted to develop a confidence interval estimate of the mean or expected value of annual sales for all restaurants located near a campus with 10,000 students. First, recall that for Armand's Pizza Parlors the expected value of annual sales is given by

$$E(y) = \beta_0 + \beta_1 x$$

So, when the student population size is 10,000, $x = 10$; hence $E(y) = \beta_0 + \beta_1(10)$.

For Armand's Pizza Parlors, the estimated regression equation was found to be $\hat{y} = 60 + 5x$. Thus, for restaurants located near a campus with 10,000 students, we would obtain $\hat{y} = 60 + 5(10) = 110$. Hence, the best estimate of the expected value of annual sales would be $110,000.

In general, the point estimate of $E(y)$ for a particular value of x is the corresponding value of $\hat{y}$ given by the estimated regression equation. We denote the particular value of x using x_p, the mean value of y at x_p using $E(y_p)$, and the estimate of $E(y_p)$ using $\hat{y}_p = b_0 + b_1 x_p$.

Since b_0 and b_1 are only estimates of β_0 and β_1, we cannot expect that the estimated value of $\hat{y}_p$ will exactly equal $E(y_p)$. For instance, in the Armand's Pizza Parlors problem, we do not expect that the mean annual sales for all restaurants located near a campus with 10,000 students to exactly equal $110,000, the estimated value. If we want to make an inference about how close $\hat{y}_p$ is to the true mean value $E(y_p)$, however, we will have to consider the variability that exists when we develop estimates based on the estimated regression equation. Statisticians have developed the following estimate of the variance of $\hat{y}_p$:

$$\text{Estimated Variance of } \hat{y}_p = s_{\hat{y}_p}^2 = s^2 \left[\frac{1}{n} + \frac{(x_p - \bar{x})^2}{\Sigma x_i^2 - (\Sigma x_i)^2/n} \right]$$

Hence, an estimate of the standard deviation of $\hat{y}_p$ is given by the square root of the variance:

$$s_{\hat{y}_p} = s \sqrt{\frac{1}{n} + \frac{(x_p - \bar{x})^2}{\Sigma x_i^2 - (\Sigma x_i)^2/n}} \qquad (13.20)$$

The confidence interval estimate of $E(y_p)$ is as follows.

> ### Confidence Interval Estimate of $E(y_p)$
>
> $$\hat{y}_p \pm t_{\alpha/2}s_{\hat{y}_p} \qquad\qquad (13.21)$$
>
> where the confidence coefficient is $1 - \alpha$ and the t value has $n - 2$ degrees of freedom.

EXAMPLE 13.1
(CONTINUED)

For Armand's Pizza Parlors, the estimated standard deviation of $\hat{y}_p$ for a restaurant located near a campus with 10,000 students is

$$s_{\hat{y}_p} = 13.829 \sqrt{\frac{1}{10} + \frac{(10 - 14)^2}{2528 - (140)^2/10}}$$

$$= 13.829\sqrt{.1282}$$

$$= 4.95$$

Thus, to develop a 95% confidence interval estimate of the expected annual sales for all Armand's restaurants located near a campus with 10,000 students, we need to find the t value from Table 2 of Appendix B corresponding to $n - 2 = 10 - 2 = 8$ degrees of freedom and $\alpha = .05$. Doing so, we find $t_{.025} = 2.306$. Hence, using equation (13.21) the resulting confidence interval is

$$[b_0 + b_1(10)] \pm 2.306s_{\hat{y}_p}$$

$$[60 + 5(10)] \pm 2.306(4.95)$$

$$110 \pm 11.415$$

In dollars, the 95% confidence interval estimate is $110,000 \pm 11,415$. Thus, we obtain $98,585 to $121,415 as a confidence interval estimate of the expected or average sales volume for all restaurants located near a campus with 10,000 students. ◀

Note that the estimated standard deviation of $\hat{y}_p$ [see equation (13.20)] is smallest when the given value of $x_p = \bar{x}$. In this case, equation (13.20) becomes

$$s_{\hat{y}_p} = s\sqrt{\frac{1}{n} + \frac{(\bar{x} - \bar{x})^2}{\Sigma x_i^2 - (\Sigma x_i)^2/n}} = s\sqrt{\frac{1}{n}}$$

which implies that we can expect to make our best estimates of $E(y_p)$ at the mean of the independent variable.

PREDICTION INTERVAL ESTIMATE FOR AN INDIVIDUAL VALUE OF y

In the preceding discussion, we showed how to develop a confidence interval estimate of the expected annual sales for all restaurants located near a campus with 10,000 students. Now we turn to the problem of developing point and interval estimates for an individual value of y corresponding to a particular value of x. Suppose we want to predict annual sales for one specific restaurant located near Talbot College, a school with 10,000 students.

The point estimate for an individual value of y is given by $\hat{y}_p = b_0 + b_1 x_p$. Hence, the point estimate of annual sales for one specific restaurant located near Talbot College is $\hat{y}_p = 60 + 5(10) = 110$. Note that this is the same as the point estimate of the mean annual sales for all restaurants located near a campus with 10,000 students.

To develop a prediction interval estimate, we must first determine the variance associated with using $\hat{y}_p$ as an estimate of a particular value of y when $x = x_p$. This variance is the sum of the following two components:

1. The variance of the individual y values about the mean $E(y_p)$, an estimate of which is given by s^2.
2. The variance associated with using $\hat{y}_p$ to estimate $E(y_p)$, an estimate of which is given by $s_{\hat{y}_p}^2$.

Statisticians have shown that an estimate of the variance of an individual value of y_p, which we denote s_{ind}^2, is given by

$$s_{ind}^2 = s^2 + s_{\hat{y}_p}^2$$

$$= s^2 + s^2 \left[\frac{1}{n} + \frac{(x_p - \bar{x})^2}{\Sigma x_i^2 - (\Sigma x_i)^2/n} \right]$$

$$= s^2 \left[1 + \frac{1}{n} + \frac{(x_p - \bar{x})^2}{\Sigma x_i^2 - (\Sigma x_i)^2/n} \right]$$

Hence, an estimate of the standard deviation of an individual value of y_p is given by

$$s_{ind} = s \sqrt{1 + \frac{1}{n} + \frac{(x_p - \bar{x})^2}{\Sigma x_i^2 - (\Sigma x_i)^2/n}} \tag{13.22}$$

The prediction interval estimate of y_p is given by equation (13.23).

Prediction Interval Estimate of y_p

$$\hat{y}_p \pm t_{\alpha/2} s_{ind} \tag{13.23}$$

where the confidence coefficient is $1 - \alpha$ and the t value has $n - 2$ degrees of freedom.

EXAMPLE 13.1 ▶
(CONTINUED)
For Armand's Pizza Parlors, the estimated standard deviation corresponding to the prediction of annual sales for one specific restaurant located near a campus with 10,000 students is computed as follows:

$$s_{ind} = 13.829 \sqrt{1 + \frac{1}{10} + \frac{(10 - 14)^2}{2528 - (140)^2/10}}$$

$$= 13.829 \sqrt{1.1282}$$

$$= 14.69$$

Thus, using equation (13.23) a 95% prediction interval for sales for one specific restaurant located near a campus with 10,000 students is

$$[b_0 + b_1(10)] \pm t_{\alpha/2}(14.69)$$

$$[60 + 5(10)] \pm 2.306(14.69)$$

$$110 \pm 33.875$$

Therefore, the 95% prediction interval for annual sales for one specific restaurant located near a campus with 10,000 students is $76,125 to $143,875. We note that this prediction interval is greater in width than the confidence interval for mean sales of all restaurants located near campuses with 10,000 students ($98,585 to $121,415). This difference simply reflects the fact that we are able to estimate the mean annual sales with more precision than we can the annual sales for any individual restaurant.

EXERCISES

METHODS

SELF TEST

30. The data from Exercise 1 are shown below:

x_i	1	2	3	4	5
y_i	3	7	5	11	14

    a. Use equation (13.20) to estimate the standard deviation of $\hat{y}_p$ when $x = 4$.
    b. Use equation (13.21) to develop a 95% confidence interval estimate of the expected value of y when $x = 4$.
    c. Use equation (13.22) to estimate the standard deviation of an individual value when $x = 4$.
    d. Use equation (13.23) to develop a 95% prediction interval for $x = 4$.

31. The data from Exercise 2 are shown below.

x_i	2	3	5	1	8
y_i	25	25	20	30	16

    a. Estimate the standard deviation of $\hat{y}_p$ when $x = 3$.
    b. Develop a 95% confidence interval estimate of the expected value of y when $x = 3$.
    c. Estimate the standard deviation of an individual value when $x = 3$.
    d. Develop a 95% prediction interval when $x = 3$.

32. The data from Exercise 3 are shown below.

x_i	2	4	5	7	8
y_i	2	3	2	6	4

Develop the 95% confidence and prediction intervals when $x = 3$. Explain why these two intervals are different.

APPLICATIONS

33. As an extension of Exercise 12, develop a 95% confidence interval for estimating the mean percentage of management jobs held by women for companies in which 60% of the employees are women.

SELF TEST

34. As an extension of Exercise 5, develop a 95% confidence interval for estimating the mean starting salary for students with a 3.0 GPA.

35. As an extension of Exercise 10, develop a 95% confidence interval for estimating the mean selling price for homes with 2200 square feet of living space.

SELF TEST   **36.** As an extension of Exercise 5, develop a 95% prediction interval for estimating the starting salary of Joe Heller, who has a GPA of 3.0.

37. As an extension of Exercise 10, develop a 95% prediction interval for the selling price of a home on Highland Terrace with 2800 square feet.

38. A study conducted by a department of transportation regarding driving speed and mileage for midsize automobiles resulted in the data shown below.

Driving Speed (mph)	Mileage (mpg)	Driving Speed (mph)	Mileage (mpg)
30	28	25	32
50	25	60	21
40	25	25	35
55	23	50	26
30	30	55	25

a. Determine the estimated regression equation that relates mileage to the driving speed.
b. At the $\alpha = .05$ level of significance, determine whether mileage and driving speed are related.
c. Did the estimated regression line provide a good fit to the data?
d. Develop a 95% confidence interval for estimating the mean mileage for cars that are driven at 50 miles per hour.
e. If we were interested in one specific car that was driven at 50 miles per hour, how would our estimate of mileage change as compared to the estimate developed in part (d)?

13.6 COMPUTER SOLUTION OF REGRESSION PROBLEMS

Performing all the computations associated with regression analysis can be quite time-consuming. In this section we discuss how the computational burden can be simplified by using a computer software package. The general procedure followed in using computer packages is for the user to input the data (x and y values for the sample) together with some instructions concerning the types of analyses that are required. The software package performs the analysis and prints the results in an output report. Before discussing the details of this approach, we discuss the use of the analysis of variance (ANOVA) table as a device for summarizing the calculations performed in regression analysis. The ANOVA table is an important component of the output report produced by most software packages.

THE ANOVA TABLE

In Chapter 12 we saw how the ANOVA table could provide a convenient summary of the computational aspects of analysis of variance. In regression analysis, a similar table can be developed. Table 13.6 shows the general form of the ANOVA table for problems involving one independent variable, and Table 13.7 shows the ANOVA table for the Armand's Pizza Parlors problem. It can be seen that the relationship that holds among the sum of squares (i.e., SST = SSR + SSE) also holds for the degrees of freedom. That is,

TABLE 13.6   **GENERAL FORM OF THE ANOVA TABLE FOR REGRESSION ANALYSIS PROBLEMS WITH ONE INDEPENDENT VARIABLE**

Source of Variation	Sum of Squares	Degrees of Freedom	Mean Square	F
Regression	SSR	1	$MSR = \dfrac{SSR}{1}$	$F = \dfrac{MSR}{MSE}$
Error	SSE	$n - 2$	$MSE = \dfrac{SSE}{n - 2}$	
Total	SST	$n - 1$		

TABLE 13.7   **ANOVA TABLE FOR ARMAND'S PIZZA PARLORS**

Source of Variation	Sum of Squares	Degrees of Freedom	Mean Square	F
Regression	14,200	1	$\dfrac{14{,}200}{1} = 14{,}200$	$\dfrac{14{,}200}{191.25} = 74.25$
Error	1,530	8	$\dfrac{1530}{8} = 191.25$	
Total	15,730	9		

the degrees of freedom for the total sum of squares is equal to the degrees of freedom for the regression sum of squares plus the degrees of freedom for the error sum of squares.

COMPUTER OUTPUT

In Figure 13.9 we show the input and the Minitab computer output for the Armand's Pizza Parlors problem. Values of the independent variable, student population, are entered into column 1 and values of the dependent variable, annual sales, are entered into column 2. After the data has been input, the variables are given names. The independent variable x is labeled POP and the dependent variable is labeled SALES. The command REGRESS C2 1 C1 is given by the user to perform the regression analysis. The C2 indicates that the dependent variable is in column 2 and the 1 C1 indicates that one independent variable, located in column 1, is to be used in the regression analysis. The subcommand PREDICT 10 is given to direct the software package to provide interval estimates of sales when the population is 10,000. The interpretation of the output is as follows:

1. Minitab prints the estimated regression equation as SALES = 60.0 + 5.00 POP.
2. A table is printed that shows the values of the coefficients b_0 and b_1, the standard deviation of each coefficient, the t value obtained by dividing each coefficient value by its standard deviation, and the p-value associated with the t test. Thus, to test H_0: $\beta_1 = 0$ versus H_a: $\beta_1 \neq 0$, we could compare 8.62 (located in the t-ratio column) to the appropriate critical value. This is the procedure described in the last part of Section 13.4. Alternatively, we could use the p-value provided by Minitab to perform the same test. Recall from Chapter 8 that the p-value is the probability of obtaining a sample result more unlikely than what is observed; if the

FIGURE 13.9 MINITAB OUTPUT FOR ARMAND'S PIZZA PARLORS

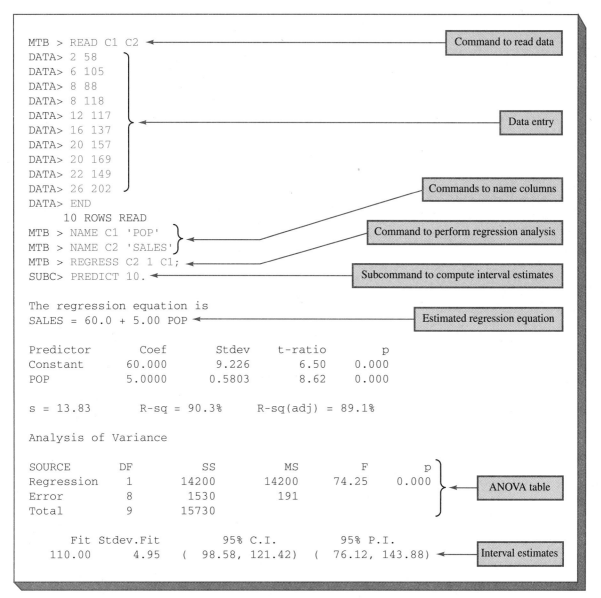

```
MTB > READ C1 C2                          ◄────────── Command to read data
DATA> 2 58
DATA> 6 105
DATA> 8 88
DATA> 8 118
DATA> 12 117                              ◄────────── Data entry
DATA> 16 137
DATA> 20 157
DATA> 20 169
DATA> 22 149
DATA> 26 202
DATA> END
      10 ROWS READ
MTB > NAME C1 'POP'                       ◄────────── Commands to name columns
MTB > NAME C2 'SALES'
MTB > REGRESS C2 1 C1;                    ◄────────── Command to perform regression analysis
SUBC> PREDICT 10.                         ◄────────── Subcommand to compute interval estimates


The regression equation is
SALES = 60.0 + 5.00 POP                   ◄────────── Estimated regression equation


Predictor      Coef        Stdev      t-ratio         p
Constant     60.000        9.226         6.50     0.000
POP          5.0000       0.5803         8.62     0.000


s = 13.83       R-sq = 90.3%     R-sq(adj) = 89.1%


Analysis of Variance


SOURCE         DF           SS           MS          F         p
Regression      1        14200        14200      74.25     0.000    ◄──────── ANOVA table
Error           8         1530          191
Total           9        15730


     Fit Stdev.Fit         95% C.I.          95% P.I.
  110.00       4.95   ( 98.58, 121.42)   ( 76.12, 143.88)   ◄──────── Interval estimates
```

p-value is less than α we reject H_0. Since the p-value in this case is zero (to three decimal places), the sample results indicate that the null hypothesis (H_0: $\beta_1 = 0$) should be rejected.

3. Minitab prints the standard error of the estimate $s = 13.83$, as well as information regarding the goodness of fit. Note that "R-sq = 90.3%" is the coefficient of determination expressed as a percentage. The output "R-Sq (adj) = 89.1%" is discussed in Chapter 14.

4. The ANOVA table is printed below the heading, Analysis of Variance. Note that DF is an abbreviation for degrees of freedom and that MSR is given as 14,200 and MSE as 191. The ratio of these two values provides the F value of 74.25; in Section 13.4 we showed how the F value can be used to determine whether there

is a significant statistical relationship between SALES and POP. Minitab also prints the *p*-value associated with this *F* test. Since the *p*-value is zero (to three decimal places), the relationship is judged statistically significant.

5. The 95% confidence interval estimate of the expected annual sales and the 95% prediction interval estimate of sales for an individual restaurant located near a campus with 10,000 students is printed below the ANOVA table. The confidence interval is (98.58, 121.42) and the prediction interval is (76.12, 143.88).

EXERCISES

METHODS

SELF TEST

39. The data from Exercise 1 are shown below.

x_i	1	2	3	4	5
y_i	3	7	5	11	14

Develop the ANOVA table for these data.

40. The data from Exercise 2 are shown below.

x_i	2	3	5	1	8
y_i	25	25	20	30	16

Develop the ANOVA table for these data.

41. The data from Exercise 3 are shown below.

x_i	2	4	5	7	8
y_i	2	3	2	6	4

Develop the ANOVA table for these data.

APPLICATIONS

SELF TEST

42. The commercial division of a real estate firm is conducting a regression analysis of the relationship between *x*, annual gross rents ($1000s), and *y*, selling price ($1000s), for apartment buildings. Data have been collected on a number of properties recently sold, and the following computer output has been obtained.

```
The regression equation is
Y = 20.0 + 7.21X

Predictor        Coef        Stdev      t-ratio
Constant       20.000       3.2213         6.21
X               7.210       1.3626         5.29


Analysis of Variance

SOURCE          DF              SS
Regression       1         41587.3
Error            7
Total            8         51984.1
```

    a. How many apartment buildings were in the sample?
    b. Write the estimated regression equation.
    c. What is the value of s_{b_1}?
    d. Use the F statistic to test the significance of the relationship at an $\alpha = .05$ level of significance.
    e. Estimate the selling price of an apartment building with gross annual rents of $50,000.

43. Shown below is a portion of the computer output for a regression analysis relating $y =$ maintenance expense (dollars per month) to $x =$ usage (hours per week) of a particular brand of computer terminal:

```
The regression equation is
Y = 6.1092 + .8951X


Predictor         Coef          Stdev
Constant          6.1092        0.9361
X                 0.8951        0.1490


Analysis of Variance


SOURCE            DF            SS              MS


Regression        1            1575.76         1575.76
Error             8             349.14           43.64
Total             9            1924.90
```

    a. Write the estimated regression equation.
    b. Test to see (use a t test) if monthly maintenance expense is related to usage at the .05 level of significance.
    c. Use the estimated regression equation to predict monthly maintenance expense for any terminal that is used 25 hours per week.

44. Regression analysis was used to relate x, the number of salespersons at a branch office, to y, the annual sales at the office ($1000s). The following computer output was obtained.

```
The regression equation is
Y = 80.0 + 50.00X


Predictor         Coef          Stdev        t-ratio
Constant          80.0          11.333          7.06
X                 50.0           5.482          9.12


Analysis of Variance


SOURCE            DF            SS              MS
Regression        1            6828.6          6828.6
Error             28           2298.8            82.1
Total             29           9127.4
```

    a. Write the estimated regression equation.
    b. How many branch offices were involved in the study?

c. Compute the F statistic, and test the significance of the relationship at an $\alpha = .05$ level of significance.

d. Predict the annual sales at the Memphis branch office, which has 12 salespersons.

 PRESCRIP   **45.** The following data show the dollar value of prescriptions for 13 pharmacies in Iowa and the population of the city served by the given pharmacy ("The Use of Categorical Variables in Data Envelopment Analysis," R. Banker and R. Morey, *Management Science,* December 1986).

Population	Value ($000s)	Population	Value ($000s)
1410	61	1070	45
1523	92	1694	183
1354	93	1910	156
822	45	1745	120
746	50	1353	75
1281	29	1016	122
1016	56		

a. Develop a scatter diagram for these data; plot population on the horizontal axis.

b. Does there appear to be any relationship between these two variables?

c. Use the computer package to develop the estimated regression line that could be used to predict the dollar value of prescriptions given the population of the city.

d. Test for the significance of the relationship at an $\alpha = .05$ level of significance.

e. Predict the dollar value for a particular city with a population of 1500 people. Use $\alpha = .05$.

 HOME1   **46.** The National Association of Home Builders compared the median home prices with the median household incomes in cities throughout the United States (*USA Today,* September 10, 1991) with 23 of the most affordable cities listed below. Both home prices and household incomes are shown in thousands of dollars.

City	Median Income	Median Home Price	City	Median Income	Median Home Price
Amarillo, Texas	36.7	69.0	Mansfield, Ohio	35.6	58.5
Brazoria, Texas	42.4	80.0	Milwaukee, Wisconsin	41.8	72.0
Canton, Ohio	34.1	66.0	Oklahoma City, Oklahoma	34.5	63.0
Davenport, Iowa	38.4	59.0	Omaha, Nebraska	38.8	65.0
Daytona Beach, Florida	31.0	63.0	Rockford, Illinois	41.6	73.5
Detroit, Michigan	44.6	77.0	Saginaw, Michigan	39.7	61.0
Fort Walton Beach, Florida	34.2	65.0	Shreveport, Louisiana	34.4	66.0
Grand Rapids, Michigan	40.3	73.0	Toledo, Ohio	39.4	65.0
Jackson, Michigan	36.8	60.0	Tulsa, Oklahoma	36.2	68.0
Kansas City, Missouri	41.1	77.0	Winter Haven, Florida	30.2	56.0
Lansing, Michigan	40.0	70.0	Youngstown, Ohio	34.9	59.0
Lorain, Ohio	38.8	72.5			

a. Develop a scatter diagram for these data; plot median income on the horizontal axis.

b. Does there appear to be any relationship between these two variables?

    c. Use a computer package to develop the estimated regression equation that could be used to predict the median home price given the median income.

    d. Test the significance of the relationship at the $\alpha = .05$ level of significance.

    e. Did the estimated regression equation provide a good fit? Explain.

    f. Predict the expected median home price for cities with a median income of $35,000.

    g. Predict the median home price for Elmira, New York, a city with a median income of $35,000.

13.7 RESIDUAL ANALYSIS: TESTING MODEL ASSUMPTIONS

For each observation in a regression analysis, there is a residual; it is the difference between the observed value of the dependent variable y_i and the value predicted by the regression equation $\hat{y}_i$. The residual for observation i, $y_i - \hat{y}_i$, is an estimate of the error resulting from using the estimated regression equation to predict the value of y_i.

The analysis of residuals plays an important role in validating the assumptions made in regression analysis. In Section 13.4 we showed how hypothesis testing can be used to determine whether a regression relationship is statistically significant. Hypothesis tests concerning regression relationships are based on the assumptions made about the regression model. If the assumptions made are not satisfied, the hypothesis tests are not valid, and the estimated regression equation should not be used. However, keep in mind that the regression model is being used only as an approximation of reality, so good judgment must be used to determine whether an assumption violation is severe enough to invalidate the model.

There are two key issues in verifying that the assumptions are satisfied in a regression model. Are the four assumptions concerning the error term ϵ satisfied, and is the form we have assumed for the model appropriate?

Regression analysis begins with an assumption concerning the appropriate form of the regression model. The simple linear regression model assumes the form

$$y = \beta_0 + \beta_1 x + \epsilon$$

With this form, y is a linear function of x plus an error term ϵ. The assumptions regarding the error term (presented in Section 13.3) are as follows:

1. $E(\epsilon) = 0$.

2. The variance of ϵ, denoted by σ^2, is the same for all values of x.

3. The values of ϵ are independent.

4. The error term ϵ is a normally distributed random variable.

Validating the assumptions concerning the error term ϵ means using the residuals to check to see whether these assumptions seem reasonable.

Recall that the first assumption concerning ϵ implies that the regression equation is

$$E(y) = \beta_0 + \beta_1 x$$

This regression equation shows a linear relationship between x and the expected value of y, $E(y)$. Validating the assumption concerning model form means satisfying ourselves that the relationship between independent and dependent variable is adequately represented by the regression equation. It is possible that the true relationship between x and y is curvilinear and/or that more independent variables should have been included (multiple regression). We will see how the statistician uses residual analysis to recognize when this assumption concerning model form might be violated.

The residuals $y_i - \hat{y}_i$ are estimates of ϵ; with n observations in a regression analysis, we have n residuals. Residual plots are graphical presentations of the residuals that help reveal patterns and thus help determine whether the assumptions concerning ϵ and the form of the regression model are satisfied. Three of the most common residual plots are

1. A plot of the residuals against the independent variable x.
2. A plot of the residuals against the predicted value of the dependent variable $\hat{y}$.
3. A standardized residual plot in which each residual is standardized by dividing the residual by its standard deviation.

RESIDUAL PLOT AGAINST x

A residual plot against the independent variable x is constructed by placing x on the horizontal axis and the residuals on the vertical axis. One point is plotted for each observation; the first coordinate of each point is x_i, and the second coordinate is the residual $y_i - \hat{y}_i$. Figure 13.10 shows some of the patterns statisticians look for when analyzing residuals. Panel A shows the type of plot to expect when the assumptions are satisfied. The patterns shown in panels B and C indicate violation of one or more assumptions.

If the assumptions regarding ϵ are valid and a linear relationship between x and the expected value of y is appropriate, the residual plot should give an overall impression of a horizontal band of points. Panel A shows the type of pattern to be expected in this case. On the other hand, if the variance of ϵ is not constant—for example, the variability about the regression line is greater for larger values of x—we would observe a pattern such as that of panel B. Finally, if we observe a residual pattern such as that of panel C, we would conclude that the model is not adequate; that is, the assumption of a linear relationship between x and y is not appropriate.

EXAMPLE 13.1 ▶
(CONTINUED)

A plot of the residuals against the independent variable x for Armand's Pizza Parlors is shown in Figure 13.11 (the residuals were computed in Table 13.3). We see that the residuals appear to follow the pattern of panel A in Figure 13.10. We thus conclude that the assumptions regarding ϵ are satisfied and that a linear relationship between x and the expected value of y is appropriate. ◀

RESIDUAL PLOT AGAINST $\hat{y}$

A residual plot against the predicted value of the dependent variable is constructed by placing the predicted value on the horizontal axis and plotting each residual directly above the corresponding value of $\hat{y}_i$. A plot of the residuals against the predicted values for Armand's Pizza Parlors (the predicted values were also computed in Table 13.3) is shown in Figure 13.12. Note that the pattern of this residual plot is the same as the pattern of the residual plot against the independent variable x. For simple linear regression, both the residual plot against x and the residual plot against $\hat{y}$ provide the same information. With multiple regression models (more than one independent variable), the residual plot against $\hat{y}$ is more widely used.

STANDARDIZED RESIDUAL PLOT

Many of the residual plots provided by computer software packages use a standardized version of the residuals. As we have seen in earlier chapters, a random variable is standardized by subtracting its mean and dividing the result by its standard deviation.

FIGURE 13.10   RESIDUAL PLOTS FROM THREE REGRESSION STUDIES

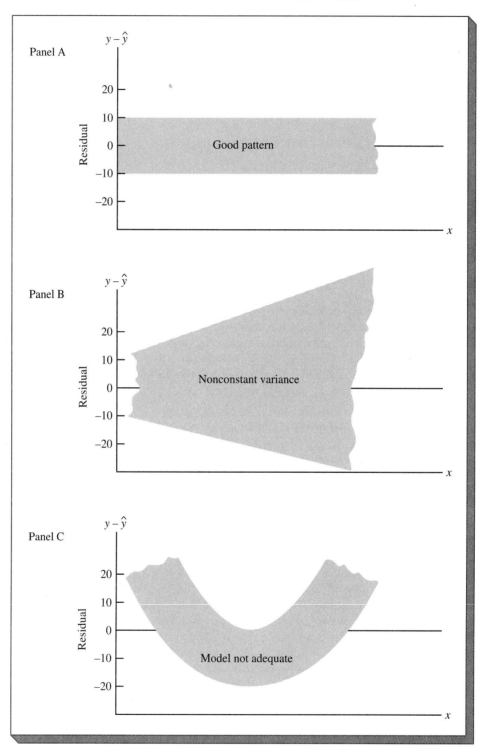

FIGURE 13.11   **PLOT OF THE RESIDUALS AGAINST THE INDEPENDENT VARIABLE** x **FOR ARMAND'S PIZZA PARLORS**

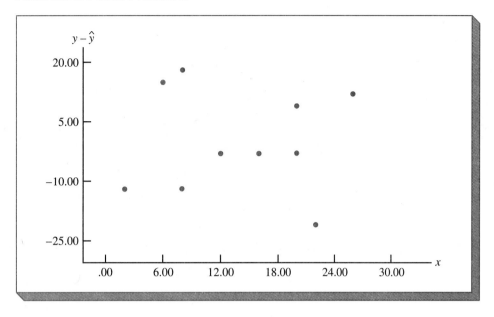

FIGURE 13.12   **PLOT OF THE RESIDUALS AGAINST THE PREDICTED VALUES** $\hat{y}$ **FOR ARMAND'S PIZZA PARLORS**

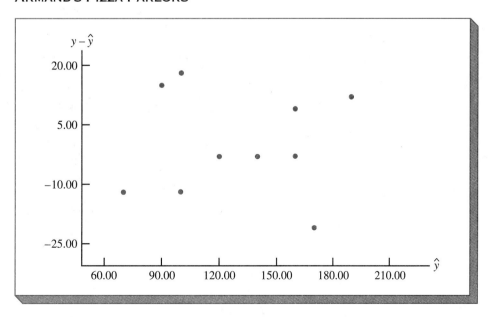

With the least squares method, the mean of the residuals is zero. Thus, simply dividing each residual by its standard deviation provides the standardized residual.

It can be shown that the standard deviation of the ith residual depends on $s = \sqrt{\text{MSE}}$ and the corresponding value of the independent variable.

Standard Deviation of ith Residual*

$$s_{y_i - \hat{y}_i} = \sqrt{s^2(1 - h_i)} \qquad (13.24)$$

where

$$s_{y_i - \hat{y}_i} = \text{standard deviation of residual } i$$

$$h_i = \frac{1}{n} + \frac{(x_i - \bar{x})^2}{\Sigma x_i^2 - (\Sigma x_i)^2/n} = \frac{1}{n} + \frac{(x_i - \bar{x})^2}{\Sigma(x_i - \bar{x})^2}$$

Note that equation (13.24) shows that residuals corresponding to different values of x have different standard deviations. Once the standard deviation of each residual is calculated, we compute the *standardized residual* by dividing each residual by its corresponding standard deviation. A standardized residual plot is then constructed by replacing the residuals in the previous residual plots with their standardized values.

EXAMPLE 13.1
(CONTINUED)

Table 13.8 shows the calculations involved for Armand's Pizza Parlors. (Recall that $s^2 = 191.25$.) Figure 13.13 is a plot of the standardized residuals against x. Note that, in this case, the standardized residual plot has the same general pattern as the original residual plot shown in Figure 13.11. ◄

TABLE 13.8 **COMPUTATION OF STANDARDIZED RESIDUALS FOR ARMAND'S PIZZA PARLORS**

Restaurant i	$x_i - \bar{x}$	$\dfrac{(x_i - \bar{x})^2}{\Sigma x_i^2 - (\Sigma x_i)^2/10}$	h_i	$s_{y_i - \hat{y}_i}$	$y_i - \hat{y}_i$	Standardized Residual
1	−12	.2535	.3535	11.1193	−12	−1.0792
2	−8	.1127	.2127	12.2709	15	1.2224
3	−6	.0634	.1634	12.6493	−12	−.9487
4	−6	.0634	.1634	12.6493	18	1.4230
5	−2	.0070	.1070	13.0682	−3	−.2296
6	2	.0070	.1070	13.0682	−3	−.2296
7	6	.0634	.1634	12.6493	−3	−.2372
8	6	.0634	.1634	12.6493	9	.7115
9	8	.1127	.2127	12.2709	−21	−1.7114
10	12	.2535	.3535	11.1193	12	1.0792

Note: From Table 13.2, we can compute $\bar{x} = 14$ and $\Sigma x_i^2 - (\Sigma x_i^2)/10 = 568$. The values of $y_i - \hat{y}_i$ are given in Table 13.3.

---

*This is actually an estimate of the standard deviation of the ith residual, since s^2 is used instead of σ^2. The value of σ^2 is never known when working with real data and is always estimated by s^2.

FIGURE 13.13   **PLOT OF THE STANDARDIZED RESIDUALS AGAINST x FOR ARMAND'S PIZZA PARLORS**

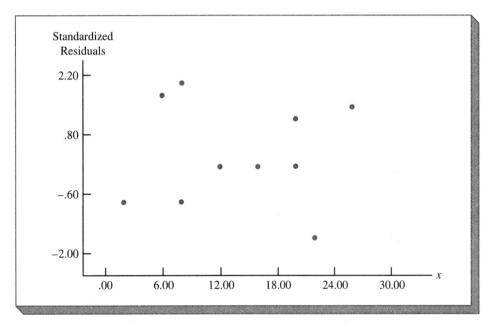

Because of the effort required to compute the estimated values of $\hat{y}$, the residuals, and the standardized residuals, most statistical packages provide these values as optional regression output. As a result, the development of residual plots such as a standardized residual plot can be easily obtained. For large problems, these packages provide the only practical means for developing the residual plots we have discussed in this section.

The standardized residual plot can provide insight concerning the normality assumption for ϵ (assumption 4). If the normality assumption is satisfied, the standardized residuals should appear to come from a standard normal probability distribution.* Thus, when looking at a standardized residual plot, we should expect to see approximately 95% of the standardized residuals between -2 and $+2$. Referring to the standardized residuals for Armand's Pizza Parlors (Table 13.8 and Figure 13.13), we see that they are all between -2 and $+2$. Thus, we conclude that the normality assumption is valid.

For simple linear regression, both the standardized residual plot against the independent variable x and the standardized residual plot against $\hat{y}$ provide the same information. With multiple regression models, the standardized residual plot against $\hat{y}$ is more widely used.

NORMAL PROBABILITY PLOT

Another approach that can be used to test whether the error terms are normally distributed involves the development of a normal probability plot. To show how the normal probability plot is developed, we need to introduce the concept of *normal scores*.

Consider an experiment in which 10 values are randomly selected from a normal probability distribution with a mean of 0 and a standard deviation of 1. Suppose this

---

*Since s^2 is substituted for σ^2 in equation (13.24), the probability distribution of the standardized residuals is not technically normal. However, in most regression studies, the sample size is large enough that a normal approximation is very good.

experiment is repeated over and over, and that for each sample the 10 values selected are ordered from the smallest to the largest. To begin with, let us consider only the smallest value in each sample. The *first-order statistic* is a random variable that represents the values of the smallest observation that would be obtained in repeated sampling. Since the value of the first-order statistic varies from sample to sample, the resulting probability distribution is called the sampling distribution of the first-order statistic.

Statisticians have shown that for samples of size 10, the expected value of the first-order statistic, referred to as a *normal score,* is approximately -1.55. In general, if we have a data set consisting of n observations, there are n-order statistics and hence n normal scores. For the case in which we have a sample of size $n = 10$, the normal scores corresponding to the order statistics from 1 to 10 are shown in Table 13.9.

EXAMPLE 13.1
(CONTINUED)

To show how the normal scores can be used to determine whether the standardized residuals for Armand's Pizza Parlors are normally distributed, we begin by ordering the 10 standardized residuals shown in Table 13.8 from smallest to largest. After rounding we obtain:

$$-1.71, -1.08, -.95, -.24, -.23, -.23, .71, 1.08, 1.22, 1.42.$$

We next form a table in which the smallest value for each standardized residual is associated with the smallest normal score, the next smallest value of each standardized residual is associated with the next smallest normal score, and so on; the results are shown in Table 13.10. If the standardized residuals came from a normal distribution, the smallest standardized residual should be close to the smallest normal score, the next smallest standardized residual should be close to the next smallest normal score, and so on. Thus, if we were to plot the standardized residuals (vertical axis) against the normal

TABLE 13.9   NORMAL SCORES FOR $n = 10$

Order Statistic	Normal Score	Order Statistic	Normal Score
1	-1.55	6	.12
2	-1.00	7	.37
3	$-.65$	8	.65
4	$-.37$	9	1.00
5	$-.12$	10	1.55

TABLE 13.10   STANDARDIZED RESIDUALS AND NORMAL SCORES

Standardized Residual	Normal Score	Standardized Residual	Normal Score
-1.71	-1.55	$-.23$	.12
-1.08	-1.00	.71	.37
$-.95$	$-.65$	1.08	.65
$-.24$	$-.37$	1.22	1.00
$-.23$	$-.12$	1.42	1.55

FIGURE 13.14 NORMAL PROBABILITY PLOT FOR ARMAND'S PIZZA PARLORS

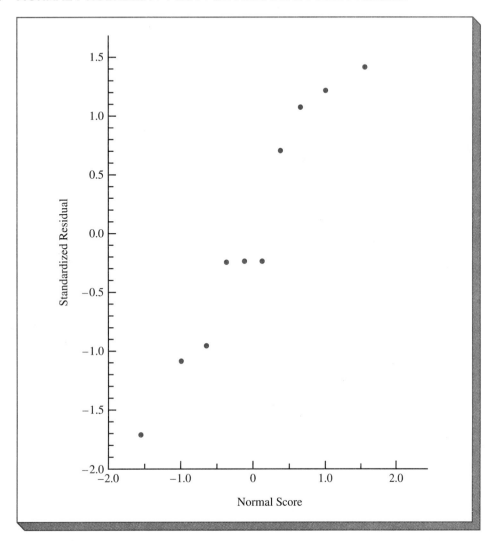

scores (horizontal axis), the points should appear to fall on a 45 line that passes through the origin. Such a plot is referred to as a *normal probability plot.*

In Figure 13.14 we show the normal probability plot for the standardized residuals for Armand's Pizza Parlors. Clearly, the points do not all fall exactly on a 45° line that passes through the origin. Given the small sample size, this is not surprising. Thus, judgment must be used to determine whether the pattern observed is different enough from what we would expect if the data were normally distributed. In this case, we feel that the normal probability plot does support the conclusion that the residuals are normally distributed. In general, the straighter the plot, the stronger the evidence supporting normality. Any curvature in the normal probability plot, however, is evidence that the residuals are not normally distributed. ◀

The computation of the normal scores for any sample size is a complex problem that involves mathematical concepts beyond the scope of this text. Fortunately, many statistical packages (e.g., Minitab) provide commands that will generate the normal scores corresponding to any set of data.

The analysis of residuals is the primary method by which statisticians verify that the assumptions are satisfied in a regression model. To validate a model, both the assumption concerning model form (simple linear in this chapter) and the assumptions on the error term are checked for possible violations. Even if no violations are found, it does not necessarily follow that the model will yield good predictions. However, if the model is statistically significant and r^2 is large, one should obtain good results.

EXERCISES

METHODS

SELF TEST

47. Given are data for two variables, x and y.

x_i	6	11	15	18	20
y_i	6	8	12	20	30

   a. Develop an estimated regression equation for these data.
   b. Compute the residuals.
   c. Develop a plot of the residuals against the independent variable x. Do the assumptions concerning the error terms seem to be satisfied?
   d. Compute the standardized residuals.
   e. Develop a plot of the standardized residuals against $\hat{y}$. What conclusions can you draw from this plot?

48. The data shown below were used in a regression study.

Observation	x_i	y_i	Observation	x_i	y_i
1	2	4	6	7	6
2	3	5	7	7	9
3	4	4	8	8	5
4	5	6	9	9	11
5	7	4			

   a. Develop an estimated regression equation for these data.
   b. Construct a plot of the residuals. Do the assumptions concerning the error terms seem to be satisfied?

APPLICATIONS

SELF TEST

49. In Exercise 7 data concerning advertising expenditures and sales at Eddie's Restaurants were given. These data (both in $1000s) are repeated here.

Advertising Expenditures	Sales	Advertising Expenditures	Sales
1.0	19.0	10.0	52.0
2.0	32.0	14.0	53.0
4.0	44.0	20.0	54.0
6.0	40.0		

a. Let x equal advertising expenditures and y equal sales. Use the least squares method to develop a straight line approximation to the relationship between the two variables.
b. Test whether sales and advertising expenditures are related at the $\alpha = .05$ level of significance.
c. Prepare a residual plot of $y - \hat{y}$ versus $\hat{y}$. Use the result of part (a) to obtain the values of $\hat{y}$.
d. What conclusions can you draw from residual analysis? Should this model be used, or should we look for a better one?

50. Refer to Exercise 9, where an estimated regression equation relating years of experience and annual sales was developed.
a. Compute the residuals and construct a residual plot for this problem.
b. Do the assumptions concerning the error terms seem reasonable in light of the residual plot?

51. The following data show the number of employees and the yearly revenues for the 10 largest wholesale bakers (*Louis Rukeyser's Business Almanac*).

Company	Employees	Revenues ($1,000,000s)
Nabisco Brands USA	9,500	1,734
Continental Baking Co.	22,400	1,600
Campbell Taggart, Inc.	19,000	1,044
Keebler Company	8,943	988
Interstate Bakeries Corp.	11,200	704
Flowers Industries, Inc.	10,200	557
Sunshine Biscuits, Inc.	5,000	490
American Bakeries Co.	6,600	461
Entenmann's Inc.	3,734	450
Kitchens of Sara Lee	1,550	405

a. Use a computer package to develop an estimated regression equation relating revenues y to the number of employees x.
b. Plot the standardized residuals against the independent variable.
c. Do the assumptions concerning the error terms and model form seem reasonable in light of the standardized residual plot?

13.8 RESIDUAL ANALYSIS: OUTLIERS AND INFLUENTIAL OBSERVATIONS

In Section 13.7 we showed how residual analysis could be used to determine when violations of assumptions concerning the regression model had occurred. In this section, we discuss how residual analysis can be used to identify observations that can be classified as outliers or as being especially influential in determining the estimated regression equation. Some steps that should be taken when such observations have been found are noted.

DETECTING OUTLIERS

Figure 13.15 shows a scatter diagram for a data set that has an outlier, a data point (observation) that does not fit the trend exhibited by the remaining data. Outliers

FIGURE 13.15   A DATA SET WITH AN OUTLIER

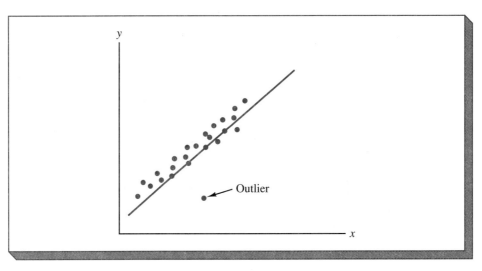

TABLE 13.11   DATA SET ILLUSTRATING THE EFFECT OF AN OUTLIER

x_i	y_i	x_i	y_i
1	45	3	45
1	55	4	30
2	50	4	35
3	75	5	25
3	40	6	15

represent observations that are suspect and warrant careful examination. They may represent erroneous data; if so, the data should be corrected. They may signal a violation of model assumptions; if so, other models should be considered. And, finally, they may simply be unusual values that have occurred by chance. In this case, they should be retained.

EXAMPLE 13.3▶ To illustrate the process of detecting outliers, consider the data set shown in Table 13.11; a scatter diagram is shown in Figure 13.16. Except for observation 4 ($x = 3$, $y = 75$), a pattern suggesting a negative linear relationship is apparent. Indeed, given the pattern of the rest of the data, we would have expected the y value for observation 4 to be much smaller and would thus identify the observation as an outlier. For the case of simple linear regression, one can usually detect outliers by simply examining the scatter diagram. ◀

The standardized residuals can also be used to identify outliers. If the value of y for a particular x is unusually large or small (does not seem to follow the trend of the rest of the data), the corresponding standardized residual will be large in absolute value. Many computer packages automatically identify observations with standardized residuals that are large in absolute value. In Figure 13.17 we show the Minitab output from a regression analysis of the data in Table 13.11. The next to last line of the output shows that the

FIGURE 13.16   SCATTER DIAGRAM FOR DATA SET OF TABLE 13.11

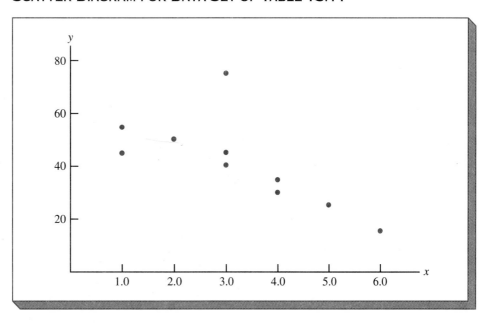

FIGURE 13.17   MINITAB OUTPUT FOR REGRESSION ANALYSIS OF DATA SET
WITH OUTLIER (TABLE 13.11)

```
The regression equation is
Y = 65.0 - 7.33 X

Predictor         Coef       Stdev       t-ratio          p
Constant        64.958       9.258          7.02      0.000
X               -7.331       2.608         -2.81      0.023

s = 12.67       R-sq = 49.7%      R-sq(adj) = 43.4%

Analysis of Variance

SOURCE          DF            SS            MS         F          p
Regression       1        1268.2        1268.2      7.90      0.023
Error            8        1284.3         160.5
Total            9        2552.5

Unusual Observations
Obs.        X             Y        Fit Stdev.Fit   Residual    St.Resid
   4     3.00         75.00      42.97      4.04      32.03       2.67R

R denotes an obs. with a large st. resid.
```

standardized residual for observation 4 is 2.67. Minitab considers a standardized residual of less than -2 or greater than $+2$ to be an outlier; in such cases, the observation is printed on a separate line with an R next to the standardized residual, as shown in Figure 13.17. Assuming normally distributed errors, standardized residuals should fall outside these limits only approximately 5% of the time.

In deciding how to handle an outlier, we should first check to see whether it is a valid observation. Perhaps an error has been made in initially recording the data or in entering the data into the computer system. For example, suppose that in checking the data for the outlier in Table 13.11, we find that an error has been made and that the correct value for observation 4 is $x = 3$, $y = 30$. Figure 13.18 shows the Minitab output obtained after correcting the value of y_4. We see that the effect of using an incorrect value for the dependent variable had a substantial effect on the goodness of fit. With the correct data, the value of r^2 has increased from 49.7 to 83.8% and the value of b_0 has decreased from 64.958 to 59.237. The slope of the line, however, has changed only from -7.331 to -6.949.

DETECTION OF INFLUENTIAL OBSERVATIONS

In regression analysis, it sometimes happens that one or more observations have a strong influence on the results obtained. Figure 13.19 shows an example of an influential observation in simple linear regression. The estimated regression line has a negative slope. But, if the influential observation is dropped from the data set, the slope of the estimated regression line would change from negative to positive, and the y-intercept would be smaller. Clearly, this one observation is much more influential in determining the estimated regression line than any of the others; dropping one of the other observations from the data set would have very little effect on the estimated regression equation.

Influential observations can be identified from a scatter diagram when only one independent variable is present. An influential observation may be an outlier (an observation with a y value that deviates substantially from the trend); it may correspond

---

FIGURE 13.18 MINITAB OUTPUT FOR REVISED DATA SET IN TABLE 13.11

```
The regression equation is
Y = 59.2 - 6.95 X


Predictor        Coef       Stdev      t-ratio         p
Constant       59.237       3.835        15.45     0.000
X              -6.949       1.080        -6.43     0.000


s = 5.248      R-sq = 83.8%      R-sq(adj) = 81.8%


Analysis of Variance


SOURCE           DF          SS           MS         F         p
Regression        1       1139.7       1139.7     41.38     0.000
Error             8        220.3         27.5
Total             9       1360.0
```

to an x value far away from its mean (e.g., see Figure 13.19); or it may be caused by a combination of the two (a somewhat off-trend y value and a somewhat extreme x value).

Since influential observations can have such a dramatic effect on the estimated regression equation, it is important that they be examined carefully. We should first check to make sure that no error has been made in collecting or recording the data. If an error has occurred, it can be corrected and a new estimated regression equation can be developed. On the other hand, if the observation is valid, we might consider ourselves fortunate to have it. Such a point, if valid, can contribute to a better understanding of the appropriate model and can lead to a better estimated regression equation. The presence of the influential observation in Figure 13.19, if valid, would suggest trying to obtain data on intermediate values of x to better understand the relationship between x and y.

Observations with extreme values for the independent variables are called high leverage points. The influential observation in Figure 13.19, caused by an extreme value of x, is a point with high leverage. The leverage of an observation is determined by how far the values of the independent variables are from their mean values. For the single-independent-variable case, the leverage of the ith observation, denoted h_i, can be computed using equation (13.25).

Leverage of Observation i

$$h_i = \frac{1}{n} + \frac{(x_i - \bar{x})^2}{\Sigma(x_i - \bar{x})^2} \qquad (13.25)$$

From the formula, it is clear that the farther x_i is from its mean $\bar{x}$, the higher the leverage of observation i.

Many computer packages automatically identify observations with high leverage as part of the standard regression output. To illustrate how the Minitab statistical package identifies points with high leverage, let us consider the data set presented in Table 13.12.

EXAMPLE 13.4 ▶ A scatter diagram for the data set in Table 13.12 is shown in Figure 13.20. From the scatter diagram, it is clear that observation 7 ($x = 70$, $y = 100$) is an observation with an

FIGURE 13.19   **A DATA SET WITH AN INFLUENTIAL OBSERVATION**

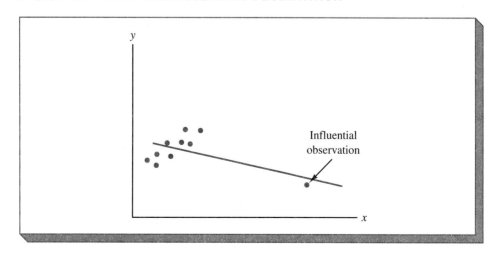

TABLE 13.12   DATA SET WITH A HIGH LEVERAGE OBSERVATION

x_i	y_i	x_i	y_i
10	125	20	120
10	130	25	110
15	120	70	100
20	115		

FIGURE 13.20   SCATTER DIAGRAM FOR THE DATA SET IN TABLE 13.12

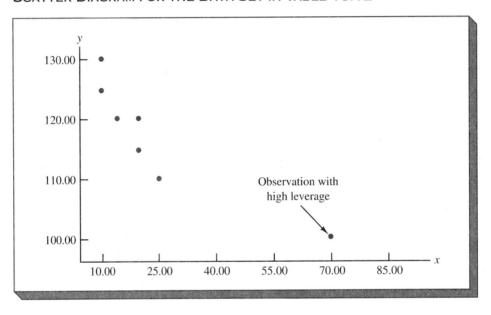

extreme value of x. Thus, we would expect it to be identified as a point with high leverage. For this observation, the leverage is computed using equation (13.25) as follows:

$$h_7 = \frac{1}{n} + \frac{(x_7 - \bar{x})^2}{\Sigma(x_i - \bar{x})^2} = \frac{1}{7} + \frac{(70 - 24.286)^2}{2621.43} = .94$$

For the case of simple linear regression, Minitab identifies observations as having high leverage if $h_i > 6/n$; for the data set in Table 13.12, $6/n = 6/7 = .86$. Since $h_7 = .94 > .86$, Minitab will identify observation 7 as a high leverage point. Figure 13.21 shows the Minitab output for a regression analysis of this data set. Observation 7 ($x = 70$, $y = 100$) is identified as having high leverage; it is printed on a separate line at the bottom, with an X in the right-hand margin. ◀

As we indicated previously, an influential observation can be caused by a combination of a large residual and a high leverage. Diagnostic procedures are available that take both into account in determining when an observation is influential. One such measure, called Cook's D statistic, will be discussed in Chapter 14.

FIGURE 13.21   MINITAB OUTPUT FOR THE DATA SET IN TABLE 13.12

```
The regression equation is
Y = 127 -0.425 X

Predictor        Coef        Stdev      t-ratio         p
Constant       127.466        2.961       43.04     0.000
X              -0.42507      0.09537      -4.46     0.007

s = 4.883        R-sq = 79.9%       R-sq(adj) = 75.9%

Analysis of Variance

SOURCE          DF            SS            MS          F          p
Regression       1         473.65        473.65      19.87     0.007
Error            5         119.21         23.84
Total            6         592.86

Unusual Observations
Obs.       X             Y        Fit Stdev.Fit   Residual    St.Resid
  7      70.0        100.00      97.71      4.73       2.29        1.91 X

X denotes an obs. whose X value gives it large influence
```

NOTES AND
COMMENTS

Once an observation has been identified as potentially influential its impact on the estimated regression equation should be evaluated. More advanced texts discuss diagnostics for doing so. However, if one is not familiar with the more advanced material, a simple procedure is to run the regression analysis with and without the observation. Although more time-consuming, this approach will reveal the influence of the observation on the results.

EXERCISES

METHODS

SELF TEST

52. Consider the following data for two variables, x and y.

x_i	135	110	130	145	175	160	120
y_i	145	100	120	120	130	130	110

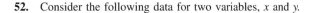

  a. Compute the standardized residuals for these data. Do there appear to be any outliers in the data? Explain.
  b. Plot the standardized residuals against $\hat{y}$. Does this plot reveal any outliers?
  c. Develop a scatter diagram for these data. Does the scatter diagram indicate any outliers in the data? In general, what implications does this have for simple linear regression?

53. Consider the following data for two variables, x and y.

x_i	4	5	7	8	10	12	12	22
y_i	12	14	16	15	18	20	24	19

a. Compute the standardized residuals for these data. Do there appear to be any outliers in the data? Explain.

b. Compute the leverage values for these data. Do there appear to be any influential observations in these data? Explain.

c. Develop a scatter diagram for these data. Does the scatter diagram indicate any influential observations? Explain.

APPLICATIONS

SELF TEST ▷ **54.** The following data show the number of golf courses and the number of paid rounds of golf (in millions) for the Myrtle Beach, South Carolina, area over a 10-year period (*Myrtle Beach Magazine,* October 1991).

Number of Golf Courses	Number of Paid Rounds of Golf	Number of Golf Courses	Number of Paid Rounds of Golf
26	1.0	35	1.6
30	1.1	38	1.8
31	1.2	43	2.0
32	1.3	57	2.5
33	1.4	67	3.0

a. Develop the estimated regression equation for these data.

b. Use residual analysis to determine if any outliers and/or influential observations are present. Briefly summarize your findings and conclusions.

 HOME2 **55.** The National Association of Home Builders compared the median home prices with the median household incomes in cities throughout the United States (*USA Today,* September 10, 1991). The 25 most affordable cities are shown below. Both home prices and household incomes are shown in thousands of dollars.

City	Median Income	Median Home Price	City	Median Income	Median Home Price
Amarillo, Texas	36.7	69.0	Milwaukee, Wisconsin	41.8	72.0
Brazoria, Texas	42.4	80.0	Minneapolis, Minnesota	48.0	91.0
Canton, Ohio	34.1	66.0	Nashua, New Hampshire	52.9	111.0
Davenport, Iowa	38.4	59.0	Oklahoma City, Oklahoma	34.5	63.0
Daytona Beach, Florida	31.0	63.0	Omaha, Nebraska	38.8	65.0
Detroit, Michigan	44.6	77.0	Rockford, Illinois	41.6	73.5
Fort Walton Beach, Florida	34.2	65.0	Saginaw, Michigan	39.7	61.0
Grand Rapids, Michigan	40.3	73.0	Shreveport, Louisiana	34.4	66.0
Jackson, Michigan	36.8	60.0	Toledo, Ohio	39.4	65.0
Kansas City, Missouri	41.1	77.0	Tulsa, Oklahoma	36.2	68.0
Lansing, Michigan	40.0	70.0	Winter Haven, Florida	30.2	56.0
Lorain, Ohio	38.8	72.5	Youngstown, Ohio	34.9	59.0
Mansfield, Ohio	35.6	58.5			

a. Develop the estimated regression equation for these data that can be used to predict the median home price given the median income.

b. Use residual analysis to determine if any outliers and/or influential observations are present. Briefly summarize your findings and conclusions.

 TEMPSC

56. The following data show the temperatures for air and water in the Myrtle Beach, South Carolina area (*Myrtle Beach and South Carolina Grand Strand,* 1992).

Month	Air	Water	Month	Air	Water
January	57	49	July	88	83
February	59	51	August	88	80
March	65	56	September	84	77
April	75	66	October	75	72
May	81	71	November	68	60
June	86	78	December	59	50

a. Develop the estimated regression equation for these data that can be used to predict the water temperature given the air temperature.

b. Use residual analysis to determine if any outliers and/or influential observations are present. Briefly summarize your findings and conclusions.

13.9 CORRELATION ANALYSIS

As we indicated in the introduction to this chapter, there are some situations in which the decision maker is not as concerned with the equation that relates two variables as in measuring the extent to which the two variables are related. In such cases, a statistical technique referred to as correlation analysis can be used to determine the strength of the relationship between the two variables.* The output of a correlation study is a number referred to as the correlation coefficient. Because of the way in which it is defined, values of the correlation coefficient are always between -1 and $+1$. A value of $+1$ indicates that x and y are perfectly related in a positive linear sense. That is, all the points lie on a straight line that has a positive slope. A value of -1 indicates that x and y are perfectly related in a negative linear sense. That is, all the points lie on a straight line that has a negative slope. Values of the correlation coefficient close to zero indicate that x and y are not linearly related.

EXAMPLE 13.5 ▶

To illustrate correlation analysis, we consider the situation of a stereo and sound-equipment store located in San Francisco. Management would like to investigate whether there is any relationship between the number of commercials shown on Friday evening television (x) and the resulting sales volume on Saturday (y) measured in hundreds of dollars. The sample data that were obtained are shown in Table 13.13.

In Figure 13.22 we show a scatter diagram of these data. The scatter diagram appears to indicate that there is a positive linear relationship between x and y. ◀

---

*In correlation analysis, it is assumed that x and y are both random variables.

TABLE 13.13   **SAMPLE DATA FOR THE STEREO AND SOUND-EQUIPMENT PROBLEM**

Store	Number of Commercials	Sales Volume ($100s)	Store	Number of Commercials	Sales Volume ($100s)
1	2	24	5	4	25
2	5	28	6	1	24
3	1	22	7	5	26
4	3	26			

FIGURE 13.22   **SCATTER DIAGRAM FOR THE STEREO AND SOUND-EQUIPMENT PROBLEM**

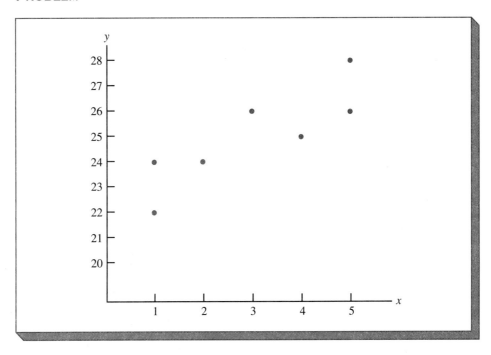

COVARIANCE

To measure the degree of linear association between these two variables, we first define a measure of linear association known as the *covariance*. *Sample covariance* is defined as follows.

Sample Covariance

$$s_{xy} = \frac{\Sigma(x_i - \bar{x})(y_i - \bar{y})}{n - 1}$$

(13.26)

In this formula, each x_i value is paired with a y_i value. We then sum the products obtained by multiplying the deviation of each x_i from its sample mean $\bar{x}$ times the deviation of the corresponding y_i from its sample mean $\bar{y}$; this sum is then divided by $n - 1$.

EXAMPLE 13.5
(CONTINUED)

To measure the strength of the linear relationship between the number of commercials x and the sales volume y in the stereo and sound-equipment problem, we can use equation (13.26) to compute the sample covariance. The calculations in Table 13.14 illustrate the computations of $\Sigma(x_i - \bar{x})(y_i - \bar{y})$. Note that $\bar{x} = 21/7 = 3$ and $\bar{y} = 175/7 = 25$. Using equation (13.26), we obtain

$$s_{xy} = \frac{\Sigma(x_i - \bar{x})(y_i - \bar{y})}{n - 1} = \frac{17}{6} = 2.8333 \qquad \blacktriangleleft$$

The formula for computing the covariance of a population of size N is similar to equation (13.26), but we use different notation to indicate that we are dealing with the entire population.

Population Covariance

$$\sigma_{xy} = \frac{\Sigma(x_i - \mu_x)(y_i - \mu_y)}{N} \qquad \text{(13.27)}$$

In equation (13.27) we use the notation μ_x for the population mean of the variable x and μ_y for the population mean of the variable y. The sample covariance s_{xy} is an estimate of the population covariance σ_{xy} based on a sample of size n.

INTERPRETATION OF THE COVARIANCE

To aid in the interpretation of the *sample covariance*, consider Figure 13.23. It is the same as the scatter diagram of Figure 13.22 with a vertical line at $x = 3$ (the value of $\bar{x}$) and a horizontal line at $y = 25$ (the value of $\bar{y}$). Four quadrants have been identified on the graph. Points that fall in quadrant I correspond to x_i values greater than $\bar{x}$ and y_i values

---

TABLE 13.14   **CALCULATIONS FOR THE SAMPLE COVARIANCE FOR THE STEREO AND SOUND-EQUIPMENT PROBLEM**

	x_i	y_i	$x_i - \bar{x}$	$y_i - \bar{y}$	$(x_i - \bar{x})(y_i - \bar{y})$
	2	24	-1	-1	1
	5	28	2	3	6
	1	22	-2	-3	6
	3	26	0	1	0
	4	25	1	0	0
	1	24	-2	-1	2
	5	26	2	1	2
Totals	21	175	0	0	17

FIGURE 13.23   QUADRANTS I, II, III, AND IV FOR THE STEREO AND SOUND-EQUIPMENT PROBLEM

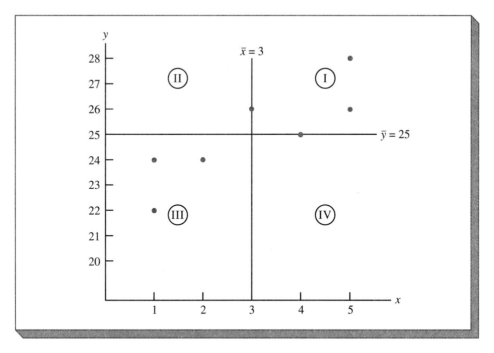

greater than $\bar{y}$; points that fall in quadrant II correspond to x_i values less than $\bar{x}$ and y_i values greater than $\bar{y}$, and so on. Thus, the value of $(x_i - \bar{x})(y_i - \bar{y})$ must be positive for points located in quadrant I, negative for points located in quadrant II, positive for points located in quadrant III, and negative for points located in quadrant IV.

   If the value of s_{xy} is positive, the points that have had the greatest effect on s_{xy} must lie in quadrants I and III. Hence, a positive value for s_{xy} is indicative of a positive linear association between x and y; that is, as the value of x increases, the value of y increases. If the value of s_{xy} is negative, however, the points that have had the greatest effect on s_{xy} lie in quadrants II and IV. Hence, a negative value for s_{xy} is indicative of a negative linear association between x and y; that is, as the value of x increases, the value of y decreases. Finally, if the points are evenly distributed across all four quadrants, the value of s_{xy} will be close to zero, indicating no linear association between x and y. Figure 13.24 shows the values of s_{xy} that can be expected with these three different types of scatter diagrams.

   From the previous discussion, it might appear that a large positive value for the covariance is indicative of a strong positive linear relationship and that a large negative value is indicative of a strong negative linear relationship. However, one problem with using covariance as a measure of the strength of the linear relationship is that the value we obtain for the covariance depends on the units of measurement for x and y. For example, suppose we were interested in the relationship between height x and weight y for individuals. If height is measured in inches, we will get much larger numerical values for $(x_i - \bar{x})$ than if it is measured in feet. Thus, with height measured in inches, we would obtain larger values for $\Sigma(x_i - \bar{x})(y_i - \bar{y})$—and hence a larger covariance—when, in fact, there is no difference in the relationship. A measure of relationship that avoids this difficulty is the *correlation coefficient*.

FIGURE 13.24   INTERPRETATION OF SAMPLE COVARIANCE

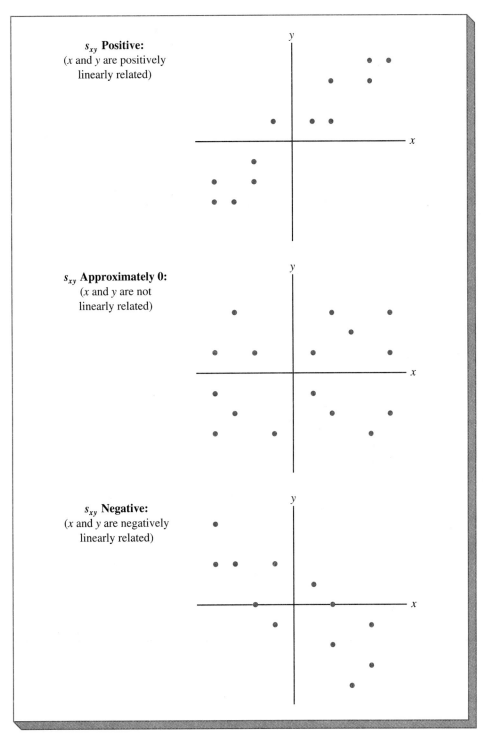

CORRELATION COEFFICIENT

For sample data, the correlation coefficient is defined as follows.

Sample Correlation Coefficient

$$r_{xy} = \frac{s_{xy}}{s_x s_y} \tag{13.28}$$

where

r_{xy} = sample correlation coefficient
s_{xy} = sample covariance
s_x = sample standard deviation of x
s_y = sample standard deviation of y

Equation (13.28) shows that the sample correlation coefficient is computed by dividing the sample covariance by the product of the standard deviation of x and the standard deviation of y. Before we consider further interpretation of the sample correlation coefficient, let us consider the use of equation (13.28) for the stereo and sound-equipment problem.

EXAMPLE 13.5 (CONTINUED)

Using the data presented in Table 13.14, we can compute the sample correlation coefficient.

$$s_x = \sqrt{\frac{\Sigma(x_i - \bar{x})^2}{n-1}} = \sqrt{\frac{18}{6}} = 1.7321$$

$$s_y = \sqrt{\frac{\Sigma(y_i - \bar{y})^2}{n-1}} = \sqrt{\frac{22}{6}} = 1.9149$$

and, since $s_{xy} = 2.8333$, we have

$$r_{xy} = \frac{s_{xy}}{s_x s_y} = \frac{2.8333}{(1.7321)(1.9149)} = .854$$

When using a calculator to compute the sample correlation coefficient, the formula given by equation (13.29) is preferred because the computation of each deviation $x_i - \bar{x}$ and $y_i - \bar{y}$ is not necessary, and thus less round-off error is introduced.

Sample Correlation Coefficient, Alternate Formula

$$r_{xy} = \frac{\Sigma x_i y_i - (\Sigma x_i \Sigma y_i)/n}{\sqrt{\Sigma x_i^2 - (\Sigma x_i)^2/n} \sqrt{\Sigma y_i^2 - (\Sigma y_i)^2/n}} \tag{13.29}$$

EXAMPLE 13.5 (CONTINUED)

Algebraically, Equations (13.28) and (13.29) are equivalent. In Table 13.15 we provide the calculations needed to use equation (13.29). Using these computations and equation (13.29), we obtain:

TABLE 13.15   COMPUTATIONS FOR USING THE ALTERNATE FORMULA FOR COMPUTING r_{xy} FOR THE STEREO AND SOUND-EQUIPMENT PROBLEM

	x_i	y_i	x_iy_i	x_i^2	y_i^2
	2	24	48	4	576
	5	28	140	25	784
	1	22	22	1	484
	3	26	78	9	676
	4	25	100	16	625
	1	24	24	1	576
	5	26	130	25	676
Totals	21	175	542	81	4397

$$r_{xy} = \frac{542 - (21)(175)/7}{\sqrt{81 - (21)^2/7}\sqrt{4397 - (175)^2/7}} = \frac{17}{19.8997} = .854$$

Thus, we see that the value obtained for r_{xy} using equation (13.29) is the same as the value obtained using equation (13.28). ◀

The formula for computing the correlation coefficient of a population, denoted by the Greek letter ρ_{xy} (rho, pronounced "row"), is as follows.

Population Correlation Coefficient

$$\rho_{xy} = \frac{\sigma_{xy}}{\sigma_x \sigma_y} \tag{13.30}$$

where

ρ_{xy} = population correlation coefficient
σ_{xy} = population covariance
σ_x = population standard deviation for x
σ_y = population standard deviation for y

The sample correlation coefficient r_{xy} is an estimate of the population correlation coefficient ρ_{xy}.

INTERPRETATION OF THE CORRELATION COEFFICIENT

First let us consider a simple example that illustrates the concept of perfect positive linear association.

EXAMPLE 13.6 ▶ The scatter diagram shown in Figure 13.25 depicts the relationship between the following $n = 3$ pairs of points.

x_i	1	2	3
y_i	10	30	50

FIGURE 13.25   **SCATTER DIAGRAM FOR EXAMPLE 13.6: A PERFECT POSITIVE LINEAR ASSOCIATION**

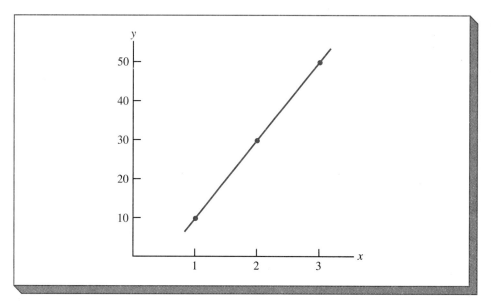

The straight line drawn through each of the three points shows that there is a perfect linear relationship between the two variables x and y. The calculations needed to compute r_{xy} are shown in Table 13.16. Using the values in this table, we obtain

$$r_{xy} = \frac{\Sigma x_i y_i - (\Sigma x_i \Sigma y_i)/n}{\sqrt{\Sigma x_i^2 - (\Sigma x_i)^2/n}\sqrt{\Sigma y_i^2 - (\Sigma y_i)^2/n}}$$

$$= \frac{220 - (6)(90)/3}{\sqrt{14 - (6)^2/3}\sqrt{3500 - (90)^2/3}} = \frac{40}{40} = 1$$

Thus, we see that the value of the sample correlation coefficient for this data set is 1. ◀

In general, it can be shown that if all the points in a data set fall on a straight line having positive slope, then the value of the sample correlation coefficient is +1; that is, a sample correlation coefficient of +1 corresponds to a perfect positive linear association between x and y. Moreover, if the points in the data set fall on a straight line having negative slope, the value of the sample correlation coefficient is −1; that is, a sample

TABLE 13.16   **CALCULATIONS FOR COMPUTING r_{xy} FOR EXAMPLE 13.6**

	x_i	y_i	$x_i y_i$	x_i^2	y_i^2
	1	10	10	1	100
	2	30	60	4	900
	3	50	150	9	2500
Totals	6	90	220	14	3500

correlation coefficient of -1 corresponds to a perfect negative linear association between x and y.

Let us now suppose that for a certain data set there is a positive linear association between x and y but that the relationship is not perfect. The value of r_{xy} will be less than 1, indicating that the points in the scatter diagram do not all fall on a straight line. As the points in a data set deviate more and more from a perfect positive linear association, the value of r_{xy} becomes smaller and smaller. A value of r_{xy} equal to zero indicates no linear relationship between x and y, and values of r_{xy} near zero indicate a weak relationship.

EXAMPLE 13.5
(CONTINUED)

Recall that for the data set involving the stereo and sound-equipment store, $r_{xy} = +.854$. Since $r_{xy} = +.854$, we conclude that there is a positive linear association between the number of commercials and Saturday sales volume. More specifically, an increase in the number of commercials is associated with an increase in sales volume. ◀

We have stated that values of r_{xy} near $+1$ indicate a strong linear association between two variables and values of r_{xy} near zero indicate little or no linear association between the variables. But we must be careful not to conclude that a value of r_{xy} near zero means there is no relationship between the variables.

EXAMPLE 13.7

The scatter diagram in Figure 13.26 shows a case where $r_{xy} = 0$ and there is no linear relationship; however, in this case, there is a perfect curvilinear relationship between the

FIGURE 13.26   SCATTER DIAGRAM FOR EXAMPLE 13.7: A PERFECT CURVILINEAR RELATIONSHIP WITH $r_{xy} = 0$

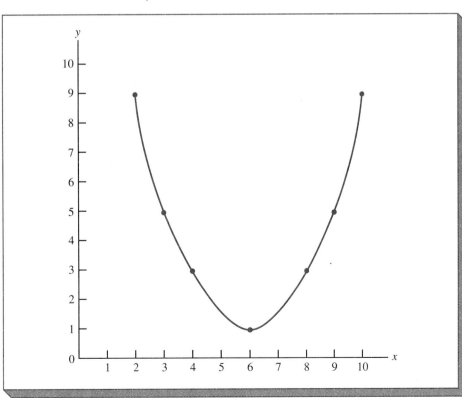

variables. Table 13.17 provides the calculations needed to compute r_{xy} for this example. Using these calculations, the computation of r_{xy} is as follows:

$$r_{xy} = \frac{210 - (42)(35)/7}{\sqrt{310 - (42)^2/7}\sqrt{231 - (35)^2/7}} = \frac{0}{56.9912} = 0$$

◄

To reiterate, Example 13.7 illustrates an important concept regarding the proper interpretation of the sample correlation coefficient. The sample correlation coefficient measures only the degree of *linear association* between the two variables. A value of r_{xy} equal to zero cannot be interpreted as implying that there is no relationship between the two variables. One should always look at the associated scatter diagram as well as the value of the sample correlation coefficient when attempting to determine if, and how, two variables are related.

In closing this part of our discussion on correlation, we caution that while a correlation coefficient near ±1 does imply a strong linear association between two variables, it does not imply a cause-and-effect relationship. Conclusions concerning cause-and-effect must be based on the judgment of the analyst.

DETERMINING THE SAMPLE CORRELATION COEFFICIENT FROM THE REGRESSION ANALYSIS OUTPUT

In this discussion, we will assume that the least squares estimated regression equation is $\hat{y} = b_0 + b_1 x$. In such cases, the sample correlation coefficient can be computed using one of the following formulas.

Sample Correlation Coefficient

$$r_{xy} = (\text{sign of } b_1)\sqrt{\text{Coefficient of Determination}} = \pm\sqrt{r^2} \qquad \textbf{(13.31)}$$

$$r_{xy} = b_1\left(\frac{s_x}{s_y}\right) \qquad \textbf{(13.32)}$$

where

$$b_1 = \text{slope of the estimated regression equation}$$
$$s_x = \text{sample standard deviation of } x$$
$$s_y = \text{sample standard deviation of } y$$

Note that the sign of the sample correlation coefficient is the same as the sign of b_1, the slope of the estimated regression equation.

EXAMPLE 13.1 ►
(CONTINUED)

For the Armand's Pizza Parlors problem presented earlier in this chapter, $b_1 = 5$, indicating a positive relationship. Thus, with $r^2 = .903$ we obtain

$$r_{xy} = \pm\sqrt{.903}$$

$$= +.95$$

◄

TABLE 13.17   CALCULATIONS FOR COMPUTING r_{xy} FOR EXAMPLE 13.7

	x_i	y_i	$x_i y_i$	x_i^2	y_i^2
	2	9	18	4	81
	3	5	15	9	25
	4	3	12	16	9
	6	1	6	36	1
	8	3	24	64	9
	9	5	45	81	25
	10	9	90	100	81
Totals	42	35	210	310	231

TESTING FOR SIGNIFICANCE

The sample correlation coefficient is a point estimator of the population correlation coefficient. With ρ_{xy} denoting the population correlation coefficient, a statistical test for the significance of a linear association between x and y can be performed by testing the following hypotheses:

$$H_0: \rho_{xy} = 0$$

$$H_a: \rho_{xy} \neq 0$$

It can be shown that testing these hypotheses is equivalent to testing the hypotheses regarding the significance of β_1, the slope of the regression equation. Recall that the appropriate hypotheses in this case are

$$H_0: \beta_1 = 0$$

$$H_a: \beta_1 \neq 0$$

Since for the Armand's Pizza Parlors problem we earlier rejected the null hypothesis $H_0: \beta_1 = 0$ (see Section 13.4), we can also reject the null hypothesis $H_0: \rho_{xy} = 0$ and conclude that x and y are correlated. Alternatively, statisticians have developed a procedure for testing the following hypothesis without performing a regression study:

$$H_0: \rho_{xy} = 0$$

$$H_a: \rho_{xy} \neq 0$$

It can be shown that if H_0 is true, then the value of

$$r_{xy} \sqrt{\frac{n-2}{1-r_{xy}^2}} \tag{13.33}$$

has a t distribution with $n - 2$ degrees of freedom.

EXAMPLE 13.1 ▶ (CONTINUED)

For Armand's Pizza Parlors with $\alpha = .05$ and $n - 2 = 10 - 2 = 8$ degrees of freedom, we see that the appropriate t value from Table 2 of Appendix B is 2.306. Thus, if the value of expression (13.33) exceeds 2.306 or is less than -2.306, we must reject the null hypothesis $H_0: \rho_{xy} = 0$.

With a sample correlation coefficient of $r_{xy} = .95$, the value of expression (13.33) is

$$.95\sqrt{\frac{8}{1 - .903}} = 8.63$$

Since 8.63 exceeds the t value of 2.306, we reject H_0 and hence conclude that x and y have a significant correlation. We note that this test yields the same result as the previous test on β_1.

EXERCISES

METHODS

SELF TEST

57. Given are five observations taken for two variables.

x_i	4	6	11	3	16
y_i	50	50	40	60	30

    a. Develop a scatter diagram with x on the horizontal axis.
    b. What does the scatter diagram developed in part (a) indicate about the relationship between the two variables?
    c. Compute and interpret the sample covariance for these data.
    d. Compute and interpret the sample correlation coefficient for these data.

58. Given are five observations taken for two variables.

x_i	6	11	15	21	27
y_i	6	9	6	17	12

    a. Develop a scatter diagram for these data.
    b. What does the scatter diagram indicate about a possible relationship between x and y?
    c. Compute and interpret the sample covariance for these data.
    d. Compute and interpret the sample correlation coefficient for these data.

59. Eight observations on two random variables are given.

x_i	2	9	6	8	4	7	5	6
y_i	11	4	6	5	9	4	9	7

    a. Compute r_{xy}.
    b. Test the hypotheses

$$H_0: \rho_{xy} = 0$$

$$H_a: \rho_{xy} \neq 0$$

at the $\alpha = .01$ level of significance.

APPLICATIONS

60. A high school guidance counselor collected the following data regarding the grade point average (GPA) and the SAT mathematics test score for six seniors.

GPA	2.7	3.5	3.7	3.3	3.6	3.0
SAT	440	560	720	520	640	480

a. Develop a scatter diagram for these data with GPA as the independent variable.

b. Does there appear to be any relationship between the GPA and the SAT mathematics test score? Explain.

c. Compute and interpret the sample covariance for these data.

d. Compute the sample correlation coefficient for these data. What does this value tell us about the relationship between the two variables?

61. A study conducted by a department of transportation regarding driving speed and mileage for midsize automobiles resulted in the data shown below.

Driving Speed	Mileage	Driving Speed	Mileage
30	28	25	32
50	25	60	21
40	25	25	35
55	23	50	26
30	30	55	25

a. Compute and interpret the sample correlation coefficient for these data.

b. Test the hypotheses

$$H_0 : \rho_{xy} = 0$$

$$H_a : \rho_{xy} \neq 0$$

at the $\alpha = .01$ level of significance.

62. A sociologist collected the following data regarding the ages of wives and husbands when they married.

Wife's Age	19	42	28	25	36
Husband's Age	20	32	31	24	33

a. Develop a scatter diagram for these data with the wife's age on the horizontal axis.

b. Does there appear to be a linear association? Explain.

c. Compute and interpret the sample correlation coefficient for these data.

63. The following estimated regression equation has been developed to estimate the relationship between x, the number of units produced per week, and y, the total weekly cost of production ($):

$$\hat{y} = 60 + 3.2x$$

The standard deviation of weekly production is 10 units, and the standard deviation of weekly cost is $35.00. Compute the sample correlation coefficient r_{xy}.

64. As more U.S. households receive cable television, the advertising revenue has continued to increase. The following data show the 1988 and 1987 expenditures for the top 10 cable television advertisers.

Advertiser	1988 Expenditure ($1,000,000s)	1987 Expenditure ($1,000,000s)
Procter & Gamble	30.2	23.7
Philip Morris	23.1	20.6
Anheuser-Busch	21.4	22.9
Time	21.2	16.4
General Mills	20.0	18.6
RJR Nabisco	14.3	14.7
Eastman Kodak	11.0	2.6
Clorox	10.1	6.9
Mars	10.0	14.9
Chrysler	9.5	6.1

a. Develop a scatter diagram for the data. Does it appear that the two variables are linearly related?

b. Compute the sample correlation coefficient for these data.

c. Test the hypotheses

$$H_0: \rho_{xy} = 0$$

$$H_a: \rho_{xy} \neq 0$$

at the $\alpha = .01$ level of significance.

 HIGHLOW   **65.** The daily high and low temperatures for 24 cities are given below (*USA Today*, April 6, 1992):

City	High	Low	City	High	Low
Tampa	80	58	Birmingham	68	32
Kansas City	69	40	Minneapolis	62	39
Boise	58	35	Portland	50	41
Los Angeles	71	57	Memphis	67	41
Philadelphia	56	35	Buffalo	44	28
Milwaukee	47	29	Cincinnati	55	29
Chicago	52	25	Charlotte	61	37
Albany	50	28	Boston	50	35
Houston	63	50	Tulsa	73	50
Salt Lake City	61	49	Washington, D.C.	56	35
Miami	79	56	Las Vegas	80	53
Cheyenne	66	35	Detroit	52	29

a. What is the correlation between the high and low temperatures?

b. Using a .05 level of significance, test for a significant correlation. What is your conclusion?

SUMMARY

In this chapter we introduced the topics of regression and correlation analysis. We discussed how regression analysis can be used to develop an equation showing how variables are related and how correlation analysis can be used to determine the strength of the relationship between two variables. Before concluding our discussion, however, we would like to reemphasize a potential misinterpretation of these studies. Regression and correlation analyses can indicate only how or to what extent the variables are associated with each other. These techniques cannot be interpreted directly as showing cause-and-effect relationships.

GLOSSARY

Note: The definitions here are all stated with the understanding that simple linear regression and correlation are being considered.

Dependent variable The variable that is being predicted or explained. It is denoted by y in the regression equation.

Independent variable The variable that is doing the predicting or explaining. It is denoted by x in the regression equation.

Simple linear regression The simplest kind of regression, involving only two variables that are related approximately by a straight line.

Scatter diagram A graph of the available data in which the independent variable appears on the horizontal axis and the dependent variable appears on the vertical axis.

Least squares method The approach used to develop the estimated regression equation that minimizes the sum of squared residuals.

Estimated regression equation The estimate of the regression equation obtained by the least squares method; that is, $\hat{y} = b_0 = b_1 x$.

Coefficient of determination (r^2) A measure of how well the estimated regression equation fits the data.

Residual The difference between the observed value of the dependent variable and the value predicted using the estimated regression equation; that is, $y_i - \hat{y}_i$.

Deterministic model A relationship between an independent variable and a dependent variable whereby specifying the value of the independent variable allows one to compute the value of the dependent variable exactly.

Probabilistic model A relationship between an independent variable and a dependent variable in which specifying the value of the independent variable is not sufficient to allow determination of the value of the dependent variable.

Regression equation The mathematical equation relating the independent variable to the expected value of the dependent variable; that is, $E(y) = \beta_0 + \beta_1 x$.

Standardized residual The value obtained by dividing the residual by its standard deviation.

Sample correlation coefficient (r_{xy}) A statistical measure of the linear association between two variables.

KEY FORMULAS

Estimated Regression Equation

$$\hat{y} = b_0 + b_1 x \tag{13.1}$$

Slope and y-Intercept for the Estimated Regression Equation

$$b_1 = \frac{\Sigma(x_i - \bar{x})(y_i - \bar{y})}{\Sigma(x_i - \bar{x})^2} = \frac{\Sigma x_i y_i - (\Sigma x_i \Sigma y_i)/n}{\Sigma x_i^2 - (\Sigma x_i)^2/n} \tag{13.3}$$

$$b_0 = \bar{y} - b_1 \bar{x} \tag{13.4}$$

Sum of Squares Due to Error

$$SSE = \Sigma(y_i - \hat{y}_i)^2 \tag{13.5}$$

Total Sum of Squares

$$SST = \Sigma(y_i - \bar{y})^2 \tag{13.6}$$

Sum of Squares Due to Regression

$$SSR = \Sigma(\hat{y}_i - \bar{y})^2 \tag{13.7}$$

Relationship Among SST, SSR, and SSE

$$SST = SSR + SSE \tag{13.8}$$

Coefficient of Determination

$$r^2 = \frac{SSR}{SST} \tag{13.9}$$

Computational Formula for SSR

$$SSR = \frac{[\Sigma x_i y_i - (\Sigma x_i \Sigma y_i)/n]^2}{\Sigma x_i^2 - (\Sigma x_i)^2/n} \tag{13.10}$$

Computational Formula for SST

$$SST = \Sigma y_i^2 - \frac{(\Sigma y_i)^2}{n} \tag{13.11}$$

Regression Model

$$y = \beta_0 + \beta_1 x + \epsilon \tag{13.12}$$

Regression Equation

$$E(y) = \beta_0 + \beta_1 x \tag{13.13}$$

Mean Square Error (Estimate of σ^2)

$$s^2 = MSE = \frac{SSE}{n - 2} \tag{13.14}$$

Standard Error of the Estimate

$$s = \sqrt{MSE} = \sqrt{\frac{\Sigma(y_i - \hat{y}_i)^2}{n - 2}} \tag{13.15}$$

Standard Deviation of b_1

$$\sigma_{b_1} = \frac{\sigma}{\sqrt{\Sigma x_i^2 - (\Sigma x_i)^2/n}} \tag{13.16}$$

Estimated Standard Deviation of b_1

$$s_{b_1} = \frac{s}{\sqrt{\Sigma x_i^2 - (\Sigma x_i)^2/n}} \tag{13.17}$$

Mean Square Due to Regression

$$\text{MSR} = \frac{\text{SSR}}{\text{Number of Independent Variables}} \tag{13.18}$$

The F Statistic

$$F = \frac{\text{MSR}}{\text{MSE}} \tag{13.19}$$

Estimated Standard Deviation of $\hat{y}_p$

$$s_{\hat{y}_p} = s\sqrt{\frac{1}{n} + \frac{(x_p - \bar{x})^2}{\Sigma x_i^2 - (\Sigma x_i)^2/n}} \tag{13.20}$$

Confidence Interval Estimate of $E(y_p)$

$$\hat{y}_p \pm t_{\alpha/2} s_{\hat{y}_p} \tag{13.21}$$

Estimated Standard Deviation When Predicting an Individual Value

$$s_{\text{ind}} = s\sqrt{1 + \frac{1}{n} + \frac{(x_p - \bar{x})^2}{\Sigma x_i^2 - (\Sigma x_i)^2/n}} \tag{13.22}$$

Prediction Interval Estimate of y_p

$$\hat{y}_p \pm t_{\alpha/2} s_{\text{ind}} \tag{13.23}$$

Standard Deviation of ith Residual

$$s_{y_i - \hat{y}_i} = \sqrt{s^2(1 - h_i)} \tag{13.24}$$

Leverage of Observation i

$$h_i = \frac{1}{n} + \frac{(x_i - \bar{x})^2}{\Sigma(x_i - \bar{x})^2} \tag{13.25}$$

Sample Covariance

$$s_{xy} = \frac{\Sigma(x_i - \bar{x})(y_i - \bar{y})}{n - 1} \tag{13.26}$$

Sample Correlation Coefficient

$$r_{xy} = \frac{s_{xy}}{s_x s_y} \tag{13.28}$$

Sample Correlation Coefficient, Alternate Formula

$$r_{xy} = \frac{\Sigma x_i y_i - (\Sigma x_i \Sigma y_i)/n}{\sqrt{\Sigma x_i^2 - (\Sigma x_i)^2/n}\sqrt{\Sigma y_i^2 - (\Sigma y_i)^2/n}} \tag{13.29}$$

Determining the Sample Correlation Coefficient from the Regression Analysis Output

$$r_{xy} = (\text{sign of } b_1)\sqrt{\text{Coefficient of Determination}} = \pm\sqrt{r^2} \tag{13.31}$$

REVIEW QUIZ

TRUE/FALSE

1. The least squares method is used to determine an estimated regression line that minimizes the squared deviations of the data values from the line.
2. The least squares method is applicable only in situations where the estimated regression line has a positive slope.
3. If the slope of the estimated regression line is positive, the correlation coefficient must be negative.
4. The slope of the estimated regression line (b_1) is a sample statistic, since, like other sample statistics, it is computed from the sample observations.
5. The coefficient of determination is the square root of the correlation coefficient.
6. The sum of squares due to regression (SSR) plus the sum of squares due to error (SSE) must equal the total sum of squares (SST).
7. A t test can be used to test whether or not there is a significant regression relationship.
8. The sampling distribution of b_1 is normal if the usual regression assumptions are satisfied.
9. An interval estimate for a particular value of the dependent variable yields a smaller interval than an interval estimate of the expected value of the dependent variable.
10. Residual analysis is only used to check the assumptions regarding ϵ.
11. If two variables are perfectly linearly related, the sample correlation coefficient must equal -1 or 1.
12. The residual is the difference between the actual value of a dependent variable and the value predicted by the estimated regression line.

MULTIPLE CHOICE

13. If two variables x and y have a significant linear relationship, then
    a. there may or may not be any causal relationship between x and y
    b. x causes y to happen
    c. y causes x to happen
14. For the estimated regression line $\hat{y} = 3 - 10x$, the correlation coefficient r_{xy}
    a. equals 0
    b. is less than 0
    c. is greater than 0
15. If the correlation coefficient for two variables is $-.9$, the coefficient of determination is
    a. .9
    b. $-.81$
    c. .81
16. If a data set has SST $= 200$ and SSE $= 150$, then the coefficient of determination is
    a. .25
    b. .50
    c. .75
    d. 50
17. Compared to the prediction interval for a particular value of y, the confidence interval for a mean value of y will be
    a. narrower
    b. wider
    c. not enough information is given
18. A sample correlation coefficient is calculated from 15 pairs of x and y observations. The t distribution used to determine whether or not this coefficient is statistically significant will have how many degrees of freedom?
    a. 13   b. 14
    c. 15   d. 28

19. Which of the following is correct?
    a. SST = SSR − SSE
    b. SSE = SSR + SSE
    c. SSR = SSE + SST
    d. SSE = SST − SSR

20. The coefficient of determination is calculated as
    a. SST/SSE
    b. SSR/SST
    c. SSR/SSE
    d. SSE/SSR

21. Which of the following is an appropriate test statistic to test the null hypothesis that there is no linear relationship between x and y?
    a. SSR/SST
    b. MSE
    c. MSR/MSE
    d. MSE/MST

22. Which of the following points are *always* on the estimated linear regression line?
    a. $x = \bar{x}, y = \bar{y}$
    b. $x = 0, y = 0$
    c. $x = 1, y = 0$
    d. $x = 0, y = 1$

▽

SUPPLEMENTARY EXERCISES

66. What is the difference between regression analysis and correlation analysis?

67. Does a high value of r^2 imply that two variables are causally related? Explain.

68. In your own words, explain the difference between an interval estimate of the mean value of y for a given x and an interval estimate for an individual value of y for a given x.

69. What is the purpose of testing whether or not $\beta_1 = 0$? If we reject $\beta_1 = 0$, does this imply a good fit?

70. In a manufacturing process, the assembly-line speed (feet per minute) was thought to affect the number of defective parts found during the inspection process. To test this theory, management devised a situation where the same batch of parts was inspected visually at a variety of line speeds. The data collected are shown below.

Line Speed	Number of Defective Parts Found	Line Speed	Number of Defective Parts Found
20	21	30	16
20	19	60	14
40	15	40	17

    a. Develop the estimated regression equation that relates line speed to the number of defective parts found.
    b. At the $\alpha = .05$ level of significance, determine whether line speed and number of defective parts found are related.
    c. Did the estimated regression equation provide a good fit to the data?
    d. Develop a 95% confidence interval to predict the mean number of defective parts for a line speed of 50 feet per minute.

71. A study was conducted by Monsanto Company to determine the relationship between the percentage of supplemental methionine used in feed and the body weight of poultry. Using the data collected in this study, regression analysis was used to develop the following estimated regression line:

$$\hat{y} = .21 + .42x$$

where

$\hat{y}$ = estimated body weight in kilograms

x = percentage of supplemental methionine used in the feed

The coefficient of determination r^2 was .78, indicating a reasonably good fit for the data. Suppose it is known that a sample size of 30 was used for the study and that SST = 45.
a. Compute SSR and SSE.
b. Test for a significant regression relationship using $\alpha = .01$.
c. What is the value of the sample correlation coefficient?

72. The PJH&D Company is in the process of deciding whether to purchase a maintenance contract for its new word-processing system. They feel that maintenance expense should be related to usage and have collected the information shown below on weekly usage (hours) and annual maintenance expense.

Weekly Usage (hours)	Annual Maintenance Expense ($100s)	Weekly Usage (hours)	Annual Maintenance Expense ($100s)
13	17.0	17	30.5
10	22.0	24	32.5
20	30.0	31	39.0
28	37.0	40	51.5
32	47.0	38	40.0

a. Develop the estimated regression equation that relates annual maintenance expense, in hundreds of dollars, to weekly usage.
b. Test the significance of the relationship in part (a) at the $\alpha = .05$ level of significance.
c. PJH&D expects to operate the word processor 30 hours per week. Develop a 95% prediction interval for the company's annual maintenance expense.
d. If the maintenance contract costs $3000 per year, would you recommend purchasing it? Why or why not?

73. A sociologist was hired by a large city hospital to investigate the relationship between the number of unauthorized days that an employee is absent per year and the distance (miles) between home and work for the employee. A sample of 10 employees was chosen, and the following data were collected.

Distance to Work (miles)	Number of Days Absent	Distance to Work (miles)	Number of Days Absent
1	8	10	3
3	5	12	5
4	8	14	2
6	7	14	4
8	6	18	2

a. Develop a scatter diagram for these data. Does a linear relationship appear reasonable? Explain.
b. Develop the least squares estimated regression equation.
c. Is there a significant relationship between the two variables? Use $\alpha = .05$.
d. Did the estimated regression equation provide a good fit? Explain.
e. Use the estimated regression equation developed in part (b) to develop a 95% confidence interval estimate of the expected number of days absent for employees living 5 miles from the company.

74. The owner of a chain of fast-food restaurants would like to investigate the relationship between the daily sales volume of a company restaurant and the number of competitor restaurants within a 1-mile radius of the firm's restaurant. The following data were collected.

Number of Competitors	Sales ($)	Number of Competitors	Sales ($)
1	3600	3	2700
1	3300	4	2500
2	3100	5	2300
3	2900	5	2000

a. Develop the least squares estimated regression equation that relates daily sales volume to the number of competitor restaurants within a 1-mile radius.
b. Is there a significant relationship between the two variables? Use $\alpha = .05$.
c. Did the estimated regression equation provide a good fit? Explain.
d. Use the estimated regression line developed in part (a) to develop a 95% interval estimate of the daily sales volume for a particular company restaurant that has four competitors within a 1-mile radius.

75. Performance data for a Century Coronado 21 with a 310-hp MerCruiser V-8 gasoline inboard engine was reported in *Boating,* September 1991. Data on how the engine revolutions per minute (rpm) affected boat speed in miles per hour (mph) are shown below.

rpm	mph	rpm	mph
1000	6.1	3500	33.6
1500	10.7	4000	37.9
2000	20.9	4500	40.2
2500	27.5	4800	40.7
3000	31.5		

a. Develop the estimated regression equation showing how boat speed is related to the engine revolutions per minute.
b. Test the significance of the relationship at the $\alpha = .05$ level of significance.
c. Develop a plot of the standardized residuals against $\hat{y}$. What conclusions can you draw from this plot?

76. The 1992 U.S. Men's Olympic Marathon Trials (Columbus, Ohio, April 11, 1992) provided marathon qualifying times and ages for 109 runners. Data below show the number of minutes that the qualifying times exceeded 2 hours and the age for a sample of eight runners.

Age	Qualifying Time	Age	Qualifying Time
33	12.1	26	14.1
33	12.6	25	14.6
31	13.1	30	15.0
26	14.0	29	15.5

   a. Develop the estimated regression equation showing how qualifying time is related to age.

   b. Test the significance of the relationship at the $\alpha = .05$ level of significance.

   c. Develop a plot of the standardized residuals against $\hat{y}$. What conclusions can you draw from this plot?

77. The regional transit authority for a major metropolitan area would like to determine if there is any relationship between the age of a bus and the annual maintenance cost. A sample of 10 buses resulted in the data shown below. Compute the sample correlation coefficient.

Age of Bus (years)	Maintenance Cost ($)	Age of Bus (years)	Maintenance Cost ($)
1	350	3	550
2	370	4	750
2	480	4	800
2	520	5	790
2	590	5	950

78. Reconsider the regional transit authority problem presented in Exercise 77.

   a. Develop the least squares estimated regression equation.

   b. Test to see if the two variables are significantly related at $\alpha = .05$.

   c. Did the least squares line provide a good fit to the observed data? Explain.

   d. Develop a 95% prediction interval for the maintenance cost for a specific bus that is 4 years old.

79. A psychology professor at Givens College is interested in the relation between hours spent studying and total points earned in the course. Data collected on 10 students who took the course last quarter are given below.

Hours Spent Studying	Total Points Earned	Hours Spent Studying	Total Points Earned
45	40	65	50
30	35	90	90
90	75	80	80
60	65	55	45
105	90	75	65

Compute the sample correlation coefficient for these data.

80. Reconsider the Givens College data in Exercise 79.

    a. Develop an estimated regression equation relating total points earned to hours spent studying.

    b. Test the significance of the model at the $\alpha = .05$ level.

    c. Predict the total points earned by Mark Sweeney. He spent 95 hours studying.

    d. Develop a 95% prediction interval for the total points earned by Mark Sweeney.

81. *USA Today* publishes college basketball computer rankings that are based on a strength rating computed for each team. The strength ratings can also be used to predict the victory margin for games. For the visiting team, the predicted score is its strength rating. For the home team, the predicted score is the strength rating plus 4½. For each game, the actual victory margin is the winning team's score minus the losing team's score. The predicted victory margin is the predicted score for the winning team minus the predicted score for the losing team.

    A sample was taken of 10 college basketball games selected from *USA Today*, December 13, 1988, to investigate the accuracy of the predicted victory margin. The following table gives the data obtained. The strength ratings are in parentheses.

Visiting Team	Score	Home Team	Score
Eastern Michigan (77.06)	57	Michigan (101.07)	80
Jackson State (65.75)	71	Iowa (96.63)	86
Georgia Southern (80.01)	80	Eastern Kentucky (61.24)	69
Seton Hall (92.66)	96	Rutgers (72.22)	70
Niagara (70.32)	78	St. Bonaventure (70.57)	81
Fairfield (63.01)	48	Connecticut (85.15)	71
S. Carolina St. (73.29)	70	Clemson (80.13)	93
Monmouth (57.43)	70	Maryland (80.53)	74
Illinois-Chicago (73.95)	74	Michigan State (83.45)	96
Oral Roberts (71.37)	75	Georgetown (87.48)	91

    a. Let x be the predicted victory margin and y be the actual victory margin. Use the strength ratings and actual game scores to compute x and y for the 10 games.

    b. Compute the correlation coefficient between the predicted victory margin and the actual victory margin.

    c. Test for a significant relationship. Use $\alpha = .01$.

    d. Develop an estimated regression equation using the predicted victory margin as the independent variable and the actual victory margin as the dependent variable.

    e. What are the values of b_0 and b_1? Comment on what the values of β_0 and β_1 should be for an ideal system of predicting victory margins.

COMPUTER EXERCISE: U.S. DEPARTMENT OF TRANSPORTATION

 SAFETY

As part of a study on transportation safety, the U.S. Department of Transportation collected data on the number of fatal accidents per 1000 licenses and the percentage of licensed drivers under the age of 21 in a sample of 42 cities. Data collected over a 1-year period are shown below. These data are available on the data disk in the file named SAFETY.

Percent Under 21	Fatal Accidents Per 1000 Licenses	Percent Under 21	Fatal Accidents Per 1000 Licenses
13	2.962	17	4.100
12	0.708	8	2.190
8	0.885	16	3.623
12	1.652	15	2.623
11	2.091	9	0.835
17	2.627	8	0.820
18	3.830	14	2.890
8	0.368	8	1.267
13	1.142	15	3.224
8	0.645	10	1.014
9	1.028	10	0.493
16	2.801	14	1.443
12	1.405	18	3.614
9	1.433	10	1.926
10	0.039	14	1.643
9	0.338	16	2.943
11	1.849	12	1.913
12	2.246	15	2.814
14	2.855	13	2.634
14	2.352	9	0.926
11	1.294	17	3.256

REPORT

1. Develop numerical and graphical summaries of the data.
2. Use regression analysis to investigate the relationship between the number of fatal accidents and the percentage of drivers under the age of 21. Discuss your findings.
3. What conclusions and/or recommendations can your derive from your analysis?

APPENDIX

CALCULUS-BASED DERIVATION OF LEAST SQUARES FORMULAS

As mentioned in the chapter, the least squares method is a procedure for determining the values of b_0 and b_1 that minimize the sum of squared residuals. The sum of squared residuals is given by

$$\Sigma(y_i - \hat{y}_i)^2$$

Substituting $\hat{y}_i = b_0 + b_1 x_i$, we get

$$\Sigma(y_i - b_0 - b_1 x_i)^2 \tag{13A.1}$$

as the expression that must be minimized.

To minimize expression (13A.1), we must take the partial derivatives with respect to b_0 and b_1, set them equal to zero, and solve. Doing so, we get

$$\frac{\partial \Sigma(y_i - b_0 - b_1 x_1)^2}{\partial b_0} = -2\Sigma(y_i - b_0 - b_1 x_i) = 0 \tag{13A.2}$$

$$\frac{\partial\Sigma(y_i - b_0 - b_1 x_i)^2}{\partial b_1} = -2\Sigma x_i(y_i - b_0 - b_1 x_i) = 0 \tag{13A.3}$$

Dividing equation (13A.2) by 2 and summing each term individually yields

$$-\Sigma y_i + \Sigma b_0 + \Sigma b_1 x_i = 0$$

Bringing Σy_i to the other side of the equal sign and noting that $\Sigma b_0 = nb_0$, we obtain

$$nb_0 + (\Sigma x_i)b_1 = \Sigma y_i \tag{13A.4}$$

Similar algebraic simplification applied to equation (13A.3) yields

$$(\Sigma x_i)b_0 + (\Sigma x_i^2)b_1 = \Sigma x_i y_i \tag{13A.5}$$

Equations (13A.4) and (13A.5) are known as the *normal equations*. Solving equation (13A.4) for b_0 yields

$$b_0 = \frac{\Sigma y_i}{n} - b_1\frac{\Sigma x_i}{n} \tag{13A.6}$$

Using equation (13A.6) to substitute for b_0 in equation (13A.5) provides

$$\frac{\Sigma x_i \Sigma y_i}{n} - \frac{(\Sigma x_i)^2}{n}b_1 + (\Sigma x_i^2)b_1 = \Sigma x_i y_i \tag{13A.7}$$

Rearranging equation (13A.7), we obtain

$$b_1 = \frac{\Sigma x_i y_i - (\Sigma x_i \Sigma y_i)/n}{\Sigma x_i^2 - (\Sigma x_i)^2/n} \tag{13A.8}$$

Since $\bar{y} = \Sigma y_i/n$ and $\bar{x} = \Sigma x_i/n$, we can rewrite equation (13A.6):

$$b_0 = \bar{y} - b_1\bar{x} \tag{13A.9}$$

Equations (13A.8) and (13A.9) are the formulas we used in the chapter to compute the coefficients in the estimated regression equation.

CHAPTER 14

MULTIPLE REGRESSION

WHAT YOU WILL LEARN IN THIS CHAPTER

- What multiple regression analysis is
- How to interpret the coefficients in a multiple regression model
- The important role computer packages play in performing multiple regression analysis
- How to use the t and F distributions to test for significant relationships in multiple regression analysis
- How to use qualitative variables in regression analysis
- How to analyze the residuals in a multiple regression model

CONTENTS

Father Versus Mother in Custody for Children of Divorce

Mothers have traditionally been awarded custody of children in divorce cases, due to the presumption by the courts that mothers were uniquely suited to raising children and also because of the powerful influence of our economic system, which has favored male employment. Thus, children tend to go with the mother, whereas the father works to provide the financial support. However, the changing roles of both men and women and the various interrelated social changes have recently caused courts to consider fathers as custodians of children in divorce cases.

Although there are many variables, or factors, that need to be evaluated in making a custody decision, one consideration is that of providing the best environment for the children's academic achievement. Researchers Frederick Shilling and Patrick Lynch have investigated the effect that custody with the father rather than the mother has on the academic achievement of eighth-grade children. Data for the Shilling-Lynch study were obtained from a sample of 3160 single-parent children in the eighth grade. In the sample, 550 children lived with their fathers, and 2610 children lived with their mothers.

The children were measured on their academic performance in reading, mathematics, and a composite of these two measures. Data were also collected on a measure of socioeconomic status based on parental occupation and education, sex of the child, sex of the single parent, residence (urban, suburban, rural) and a measure of the student's perception of parental interest in school. Using multiple regression, the research-

Divorce courts have historically favored the mother in custody battles over children.

ers developed an equation that used the socioeconomic status, sex of child, sex of parent, residence, and parental interest variables to predict academic performance scores. In terms of the sex of the single parent, the researchers found that verbal, mathematical, and overall academic achievement scores were better for children living with their mothers, other things being equal. The study can be construed as an encouragement to courts to continue assigning young children of broken families to the mothers instead of to the fathers, especially when academic achievement is an important factor in determining the best placement for the child.

This *Statistics in Practice* is based on "Father Versus Mother Custody and Academic Achievement of Eighth Grade Children," by Frederick Shilling and Patrick D. Lynch, *Journal of Research and Development in Education* 18 (1985).

14.1 THE MULTIPLE REGRESSION MODEL AND ITS ASSUMPTIONS

To introduce multiple regression analysis, let us consider a simple example to which the techniques of multiple regression analysis can be applied.

EXAMPLE 14.1 ▶ Researchers studied the relationship of beer consumption, state alcohol policies, and motor vehicle regulations to the number of fatal automobile accidents. One objective of the study was to develop an equation that could be used to predict the number of fatal accidents given the driving age (percent of drivers under 21) and the number of outlets selling alcohol for on-premises consumption. The number of fatal accidents would be the dependent variable, and the driving age and the number of outlets would be the independent variables.   ◀

In Example 14.1 we see that there are two independent variables, which can be denoted as

$$x_1 = \text{driving age}$$

$$x_2 = \text{number of outlets}$$

The advantage of this notation is that with multiple regression problems involving more than two independent variables, we can continue to refer to each independent variable as *x* with an appropriate subscript. As in Chapter 13, we refer to the dependent variable with the letter *y*. Thus, in Example 14.1 we denote the dependent variable as *y* = *number of fatal accidents.*

The probabilistic model for multiple regression analysis is a direct extension of the one introduced in the previous chapter for simple linear regression. For the case of two independent variables, the multiple regression model is given by

$$y = \beta_0 + \beta_1 x_1 + \beta_2 x_2 + \epsilon \qquad (14.1)$$

The relationship shown in equation (14.1) is a *multiple regression model* involving two independent variables. Note that if $\beta_2 = 0$, then x_2 is not related to *y*, and hence, the multiple regression model reduces to the single-independent-variable model discussed in Chapter 13; that is, $y = \beta_0 + \beta_1 x_1 + \epsilon$.

The multiple regression model of equation (14.1) can be extended to the case of *p* independent variables by simply adding more terms. Equation (14.2) shows the general case.

Multiple Regression Model

$$y = \beta_0 + \beta_1 x_1 + \beta_2 x_2 + \cdots + \beta_p x_p + \epsilon \qquad (14.2)$$

Note that if $\beta_3, \beta_4, \ldots, \beta_p$ all equal zero, equation (14.2) reduces to the two-independent-variable multiple regression model of equation (14.1).

The assumptions made about the error term ϵ in Chapter 13 also apply in multiple regression analysis.

**Assumptions About the Error Term ϵ
in the Regression Model $y = \beta_0 + \beta_1 x_1 + \cdots + \beta_p x_p + \epsilon$**

1. The error ϵ is a random variable with mean or expected value of zero; that is, $E(\epsilon) = 0$.
   Implication: For given values of $x_1, x_2, \ldots, x_p$, the expected, or average, value of *y* is given by

$$E(y) = \beta_0 + \beta_1 x_1 + \beta_2 x_2 + \cdots + \beta_p x_p \qquad (14.3)$$

> Equation (14.3) is referred to as the *multiple regression equation*. In this equation, $E(y)$ represents the average of all possible values of y that might occur for the given values of $x_1, x_2, \ldots, x_p$.
>
> 2. The variance of ϵ is denoted by σ^2 and is the same for all values of the independent variables $x_1, x_2, \ldots, x_p$.
>   *Implication:* The variance of y equals σ^2 and is the same for all values of $x_1, x_2, \ldots, x_p$.
> 3. The values of ϵ are independent.
>   *Implication:* The size of the error for a particular set of values for the independent variables is not related to the size of the error for any other set of values.
> 4. The error ϵ is a normally distributed random variable reflecting the deviation between the y value and the expected value of y given by $\beta_0 + \beta_1 x_1 + \beta_2 x_2 + \cdots + \beta_p x_p$.
>   *Implication:* Since $\beta_0, \beta_1, \ldots, \beta_p$ are constants, for the given values of $x_1, x_2, \ldots, x_p$, the dependent variable y is also a normally distributed random variable.

To obtain more insight into the form of the relationship given by equation (14.3), consider for the moment the following two-independent-variable multiple regression equation:

$$E(y) = \beta_0 + \beta_1 x_1 + \beta_2 x_2 \tag{14.4}$$

The graph of this equation is a plane in three-dimensional space. Figure 14.1 shows such a graph with x_1 and x_2 on the horizontal axis and y on the vertical axis. Note that ϵ is

---

FIGURE 14.1   **GRAPH OF THE REGRESSION EQUATION FOR MULTIPLE REGRESSION ANALYSIS WITH TWO INDEPENDENT VARIABLES**

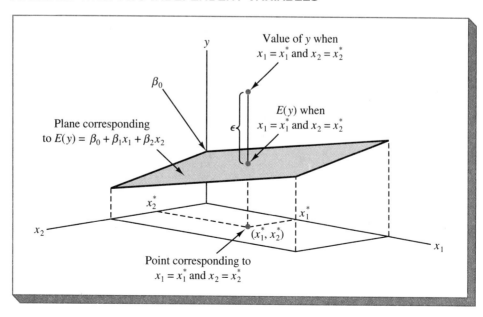

shown as the difference between the actual y value and the expected value of y, $E(y)$, when $x_1 = x_1^*$ and $x_2 = x_2^*$.

In regression analysis, the term *response variable* is often used in place of the term *dependent variable*. Furthermore, since the multiple regression equation generates a plane or surface, its graph is referred to as a response surface.

In the previous chapter the least squares method was used to develop estimates of β_0 and β_1 for the simple linear regression model. In multiple regression analysis, the least squares method is used in an analogous manner to develop estimates of the parameters $\beta_0, \beta_1, \beta_2, \ldots, \beta_p$. These estimates are denoted $b_0, b_1, b_2, \ldots, b_p$; the corresponding estimated regression equation is written as follows.

Estimated Regression Equation

$$\hat{y} = b_0 + b_1 x_1 + \cdots + b_p x_p \qquad (14.5)$$

At this point, we can begin to see the similarity between the concepts of multiple regression analysis and those of the previous chapter. The concepts of simple linear regression have simply been extended to the case involving more than one independent variable.

14.2 DEVELOPING THE ESTIMATED REGRESSION EQUATION

To show how an estimated regression equation can be developed in situations that involve two or more independent variables, let us consider the problem facing Butler Trucking.

EXAMPLE 14.2▶ Butler Trucking is an independent trucking company located in southern California. A major portion of Butler's business involves deliveries throughout its local area. To develop better work schedules, Butler Trucking would like to use an estimated regression equation to help predict total daily travel time.

Initially, Butler Trucking felt that travel time should be closely related to miles traveled. A random sample of 10 days of operation was taken; the data obtained are presented in Table 14.1 with the corresponding scatter diagram shown in Figure 14.2. The scatter diagram indicates that the number of miles traveled x_1 and the travel time y are

---

TABLE 14.1   PRELIMINARY DATA FOR BUTLER TRUCKING

Day	x_1 = Miles Traveled	y = Travel Time (hours)	Day	x_1 = Miles Traveled	y = Travel Time (hours)
1	100	9.3	6	80	6.2
2	50	4.8	7	75	7.4
3	100	8.9	8	65	6.0
4	100	6.5	9	90	7.6
5	50	4.2	10	90	6.1

FIGURE 14.2   SCATTER DIAGRAM OF PRELIMINARY DATA FOR BUTLER TRUCKING

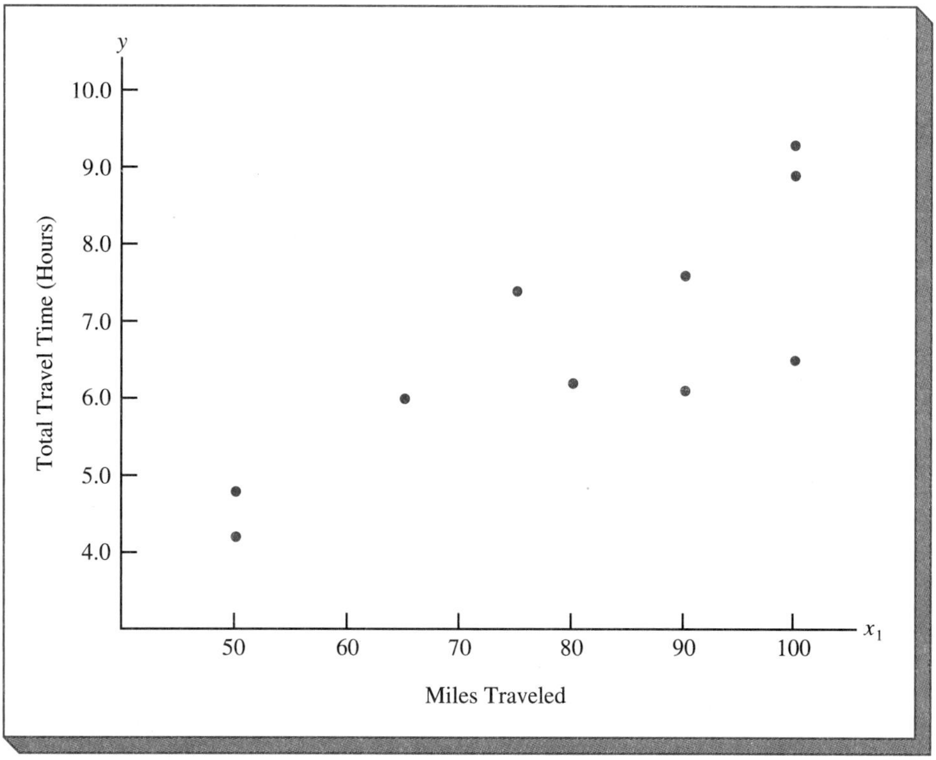

positively related; that is, as x_1 increases, y increases. As a result, the following regression model was hypothesized.

$$y = \beta_0 + \beta_1 x_1 + \epsilon \qquad (14.6)$$

Note that this is nothing more than the simple linear regression model with x_1 replacing x.

In Figure 14.3 we show the Minitab computer output corresponding to the data in Table 14.1. The resulting estimated regression equation is shown on the computer output as

$$\text{TIME} = 1.27 + 0.0678 \text{ MILES}$$

where

$$\text{TIME} = \text{travel time in hours}$$
$$\text{MILES} = \text{miles traveled}$$

In other words,

$$\hat{y} = 1.27 + 0.0678 x_1$$

The relationship between x_1 and y does appear to be statistically significant. The p-value for the t test and the F test is .004. Moreover, with a coefficient of determination of $r^2 = 0.664$, we see that 66.4% of the variability in travel time has been explained by the relationship with miles traveled. However, before reaching any final conclusions regarding the usefulness of the estimated regression equation, let us consider the standardized residual plot shown in Figure 14.4.

FIGURE 14.3   MINITAB OUTPUT FOR BUTLER TRUCKING WITH ONE INDEPENDENT VARIABLE

```
The regression equation is
TIME = 1.27 + 0.0678 MILES

Predictor        Coef        Stdev      t-ratio          p
Constant        1.274       1.401         0.91      0.390
MILES         0.06783     0.01706         3.98      0.004

s = 1.002       R-sq = 66.4%      R-sq(adj) = 62.2%

Analysis of Variance

SOURCE         DF          SS          MS          F          p
Regression      1       15.871      15.871       15.81      0.004
Error           8        8.029       1.004
Total           9       23.900
```

FIGURE 14.4   STANDARDIZED RESIDUAL PLOT FOR BUTLER TRUCKING WITH ONE INDEPENDENT VARIABLE

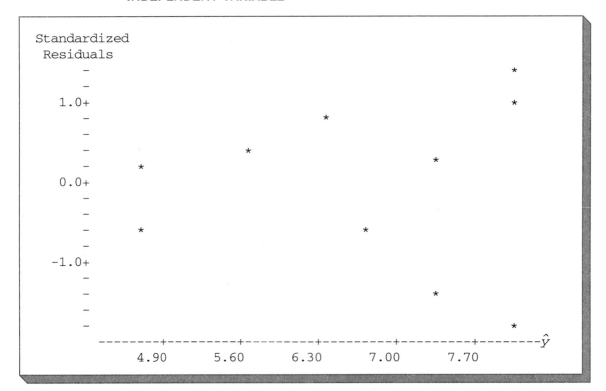

In Figure 14.4 the standardized residuals are shown on the vertical axis and the predicted values $\hat{y}$ are shown on the horizontal axis. The standardized residual plot indicates that the variability of the standardized residuals is increasing as $\hat{y}$ increases and that the assumption of constant variance for the error term (assumption 2) does not appear to be satisfied. Since the tests for statistical significance in regression analysis are based on the assumptions regarding the error term, our previous conclusion regarding the statistical significance of the relationship must be questioned. In addition, even though 66.4% of the variability has been explained using just miles traveled, 33.6% of the variability is still unexplained; the goodness of fit can perhaps be improved. Given the problem we observed with the standardized residual plot, we should doubt the adequacy of the existing model.

Looking for an alternative, it was suggested that perhaps the number of deliveries made could also be used to help predict travel time and hence improve the regression model. The data, with the addition of the number of deliveries, are shown in Table 14.2, where x_2 denotes the number of deliveries.

MULTIPLE REGRESSION AND THE LEAST SQUARES CRITERION

In Chapter 13 we showed how the least squares method can be used to find the straight line that provides the best approximation for the relationship between the independent and dependent variables. The criterion for the least squares method is given by equation (14.7).

Least Squares Criterion

$$\text{Min } \Sigma(y_i - \hat{y}_i)^2 \tag{14.7}$$

where

y_i = observed value of the dependent variable for the ith observation
$\hat{y}_i$ = estimated value of the dependent variable for the ith observation

---

TABLE 14.2   **DATA FOR BUTLER TRUCKING WITH MILES TRAVELED (x_1) AND NUMBER OF DELIVERIES (x_2) AS THE INDEPENDENT VARIABLES**

Day	x_1 = Miles Traveled	x_2 = Number of Deliveries	y = Travel Time (hours)
1	100	4	9.3
2	50	3	4.8
3	100	4	8.9
4	100	2	6.5
5	50	2	4.2
6	80	2	6.2
7	75	3	7.4
8	65	4	6.0
9	90	3	7.6
10	90	2	6.1

BUTLER

In multiple regression, this same criterion applies. In the Butler Trucking problem, the estimated regression equation that includes the effect of the miles traveled (x_1) as well as the number of deliveries (x_2) is

$$\hat{y} = b_0 + b_1x_1 + b_2x_2$$

Thus, the predicted value for the ith observation is

$$\hat{y}_i = b_0 + b_1x_{1i} + b_2x_{2i}$$

where

$$x_{1i} = \text{value of } x_1 \text{ for the } i\text{th observation}$$
$$x_{2i} = \text{value of } x_2 \text{ for the } i\text{th observation}$$

The least squares method determines the values of b_0, b_1, and b_2 that satisfy the following criterion:

$$\text{Min } \Sigma(y_i - \hat{y}_i)^2 = \text{Min } \Sigma(y_i - b_0 - b_1x_{1i} - b_2x_{2i})^2$$

In Chapter 13 we presented formulas for computing the values of b_0 and b_1 for problems involving one independent variable. In the multiple regression case, the usual presentation of the formulas for the coefficients of the estimated regression equation involves the use of matrix algebra and is beyond the scope of this text.* Thus, we will focus our attention on how computer software packages can be used to obtain the estimated regression equation. In subsequent sections we will show how computer output can be used to determine the goodness of fit, test for significance, and develop interval estimates.

COMPUTER SOLUTION

To illustrate how computer software packages can be used to obtain the estimated regression equation in multiple regression, recall the situation facing Butler Trucking in Example 14.2.

EXAMPLE 14.2
(CONTINUED)

In Figure 14.5 we show the output from the Minitab computer package for the version of the Butler Trucking problem with two independent variables, miles traveled and the number of deliveries. The estimated regression equation is

$$\text{TIME} = -0.869 + 0.0611 \text{ MILES} + 0.923 \text{ DELIV}$$

where

$$\text{TIME} = \text{travel time (hours)}$$
$$\text{MILES} = \text{miles traveled}$$
$$\text{DELIV} = \text{number of deliveries}$$

In other words,

$$\hat{y} = -0.869 + 0.0611x_1 + 0.923x_2$$

The least squares estimates $b_0 = -0.869$, $b_1 = 0.0611$, and $b_2 = 0.923$ are also shown in the output in the column labeled Coef; however, in this column the estimates are shown with more significant digits.

---

*In the appendix to this chapter we show how the least squares estimates can be obtained algebraically for the Butler Trucking problem with two independent variables.

FIGURE 14.5   MINITAB OUTPUT FOR BUTLER TRUCKING WITH TWO INDEPENDENT VARIABLES

```
The regression equation is
TIME = - 0.869 + 0.0611 MILES + 0.923 DELIV

Predictor          Coef          Stdev      t-ratio          p
Constant        -0.8687        0.9515        -0.91      0.392
MILES          0.061135      0.009888         6.18      0.000
DELIV            0.9234        0.2211         4.18      0.004

s = 0.5731        R-sq = 90.4%       R-sq(adj) = 87.6%

Analysis of Variance

SOURCE          DF           SS           MS           F          p
Regression       2        21.601       10.800       32.88      0.000
Error            7         2.299        0.328
Total            9        23.900
```

Note that the sum of squares due to regression (SSR = 21.601) plus the sum of squares due to error (SSE = 2.299) is equal to the total sum of squares (SST = 23.9); that is, SST = SSR + SSE. In addition, we see that the degrees of freedom associated with the regression sum of squares (2) plus the degrees of freedom associated with the error sum of squares (7) are equal to the degrees of freedom associated with the total sum of squares (9). In general, the degrees of freedom for the regression sum of squares is the number of independent variables p. The degrees of freedom for the total sum of squares is the number of observations minus one, $n - 1$. So, the degrees of freedom for the error sum of squares is $n - p - 1$. Note also that the mean square due to regression (MSR = 10.8) is the regression sum of squares divided by the regression degrees of freedom, and that the mean square due to error (MSE = 0.328) is the error sum of squares divided by the error degrees of freedom. The F value (F = 32.88) is MSR/MSE. ◀

Let us compare the computer output in Figure 14.3 for the model involving just one independent variable, miles traveled, with the output in Figure 14.5. Note that the total sum of squares (SST = 23.9) is the same in both cases. To better understand the significance of this result, recall that the total sum of squares is

$$\text{SST} = \Sigma(y_i - \bar{y})^2 \tag{14.8}$$

Since SST depends only on the observed values of the dependent variable, its value does not depend on how many independent variables we have in the regression model. Recognizing that the total sum of squares remains the same for every model is a key point in understanding what happens when we add additional independent variables to the model; that is, since SST = SSR + SSE, whenever we add additional independent variables to the model, we will see a decrease in the error sum of squares and a corresponding increase in the regression sum of squares.

In the following sections we will discuss these relationships among the sources of variation in more detail. Before doing so, however, let us consider more carefully the interpretation of the coefficients in the estimated multiple regression equation.

A NOTE ON INTERPRETATION OF COEFFICIENTS

One observation can be made at this time concerning the relationship between the estimated regression equation with only the miles traveled as an independent variable and the equation that includes the number of deliveries as a second independent variable. The value of b_1 is not the same in both cases. In simple linear regression, we interpret b_1 as an estimate of the amount of change in y for a 1-unit change in the independent variable. In multiple regression analysis, this interpretation must be modified somewhat. That is, in multiple regression analysis, we interpret each regression coefficient as follows: b_i represents an estimate of the change in y corresponding to a 1-unit change in x_i, when all other independent variables are held constant.

EXAMPLE 14.2
(CONTINUED)

Recall that in the Butler Trucking problem involving two independent variables, $b_1 = .0611$. Thus, .0611 hours is an estimate of the expected increase in travel time corresponding to an increase of 1 mile in the distance traveled when the number of deliveries is held constant. Similarly, since $b_2 = .923$, an estimate of the expected increase in travel time corresponding to an increase of 1 delivery when the number of miles traveled is held constant is .923 hours.   ◀

EXERCISES

Note to student: The exercises involving data in this and subsequent sections were designed to be solved using a computer software package.

METHODS

1. Shown below is the estimated regression equation for a model involving two independent variables and 10 observations.

$$\hat{y} = 29.1270 + .5906x_1 + .4980x_2$$

   a. Interpret b_1 and b_2 in this estimated regression equation.
   b. Estimate y when $x_1 = 180$ and $x_2 = 310$.

 SELF TEST

2. Consider the following data involving a dependent variable y and two independent variables, x_1 and x_2.

y	x_1	x_2	y	x_1	x_2
94	30	12	175	51	19
108	47	10	170	74	7
112	25	17	117	36	12
178	51	16	142	59	13
94	40	5	211	76	16

a. Using these data, develop an estimated regression equation relating y to x_1. Estimate y if $x_1 = 45$.

b. Using these data, develop an estimated regression equation relating y to x_2. Estimate y if $x_2 = 15$.

c. Using these data, develop an estimated regression equation relating y to x_1 and x_2. Estimate y if $x_1 = 45$ and $x_2 = 15$.

3. In a regression analysis involving 30 observations, the following estimated regression equation was obtained.

$$\hat{y} = 17.6 + 3.8x_1 - 2.3x_2 + 7.6x_3 + 2.7x_4$$

a. Interpret b_1, b_2, and b_3 in this estimated regression equation.

b. Estimate y when $x_1 = 10$, $x_2 = 5$, $x_3 = 1$, and $x_4 = 2$.

APPLICATIONS

4. A shoe store has developed the following estimated regression equation relating sales to inventory investment and advertising expenditures:

$$\hat{y} = 25 + 10x_1 + 8x_2$$

where

$$x_1 = \text{inventory investment (\$1000s)}$$
$$x_2 = \text{advertising expenditures (\$1000s)}$$
$$y = \text{sales (\$1000s)}$$

a. Estimate sales if there is a $15,000 investment in inventory and an advertising budget of $10,000.

b. Interpret b_1 and b_2 in this estimated regression equation.

5. The owner of TAI Movie Theaters, Inc., would like to investigate the effect of television advertising on weekly gross revenue. The following historical data (both in $1000s) were collected.

Weekly Gross Revenue	Television Advertising	Weekly Gross Revenue	Television Advertising
96	5.0	95	3.0
90	2.0	94	3.5
95	4.0	94	2.5
92	2.5	94	3.0

a. Using these data, develop an estimated regression equation relating weekly gross revenue to television advertising expenditure.

b. Estimate the weekly gross revenue in a week in which $3500 is spent on television advertising.

SELF TEST ▷ 6. As an extension of Exercise 5, consider the possibility of incorporating the effects of both newspaper and television advertising on weekly gross revenue. The following data (in $1000s) were developed from historical records.

MOVIE

Weekly Gross Revenue	Television Advertising	Newspaper Advertising
96	5.0	1.5
90	2.0	2.0
95	4.0	1.5
92	2.5	2.5
95	3.0	3.3
94	3.5	2.3
94	2.5	4.2
94	3.0	2.5

a. Find an estimated regression equation relating weekly gross revenue to television and newspaper advertising.

b. Is the coefficient for television advertising expenditures the same in Exercise 5(a) and Exercise 6(a)? Interpret this coefficient in each case.

MOWER

7. Heller Company manufactures lawn mowers and related lawn equipment. They believe that the quantity of lawn mowers sold depends on the price of their mower and the price of a competitor's mower. Let

$$y = \text{quantity sold (1000s)}$$
$$x_1 = \text{price of competitor's mower (\$)}$$
$$x_2 = \text{price of Heller's mower (\$)}$$

Management would like an estimated regression equation that relates quantity sold to the price of the Heller mower and the competitor's mower. The following data are available concerning prices in 10 different cities.

Competitor's Price (x_1)	Heller's Price (x_2)	Quantity Sold (y)
120	100	102
140	110	100
190	90	120
130	150	77
155	210	46
175	150	93
125	250	26
145	270	69
180	300	65
150	250	85

a. Determine the estimated regression equation that can be used to predict the quantity sold given the competitor's price and Heller's price.

b. Interpret b_1 and b_2.

c. Predict the quantity sold in a city where Heller prices its mower at $160 and the competitor prices its mower at $170.

PHARMACY

8. The following data show the value of prescriptions sold for 13 pharmacies in Iowa, the population of the city served by the given pharmacy, and the average prescription inventory value. ("The Use of Categorical Variables in Data Envelopment Analysis," R. Banker and R. Morey, *Management Science,* December 1986).

Value ($1000s)	Population	Average Inventory Value ($)	Value ($1000s)	Population	Average Inventory Value ($)
61	1,410	8,000	45	1,070	5,000
92	1,523	9,000	183	1,694	27,000
93	1,354	13,694	156	1,910	21,560
45	822	4,250	120	1,745	15,000
50	746	6,500	75	1,353	8,500
29	1,281	7,000	122	1,016	18,000
56	1,016	4,500			

a. Determine the estimated regression equation that can be used to predict the dollar value of prescriptions y given the population size x_1 and the average inventory value x_2.

b. What other variables do you think might be useful in predicting y?

 SCHOOLS1

9. Two experts provided subjective lists of school districts that they think are among the best in the country. For each school district the average class size, the combined SAT score, and the percentage of students who attended a 4-year college were provided (*The Wall Street Journal,* March 31, 1989).

District	Average Class Size	Combined SAT Score	% Attend 4-Year College
Blue Springs, Mo.	25	1083	74
Garden City, N.Y.	18	997	77
Indianapolis, Ind.	30	716	40
Newport Beach, Calif.	26	977	51
Novi, Mich.	20	980	53
Piedmont, Calif.	28	1042	75
Pittsburgh, Pa.	21	983	66
Scarsdale, N.Y.	20	1110	87
Wayne, Pa.	22	1040	85
Weston, Mass.	21	1031	89
Farmingdale, N.Y.	22	947	81
Mamaroneck, N.Y.	20	1000	69
Mayfield, Ohio	24	1003	48
Morristown, N.J.	22	972	64
New Rochelle, N.Y.	23	1039	55
Newtown Square, Pa.	17	963	79
Omaha, Neb.	23	1059	81
Shaker Heights, Ohio	23	940	82

a. Using these data, develop an estimated regression equation relating the percentage of students who attend a 4-year college to the average class size and the combined SAT score.

b. Estimate the percentage of students who attend a 4-year college if the average class size is 20 and the combined SAT score is 1000.

 HOUSING

10. Data on housing markets were provided for 100 different cities in the United States (*U.S. News & World Report,* April 6, 1992). A sample of 16 cities with data on the median cost of a new home ($1000s), the number of new housing starts during 1991–1992 (1000s), and the average household income ($1000s) are shown at the top of page 596.

City	Cost	Housing Starts	Household Income
Chicago	181.8	12.9	61.0
Dayton, Ohio	107.8	3.8	48.4
Atlanta	100.6	24.2	54.7
Oklahoma City	68.9	3.3	53.2
Columbia, S.C.	90.3	3.1	57.4
Tacoma, Wash.	96.1	4.1	51.1
Mobile, Ala.	68.5	1.0	41.0
Baltimore	121.8	11.1	62.8
West Palm Beach	130.4	8.9	58.1
San Antonio	72.5	1.5	57.0
Pittsburgh	79.5	4.9	49.2
Jacksonville	82.1	8.0	47.5
Cleveland	122.9	5.4	54.0
Gary, Ind.	98.2	3.2	45.7
Scranton, Pa.	81.6	2.5	44.8
Richmond, Va.	102.8	6.0	64.5

a. Using these data, develop the estimated regression equation relating cost to the number of housing starts and the household income.

b. Estimate the median cost for a city with 8000 housing starts and an average household income of $50,000.

14.3   DETERMINING THE GOODNESS OF FIT

In Chapter 13 we used the coefficient of determination (r^2) to evaluate the goodness of fit for the regression relationship. Recall that r^2 was computed as

$$r^2 = \frac{\text{SSR}}{\text{SST}}$$

In multiple regression analysis, we compute a similar quantity, called the *multiple coefficient of determination.*

Multiple Coefficient of Determination

$$R^2 = \frac{\text{SSR}}{\text{SST}} \qquad\qquad (14.9)$$

When multiplied by 100, the multiple coefficient of determination represents the percentage of variability in y that is explained by the estimated regression equation.

EXAMPLE 14.2 ▶
(CONTINUED)

In the case of Butler Trucking (refer to Figure 14.5 for SSR and SST), we find

$$R^2 = \frac{21.601}{23.900} = .9038$$

Therefore, 90.38% of the variability in y is explained by the relationship with miles traveled and number of deliveries. In Figure 14.5, this is rounded to show R-sq = 90.4%.

Refer now to Section 14.2. Note that the regression model with only miles traveled as the independent variable had $r^2 = .664$. Therefore, the percentage of variability explained has increased from 66.4% to 90.38%. ◀

In general, it is always true that R^2 will increase as more independent variables are added to the regression model because adding variables to the model causes the prediction errors to be smaller, hence reducing SSE. Since SST = SSR + SSE, when SSE gets smaller, SSR must get larger, causing $R^2 = $ SSR/SST to increase.

Many analysts recommend adjusting R^2 for the number of independent variables to avoid overestimating the impact of adding an independent variable on the amount of explained variability. This *adjusted multiple coefficient of determination* is computed as follows.

> **Adjusted Multiple Coefficient of Determination**
>
> $$R_a^2 = 1 - (1 - R^2)\frac{n - 1}{n - p - 1} \qquad (14.10)$$

EXAMPLE 14.2 ▶
(CONTINUED)

For Butler Trucking, we obtain

$$R_a^2 = 1 - (1 - 0.9038)\frac{10 - 1}{10 - 2 - 1}$$

$$= 1 - (0.0962)(1.2857)$$

$$= 0.8763$$

Thus, after adjusting for the number of independent variables in the model, 87.63% of the variability in y has been accounted for. Note that both the value of R^2 and the value of R_a^2 are provided by the Minitab output shown in Figure 14.5 where $R_a^2 = 0.8763$ is rounded to show R-sq(adj) = 87.6%. ◀

NOTES AND
COMMENTS

If the value of R^2 is small and the model contains a large number of independent variables, the adjusted coefficient of determination can take on a negative value; in such cases, Minitab sets the adjusted coefficient of determination to zero.

▷

EXERCISES

METHODS

11. In Exercise 1 the following estimated regression equation based on 10 observations was presented:

$$\hat{y} = 29.1270 + .5906x_1 + .4980x_2$$

The values of SST and SSR are 6724.125 and 6216.375, respectively.
   a. Find SSE.   b. Compute R^2.   c. Compute R_a^2.
   d. Comment on the goodness of fit.

SELF TEST

12. In Exercise 2, 10 observations were provided for a dependent variable y and two independent variables x_1 and x_2; for these data $SST = 15,182.9$, and SSR $= 14.052.2$.
   a. Compute R^2.
   b. Compute R_a^2.
   c. Does the estimated regression equation explain a large amount of the variability in the data? Explain.

13. In Exercise 3 the following estimated regression equation based on 30 observations was developed.

$$\hat{y} = 17.6 + 3.8x_1 - 2.3x_2 + 7.6x_3 + 2.7x_4$$

The values of SST and SSR are 1805 and 1760, respectively.
   a. Compute R^2.
   b. Compute R_a^2.
   c. Comment on the goodness of fit.

APPLICATIONS

SELF TEST

14. In Exercise 4 the following estimated regression equation relating sales to inventory investment and advertising expenditures was given:

$$\hat{y} = 25 + 10x_1 + 8x_2$$

The data used to develop the model came from a survey of 10 stores; for these data, $SST = 16,000$ and $SSR = 12,000$.
   a. For the estimated regression equation given, compute R^2.
   b. Compute R_a^2.
   c. Does the model appear to explain a large amount of variability in the data? Explain.

15. Refer to Exercise 9.
   a. What are R^2 and R_a^2 for this problem?
   b. Does the estimated regression equation provide a good fit to the data? Explain.

16. Refer to Exercise 10.
   a. What are R^2 and R_a^2 for this problem?
   b. Does the estimated regression equation provide a good fit to the data? Explain.

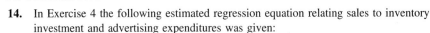

14.4 TEST FOR A SIGNIFICANT RELATIONSHIP

The regression equation that we have assumed for the Butler Trucking problem with two independent variables is

$$E(y) = \beta_0 + \beta_1 x_1 + \beta_2 x_2$$

Therefore, the appropriate test for determining whether there is a significant relationship among x_1, x_2, and y is as follows:

$$H_0: \beta_1 = \beta_2 = 0$$

H_a: One or more of the parameters (β_i's) is not equal to zero

If we reject H_0, we conclude that there is a significant relationship among x_1, x_2, and y.

The test used to determine whether there is a significant relationship in the multiple regression case is an F test very similar to the one introduced in Chapter 13 for models involving one independent variable. First, we will show how it is used by applying it in the Butler Trucking problem. Then we will generalize the application of the test to models involving p independent variables.

THE F TEST

It can be shown that if the null hypothesis (H_0: $\beta_1 = \beta_2 = 0$) is true and the four underlying regression model assumptions are valid, then the sampling distribution of MSR/MSE follows an F distribution. The number of numerator degrees of freedom is equal to the degrees of freedom associated with the sum of squares due to regression, and the denominator degrees of freedom is equal to the degrees of freedom associated with the sum of squares due to error. Recall that the sum of squares due to regression (SSR) measures the amount of the variability in y explained by the regression model. Thus, we would expect large values of F = MSR/MSE to cast doubt on the null hypothesis (no relationship between the dependent and independent variables). Indeed, this is true; small values of F do not permit us to reject H_0, and large values lead to the rejection of H_0.

EXAMPLE 14.2
(CONTINUED)

Let us return to the Butler Trucking problem and test the significance of the multiple regression model. Refer to Figure 14.5. We see that

$$F = \text{MSR}/\text{MSE}$$

$$= 10.8/.328$$

$$= 32.88$$

To test the hypothesis that $\beta_1 = \beta_2 = 0$, we must determine whether 32.88 is a value that appears likely when random sampling from an F distribution with 2 numerator and 7 denominator degrees of freedom.

Suppose that the level of significance is $\alpha = .05$. The critical value from the F distribution table (Table 4 of Appendix B) is 4.74. That is, if we are sampling randomly from an F distribution with numerator and denominator degrees of freedom equal to 2 and 7, respectively (as we would be if H_0 were true), then only 5% of the time would we get a value larger than 4.74. Figure 14.6 illustrates the determination of the critical region.

We can use the above analysis to formulate the following rejection rule at the .05 significance level.

$$\text{Reject } H_0 \text{ if } \frac{\text{MSR}}{\text{MSE}} > 4.74$$

FIGURE 14.6   **DETERMINATION OF THE CRITICAL VALUE AND REJECTION REGION USING $\alpha = .05$**

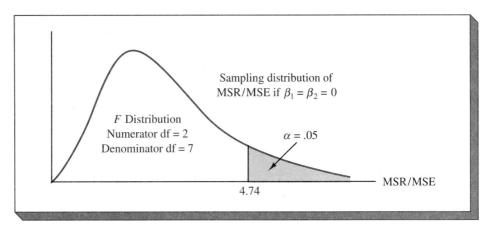

Since the value of MSR/MSE was 32.88, we can reject H_0. Hence, we conclude that there is a significant relationship between total travel time and the two independent variables. Thus, the estimated regression equation should be useful in predicting y for values of the independent variables within the range of those included in the sample.   ◀

The p-value provided by Minitab (see Figure 14.5) provides a convenient way to perform this same test. Since the p-value provides the probability of obtaining a sample result more unlikely than what is observed, the p-value of 0.000 indicates that for a level of significance of $\alpha = .05$, the null hypothesis that $\beta_1 = \beta_2 = 0$ should be rejected. In practice, the ease of use of the p-value approach to testing has made it the preferred method for performing hypothesis tests in regression analysis.

THE GENERAL ANOVA TABLE AND THE F TEST

Now that we know how the F test can be applied for a multiple regression model with two independent variables, let us generalize the test to the case involving a model with p independent variables. The appropriate hypothesis test to determine whether there is a significant relationship is as follows:

$$H_0: \beta_1 = \beta_2 = \cdots = \beta_p = 0$$

H_a: One or more of the parameters is not equal to zero

Again, if we reject H_0, we can conclude that there is a significant relationship and that the estimated regression equation is useful for predicting or explaining the dependent variable y.

The general form of the ANOVA table for the multiple regression case involving p independent variables is shown in Table 14.3. The only change from the two-variable case is the degrees of freedom corresponding to SSR and SSE. Here the sum of squares due to regression has p degrees of freedom corresponding to the p independent variables; hence

$$\text{MSR} = \frac{\text{SSR}}{p} \tag{14.11}$$

In addition, the sum of squares due to error has $n - p - 1$ degrees of freedom; thus

$$\text{MSE} = \frac{\text{SSE}}{n - p - 1} \tag{14.12}$$

Hence, the F statistic for the case of p independent variables is computed as follows:

TABLE 14.3   **ANOVA TABLE FOR A MULTIPLE REGRESSION MODEL WITH p INDEPENDENT VARIABLES**

Source	Sum of Squares	Degrees of Freedom	Mean Square	F
Regression	SSR	p	$\text{MSR} = \dfrac{\text{SSR}}{p}$	$F = \dfrac{\text{MSR}}{\text{MSE}}$
Error	SSE	$n - p - 1$	$\text{MSE} = \dfrac{\text{SSE}}{n - p - 1}$	
Total	SST	$n - 1$		

$$F = \frac{\text{MSR}}{\text{MSE}} = \frac{\text{SSR}/p}{\text{SSE}/(n - p - 1)} \tag{14.13}$$

When looking up the critical value from the F distribution table, the numerator degrees of freedom are p and the denominator degrees of freedom are $n - p - 1$. As we mentioned earlier, the decision to reject or not to reject H_0 can be made by comparing the computed F statistic with the critical value.

t TEST FOR SIGNIFICANCE OF INDIVIDUAL PARAMETERS

If after using the F test we conclude that the multiple regression relationship is significant (i.e., we conclude that at least one of the $\beta_i \neq 0$), it is often of interest to conduct further tests to see which individual parameters β_i are significant. The t test is a statistical method for testing the significance of the individual parameters.

The hypothesis test we wish to conduct is the same for the coefficient of each independent variable. It is stated as follows:

$$H_0: \beta_i = 0$$

$$H_a: \beta_i \neq 0$$

Recall that in Chapter 13 we learned how to conduct such a test for the case where there is only one independent variable. The hypotheses were

$$H_0: \beta_1 = 0$$

$$H_a: \beta_1 \neq 0$$

To test these hypotheses, we computed the sample statistic b_1/s_{b_1}, where b_1 was the least squares estimate of β_1 and s_{b_1} was an estimate of the standard deviation of the sampling distribution of b_1. We learned that the sampling distribution of b_1/s_{b_1}, follows a t distribution with $n - 2$ degrees of freedom. Thus, to conduct the hypothesis test, we used the following rejection rule:

$$\text{Reject } H_0 \text{ if } \frac{b_1}{s_{b_1}} > t_{\alpha/2}$$

$$\text{or if } \frac{b_1}{s_{b_1}} < -t_{\alpha/2}$$

The procedure for testing individual parameters in the multiple regression case is essentially the same. The only differences are in the number of degrees of freedom for the appropriate t distribution and in the formula for computing s_{b_i}. The number of degrees of freedom is the same as for the sum of squares due to error. Thus, we use $n - p - 1$ degrees of freedom, where p is the number of independent variables. (Note that for the case of one independent variable, this reduces to the $n - 2$ degrees of freedom used in Chapter 13.) The formula for s_{b_i} is more involved, and we do not present it here; however, s_{b_i} is calculated and printed by most computer software packages for multiple regression analysis.

EXAMPLE 14.2 ▶
(CONTINUED)

Let us return now to the Butler Trucking problem to test the significance of the parameters β_1 and β_2. Note that in the Minitab printout (Figure 14.5) the values of b_1, b_2, s_{b_1}, and s_{b_2} were given as

$$b_1 = 0.061135 \qquad s_{b_1} = 0.009888$$

$$b_2 = 0.9234 \qquad s_{b_2} = 0.2211$$

Therefore, for the parameters β_1 and β_2 we obtain

$$\frac{b_1}{s_{b_1}} = \frac{0.061135}{0.009888} = 6.18$$

$$\frac{b_2}{s_{b_2}} = \frac{0.9234}{0.2211} = 4.18$$

Note that both of these values were provided by the Minitab output of Figure 14.5 under the column labeled t-ratio. Using $\alpha = .05$ and $10 - 2 - 1 = 7$ degrees of freedom, we can find the appropriate t value for our hypothesis tests in Table 2 of Appendix B. We obtain

$$t_{.025} = 2.365$$

Now, since $b_1/s_{b_1} = 6.18 > 2.365$, we reject the hypothesis that $\beta_1 = 0$. Furthermore, since $b_2/s_{b_2} = 4.18 > 2.365$, we reject the hypothesis that $\beta_2 = 0$. Note also that the p-values provided by the Minitab outputs permit us to perform these tests very easily. For instance, the p-value of 0.004 for DELIV indicates that the null hypothesis H_0: $\beta_2 = 0$ should be rejected for all values of α larger than .004.   ◄

MULTICOLLINEARITY

We have used the term *independent variable* in regression analysis to refer to any variable being used to predict or explain the value of the dependent variable. The term does not mean, however, that the independent variables themselves are independent in any statistical sense. Quite the contrary, most independent variables in a multiple regression problem are correlated to some degree with one another. For example, in the Butler Trucking problem involving the two independent variables x_1 (miles traveled) and x_2 (number of deliveries), we could treat the miles traveled as the dependent variable and the number of deliveries as the independent variable to determine whether these two variables are themselves related. We could then compute the sample correlation coefficient $r_{x_1 x_2}$ to determine the extent to which these variables are related. Doing so yields $r_{x_1 x_2} = .28$. Thus, there is some degree of linear association between the two independent variables. In multiple regression analysis, we use the term *multicollinearity* to refer to the correlation among the independent variables.

To provide a better perspective of the potential problems of multicollinearity, let us consider a modification of the Butler Trucking problem.

EXAMPLE 14.2
(CONTINUED)

Instead of x_2 being the number of deliveries, let x_2 denote the number of gallons of gasoline consumed. Clearly, x_1 (the miles traveled) and x_2 are related; that is, we know that the number of gallons of gasoline used depends on the number of miles traveled. Thus, we would conclude logically that x_1 and x_2 are highly correlated independent variables.

Assume that we obtain the equation $\hat{y} = b_0 + b_1 x_1 + b_2 x_2$ and find that the F test shows that the regression is significant. Then suppose that we were to conduct a t test on β_1 to determine if $\beta_1 \ne 0$, and we cannot reject H_0: $\beta_1 = 0$. Does this mean that travel time is not related to miles traveled? Not necessarily. What it probably means is that with x_2 already in the model, x_1 does not make a significant contribution toward determining the value of y. This would seem to make sense in our example, since if we know the amount of gasoline consumed, we do not gain much additional information useful in predicting y by knowing the miles traveled. Similarly, a t test might lead us to conclude $\beta_2 = 0$ on the grounds that, with x_1 in the model, knowledge of the amount of gasoline consumed does not add much. ◀

To summarize, the difficulty caused by multicollinearity in conducting t tests for the significance of individual parameters is that it is possible to conclude that none of the individual parameters are significantly different from zero when an F test on the overall multiple regression equation indicates a significant relationship. This problem is avoided when there is very little correlation among the independent variables.

Ordinarily, multicollinearity does not affect the way in which we perform our regression analysis or interpret the output from a study. However, when multicollinearity is severe—that is, when two or more of the independent variables are highly correlated with one another—we can run into difficulties interpreting the results of t tests on the individual parameters. In addition to the type of problem illustrated above, severe cases of multicollinearity have been shown to result in least squares estimates that even have the wrong sign. That is, in simulated studies where researchers created the underlying regression model and then applied the least squares technique to develop estimates of $\beta_0, \beta_1, \beta_2$, and so on, it has been shown that under conditions of high multicollinearity the least squares estimates can even have a sign opposite to that of the parameter being estimated. For example, β_2 might actually be $+10$ and b_2, its estimate, might turn out to be -2. Thus, little faith can be placed in the individual coefficients themselves if multicollinearity is present to a high degree.

Statisticians have developed several tests for determining whether multicollinearity is high enough to cause these types of problems. One simple test, referred to as the rule of thumb test, says that multicollinearity is a potential problem if the absolute value of the sample correlation coefficient exceeds .7 for any two of the independent variables. The other types of tests are more advanced and beyond the scope of this text.

If possible, every attempt should be made to avoid including independent variables that are highly correlated. In practice, however, it is rarely possible to adhere to this policy strictly. Thus, the decision maker should be warned that when there is reason to believe that substantial multicollinearity is present, it is difficult to separate out the effect of the individual independent variables on the dependent variable.

EXERCISES

METHODS

SELF TEST

17. In Exercise 1 the following estimated regression equation based on 10 observations was presented:

$$\hat{y} = 29.1270 + .5906x_1 + .4980x_2$$

Here SST $= 6724.125$, SSR $= 6216.375$, $s_{b_1} = .0813$, and $s_{b_2} = .0567$.
a. Compute MSR and MSE.
b. Compute F and perform the appropriate F test. Use $\alpha = .05$.
c. Perform a t test for the significance of β_1. Use $\alpha = .05$.
d. Perform a t test for the significance of β_2. Use $\alpha = .05$

18. Refer to the data presented in Exercise 2. The estimated regression equation for these data is

$$\hat{y} = -18.4 + 2.01x_1 + 4.74x_2$$

  a. Test for a significant relationship among x_1, x_2, and y. Use $\alpha = .05$.
  b. Is β_1 significant? Use $\alpha = .05$.
  c. Is β_2 significant? Use $\alpha = .05$.

19. The following estimated regression equation was developed for a model involving two independent variables:

$$\hat{y} = 40.7 + 8.63x_1 + 2.71x_2$$

After dropping x_2 from the model, the least squares method was used again to obtain an estimated regression equation involving only x_1 as an independent variable:

$$\hat{y} = 42.0 + 9.01x_1$$

  a. Give an interpretation of the coefficient of x_1 in both models.
  b. Could multicollinearity explain why the coefficient of x_1 differs in the two models? If so, how?

APPLICATIONS

SELF TEST

20. In Exercise 4 the following estimated regression equation for relating sales to inventory investment and advertising expenditures was given:

$$\hat{y} = 25 + 10x_1 + 8x_2$$

The data used to develop the model came from a survey of 10 stores; for these data SST = 16,000 and SSR = 12,000.
  a. Compute SSE, MSE, and MSR.
  b. Use an F test and an $\alpha = .05$ level of significance to determine if there is a relationship among the variables.

21. Refer to Exercise 6.
  a. Use a level of significance of $\alpha = .01$ to test the hypotheses

$$H_0: \beta_1 = \beta_2 = 0$$

$$H_a: \beta_1 \text{ or } \beta_2 \text{ is not equal to zero}$$

for the model $y = \beta_0 + \beta_1 x_1 + \beta_2 x_2 + \epsilon$, where

$$x_1 = \text{television advertising (\$1000s)}$$
$$x_2 = \text{newspaper advertising (\$1000s)}$$

  b. Use a level of significance of $\alpha = .05$ to test the significance of β_1. Should x_1 be dropped from the model?
  c. Use a level of significance of $\alpha = .05$ to test the significance of β_2. Should x_2 be dropped from the model?

22. Refer to Exercise 7 involving the Heller Company. Test the significance of the overall model at $\alpha = .05$.

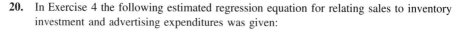
FORBES1

23. The following data show the price/earnings (P/E) ratio, the profit margin, and the growth rate for 19 companies listed in "The *Forbes* 500s on Wall Street" (*Forbes*, May 1, 1989).

Firm	P/E Ratio	Profit Margin (%)	Growth Rate (%)
Exxon	11.3	6.5	10
Chevron	10.0	7.0	5
Texaco	9.9	3.9	5
Mobil	9.7	4.3	7
Amoco	10.0	9.8	8
Pfizer	11.9	14.7	12
Bristol Meyers	16.2	13.9	14
Merck	21.0	20.3	16
American Home Products	13.3	16.9	11
Abbott Laboratories	15.5	15.2	18
Eli Lilly	18.9	18.7	11
Upjohn	14.6	12.8	10
Warner-Lambert	16.0	8.7	7
Amdahl	8.4	11.9	4
Digital	10.4	9.8	19
Hewlett-Packard	14.8	8.1	18
NCR	10.1	7.3	6
Unisys	7.0	6.9	6
IBM	11.8	9.2	6

a. Determine the estimated regression equation that can be used to predict the price/earnings ratio given the profit margin and the growth rate.

b. At the $\alpha = .05$ level of significance, determine if there is a relationship among the variables.

c. Does there appear to be any multicollinearity present in the data? Explain.

24. Refer to the data in Exercise 10. The estimated regression equation for these data is

$$\hat{y} = -5.7 + 1.54x_1 + 1.81x_2$$

where

$$\hat{y} = \text{estimated cost}$$
$$x_1 = \text{number of housing starts}$$
$$x_2 = \text{household income}$$

a. Test for a significant relationship among $x_1, x_2,$ and y. Use $\alpha = .05$.

b. Is β_1 significant? Use $\alpha = .05$.

c. Is β_2 significant? Use $\alpha = .05$.

d. Briefly discuss the results obtained.

14.5 ESTIMATION AND PREDICTION

Estimating the mean value of y and predicting an individual value of y in multiple regression is similar to that for the case of regression analysis involving one independent variable. First, recall that in Chapter 13 we showed that the point estimate of the expected value of y for a given value of x is the same as the point estimate of an individual value of y. In both cases, we used $\hat{y} = b_0 + b_1 x$ as the point estimate.

In multiple regression we use the same procedure. That is, we substitute the given values of $x_1, x_2, \ldots, x_p$ into the estimated regression equation and use the corresponding value of $\hat{y}$ as the point estimate.

EXAMPLE 14.2 ▶
(CONTINUED)

Suppose that for the Butler Trucking problem, we wanted to use the estimated regression equation involving x_1 (miles traveled) and x_2 (number of deliveries) to do the following:

1. Develop a *confidence interval estimate* of the mean travel time for all trucks that travel 100 miles and make two deliveries.
2. Develop a *prediction interval estimate* of the travel time for *one specific* truck that travels 100 miles and makes two deliveries.

Using the estimated regression equation $\hat{y} = -.869 + .0611x_1 + .923x_2$ with $x_1 = 100$ and $x_2 = 2$, we obtain the following value of $\hat{y}$.

$$\hat{y} = -.869 + .0611(100) + .923(2) = 7.09$$

Hence, the point estimate of travel time in both cases is approximately 7 hours.

To develop interval estimates for the mean value of y and for an individual value of y, we use a procedure similar to that for the case of regression analysis involving one independent variable. The formulas required, however, are beyond the scope of this text. But, computer packages for multiple regression analysis will often provide confidence intervals once the values of $x_1, x_2, \ldots, x_p$ are specified by the user. In Table 14.4 we show 95% confidence and prediction interval estimates for the Butler Trucking problem for selected values of x_1 and x_2; these values were obtained using Minitab. Note that the interval estimate for an individual value of y is wider than the interval estimate for the expected value of y. This simply reflects the fact that for given values of x_1 and x_2 we can predict the mean travel time for all trucks with more precision than we can the travel time for one specific truck.   ◀

EXERCISES

METHODS

25. In Exercise 1 the following estimated regression equation based on 10 observations was presented:

$$\hat{y} = 29.1270 + .5906x_1 + .4980x_2$$

TABLE 14.4   **THE 95% CONFIDENCE AND PREDICTION INTERVAL ESTIMATES FOR BUTLER TRUCKING**

Value of x_1	Value of x_2	Confidence Interval Lower Limit	Confidence Interval Upper Limit	Prediction Interval Lower Limit	Prediction Interval Upper Limit
50	2	3.146	4.924	2.414	5.656
50	3	4.127	5.789	3.368	6.548
50	4	4.815	6.948	4.157	7.607
100	2	6.258	7.926	5.500	8.683
100	3	7.385	8.645	6.520	9.510
100	4	8.135	9.742	7.362	10.515

a. Develop a point estimate of the mean value of y when $x_1 = 180$ and $x_2 = 310$.

b. Develop a point estimate for an individual value of y when $x_1 = 180$ and $x_2 = 310$.

26. Refer to the data in Exercise 2. The estimated regression equation for these data is

$$\hat{y} = 18.4 + 2.01x_1 + 4.74x_2$$

a. Develop a 95% confidence interval estimate of the mean value of y when $x_1 = 45$ and $x_2 = 15$.

b. Develop a 95% prediction interval estimate of y when $x_1 = 45$ and $x_2 = 15$.

APPLICATIONS

27. The following estimated regression equation has been developed to predict annual sales for account executives:

$$\hat{y} = 160 + 8x_1 + 15x_2$$

where

$$\hat{y} = \text{sales (\$1000s)}$$
$$x_1 = \text{years of experience}$$
$$x_2 = \text{number of sales training programs attended}$$

a. Estimate expected annual sales for an employee with 3 years of experience who did not attend any sales training program.

b. Estimate annual sales for a given employee with 2 years of experience who attended one sales training program.

c. What is the expected increase in sales as a result of attending one sales training program?

28. Refer to Exercise 7.

a. Develop a 95% confidence interval estimate of the mean quantity sold if both the competitor's price and Heller's price are $175.

b. Would a prediction interval estimate have any meaning in this situation? Explain.

29. Refer to exercise 9.

a. Develop a 95% confidence interval estimate of the mean percentage of students who attend a 4-year college for a school district that has an average class size of 25 and whose students have a combined SAT score of 1000.

b. Suppose that a school district in Conway, South Carolina, has an average class size of 25 and a combined SAT score of 950. Develop a 95% prediction interval estimate of the percentage of students who attend a 4-year college.

14.6 ▽ THE USE OF QUALITATIVE VARIABLES

So far, the variables that we have used to build regression models have been quantitative variables; that is, variables that are measured in terms of how much or how many. Frequently, however, we will need to use independent variables that are not measured in these terms. We refer to such variables as *qualitative variables*. For instance, suppose that we were interested in predicting sales for a product that was available in either bottles or cans. Clearly the independent variable "container type" could influence the dependent variable "sales"; but, container type is a qualitative, not a quantitative, variable. The distinguishing feature of qualitative variables is that there is no natural measure of how much or how many; these variables are used to represent attributes that are either present or not present.

EXAMPLE 14.3 ▶   To illustrate how qualitative variables are used in regression analysis, let us consider the problem currently being faced by Johnson Filtration, Inc., a firm that provides sales and service for industrial water-filtration systems throughout southern Florida. To investigate the relationship between the service time to repair a particular type of system and the number of months since the machine was serviced, Johnson Filtration collected the data shown in Table 14.5; a scatter diagram for these data is shown in Figure 14.7.

TABLE 14.5   DATA FOR JOHNSON FILTRATION

Repair Time (hours)	Months Since Last Service Call	Repair Time (hours)	Months Since Last Service Call
2.9	2	4.9	7
3.0	6	4.2	9
4.8	8	4.8	8
1.8	3	4.4	4
2.9	2	4.5	6

FIGURE 14.7   SCATTER DIAGRAM FOR JOHNSON FILTRATION

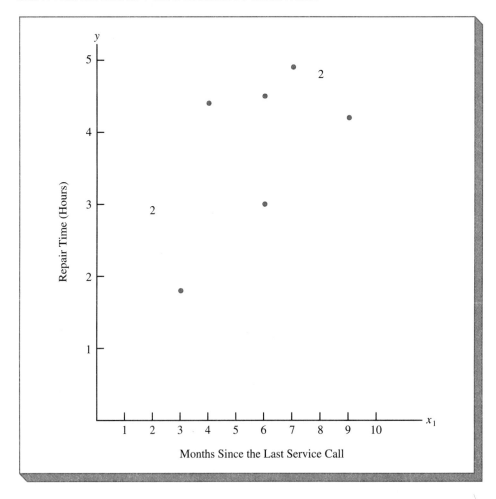

The scatter diagram indicates that there may be a linear relationship between the number of months since the last service call and the repair time in hours. As a result, the following regression model was proposed:

$$y = \beta_0 + \beta_1 x_1 + \epsilon$$

where

y = the repair time in hours
x_1 = the number of months since the last service call

The corresponding estimated regression equation is

$$\hat{y} = b_0 + b_1 x_1$$

In Figure 14.8 we show the computer output obtained using the Minitab statistical system. The estimated regression equation is

$$\text{HOURS} = 2.15 + 0.304 \text{ MONTHS}$$

where the estimated repair time is denoted as HOURS and the number of months since the last service call is denoted as MONTHS. To test for the significance of the number of months since the last service call, the appropriate hypotheses are

$$H_0: \beta_1 = 0$$

$$H_a: \beta_1 \neq 0$$

Suppose we perform the test at the .05 level of significance. Looking at the p-value for either the F test or the t test (since they are equivalent procedures for simple linear regression), we see that the p-value of .016 is less than the level of significance; thus, we must reject H_0: $\beta_1 = 0$ and conclude that the number of months since the last service call is a significant factor in predicting the repair time. However, we also note that this variable only explains 53.4% of the variability in repair time.

FIGURE 14.8   MINITAB OUTPUT FOR JOHNSON FILTRATION

```
The regression equation is
HOURS = 2.15 + 0.304 MONTHS

Predictor        Coef        Stdev      t-ratio          p
Constant       2.1473      0.6050         3.55      0.008
MONTHS         0.3041      0.1004         3.03      0.016

s = 0.7810       R-sq = 53.4%       R-sq(adj) = 47.6%

Analysis of Variance

SOURCE         DF            SS           MS          F          p
Regression      1        5.5960       5.5960       9.17      0.016
Error           8        4.8800       0.6100
Total           9       10.4760
```

After further consideration, management felt that the type of failure that occurred, mechanical or electrical, should also be considered in attempting to predict the repair time. Note that "type of repair" is an example of a qualitative variable. To incorporate the effect of type of failure into a regression model to predict repair time, we define the following variable:

$$x_2 = \begin{cases} 0 \text{ if the type of failure is mechanical} \\ 1 \text{ if the type of failure is electrical} \end{cases}$$

In regression analysis this way of representing a variable provides what is commonly referred to as a *dummy,* or *indicator* variable. Using this dummy variable, we can now write the following multiple regression equation to account for the effect of both the number of months since the last service call and the type of failure that occurred:

$$E(y) = \beta_0 + \beta_1 x_1 + \beta_2 x_2$$

To interpret the parameters in this equation, we begin by considering what the expected value of y is when the type of failure is mechanical.

INTERPRETING THE PARAMETERS

Since a mechanical failure corresponds to $x_2 = 0$, we obtain

$$E(y \mid \text{mechanical failure}) = \beta_0 + \beta_1 x_1 + \beta_2 (0)$$

or

$$E(y \mid \text{mechanical failure}) = \beta_0 + \beta_1 x_1 \qquad (14.14)$$

The vertical line separating "y" and "mechanical failure" stands for the word "given"; thus, the expression $E(y \mid \text{mechanical failure})$ denotes the "expected value of the repair time given a mechanical failure." Equation (14.14) shows that the expected repair time for a mechanical failure is a linear function of x_1, the time since the previous service call; note that the y-intercept of this equation is β_0 and the slope is β_1.

Similarly, since an electrical failure corresponds to $x_2 = 1$, the expression for the expected value of y corresponding to an electrical failure is

$$E(y \mid \text{electrical failure}) = \beta_0 + \beta_1 x_1 + \beta_2 (1)$$

or

$$E(y \mid \text{electrical failure}) = \beta_0 + \beta_1 x_1 + \beta_2 \qquad (14.15)$$

Note that equation (14.15) can be rewritten as

$$E(y \mid \text{electrical failure}) = (\beta_0 + \beta_2) + \beta_1 x_1 \qquad (14.16)$$

Equation (14.16) shows that the expected repair time for an electrical failure is also a linear function of x_1; although the slope of this equation is still β_1, the y-intercept has changed from β_0 to $\beta_0 + \beta_2$. Thus, for any value of x_1, β_2 is the difference between the expected repair time for a mechanical failure and the expected repair time for an electrical failure.

If β_2 is positive, the expected repair time for an electrical failure will be greater than the expected repair time for a mechanical failure; however, if β_2 is negative, the expected repair time for a mechanical failure will be greater. Finally, if β_2 is zero, there will be no difference between the expected repair time between a mechanical and an electrical failure.

We can also interpret β_2 by subtracting equation (14.14), the expected repair time for a mechanical failure, from equation (14.15), the expected repair time for an electrical failure; the result is

$$(\beta_0 + \beta_1 x_1 + \beta_2) - (\beta_0 + \beta_1 x_1) = \beta_2$$

This also shows that β_2 can be interpreted as the difference in the expected repair time between an electrical failure and a mechanical failure.

COMPUTER SOLUTION

Table 14.6 shows the data for the Johnson Filtration problem after classifying each observation as to the type of failure. Table 14.6 shows that $x_2 = 0$ if an observation corresponded to a mechanical failure, and $x_2 = 1$ if an observation corresponded to an electrical failure. Using these data, we can now fit the following estimated regression equation:

$$\hat{y} = b_0 + b_1 x_1 + b_2 x_2$$

In Figure 14.9 we show the computer output obtained using the Minitab statistical software package; the estimated regression equation is

$$\text{HOURS} = 0.930 + 0.388 \text{ MONTHS} + 1.26 \text{ TYPE}$$

where TYPE is the name for the dummy variable x_2. Hence, the best estimate of β_2, the difference between the expected repair time for an electrical failure (TYPE = 1) and the expected repair time for a mechanical failure (TYPE = 0), is 1.26 hours.

To test for the significance of x_2 given the effect of x_1, the appropriate hypotheses are

$$H_0: \beta_2 = 0$$

$$H_a: \beta_2 \neq 0$$

Using a .05 level of significance, the p-value of .005 (corresponding to the t test for TYPE) shows that for $\alpha = .05$, we can reject H_0 and conclude that type of failure is a significant factor in predicting the repair time once the effect of the time since the last service call has been accounted for. Note also that R_a^2 of 81.9% shows that a good fit (much better than before) has been obtained using these two predictors.

TABLE 14.6 **JOHNSON FILTRATION DATA CLASSIFIED FOR TYPE OF FAILURE**

Repair Time (Hours)	Months Since Last Service Call	Type of Failure*	Repair Time (Hours)	Months Since Last Service Call	Type of Failure*
2.9	2	1	4.9	7	1
3.0	6	0	4.2	9	0
4.8	8	1	4.8	8	0
1.8	3	0	4.4	4	1
2.9	2	1	4.5	6	1

*Type of failure is 0 if the failure is mechanical and 1 if the failure is electrical.

FIGURE 14.9   MINITAB OUTPUT FOR DATA IN TABLE 14.6

```
The regression equation is
HOURS = 0.930 + 0.388 MONTHS + 1.26 TYPE

Predictor        Coef        Stdev      t-ratio          p
Constant       0.9305      0.4670         1.99      0.087
MONTHS         0.38762     0.06257        6.20      0.000
TYPE           1.2627      0.3141         4.02      0.005


s = 0.4590      R-sq = 85.9%      R-sq(adj) = 81.9%

Analysis of Variance

SOURCE         DF            SS            MS           F          p
Regression      2         9.0009        4.5005      21.36      0.001
Error           7         1.4751        0.2107
Total           9        10.4760
```

To provide additional insight into the role that qualitative variables play in regression analysis, in Figure 14.10 we have shown a scatter diagram for these data in which we have identified each point with the letter M for a mechanical failure, and the letter E for an electrical failure. Figure 14.10 suggests that there may be one underlying function relating y and x_1 for all the M points and another function relating y and x_1 for all the E points. Note that the estimated regression equation corresponding to a mechanical failure (TYPE = 0) is

$$\text{HOURS} = 0.930 + 0.388 \text{ MONTHS}$$

and the estimated regression equation corresponding to a mechanical failure (TYPE = 1) is

$$\text{HOURS} = 0.930 + 0.388 \text{ MONTHS} + 1.26$$

or

$$\text{HOURS} = 2.19 + 0.388 \text{ MONTHS}$$

Thus, by using a dummy variable, we have been able to develop two equations for predicting the repair time in hours; one corresponding to mechanical failures and one corresponding to electrical failures. ◀

MORE COMPLEX PROBLEMS

Let us now consider a problem involving a qualitative variable with more than two levels.

EXAMPLE 14.4▶   Suppose a manufacturer of copy machines has organized the sales territories for a particular state into three regions: A, B, and C. Management would like to investigate the relationship between the number of units sold each week y and the region. Since region

FIGURE 14.10   SCATTER DIAGRAM FOR DATA IN TABLE 14.6

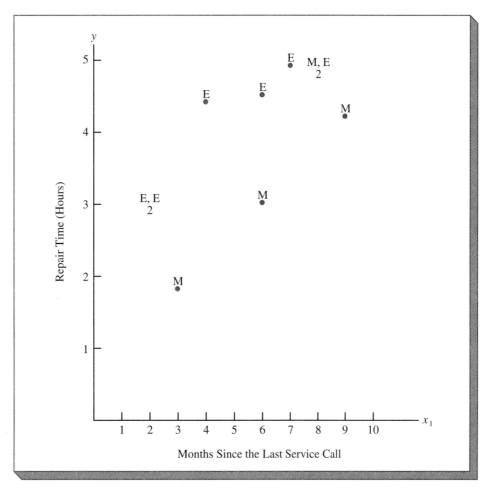

is a qualitative variable that takes on three values, we need two dummy variables to represent it. One possible specification is

$$D_1 = \begin{cases} 1 & \text{if region B} \\ 0 & \text{otherwise} \end{cases}$$

$$D_2 = \begin{cases} 1 & \text{if region C} \\ 0 & \text{otherwise} \end{cases}$$

With this definition of dummy variables, the values of D_1 and D_2 for each of the three regions are given in Table 14.7. We have defined two dummy variables D_1 and D_2 to represent the single qualitative variable, region. Observations corresponding to region A would be coded as $D_1 = 0$ and $D_2 = 0$, observations corresponding to region B would be coded as $D_1 = 1$ and $D_2 = 0$, and observations corresponding to region C would be coded as $D_1 = 0$ and $D_2 = 1$. Using these two dummy variables, then, we can write the regression equation relating the number of units sold to the region in which the sales take place as follows:

$$E(y) = \beta_0 + \beta_1 D_1 + \beta_2 D_2$$

| TABLE 14.7 | **CODING SYSTEM 1** |

D_1	D_2	Region
0	0	A
1	0	B
0	1	C

| TABLE 14.8 | **CODING SYSTEM 2** |

D_1	D_2	Region
0	0	B
1	0	A
0	1	C

| TABLE 14.9 | **CODING SYSTEM 3** |

D_1	D_2	Region
0	0	C
1	0	A
0	1	B

Thus,

$$E(y \mid \text{region A}) = \beta_0 + \beta_1(0) + \beta_2(0) = \beta_0 \qquad (14.17)$$

$$E(y \mid \text{region B}) = \beta_0 + \beta_1(1) + \beta_2(0) = \beta_0 + \beta_1 \qquad (14.18)$$

$$E(y \mid \text{region C}) = \beta_0 + \beta_1(0) + \beta_2(1) = \beta_0 + \beta_2 \qquad (14.19)$$

To interpret β_1 we subtract equation (14.17) from equation (14.18); hence, β_1 is the difference between the expected number of units sold in region B and the expected number of units sold in region A. Similarly, by subtracting equation (14.17) from equation (14.19), we see that β_2 is the difference between the expected number of units sold in region C and the expected number of units sold in region A.

In general, whenever we want to incorporate the effect of a qualitative variable that takes on k values, we use $k - 1$ dummy variables. Note that there are several possible ways in which we can code the $k - 1$ dummy variables. For example, we could have coded the data in our sales problem as shown in Table 14.8.

Using this coding system, observations corresponding to region B would be coded as $D_1 = 0$ and $D_2 = 0$, observations corresponding to region A would be coded as $D_1 = 1$ and $D_2 = 0$, and observations corresponding to region C would be coded as $D_1 = 0$ and $D_2 = 1$. Hence,

$$E(y \mid \text{region B}) = \beta_0 + \beta_1(0) + \beta_2(0) = \beta_0$$

$$E(y \mid \text{region A}) = \beta_0 + \beta_1(1) + \beta_2(0) = \beta_0 + \beta_1$$

$$E(y \mid \text{region C}) = \beta_0 + \beta_1(0) + \beta_2(1) = \beta_0 + \beta_2$$

Thus, β_1 denotes the difference between the expected number of units sold in region A and the expected number of units sold in region B, and β_2 denotes the difference between the expected number of units sold in region C and the expected number of units sold in region B. Note that the interpretations of β_1 and β_2 are different from the interpretation obtained using the previous coding system.

A third possibility would have been to code the data as shown in Table 14.9. It does not matter which coding system we use as long as we are consistent in defining both the data that will be analyzed and interpreting the results.

A CAUTIONARY NOTE WHEN USING DUMMY VARIABLES

Suppose that we had a fourth region in the sales territory illustration, identified as region D. Since we have four levels or regions, we would need to use three dummy variables. A common mistake is attempting to use only two dummy variables by setting both D_1 and D_2 equal to 1 if the observation corresponds to region D. Note that if this approach is followed, the expected sales for region D is

$$E(y \mid \text{region D}) = \beta_0 + \beta_1(1) + \beta_2(1) = \beta_0 + \beta_1 + \beta_2 \qquad (14.20)$$

If we subtract equation (14.17) from equation (14.20) we see that $\beta_1 + \beta_2$ is the difference between the expected sales in region D and region A. However, since β_1 is the difference between the expected sales in regions B and A, and β_2 is the difference between the expected sales in regions C and A, this approach forces the difference between D and A to be a linear combination of the other two differences. In general, using a coding system that forces any conclusions on the data is not warranted. Instead, the proper approach is to let the data tell us what the relationship is. When using dummy variables, we can ensure a correct interpretation by always following the rule of using $k - 1$ dummy variables whenever the qualitative variable takes on k values.

EXERCISES

METHODS

SELF TEST

30. Consider a regression study involving a dependent variable y, a quantitative independent variable x_1, and a qualitative variable with two levels (level 1 and level 2).
    a. Write a multiple regression equation relating x_1 and the qualitative variable to y.
    b. What is the expected value of y corresponding to level 1 of the qualitative variable?
    c. What is the expected value of y corresponding to level 2 of the qualitative variable?
    d. Interpret the parameters in your regression equation.

31. Consider a regression study involving a dependent variable y, a quantitative independent variable x_1, and a qualitative independent variable with three possible levels (level 1, level 2, and level 3).
    a. How many dummy variables are required to represent the qualitative variable?
    b. Write a multiple regression equation relating x_1 and the qualitative variable to y.
    c. Interpret the parameters in your regression equation.

APPLICATIONS

SELF TEST

32. The following regression model has been proposed to predict sales at a fast-food outlet:

$$y = \beta_0 + \beta_1 x_1 + \beta_2 x_2 + \beta_3 x_3 + \epsilon$$

where

$$x_1 = \text{number of competitors within 1 mile}$$
$$x_2 = \text{population within 1 mile (1000s)}$$
$$x_3 = \begin{cases} 1 & \text{if drive-up window present} \\ 0 & \text{otherwise} \end{cases}$$
$$y = \text{sales (\$1000s)}$$

The following estimated regression equation was developed after 20 outlets were surveyed:

$$\hat{y} = 10.1 - 4.2x_1 + 6.8x_2 + 15.3x_3$$

a. What is the expected amount of sales attributable to the drive-up window?

b. Predict sales for a store with two competitors, a population of 8000 within 1 mile, and no drive-up window.

c. Predict sales for a store with one competitor, a population of 3000 within 1 mile, and a drive-up window.

 REPAIR

33. Refer to the Johnson Filtration problem, Example 14.3, introduced in this section. Suppose that in addition the data already mentioned (i.e., the number of months since the machine was serviced and whether a mechanical or an electrical failure had occurred), management also obtained data showing which repairperson had performed the service. The revised data are shown below.

Repair Time (hours)	Months Since Previous Service Call (months)	Type of Failure	Repairperson
2.9	2	Electrical	Dave Newton
3.0	6	Mechanical	Dave Newton
4.8	8	Electrical	Bob Jones
1.8	3	Mechanical	Dave Newton
2.9	2	Electrical	Dave Newton
4.9	7	Electrical	Bob Jones
4.2	9	Mechanical	Bob Jones
4.8	8	Mechanical	Bob Jones
4.4	4	Electrical	Bob Jones
4.5	6	Electrical	Dave Newton

a. Ignore for now the months since the previous service call (x_1) and the repairperson who performed the service. Develop the estimated simple linear regression equation to predict the repair time (y) given the type of failure (x_2). Recall that $x_2 = 0$ if the failure is mechanical and 1 if the failure is electrical.

b. Does the equation that you developed in part (a) provide a good fit for the observed data? Explain.

c. Ignore for now the months since the previous service call and the type of failure associated with the machine. Develop the estimated simple linear regression equation to predict the repair time given the repairperson who performed the service. Let $x_3 = 0$ if Bob Jones performed the service and $x_3 = 1$ if Dave Newton performed the service.

d. Does the equation that you developed in part (c) provide a good fit for the observed data? Explain.

34. This problem is an extension of the situation described in Exercise 33.

a. Develop the estimated regression equation to predict the repair time given the number of months since the previous service call, the type of failure, and the repairperson who performed the service.

b. At the $\alpha = .05$ level of significance, test whether the estimated regression equation developed in part (a) represents a significant relationship between the independent variables and the dependent variable.

c. Is the addition of the independent variable x_3, the repairperson who performed the service, statistically significant? Use $\alpha = .05$. What explanation can you give for the results observed?

FORBES

35. The following data show the price/earnings ratio, the profit margin, and the growth rate for 19 companies listed in "The *Forbes* 500s on Wall Street" (*Forbes,* May 1, 1989). The data in the column labeled "Industry" are simply codes used to define the industry for each company: 1 = energy-international oil, 2 = health-drugs, and 3 = electronics-computers.

Firm	P/E Ratio	Profit Martin	Growth Rate	Industry
Exxon	11.3	6.5	10	1
Chevron	10.0	7.0	5	1
Texaco	9.9	3.9	5	1
Mobil	9.7	4.3	7	1
Amoco	10.0	9.8	8	1
Pfizer	11.9	14.7	12	2
Bristol Meyers	16.2	13.9	14	2
Merck	21.0	20.3	16	2
American Home Products	13.3	16.9	11	2
Abbott Laboratories	15.5	15.2	18	2
Eli Lilly	18.9	18.7	11	2
Upjohn	14.6	12.8	10	2
Warner-Lambert	16.0	8.7	7	2
Amdahl	8.4	11.9	4	3
Digital	10.4	9.8	19	3
Hewlett-Packard	14.8	8.1	18	3
NCR	10.1	7.3	6	3
Unisys	7.0	6.9	6	3
IBM	11.8	9.2	6	3

Develop the estimated regression equation that can be used to predict the price/earnings ratio given the profit margin, growth rate, and type of industry.

 STROKE

36. The American Heart Association collects data on the risk of strokes. A 10-year study provided data on how age, blood pressure, and smoking relate to the risk of strokes (*U.S. News & World Report,* April 13, 1992). Assume that data from a portion of this study are shown below. Risk is interpreted as the probability (times 100) that the patient will have a stroke over the next 10-year period. For the smoking variable, a 1 indicates a smoker and a 0 indicates a nonsmoker.

Risk	Age	Pressure	Smoker	Risk	Age	Pressure	Smoker
12	57	152	0	22	71	152	0
24	67	163	0	36	70	173	1
13	58	155	0	15	67	135	1
56	86	177	1	48	77	209	1
28	59	196	0	15	60	199	0
51	76	189	1	36	82	119	1
18	56	155	1	8	66	166	0
31	78	120	0	34	80	125	1
37	80	135	1	3	62	117	0
15	78	98	0	37	59	207	1

a. Using these data, develop an estimated regression equation that relates risk of a stroke to the person's age, blood pressure, and whether the person is a smoker.

b. Is smoking a significant factor in the risk of a stroke? Explain. Use $\alpha = .05$.

c. What is the probability of a stroke over the next 10 years for Art Speen, a 68-year-old smoker who has a blood pressure of 175? What action might the physician recommend for this patient?

14.7 RESIDUAL ANALYSIS

In Chapter 13 we showed how residual plots could be used to determine whether the assumptions made regarding the error term and model form are valid. In the multiple regression case, these same methods can be used to validate the assumptions made for the proposed model and provide additional information concerning the adequacy of the fitted least squares equation.

The statistical tests that we discussed in Section 14.4 are based on the assumptions presented in Section 14.1 for the error term ϵ and the assumption that the population regression equation has the linear form given by $E(y) = \beta_0 + \beta_1 x_1 + \cdots + \beta_p x_p$. Residual analysis will allow us to make a judgment about whether the model assumptions appear to be satisfied. In addition, residual analysis can often provide insight as to whether a different type of model—for example, one involving more variables or a different functional form—might better describe the observed relationship.

In simple linear regression, we showed how a residual plot against the independent variable x and a residual plot against the predicted value $\hat{y}$ could be used to determine whether the assumptions regarding the error term are appropriate. In multiple regression analysis, these plots can be developed and interpreted similarly; the only difference in multiple regression is that plots against more than one independent variable can be developed. Statisticians usually look first at the residual plot against $\hat{y}$ and then review the plots against the independent variables if necessary.

In Chapter 13 we indicated that many of the residual plots provided by statistical software packages use a standardized version of the residuals. For simple linear regression, the ith standardized residual is defined as follows.

Standardized Residual

$$\frac{y_i - \hat{y}_i}{s\sqrt{1 - h_i}} \tag{14.21}$$

where

$$h_i = \frac{1}{n} + \frac{(x_i - \bar{x})^2}{\Sigma(x_i - \bar{x})^2} \tag{14.22}$$

The comments that we made in the previous paragraph apply equally well for residual plots developed using the standardized residuals.

The computation of the predicted values, the residuals, and the standardized residuals is too complex to do by hand. Using a statistical software package such as Minitab, however, these values can be simply obtained as part of the standard regression analysis output.

EXAMPLE 14.2
(CONTINUED)

In Table 14.10, we show the predicted values, residuals, and the standardized residuals for the Butler Trucking problem; these values were obtained using the Minitab statistical software package. Using the values in Table 14.10, the appropriate residual plots can be simply constructed. Our preference is to initially plot the standardized residuals versus the predicted values. In reviewing the subsequent standardized residual plot, we will look for the same patterns that we looked for when analyzing residuals in simple linear regression. ◄

Figure 14.11 shows three forms of standardized residual plots versus $\hat{y}$, the value of y predicted by the regression equation. The plot in panel A shows the type of pattern to expect when the model assumptions are satisfied. The plot in panel B shows a pattern that can be expected when the constant variance assumption is not satisfied. The error term gets larger as the value of $\hat{y}$ increases. Adding another independent variable will sometimes correct this problem. It was this type of pattern that suggested that adding a variable might be helpful in the Butler Trucking problem. Finally, the plot in panel C shows a case where the model is not adequate. When $\hat{y}$ is small, the error term is positive; when $\hat{y}$ assumes intermediate values, the error term is negative; and when $\hat{y}$ is large, the error term is again positive. Often a curvilinear model is needed in this case.

EXAMPLE 14.2
(CONTINUED)

In Figure 14.12 we show a standardized residual plot against the predicted values $\hat{y}$ for the Butler Trucking problem involving two independent variables. Note that this residual plot does not indicate any abnormality; thus, we conclude that the model assumptions are reasonable. ◄

TABLE 14.10   RESIDUAL ANALYSIS FOR BUTLER TRUCKING

Miles Traveled (x_1)	Deliveries (x_2)	Travel Time (y)	Predicted Value ($\hat{y}$)	Residual ($y - \hat{y}$)	Standardized Residual
100	4	9.3	8.93846	0.361541	0.78344
50	3	4.8	4.95830	−0.158304	−0.34962
100	4	8.9	8.93846	−0.038460	−0.08334
100	2	6.5	7.09161	−0.591609	−1.30929
50	2	4.2	4.03488	0.165121	0.38167
80	2	6.2	5.86892	0.331083	0.65431
75	3	7.4	6.48667	0.913331	1.68917
65	4	6.0	6.79875	−0.798749	−1.77372
90	3	7.6	7.40369	0.196311	0.36703
90	2	6.1	6.48026	−0.380263	−0.77639

FIGURE 14.11   POSSIBLE RESIDUAL PATTERNS AND THEIR CAUSES

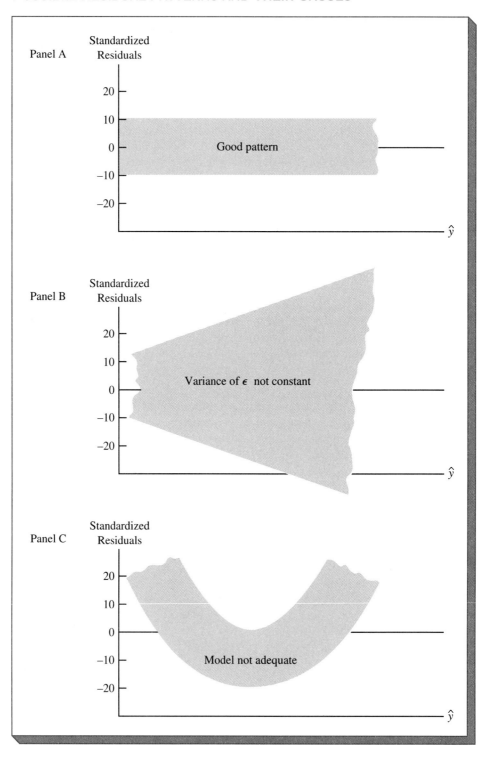

FIGURE 14.12   STANDARDIZED RESIDUAL PLOT FOR BUTLER TRUCKING WITH TWO
INDEPENDENT VARIABLES

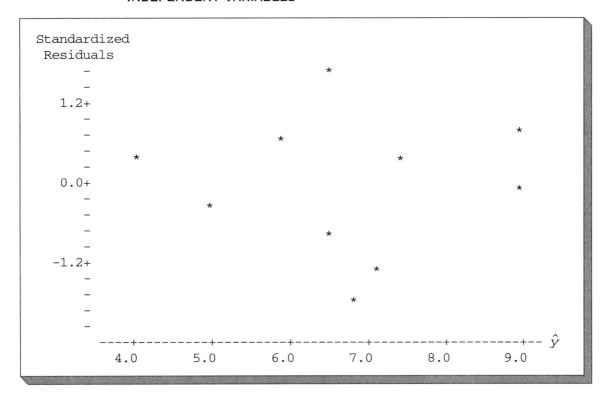

OUTLIERS

An *outlier* is a data point (observation) that does not follow the trend exhibited by the remaining data. In Chapter 13 we showed how standardized residuals can be used to identify outliers in a data set. If outliers are present, the fit provided by the estimated regression equation will not be as good as it would be if there were no outliers, and the value of *s*, the standard error of the estimate, will be larger. Since *s* appears in the denominator of equation (14.21), the size of the standardized residuals will be decreased as *s* increases. Thus, even though a residual is unusually large, it can be difficult to identify it as an outlier using the standardized residuals. An approach that can circumvent this type of problem uses a form of the residuals that are referred to as *studentized deleted residuals*.

USE OF STUDENTIZED DELETED RESIDUALS FOR IDENTIFYING OUTLIERS

Suppose that the *i*th observation was deleted from the data set and a new estimated regression equation was developed. Let $s_{(i)}$ denote the standard error of the estimate based on the data set with the *i*th observation deleted. If we compute the standardized residuals for the original data set using $s_{(i)}$ instead of *s*, the resulting values are referred to as the *studentized deleted residuals*.

Studentized Deleted Residual

$$\frac{y_i - \hat{y}_i}{s_{(i)}\sqrt{1 - h_i}} \qquad (14.23)$$

If the ith observation is an outlier, deleting this observation will result in a smaller value for the standard error of the estimate; that is, $s_{(i)} < s$. Since the ith studentized deleted residual is computed using $s_{(i)}$, the absolute value of the ith studentized residual will be larger than the absolute value for the standardized residual.

Many statistical software packages provide options for obtaining the studentized deleted residuals as part of the regression analysis output. For example, in the Minitab statistical system, the studentized deleted residuals can be obtained using the REGRESS command with the subcommand TRESID.

EXAMPLE 14.2 ▶
(CONTINUED)

In Table 14.11 we show the values of the studentized deleted residuals for the Butler Trucking problem involving 10 observations and two independent variables. To identify outliers in the data set, we look for large absolute values for the studentized deleted residuals.

The t distribution can be used to determine how large the ith studentized deleted residual must be to conclude that the ith observation is an outlier. Recall that p denotes the number of independent variables in the model and n denotes the number of observations in the original data set; thus, if we delete the ith observation, the total number of observations drops to $n - 1$ and the error sum of squares has $n - p - 2$ degrees of freedom. For the Butler Trucking problem involving two independent variables, the error degrees of freedom corresponding to the fitted model in which we delete the ith observation is $10 - 2 - 2 = 6$. At the .05 level of significance, the t distribution table shows that for 6 degrees of freedom, $t_{.025} = 2.447$. Thus, if the value of the ith studentized deleted residual is less than -2.447 or greater than $+2.447$, we conclude that at the .05 level of significance, the ith observation is an outlier. Since none

TABLE 14.11   STUDENTIZED DELETED RESIDUALS FOR BUTLER TRUCKING

Miles Traveled (x_1)	Deliveries (x_2)	Travel Time (y)	Standardized Residual	Studentized Deleted Residual
100	4	9.3	0.78344	0.75939
50	3	4.8	−0.34962	−0.32654
100	4	8.9	−0.08334	−0.07720
100	2	6.5	−1.30929	−1.39494
50	2	4.2	0.38167	0.35709
80	2	6.2	0.65431	0.62519
75	3	7.4	1.68917	2.03187
65	4	6.0	−1.77372	−2.21314
90	3	7.6	0.36703	0.34312
90	2	6.1	−0.77639	−0.75190

of the studentized deleted residuals in table 14.11 exceeds these limits, we conclude that there are no outliers in the data. ◀

INFLUENTIAL OBSERVATIONS

In Chapter 13 we discussed how leverage can be used to identify observations for which the value of the independent variable has a strong influence on the results obtained. Recall that the leverage of the ith observation was defined for the case of simple linear regression as follows.

> **Leverage of Observation i**
>
> $$h_i = \frac{1}{n} + \frac{(x_i - \bar{x})^2}{\Sigma(x_i - \bar{x})^2}$$ (14.24)

In multiple regression analysis, we also use leverage to identify observations that have a strong influence on the coefficients. In the Minitab statistical computing system, an observation is considered *influential* if $h_i > 3(p + 1)/n$, where p denotes the number of independent variables and n is the number of observations.

EXAMPLE 14.2 ▶
(CONTINUED)

As an illustration, let us consider the Butler Trucking problem involving $n = 10$ observations and $p = 2$ independent variables. The critical value for leverage is

$$\frac{3(p + 1)}{n} = \frac{3(2 + 1)}{10} = \frac{9}{10} = .9$$

The values of h_i, obtained using the Minitab statistical computing system, are shown in Table 14.12. Since none of the values exceed .9, we conclude that there are no influential observations in the data. ◀

TABLE 14.12   LEVERAGE VALUES FOR BUTLER TRUCKING

Miles Traveled (x_1)	Deliveries (x_2)	Travel Time (y)	Leverage (h_i)
100	4	9.3	0.351704
50	3	4.8	0.375863
100	4	8.9	0.351704
100	2	6.5	0.378451
50	2	4.2	0.430220
80	2	6.2	0.220557
75	3	7.4	0.110009
65	4	6.0	0.382657
90	3	7.6	0.129098
90	2	6.1	0.269737

USE OF COOK'S DISTANCE MEASURE TO IDENTIFY INFLUENTIAL OBSERVATIONS

A problem that can arise when using leverage to identify influential observations is that an observation can be identified as having high leverage and not necessarily be influential in terms of the resulting estimated regression equation.

EXAMPLE 14.5 ▶ Table 14.13 shows a data set consisting of eight observations and their corresponding leverage values (obtained using Minitab). Since the leverage for the eighth observation is .91 > .75 (the critical leverage value when $p = 1$ and $n = 8$), this observation is identified as an influential observation. Before reaching any final conclusions, however, let us consider the situation from a different perspective.

In Figure 14.13 we show the scatter diagram and the estimated regression equation corresponding to the data set in Table 14.13. Using Minitab, the following estimated regression equation was developed for these data:

$$\hat{y} = 18.2 + 1.39x$$

The straight line shown in Figure 14.13 is the graph of this equation. Now, let's delete the observation $x = 15$, $y = 39$ from the data set and fit a new estimated regression equation to the remaining seven observations; the new estimated regression equation is

$$\hat{y} = 18.1 + 1.42x$$

We note that the y-intercept and slope of the new estimated regression equation have changed very little from the values obtained using all the data. Thus, although the eighth observation was identified as an influential observation using the leverage criterion, this observation clearly had little influence on the results obtained. Thus, in some situations using only leverage to identify influential observations can lead to wrong conclusions. ◀

Cook's distance measure uses both the values of the residuals and the leverage values to determine whether an observation is influential.

Cook's Distance Measure

$$D_i = \frac{(y_i - \hat{y}_i)^2}{(p + 1)s^2}\left[\frac{h_i}{(1 - h_i)^2}\right]$$

(14.25)

where

p = number of independent variables
s^2 = variance of the estimate (developed using all n observations)
h_i = leverage for the ith observation (developed using all n observations)

Note that the value of D_i depends on the value of the ith residual $y_i - \hat{y}_i$ and the leverage of the ith observation h_i. Thus, the ith observation can be influential if it has (1) a large

TABLE 14.13   DATA SET ILLUSTRATING POTENTIAL PROBLEM USING THE LEVERAGE CRITERION

x_i	y_i	Leverage h_i	x_i	y_i	Leverage h_i
1	18	0.204170	4	23	0.125977
1	21	0.204170	4	24	0.125977
2	22	0.164205	5	26	0.127715
3	21	0.138141	15	39	0.909644

FIGURE 14.13   SCATTER DIAGRAM FOR THE DATA SET IN TABLE 14.13

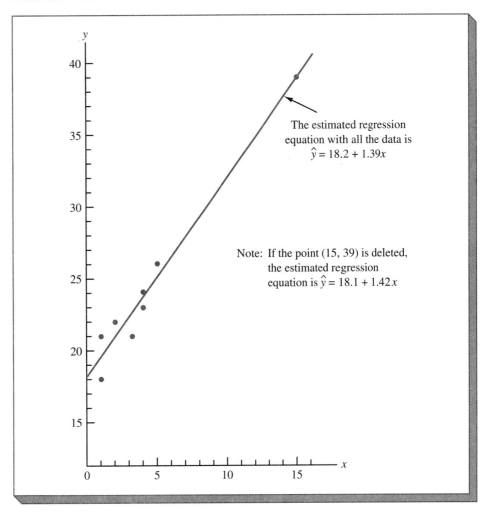

The estimated regression equation with all the data is $\hat{y} = 18.2 + 1.39x$

Note: If the point (15, 39) is deleted, the estimated regression equation is $\hat{y} = 18.1 + 1.42x$

TABLE 14.14   COOK'S DISTANCE MEASURE

x_i	y_i	Cook's Distance Measure D_i	x_i	y_i	Cook's Distance Measure D_i
1	18	0.265344	4	23	0.029850
1	21	0.230628	4	24	0.004164
2	22	0.089443	5	26	0.044397
3	21	0.113389	15	39	0.055598

---

residual and a small leverage value, (2) a small residual and a large leverage value, or (3) a large residual and a large leverage value. As a rule of thumb, values of $D_i > 1$ are considered large enough to conclude that the ith observation is influential and thus should be studied further.

EXAMPLE 14.5
(CONTINUED)

In Table 14.14 we show the values of Cook's distance measure for the data set shown in Table 14.13. Since none of the values exceeds 1, we conclude that the data set does not contain any influential observations. Recall that when we just used leverage to identify influential observations, we erroneously concluded that observation 8 was influential. ◄

NOTES AND
COMMENTS

To determine whether the value of Cook's distance measure D_i is large enough to conclude that the ith observation is influential, we can also compare the value of D_i to the 50th percentile of an F distribution (denoted $F_{.50}$) with $p + 1$ numerator degrees of freedom and $n - p - 1$ denominator degrees of freedom. The F tables corresponding to a 50% level of significance must be available to carry out the test. The rule of thumb we provided ($D_i > 1$) is based on the fact that the table value is very close to 1 for a wide variety of cases.

EXERCISES

METHODS

SELF TEST

37.   Shown below are data for two variables, x and y.

x_i	1	2	3	4	5
y_i	3	7	5	11	14

   a.   Develop the estimated regression equation for these data.
   b.   Plot the standardized residuals versus $\hat{y}$. Do there appear to be any outliers in these data? Explain.
   c.   Compute the studentized deleted residuals for these data. At the $\alpha = .05$ level of significance, can any of these observations be classified as an outlier? Explain.

38. Shown below are data for two variables, x and y.

x_i	22	24	26	28	40
y_i	12	21	31	35	70

    a. Develop the estimated regression equation for these data.
    b. Compute the studentized deleted residuals for these data. At the $\alpha = .05$ level of significance, can any of these observations be classified as an outlier? Explain.
    c. Compute the leverage values for these data. Do there appear to be any influential observations in these data? Explain.
    d. Compute Cook's distance measure for these data. Are any observations influential? Explain.

APPLICATIONS

SELF TEST

39. In Exercise 6 data (in $1000s) were presented showing the weekly gross revenue, the television advertising, and the newspaper advertising for TAI Movie Theaters; these data are repeated below.

MOVIE

Weekly Gross Revenue ($1000s)	Television Advertising ($1000s)	Newspaper Advertising ($1000s)
96	5.0	1.5
90	2.0	2.0
95	4.0	1.5
92	2.5	2.5
95	3.0	3.3
94	3.5	2.3
94	2.5	4.2
94	3.0	2.5

    a. Find an estimated regression equation relating weekly gross revenue to television and newspaper advertising.
    b. Plot the standardized residuals against $\hat{y}$. Does the residual plot support the assumptions regarding ϵ? Explain.
    c. Check for any outliers in these data. What are your conclusions?
    d. Are there any influential observations? Explain.

40. Refer to the data in Exercise 23.
    a. Plot the standardized residuals against $\hat{y}$. Does the residual plot support the assumptions regarding ϵ? Explain.
    b. Check for any outliers in these data. What are your conclusions?
    c. Are there any influential observations? Explain.

41. Refer to the data in Exercise 9. Let SIZE denote the average class size, SAT the combined SAT score, and %COLLEGE the percentage of students who attended a 4-year college.
    a. Develop an estimated regression equation that can be used to predict %COLLEGE given SAT.
    b. Based on the estimated regression equation developed in part (a), do there appear to be any outliers and/or influential observations in these data? Explain.
    c. Develop an estimated regression equation that can be used to predict %COLLEGE given SIZE and SAT.
    d. Based on the estimated regression equation developed in part (c), do there appear to be any outliers and/or influential observations? Explain.

▽
SUMMARY

In this chapter we showed how extensions of the concepts of simple linear regression can be used to develop an estimated regression equation for predicting y that involves several independent variables. We noted that the interpretation of the coefficients had to be modified somewhat for this case. That is, we interpreted b_i as an estimate of the change in the dependent variable y that would result from a 1-unit change in independent variable x_i when the other independent variables do not change.

A key part of any multiple regression study is the use of a computer software package for carrying out the computational work. Many excellent packages exist and, after a short learning period, can be used to develop the estimated regression equation, conduct the appropriate significance tests, and prepare residual plots.

In Section 14.6, we discussed how dummy variables can be used to account for the effect of qualitative variables. Then, in Section 14.7 we extended the application of residual analysis by showing additional procedures for identifying outliers and influential observations.

▽
GLOSSARY

Multiple regression model A regression model in which more than one independent variable is used to predict the dependent variable.

Response variable Another term for dependent variable.

Multiple coefficient of determination (R^2) A measure of the goodness of fit for the estimated regression equation.

Adjusted multiple coefficient of determination (R_a^2) A measure of the goodness of fit for the estimated regression equation that accounts for the number of independent variables in the model.

Multicollinearity A term used to describe the case when the independent variables in a multiple regression model are correlated.

Qualitative variable A variable that is not measured in terms of how much or how many, but instead is assigned values to represent categories.

Dummy variable A variable that takes on the values 0 or 1 and is used to incorporate the effects of qualitative variables in a regression model.

Outlier An observation with a residual that is far greater in magnitude than the rest of the residual values.

Studentized deleted residuals A form of the standardized residuals computed when the ith observation is deleted from the data set and a new estimated regression equation is developed.

Influential observation An observation that has a great deal of influence in determining the estimated regression equation.

▽
KEY FORMULAS

Multiple Regression Model

$$y = \beta_0 + \beta_1 x_1 + B_2 x_2 + \cdots + \beta_p x_p + \epsilon \tag{14.2}$$

Multiple Regression Equation

$$E(y) = \beta_0 + \beta_1 x_1 + \cdots + \beta_p x_p \tag{14.3}$$

Estimated Regression Equation

$$\hat{y} = b_0 + b_1 x_1 + \cdots + b_p x_p \qquad \text{(14.5)}$$

Least Squares Criterion

$$\min \Sigma (y_i - \hat{y}_i)^2 \qquad \text{(14.7)}$$

Total Sum of Squares

$$\text{SST} = \Sigma (y_i - \bar{y})^2 \qquad \text{(14.8)}$$

Multiple Coefficient of Determination

$$R^2 = \frac{\text{SSR}}{\text{SST}} \qquad \text{(14.9)}$$

Adjusted Multiple Coefficient of Determination

$$R_a^2 = 1 - (1 - R^2)\frac{n-1}{n-p-1} \qquad \text{(14.10)}$$

Mean Square Due to Regression

$$\text{MSR} = \frac{\text{SSR}}{p} \qquad \text{(14.11)}$$

Mean Square Due to Error

$$\text{MSE} = \frac{\text{SSE}}{n-p-1} \qquad \text{(14.12)}$$

The F Statistic

$$F = \frac{\text{MSR}}{\text{MSE}} = \frac{\text{SSR}/p}{\text{SSE}/(n-p-1)} \qquad \text{(14.13)}$$

Standardized Residual

$$\frac{y_i - \hat{y}_i}{s\sqrt{1 - h_i}} \qquad \text{(14.21)}$$

Studentized Deleted Residual

$$\frac{y_i - \hat{y}_i}{s_{(i)}\sqrt{1 - h_i}} \qquad \text{(14.23)}$$

Leverage of Observation i

$$h_i = \frac{1}{n} + \frac{(x_i - \bar{x})^2}{\Sigma(x_i - \bar{x})^2} \qquad \text{(14.24)}$$

Cook's Distance Measure

$$D_i = \frac{(y_i - \hat{y}_i)^2}{(p+1)s^2}\left[\frac{h_i}{(1-h_i)^2}\right] \qquad \text{(14.25)}$$

▽

REVIEW QUIZ

TRUE/FALSE

1. In multiple regression analysis, there are two or more dependent variables.
2. The least squares method cannot be used to develop the coefficients of the estimated regression equation for multiple regression analysis.
3. The coefficient of x_1 in an estimated regression equation will be the same no matter how many independent variables are involved.
4. The computer solution of a multiple regression problem provides all the coefficients for the estimated regression equation.
5. An F test can be used to test for a significant relationship in multiple regression analysis.
6. The number of degrees of freedom for the error sum of squares in multiple regression analysis decreases as the number of independent variables increases.
7. The number of degrees of freedom in the t test for the significance of individual parameters in multiple regression analysis is always $n - 3$.
8. The adjusted multiple coefficient of determination will always be smaller than the unadjusted multiple coefficient of determination.

MULTIPLE CHOICE

For Questions 9–12, consider a regression model in which two independent variables, x_1 and x_2, are used to explain the dependent variable, y.

9. In the test of the hypotheses H_0: $\beta_1 = \beta_2 = 0$ and H_a: either β_1 or β_2 or both $\neq 0$, the test statistic MSR/MSE has a sampling distribution that is the
   a. t distribution
   b. F distribution
   c. normal distribution
   d. chi-square distribution
10. The degrees of freedom for the sampling distribution in Question 9 are
    a. $n - 2$ b. $n - 3$
    c. 3 and $n - 2$ d. 2 and $n - 3$
11. Which sampling distribution is used to test H_0: $\beta_2 = 0$, H_a: $\beta_2 \neq 0$?
    a. t distribution
    b. F distribution
    c. normal distribution
    d. chi-square distribution
12. The degrees of freedom for the distribution in Question 11 are
    a. $n - 2$ b. $n - 3$
    c. 3 and $n - 2$ d. 2 and $n - 3$

Questions 13–15 involve the following problem. A car-rental company wanted to predict the annual operating cost (y) of its cars using both the number of miles a car was driven (x_1) and the size of the car (subcompact, compact, midsize, and full-size). To incorporate the effect of the size of the car, the following dummy variables were defined.

x_2	x_3	x_4	Car Size
0	0	0	subcompact
1	0	0	compact
0	1	0	midsize
0	0	1	full-size

13. The mean annual operating cost for a compact car is
    a. β_0
    b. $\beta_0 + \beta_1 x_1 + \beta_2 x_2$
    c. $\beta_0 + \beta_1 x_1 + \beta_2$
    d. $\beta_0 + \beta_1 x_1 + \beta_3$
    e. $\beta_0 + \beta_1 x_1 + \beta_4$
14. The mean annual operating cost for a subcompact car is
    a. β_0
    b. $\beta_0 + \beta_1 x_1$
    c. $\beta_0 + \beta_1 x_1 + \beta_2$
    d. $\beta_0 + \beta_1 x_1 + \beta_3$
    e. $\beta_0 + \beta_1 x_1 + \beta_4$
15. The difference in mean annual operating costs between a full-size car and a subcompact car is
    a. β_0
    b. β_4
    c. $\beta_0 + \beta_4$
    d. $\beta_0 + \beta_1 x_1 + \beta_4$

SUPPLEMENTARY EXERCISES

42. The admissions officer for Clearwater College developed the following estimated regression equation relating the final college GPA to the students' high-school GPA and their SAT mathematics score.

    $$\hat{y} = -1.41 + .0235x_1 + .00486x_2$$

    where

    $$x_1 = \text{high school grade point average}$$
    $$x_2 = \text{SAT mathematics score}$$
    $$y = \text{final college grade point average}$$

    a. Interpret the coefficients in this estimated regression equation.
    b. Estimate the final college GPA for a student who has a high school average of 84 and a score of 540 on the SAT mathematics test.

43. The personnel director for Electronics Associates developed the following estimated regression equation relating an employee's score on a job satisfaction test to his or her length of service and wage rate:

    $$\hat{y} = 14.4 - 8.69x_1 + 13.5x_2$$

    where

    $$x_1 = \text{length of service (years)}$$
    $$x_2 = \text{wage rate (dollars)}$$
    $$y = \text{job satisfaction test score (higher scores}$$
    $$\text{indicate more job satisfaction)}$$

    a. Interpret the coefficients in this estimated regression equation.
    b. Develop an estimate of the job satisfaction test score for an employee who has 4 years of service and makes $6.50 per hour.

44. In a regression analysis involving 18 observations and four independent variables, it was determined that SSR = 18,051.63 and SSE = 1014.3.
    a. Determine R^2 and R_a^2.
    b. Test for the significance of the relationship at the $\alpha = .01$ level of significance.

45. The following estimated regression equation involving three independent variables has been developed:

$$\hat{y} = 18.31 + 8.12x_1 + 17.9x_2 - 3.6x_3$$

Computer output indicates that $s_{b_1} = 2.1$, $s_{b_2} = 9.72$, and $s_{b_3} = .71$. There were 15 observations in the study.

a. Test H_0: $\beta_1 = 0$ at $\alpha = .05$.
b. Test H_0: $\beta_2 = 0$ at $\alpha = .05$.
c. Test H_0: $\beta_3 = 0$ at $\alpha = .05$.
d. Would you recommend dropping any of the independent variables from the model?

46. Shown is a partial computer output from a regression analysis.

```
The regression equation is
Y = 8.103 + 7.602 X1 + 3.111 X2


Predictor      Coef      Stdev      t-ratio
Constant     _____     2.667     _____
X1           _____     2.105     _____
X2           _____     0.613     _____


s = 3.35          R-sq = 92.3%          R-sq(adj) = _____

Analysis of Variance

SOURCE             DF             SS          MS           F
Regression      _____         1612       _____      _____
Error              12          _____   .  _____
Total           _____       _____
```

a. Compute the appropriate t-ratios.
b. Test for the significance of β_1 and β_2 at $\alpha = .05$.
c. Compute the entries in the DF, SS, and MS = SS/DF columns.
d. Compute R_a^2.

47. Recall that in Exercise 42, the admissions officer for Clearwater College developed the following estimated regression equation relating the final college GPA (y) to the student's high-school GPA (x_1) and their SAT mathematics score (x_2).

$$\hat{y} = -1.41 + .0235x_1 + .00486x_2$$

A portion of the Minitab computer output is shown.

```
The regression equation is
Y = -1.41 + .0235 X1 + .00486 X2


Predictor          Coef          Stdev      t-ratio
Constant         -1.4053        0.4848     _____
X1               0.023467       0.008666   _____
X2             _____      0.001077   _____
```

—*This Minitab computer output continues at the top of page 633*

```
s = 0.1298      R-sq = _____       R-sq(adj) = _____
```

Analysis of Variance

```
SOURCE            DF          SS          MS          F
Regression      _____      1.76209     _____     _____
Error           _____      _____     _____
Total             9         1.88000
```

a. Complete the missing entries in this output.
b. Compute F and test at the $\alpha = .05$ level to see whether a significant relationship exists.
c. Did the estimated regression equation provide a good fit to the data? Explain.
d. Use the t test and $\alpha = .05$ to test H_0: $\beta_1 = 0$ and H_0: $\beta_2 = 0$.

48. Recall that in Exercise 43 the personnel director for Electronics Associates developed the following estimated regression equation relating an employee's score on a job satisfaction test (y) to their length of service (x_1) and wage rate (x_2):

$$\hat{y} = 14.4 - 8.69x_1 + 13.5x_2$$

A portion of the Minitab computer output is shown.

```
The regression equation is
Y = 14.4 - 8.69 X1 + 13.52 X2

Predictor      Coef       Stdev      t-ratio
Constant      14.448      8.191        1.76
X1            _____     1.555      _____
X2            13.517      2.085      _____

s = 3.773      R-sq = _____      R-sq(adj) = _____
```

Analysis of Variance

```
SOURCE            DF          SS          MS          F
Regression        2        _____     _____     _____
Error           _____     71.17       _____
Total             7         720.0
```

a. Complete the missing entries in this output.
b. Compute F and test at the $\alpha = .05$ level to see whether a significant relationship exists.
c. Did the estimated regression equation provide a good fit to the data? Explain.
d. Use the t test and $\alpha = .05$ to test H_0: $\beta_1 = 0$ and H_0: $\beta_2 = 0$.

49. Bauman Construction Company makes bids on a variety of projects. In an effort to estimate the bid to be made by one of its competitors, Bauman has obtained data on 15 previous bids and developed the following estimated regression equation:

$$\hat{y} = 80 + 45x_1 - 3x_2$$

where
$\hat{y}$ = competitor's bid ($1000s)
x_1 = square feet (1000s)
x_2 = local index of construction activity

a. Estimate the competitor's bid on a project involving 50,000 square feet and an index of construction activity of 70.

b. If SSR = 19,780 and SST = 21,533, test the significance of the relationship at α = .01.

 JOBS

50. The following data set, reported in *Louis Rukeyser's Business Almanac* (1988, Simon and Schuster, p. 47), shows the percentage of management jobs held by women in various companies and the percentage of women in each company.

Industry/Company	Management Jobs Held by Women (%)	Women Employees (%)
Industrial		
Du Pont	7	22
Exxon	8	27
General Motors	6	19
Goodyear Tire and Rubber	25	39
Technology		
AT&T	32	48
General Electric	6	26
IBM	16	28
Xerox	23	38
Consumer products		
Johnson & Johnson	18	47
PepsiCo	28	46
Phillip Morris (excluding General Foods)	14	31
Proctor & Gamble	17	28
Retailing and trade		
Federated Department Stores	61	72
Kroger	16	47
Marriott	32	51
McDonald's	46	57
Sears, Roebuck	36	55
Media		
ABC (excluding Capital Cities)	36	43
Time	46	54
Times Mirror	27	37
Financial Services		
American Express	37	57
BankAmerica	64	72
Chemical Bank	34	57
Prudential Life Insurance	32	53
Wells Fargo Bank	58	71

a. Fit a simple linear model that can be used to predict the percentage of management jobs held by women given the percentage of women employed by the company.

b. Did the model developed in part (a) provide a good fit to the data? Explain.

c. Use dummy variables to develop a model that relates the percentage of management jobs held by women to the type of industry (industrial, technology, etc.).

d. What conclusions can you reach based on the model developed in part (c)?

e. Develop a model that can be used to predict the percentage of management jobs held by women using the percentage of women employed by the company and the type of industry.

f. What final conclusions can be made regarding the percentage of management jobs held by women based on your analyses?

51. Refer to Exercise 50.

    a. Develop a standardized residual plot for the estimated regression equation developed in part (a) of Exercise 50. Does the pattern of the residual plot appear acceptable? Explain.

    b. Are there any outliers? Explain.

    c. Are there any influential observations? If so, what effect do they have on the model?

FOOTBALL   **52.** The following table shows some of the data available for 14 teams in the National Football League at the end of week 15 for the 1988 season.

| | | | | | Interceptions | |
| | | | | | Made by Team | Made by Opponent |
Team	Won–Lost	Points Scored	Rushing Yards	Passing Yards		
Atlanta	5–10	305	1907	2473	19	23
Chicago	12–3	187	2134	2718	14	24
Dallas	3–12	358	1858	3386	24	10
Detroit	4–11	292	1184	1971	15	12
Green Bay	3–12	298	1274	3046	22	20
L.A. Rams	9–6	277	1882	3604	17	22
Minnesota	10–5	206	1744	3633	16	35
New Orleans	9–6	274	1843	2963	15	17
N.Y. Giants	10–5	277	1492	3096	14	15
Philadelphia	9–6	312	1812	3247	17	29
Phoenix	7–8	372	1909	3633	19	14
San Francisco	10–5	256	2453	3131	14	21
Tampa Bay	4–11	340	1650	3169	33	18
Washington	7–8	367	1377	3930	24	14

    a. Develop an estimated regression equation that can be used to predict the number of points scored given the number of interceptions made by the team.

    b. Develop a standardized residual plot for the model developed in part (a). Does the pattern of the residual plot appear acceptable? Explain.

    c. Are there any outliers? Explain.

    d. Are there any influential observations? If so, what effect do they have on the model?

53. Refer to the data set in Exercise 52.

    a. Develop an estimated regression equation that can be used to predict the number of points scored given the number of interceptions made by the opponents.

    b. Develop a standardized residual plot for the model developed in part (a). Does the pattern of the residual plot appear acceptable? Explain.

    c. Are there any outliers? Explain.

    d. Are there any influential observations? If so, what effect do they have on the model?

COMPUTER CASE: CONSUMER RESEARCH, INC.

CONSUMER   Consumer Research, Inc., is an independent agency that conducts research on consumer attitudes and behaviors for a variety of firms. In one study, a client asked for an investigation of the consumer characteristics that can be used to predict the amount charged by credit-card users. Data were collected on the annual income, household size, and annual credit-card charges for a sample of 50 consumers. The following data are also on the data disk in the file named CONSUMER.

Income ($1000s)	Household Size	Amount Charged ($)	Income ($1000s)	Household Size	Amount Charged ($)
54	3	4016	54	6	5573
30	2	3159	30	1	2583
32	4	5100	48	2	3866
50	5	4742	34	5	3586
31	2	1864	67	4	5037
55	2	4070	50	2	3605
37	1	2731	67	5	5345
40	2	3348	55	6	5370
66	4	4764	52	2	3890
51	3	4110	62	3	4705
25	3	4208	64	2	4157
48	4	4219	22	3	3579
27	1	2477	29	4	3890
33	2	2514	39	2	2972
65	3	4214	35	1	3121
63	4	4965	39	4	4183
42	6	4412	54	3	3730
21	2	2448	23	6	4127
44	1	2995	27	2	2921
37	5	4171	26	7	4603
62	6	5678	61	2	4273
21	3	3623	30	2	3067
55	7	5301	22	4	3074
42	2	3020	46	5	4820
41	7	4828	66	4	5149

REPORT

1. Use methods of descriptive statistics to summarize the data. Comment on the findings.

2. Develop estimated regression equations first using annual income as the independent variable and then using household size as the independent variable. Which variable is the better predictor of annual credit-card charges? Discuss your findings.

3. Develop an estimated regression equation with annual income and household size as the independent variables. Discuss your findings.

4. What is the predicted annual credit-card charge for a three-person household with an annual income of $40,000.

5. Discuss the need for additional independent variables that could be added to the model. What additional variables might be helpful?

APPENDIX

CALCULUS-BASED DERIVATION AND SOLUTION OF MULTIPLE REGRESSION PROBLEMS WITH TWO INDEPENDENT VARIABLES

For the multiple regression analysis case involving two independent variables, the least squares criterion calls for the minimization of

$$\text{SSE} = \Sigma(y_i - b_0 - b_1 x_{1i} - b_2 x_{2i})^2 \tag{14A.1}$$

To minimize equation (14A.1), we must take the partial derivatives of SSE with respect to b_0, b_1, and b_2. We can then set the partial derivatives equal to zero and solve for the estimated regression coefficients b_0, b_1, and b_2. Taking the partial derivatives and setting them equal to zero provides

$$\frac{\partial \text{SSE}}{\partial b_0} = -2\Sigma(y_i - b_0 - b_1 x_{1i} - b_2 x_{2i}) = 0 \tag{14A.2}$$

$$\frac{\partial \text{SSE}}{\partial b_1} = -2\Sigma x_{1i}(y_i - b_0 - b_1 x_{1i} - b_2 x_{2i}) = 0 \tag{14A.3}$$

$$\frac{\partial \text{SSE}}{\partial b_2} = -2\Sigma x_{2i}(y_i - b_0 - b_1 x_{1i} - b_2 x_{2i}) = 0 \tag{14A.4}$$

Dividing equation (14A.2) by 2 and summing the terms individually yields

$$-\Sigma y_i + \Sigma b_0 + \Sigma b_1 x_{1i} + \Sigma b_2 x_{2i} = 0$$

Bringing Σy_i to the right-hand side of the equation and noting that $\Sigma b_0 = n b_0$, we obtain

$$n b_0 + (\Sigma x_{1i})b_1 + (\Sigma x_{2i})b_2 = \Sigma y_i \tag{14A.5}$$

Similar algebraic simplification applied to equations (14A.3) and (14A.4) leads to the equations:

$$(\Sigma x_{1i})b_0 + (\Sigma x_{1i}^2)b_1 + (\Sigma x_{1i} x_{2i})b_2 = \Sigma x_{1i} y_i \tag{14A.6}$$

$$(\Sigma x_{2i})b_0 + (\Sigma x_{1i} x_{2i})b_1 + (\Sigma x_{2i}^2)b_2 = \Sigma x_{2i} y_i \tag{14A.7}$$

Equations (14A.5) through (14A.7) are referred to as the *normal equations*. Application of these procedures to a regression model involving p independent variables would lead to $p + 1$ normal equations of this type. However, matrix algebra is usually used for that type of derivation.

SOLVING THE NORMAL EQUATIONS FOR BUTLER TRUCKING

Refer to Table 14A.1. Substituting the values in Table 14A.1 results in the following normal equations:

$$10 b_0 + 800 b_1 + 29 b_2 = 67.0 \tag{14A.8}$$

$$800 b_0 + 67{,}450 b_1 + 2345 b_2 = 5594.0 \tag{14A.9}$$

$$29 b_0 + 2345 b_1 + 91 b_2 = 202.2 \tag{14A.10}$$

TABLE 14A.1   CALCULATION OF COEFFICIENTS FOR NORMAL EQUATIONS

y_i	x_{1i}	x_{2i}	x_{1i}^2	x_{2i}^2	$x_{1i}x_{2i}$	$x_{1i}y_i$	$x_{2i}y_i$
9.3	100	4	10,000	16	400	930	37.2
4.8	50	3	2,500	9	150	240	14.4
8.9	100	4	10,000	16	400	890	35.6
6.5	100	2	10,000	4	200	650	13.0
4.2	50	2	2,500	4	100	210	8.4
6.2	80	2	6,400	4	160	496	12.4
7.4	75	3	5,625	9	225	555	22.2
6.0	65	4	4,225	16	260	390	24.0
7.6	90	3	8,100	9	270	684	22.8
6.1	90	2	8,100	4	180	549	12.2
67.0	800	29	67,450	91	2,345	5,594	202.2

By multiplying equation (14A.8) by 80 and subtracting the result from equation (14A.9), we can eliminate b_0 and obtain an equation involving only b_1 and b_2.

$$\begin{array}{rl} 800b_0 + 67,450.0b_1 + 2345b_2 = & 5594.0 \\ -800b_0 - 64,000.5b_1 - 2320b_2 = & -5360 \\ \hline 3450b_1 + \quad 25b_2 = & 234.0 \end{array}$$ (14A.11)

Now multiply equation (14A.8) by 2.9 and subtract the result from equation (14A.10). This manipulation yields a second equation involving only b_1 and b_2.

$$\begin{array}{rl} 29b_0 + 2345b_1 + 91.0b_2 = & 202.2 \\ -29b_0 - 2320b_1 - 84.1b_2 = & -194.3 \\ \hline 25b_1 + \quad 6.9b_2 = & 7.9 \end{array}$$ (14A.12)

With equations (14A.11) and (14A.12), we can solve simultaneously for b_1 and b_2. Multiplying equation (14A.12) by 25/6.9 and subtracting the result from equation (14A.11) gives us an equation involving only b_1.

$$\begin{array}{rl} 3,450.0000b_1 + 25b_2 = & 234.0000 \\ - \quad 90.5797b_1 - 25b_2 = & -28.6232 \\ \hline 3,359.4203b_1 \qquad\quad = & 205.3768 \end{array}$$ (14A.13)

Using equation (14A.13) to solve for b_1, we get

$$b_1 = \frac{205.3768}{3,359.4203} = .061135$$

Using this value for b_1, we can substitute into equation (14A.12) to solve for b_2:

$$25(.061135) + 6.9b_2 = 7.9$$

$$1.528375 + 6.9b_2 = 7.9$$

$$6.9b_2 = 6.371625$$

$$b_2 = .923424$$

Now we can substitute the values obtained for b_1 and b_2 into equation (14A.8), thus obtaining b_0:

$$10b_0 + 800(.061135) + 29(.923424) = 67.0$$

$$10b_0 + 48.90800 \quad\quad + 26.779296 \quad = 67.0$$

$$10b_0 \quad\quad\quad\quad\quad\quad\quad\quad\quad\quad = -8.687296$$

$$b_0 \quad\quad\quad\quad\quad\quad\quad\quad\quad\quad = -.8687296$$

Rounding, we obtain the following estimated regression for Butler Trucking:

$$\hat{y} = -.8687 + .0611x_1 + .9234x_2 \tag{14A.14}$$

REGRESSION ANALYSIS: MODEL BUILDING

CONTENTS

Working Wives and Expenditures on Time-Saving Services

The dramatic rise of married women in the labor force has increased the number of families where both spouses are wage earners. Since working wives have fewer hours per week to devote to housework than do full-time homemakers, how does the housework get done when both husband and wife are in the work force? Perhaps the husband or other members of the family take on more of the household duties. However, research has shown that, in general, this is not the case. Thus, a working wife with less time at home must find other ways to successfully balance her role in the home with her role at the office.

Don Bellante and Ann Foster, researchers at Auburn University, have studied how much working wives spend on time-saving goods and services, such as the expenditure on food away from home, child care, domestic services, clothing care, and personal care, in order to become more efficient in meeting their household responsibilities. Specifically, the researchers used regression analysis to study the relationship between these time-saving services and factors such as family income, weeks worked per year, hours worked per week, family size, the number of children under the age of six, wife's education, and homeownership status.

The research findings showed that the more income the family has the more the family will spend on time-saving services. In addition, it was found that spending increases when the number of hours worked per week increases, when there are more children under the age of six, when the

Time-saving services help working wives balance their home, family, and career responsibilities.

wife has more education, and when the family owns a home. The research supports the notion that when the job at home gets bigger, the families with working wives tend to spend more on time-saving services. The challenge of balancing a career, a home, and a family is an everyday battle for the modern working wife.

This *Statistics in Practice* is based on "Working Wives and Expenditures on Services," by Don Bellante and Ann C. Foster, *Journal of Consumer Research*, September, 1984.

Model building in regression analysis is the process of developing a regression model that best describes the relationship among the independent and dependent variables. The major issues are finding the proper form (e.g., linear or curvilinear) of the relationship and selecting the variables. Variable selection involves determining which of a candidate set of independent variables should be included in the regression model.

In Chapters 13 and 14 we introduced and worked with two regression models:

Simple Linear Regression: $\qquad y = \beta_0 + \beta_1 x + \epsilon$

Multiple Regression: $\qquad y = \beta_0 + \beta_1 x_1 + \beta_2 x_2 + \cdots + \beta_p x_p + \epsilon$

Both of these models specify a linear relationship between the independent variable(s) and the dependent variable.

The primary method introduced in Chapters 13 and 14 to determine the adequacy of a regression model was residual analysis. Essentially, if the residual plot looked like a horizontal band, we concluded that the assumptions concerning model form and the error term were satisfied. Otherwise, we concluded that the model was inadequate. When the regression model is judged inadequate, it may be because we have chosen the wrong functional form for the model (e.g., the actual relationship is curvilinear) and/or the proper independent variables have not been included. In Chapter 14, we saw that when the residual plot showed a nonconstant variance, a model that corrected this deficiency was created by adding a second independent variable.

In this chapter we focus on the model-building issues of proper model form identification and variable selection. Section 15.1, which establishes the framework for model building, introduces the concept of a *general linear model*. Surprisingly, the general linear model makes it possible for us to accommodate curvilinear relationships between the independent variables and the dependent variable with no more computational difficulty than that involved in multiple regression with linear relationships.

Section 15.2 provides the foundation for the more sophisticated computer-based procedures for variable selection. The issue of when one variable or a group of variables should be added to a regression model is examined. We show that the general approach to determining when to add or delete variables is based on the value of an F statistic. In Section 15.3 a larger problem involving 25 observations on eight independent variables is introduced. This larger problem is used to illustrate computer-based variable-selection procedures in Section 15.4. The stepwise-regression, forward-selection, backward-elimination, and best-subsets regression procedures are explained.

In Section 15.5 we show how the Durbin-Watson test is used to detect serial or autocorrelation. The chapter concludes with a discussion of how regression analysis can be used to solve analysis of variance and experimental design problems.

15.1   THE GENERAL LINEAR MODEL

Suppose that we have collected data for one dependent variable y and k independent variables $x_1, x_2, \ldots, x_k$. The objective is to use these data to develop an estimated regression equation that provides the best relationship between the dependent and independent variables. To provide a general framework for developing more complex relationships among the independent variables, we introduce the concept of a general linear model involving p independent variables.

General Linear Model

$$y = \beta_0 + \beta_1 z_1 + \beta_2 z_2 + \cdots + \beta_p z_p + \epsilon \qquad \textbf{(15.1)}$$

In equation (15.1) each of the independent variables z_j (where $j = 1, 2, \ldots, p$) is a function of $x_1, x_2, \ldots, x_k$ (the variables for which data have been collected). In some cases, each z_j may be a function of only one x variable. The simplest case occurs in which we have collected data for just one variable x_1 and we want to estimate y using a straight-line relationship. In this case, $z_1 = x_1$ and equation (15.1) becomes

$$y = \beta_0 + \beta_1 x_1 + \epsilon \tag{15.2}$$

Note that this is just the simple linear regression model introduced in Chapter 13 with the exception that the independent variable is labeled x_1 instead of x. In the statistical literature, this model is referred to as a *simple first-order model with one predictor variable.*

MODELING CURVILINEAR RELATIONSHIPS

More complex types of relationships can be modeled with equation (15.1). To illustrate how this is done, let us consider the problem facing Reynolds, Inc., a manufacturer of industrial scales and laboratory equipment.

EXAMPLE 15.1 ▶

 REYNOLDS

Reynolds, Inc., would like to investigate the relationship between length of employment of their salespeople and the number of their electronic laboratory scales sold. Table 15.1 gives the number of scales sold by 15 randomly selected salespeople for the most recent sales period and the number of months each salesperson has been employed by the firm. Figure 15.1 shows the scatter diagram for these data. The scatter diagram indicates a possible curvilinear relationship between the length of time employed and the number of units sold.

Before considering how to develop a curvilinear relationship for Reynolds, let us consider the Minitab output shown in Figure 15.2 corresponding to a simple first-order model; the estimated regression equation is

$$\text{SALES} = 111 + 2.38 \text{ MONTHS}$$

where

$$\text{SALES} = \text{number of electronic laboratory scales sold}$$
$$\text{MONTHS} = \text{the number of months the salesperson has been employed}$$

Figure 15.3 shows the corresponding standardized residual plot. Although the computer output shows that a linear relationship explains a high percentage of the variability in

---

TABLE 15.1   **DATA FOR THE REYNOLDS PROBLEM**

Scales Sold	Months Employed	Scales Sold	Months Employed
275	41	189	40
296	106	235	51
317	76	83	9
376	104	112	12
162	22	67	6
150	12	325	56
367	85	189	19
308	111		

FIGURE 15.1   SCATTER DIAGRAM FOR THE REYNOLDS PROBLEM

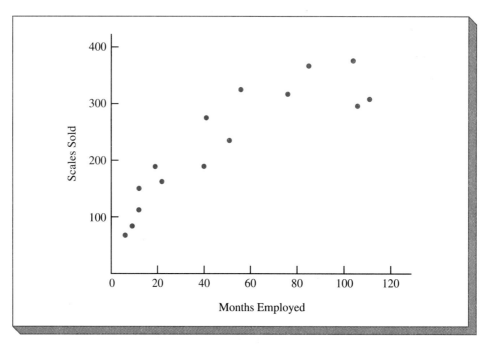

FIGURE 15.2   MINITAB OUTPUT FOR THE REYNOLDS PROBLEM: FIRST-ORDER MODEL

```
The regression equation is
SALES = 111 + 2.38 MONTHS

Predictor        Coef        Stdev      t-ratio          p
Constant        111.23       21.63        5.14      0.000
MONTHS          2.3768       0.3489       6.81      0.000

s = 49.52        R-sq = 78.1%      R-sq(adj) = 76.4%

Analysis of Variance

SOURCE          DF            SS          MS          F          p
Regression       1         113783      113783      46.41      0.000
Error           13          31874        2452
Total           14         145657
```

sales (R-sq = 78.1%), the residual plot also suggests that some type of curvilinear relationship is needed. The residuals are negative for small values of MONTHS, positive for intermediate values, and negative for large values of MONTHS.

To account for the curvilinear relationship we see in the data, we set $z_1 = x_1$ and $z_2 = x_1^2$ in equation (15.1) to obtain the model

**FIGURE 15.3   STANDARDIZED RESIDUAL PLOT FOR THE REYNOLDS PROBLEM:
FIRST-ORDER MODEL**

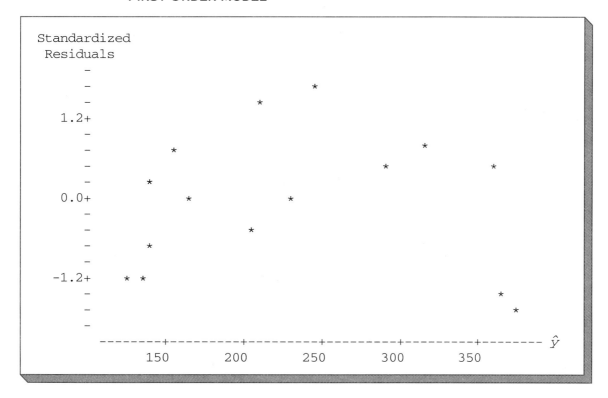

$$y = \beta_0 + \beta_1 x_1 + \beta_2 x_1^2 + \epsilon \qquad (15.3)$$

This model is referred to as a *second-order model with one predictor variable.* To develop
an estimated regression equation corresponding to this second-order model, we need to
provide the statistical software package we are using with the original data in Table 15.1,
as well as that data corresponding to adding a second independent variable, which is the
square of the number of months the employee has been with the firm. In Figure 15.4 we
show the Minitab output corresponding to the second-order model; the estimated
regression equation is

SALES = 45.3 + 6.34 MONTHS − 0.0345 MONTHSQ

where

MONTHSQ = the square of the number of months the
salesperson has been employed

The corresponding standardized residual plot is shown in Figure 15.5. The residual plot
shows that the previous curvilinear pattern has been removed. At the .05 level of
significance, the computer output shows that the overall model is significant (p-value for
the F test is 0.000); note also that the p-values corresponding to the t-ratios for both
MONTHS and MONTHSQ are less than .05, and hence, we can conclude that adding
MONTHSQ to the model involving MONTHS is significant. With an R-sq(adj) value of
88.6%, we should be pleased with the fit provided by this second-order model. More

FIGURE 15.4   MINITAB OUTPUT FOR THE REYNOLDS PROBLEM: SECOND-ORDER MODEL

```
The regression equation is
SALES = 45.3 + 6.34 MONTHS - 0.0345 MONTHSQ

Predictor        Coef        Stdev      t-ratio          p
Constant        45.35        22.77         1.99      0.070
MONTHS          6.345        1.058         6.00      0.000
MONTHSQ     -0.034486     0.008948        -3.85      0.002

s = 34.45        R-sq = 90.2%      R-sq(adj) = 88.6%

Analysis of Variance

SOURCE       DF          SS           MS          F          p
Regression    2       131413        65707      55.36      0.000
Error        12        14244         1187
Total        14       145657
```

FIGURE 15.5   STANDARDIZED RESIDUAL PLOT FOR THE REYNOLDS PROBLEM: SECOND-ORDER MODEL

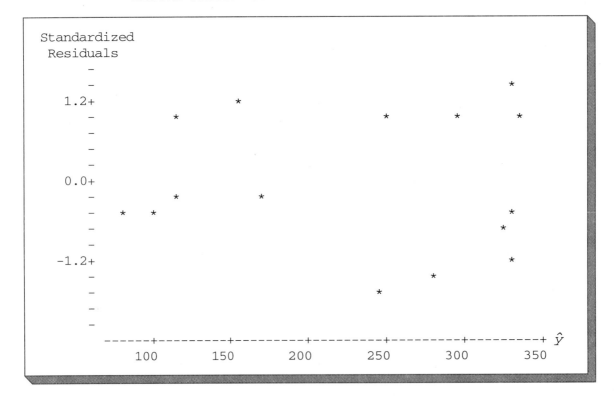

importantly, however, we now see how easy it is to handle curvilinear relationships in regression analysis. ◀

It should now be apparent that many types of relationships can be modeled using equation (15.1). Thus, the regression techniques with which we have been working are definitely not limited to linear, or straight-line, relationships. In multiple regression analysis, the word *linear* in the term "general linear model" refers only to the fact that $\beta_0, \beta_1, \ldots, \beta_p$ all have exponents of 1; it does not imply that the relationship between y and the x_i's is linear. Indeed, in this section we have seen one example where equation (15.1) can be used to model a curvilinear relationship.

INTERACTION

If the original data set consists of observations for y and two independent variables x_1 and x_2, we can develop a complete second-order model with two predictor variables by setting $z_1 = x_1$, $z_2 = x_2$, $z_3 = x_1^2$, $z_4 = x_2^2$, and $z_5 = x_1 x_2$ in the general linear model of equation (15.1). The model obtained is

$$y = \beta_0 + \beta_1 x_1 + \beta_2 x_2 + \beta_3 x_1^2 + \beta_4 x_2^2 + \beta_5 x_1 x_2 + \epsilon \qquad (15.4)$$

In this second-order model, the variable $z_5 = x_1 x_2$ is added to account for the potential effects of the two variables acting together. This type of effect is called *interaction*.

EXAMPLE 15.2 ▶ To illustrate interaction and what it means, let us review the regression study conducted by Tyler Personal Care for one of its new shampoo products. Two factors believed to have the most influence on sales were unit selling price and advertising expenditure. To investigate the effects of these two variables on sales, prices of $2.00, $2.50, and $3.00 were paired with advertising expenditures of $50,000 and $100,000 in 24 test markets. The unit sales that were observed (in 1000s) are shown in Table 15.2.

TABLE 15.2   **DATA FOR THE TYLER PERSONAL CARE PROBLEM**

 TYLER

Price	Advertising Expenditure ($1000s)	Unit Sales (1000s)	Price	Advertising Expenditure ($1000s)	Unit Sales (1000s)
$2.00	50	478	$2.00	100	810
$2.50	50	373	$2.50	100	653
$3.00	50	335	$3.00	100	345
$2.00	50	473	$2.00	100	832
$2.50	50	358	$2.50	100	641
$3.00	50	329	$3.00	100	372
$2.00	50	456	$2.00	100	800
$2.50	50	360	$2.50	100	620
$3.00	50	322	$3.00	100	390
$2.00	50	437	$2.00	100	790
$2.50	50	365	$2.50	100	670
$3.00	50	342	$3.00	100	393

Table 15.3 provides a summary of these data. Note that the mean unit sales corresponding to a price of $2.00 and an advertising expenditure of $50,000 is 461,000, and the mean unit sales corresponding to a price of $2.00 and an advertising expenditure of $100,000 is 808,000; thus, holding price constant at $2.00, the difference in mean unit sales between an advertising expenditure of $50,000 and $100,000 is 808,000 − 461,000 = 347,000. When the price of the product is $2.50, the difference in mean unit sales is 646,000 − 364,000 = 282,000. Finally, when the price is $3.00, the difference in mean unit sales is 375,000 − 332,000 = 43,000. Clearly, the difference in mean unit sales between an advertising expenditure of $50,000 and $100,000 depends on the price of the product. In other words, at higher selling prices, we see that the effect of increased advertising expenditure diminishes. These observations provide evidence of interaction between the price and advertising expenditure variables.

To provide another perspective of interaction, Figure 15.6 shows the mean unit sales for the six price–advertising expenditure combinations. This graph also shows that the effect of advertising expenditure on mean unit sales depends on the level of the price of the product; we again see the effect of interaction. When interaction between two variables is present, we cannot study the effect of one variable on the response y independently of the other variable. In other words, meaningful conclusions can only be developed if we consider the joint effect that both variables have on the response.

To account for the effect of interaction, we will use the following regression model:

$$y = \beta_0 + \beta_1 x_1 + \beta_2 x_2 + \beta_3 x_1 x_2 + \epsilon \qquad \textbf{(15.5)}$$

where

$$y = \text{unit sales (1000s)}$$
$$x_1 = \text{selling price (\$)}$$
$$x_2 = \text{advertising expenditure (\$1000s)}$$

Note that this model is simply equation (15.4) with the curvilinear effects of x_1^2 and x_2^2 dropped from the model; it reflects Tyler's belief that the number of units sold depends linearly on selling price and advertising expenditures (accounted for by the $\beta_1 x_1$ and $\beta_2 x_2$ terms), and that there is interaction between the two variables (accounted for by the $\beta_3 x_1 x_2$ term).

To develop an estimated regression equation, a general linear model involving three independent variables ($z_1, z_2,$ and z_3) was used.

TABLE 15.3   MEAN UNIT SALES FOR THE TYLER PERSONAL CARE PROBLEM

Advertising Expenditure	Price		
	$2.00	$2.50	$3.00
$50,000	461	364	332
$100,000	808	646	375

Mean unit sales of 808,000 when price = $2.00 and advertising expenditure = $100,000

FIGURE 15.6   **MEAN UNIT SALES AS A FUNCTION OF SELLING PRICE AND ADVERTISING EXPENDITURE FOR THE TYLER PERSONAL CARE PROBLEM**

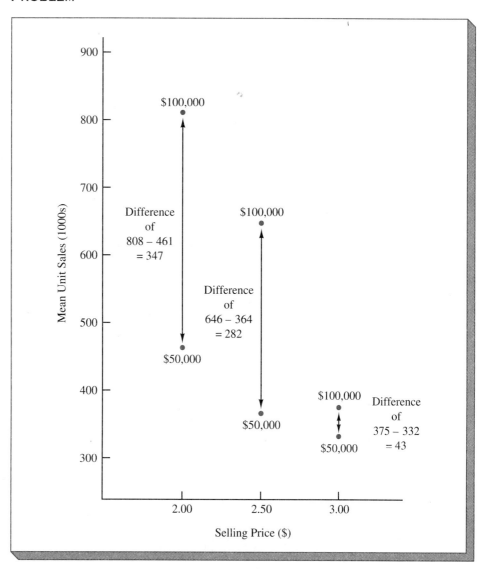

$$y = \beta_0 + \beta_1 z_1 + \beta_2 z_2 + \beta_3 z_3 + \epsilon \tag{15.6}$$

where

$$z_1 = x_1$$
$$z_2 = x_2$$
$$z_3 = x_1 x_2$$

In Figure 15.7 we show the Minitab output corresponding to the interaction model for the Tyler Personal Care problem. The resulting estimated regression equation is

$$\text{SALES} = -276 + 175 \text{ PRICE} + 19.7 \text{ ADVER} - 6.08 \text{ PRICEADV}$$

FIGURE 15.7   MINITAB OUTPUT FOR THE TYLER PERSONAL CARE PROBLEM

```
The regression equation is
SALES = - 276 + 175 PRICE + 19.7 ADVER - 6.08 PRICEADV

Predictor       Coef       Stdev      t-ratio         p
Constant      -275.8       112.8        -2.44      0.024
PRICE         175.00       44.55         3.93      0.001
ADVER         19.680       1.427        13.79      0.000
PRICEADV     -6.0800      0.5635       -10.79      0.000


s = 28.17       R-sq = 97.8%       R-sq(adj) = 97.5%

Analysis of Variance

SOURCE        DF          SS           MS          F         p
Regression     3       709316       236439     297.87     0.000
Error         20        15875          794
Total         23       725191
```

where

$$\text{SALES} = \text{unit sales (1000s)}$$
$$\text{PRICE} = \text{price of the product (\$)}$$
$$\text{ADVER} = \text{advertising expenditure (\$1000s)}$$
$$\text{PRICEADV} = \text{interaction term (PRICE times ADVER)}$$

Since the p-value corresponding to the t test for PRICEADV is 0.000, we conclude that interaction is significant given the linear effect of the price of the product and the advertising expenditure. Thus, the regression results do show that the effect of advertising expenditure on unit sales depends on the level of the selling price. ◀

TRANSFORMATIONS INVOLVING THE DEPENDENT VARIABLE

In showing how the general linear model can be used to model a variety of possible relationships between the independent variables and the dependent variable, we have focused our attention on transformations involving one or more of the independent variables. It is often worthwhile to consider transformations involving the dependent variable y.

EXAMPLE 15.3 ▶ To provide an illustration of when we might want to transform the dependent variable, consider the data shown in Table 15.4. The data show the miles per gallon ratings and weights for 12 automobiles. The scatter diagram in Figure 15.8 shows a negative linear relationship between these two variables. As a result, we will use a simple first-order model to relate these two variables. The Minitab output that we obtained is shown in Figure 15.9; the resulting estimated regression equation is

MPG

$$\text{MPG} = 56.1 - 0.0116 \text{ WEIGHT}$$

TABLE 15.4   MILES PER GALLON RATINGS AND WEIGHTS FOR 12 AUTOMOBILES

Miles Per Gallon	Weight	Miles Per Gallon	Weight
28.7	2289	23.9	2657
29.2	2113	30.5	2106
34.2	2180	18.1	3226
27.9	2448	19.5	3213
33.3	2026	14.3	3607
26.4	2702	20.9	2888

FIGURE 15.8   SCATTER DIAGRAM FOR THE MILES PER GALLON PROBLEM

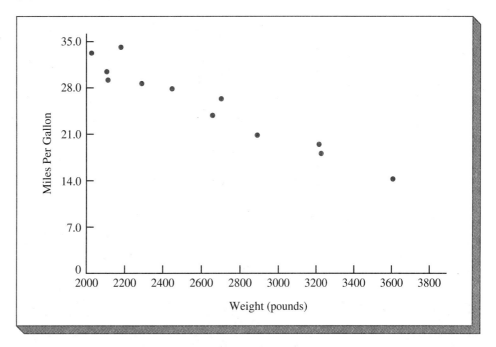

where

$$MPG = \text{miles per gallon rating}$$
$$WEIGHT = \text{weight of the car in pounds}$$

The model is significant (p-value for the F test is 0.000) and the fit is very good (R-sq = 93.5%). However, we note that observation 3 has been identified as having a large standardized residual.

In Figure 15.10 we show the standardized residual plot corresponding to the first-order model. The pattern that we observe does not look like the horizontal band we should expect to observe if the assumptions regarding the error term are valid. Instead, it appears that the variability in the residuals increases as the predicted value of $\hat{y}$ increases. In other words, we have the wedge-shaped pattern referred to in Chapters 13 and 14 as being indicative of nonconstant variance. Thus, we are not justified in reaching any conclusions

FIGURE 15.9   MINITAB OUTPUT FOR THE MILES PER GALLON PROBLEM

```
The regression equation is
MPG = 56.1 - 0.0116 WEIGHT


Predictor        Coef        Stdev      t-ratio          p
Constant       56.096        2.582        21.72      0.000
WEIGHT     -0.0116436    0.0009677       -12.03      0.000


s = 1.671       R-sq = 93.5%      R-sq(adj) = 92.9%

Analysis of Variance

SOURCE         DF           SS          MS          F          p
Regression      1       403.98      403.98     144.76      0.000
Error          10        27.91        2.79
Total          11       431.88


Unusual Observations
Obs.   WEIGHT       MPG       Fit Stdev.Fit   Residual    St.Resid
  3      2180    34.200    30.713     0.644      3.487        2.26R

R denotes an obs. with a large st. resid.
```

FIGURE 15.10   STANDARDIZED RESIDUAL PLOT FOR THE MILES PER GALLON
                PROBLEM

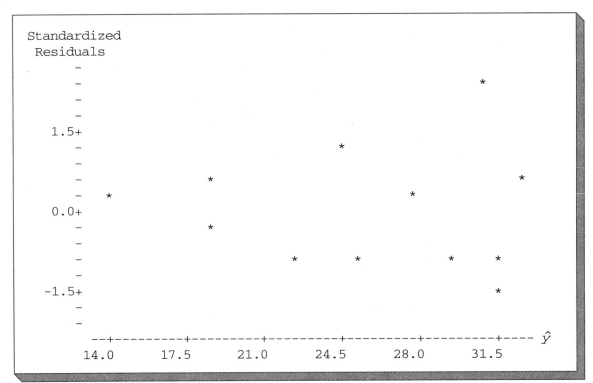

regarding the statistical significance of our resulting model since the underlying assumptions for the tests of significance do not appear to be satisfied. ◀

The problem of nonconstant variance can often be corrected by transforming the dependent variable to a different scale. For instance, if we work with the logarithm of the dependent variable instead of the original dependent variable, the effect will be to compress the values of the dependent variable and thus diminish the effects of nonconstant variance in the data. Most statistical packages provide the ability to apply logarithmic transformations using either the base 10 (common logarithm) or the base $e = 2.71828 \ldots$ (natural logarithm).

EXAMPLE 15.3 ▶
(CONTINUED)

We applied a natural logarithmic transformation to the miles per gallon data and developed the estimated regression equation relating weight to the natural logarithm of miles per gallon. The regression results we obtained using the natural logarithm of miles per gallon as the dependent variable, labeled LOGEMPG in the output, are shown in Figure 15.11; the corresponding standardized residual plot is shown in Figure 15.12.

Looking at the residual plot in Figure 15.12, we see that the wedge-shaped pattern has now disappeared. Moreover, none of the observations have been identified as having a large standardized residual. The model using the logarithm of miles per gallon as the dependent variable is statistically significant and provides an excellent fit to the observed data. Thus, we would recommend using the estimated regression equation

$$\text{LOGEMPG} = 4.52 - 0.000501 \text{ WEIGHT}$$

To estimate the miles per gallon rating for a car that weighs 2500 pounds, we first develop an estimate of the logarithm of the miles per gallon rating.

$$\text{LOGEMPG} = 4.52 - 0.000501(2500) = 3.2675$$

The miles per gallon estimate is obtained by finding the number whose natural logarithm

FIGURE 15.11   MINITAB OUTPUT FOR THE MILES PER GALLON PROBLEM: LOGARITHMIC TRANSFORMATION

```
The regression equation is
LOGEMPG = 4.52 -0.000501 WEIGHT

Predictor        Coef        Stdev      t-ratio          p
Constant      4.52423     0.09932        45.55      0.000
WEIGHT     -0.00050110  0.00003722       -13.46      0.000

s = 0.06425      R-sq = 94.8%      R-sq(adj) = 94.2%

Analysis of Variance

SOURCE          DF          SS          MS          F          p
Regression       1      0.74822     0.74822     181.22      0.000
Error           10      0.04129     0.00413
Total           11      0.78950
```

FIGURE 15.12   STANDARDIZED RESIDUAL PLOT FOR THE MILES PER GALLON PROBLEM: LOGARITHMIC TRANSFORMATION

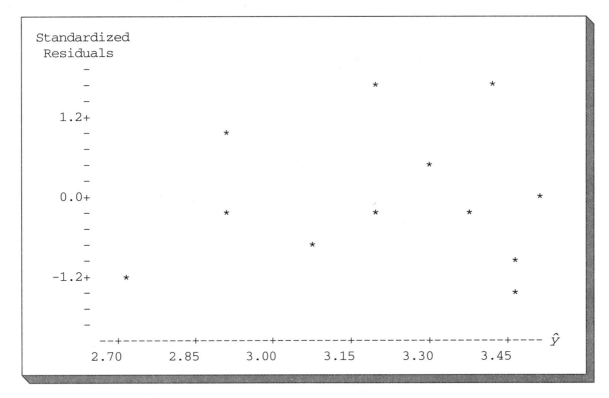

is 3.2675. Using a calculator with an exponential function, or raising e to the power 3.2675, we obtain 26.2 miles per gallon. ◀

Another approach to dealing with problems of nonconstant variance involves using $1/y$ as the dependent variable instead of y. This type of transformation is called a *reciprocal transformation*. For instance, if the dependent variable is measured in miles per gallon, the reciprocal transformation would result in a new dependent variable whose units would be 1/(miles per gallon) or gallons per mile. In general, there is no way to determine whether a logarithmic transformation or a reciprocal transformation will perform best without actually trying each transformation.

NONLINEAR MODELS THAT ARE INTRINSICALLY LINEAR

Models in which the parameters $(\beta_0, \beta_1, \ldots, \beta_p)$ have exponents other than 1 are referred to as nonlinear models. However, for the case of the exponential model, it is possible to perform a transformation of variables that will permit us to perform regression analysis using equation (15.1), the general linear model. To show how this is done, consider the following regression equation:

$$E(y) = \beta_0\beta_1^x \tag{15.7}$$

This regression equation is appropriate in cases where the dependent variable y increases or decreases by a constant percentage, instead of by a fixed amount, as x increases.

EXAMPLE 15.4 As an example, suppose that sales for a product y were related to advertising expenditure x (in \$1000s) according to the following regression equation:

$$E(y) = 500(1.2)^x$$

Thus, for $x = 1$, $E(y) = 500(1.2)^1 = 600$; for $x = 2$, $E(y) = 500(1.2)^2 = 720$; and for $x = 3$, $E(y) = 500(1.2)^3 = 864$. Note that $E(y)$ is not increasing by a constant amount in this case, but by a constant percentage; the percentage increase is 20%.

We can transform this nonlinear regression equation to a linear regression equation by taking the logarithm of both sides of equation (15.7):

$$\log E(y) = \log \beta_0 + x \log \beta_1 \qquad (15.8)$$

Now if we let $y' = \log E(y)$, $\beta'_0 = \log \beta_0$, and $\beta'_1 = \log \beta_1$, we can rewrite equation (15.8) as

$$y' = \beta'_0 + \beta'_1 x$$

It is clear that the formulas for simple linear regression can now be used to develop estimates of β'_0 and β'_1. Denoting the estimates as b'_0 and b'_1 leads to the following estimated regression equation:

$$\hat{y}' = b'_0 + b'_1 x \qquad (15.9)$$

To obtain predictions of the original dependent variable y given a value of x, we would first substitute the value of x into equation (15.9) and compute $\hat{y}'$. The antilog of $\hat{y}'$ would be our prediction of y, or the expected value of y.

We should make it clear that there are many nonlinear models that cannot be transformed into an equivalent linear model. However, such models have had limited use in business and economic applications. Furthermore, the mathematical background needed for study of such models is beyond the scope of this text.

EXERCISES

METHODS

SELF TEST

1. Consider the following data for two variables, x and y.

x	22	24	26	30	35	40
y	12	21	33	35	40	36

    a. Develop an estimated regression equation for these data of the form $\hat{y} = b_0 + b_1 x$.
    b. Using the results from part (a), test for a significant relationship between x and y; use $\alpha = .05$.
    c. Develop a scatter diagram for these data. Does the scatter diagram suggest an estimated regression equation of the form $\hat{y} = b_0 + b_1 x + b_2 x^2$? Explain.
    d. Develop an estimated regression equation for these data of the form $\hat{y} = b_0 + b_1 x + b_2 x^2$.
    e. Refer to part (d). Is the relationship between x, x^2, and y significant? Use $\alpha = .05$.
    f. Predict the value of y when $x = 25$.

2. Consider the following data for two variables, x and y.

x	9	32	18	15	26
y	10	20	21	16	22

   a. Develop an estimated regression equation for these data of the form $\hat{y} = b_0 + b_1 x$. Comment on the adequacy of this equation for predicting y.

   b. Develop an estimated regression equation for these data of the form $\hat{y} = b_0 + b_1 x + b_2 x^2$. Comment on the adequacy of this equation for predicting y.

   c. Predict the value of y when $x = 20$.

3. Consider the following data for two variables, x and y.

x	2	3	4	5	7	7	7	8	9
y	4	5	4	6	4	6	9	5	11

   a. Does there appear to be a linear relationship between x and y? Explain.

   b. Develop the estimated regression equation relating x and y.

   c. Plot the standardized residuals versus $\hat{y}$ for the estimated regression equation developed in part (b). Do the model assumptions appear to be satisfied? Explain.

   d. Perform a logarithmic transformation on the dependent variable y. Develop an estimated regression equation using the transformed dependent variable. Do the model assumptions appear to be satisfied using the transformed dependent variable? Does a reciprocal transformation work better in this case? Explain.

APPLICATIONS

4. The highway department is doing a study on the relationship between traffic flow and speed. The following model has been hypothesized:

$$y = \beta_0 + \beta_1 x + \epsilon$$

where

$$y = \text{traffic flow in vehicles per hour}$$
$$x = \text{vehicle speed in miles per hour}$$

The data shown below have been collected during rush hour for six highways leading out of the city.

Traffic Flow (y)	Vehicle Speed (x)	Traffic Flow (y)	Vehicle Speed (x)
1256	35	1335	45
1329	40	1349	50
1226	30	1124	25

   a. Develop an estimated regression equation for these data.

   b. Using $\alpha = .01$, test for a significant relationship.

 SELF TEST

5. In working further with the problem of Exercise 4, statisticians suggested the use of the following curvilinear estimated regression equation:

$$\hat{y} = b_0 + b_1 x + b_2 x^2$$

   a. Use the data of Exercise 4 to estimate the parameters of this estimated regression equation.

   b. Using $\alpha = .01$, test for a significant relationship.

   c. Estimate the traffic flow in vehicles per hour at speeds of 38 miles per hour.

6. A study of emergency service facilities investigated the relationship between the number of facilities and the average distance traveled to provide the emergency service (*Management Science*, July 1988). The data collected are shown below.

Number of Facilities	Average Travel Distance (miles)	Number of Facilities	Average Travel Distance (miles)
9	1.66	21	.62
11	1.12	27	.51
16	.83	30	.47

   a. Develop a scatter diagram for these data treating average travel distance as the dependent variable.
   b. Does a simple linear model appear to be appropriate? Explain.
   c. Fit a model to the data that you believe will best explain the relationship between these two variables.

7. Performance data for a Century Coronado 21 with a 310-hp MerCruiser V-8 gasoline inboard engine was reported in *Boating,* September 1991. Data on how the engine speed in revolutions per minute (rpm) affected boat speed in miles per hour (mph) are shown below.

Engine Speed (rpm)	Boat Speed (mph)	Engine Speed (rpm)	Boat Speed (mph)
1000	6.1	3500	33.6
1500	10.7	4000	37.9
2000	20.9	4500	40.2
2500	27.5	4800	40.7
3000	31.5		

   a. Develop an estimated regression equation that shows how boat speed is related to engine speed.
   b. Estimate the boat speed for an engine being run at 2800 rpm.

 POLAROID

8. A study by Polaroid of the relationship between film speed and the age of Polaroid extended range, color, professional quality print film resulted in the following data.

Film Age (months)	Change in Film Speed	Film Age (months)	Change in Film Speed
0	0.0	8	−85.1
1	−20.5	9	−85.6
2	−36.8	10	−99.3
3	−50.8	11	−101.4
4	−56.1	12	−100.4
5	−67.4	13	−111.5
6	−72.3		
7	−82.2		

a. Develop a scatter diagram for these data treating change in film speed as the dependent variable.

b. Does a simple linear model appear to be appropriate?

c. Fit a model to these data that you believe will best explain the relationship between these two variables.

9. Data on housing markets was provided for 100 different cities in the United States (*U.S. News & World Report,* April 6, 1992). A sample of 16 cities with data (in 1000s) on the median cost of a new home, the number of new housing starts during 1991–1992 and the average household income ($1000s) are shown below.

City	Cost	Housing Starts	Household Income
Chicago	181.8	12.9	61.0
Dayton, Ohio	107.8	3.8	48.4
Atlanta	100.6	24.2	54.7
Oklahoma City	68.9	3.3	53.2
Columbia, S.C.	90.3	3.1	57.4
Tacoma, Wash.	96.1	4.1	51.1
Mobile, Ala.	68.5	1.0	41.0
Baltimore	121.8	11.1	62.8
West Palm Beach	130.4	8.9	58.1
San Antonio	72.5	1.5	57.0
Pittsburgh	79.5	4.9	49.2
Jacksonville	82.1	8.0	47.5
Cleveland	122.9	5.4	54.0
Gary, Ind.	98.2	3.2	45.7
Scranton, Pa.	81.6	2.5	44.8
Richmond, Va.	102.8	6.0	64.5

 HOUSING

a. Using these data, develop an estimated regression equation that can be used to predict the median cost of a new home.

b. Use the estimated regression equation developed in part (a) to predict the median cost of a new home in a city with 8000 housing starts and a household income of $45,000.

15.2 DETERMINING WHEN TO ADD OR DELETE VARIABLES

In this section we will show how an F test can be used to determine whether it is advantageous to add one variable—or a group of variables—to a multiple regression model. This test is based on a determination of the amount of reduction in the error sum of squares resulting from adding one or more independent variables to the model. We will first illustrate how the test might be used in the context of the Butler Trucking problem introduced in Example 14.2.

EXAMPLE 15.5 ▶ Recall that Butler Trucking wanted to develop an estimated regression equation to predict total daily travel time for their trucks using two independent variables: miles traveled and number of deliveries. With miles traveled x_1 as the only independent variable, the least squares procedure provided the following estimated regression equation:

$$\hat{y} = 1.27 + 0.0678x_1$$

In Chapter 14 we showed that the error sum of squares for this model was SSE = 8.029. When x_2, the number of deliveries, was added as a second independent variable, we obtained the following estimated regression equation:

$$\hat{y} = -.869 + 0.0611x_1 + 0.923x_2$$

The error sum of squares for this model was 2.299. Clearly, adding x_2 resulted in a reduction of SSE. The question we want to answer is, Does adding the variable x_2 lead to a *significant* reduction in SSE?

We will use the notation $SSE(x_1)$ to denote the error sum of squares when x_1 is the only independent variable in the model, $SSE(x_1, x_2)$ to denote the error sum of squares when x_1 and x_2 are both in the model, and so on. Hence, the reduction in SSE resulting from adding x_2 to the model involving just x_1 is

$$SSE(x_1) - SSE(x_1, x_2) = 8.029 - 2.299 = 5.730$$

An *F* test is conducted to determine whether this reduction is significant.

The numerator of the *F* statistic is the reduction in SSE divided by the number of variables added to the original model. Here only one variable, x_2, has been added; thus, the numerator of the *F* statistic is

$$\frac{SSE(x_1) - SSE(x_1, x_2)}{1} = 5.730$$

The result is a measure of the reduction in SSE per variable added to the model. The denominator of the *F* statistic is the mean square error for the model that includes all of the independent variables. For Butler Trucking, this corresponds to the model containing both x_1 and x_2; thus, $p = 2$ and hence

$$MSE = \frac{SSE(x_1, x_2)}{n - p - 1} = \frac{2.299}{7} = .3284$$

The following *F* statistic provides the basis for testing whether the addition of x_2 is statistically significant:

$$F = \frac{\dfrac{SSE(x_1) - SSE(x_1, x_2)}{1}}{\dfrac{SSE(x_1, x_2)}{n - p - 1}} \qquad\qquad (15.10)$$

In general, the numerator degrees of freedom for this *F* test is equal to the number of variables added to the model, and the denominator degrees of freedom is equal to $n - p - 1$.

For the Butler Trucking problem, we obtain

$$F = \frac{\dfrac{5.730}{1}}{\dfrac{2.299}{7}} = \frac{5.730}{.3284} = 17.45$$

Refer to Table 4 of Appendix B. We find that for a level of significance of $\alpha = .05$, $F_{.05} = 5.59$. Since $F = 17.45 > F_{.05} = 5.59$, we reject the null hypothesis that x_2 is not statistically significant; in other words, adding x_2 to the model involving only x_1 results in a significant reduction in the error sum of squares. ◄

When we want to test for the significance of adding only one additional independent variable to an existing model, the result found with the F test just described could also be obtained by using the t test for the significance of an individual parameter (described in Section 14.4). Indeed, the F statistic we just computed is the square of the t statistic used to test the significance of an individual parameter.

Since the t test is equivalent to the F test when only one variable is being added to the model, we can now further clarify the proper use of the t test for testing the significance of an individual parameter. If an individual parameter is not significant, the corresponding variable can be dropped from the model. However, no more than one variable can ever be dropped from a model on the basis of a t test; if one variable is dropped, a second variable that was not significant initially might become significant.

We now turn to a consideration of whether the addition of more than one variable—as a set—results in a significant reduction in the error sum of squares.

THE GENERAL CASE

Consider the following multiple regression model involving q independent variables, where $q < p$:

$$y = \beta_0 + \beta_1 x_1 + \beta_2 x_2 + \cdots + \beta_q x_q + \epsilon \tag{15.11}$$

If we add variables $x_{q+1}, x_{q+2}, \ldots, x_p$ to this model, we obtain a model involving p independent variables:

$$y = \beta_0 + \beta_1 x_1 + \beta_2 x_2 + \cdots + \beta_q x_q \tag{15.12}$$
$$+ \beta_{q+1} x_{q+1} + \beta_{q+2} x_{q+2} + \cdots + \beta_p x_p + \epsilon$$

To test whether the addition of $x_{q+1}, x_{q+2}, \ldots, x_p$ is statistically significant, the null and alternative hypotheses can be stated as follows:

$$H_0: \beta_{q+1} = \beta_{q+2} = \cdots = \beta_p = 0$$

$$H_a: \text{One or more of the parameters } (\beta_i, i = q + 1, q + 2, \cdots, p)$$
$$\text{is not equal to zero}$$

The following F statistic provides the basis for testing whether the additional variables are statistically significant:

$$F = \frac{\dfrac{\text{SSE}(x_1, x_2, \ldots, x_q) - \text{SSE}(x_1, x_2, \ldots, x_q, x_{q+1}, \ldots, x_p)}{p - q}}{\dfrac{\text{SSE}(x_1, x_2, \ldots, x_q, x_{q+1}, \ldots, x_p)}{n - p - 1}} \tag{15.13}$$

This computed F value is then compared with F_α, the table value with $p - q$ numerator degrees of freedom and $n - p - 1$ denominator degrees of freedom. If $F > F_\alpha$, we reject H_0 and conclude that the set of additional variables is statistically significant. Note that for the special case where $q = 1$ and $p = 2$, equation (15.13) reduces to equation (15.10).

Equation (15.13) is somewhat complex. To provide a simpler description of this F ratio, we can refer to the model with the fewer number of independent variables as the reduced model and the model with the greater number of independent variables as the full model. If we let SSE(reduced) denote the error sum of squares for the reduced model and

SSE(full) denote the error sum of squares for the full model, we can write the numerator of equation (15.13) as

$$\frac{\text{SSE(reduced)} - \text{SSE(full)}}{\text{number of extra terms}} \tag{15.14}$$

Note that "number of extra terms" denotes the difference between the number of independent variables in the full model and the number of independent variables in the reduced model. The denominator of equation (15.13) is the error sum of squares for the full model divided by the corresponding degrees of freedom; in other words, the denominator is the mean square error for the full model. Denoting the mean square error for the full model as MSE(full) enables us to write equation (15.13) as

$$F = \frac{\dfrac{\text{SSE(reduced)} - \text{SSE(full)}}{\text{number of extra terms}}}{\text{MSE(full)}} \tag{15.15}$$

EXAMPLE 15.6 ▶ To illustrate the use of this F statistic, suppose that we had a regression problem involving 30 observations. One model involving the independent variables x_1, x_2, and x_3 had an error sum of squares of 150, and a second model involving the independent variables x_1, x_2, x_3, x_4, and x_5 had an error sum of squares of 100. Did the addition of the two independent variables x_4 and x_5 result in a significant reduction in the error sum of squares?

First, note that the degrees of freedom for the error sum of squares for the full model is $n - p - 1 = 30 - 5 - 1 = 24$, and hence MSE(full) $= 100/24 = 4.17$. Thus, the F statistic is

$$F = \frac{\dfrac{150 - 100}{2}}{4.17} = 6.00$$

This computed F value is compared with the table F value with 2 numerator and 24 denominator degrees of freedom. At the .05 level of significance, Table 4 of Appendix B shows $F_{.05} = 3.40$. Since $F = 6.00$ is greater than 3.40, we conclude that the addition of variables x_4 and x_5 is statistically significant. ◀

NOTES AND COMMENTS

The F statistic can also be computed based on the difference in the regression sums of squares. To show this form of the F statistic, we first note that

$$\text{SSE(reduced)} = \text{SST} - \text{SSR(reduced)}$$

$$\text{SSE(full)} = \text{SST} - \text{SSR(full)}$$

Hence

$$\text{SSE(reduced)} - \text{SSE(full)} = [\text{SST} - \text{SSR(reduced)}] - [\text{SST} - \text{SSR(full)}]$$

$$= \text{SSR(full)} - \text{SSR(reduced)}$$

Thus,

$$F = \frac{\dfrac{\text{SSR(full)} - \text{SSR(reduced)}}{\text{number of extra terms}}}{\text{MSE(full)}}$$

EXERCISES

METHODS

10. In a regression analysis involving 27 observations, the following estimated regression equation was developed:

$$\hat{y} = 16.3 + 2.3x_1 + 12.1x_2 - 5.8x_3$$

Also, the following standard errors were obtained:

$$s_{b_1} = .53 \qquad s_{b_2} = 8.15 \qquad s_{b_3} = 1.30$$

At an $\alpha = .05$ level of significance, conduct the following hypothesis tests:
a. $H_0: \beta_1 = 0$ versus $H_a: \beta_1 \neq 0$.
b. $H_0: \beta_2 = 0$ versus $H_a: \beta_2 \neq 0$.
c. $H_0: \beta_3 = 0$ versus $H_a: \beta_3 \neq 0$.
d. Can any of the variables be dropped from the model? Why or why not?

SELF TEST

11. In a regression analysis involving 30 observations, the following estimated regression equation was obtained:

$$\hat{y} = 17.6 + 3.8x_1 - 2.3x_2 + 7.6x_3 + 2.7x_4$$

For this model, SST = 1805 and SSR = 1760.
a. At $\alpha = .05$, test the significance of the relationship among the variables.

Suppose that variables x_1 and x_4 were dropped from the model, and the following estimated regression equation was obtained:

$$\hat{y} = 11.1 - 3.6x_2 + 8.1x_3$$

For this model, SST = 1805 and SSR = 1705.
b. Compute SSE(x_1, x_2, x_3, x_4).
c. Compute SSE(x_2, x_3).
d. Using $\alpha = .05$, test if x_1 and x_4 contribute significantly to the model.

APPLICATIONS

SELF TEST

12. The following table shows some of the data available for 14 teams in the National Football League at the end of week 15 for the 1988 season.

FOOTBALL

| | | | | | Interceptions | |
Team	Won–Lost	Total Points	Rushing Yards	Passing Yards	Made by Team	Made by Opponent
Atlanta	5–10	305	1907	2473	19	23
Chicago	12–3	187	2134	2718	14	24
Dallas	3–12	358	1858	3386	24	10
Detroit	4–11	292	1184	1971	15	12
Green Bay	3–12	298	1274	3046	22	20
L.A. Rams	9–6	277	1882	3604	17	22
Minnesota	10–5	206	1744	3633	16	35
New Orleans	9–6	274	1843	2963	15	17
N.Y. Giants	10–5	277	1492	3096	14	15

—Table continues on next page

| | | | | Interceptions | |
| | | Total | Rushing | Passing | Made by | Made by |
Team	Won–Lost	Points	Yards	Yards	Team	Opponent
Philadelphia	9–6	312	1812	3247	17	29
Phoenix	7–8	372	1909	3633	19	14
San Francisco	10–5	256	2453	3131	14	21
Tampa Bay	4–11	340	1650	3169	33	18
Washington	7–8	367	1377	3930	24	14

a. Develop an estimated regression equation that can be used to predict the total points scored given the number of interceptions made by the team.

b. Develop an estimated regression equation that can be used to predict the total points scored given the number of interceptions made by the team, the number of rushing yards, and the number of interceptions made by the opponents.

c. At the $\alpha = .05$ level of significance, test to see if the addition of the number of rushing yards and the number of interceptions made by the opponents contribute significantly to the estimated regression equation developed in part (a). Explain.

13. Refer to Exercise 12.

a. Develop an estimated regression equation that relates the total points to the number of passing yards, interceptions made by the team, and interceptions made by the opponents.

b. Develop an estimated regression equation using the independent variables in part (a) and the number of rushing yards.

c. At the $\alpha = .05$ level of significance, did the number of rushing yards contribute significantly to the estimated regression equation developed in part (a)? Explain.

14. The American Heart Association collects data on the risk of strokes. A 10-year study provided data on how age, blood pressure, and smoking relate to the risk of strokes (*U.S. News & World Report,* April 13, 1992). Assume that data from a portion of this study are shown below. Risk is interpreted as the probability (times 100) that the patient will have a stroke over the next 10-year period. For the smoking variable, a 1 indicates a smoker and a 0 indicates a nonsmoker.

 STROKE

Risk	Age	Blood Pressure	Smoker	Risk	Age	Blood Pressure	Smoker
12	57	152	0	22	71	152	0
24	67	163	0	36	70	173	1
13	58	155	0	15	67	135	1
56	86	177	1	48	77	209	1
28	59	196	0	15	60	199	0
51	76	189	1	36	82	119	1
18	56	155	1	8	66	166	0
31	78	120	0	34	80	125	1
37	80	135	1	3	62	117	0
15	78	98	0	37	59	207	1

a. Develop an estimated regression equation that can be used to predict the risk of stroke given the age and blood-pressure level.

b. Consider adding two independent variables to the model developed in part (a), one involving the interaction between age and blood-pressure level and the other involving whether the person is a smoker. Develop an estimated regression equation using all four independent variables.

c. At the $\alpha = .05$ level of significance, test to see if the addition of the interaction term and whether a person is a smoker contribute significantly to the estimated regression equation developed in part (a).

15.3 FIRST STEPS IN THE ANALYSIS OF A LARGER PROBLEM: THE CRAVENS DATA

In introducing multiple regression analysis, we utilized the Butler Trucking problem extensively. Although the small size of this problem was an advantage when exploring introductory concepts, its limited size makes it difficult to illustrate some of the variable-selection issues involved in model building. To provide an illustration of the *variable-selection procedures* discussed in the next section, we now introduce a data set consisting of 25 observations on eight independent variables. Permission to use these data was provided by Dr. David W. Cravens of the Department of Marketing at Texas Christian University. Consequently, we refer to the data set as the Cravens data.*

EXAMPLE 15.7 ▶ The Cravens data involve a company that sells products in a number of sales territories, each of which is assigned to a single sales representative. A regression analysis was conducted to determine whether sales in each territory could be explained using a variety of predictor (independent) variables. A random sample of 25 sales territories resulted in the data shown in Table 15.5; the variable definitions are shown in Table 15.6.

As a preliminary step, let us consider the sample correlation coefficients between each pair of variables. Figure 15.13 shows the correlation matrix obtained using the Minitab correlation command. Note that the sample correlation coefficient between SALES and TIME is .623, between SALES and POTEN is .598, and so on.

Looking at the sample correlation coefficients between the independent variables, we see that the correlation between TIME and ACCTS is .758; thus, if ACCTS is used as an independent variable, TIME would not provide much more explanatory power to the model. Recall from the discussion of multicollinearity in Section 14.4 that the rule-of-thumb test says that multicollinearity can cause problems if the absolute value of the sample correlation coefficient exceeds .7 for any two of the independent variables. If possible, then, we should avoid including both TIME and ACCTS in the same regression model. The sample correlation coefficient of .549 between CHANGE and RATING is also quite high and may warrant further consideration.

Looking at the sample correlation coefficients between SALES and each of the independent variables can provide us with a quick indication of which independent

---

*For details see David W. Cravens, Robert B. Woodruff, and Joe C. Stamper, "An Analytical Approach for Evaluating Sales Territory Performance," *Journal of Marketing* 36 (January 1972): 31–37.

TABLE 15.5   THE CRAVENS DATA

SALES	TIME	POTEN	ADV	SHARE	CHANGE	ACCTS	WORK	RATING
3,669.88	43.10	74,065.1	4,582.9	2.51	0.34	74.86	15.05	4.9
3,473.95	108.13	58,117.3	5,539.8	5.51	0.15	107.32	19.97	5.1
2,295.10	13.82	21,118.5	2,950.4	10.91	−0.72	96.75	17.34	2.9
4,675.56	186.18	68,521.3	2,243.1	8.27	0.17	195.12	13.40	3.4
6,125.96	161.79	57,805.1	7,747.1	9.15	0.50	180.44	17.64	4.6
2,134.94	8.94	37,806.9	402.4	5.51	0.15	104.88	16.22	4.5
5,031.66	365.04	50,935.3	3,140.6	8.54	0.55	256.10	18.80	4.6
3,367.45	220.32	35,602.1	2,086.2	7.07	−0.49	126.83	19.86	2.3
6,519.45	127.64	46,176.8	8,846.2	12.54	1.24	203.25	17.42	4.9
4,876.37	105.69	42,053.2	5,673.1	8.85	0.31	119.51	21.41	2.8
2,468.27	57.72	36,829.7	2,761.8	5.38	0.37	116.26	16.32	3.1
2,533.31	23.58	33,612.7	1,991.8	5.43	−0.65	142.28	14.51	4.2
2,408.11	13.82	21,412.8	1,971.5	8.48	0.64	89.43	19.35	4.3
2,337.38	13.82	20,416.9	1,737.4	7.80	1.01	84.55	20.02	4.2
4,586.95	86.99	36,272.0	10,694.2	10.34	0.11	119.51	15.26	5.5
2,729.24	165.85	23,093.3	8,618.6	5.15	0.04	80.49	15.87	3.6
3,289.40	116.26	26,878.6	7,747.9	6.64	0.68	136.58	7.81	3.4
2,800.78	42.28	39,572.0	4,565.8	5.45	0.66	78.86	16.00	4.2
3,264.20	52.84	51,866.1	6,022.7	6.31	−0.10	136.58	17.44	3.6
3,453.62	165.04	58,749.8	3,721.1	6.35	−0.03	138.21	17.98	3.1
1,741.45	10.57	23,990.8	861.0	7.37	−1.63	75.61	20.99	1.6
2,035.75	13.82	25,694.9	3,571.5	8.39	−0.43	102.44	21.66	3.4
1,578.00	8.13	23,736.3	2,845.5	5.15	0.04	76.42	21.46	2.7
4,167.44	58.44	34,314.3	5,060.1	12.88	0.22	136.58	24.78	2.8
2,799.97	21.14	22,809.5	3,552.0	9.14	−0.74	88.62	24.96	3.9

CRAVENS

TABLE 15.6   MINITAB VARIABLE DEFINITIONS FOR THE CRAVENS DATA

Variable	Definition
SALES	Total sales credited to the sales representative.
TIME	Length of time employed in months.
POTEN	Market potential; total industry sales in units for the sales territory.*
ADV	Advertising expenditure in the sales territory.
SHARE	Market share; weighted average for the past 4 years.
CHANGE	Change in the market share over the previous 4 years.
ACCTS	Number of accounts assigned to the sales representative.*
WORK	Work load; a weighted index based on annual purchases and concentrations of accounts.
RATING	Sales representative overall rating on eight performance dimensions; an aggregate rating on a 1–7 scale.

*These data were coded to preserve confidentiality.

FIGURE 15.13   SAMPLE CORRELATION COEFFICIENTS FOR THE CRAVENS DATA (AS PRINTED BY MINITAB)

	SALES	TIME	POTEN	ADV	SHARE	CHANGE	ACCTS	WORK
TIME	0.623							
POTEN	0.598	0.454						
ADV	0.596	0.249	0.174					
SHARE	0.484	0.106	-0.211	0.264				
CHANGE	0.489	0.251	0.268	0.377	0.085			
ACCTS	0.754	0.758	0.479	0.200	0.403	0.327		
WORK	-0.117	-0.179	-0.259	-0.272	0.349	-0.288	-0.199	
RATING	0.402	0.101	0.359	0.411	-0.024	0.549	0.229	-0.277

variables are, by themselves, good predictors. We see that the single best predictor of SALES is ACCTS, since it has the highest sample correlation coefficient (.754). Recall that for the case of one independent variable, the square of the sample correlation coefficient is the coefficient of determination. Thus, ACCTS can explain $(.754)^2(100) = 56.85\%$ of the variability in SALES. The next most important independent variables are TIME, POTEN, and ADV, each with a sample correlation coefficient of approximately .6.

Although there are potential multicollinearity problems, let us for the moment consider developing an estimated regression equation using all eight independent variables. Using the Minitab computer package provided the results shown in Figure 15.14. The eight-variable multiple regression model has an adjusted coefficient of determination of 88.3%, which is very high for real data. Note, however, that the p column (the p-values for the t tests of individual parameters) shows that only POTEN, ADV, and SHARE are significant at the $\alpha = .05$ level, given the effect of all the others. Thus, we might be inclined to investigate the results that would be obtained if we used just these three variables. Figure 15.15 shows the Minitab results obtained for the model that uses just these three variables. We see that the model using just three independent variables has an adjusted coefficient of determination of 82.7%, which, although not quite as good as for the eight-independent-variable model, is still very high.

How can we find a model that will do the best job given the data available? One approach sometimes advocated for determining the best model is to compute all possible regressions. That is, one could develop 8 one-variable models (each of which corresponds to one of the independent variables), 28 two-variable models (the number of combinations of 8 variables taken 2 at a time), and so on. In all, for the Cravens data, there are 255 different models involving one or more independent variables that would have to be fitted to the data.

With the excellent computer packages available today, it is possible to compute all possible regressions. But, doing so involves a great deal of computation and requires the model builder to review a great deal of computer output, much of which is associated with obviously poor models. Statisticians often prefer a more systematic approach to selecting the subset of independent variables providing the best model. In the next section, we introduce some of the more popular approaches.

FIGURE 15.14 MINITAB OUTPUT FOR THE MODEL INVOLVING ALL EIGHT INDEPENDENT VARIABLES

```
The regression equation is
SALES = - 1508 + 2.01 TIME + 0.0372 POTEN + 0.151 ADV + 199 SHARE
           + 291 CHANGE + 5.55 ACCTS + 19.8 WORK + 8 RATING
```

Predictor	Coef	Stdev	t-ratio	p
Constant	1507.8	778.6	-1.94	0.071
TIME	2.010	1.931	1.04	0.313
POTEN	0.037205	0.008202	4.54	0.000
ADV	0.15099	0.04711	3.21	0.006
SHARE	199.02	67.03	2.97	0.009
CHANGE	290.9	186.8	1.56	0.139
ACCTS	5.551	4.776	1.16	0.262
WORK	19.79	33.68	0.59	0.565
RATING	8.2	128.5	0.06	0.950

```
s = 449.0       R-sq = 92.2%      R-sq(adj) = 88.3%
```

Analysis of Variance

SOURCE	DF	SS	MS	F	p
Regression	8	38153568	4769196	23.65	0.000
Error	16	3225984	201624		
Total	24	41379552			

FIGURE 15.15 MINITAB OUTPUT FOR THE MODEL INVOLVING POTEN, ADV, AND SHARE

```
The regression equation is
SALES = - 1604 + 0.0543 POTEN + 0.167 ADV + 283 SHARE
```

Predictor	Coef	Stdev	t-ratio	p
Constant	-1603.6	505.6	-3.17	0.005
POTEN	0.054286	0.007474	7.26	0.000
ADV	0.16748	0.04427	3.78	0.001
SHARE	282.75	48.76	5.80	0.000

```
s = 545.5       R-sq = 84.9%      R-sq(adj) = 82.7%
```

Analysis of Variance

SOURCE	DF	SS	MS	F	p
Regression	3	35130240	11710080	39.35	0.000
Error	21	6249310	297586		
Total	24	41379552			

15.4   VARIABLE-SELECTION PROCEDURES

In this section, we discuss four computer-based methods for selecting the independent variables in a regression model: stepwise regression, forward selection, backward elimination, and best-subsets regression. Given a data set involving several possible independent variables, these methods can be used to identify which independent variables provide the best model. The first three methods are iterative; at each step, a single variable is added or deleted, and the new model is evaluated. The process continues until a stopping criterion indicates that the procedure cannot find a better model. The last method (best subsets) is not a one-variable-at-a-time method; it evaluates regression models involving different subsets of the independent variables.

The criterion for selecting an independent variable to add or delete from the model at each step is based on the F statistic introduced in Section 15.2. Suppose, for instance, that we were considering adding x_3 to a model involving x_1 or deleting x_3 from a model involving x_1 and x_3. In Section 15.2 we showed that

$$F = \frac{\dfrac{\text{SSE}(x_1) - \text{SSE}(x_1, x_3)}{1}}{\dfrac{\text{SSE}(x_1, x_3)}{n - 2 - 1}}$$

can be used as a criterion for determining whether the presence of x_3 in the model causes a significant reduction in the error sum of squares. The value of this F statistic is the criterion used by the first three methods to determine whether a variable should be added to or deleted from the regression model at each step. It is also used to indicate when the iterative procedure should stop. The first three procedures stop when no more significant reduction in the error sum of squares can be obtained. As also noted in Section 15.2, when only one variable at a time is to be added or deleted, the t statistic (recall that $t^2 = F$) provides the same criterion.

With the stepwise-regression procedure, a variable may be added or deleted at each step. The procedure stops when no more improvement can be obtained by adding or deleting a variable. With the forward-selection procedure, a variable is added at each step, but variables are never deleted. The procedure stops when no more improvement can be obtained by adding a variable. With the backward-elimination procedure, the procedure starts with a model involving all the possible independent variables. At each step, a variable is eliminated. The procedure stops when no more improvement can be obtained by deleting a variable.

STEPWISE REGRESSION

We will illustrate the stepwise-regression procedure using the Cravens data. To see how a step of the procedure is performed, suppose that after three steps the following three independent variables have been selected: ACCTS, ADV, and POTEN. At the next step, the procedure first determines whether any of the variables *already in the model* should be deleted. It does so by first determining which of the three variables is the least significant addition in moving from a two- to three-independent-variable model. To determine this, an F statistic is computed for each of the three variables. The F statistic for ACCTS enables us to test whether adding ACCTS to a model that already includes ADV and

POTEN leads to a significant reduction in SSE. If not, the stepwise procedure will consider dropping ACCTS from the model. Before doing so, however, the same F statistic will be computed for ADV and POTEN. The variable with the smallest F statistic makes the least significant addition in moving from a two- to three-independent-variable regression model and becomes a candidate for deletion. If any variable is to be deleted, it will be the one.

We will denote by FMIN the smallest of the F statistics for all variables in the regression model at the beginning of a new step. The variable with the smallest F statistic is the least significant addition to the model. If the value of FMIN is too small to be significant, the corresponding variable is deleted from the model. On the other hand, if FMIN is large enough to be significant, none of the variables are deleted from the model (none of the other variables can have smaller F statistics).

The user of a computer-based stepwise-regression procedure must specify a cutoff value for the F statistic so that the method can determine when FMIN is large enough to be significant. With the Minitab package, the smallest significant F value is denoted by FREMOVE. If the user does not specify a value for FREMOVE, it is automatically set equal to 4 by Minitab. Anytime FMIN < FREMOVE, the stepwise procedure of Minitab will delete the corresponding variable from the model. If FMIN ≥ FREMOVE, no variable is deleted at that step of the procedure.

If no variable can be removed from the model, the stepwise procedure next checks to see whether adding a variable can improve the model. For each variable *not in the model,* an F statistic is computed. The largest of these F statistics corresponds to the variable that will cause the largest reduction in SSE. This variable then becomes a candidate for inclusion in the model. We will denote the largest F statistic for variables not currently in the model by FMAX. Again, a cutoff value for the F statistic must be used to determine whether FMAX is large enough for the corresponding variable to make a significant improvement in the model.

The cutoff value for determining when to add a variable is denoted by FENTER in the Minitab computer package. The user of the package may specify a cutoff value for FENTER; if the user does not, Minitab will automatically set FENTER = 4. If FMAX > FENTER, the corresponding variable is added to the model, and the stepwise-regression procedure goes on to the next step. The procedure stops when no variables can be deleted and no variables can be added.

In summary, at each step of the stepwise-regression procedure, the first consideration is to see whether any variable can be removed. If none of the variables can be removed, the procedure then checks to see if any variables can be added. Because of the nature of the stepwise procedure, it is possible for a variable to enter the model at one step, be deleted at a subsequent step, and then reenter the model at a later step. The procedure stops when FMIN ≥ FREMOVE (no variables can be deleted) and FMAX ≤ FENTER (no variables can be added).

EXAMPLE 15.7
(**CONTINUED**) ▶

Figure 15.16 shows the result of using the Minitab stepwise-regression procedure for the Cravens data. As we noted in Section 15.2, when only one variable is being added, the t statistic provides the same criterion as the F statistic. (One can show that $F = t^2$.) The entries in the T-RATIO row are the t statistics. The values of FREMOVE and FENTER were both automatically set equal to 4. At step 1, there are no variables to consider for deletion. The variable providing the largest value for the F statistic is ACCTS, with $F = t^2 = (5.5)^2 = 30.25$. Since 30.25 > 4, ACCTS is added to the model. On the next three steps, ADV, POTEN, and SHARE are added to the model. After step 4, an F statistic

FIGURE 15.16 MINITAB OUTPUT USING STEPWISE REGRESSION FOR THE CRAVENS DATA

```
STEPWISE REGRESSION OF SALES ON 8 PREDICTORS, WITH N = 25

       STEP        1        2        3        4
   CONSTANT    709.32    50.30  -327.23 -1441.93

   ACCTS         21.7     19.0     15.6      9.2
   T-RATIO       5.50     6.41     5.19     3.22

   ADV                   0.227    0.216    0.195
   T-RATIO               4.50     4.77     4.74

   POTEN                          0.0219   0.0382
   T-RATIO                        2.53     4.79

   SHARE                                    190
   T-RATIO                                  3.82

   S              881      650      583      454
   R-SQ         56.85    77.51    82.77    90.04
```

was computed for each of the four variables in the model. The values of the F statistics were $t^2 = (3.22)^2 = 10.37$, $t^2 = 22.47$, $t^2 = 22.94$, and $t^2 = 14.59$ for ACCTS, ADV, POTEN, and SHARE, respectively. Thus, FMIN = 10.37, and the corresponding variable is ACCTS. Since $10.37 > 4$, no variable is dropped from the model.

An F statistic was then computed for each of the other four variables not in the model. Since all of these F statistics were less than 4, no variables were added to the model. The stepwise procedure stopped at this point; no variables could be deleted, and none could be added to improve the model. The results shown in Figure 15.16 were printed at this point. The estimated regression equation identified by the Minitab stepwise-regression procedure is

$$\hat{y} = -1441.93 + 9.2 \text{ ACCTS} + .195 \text{ ADV} + .0382 \text{ POTEN} + 190 \text{ SHARE}$$

Note also in Figure 15.16 that, with the error sum of squares being reduced at each step, $s = \sqrt{\text{MSE}}$ has been reduced from 881 with the best one-variable model to 454 after four steps. The value of R-sq has been increased from 56.85% to 90.04%. ◀

FORWARD SELECTION

Forward selection is another computer-based procedure for variable selection. It is similar to the stepwise-regression procedure except that it does not permit a variable to be deleted from the model once it has been added. The forward-selection procedure starts out with no independent variables. Then it adds variables, one at a time, as long as a significant reduction in the error sum of squares (SSE) can be achieved. When no variable can be added that will cause a further significant reduction in SSE, the procedure stops and prints out the results. For the Cravens data, the stepwise-regression procedure added one variable at each step and did not delete any variables. Thus, for the Cravens data, the forward-selection procedure leads to the same model as that provided by the stepwise procedure.

BACKWARD ELIMINATION

The backward-elimination procedure begins with a model including all the independent variables the model builder wants considered. (Figure 15.14 shows a regression model involving all eight independent variables for the Cravens data.) It then deletes one variable at a time using the same criterion as that for removing variables using the stepwise-regression procedure. The variable with the smallest F statistic is deleted, provided F is less than the preestablished cutoff criterion (FREMOVE for Minitab). The major difference between the backward-elimination procedure and the stepwise procedure is that once a variable has been removed from the model, it cannot reenter at a later step.

BEST-SUBSETS REGRESSION

Stepwise regression, forward selection, and backward elimination are approaches to choosing the regression model by adding or deleting independent variables one at a time. As such, there is no guarantee that the best model for a given number of variables will be found. Thus, these one-variable-at-a-time methods are properly viewed as heuristics for selecting a good regression model.

Some software packages have a procedure called best-subsets regression that permits the user to find, given a specified number of independent variables, the best regression model. Minitab has such a procedure.

EXAMPLE 15.7
(CONTINUED) ▶ In Figure 15.17, we show a portion of the computer output obtained using the best-subsets procedure for the Cravens data set.

This output identifies the two best one-variable estimated regression equations, the two best two-variable equations, the two best three-variable equations, and so on. The criterion used in determining which estimated regression equations are best for any number of predictors is the value of the coefficient of determination (R-sq). For instance, ACCTS, with an R-sq = 56.8%, provides the best estimated regression equation using only one independent variable; ADV and ACCTS, with an R-sq = 77.5%, provides the best estimated regression equation using two independent variables; and POTEN, ADV, and SHARE, with an R-sq = 84.9%, provides the best estimated regression equation with three independent variables. For the Cravens data, the adjusted coefficient of determination (Adj. R-sq) is largest for the model with six independent variables: TIME, POTEN, ADV, SHARE, HANGE, and ACCTS. However, the best model with four independent variables (POTEN, ADV, SHARE, ACCTS) has an adjusted coefficient of determination almost as high (88.1%). All other things being equal, a simpler model with fewer variables is usually preferred. ◀

MAKING THE FINAL CHOICE

The analysis performed on the Cravens data to this point is good preparation for choosing a final model, but more analysis should be conducted before making the final choice. As

FIGURE 15.17   PORTION OF MINITAB OUTPUT USING BEST-SUBSETS REGRESSION

```
                                                     R
                                      P   S H A      A
                                    T O   H A C W T
                                    I T A A N C O I
                    Adj.            M E D R G T R N
Vars   R-sq   R-sq        s         E N V E E S K G

  1    56.8   55.0    881.09                    X
  1    38.8   36.1    1049.3    X
  2    77.5   75.5    650.40          X       X
  2    74.6   72.3    691.11       X  X
  3    84.9   82.7    545.53       X  X X
  3    82.8   80.3    582.66       X  X        X
  4    90.0   88.1    453.86       X  X X      X
  4    89.6   87.5    463.95    X  X  X X
  5    91.5   89.3    430.21    X  X  X X X
  5    91.2   88.9    436.75       X  X X X X
  6    92.0   89.4    428.00    X  X  X X X X
  6    91.6   88.9    438.20       X  X X X X X
  7    92.2   89.0    435.67    X  X  X X X X X
  7    92.0   88.8    440.29    X  X  X X X X   X
  8    92.2   88.3    449.02    X  X  X X X X X X
```

TABLE 15.7   SELECTED MODELS INVOLVING ACCTS, ADV, POTEN, AND SHARE

Model	Independent Variables	Adj. R-sq
1	ACCTS	55.0
2	ACCTS, ADV	75.5
3	POTEN, SHARE	72.3
4	ACCTS, ADV, POTEN	80.3
5	ADV, POTEN, SHARE	82.7
6	ACCTS, ADV, POTEN, SHARE	88.1

we have noted in Chapters 13 and 14, a careful analysis of the residuals should be made. We want the residual plot for the model chosen to resemble approximately a horizontal band. For now, let us assume that there is no difficulty with the residuals.

EXAMPLE 15.7
(CONTINUED)   The best-subsets procedure has shown us that the best four-variable model utilizes the independent variables ACCTS, ADV, POTEN, and SHARE. This also happens to be the four-variable model identified with the stepwise-regression procedure. Table 15.7 is helpful in making the final choice. It shows several possible models consisting of some or all of these four independent variables.

From Table 15.7, we see that the model involving just ACCTS and ADV is pretty good. The adjusted coefficient of determination is 75.5%, whereas the model using all four variables only provides a 12.6% increase. The simpler two-variable model might be preferred, for instance, if it is difficult to measure market potential (POTEN). On the other hand, if the data are readily available, the model builder would clearly prefer the model involving all four variables if highly accurate predictions of sales are needed.

NOTES AND COMMENTS

1. In the stepwise procedure, FENTER cannot be set smaller than FREMOVE. To see why, suppose this condition was not satisfied. For instance, suppose a model builder set FENTER = 2 and FREMOVE = 4 and at some step of the procedure, a variable with an F statistic of 3 was entered into the model. At the next step, any variable with an F statistic of 3 would be a candidate for removal from the model (since 3 < FREMOVE = 4). If the stepwise procedure deleted it, then at the very next step the variable would enter again, it would be deleted again, and so on. Thus, the procedure would cycle forever. To avoid this, the stepwise procedure requires that FENTER be greater than or equal to FREMOVE.

2. Functions of the independent variables may be used to create new independent variables for use with any of the procedures of this section. For instance, if we desired $x_1 x_2$ in the model to account for interaction, the data for x_1 and x_2 would be used to create the data for $z_i = x_{1i} x_{2i}$.

3. None of the procedures that add or delete variables one at a time can be guaranteed to identify the best regression model. But they are excellent approaches to finding good models—especially when there is not much multicollinearity present.

EXERCISES

APPLICATIONS

SELF TEST

15. Two experts provided subjective lists of school districts that they think are among the best in the country. For each school district, the following data were obtained: average class size, instructional spending per student, average teacher salary, combined SAT score, percent of students taking the SAT, and the percentage of students who attended a 4-year college (*The Wall Street Journal,* March 31, 1989).

City	Average Class Size	Instructional Spending Per Student ($)	Average Teacher Salary ($)	Combined SAT Score/ (% Taking Test)	Attend 4-Year College (%)
Blue Springs, Mo.	25	3,060	29,359	1083/(8)	74
Garden City, N.Y.	18	9,700	51,000	997/(99)	77
Indianapolis, Ind.	30	3,222	30,482	716/(42)	40
Newport Beach, Calif. (Newport-Mesa)	26	4,028	37,043	977/(46)	51

—Table continues on next page

City	Average Class Size	Instructional Spending Per Student ($)	Average Teacher Salary ($)	Combined SAT Score/ (% Taking Test)	Attend 4-Year College (%)
Novi, Mich.	20	3,067	39,797	980/(15)	53
Piedmont, Calif. (Piedmont City)	28	4,208	37,274	1,042/(91)	75
Pittsburgh, Pa. (Fox Chapel area)	21	4,884	37,156	983/(80)	66
Scarsdale, N.Y. (Edgemont)	20	9,853	31,555	1,110/(98)	87
Wayne, Pa. (Radnor Township)	22	5,022	40,406	1,040/(95)	85
Weston, Mass.	21	4,680	39,800	1,031/(99)	89
Farmingdale, N.Y.	22	6,729	45,846	947/(75)	81
Mamaroneck, N.Y.	20	10,405	49,625	1,000/(90)	69
Mayfield, Ohio	24	5,881	36,228	1,003/(25)	48
Morristown, N.J.	22	6,300	37,000	972/(80)	64
New Rochelle, N.Y.	23	8,875	41,650	1,039/(80)	55
Newton Square, Pa. (Marple-Newton)	17	5,313	38,000	963/(75)	79
Omaha, Neb. (Westside)	23	4,815	32,500	1,059/(31)	81
Shaker Heights, Ohio	23	4,370	38,639	940/(56)	82

 SCHOOLS2

Let the dependent variable be the percentage of students who attend a 4-year college.
a. Develop the best one-variable model.
b. Use the stepwise procedure to develop the best model.
c. Use the backward-elimination procedure to develop the best model.
d. Use the best-subsets regression procedure to develop the best model.

16. Refer to Exercise 15. Let the dependent variable be the combined SAT score.
a. Develop the best one-variable model.
b. Use the stepwise procedure to develop the best model.
c. Use the backward-elimination procedure to develop the best model.
d. Use the best-subsets regression procedure to develop the best model.

17. Refer to the data in Exercise 12. Let the dependent variable be the number of wins.
a. Develop the best one-variable model.
b. Use the stepwise procedure to develop the best model.
c. Use the backward-elimination procedure to develop the best model.
d. Use the best-subsets regression procedure to develop the best model.

18. Refer to the data in Exercise 12. Let the dependent variable be the total points scored.
a. Develop the best one-variable model.
b. Use the stepwise procedure to develop the best model.
c. Use the backward-elimination procedure to develop the best model.
d. Use the best-subsets regression procedure to develop the best model.

19. The following data show the price/earnings ratio, the net profit margin, and the growth rate for 19 companies listed in "The *Forbes* 500s on Wall Street," (*Forbes,* May 1, 1989) The data in the column labeled "Industry" are simply codes used to define the industry for each company: 1 = energy-international oil; 2 = health-drugs, and 3 = electronics-computers.

Firm	P/E Ratio	Profit Margin	Growth Rate	Industry
Exxon	11.3	6.5	10	1
Chevron	10.0	7.0	5	1
Texaco	9.9	3.9	5	1
Mobil	9.7	4.3	7	1
Amoco	10.0	9.8	8	1
Pfizer	11.9	14.7	12	2
Bristol Meyers	16.2	13.9	14	2
Merck	21.0	20.3	16	2
American Home Products	13.3	16.9	11	2
Abbott Laboratories	15.5	15.2	18	2
Eli Lilly	18.9	18.7	11	2
Upjohn	14.6	12.8	10	2
Warner-Lambert	16.0	8.7	7	2
Amdahl	8.4	11.9	4	3
Digital	10.4	9.8	19	3
Hewlett-Packard	14.8	8.1	18	3
NCR	10.1	7.3	6	3
Unisys	7.0	6.9	6	3
IBM	11.8	9.2	6	3

 FORBES2

Develop the best model that can be used to predict price/earnings ratio. Briefly describe the process you used to develop a recommended estimated regression equation for these data.

20. Refer to Exercise 14. Develop a model that can be used to predict risk using age, blood pressure, whether a person is a smoker, and any interaction involving these variables. Briefly describe the process you used to develop an estimated regression equation for these data.

15.5 RESIDUAL ANALYSIS

In Chapters 13 and 14 we showed how residual plots can be used to determine when violations of assumptions concerning the regression model occur. We looked for violations of assumptions concerning the error term ϵ and the assumed functional form of the model. Some of the actions that can be taken when such violations are detected have been discussed in this chapter. When a different functional form is needed, curvilinear and interaction terms can be included through the use of the general linear model. When other (or more) variables need to be considered, some of the variable-selection procedures of the preceding section may be appropriate.

In Chapters 13 and 14 we also discussed how residual analysis could be used to identify observations that can be classified as outliers or as being especially influential in determining the estimated regression equation. Some steps that should be taken when such observations are found were noted. In many regression studies involving economic data, a special type of correlation involving the error terms can cause problems; it is called *serial correlation,* or *autocorrelation.* In this section we show how residual analysis using the *Durbin-Watson test* can be used to determine when autocorrelation is a problem.

AUTOCORRELATION AND THE DURBIN-WATSON TEST

The data used for regression studies are often collected over time. In such cases, it is not uncommon for the value of y at time t, denoted by y_t, to be related to the value of y at previous time periods. When this occurs, we say autocorrelation (also called serial correlation) is present in the data. If the value of y in time period t is related to its value in time period $t - 1$, we say first-order autocorrelation is present. If the value of y in time period t is related to the value of y in time period $t - 2$, we say second-order autocorrelation is present, and so on.

When autocorrelation is present, one of the assumptions of the regression model is violated: The error terms are not independent. In the case of first-order autocorrelation, the error at time t, denoted by ϵ_t, will be related to the error at time period $t - 1$, denoted by ϵ_{t-1}. Two cases of first-order autocorrelation are shown in Figure 15.18. Panel A illustrates the case of positive autocorrelation; panel B illustrates the case of negative autocorrelation. With positive autocorrelation, we expect a positive residual in one period to be followed by a positive residual in the next period, a negative residual in one period to be followed by a negative residual in the next period, and so on. With negative autocorrelation, we expect a positive residual in one period to be followed by a negative residual in the next period, then a positive residual, and so on.

When autocorrelation is present, serious errors can be made in statistical inferences about the regression model. Thus, it is important to be able to detect autocorrelation and take corrective action. We will show how the Durbin-Watson statistic can be used to detect first-order autocorrelation.

Suppose that the values of ϵ are not independent but are related in the following manner:

$$\epsilon_t = \rho\epsilon_{t-1} + z_t \qquad (15.16)$$

where ρ is a parameter with an absolute value less than 1 and z_t is a normally and independently distributed random variable with mean 0 and variance σ^2. From equation (15.16) we see that if $\rho = 0$, then the error terms are not related, and each has a mean of 0 and a variance of σ^2. In this case, there is no autocorrelation and the regression

---

FIGURE 15.18   **TWO DATA SETS WITH FIRST-ORDER AUTOCORRELATION**

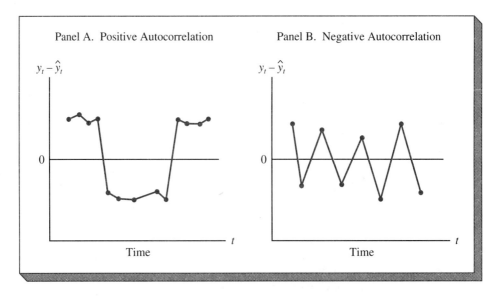

assumptions are satisfied. If $\rho > 0$, we have positive autocorrelation; if $\rho < 0$, we have negative autocorrelation. In either of these cases, the regression assumptions concerning the error term are violated.

The Durbin-Watson test for autocorrelation uses the residuals to determine whether $\rho = 0$. To simplify the notation for the Durbin-Watson statistic, we shall denote the ith residual by $e_i = y_i - \hat{y}_i$. The Durbin-Watson statistic denoted d is given by the following expression.

Durbin-Watson Statistic

$$d = \frac{\sum_{t=2}^{n} (e_t - e_{t-1})^2}{\sum_{t=1}^{n} e_t^2} \qquad (15.17)$$

If successive values of the residuals are close together (positive autocorrelation), the Durbin-Watson statistic will be small. If successive values of the residuals are far apart (negative autocorrelation), the Durbin-Watson statistic will tend to be large.

The Durbin-Watson statistic ranges in value between 0 and 4, with a value of 2 indicating no autocorrelation is present. Durbin and Watson have developed tables that can be used to determine when their test statistic indicates the presence of autocorrelation. Table 15.8 shows lower and upper bounds (d_L and d_U) for hypothesis tests using $\alpha = .05$, $\alpha = .025$, and $\alpha = .01$; n denotes the number of observations, and k is the number of independent variables in the model. The null hypothesis to be tested is always taken to be one of no autocorrelation:

$$H_0: \rho = 0$$

The alternative hypothesis to test for positive autocorrelation is

$$H_a: \rho > 0$$

The alternative hypothesis to test for negative autocorrelation is

$$H_a: \rho < 0$$

A two-sided test is also possible. In this case, the alternative hypothesis is

$$H_a: \rho \neq 0$$

Figure 15.19 shows how the values of d_L and d_U in Table 15.8 are to be used to test for autocorrelation. Panel A illustrates the test for positive autocorrelation. If $d < d_L$, we conclude positive autocorrelation is present. If $d_L \leq d \leq d_U$, we say the test is inconclusive. If $d > d_U$, we conclude there is no evidence of positive autocorrelation.

Panel B illustrates the test for negative autocorrelation. If $d > 4 - d_L$, we conclude negative autocorrelation is present. If $4 - d_U \leq d \leq 4 - d_L$, we say the test is inconclusive. If $d < 4 - d_U$, we conclude there is no evidence of negative autocorrelation.

Panel C illustrates the two-sided test. If $d < d_L$ or $d > 4 - d_L$, we reject H_0 and conclude autocorrelation is present. If $d_L \leq d \leq d_U$ or $4 - d_U \leq d \leq 4 - d_L$, we say the test is inconclusive. If $d_U < d < 4 - d_U$, we conclude there is no evidence of autocorrelation.

TABLE 15.8   **CRITICAL VALUES FOR THE DURBIN-WATSON TEST FOR AUTOCORRELATION**

NOTE: ENTRIES IN THE TABLE GIVE THE CRITICAL VALUES FOR A ONE-TAILED DURBIN-WATSON TEST FOR AUTOCORRELATION. FOR A TWO-TAILED TEST, THE LEVEL OF SIGNIFICANCE IS DOUBLED.

Significance Points of d_L and d_U: $\alpha = .025$
Number of Independent Variables

k	1		2		3		4		5	
n	d_L	d_U	d_L	d_U	d_L	d_U	d_L	d_U	d_L	d_U
15	1.08	1.36	0.95	1.54	0.82	1.75	0.69	1.97	0.56	2.21
16	1.10	1.37	0.98	1.54	0.86	1.73	0.74	1.93	0.62	2.15
17	1.13	1.38	1.02	1.54	0.90	1.71	0.78	1.90	0.67	2.10
18	1.16	1.39	1.05	1.53	0.93	1.69	0.82	1.87	0.71	2.06
19	1.18	1.40	1.08	1.53	0.97	1.68	0.86	1.85	0.75	2.02
20	1.20	1.41	1.10	1.54	1.00	1.68	0.90	1.83	0.79	1.99
21	1.22	1.42	1.13	1.54	1.03	1.67	0.93	1.81	0.83	1.96
22	1.24	1.43	1.15	1.54	1.05	1.66	0.96	1.80	0.86	1.94
23	1.26	1.44	1.17	1.54	1.08	1.66	0.99	1.79	0.90	1.92
24	1.27	1.45	1.19	1.55	1.10	1.66	1.01	1.78	0.93	1.90
25	1.29	1.45	1.21	1.55	1.12	1.66	1.04	1.77	0.95	1.89
26	1.30	1.46	1.22	1.55	1.14	1.65	1.06	1.76	0.98	1.88
27	1.32	1.47	1.24	1.56	1.16	1.65	1.08	1.76	1.01	1.86
28	1.33	1.48	1.26	1.56	1.18	1.65	1.10	1.75	1.03	1.85
29	1.34	1.48	1.27	1.56	1.20	1.65	1.12	1.74	1.05	1.84
30	1.35	1.49	1.28	1.57	1.21	1.65	1.14	1.74	1.07	1.83
31	1.36	1.50	1.30	1.57	1.23	1.65	1.16	1.74	1.09	1.83
32	1.37	1.50	1.31	1.57	1.24	1.65	1.18	1.73	1.11	1.82
33	1.38	1.51	1.32	1.58	1.26	1.65	1.19	1.73	1.13	1.81
34	1.39	1.51	1.33	1.58	1.27	1.65	1.21	1.73	1.15	1.81
35	1.40	1.52	1.34	1.58	1.28	1.65	1.22	1.73	1.16	1.80
36	1.41	1.52	1.35	1.59	1.29	1.65	1.24	1.73	1.18	1.80
37	1.42	1.53	1.36	1.59	1.31	1.66	1.25	1.72	1.19	1.80
38	1.43	1.54	1.37	1.59	1.32	1.66	1.26	1.72	1.21	1.79
39	1.43	1.54	1.38	1.60	1.33	1.66	1.27	1.72	1.22	1.79
40	1.44	1.54	1.39	1.60	1.34	1.66	1.29	1.72	1.23	1.79
45	1.48	1.57	1.43	1.62	1.38	1.67	1.34	1.72	1.29	1.78
50	1.50	1.59	1.46	1.63	1.42	1.67	1.38	1.72	1.34	1.77
55	1.53	1.60	1.49	1.64	1.45	1.68	1.41	1.72	1.38	1.77
60	1.55	1.62	1.51	1.65	1.48	1.69	1.44	1.73	1.41	1.77
65	1.57	1.63	1.54	1.66	1.50	1.70	1.47	1.73	1.44	1.77
70	1.58	1.64	1.55	1.67	1.52	1.70	1.49	1.74	1.46	1.77
75	1.60	1.65	1.57	1.68	1.54	1.71	1.51	1.74	1.49	1.77
80	1.61	1.66	1.59	1.69	1.56	1.72	1.53	1.74	1.51	1.77
85	1.62	1.67	1.60	1.70	1.57	1.72	1.55	1.75	1.52	1.77
90	1.63	1.68	1.61	1.70	1.59	1.73	1.57	1.75	1.54	1.78
95	1.64	1.69	1.62	1.71	1.60	1.73	1.58	1.75	1.56	1.78
100	1.65	1.69	1.63	1.72	1.61	1.74	1.59	1.76	1.57	1.78

—Table continues on next page

Source: J. Durbin and G. S. Watson, "Testing for serial correlation in least square regression II," *Biometrika,* 38, 1951, 159–178.

TABLE 15.8   (CONTINUED)

Significance Points of d_L and d_U: $\alpha = .025$

Number of Independent Variables

k	1		2		3		4		5	
n	d_L	d_U	d_L	d_U	d_L	d_U	d_L	d_U	d_L	d_U
15	0.95	1.23	0.83	1.40	0.71	1.61	0.59	1.84	0.48	2.09
16	0.98	1.24	0.86	1.40	0.75	1.59	0.64	1.80	0.53	2.03
17	1.01	1.25	0.90	1.40	0.79	1.58	0.68	1.77	0.57	1.98
18	1.03	1.26	0.93	1.40	0.82	1.56	0.72	1.74	0.62	1.93
19	1.06	1.28	0.96	1.41	0.86	1.55	0.76	1.72	0.66	1.90
20	1.08	1.28	0.99	1.41	0.89	1.55	0.79	1.70	0.70	1.87
21	1.10	1.30	1.01	1.41	0.92	1.54	0.83	1.69	0.73	1.84
22	1.12	1.31	1.04	1.42	0.95	1.54	0.86	1.68	0.77	1.82
23	1.14	1.32	1.06	1.42	0.97	1.54	0.89	1.67	0.80	1.80
24	1.16	1.33	1.08	1.43	1.00	1.54	0.91	1.66	0.83	1.79
25	1.18	1.34	1.10	1.43	1.02	1.54	0.94	1.65	0.86	1.77
26	1.19	1.35	1.12	1.44	1.04	1.54	0.96	1.65	0.88	1.76
27	1.21	1.36	1.13	1.44	1.06	1.54	0.99	1.64	0.91	1.75
28	1.22	1.37	1.15	1.45	1.08	1.54	1.01	1.64	0.93	1.74
29	1.24	1.38	1.17	1.45	1.10	1.54	1.03	1.63	0.96	1.73
30	1.25	1.38	1.18	1.46	1.12	1.54	1.05	1.63	0.98	1.73
31	1.26	1.39	1.20	1.47	1.13	1.55	1.07	1.63	1.00	1.72
32	1.27	1.40	1.21	1.47	1.15	1.55	1.08	1.63	1.02	1.71
33	1.28	1.41	1.22	1.48	1.16	1.55	1.10	1.63	1.04	1.71
34	1.29	1.41	1.24	1.48	1.17	1.55	1.12	1.63	1.06	1.70
35	1.30	1.42	1.25	1.48	1.19	1.55	1.13	1.63	1.07	1.70
36	1.31	1.43	1.26	1.49	1.20	1.56	1.15	1.63	1.09	1.70
37	1.32	1.43	1.27	1.49	1.21	1.56	1.16	1.62	1.10	1.70
38	1.33	1.44	1.28	1.50	1.23	1.56	1.17	1.62	1.12	1.70
39	1.34	1.44	1.29	1.50	1.24	1.56	1.19	1.63	1.13	1.69
40	1.35	1.45	1.30	1.51	1.25	1.57	1.20	1.63	1.15	1.69
45	1.39	1.48	1.34	1.53	1.30	1.58	1.25	1.63	1.21	1.69
50	1.42	1.50	1.38	1.54	1.34	1.59	1.30	1.64	1.26	1.69
55	1.45	1.52	1.41	1.56	1.37	1.60	1.33	1.64	1.30	1.69
60	1.47	1.54	1.44	1.57	1.40	1.61	1.37	1.65	1.33	1.69
65	1.49	1.55	1.46	1.59	1.43	1.62	1.40	1.66	1.36	1.69
70	1.51	1.57	1.48	1.60	1.45	1.63	1.42	1.66	1.39	1.70
75	1.53	1.58	1.50	1.61	1.47	1.64	1.45	1.67	1.42	1.70
80	1.54	1.59	1.52	1.62	1.49	1.65	1.47	1.67	1.44	1.70
85	1.56	1.60	1.53	1.63	1.51	1.65	1.49	1.68	1.46	1.71
90	1.57	1.61	1.55	1.64	1.53	1.66	1.50	1.69	1.48	1.71
95	1.58	1.62	1.56	1.65	1.54	1.67	1.52	1.69	1.50	1.71
100	1.59	1.63	1.57	1.65	1.55	1.67	1.53	1.70	1.51	1.72

—Table continues on next page

If significant autocorrelation is identified, we should investigate whether we have omitted one or more key independent variables that have time-ordered effects on the dependent variable. If no such variables can be identified, including an independent variable that measures the time of the observation (for instance, the value of this variable

TABLE 15.8   (CONTINUED)

Significance Points of d_L and d_U: $\alpha = .01$

Number of Independent Variables

	k	1		2		3		4		5	
n	d_L	d_U	d_L	d_U	d_L	d_U	d_L	d_U	d_L	d_U
15	0.81	1.07	0.70	1.25	0.59	1.46	0.49	1.70	0.39	1.96
16	0.84	1.09	0.74	1.25	0.63	1.44	0.53	1.66	0.44	1.90
17	0.87	1.10	0.77	1.25	0.67	1.43	0.57	1.63	0.48	1.85
18	0.90	1.12	0.80	1.26	0.71	1.42	0.61	1.60	0.52	1.80
19	0.93	1.13	0.83	1.26	0.74	1.41	0.65	1.58	0.56	1.77
20	0.95	1.15	0.86	1.27	0.77	1.41	0.68	1.57	0.60	1.74
21	0.97	1.16	0.89	1.27	0.80	1.41	0.72	1.55	0.63	1.71
22	1.00	1.17	0.91	1.28	0.83	1.40	0.75	1.54	0.66	1.69
23	1.02	1.19	0.94	1.29	0.86	1.40	0.77	1.53	0.70	1.67
24	1.04	1.20	0.96	1.30	0.88	1.41	0.80	1.53	0.72	1.66
25	1.05	1.21	0.98	1.30	0.90	1.41	0.83	1.52	0.75	1.65
26	1.07	1.22	1.00	1.31	0.93	1.41	0.85	1.52	0.78	1.64
27	1.09	1.23	1.02	1.32	0.95	1.41	0.88	1.51	0.81	1.63
28	1.10	1.24	1.04	1.32	0.97	1.41	0.90	1.51	0.83	1.62
29	1.12	1.25	1.05	1.33	0.99	1.42	0.92	1.51	0.85	1.61
30	1.13	1.26	1.07	1.34	1.01	1.42	0.94	1.51	0.88	1.61
31	1.15	1.27	1.08	1.34	1.02	1.42	0.96	1.51	0.90	1.60
32	1.16	1.28	1.10	1.35	1.04	1.43	0.98	1.51	0.92	1.60
33	1.17	1.29	1.11	1.36	1.05	1.43	1.00	1.51	0.94	1.59
34	1.18	1.30	1.13	1.36	1.07	1.43	1.01	1.51	0.95	1.59
35	1.19	1.31	1.14	1.37	1.08	1.44	1.03	1.51	0.97	1.59
36	1.21	1.32	1.15	1.38	1.10	1.44	1.04	1.51	0.99	1.59
37	1.22	1.32	1.16	1.38	1.11	1.45	1.06	1.51	1.00	1.59
38	1.23	1.33	1.18	1.39	1.12	1.45	1.07	1.52	1.02	1.58
39	1.24	1.34	1.19	1.39	1.14	1.45	1.09	1.52	1.03	1.58
40	1.25	1.34	1.20	1.40	1.15	1.46	1.10	1.52	1.05	1.58
45	1.29	1.38	1.24	1.42	1.20	1.48	1.16	1.53	1.11	1.58
50	1.32	1.40	1.28	1.45	1.24	1.49	1.20	1.54	1.16	1.59
55	1.36	1.43	1.32	1.47	1.28	1.51	1.25	1.55	1.21	1.59
60	1.38	1.45	1.35	1.48	1.32	1.52	1.28	1.56	1.25	1.60
65	1.41	1.47	1.38	1.50	1.35	1.53	1.31	1.57	1.28	1.61
70	1.43	1.49	1.40	1.52	1.37	1.55	1.34	1.58	1.31	1.61
75	1.45	1.50	1.42	1.53	1.39	1.56	1.37	1.59	1.34	1.62
80	1.47	1.52	1.44	1.54	1.42	1.57	1.39	1.60	1.36	1.62
85	1.48	1.53	1.46	1.55	1.43	1.58	1.41	1.60	1.39	1.63
90	1.50	1.54	1.47	1.56	1.45	1.59	1.43	1.61	1.41	1.64
95	1.51	1.55	1.49	1.57	1.47	1.60	1.45	1.62	1.42	1.64
100	1.52	1.56	1.50	1.58	1.48	1.60	1.46	1.63	1.44	1.65

could be 1 for the first observation, 2 for the second observation, etc.) will sometimes eliminate or reduce the autocorrelation. When these attempts to reduce or remove autocorrelation do not work, transformations on the dependent or independent variables can prove helpful; a discussion of such transformations can be found in more advanced texts on regression analysis.

FIGURE 15.19   HYPOTHESIS TEST FOR AUTOCORRELATION USING THE DURBIN-WATSON TEST

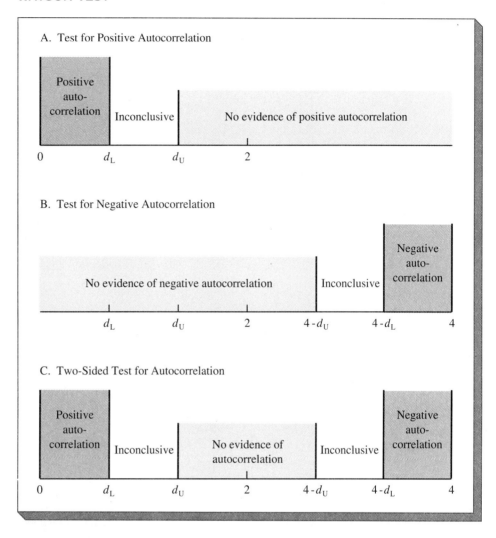

In closing this section, we note that the Durbin-Watson tables list the smallest sample size as 15. The reason for this is that the test is generally inconclusive for smaller sample sizes; in fact, many statisticians believe that the sample size should be at least 50 for the test to produce worthwhile results.

EXERCISES

APPLICATIONS

SELF TEST

21. Consider the data set presented in Exercise 19.
   a. Develop the estimated regression equation that can be used to predict the price/earnings ratio given the profit margin.
   b. Plot the residuals obtained from the model developed in part (a) as a function of the order in which the data are presented. Does there appear to be any autocorrelation present in the data? Explain.
   c. At the .05 level of significance, test for any positive autocorrelation in the data.

22. Refer to the Cravens data set presented in Table 15.5. In Section 15.4 we showed that the model involving ACCTS, ADV, POTEN, and SHARE had an adjusted coefficient of determination of 88.1%. At the .05 level of significance, use the Durbin-Watson test to determine if positive autocorrelation is present.

15.6 MULTIPLE REGRESSION APPROACH TO ANALYSIS OF VARIANCE AND EXPERIMENTAL DESIGN

In Section 14.6 we discussed the use of dummy variables in multiple regression analysis. In this section we show how the use of dummy variables in a multiple regression equation can provide another approach to solving analysis of variance and experimental design problems. We will demonstrate the multiple regression approach to analysis of variance by applying it to the National Computer Products, Inc. (NCP), problem introduced in Chapter 12.

EXAMPLE 15.8 ▶ Recall that NCP manufactures printers and fax machines at plants located in Charlotte, Houston, and San Diego. To measure how much their employees know about total quality management, a random sample of six employees was selected from each plant and given a quality-awareness exam. NCP would like to use the exam scores obtained for these 18 employees to determine whether the mean examination score is the same at each plant.

We begin the regression approach to this problem by defining two dummy variables that will be used to indicate the plant from which each sample observation was selected. Since there are three plants or populations in the NCP problem, we need two dummy variables. In general, if the factor being investigated involves k distinct levels or populations, we need to define $k - 1$ dummy variables. For the NCP problem, we define x_1 and x_2 as shown in Table 15.9.

We can use the dummy variables x_1 and x_2 to relate the score on the quality-awareness examination y to the plant at which the employee works:

$$E(y) = \text{Expected value of the score on the quality-awareness examination}$$
$$= \beta_0 + \beta_1 x_1 + \beta_2 x_2$$

Thus, if we are interested in the expected value of the examination score for an employee who works at the Charlotte plant, our procedure for assigning numerical values to the dummy variables x_1 and x_2 would result in setting $x_1 = x_2 = 0$. The multiple regression equation then reduces to

$$E(y) = \beta_0 + \beta_1(0) + \beta_2(0) = \beta_0$$

Thus, we can interpret β_0 as the expected value of the examination score for employees who work at the Charlotte plant.

TABLE 15.9　NCP PROBLEM WITH DUMMY VARIABLES

x_1	x_2	
0	0	Observation is associated with the Charlotte plant
1	0	Observation is associated with the Houston plant
0	1	Observation is associated with the San Diego plant

Next, let us consider the forms of the multiple regression equation for each of the other plants. For the Houston plant, $x_1 = 1$ and $x_2 = 0$, and

$$E(y) = \beta_0 + \beta_1(1) + \beta_2(0) = \beta_0 + \beta_1$$

For the San Diego plant, $x_1 = 0$ and $x_2 = 1$, and

$$E(y) = \beta_0 + \beta_1(0) + \beta_2(1) = \beta_0 + \beta_2$$

We see that $\beta_0 + \beta_1$ represents the expected value of the examination score for employees who work at the Houston plant, and $\beta_0 + \beta_2$ represents the expected value of the examination score for employees at the San Diego plant.

We now want to estimate the coefficients β_0, β_1, and β_2 and hence develop estimates of the expected value of the examination score for each plant. The sample data consisting of 18 observations of x_1, x_2, and y were entered into Minitab. The actual input data and the output from Minitab are shown in Table 15.10 and Figure 15.20, respectively.

Refer to Figure 15.20. We see that the estimates of β_0, β_1, and β_2 are $b_0 = 79$, $b_1 = -5$, and $b_2 = -13$. Thus, the best estimate of the expected value of the examination score for each plant is as follows.

Plant	Estimate of $E(y)$
Charlotte	$b_0 = 79$
Houston	$b_0 + b_1 = 79 - 5 = 74$
San Diego	$b_0 + b_2 = 79 - 13 = 66$

Note that the best estimate of the expected value of the examination score for each plant obtained from the regression analysis is the same as the sample means found earlier when applying the ANOVA procedure. That is, $\bar{x}_1 = 79$, $\bar{x}_2 = 74$, and $\bar{x}_3 = 66$.

TABLE 15.10   INPUT DATA FOR THE NCP PROBLEM

Observations Correspond to	x_1	x_2	y	Observations Correspond to	x_1	x_2	y
Charlotte Plant	0	0	85	San Diego Plant	0	1	59
	0	0	75		0	1	64
	0	0	82		0	1	62
	0	0	76		0	1	69
	0	0	71		0	1	75
	0	0	85		0	1	67
Houston Plant	1	0	71				
	1	0	75				
	1	0	73				
	1	0	74				
	1	0	69				
	1	0	82				

FIGURE 15.20   MULTIPLE REGRESSION OUTPUT FOR THE NCP PROBLEM

```
The regression equation is
Y = 79.0 - 5.00 X1 - 13.0 X2

Predictor          Coef        Stdev      t-ratio          p
Constant         79.000        2.186        36.14      0.000
X1               -5.000        3.091        -1.62      0.127
X2              -13.000        3.091        -4.21      0.001

s = 5.354        R-sq = 54.5%       R-sq(adj) = 48.5%

Analysis of Variance

SOURCE          DF           SS           MS          F          p
Regression       2        516.00       258.00       9.00      0.003
Error           15        430.00        28.67
Total           17        946.00
```

Now let us see how we can use the output from the multiple regression package to perform the ANOVA test on the difference in the means for the three plants. First, we observe that if there is no difference in the means, then

$$E(y) \text{ for the Houston plant} - E(y) \text{ for the Charlotte plant} = 0$$

$$E(y) \text{ for the San Diego plant} - E(y) \text{ for the Charlotte plant} = 0$$

Since β_0 equals $E(y)$ for the Charlotte plant and $\beta_0 + \beta_1$ equals $E(y)$ for the Houston plant, the first difference above is equal to $(\beta_0 + \beta_1) - \beta_0 = \beta_1$. Moreover, since $\beta_0 + \beta_2$ equals $E(y)$ for the San Diego plant, the second difference above is equal to $(\beta_0 + \beta_2) - \beta_0 = \beta_2$. Hence, we would conclude that there is no difference in the three means if $\beta_1 = 0$ and $\beta_2 = 0$. Thus, the null hypothesis for a test for difference of means can be stated as

$$H_0: \beta_1 = \beta_2 = 0$$

Recall that to test this type of null hypothesis about the significance of the regression relationship, we must compare the value of MSR/MSE to the critical value from an F distribution with numerator and denominator degrees of freedom equal to the degrees of freedom for the regression sum of squares and the error sum of squares, respectively. In the current problem, the regression sum of squares has 2 degrees of freedom and the error sum of squares has 15 degrees of freedom. Thus, the values for MSR and MSE are

$$\text{MSR} = \frac{\text{SSR}}{2} = \frac{516}{2} = 258$$

$$\text{MSE} = \frac{\text{SSE}}{15} = \frac{430}{15} = 28.67$$

Hence, the computed F value is

$$F = \frac{\text{MSR}}{\text{MSE}} = \frac{258}{28.67} = 9.00$$

At the $\alpha = .05$ level of significance, the critical value of F with 2 numerator and 15 denominator degrees of freedom is 3.68. Since the observed value of F is greater than the critical value of 3.68, we reject the null hypothesis H_0: $\beta_1 = \beta_2 = 0$. Hence, we conclude that the means for the three plants are not all equal.

EXERCISES

METHODS

SELF TEST

23. Consider a completely randomized design involving four treatments: A, B, C, and D. Write a multiple regression equation that can be used to analyze these data. Define all variables.

24. Write a multiple regression equation that can be used to analyze the data for a randomized block design involving three treatments and two blocks. Define all variables.

25. Write a multiple regression equation that can be used to analyze the data for a two-factor factorial design with two levels for factor A and three levels for factor B. Define all variables.

APPLICATIONS

SELF TEST

26. The Jacobs Chemical Company wants to estimate the mean time (minutes) required to mix a batch of material on machines produced by three different manufacturers. To limit the cost of testing, four batches of material were mixed on machines produced by each of the three manufacturers. The times needed to mix the material were recorded and are as follows.

Manufacturer 1	Manufacturer 2	Manufacturer 3
20	28	20
26	26	19
24	31	23
22	27	22

a. Write a multiple regression equation that can be used to analyze the data.
b. What are the best estimates of the coefficients in your regression equation?
c. In terms of the regression equation coefficients, what hypotheses do we have to test to see if the mean time to mix a batch of material is the same for each manufacturer?
d. For the $\alpha = .05$ level of significance, what conclusion should be drawn?

27. Four different paints are advertised as having the same drying time. To check the manufacturers' claims, five paint samples were tested for each brand of paint. The time in minutes until the paint was dry enough for a second coat to be applied was recorded for each sample. The data obtained are shown below.

Paint 1	Paint 2	Paint 3	Paint 4
128	144	133	150
137	133	143	142
135	142	137	135
124	146	136	140
141	130	131	153

a. Using $\alpha = .05$, test for any significant differences in mean drying times among the paints.

b. What is your estimate of mean drying time for paint 2? How is it obtained from the computer output?

28. An automobile dealer conducted a test to determine if the time needed to complete a minor engine tuneup depends on whether a computerized engine analyzer or an electronic analyzer is used. Because tuneup time varies among compact, intermediate, and full-sized cars, the three types of cars were used as blocks in the experiment. The data (time in minutes) obtained are shown below.

		Car		
		Compact	Intermediate	Full-size
Analyzer	Computerized	50	55	63
	Electronic	42	44	46

Using $\alpha = .05$, test for any significant differences.

29. A mail-order catalog firm designed a factorial experiment to test the effect of the size of a magazine advertisement and the advertisement design on the number of catalog requests received (1000s). Three advertising designs and two sizes of advertisements were considered. The data are shown below. Test for any significant effects due to type of design, size of advertisement, or interaction. Use $\alpha = .05$.

		Size of Advertisement	
		Small	Large
Design	A	8 12	12 8
	B	22 14	26 30
	C	10 18	18 14

SUMMARY

In this chapter we discussed several concepts used by model builders to identify the best estimated regression equation. First, we introduced the concept of a general linear model to show how the methods discussed in Chapters 13 and 14 could be extended to handle curvilinear relationships and interaction effects. Then, we discussed how transformations involving the dependent variable could be used in model building to account for problems such as nonconstant variance in the error terms.

In applications of regression analysis to real problems, there are usually a large number of potential independent variables to consider. We presented a general approach, based on an F statistic, for adding or deleting variables from a regression model. We then introduced a larger problem involving 25 observations and eight independent variables. We saw that one issue encountered when solving larger problems is finding the best subset of the possible independent variables. To help in this regard, we discussed four variable-selection procedures: stepwise regression, forward selection, backward elimination, and best-subsets regression.

In Section 15.5, we extended the applications of residual analysis to show the Durbin-Watson test for autocorrelation. The chapter concluded with a discussion of how multiple regression models could be developed to provide another approach for solving analysis of variance and experimental design problems.

GLOSSARY

General linear model A model of the form $y = \beta_0 + \beta_1 z_1 + \beta_2 z_2 + \cdots + \beta_p z_p + \epsilon$, where each of the independent variables $z_j, j = 1, 2, \ldots, p,$ is a function of $x_1, x_2, \ldots, x_k,$ the variables for which data have been collected.

Interaction The joint effect of two variables acting together.

Variable-selection procedures Computer-based methods for selecting a subset of the potential independent variables for a regression model.

Autocorrelation Correlation in the errors that arises when the error terms at successive points in time are related.

Serial correlation Same as autocorrelation.

Durbin-Watson test A test to determine whether first-order autocorrelation is present.

KEY FORMULAS

General Linear Model

$$y = \beta_0 + \beta_1 z_1 + \beta_2 z_2 + \cdots + \beta_p z_p + \epsilon \tag{15.1}$$

General F Test for Adding or Deleting $p - q$ Variables

$$F = \dfrac{\dfrac{\text{SSE}(x_1, x_2, \ldots, x_q) - \text{SSE}(x_1, x_2, \ldots, x_q, x_{q+1}, \ldots, x_p)}{p - q}}{\dfrac{\text{SSE}(x_1, x_2, \ldots, x_q, x_{q+1}, \ldots, x_p)}{n - p - 1}} \tag{15.13}$$

Autocorrelated Error Terms

$$\epsilon_t = \rho \epsilon_{t-1} + z_t \tag{15.16}$$

Durbin-Watson Statistic

$$d = \frac{\displaystyle\sum_{t=2}^{n} (e_t - e_{t-1})^2}{\displaystyle\sum_{t=1}^{n} e_t^2} \tag{15.17}$$

REVIEW QUIZ

TRUE/FALSE

1. Model building is solely concerned with identifying which independent variables should be included in the model.
2. Since the "truly correct model" can never be found for real data, the model builder's goal is to find the best model from a set of acceptable models.

3. Models in which the independent variables have exponents other than 1 are referred to as nonlinear models.
4. All nonlinear models can be transformed into equivalent linear models using the logarithmic transformation.
5. Interaction (and the method for dealing with it) does not apply when working with qualitative variables.
6. The t test for determining the significance of adding only one additional independent variable to an existing model is equivalent to the F test.
7. With the stepwise regression procedure, a variable is added at each step, but variables are never deleted.
8. At each step of the stepwise regression procedure, the first consideration is to see whether any variable can be added.
9. In the stepwise procedure, FENTER cannot be set greater than FREMOVE.
10. When autocorrelation is present, at least one of the assumptions of the regression model is violated.
11. The Durbin-Watson test is generally inconclusive for small sample sizes.

MULTIPLE CHOICE

12. The model $y = \beta_0 + \beta_1 x_1 + \beta_2 x_1^2 + \epsilon$ is called a
     a. simple nonlinear model with one predictor variable
     b. simple first-order model with one predictor variable
     c. second-order model with one predictor variable
     d. second-order model with two predictor variables
13. Interaction terms are added to a regression model in order to
     a. study the joint effect of quantitative and qualitative variables
     b. account for curvilinear effects in the data
     c. transform a nonlinear model into an equivalent linear model
     d. account for the joint effect of two independent variables
14. The F test for determining whether to add two variables to an existing model is based on a determination of
     a. the increase in SSE resulting from adding the two variables
     b. the decrease in SSR resulting from adding the two variables
     c. the change in SST resulting from adding the two variables
     d. the decrease in SSE resulting from adding the two variables
     e. none of the above
15. The forward-selection procedure
     a. starts out with all the independent variables
     b. adds variables, one at a time, until all the independent variables have entered the model
     c. allows a variable that entered at a previous step to be deleted at a later step
     d. none of the above
16. In the stepwise procedure, which of the following are acceptable values for FENTER AND FREMOVE?
     a. FENTER = 2 and FREMOVE = 3
     b. FENTER = 3 AND FREMOVE = 2
     c. none of the above combinations are acceptable

SUPPLEMENTARY EXERCISES

30. Refer to the Cravens data set presented in Table 15.5.
     a. Develop a scatter diagram showing SALES as a function of TIME.
     b. Does a linear relationship between SALES and TIME appear to be appropriate? Explain.
     c. Develop a model that can be used to predict SALES using just TIME or some appropriate function of TIME.

31. A study reported in the *Journal of Accounting Research* (Vol. 2, No. 2 Autumn 1987) investigated the relationship between audit delay (AUDELAY), the length of time from a company's fiscal year-end to the date of the auditor's report, and variables that describe the client and the auditor. Some of the independent variables that were included in this study were

INDUS	A dummy variable that was coded as 1 if the firm was an industrial company or 0 if the firm was a bank, savings and loan, or insurance company.
PUBLIC	A dummy variable coded as 1 if the company was traded on an organized exchange or over the counter; otherwise coded 0.
ICQUAL	A measure of overall quality of internal controls, as judged by the auditor, using a five-point scale ranging from "virtually none" (1) to "excellent" (5).
INTFIN	A measure ranging from 1 to 4, as judged by the auditor, where 1 indicates "all work performed subsequent to year-end" and 4 indicates "most work performed prior to year-end."

Suppose that in a similar study a sample of 40 companies provided the following data.

AUDELAY	INDUS	PUBLIC	ICQUAL	INTFIN
62	0	0	3	1
45	0	1	3	3
54	0	0	2	2
71	0	1	1	2
91	0	0	1	1
62	0	0	4	4
61	0	0	3	2
69	0	1	5	2
80	0	0	1	1
52	0	0	5	3
47	0	0	3	2
65	0	1	2	3
60	0	0	1	3
81	1	0	1	2
73	1	0	2	2
89	1	0	2	1
71	1	0	5	4
76	1	0	2	2
68	1	0	1	2
68	1	0	5	2
86	1	0	2	2
76	1	1	3	1
67	1	0	2	3
57	1	0	4	2
55	1	1	3	2
54	1	0	5	2
69	1	0	3	3
82	1	0	5	1
94	1	0	1	1
74	1	1	5	2
75	1	1	4	3
69	1	0	2	2
71	1	0	4	4

—Table continues on next page

AUDIT

AUDELAY	INDUS	PUBLIC	ICQUAL	INTFIN
79	1	0	5	2
80	1	0	1	4
91	1	0	4	1
92	1	0	1	4
46	1	1	4	3
72	1	0	5	2
85	1	0	5	1

   a. Develop the estimated regression equation using all of the independent variables.

   b. Did the model developed in part (a) provide a good fit? Explain.

   c. Develop a scatter diagram that shows AUDELAY as a function of INTFIN. What does this scatter diagram indicate about the relationship between AUDELAY and INTFIN?

   d. Based on your observations regarding the relationship between AUDELAY and INTFIN, develop an alternative model to the one developed in part (a) to explain as much of the variability in AUDELAY as possible.

32. The following data set, reported in *Louis Rukeyser's Business Almanac* (1988, Simon & Schuster, p. 47), shows the percentage of management jobs held by women in various companies and the percentage of women in each company.

JOBS

Industry/Company	Management Jobs Held by Women (%)	Women Employees (%)
Industrial		
Du Pont	7	22
Exxon	8	27
General Motors	6	19
Goodyear Tire and Rubber	25	39
Technology		
AT&T	32	48
General Electric	6	26
IBM	16	28
Xerox	23	38
Consumer products		
Johnson & Johnson	18	47
PepsiCo	28	46
Phillip Morris (excluding General Foods)	14	31
Proctor & Gamble	17	28
Retailing and trade		
Federated Department Stores	61	72
Kroger	16	47
Marriott	32	51
McDonald's	46	57
Sears, Roebuck	36	55
Media		
ABC (excluding Capital Cities)	36	43
Time	46	54
Times Mirror	27	37

—Table continues

Industry/Company	Management Jobs Held by Women (%)	Women Employees (%)
Financial Services		
American Express	37	57
BankAmerica	64	72
Chemical Bank	34	57
Prudential Life Insurance	32	53
Wells Fargo Bank	58	71

In addition to the percentage of women employed in the company, create additional independent variables by using dummy variables to account for the type of industry.

a. Develop the best one-variable model that can be used to predict the percentage of management jobs held by women.

b. Use the stepwise procedure to develop the best model.

c. Use the backward-elimination procedure to develop the best model.

33. Refer to the data in Exercise 31. Consider a model in which only INDUS is used to predict AUDELAY. At the $\alpha = .01$ level of significance, test for any positive autocorrelation in the data.

34. Refer to the data in Exercise 31.

a. Develop an estimated regression equation that can be used to predict AUDELAY using INDUS and ICQUAL.

b. Plot the residuals obtained from the model developed in part (a) as a function of the order in which the data are presented. Does there appear to be any autocorrelation present in the data? Explain.

c. At the .05 level of significance, test for any positive autocorrelation in the data.

35. Refer to the data in Exercise 32.

a. Develop an estimated regression equation that can be used to predict the percentage of management jobs held by women given the percentage of women employees in the company.

b. Plot the residuals obtained from the model developed in part (a) as a function of the order in which the data are presented. Does there appear to be any autocorrelation present in the data? Explain.

c. At the .05 level of significance, test for any positive autocorrelation in the data.

36. A study was conducted to investigate the browsing activity by shoppers (*Journal of the Academy of Marketing Science,* Winter 1989). Shoppers were classified as nonbrowsers, light browsers, and heavy browsers. For each shopper in the study, a measure was obtained to determine how comfortable the shopper was in the store. Higher scores indicated greater comfort. Assume that the following data are from this study. Use a .05 level of significance to test for differences in comfort levels among the three types of browsers.

	Browser			Browser	
Nonbrowser	Light	Heavy	**Nonbrowser**	Light	Heavy
4	5	5	3	7	4
5	6	7	4	4	6
6	5	5	5	6	5
3	4	7	4	5	7

COMPUTER CASE: UNEMPLOYMENT STUDY

Layoffs and unemployment have affected a substantial number of workers in recent years. A study reported in *Industrial and Labor Relations Review* (April 1988) collected data on variables that may be related to the number of weeks a manufacturing worker has been jobless. The dependent variable in the study (WEEKS) was defined as the number of weeks a worker has been jobless due to a layoff. The independent variables in the study were as follows:

AGE   The age of the worker.
EDUC   The number of years of education.
MARRIED   A dummy variable; 1 if married, 0 otherwise.
HEAD   A dummy variable; 1 if the head-of-household, 0 otherwise.
TENURE   The number of years on the old job.
MGT   A dummy variable; 1 if management occupation, 0 otherwise.
SALES   A dummy variable; 1 if sales occupation, 0 otherwise.

 LAYOFFS

Assume the following data were collected for 50 displaced workers. These data are available on the data disk in the file named LAYOFFS.

WEEKS	AGE	EDUC	MARRIED	HEAD	TENURE	MGT	SALES
37	30	14	1	1	1	0	0
62	27	14	1	0	6	0	0
49	32	10	0	1	11	0	0
73	44	11	1	0	2	0	0
8	21	14	1	1	2	0	0
15	26	13	1	0	7	1	0
52	26	15	1	0	6	0	0
72	33	13	0	1	6	0	0
11	27	12	1	1	8	0	0
13	33	12	0	1	2	0	0
39	20	11	1	0	1	0	0
59	35	7	1	1	6	0	0
39	36	17	0	1	9	1	0
44	26	12	1	1	8	0	0
56	36	15	0	1	8	0	0
31	38	16	1	1	11	0	1
62	34	13	0	1	13	0	0
25	27	19	1	0	8	0	0
72	44	13	1	0	22	0	0
65	45	15	1	1	6	0	0
44	28	17	0	1	3	0	1
49	25	10	1	1	1	0	0
80	31	15	1	0	12	0	0
7	23	15	1	0	2	0	0
14	24	13	1	1	7	0	0
94	62	13	0	1	8	0	0
48	31	16	1	0	11	0	0
82	48	18	0	1	30	0	0

—Table continues

WEEKS	AGE	EDUC	MARRIED	HEAD	TENURE	MGT	SALES
50	35	18	1	1	5	0	0
37	33	14	0	1	6	0	1
62	46	15	0	1	6	0	0
37	35	8	0	1	6	0	0
40	32	9	1	1	13	0	0
16	40	17	1	0	8	1	0
34	23	12	1	1	1	0	0
4	36	16	0	1	8	0	1
55	33	12	1	0	10	0	1
39	32	16	0	1	11	0	0
80	62	15	1	0	16	0	1
19	29	14	1	1	12	0	0
98	45	12	1	0	17	0	0
30	38	15	0	1	6	0	1
22	40	8	1	1	16	0	1
57	42	13	1	0	2	1	0
64	45	16	1	1	22	0	0
22	39	11	1	1	4	0	0
27	27	15	1	0	10	0	1
20	42	14	1	1	6	1	0
30	31	10	1	1	8	0	0
23	33	13	1	1	8	0	0

REPORT

Use the methods presented in this and previous chapters to analyze this data set. Prepare a report which summarizes your analysis, including key statistical results, conclusions, and recommendations. Include any technical material that you feel is appropriate (computer output, residual plots, etc.) in an appendix.

CHAPTER 16

TESTS OF GOODNESS OF FIT AND INDEPENDENCE

WHAT YOU WILL LEARN IN THIS CHAPTER

- What a goodness of fit test is
- What a contingency table is
- How to use the chi-square distribution to conduct tests of goodness of fit and independence

CONTENTS

Do Soccer Injuries Differ Between Game and Training Sessions?

A study of injuries occurring in three Swedish semiprofessional soccer teams showed that 75% of the players sustained one or more injuries each year. The study showed an average of 13 injuries per 1000 hours of game conditions and 3 injuries per 1000 hours during training. Each injury was classified as minor, moderate, or major. Minor injuries were those in which the player was absent from training and/or games for less than 1 week, moderate injuries were those in which the player was absent from 1 week to 1 month, and major injuries resulted in absences of more than 1 month. To better understand the relationship between the severity of an injury and when it occurred, training versus game conditions, the researchers collected the data shown below.

The risk of injury is high for today's soccer players.

	Minor	Moderate	Major	Totals
Game	15	13	12	40
Training	6	9	8	23
Totals	21	22	20	63

These data show that 40/63 or 63.5% of the injuries occurred during game conditions and 36.5% of the injuries occurred during training. Moreover, for injuries sustained during a game, 15/40 or 38% of the injuries were classified as minor, 13/40 or 33% were classified as moderate, and 12/40 or 30% were classified as major. Similarly, 26% of the injuries sustained during training were minor, 39% were moderate, and 35% were major. Based on these data, can we conclude that the severity of an injury is independent of the playing conditions under which the injury occurred? Using a statistical test for independence, it was concluded that the severity of the injury is independent of the playing conditions.

The results from this study also showed that the position of the player within the team did not influence the injury rate, and that there was no significant difference between ankle sprain rates in players using ankle tape versus players not using ankle tape. The fact that 20/63 or 32% of the injuries sustained were major indicates that soccer at this level is a relatively dangerous sport.

This *Statistics in Practice* is based on "Does a major knee injury definitely sideline an elite soccer player?" by Björn Engström, Magnus Forssblad, Christer Johansson, and Hans Törnkvist, *The American Journal of Sports Medicine* (January 1990)

In Chapter 11 we introduced the chi-square distribution and illustrated how it could be used in interval estimation and hypothesis testing concerning population variances. In this chapter we introduce two more hypothesis-testing procedures that are based on the use of the chi-square distribution. As with other hypothesis-testing procedures, these tests compare observed sample results with those that are expected when the null hypothesis is true. The decision of whether to reject the null hypothesis is based on how "close" the sample or observed results are to the expected results.

In the following section we introduce a *goodness of fit test* involving a multinomial population. Later we discuss a test for independence using contingency tables.

16.1 GOODNESS OF FIT TEST—A MULTINOMIAL POPULATION

In this section we consider the situation in which each element of a population is assigned to one and only one of several classes or categories. Such a population is described by a *multinomial probability distribution.* Example 16.1 describes a multinomial distribution with three classes, or categories.

EXAMPLE 16.1 ▶ Patients that arrive for treatment at the emergency room of a large metropolitan hospital are assigned to one of the following three categories based on the seriousness of their condition.

Category 1: Patient condition is stable; immediate treatment by a physician is not required.

Category 2: Patient condition is serious; immediate treatment is not required, but patient should be monitored for vital signs until a physician is available.

Category 3: Patient condition is critical; the patient's life will be endangered without immediate treatment.

In this example, the population of interest is a multinomial population since the condition of each patient is classified into one and only one of three classifications or categories: stable, serious, and critical. ◀

In general, a multinomial population involves k categories. A goodness of fit test can be used to determine whether a hypothesized probability distribution provides a good description of a particular population of interest. To perform such a test, we must first formulate a null hypothesis concerning the particular multinomial probability distribution.

EXAMPLE 16.1 ▶
(CONTINUED) Over the past year, the hospital's records show that 50% of the patients who arrived for treatment were classified as stable, 30% were classified as serious, and 20% were classified as critical. The hospital's reputation for providing superior emergency room treatment, however, has resulted in an increased volume for the emergency room. The director is concerned that the percentage of patients classified as having stable, serious, or critical conditions may have also changed. Let us define the following notation:

$$p_1 = \text{fraction of patients classified as stable}$$

$$p_2 = \text{fraction of patients classified as serious}$$

$$p_3 = \text{fraction of patients classified as critical}$$

Based on the assumption that the increase in volume for the emergency room has not altered the distribution of patients among the categories, the null and alternative hypotheses would be stated as follows:

$$H_0\text{: } p_1 = .50, p_2 = .30, \text{ and } p_3 = .20$$

$$H_a\text{: The population proportions are not}$$
$$p_1 = .50, p_2 = .30, \text{ and } p_3 = .20 \quad ◀$$

Once the hypotheses have been formulated, we must obtain a simple random sample of n items from the population in order to conduct the test. Using the sample of n items, we record the observed frequencies for each of the k classes or categories. Then, given the usual hypothesis-testing assumption that the null hypothesis is true, we determine the expected frequency for each category. The expected frequency for each category is found by multiplying the sample size n by the proportion assumed to be in that category under the null hypothesis.

EXAMPLE 16.1
(CONTINUED)

▶ Suppose the hospital has selected a sample of 200 patients who have been treated since the volume increased in the emergency room. The observed frequencies for this group are summarized as follows.

Stable Condition	Serious Condition	Critical Condition
98	48	54

The next step is to compute the expected frequencies for the 200 patients under the null hypothesis assumption that $p_1 = .50$, $p_2 = .30$, and $p_3 = .20$. Doing so provides the following expected frequencies.

Stable Condition	Serious Condition	Critical Condition
200(.50) = 100	200(.30) = 60	200(.20) = 40

Note that the expected frequency for each category is found by multiplying the sample size of 200 by the hypothesized proportion for the category. ◀

The goodness of fit test now focuses on the differences between the observed frequencies and the expected frequencies. Large differences between observed and expected frequencies cast doubt on the assumption that the hypothesized proportions are correct. Whether the differences between the observed and expected frequencies are "large" or "small" is a question answered with the aid of the following chi-square test statistic.

Test Statistic for Goodness of Fit

$$\chi^2 = \sum_{i=1}^{k} \frac{(f_i - e_i)^2}{e_i} \tag{16.1}$$

where

f_i = observed frequency for category i
e_i = expected frequency for category i based on the assumption that H_0 is true
k = the number of categories

Note: The test statistic has a chi-square distribution with $k - 1$ degrees of freedom, provided the expected frequencies are 5 *or more* for all categories.

From the numerator of equation (16.1), we see that larger differences between observed and expected frequencies result in larger values for χ^2. Therefore, large values of χ^2 lead to the rejection of H_0. How large χ^2 must be before rejecting H_0 depends on the level of significance chosen for the hypothesis test.

EXAMPLE 16.1 ▶
(CONTINUED)

With expected frequencies $e_i \geq 5$ for all three categories, the sample-size requirement is satisfied, and we can proceed with the computation of the chi-square (χ^2) value in equation (16.1), as follows:

$$\chi^2 = \frac{(98 - 100)^2}{100} + \frac{(48 - 60)^2}{60} + \frac{(54 - 40)^2}{40}$$

$$= \frac{4}{100} + \frac{144}{60} + \frac{196}{40}$$

$$= .04 + 2.40 + 4.90 = 7.34$$

We will reject the null hypothesis only if the differences between observed and expected cell frequencies are *large*. Thus, the larger the value of χ^2, the more likely it is we will reject the null hypothesis. Using $\alpha = .05$, we will place a rejection area of .05 in the upper tail of the chi-square distribution. Checking the chi-square distribution table (Table 3 of Appendix B), we find that with $k - 1 = 3 - 1 = 2$ degrees of freedom $\chi^2_{.05} = 5.99147$. Thus, as with similar one-tailed tests, we will reject H_0 if the computed chi-square value exceeds the critical value of 5.99147.

Since $\chi^2 = 7.34$ is larger than the critical value of 5.99147, we reject the null hypothesis. In rejecting H_0, we conclude that the increase in volume for the emergency room has altered the percentages of patients whose conditions are stable, serious, or critical. While the goodness of fit test itself permits no further conclusions, we can informally compare the observed and expected frequencies to obtain an idea of where the significant differences are. For instance, considering the critical-condition category, we find that the observed frequency of 54 is larger than the expected frequency of 40. Since the expected frequency was based on the historical percentage observed, the larger observed frequency suggests that associated with the increase in volume for the emergency room has been an increase in the percentage of patients whose conditions are classified as critical. Similarly, there has been a decrease in the percentage of patients whose conditions are classified as serious. ◀

As illustrated in Example 16.1, the goodness of fit test uses the chi-square distribution to determine whether a hypothesized probability distribution for a population provides a good fit. Acceptance or rejection of the hypothesized probability distribution depends on differences between the observed frequencies in a sample and the expected frequencies based on the assumed probability distribution. Let us outline the general steps that can be used to conduct a goodness of fit test for any hypothesized probability distribution:

1. Formulate a null hypothesis indicating a hypothesized distribution for k classes or categories of a population.
2. Select a simple random sample of n items, and record the observed frequencies for each of the k classes or categories.
3. Based on the assumption that the null hypothesis is true, determine the expected frequencies for each category.
4. Use the observed and expected frequencies and equation (16.1) to compute a value of χ^2 for the test.

5. Rejection rule:

$$\text{Reject } H_0 \text{ if } \chi^2 > \chi^2_\alpha$$

where α is the level of significance for the test.

NOTES AND COMMENTS

1. The multinomial probability distribution is an extension of the binomial distribution to the case of three or more categories of outcomes.
2. The goodness of fit test is not restricted to the multinomial probability distribution. For instance, it can be applied to the Poisson and Normal distributions.

EXERCISES

METHODS

SELF TEST

1. Conduct a test of the following hypotheses using the χ^2 goodness of fit test.

$$H_0:\ p_A = .40, p_B = .40, \text{ and } p_C = .20$$

$$H_a:\ \text{The population proportions are not}$$
$$p_A = .40,\, p_B = .40, \text{ and } p_C = .20$$

A sample of size 200 yielded 60 in category A, 120 in category B, and 20 in category C. Use $\alpha = .01$ and test to see if the proportions are as stated in H_0.

2. Suppose we have a multinomial population with four categories: A, B, C, and D. The null hypothesis is that the proportion of items in each category is the same.

$$H_0:\ p_A = p_B = p_C = p_D = .25$$

A sample of size 300 yielded the following numbers in each category:

A: 85, B: 95, C: 50, D: 70.

Use $\alpha = .05$ and the χ^2 test to determine if H_0 should be rejected.

APPLICATIONS

SELF TEST

3. During the first 13 weeks of the television season, the Saturday evening 8:00 P.M. to 9:00 P.M. audience proportions were recorded as ABC, 29%; CBS, 28%; NBC, 25%; and independents, 18%. A sample of 300 homes 2 weeks after a Saturday night schedule revision showed the following viewing audience data: ABC, 95 homes; CBS, 70 homes; NBC, 89 homes; and independents, 46 homes. Test with $\alpha = .05$ to determine if the viewing audience proportions have changed.

4. Where do America's millionaires live? Assume a sample of 300 millionaires showed 63 living in the Northeast, 62 living in the Midwest, 100 living in the South, and 75 living in the West. The sample results are based on data from *Louis Rukeyser's Business Almanac* (1988, Simon and Schuster). Conduct a hypothesis test for $H_0:\ p_1 = p_2 = p_3 = p_4 = .25$ with $\alpha = .05$. What is your conclusion about where America's millionaires live?

5. The four major competitors in the computer-workstation market are Sun Microsystems (29%), Hewlett-Packard (18.8%), IBM (16%), and Digital Equipment (11.6%) with other manufacturers holding 24.6% of the market (*USA Today*, February 13, 1992). Assume that 1 year later a survey of 400 computer workstations found 106 Sun, 72 Hewlett-Packard, 80 IBM, 48 Digital, and 94 other systems in use. Do the data suggest any changes have occurred during the 1-year period? Test at a .05 level of significance.

6. A new container design has been adopted by a manufacturer. Color preferences indicated in a sample of 150 individuals are as follows.

Red	Blue	Green
40	64	46

Test using $\alpha = .10$ to see if the color preferences are different. (*Hint:* Formulate the null hypothesis as H_0: $p_1 = p_2 = p_3 = 1/3$.)

7. Consumer panel preferences for three proposed store displays are as follows.

Display A	Display B	Display C
43	53	39

Use $\alpha = .05$ and test to see if there is a difference in preference among the three display designs.

8. Grade-distribution guidelines for a statistics course at a major university are as follows: 10% A, 30% B, 40% C, 15% D, and 5% F. A sample of 120 statistics grades at the end of a semester showed 18 A's, 30 B's, 40 C's, 22 D's, and 10 F's. Use $\alpha = .05$, and test to see if the actual grades deviate significantly from the grade-distribution guidelines.

9. An October 1988 poll sponsored by the *Cincinnati Post* used a random sample of registered voters throughout the state of Ohio to determine how Ohioans rate their local schools. The following rating categories and results were reported.

School Rating Category	Frequency
Excellent	155
Good	234
Fair	129
Poor	98

Assume Ohio school administrators had expected rating percentages of 20% excellent, 40% good, 25% fair, and 15% poor for the population of Ohio registered voters. Use the chi-square test and the actual survey data to determine whether the survey is consistent with the administrators expectations.

10. At Ontario University, entering freshmen have historically selected the following colleges.

College	Percentage
Business	15
Education	20
Engineering	30
Liberal Arts	25
Science	10

Data obtained for the most recent class show that 73 students selected business, 105 selected education, 150 selected engineering, 124 chose liberal arts, and 47 selected science. Use $\alpha = .10$, and test whether or not the historical percentages have changed.

16.2 TEST OF INDEPENDENCE—CONTINGENCY TABLES

Another important application of the chi-square distribution involves using sample data to test for the independence of two variables. To illustrate the test of independence, we consider the study conducted by Alber's Brewery.

EXAMPLE 16.2 ▶ Alber's Brewery manufactures and distributes three types of beers: a low-calorie light beer, a regular beer, and a dark beer. In an analysis of the market segments for the three beers, the firm's market research group has raised the question of whether preferences for the three beers differ between male and female beer drinkers. If beer preference is independent of the sex of the beer drinker, one advertising campaign will be initiated for all of Alber's beers. However, if beer preference depends on the sex of the beer drinker, the firm will tailor its promotions toward different target markets.

A test of independence addresses the question of whether the beer preference (light, regular, or dark) is independent of the sex of the beer drinker (male, female). The hypotheses for this test of independence are as follows.

H_0: Beer preference is independent of the sex of the beer drinker

H_a: Beer preference is not independent of the sex of the beer drinker (i.e., males and females differ in their preferences)

Table 16.1 can be used to describe the situation being studied. By identifying the population as all male and female beer drinkers, a sample can be selected and each individual asked to state his or her preference for the three Alber's beers. Every individual in the sample will be placed in one of the six cells in the table. For example, an individual may be a male preferring regular beer (cell 1, 2), a female preferring light beer (cell 2, 1), a female preferring dark beer (cell 2, 3), and so on. Since we have listed all possible combinations of beer preference and sex—or, in other words, listed all possible contingencies—Table 16.1 is called a *contingency table*. The test of independence makes use of the contingency table format and for this reason is sometimes referred to as a *contingency table test*.

Suppose that a simple random sample of 150 beer drinkers has been selected. After taste-testing the three beers, the individuals in the sample are asked to state their preference, or first choice. The contingency table in Table 16.2 summarizes the responses

TABLE 16.1   CONTINGENCY TABLE—BEER PREFERENCE AND SEX OF BEER DRINKERS

		Beer Preference		
		Light	*Regular*	*Dark*
Sex	*Male*	(Cell 1, 1)	(Cell 1, 2)	(Cell 1, 3)
	Female	(Cell 2, 1)	(Cell 2, 2)	(Cell 2, 3)

TABLE 16.2   **SAMPLE RESULTS OF BEER PREFERENCES FOR MALE AND FEMALE BEER DRINKERS (OBSERVED FREQUENCIES)**

		Beer Preference			
		Light	Regular	Dark	Totals
Sex	Male	20	40	20	80
	Female	30	30	10	70
	Totals	50	70	30	150

TABLE 16.3   **EXPECTED FREQUENCIES IF BEER PREFERENCE IS INDEPENDENT OF THE SEX OF THE BEER DRINKER**

		Beer Preference			
		Light	Regular	Dark	Totals
Sex	Male	26.67	37.33	16.00	80
	Female	23.33	32.67	14.00	70
	Totals	50	70	30	150

to the study. As we see in the contingency table, the data for the test of independence are collected in terms of counts, or frequencies, for each cell or category. Thus, of the 150 individuals in the sample, 20 were men who favored light beer, 40 were men who favored regular beer, 20 were men who favored dark beer, and so on.

The data in Table 16.2 contain the sample, or observed, frequencies for each of six classes or categories. If we can determine the expected frequencies under the assumption of independence between beer preference and sex of the beer drinker, we can use the chi-square distribution, just as we did in the previous section, to determine whether or not there is a significant difference between observed and expected frequencies.

Expected frequencies for the cells of the contingency table are based on the following rationale: First, we assume that the null hypothesis of independence between beer preference and sex of the beer drinker is true. Then we note that the sample of 150 beer drinkers showed a total of 50 preferring light beer, 70 preferring regular beer, and 30 preferring dark beer. In terms of fractions, we conclude that $50/150 = 1/3$ of the beer drinkers prefer light beer, $70/150 = 7/15$ prefer regular beer, and $30/150 = 1/5$ prefer dark beer. If the *independence* assumption is valid, we argue that these same fractions must be applicable to both male and female beer drinkers. Thus, under the assumption of independence, we would expect the 80 male beer drinkers to show that $(1/3)80 = 26.67$ prefer light beer, $(7/15)80 = 37.33$ prefer regular beer, and $(1/5)80 = 16$ prefer dark beer. Application of these same fractions to the 70 female beer drinkers provides the expected frequencies as shown in Table 16.3.

Let e_{ij} denote the expected frequency for the contingency table category in row i and column j. With this notation, let us reconsider the expected frequency calculation for males (row $i = 1$) who prefer regular beer (column $j = 2$)—that is, expected frequency e_{12}. Following our previous argument for the computation of expected frequencies, we showed that

$$e_{12} = \left(\frac{7}{15}\right)80 = 37.33$$

Writing this slightly differently, we find

$$e_{12} = \left(\frac{7}{15}\right)80 = \left(\frac{70}{150}\right)80 = \frac{(80)(70)}{150} = 37.33$$

Note that 80 in the above expression is the total number of males (row 1 total), 70 is the total number preferring regular beer (column 2 total), and 150 is the total sample size. Thus, we see that

$$e_{12} = \frac{(\text{Row 1 Total})(\text{Column 2 Total})}{\text{Sample Size}} \qquad \blacktriangleleft$$

Generalization of the last expression in Example 16.2 shows that the following formula provides the expected frequencies for a contingency table in the test for independence.

Expected Frequencies for Contingency Tables Under the Assumption of Independence

$$e_{ij} = \frac{(\text{Row } i \text{ Total})(\text{Column } j \text{ Total})}{\text{Sample Size}} \qquad \textbf{(16.2)}$$

The test procedure for comparing the observed frequencies of Table 16.2 with the expected frequencies of Table 16.3 is similar to the goodness of fit calculations made in the previous section. Specifically, the value of χ^2, based on the differences between the observed and expected frequencies, is computed as follows.

Contingency Table Test Statistic

$$\chi^2 = \sum_i \sum_j \frac{(f_{ij} - e_{ij})^2}{e_{ij}} \qquad \textbf{(16.3)}$$

where

f_{ij} = observed frequency for contingency table category in row i and column j

e_{ij} = expected frequency for contingency table category in row i and column j based on the assumption of independence

Note: With n rows and m columns in the contingency table, the test statistic has a chi-square distribution with $(n - 1)(m - 1)$ degrees of freedom provided the expected frequencies are 5 *or more* for all categories.

The double summation in equation (16.3) indicates that the calculation must be summed for each cell in the contingency table.

Before applying equation (16.3), we must check to see that the expected frequencies in each cell are at least 5. This is the same check that we used in the previous section to determine whether the sample size was large enough for the chi-square distribution assumption to be made.

EXAMPLE 16.2 ▶
(CONTINUED)

Since all expected frequencies in Table 16.3 are at least 5, we can conclude that the sample size is adequate. The resulting value of χ^2 is found as follows:

$$\chi^2 = \frac{(20 - 26.67)^2}{26.67} + \frac{(40 - 37.33)^2}{37.33} + \cdots + \frac{(10 - 14.00)^2}{14.00}$$

$$= 1.67 + .19 + \cdots + 1.14 = 6.13$$

The number of degrees of freedom for the appropriate chi-square distribution is computed by multiplying the *number of rows minus* 1 times the *number of columns minus* 1. With two rows and three columns, we have $(2 - 1)(3 - 1) = (1)(2) = 2$ degrees of freedom for the test of independence of beer preference and the sex of the beer drinker. Using $\alpha = .05$ for the level of significance of the test, Table 3 of Appendix B shows an upper-tail value of $\chi^2_{.05} = 5.99147$. Again, we are using the upper-tail value because we will reject the null hypothesis only if the differences between observed and expected frequencies provide a large value of χ^2. Since $\chi^2 = 6.13$ is greater than the critical value of $\chi^2_{.05} = 5.99147$, we reject the null hypothesis of independence and conclude that the preference for the beers is not independent of the sex of the beer drinkers.

Although the test for independence allows only the above conclusion, we can again informally compare the observed and expected frequencies to obtain an idea of how the dependence between the beer preference and sex of the beer drinker comes about. (Refer to Tables 16.2 and 16.3.) We see that male beer drinkers have higher observed than expected frequencies for both regular and dark beers, while female beer drinkers have a higher observed than expected frequency for only the light beer. These observations give us an insight into the differing preferences between male and female beer drinkers. This information can be used by the company in targeting its promotions for the different beers. ◀

EXERCISES

METHODS

SELF TEST ▷

11. Shown in the table below is a 2 × 3 contingency table with observed frequencies for a sample of 200. Test for independence of the row and column factors using the χ^2 test with $\alpha = .025$.

	Column Factor		
Row Factor	A	B	C
P	20	44	50
Q	30	26	30

12. Shown below is a 3×3 contingency table with observed frequencies from a sample of 240. Test for independence of the row and column factors using the χ^2 test with $\alpha = .05$.

	Column Factor		
Row Factor	A	B	C
P	20	30	20
Q	30	60	25
R	10	15	30

APPLICATIONS

SELF TEST

13. The 1992 NCAA basketball championship final four teams were Duke, Michigan, Indiana, and Cincinnati. Data below show the season 3-point shooting records (*NCAA Final Four Program,* April 1992) for the four teams. At the .05 level of significance, is there a difference in the 3-point shooting abilities among the four teams? What is your conclusion?

3-Point Shooting	**Duke**	**Michigan**	**Indiana**	**Cincinnati**
Made	160	113	154	202
Missed	214	228	215	331

14. The number of units sold by three salespersons over a 3-month period are shown in the table below. Use $\alpha = .05$, and test for the independence of salesperson and type of product. What is your conclusion?

	Product		
Salesperson	A	B	C
Troutman	14	12	4
Kempton	21	16	8
McChristian	15	5	10

15. Starting positions for chemistry and engineering graduates are classified by industry as shown below.

	Industry			
Degree Major	Oil	Chemical	Electrical	Computer
Chemistry	30	15	15	40
Engineering	30	30	20	20

Use $\alpha = .01$, and test for independence of degree major and industry type.

16. A *CBS News/New York Times* poll (February 24, 1991) asked individuals a series of questions about the involvement of the United States in the Persian Gulf war. One question asked men and women the following: Do you think the United States did the right thing in starting the ground war against Iraq, or should the United States have waited longer to see if bombing from the air worked? Assume the responses for men and women were summarized in the contingency table shown below. Use the chi-square test of independence to analyze these data. What is your conclusion at a .05 level of significance?

Response	Men	Women
Right thing	243	207
Waited longer	48	66
Not sure	9	27

17. Medical researchers at Harvard and Boston University randomly assigned 227 General Electric employees with alcohol problems to one of three alcohol-treatment groups: a 28-day hospitalization followed by Alcoholics Anonymous (AA) meetings, AA meetings only with no hospitalization, or a choice of the two programs (*USA Today,* September 12, 1991). Two years later, the researchers identified the patients who had remained sober after completing the program. Assume that the data are as shown in the following contingency table.

Status/Program	Hospitalization	AA only	Choice
Remained sober	28	13	12
Did not remain sober	48	63	63

Use the chi-square test of independence with a .01 level of significance to analyze the data. What is your conclusion and recommendation?

18. The *GMAC Occasional Papers* (March 1988) provided data on the primary reason for application to an MBA program by full-time and part-time students. Do the data suggest that full-time and part-time students differ in their reason for applying to MBA programs? Explain. Use $\alpha = .01$.

Student Status	Primary Reason for Application		
	Program Quality	Convenience/Cost	Other
Full-time	421	393	76
Part-time	400	593	46

19. A sport preference poll shows the following data for men and women.

Sex	Favorite Sport		
	Baseball	Basketball	Football
Men	19	15	24
Women	16	18	16

Use $\alpha = .05$, and test for similar sport preferences by men and women. What is your conclusion?

20. Three suppliers provide the following data on defective parts.

	Part Quality		
Supplier	Good	Minor Defect	Major Defect
A	90	3	7
B	170	18	7
C	135	6	9

Use $\alpha = .05$, and test for independence between supplier and part quality. What does the result of your analysis tell the purchasing department?

21. A study of educational levels of voters and their political party affiliations showed the following results.

	Party Affiliation		
Educational Level	Democratic	Republican	Independent
Did not complete high school	40	20	10
High school degree	30	35	15
College degree	30	45	25

Use $\alpha = .01$, and test if party affiliation is independent of the educational level of the voters.

22. *Personnel Administrator* (January 1984) provided the following data as an example of selection among 40 male and 40 female applicants for 12 positions.

Applicant	**Selected**	**Not Selected**	**Total**
Male	7	33	40
Female	5	35	40

a. The chi-square test of independence was suggested as a way of determining if the decision to hire seven males and five females should be interpreted as having a selection bias in favor of males. Conduct the test of independence using $\alpha = .10$. What is your conclusion?
b. Using the same test, would the decision to hire eight males and four females suggest concern for a selection bias?
c. How many males could be hired for the 12 positions before the procedure would suggest concern for a selection bias?

SUMMARY

In this chapter we introduced the goodness of fit test and the test of independence. Both are based on the chi-square distribution. The purpose of the goodness of fit test is to determine whether or not a hypothesized probability distribution provides a good description of a particular population of interest. The computations for conducting the goodness of fit test involve comparing observed frequencies from a sample with expected frequencies when the hypothesized probability distribution is assumed true. A chi-square distribution is used to determine whether the differences in observed and expected frequencies are large enough to reject the hypothesized probability distribution. We illustrated the goodness of fit test for an assumed multinomial probability distribution.

A test of independence for two variables is a straightforward extension of the methodology employed in the goodness of fit test for a multinomial population. A contingency table is used to display the observed and expected frequencies. Then a chi-square value is computed. Large chi-square values, caused by large differences between observed and expected frequencies, lead to the rejection of the null hypothesis of independence.

GLOSSARY

Goodness of fit test A statistical test conducted to determine whether to accept or reject a hypothesized probability distribution as a description of a population.

Contingency table A table used to summarize observed and expected frequencies for a test of independence of two variables associated with a population.

KEY FORMULAS

Goodness of Fit Test

$$\chi^2 = \sum_{i=1}^{k} \frac{(f_i - e_i)^2}{e_i} \tag{16.1}$$

Expected Frequencies for Contingency Tables Under the Assumption of Independence

$$e_{ij} = \frac{(\text{Row } i \text{ Total})(\text{Column } j \text{ Total})}{\text{Sample Size}} \tag{16.2}$$

Contingency Table Test Statistic

$$\chi^2 = \sum_i \sum_j \frac{(f_{ij} - e_{ij})^2}{e_{ij}} \tag{16.3}$$

REVIEW QUIZ

TRUE/FALSE

1. A goodness of fit test can be used to determine whether a particular multinomial distribution provides a good description of a population.
2. The chi-square probability distribution should not be used for a goodness of fit test.
3. In computing the expected frequencies for a goodness of fit test, it is not proper to assume the null hypothesis is true.
4. In conducting a goodness of fit test, the expected frequency for each category must be greater than or equal to 10.
5. In conducting a goodness of fit test, the observed frequency for one or more categories may be less than 5.

6. The chi-square distribution is used in conducting a contingency table test.
7. In a contingency table test of independence, the number of rows must be equal to the number of columns.
8. In conducting a contingency table test for independence, the expected frequency for each cell must be at least 5.
9. In conducting either a goodness of fit or contingency table test, the larger the differences between the observed and expected frequencies, the more likely it is that the null hypothesis will be rejected.
10. The appropriate number of degrees of freedom for a contingency table test is given by the product of the number of rows times the number of columns.

MULTIPLE CHOICE

11. The sampling distribution for a goodness of fit test is the
    a. chi-square distribution
    b. t distribution
    c. F distribution
    d. normal distribution
    Use the following information for Questions 12–15. A firm that manufactures kitchen appliances takes a random sample of 300 families to check whether there is any significant difference in color preferences for appliances. The results are as follows.

Color Preferred	Number of Families
White	65
Coppertone	89
Avocado	72
Harvest gold	74
Total	300

12. The critical value for the multinomial goodness of fit test with a .01 level of significance is
    a. 6.25139
    b. 7.77944
    c. 11.3449
    d. 13.2767
13. The expected frequency is
    a. 4 for each color
    b. 50 for each color
    c. 75 for each color
    d. different for each color
14. The calculated value of the test statistic for this chi-square goodness of fit test is
    a. 4.08
    b. 8.52
    c. 11.3
    d. 306
15. The result of the test at the .01 level of significance is which of the following?
    a. There is a significant difference in color preference.
    b. There is no significant difference in color preference.
16. The value of χ^2 that cuts off an area in the upper tail of .01 with 8 degrees of freedom is closest to
    a. 14
    b. 19
    c. 22
    d. 25

17. The value of χ^2 that cuts off an area in the upper tail of .025 with 70 degrees of freedom is closest to
  a. 20
  b. 50
  c. 75
  d. 100

SUPPLEMENTARY EXERCISES

23. In setting sales quotas, the marketing manager makes the assumption that order potentials are the same in each of four sales territories. A sample of 200 sales shows the following number of orders from each region.

Sales Territories			
I	II	III	IV
60	45	59	36

Should the manager's assumption be rejected? Use $\alpha = .05$.

24. A regional transit authority was concerned about the number of riders on one of their routes. In setting up this route, the assumption was that the number of riders was uniformly distributed from Monday through Friday. The following historical data were obtained.

Day	Number of Riders
Monday	13
Tuesday	16
Wednesday	28
Thursday	17
Friday	16

Test using $\alpha = .05$ to determine if the transit authority's assumption appears to be justified.

25. An automobile dealer sells three models of a certain make of pickup truck. Over the most recent sales period, the dealer sold 27 units of model 1, 39 units of model 2, and 30 units of model 3. Using $\alpha = .05$, test whether consumer preferences vary among the three models.

26. A park is soon to be opened in a community. A sample of 140 individuals were asked to state their preference as to when they would most like to visit the park. The sample results are shown.

Week Day	Saturday	Sunday	Holiday
20	20	40	60

In developing a staffing plan, should the park manager plan on the same number visiting the park each day. Support your conclusion with a statistical test using $\alpha = .05$.

27. A sample of parts provided the following contingency table data concerning part quality and production shift.

Shift	Number Good	Number Defective
First	368	32
Second	285	15
Third	176	24

Use $\alpha = .05$, and test the hypothesis that part quality is independent of the production shift. What is your conclusion?

28. An analysis of attendance records and performance on the final examination was made for a first-year mathematics course. The following results were obtained.

Number of Classes Missed	Grade on Final		
	80 or Above	70s	Below 70
None	18	11	6
1–5	14	12	6
More than 5	3	9	20

Use $\alpha = .05$, and test for independence between number of classes missed and the grade on the final examination. What is your conclusion?

29. As part of the standard course evaluation, students are asked to rate the course as either poor, good, or excellent. The course evaluation form also asks students to indicate whether the course taken was a required part of their academic program or was taken as an elective. The dean of the college is interested in determining if the rating of the course is independent of the reason for taking the course. The following results were obtained.

Reason for Taking the Course	Rating		
	Poor	Good	Excellent
Required	16	38	16
Elective	4	10	16

How would you respond to the dean? Use $\alpha = .01$.

30. The following data were collected on the number of emergency ambulance calls for an urban county and a rural county in Virginia (*Journal of The Operational Research Society,* November 1986).

| | | Day of Week | | | | | | | Total |
		Sun	Mon	Tue	Wed	Thur	Fri	Sat	
County	Urban	61	48	50	55	63	73	43	393
	Rural	7	9	16	13	9	14	10	78
Total		68	57	66	68	72	87	53	471

Conduct a test for independence using $\alpha = .05$. What is your conclusion?

31. Office occupancy rates were reported for 1991 in four California metropolitan areas (*Business Week,* 1991). Do the following data suggest that the office vacancies are independent of metropolitan area? Use a .05 level of significance. What is your conclusion?

Occupancy Status	Los Angeles	San Diego	San Francisco	San Jose
Occupied	160	116	192	174
Vacant	40	34	33	26

32. The jury in the well-publicized Rodney King trial found two of the four police officers charged with violating King's civil rights guilty. A *USA Today*/CNN/Gallup poll of 300 blacks and 307 whites provided the following opinions concerning the verdict (*USA Today,* April 19, 1993).

	Right	Too few Convicted	Too many Convicted	Other
Blacks	99	165	0	36
Whites	168	64	37	38

Conduct a statistical test to determine if opinions concerning the verdict are independent of race. Use $\alpha = .05$.

CHAPTER 17

NONPARAMETRIC
METHODS

WHAT YOU WILL LEARN IN THIS CHAPTER

- When nonparametric methods are applicable
- The advantages of nonparametric methods
- How to use the sign test
- How to use the Wilcoxon signed-rank test
- How to use the Mann-Whitney-Wilcoxon test for differences between two populations
- How to use the Kruskal-Wallis test for differences among k populations
- How to compute the Spearman rank-correlation coefficient

CONTENTS

Physical Contact in the Family

Physical contact is an important means of connecting people with one another. The touch, the handshake, the hug, the embrace, and the kiss reflect different levels of physical intimacy and commitment. Physical contact is a nonverbal means of interpersonal bonding and an important aspect of family activity. Children require the physical contact of parents, especially during infancy, if they are to thrive. Although physical contact is an important aspect of close relationships, it has been analyzed infrequently by social scientists because of its delicacy and related measurement difficulties.

A study by Oscar Grusky, Phillip Bonacich, and Mark Peyrot attempts to measure and test hypotheses about physical contact within the family unit. Using 48 two-parent, two-child families, the researchers asked each family to pose for a family photograph. In this setting, the researchers observed who touched whom and the frequency of the physical contacts.

The researchers hypothesized that higher-status persons in the family unit would touch those of lower status more often than vice versa. The touches were expected to come from father to mother, from parent to child, and/or from older child to younger child. Using a technique from nonparametric statistics known as the sign test, the researchers found support for this hypothesis. They noted that fathers were significantly more likely to touch mothers than vice versa and that parents were more likely to touch children than vice versa.

The researchers found that the children were more likely to be touched than parents and that

Physical contact among father, mother, and children helps provide a close family relationship.

mothers were significantly more likely to be touched than fathers. Younger children were also more likely to be touched than older children. Fathers were touched less than any other member of the family and experienced the least physical contact of any family member.

The researchers concluded by noting that family physical contact is a complex multidimensional phenomenon. On one level, it signifies status and power differential between members; on another level, it reflects warmth, closeness, support, and expressiveness. Clearly, touching and physical contact are important aspects in the health and well-being of a family unit.

This *Statistics in Practice* is based on "Physical Contact in the Family," by Oscar Grusky, Phillip Bonacich, and Mark Peyrot, *Journal of Marriage and the Family* (August 1984).

The statistical methods presented in earlier chapters are generally referred to as *parametric methods*. In this chapter we introduce several statistical methods that are referred to as *nonparametric methods*. Nonparametric methods are often applicable in situations where the parametric methods of the preceding chapters are not. They typically require less restrictive assumptions concerning the level of data measurement and/or fewer assumptions concerning the form of the probability distributions generating the data.

One consideration used to determine whether a parametric or a nonparametric method should be used is the scale of measurement for the data. As noted in Chapter 1, there are four scales of measurement: nominal, ordinal, interval, and ratio. All data are generated by one of these four scales; thus, all statistical analyses are conducted with either nominal, ordinal, interval, or ratio data. Let us briefly review the four scales of measurement.

1. *Nominal scale.* The scale of measurement is nominal when the data are simply a label used to identify an attribute of the item.
2. *Ordinal scale.* The scale of measurement is ordinal when the data have all the properties of nominal data and the data can be ranked or ordered with respect to some criterion.
3. *Interval scale.* The scale of measurement is interval when the data have all the properties of ordinal data and the distance between values can be expressed in terms of a fixed unit of measure.
4. *Ratio scale.* The scale of measurement is ratio when the data have all the properties of interval data and a zero value is inherently defined for the measurement scale. With this scale, the ratio of data values is meaningful.

Most of the statistical methods referred to as parametric methods require the use of interval- or ratio-scaled data. With these scales of measurement, means, variances, standard deviations, and so on, can be computed, interpreted, and used in the analysis. Generally, with nominal or ordinal data, parametric methods cannot be used. Thus, whenever the data are nominal or ordinal, nonparametric methods are often the only way to analyze the data and draw statistical conclusions.

Another consideration used to determine whether a parametric method or a nonparametric method should be employed is the assumption concerning the population generating the data. For example, a parametric procedure for testing a hypothesis about the difference between the means of two populations was presented in Chapter 10. In the small-sample case, the t distribution can be used for this test provided we are willing to assume that both populations are normally distributed with equal variances. If this assumption about the populations is not appropriate, the parametric method based on the use of the t distribution should not be used. However, nonparametric methods, which require no assumptions about the population probability distributions, are available for testing for differences between two populations. Because of this, and other cases in which no population assumptions are required, nonparametric methods are often referred to as *distribution-free methods.*

In general, for a statistical method to be classified as nonparametric, it must satisfy at least one of the following conditions:*

1. The method may be used with nominal data.
2. The method may be used with ordinal data.
3. The method may be used with interval or ratio data when no assumption can be made about the population probability distribution.

If the scale of measurement is interval or ratio and if the necessary probability distribution assumptions for the population are appropriate, parametric methods provide a more powerful or more discerning statistical procedure. In many cases where a nonparametric method as well as a parametric method can be applied, the nonparametric method is almost as good as or almost as powerful as the parametric method. In cases where the assumptions required by parametric methods are inappropriate, only nonpara-

---

*See W. J. Conover, *Practical Nonparametric Statistics,* 2nd ed. (New York: Wiley, 1980).

metric methods are available. Because of the less restrictive data-measurement requirements and the fewer assumptions needed concerning the population distribution, nonparametric methods are regarded as more generally applicable than parametric methods. The
sign test, the Wilcoxon signed-rank test, the Mann-Whitney-Wilcoxon test, the Kruskal-
Wallis test, and Spearman rank correlation are the nonparametric methods that are
presented in this chapter.

17.1 ▽ SIGN TEST

The *sign test,* one of the oldest nonparametric methods, can be used with either nominal
or ordinal data. A common market research application of the sign test involves using a
sample of n potential customers to identify preferences for two brands of a product such
as coffee, soft drinks, detergents, and so on. Given these data, the objective is to
determine whether a difference in preference exists. As we will see, the sign test is a
nonparametric statistical procedure that can be used to help answer this question.

SMALL-SAMPLE CASE

The small-sample case for the sign test is appropriate whenever $n \leq 20$. Let us illustrate
the use of the sign test for the small-sample case by considering a study conducted for
Sun Coast Farms.

EXAMPLE 17.1 ▶ Sun Coast Farms produces an orange juice product marketed under the name "Citrus
Valley." A competitor of Sun Coast Farms has begun producing a new orange juice
product known as "Tropical Orange." In a study of consumer preferences for the two
brands, 12 individuals were given unmarked samples of the two brands of orange juice.
The brand each individual tasted first was randomly selected. After tasting the two
products, the individuals were asked to state a preference for one of the two brands. The
purpose of the study is to determine whether preferences for the two products are equal.
Letting p indicate the proportion of the population of consumers favoring Citrus Valley,
Sun Coast Farms would like to test the following hypotheses.

Hypothesis	Conclusion
H_0: $p = .50$	No difference in preference for one product over the other exists.
H_a: $p \neq .50$	A difference in preference for one product over the other exists

In recording the preference data for the 12 individuals participating in the study, a "$+$"
sign will be recorded if the individual expresses a preference for Citrus Valley and a "$-$"
sign will be recorded if the individual expresses a preference for Tropical Orange. Using
this procedure, the data will be recorded in terms of the $+$ or $-$ signs; thus, the
nonparametric test is referred to as the sign test.

Under the assumption that H_0 is true ($p = .50$), the number of $+$ signs follows a
binomial probability distribution with $p = .50$. With a sample size of $n = 12$, Table 5 in

Appendix B shows the following probabilities for the binomial probability distribution with $p = .50$.

Number of + Signs	Binomial Probability	Number of + Signs	Binomial Probability
0	.0002	7	.1934
1	.0029	8	.1208
2	.0161	9	.0537
3	.0537	10	.0161
4	.1208	11	.0029
5	.1934	12	.0002
6	.2256		

A graphical representation of this binomial probability distribution is shown in Figure 17.1. This probability distribution shows the probability of the number of + signs under the assumption that H_0 is true. We will use this sampling distribution to determine a rejection rule. The approach will be similar to the method we used to develop rejection rules for hypothesis tests in previous chapters. For example, using $\alpha = .05$, we would place a rejection region or area of approximately .025 in each tail of the distribution

FIGURE 17.1 **BINOMIAL PROBABILITIES FOR THE NUMBER OF + SIGNS WHEN $n = 12$ AND $p = .50$**

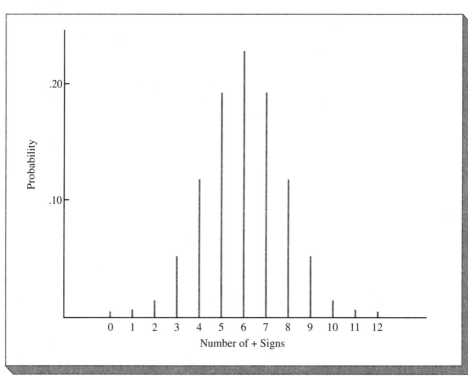

shown in Figure 17.1. Starting at the lower end of the distribution, we see that the probability of obtaining 0, 1, or 2 + signs is .0002 + .0029 + .0161 = .0192. Note that we stop at 2 + signs because adding the probability of 3 + signs would make the area in the lower tail equal to .0192 + .0537 = .0729, which substantially exceeds the desired area of .025. At the upper end of the distribution, we find the same probability of .0192 corresponding to 10, 11, or 12 + signs. Thus, the closest we can come to $\alpha = .05$ is .0192 + .0192 = .0384. As a result, we adopt the following rejection rule:

Reject H_0 if the number of + signs is less than 3 or greater than 9

The preference data that were obtained for the Sun Coast Farms example are shown in Table 17.1. Since only 2 + signs were observed, the null hypothesis is rejected. There is evidence from this study that consumers' preference differs for the two brands of orange juice. We would advise Sun Coast Farms that consumers appear to prefer the competitor's Tropical Orange brand.

In the Sun Coast Farm example, all 12 individuals in the study were able to state a preference. In many situations, one or more individuals in the sample may not be able to state a definite preference. In such cases, the individual's response of no preference can be removed from the study and the analysis conducted with a smaller sample size.

The binomial probability distribution shown in Table 5 of Appendix B can be used to determine the rejection rule for a sign test up to a sample size of $n = 20$. In addition, by considering the probabilities in only the lower or upper tail of the binomial probability distribution, rejection rules can also be developed for one-tailed tests. Appendix B does not provide binomial probability distribution tables for sample sizes greater than 20. In these cases, we can use the large-sample normal approximation of binomial probabilities to determine the appropriate rejection rule for the sign test.

LARGE-SAMPLE CASE

Using the null hypothesis H_0: $p = .50$ and a sample size of $n > 20$, the normal approximation of the sampling distribution for the number of + signs is as follows.

TABLE 17.1 **PREFERENCE DATA FOR THE SUN COAST FARMS' TASTE TEST**

Individual	Brand Preference	Recorded Data
1	Tropical Orange	−
2	Tropical Orange	−
3	Citrus Valley	+
4	Tropical Orange	−
5	Tropical Orange	−
6	Tropical Orange	−
7	Tropical Orange	−
8	Tropical Orange	−
9	Citrus Valley	+
10	Tropical Orange	−
11	Tropical Orange	−
12	Tropical Orange	−

> **Normal Approximation of the Sampling Distribution of the Number of Plus Signs When No Preference Exists**
>
> Mean: $\mu = .50n$ (17.1)
>
> Standard Deviation: $\sigma = \sqrt{.25n}$ (17.2)
>
> Distribution Form: Approximately normal, provided $n > 20$

EXAMPLE 17.2▶ A poll taken during a recent presidential election campaign asked 200 registered voters to rate the Democratic and Republican candidates in terms of best overall foreign policy. Results of the poll showed that 72 rated the Democratic candidate higher, 103 rated the Republican candidate higher, and 25 indicated no difference between the candidates. Does the poll indicate that there is a significant difference in the public's perception of the foreign policies of the two candidates?

Using the sign test, we see that $n = 200 - 25 = 175$ individuals were able to indicate the candidate they believed had the best overall foreign policy. Using equations (17.1) and (17.2), the sampling distribution of the number of plus signs has the following properties.

$$\mu = .50n = .50(175) = 87.5$$

$$\sigma = \sqrt{.25n} = \sqrt{.25(175)} = 6.6$$

In addition, with $n = 175$ we can assume that the sampling distribution is approximately normal. This distribution is shown in Figure 17.2. Since the distribution is approximately normal, we can use the table of areas for the standard normal probability distribution to develop the rejection rule for the sign test. Using Table 1 of Appendix B, we find that for a two-tailed test with $\alpha = .05$, the area in each tail is .025; the corresponding z values are -1.96 and $+1.96$. Thus, the rejection rule for this sign test can be written as follows:

$$\text{Reject } H_0 \text{ if } z < -1.96 \text{ or if } z > +1.96$$

Using the number of times the Democratic candidate received the higher foreign policy rating as the number of $+$ signs (72), we have the following value of the test statistic:

$$z = \frac{72 - 87.5}{6.6} = -2.35$$

Note that $z = -2.35$ is less than -1.96. Thus, the hypothesis of no difference in foreign policy for the two candidates should be rejected at the .05 level of significance. Based on this study, the Republican candidate is perceived to have the higher-rated foreign policy. ◀

HYPOTHESIS TESTS ABOUT A MEDIAN

In earlier chapters, we described statistical tests that can be used to make inferences about population means. We now show how the sign test can be used to conduct hypothesis tests about the value of a population median. Recall that the median splits a population such that 50% of the values fall at the median or above and 50% fall at the median or

FIGURE 17.2   NORMAL PROBABILITY DISTRIBUTION OF THE NUMBER OF + SIGNS
FOR A SIGN TEST WITH $n = 175$

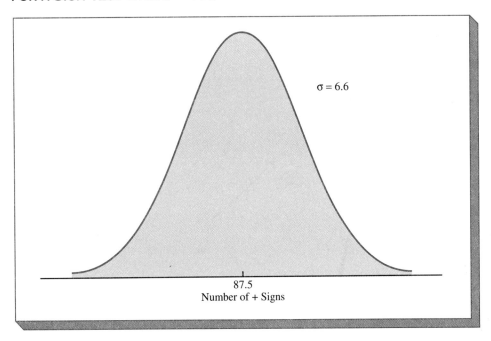

$\sigma = 6.6$

87.5
Number of + Signs

below. We can apply the sign test to conduct a hypothesis test about the value of a median by using a + sign whenever the data value in the sample is above the median and a − sign whenever the data value in the sample is below the median. Any data value exactly equal to the hypothesized value of the median should be discarded. The computations for the sign test are done in exactly the same manner as before.

EXAMPLE 17.3 ▶   The following hypothesis test is being conducted on the median price of new homes in St. Louis, Missouri.

$$H_0: \text{Median} = \$75{,}000$$

$$H_a: \text{Median} \neq \$75{,}000$$

From a sample of 62 sales of new homes, 37 had prices above \$75,000, 23 had prices below \$75,000, and 2 had prices of exactly \$75,000.

Using equations (17.1) and (17.2) for the $n = 60$ homes with prices different than \$75,000, we have

$$\mu = .50(60) = 30$$

$$\sigma = \sqrt{.25(60)} = 3.87$$

Using 37 as the number of + signs, the value of z becomes

$$z = \frac{37 - 30}{3.87} = 1.81$$

With a level of significance of $\alpha = .05$, we reject H_0 if $z < -1.96$ or if $z > +1.96$. Since $z = 1.81$, we are unable to reject H_0.

NOTES AND
COMMENTS

The number of + signs was used in the calculations to determine whether to reject the null hypothesis that $p = .50$. One could just as easily use the number of − signs; the test result would be the same.

EXERCISES

METHODS

SELF TEST

1. The following data show the preferences indicated by 10 individuals in taste tests involving two brands of a product.

Individual	Brand A Versus Brand B	Individual	Brand A Versus Brand B
1	+	6	+
2	+	7	−
3	+	8	+
4	−	9	−
5	+	10	+

With $\alpha = .05$, test for a significant difference in the preferences for the two brands. A + indicates a preference for brand A over brand B.

SELF TEST

2. The following hypothesis test is to be conducted:

$$H_0: \text{Median} \leq 150$$

$$H_a: \text{Median} > 150$$

A sample of size 30 yields 22 cases in which a value greater than 150 is obtained, 3 cases in which a value of exactly 150 is obtained, and 5 cases in which a value less than 150 is obtained. Use $\alpha = .01$ and conduct the hypothesis test.

APPLICATIONS

3. The chapter-opening *Statistics in Practice* described a case in which researchers studied physical contact between parents and children in the same family (*Journal of Marriage and the Family*, August 1984). Assume that a sample of interactions for 20 different families with two children showed 14 cases where the mother was touched more than the father, 4 cases where the father was touched more than the mother, and 2 cases where the touching was judged equal. Suppose a sign test is to be used to see if there is any difference in touching between mothers and fathers.
   a. What are the null and alternative hypotheses for the sign test?
   b. Using $\alpha = .05$, what is the rejection rule?
   c. What is your conclusion?

SELF TEST

4. A Louis Harris poll asked 1253 adults a series of questions about the state of the economy and their children's future (*Business Week*, April 6, 1992). One question was, "Do you expect your children to have a better life than you have had, a worse life, or a life about as good as yours?" The responses were 34% better, 29% worse, 33% about the same, and 4% not sure. Use the sign test and a .05 level of significance to determine if more adults feel their children will have a better future than feel their children will have a worse future. What is your conclusion?

5. In a television preference poll, a sample of 180 individuals was asked to state a preference for one of the two shows aired at the same time on Friday evenings. "Big Town Detective" was favored by 100; 65 favored "The Friday Variety Special"; and 15 were unable to state a preference for one over the other. Is there evidence of a significant difference in the preferences for the two shows? Use $\alpha = .05$ for the test.

6. Menu planning at the Hampshire House Restaurant involves the question of customer preferences for steak and seafood. A sample of 250 customers was asked to state a preference for the two menu items. A preference for steak was stated by 140, and 110 stated a preference for seafood. Use $\alpha = .05$, and test for a difference in the preference for the two menu items.

7. The nationwide median hourly wage for a particular labor group is $14.50 per hour. A sample of 200 individuals in this labor group was taken in one city: 134 individuals had a wage rate less than $14.50 per hour; 54 individuals had a wage rate greater than $14.50 per hour; 12 individuals had a wage rate of $14.50. Test the null hypothesis that the median hourly wage in this city is the same as the nationwide median hourly wage. Use a .02 level of significance.

8. In a sample of 150 college basketball games, it was found that the home team won 98 games. Test to see if these data support the claim that there is a home-team advantage in college basketball. Use a .05 level of significance. What is your conclusion?

9. The median number of part-time employees at fast-food restaurants in a particular city was known to be 15 last year. City officials think the use of part-time employees may be increasing. A sample of nine fast-foot restaurants showed that there were more than 15 part-time employees at seven of the restaurants, one restaurant had exactly 15 part-time employees, and one had fewer than 15 part-time employees. Test at $\alpha = .05$ whether it can be concluded that there has been an increase in the median number of part-time employees.

10. *The Wall Street Journal,* October 22, 1988, reported that the median age at first marriage is 25.9 years for men and 23.6 years for women. Suppose a sample of 225 first marriages in a certain Ohio county showed 122 cases where men were less than 25.9 years of age and 103 cases where men were more than 25.9 years of age. Test the hypothesis that the median age at first marriage for men in the sampled county is the same as the reported 25.9 years. Use $\alpha = .05$. What is your conclusion?

11. The median hourly wage for the population of blue- and white-collar workers nationwide is $9.00 per hour. A sample of workers was selected in the Los Angeles area (*Newsweek,* February 17, 1992). Use the sample data in the table below to test the hypotheses H_0: median $\leq 9, H_a$: median > 9 for the population of workers in Los Angeles. Using a .05 level of significance, what is your conclusion?

11.50	8.40	11.75
10.05	10.25	8.00
13.65	7.05	9.05
11.90	9.90	6.85
15.35	11.10	14.70
13.15	13.10	6.65
13.10	9.20	9.15
12.05	8.45	5.85
9.80		

17.2 WILCOXON SIGNED-RANK TEST

The *Wilcoxon signed-rank test* is the nonparametric alternative to the parametric matched-sample test presented in Chapter 10. In the matched-sample situation, each experimental unit generates two paired or matched observations, one from population 1 and one from population 2. The differences between the matched observations provide insight concerning the differences between the two populations.

The methodology of the parametric matched-sample analysis (the t test for paired differences) requires interval data and the assumption that the population of differences between the pairs of observations is *normally distributed*. With this assumption, the t distribution can be used to test the hypothesis of no difference between population means. If some question exists concerning the appropriateness of the assumption of normally distributed differences, the nonparametric Wilcoxon signed-rank test can be used. We illustrate this nonparametric test for data used to compare the effectiveness of two production methods.

EXAMPLE 17.4 ▶

A manufacturing firm is attempting to determine whether a difference between task-completion times exists for two production methods. A sample of 11 workers was selected, and each worker completed a production task using both production methods. The production method that each worker used first was selected randomly. Thus, each worker in the sample provided the pair of observations shown in Table 17.2. A positive difference in task-completion times indicates that method 1 required more time and a negative difference in times indicates that method 2 required more time. Do the data indicate that the methods are significantly different in terms of task-completion times?

The question raised is whether or not the two methods provide differences in task-completion times. In effect, we have two populations of task-completion times, one population associated with each method. The hypotheses that will be tested are

$$H_0: \text{ The populations are identical}$$

$$H_a: \text{ The populations are not identical}$$

TABLE 17.2   PRODUCTION TASK-COMPLETION TIMES (MINUTES)

Worker	Method 1	Method 2	Difference
1	10.2	9.5	.7
2	9.6	9.8	−.2
3	9.2	8.8	.4
4	10.6	10.1	.5
5	9.9	10.3	−.4
6	10.2	9.3	.9
7	10.6	10.5	.1
8	10.0	10.0	.0
9	11.2	10.6	.6
10	10.7	10.2	.5
11	10.6	9.8	.8

TABLE 17.3   RANKING OF ABSOLUTE DIFFERENCES FOR THE PRODUCTION TASK-COMPLETION TIME PROBLEM

Worker	Difference	Absolute Value of Difference	Rank	Signed Rank
1	.7	.7	8	+ 8
2	−.2	.2	2	− 2
3	.4	.4	3.5	+ 3.5
4	.5	.5	5.5	+ 5.5
5	−.4	.4	3.5	− 3.5
6	.9	.9	10	+10
7	.1	.1	1	+ 1
8	0	0	—	—
9	.6	.6	7	+ 7
10	.5	.5	5.5	+ 5.5
11	.8	.8	9	+ 9
		Sum of signed ranks		+44

If H_0 can be rejected, we will conclude that the two methods differ in terms of task-completion times.

The first step of the Wilcoxon signed-rank test is to rank the *absolute value* of the differences between the two methods. To do this, we first discard any differences of zero and then rank the remaining absolute differences from lowest to highest. Tied differences are assigned average rank values. The ranking of the absolute values of differences is shown in the fourth column of Table 17.3. Note that the difference of 0 for worker 8 is discarded from the rankings; then the smallest absolute difference of .1 is assigned the rank of 1. This ranking of absolute differences continues with the largest absolute difference of .9 assigned the rank of 10. The absolute differences of .4 for workers 3 and 5 are assigned the average rank of 3.5, while the absolute differences of .5 for workers 4 and 10 are assigned the average rank of 5.5.

Once the ranks of the absolute differences have been determined, the ranks are given the sign of the original difference in the data. For example, the .1 difference for worker 7, which was assigned the rank of 1, is given the value of +1 because the observed difference between the two methods was positive. The .2 difference, which was assigned the rank of 2, is given the value of −2 because the observed difference between the two methods was negative. The complete list of signed ranks, together with their sum, is shown in the last column of Table 17.3.

Let us return to the original hypothesis of identical population task-completion times for the two methods. If the populations representing task-completion times for each of the two methods are identical, we would expect the positive ranks and the negative ranks to cancel each other, so that the sum of the signed rank values would be approximately zero. Thus, the test for significance under the Wilcoxon signed-rank test involves determining whether the computed sum of signed ranks (+44 in our example) is significantly different from zero. ◀

Let T denote the sum of the signed-rank values in a Wilcoxon signed-rank test. It can be shown that if the two populations are identical and the number of matched pairs of data is 10 or more, the sampling distribution of T can be approximated as follows.

FIGURE 17.3   SAMPLING DISTRIBUTION OF THE WILCOXON T FOR THE PRODUCTION
TASK-COMPLETION TIME PROBLEM

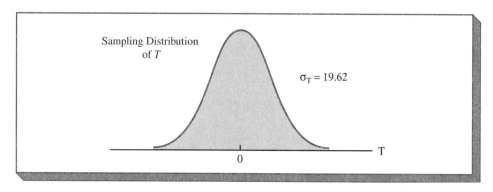

Sampling Distribution
of T

$\sigma_T = 19.62$

T

0

Sampling Distribution of T for Identical Populations

Mean: $\mu_T = 0$ (17.3)

Standard Deviation: $\sigma_T = \sqrt{\dfrac{n(n+1)(2n+1)}{6}}$ (17.4)

Distribution Form: Approximately normal, provided $n \geq 10$

EXAMPLE 17.4
(CONTINUED)

Since we discarded the observation with the difference of zero (worker 8), $n = 10$. Thus, using equation (17.4), we have

$$\sigma_T = \sqrt{\frac{10(11)(21)}{6}} = 19.62$$

The sampling distribution of T under the assumption of identical populations is shown in Figure 17.3.

The value of z is as follows:

SELF TEST

$$z = \frac{T - \mu_T}{\sigma_T} = \frac{44 - 0}{19.62} = 2.24$$

Testing the null hypothesis of no difference using a level of significance of $\alpha = .05$ the rejection region will be $z < -1.96$ or $z > +1.96$. With the value of $z = 2.24$, we reject H_0 and conclude that the two populations are not identical in terms of task-completion times. Although we have determined that a difference between populations exists, the Wilcoxon signed-rank test does not enable us to conclude in what ways the populations differ. However, the fact that method 2 showed the shorter completion times for 8 of the 11 workers would lead us to conclude that differences between the two populations indicate method 2 to be the better production method.   ◀

EXERCISES

APPLICATIONS

12.   Two fuel additives are being tested to determine their effect on miles per gallon for passenger cars. Test results using 12 cars are shown below; each car was

tested with both fuel additives. Use $\alpha = .05$ and the Wilcoxon signed-rank test to see if there is a significant difference in the additives.

Car	Additive 1	Additive 2	Car	Additive 1	Additive 2
1	20.12	18.05	7	16.16	17.20
2	23.56	21.77	8	18.55	14.98
3	22.03	22.57	9	21.87	20.03
4	19.15	17.06	10	24.23	21.15
5	21.23	21.22	11	23.21	22.78
6	24.77	23.80	12	25.02	23.70

SELF TEST

13. A sample of 10 individuals was used in a study to test the effects of a relaxant on the time required to fall asleep for male adults. Data for 10 subjects showing the number of minutes required to fall asleep with and without the relaxant are given in the following table. Use a .05 level of significance to determine if the relaxant reduces the time required to fall asleep. What is your conclusion?

Subject	Without Relaxant	With Relaxant	Subject	Without Relaxant	With Relaxant
1	15	10	6	7	5
2	12	10	7	8	10
3	22	12	8	10	7
4	8	11	9	14	11
5	10	9	10	9	6

14. Shown below are the number of baggage-related complaints per 1000 passengers for 10 airlines during the months of December 1988 and January 1989 (*U.S. Department of Transportation*, March 1989). Use $\alpha = .05$ and the Wilcoxon signed-rank test to determine if the data indicate the number of baggage-related complaints for the airline industry has *decreased* over the two months studied. What is your conclusion?

Airline	December Complaints	January Complaints
American	8.9	8.0
Delta	8.2	7.9
Continental	7.9	8.2
Eastern	7.5	7.8
Northwest	9.6	6.5
Pan American	5.0	5.1
Piedmont	12.3	11.0
TWA	11.2	10.9
United	7.7	7.4
USAir	8.6	7.9

15. A test was conducted of two overnight mail-delivery services. Two samples of identical deliveries were set up such that both delivery services were notified of the need for a delivery at the same time. The number of hours required to make the delivery is shown in the table below for each service. Do the data shown suggest a difference in the delivery times for the two services? Use a .05 level of significance for the test.

Delivery	Service 1	Service 2	Delivery	Service 1	Service 2
1	24.5	28.0	7	25.0	29.0
2	26.0	25.5	8	21.0	22.0
3	28.0	32.0	9	24.0	23.5
4	21.0	20.0	10	26.0	29.5
5	18.0	19.5	11	31.0	30.0
6	36.0	28.0			

16. Elsbernd Investors, Inc., provides a 6-week training program for newly hired management trainees. As part of the program-evaluation procedure, the firm gives each trainee a pretest and posttest. Use a one-tailed test with $\alpha = .05$, and analyze the following data as part of the evaluation of the firm's management training program. What is your conclusion?

Trainee	Pretest Score	Posttest Score	Trainee	Pretest Score	Posttest Score
1	45	65	7	57	70
2	60	70	8	70	65
3	65	63	9	72	80
4	60	67	10	66	88
5	52	60	11	78	74
6	62	58			

17. Ten test-market cities were selected as part of a market research study designed to evaluate the effectiveness of a particular advertising campaign. The sales dollars for each city were recorded for the week prior to the promotional program. Then the campaign was conducted for 2 weeks, with new sales data collected for the week immediately following the campaign. The resulting data with sales in thousands of dollars are shown below.

City	Sales Precampaign	Postcampaign	City	Sales Precampaign	Postcampaign
Kansas City	130	160	Indianapolis	80	82
Dayton	100	105	Louisville	65	55
Cincinnati	120	140	St. Louis	90	105
Columbus	95	90	Pittsburgh	140	152
Cleveland	140	130	Peoria	125	140

Use $\alpha = .05$. What conclusion would you draw concerning the value of the advertising program?

17.3 ▽ MANN-WHITNEY-WILCOXON TEST

In this section we present another nonparametric method that can be used to determine whether there is a difference between two populations. This test, unlike the signed-rank test, is not based on a matched sample. Two independent samples, one from each population, are used. This test, developed jointly by Mann, Whitney, and Wilcoxon, is sometimes referred to as the *Mann-Whitney test* and is sometimes referred to as the *Wilcoxon rank-sum test*. Both the Mann-Whitney and Wilcoxon versions of this test are equivalent; thus, we refer to it as the *Mann-Whitney-Wilcoxon (MWW) test*.

The MWW test is based on independent random samples from each population. Recall that in Chapter 10 we conducted a parametric test for the difference between the means of two populations. The hypotheses tested were as follows:

$$H_0: \ \mu_1 - \mu_2 = 0$$

$$H_a: \ \mu_1 - \mu_2 \neq 0$$

In the small-sample case, the parametric method used was based on two assumptions:

1. Both populations are normally distributed.
2. The variances of the two populations are equal.

The nonparametric MWW test does not require either of the above assumptions. The only requirement of the MWW test is that the measurement scale for the data generated by the two independent random samples is at least ordinal. Instead of testing for the difference between the means of the two populations, the MWW test is to determine whether the two populations are identical. The hypotheses for the Mann-Whitney-Wilcoxon test are as follows:

$$H_0: \ \text{The two populations are identical}$$

$$H_a: \ \text{The two populations are not identical}$$

We first demonstrate how the MWW test can be applied by showing an application for the small-sample case.

SMALL-SAMPLE CASE

The small-sample size case for the MWW test is appropriate whenever the sample sizes for both populations are less than or equal to 10. We illustrate the use of the MWW test for the small-sample case by considering the academic achievement of students attending Johnston High School.

EXAMPLE 17.5 ▷ The majority of students attending Johnston High School previously attended either Garfield Junior High School or Mulberry Junior High School. The question raised by school administrators was whether the population of students who attended Garfield was identical to the population of students who attended Mulberry in terms of academic potential. The hypotheses under consideration were expressed as follows:

$$H_0: \ \text{The two populations are identical in terms} \\ \text{of academic potential}$$

$$H_a: \ \text{The two populations are not identical in terms} \\ \text{of academic potential}$$

Using high school records, Johnston High School administrators selected a random sample of four high school students who had attended Garfield Junior High and another random sample of five students who had attended Mulberry Junior High. The current high school class standing was recorded for each of the nine students used in the study. The ordinal class standings for the nine students are shown in Table 17.4.

The first step in the MWW procedure is to rank the *combined* data from the two samples from low to high. The lowest value (class standing 8) receives a rank of 1 and the highest value (class standing 202) receives a rank of 9. The complete ranking of the nine students is as follows.

Student	Class Standing	Combined Sample Rank
Fields	8	1
Tibbs	21	2
Clark	52	3
Hart	70	4
Jones	112	5
Kirkwood	144	6
Guest	146	7
Abbott	175	8
Phipps	202	9

The next step is to sum the ranks for each sample separately. This calculation is shown in Table 17.5. The MWW procedure may utilize the sum of the ranks for either sample. In the following discussion, we use the sum of the ranks for the sample of four students from Garfield. We denote this sum by the symbol T. Thus, for our example, $T = 11$.

Let us consider for a moment the properties of the sum of the ranks for the Garfield sample. Since there are four students in the sample, Garfield could have the four top-ranking students in the study. If this were the case, $T = 1 + 2 + 3 + 4 = 10$ would be the smallest value possible for the rank sum T. On the other hand, Garfield could have the four bottom-ranking students, in which case $T = 6 + 7 + 8 + 9 = 30$ would be the largest value possible for T. Thus, T for the Garfield sample must take on a value between 10 and 30.

Note that values of T near 10 imply Garfield has the significantly better, or higher-ranking, students, whereas values of T near 30 imply Garfield has the significantly

---

TABLE 17.4   HIGH SCHOOL CLASS-STANDING DATA

Garfield Students		Mulberry Students	
Student	Class Standing	Student	Class Standing
Fields	8	Hart	70
Clark	52	Phipps	202
Jones	112	Kirkwood	144
Tibbs	21	Abbott	175
		Guest	146

TABLE 17.5   RANK SUMS FOR HIGH SCHOOL STUDENTS FROM EACH JUNIOR HIGH SCHOOL

	Garfield Students			Mulberry Students	
Student	Class Standing	Sample Rank	**Student**	Class Standing	Sample Rank
Fields	8	1	Hart	70	4
Clark	52	3	Phipps	202	9
Jones	112	5	Kirkwood	144	6
Tibbs	21	2	Abbott	175	8
			Guest	146	7
	Sum of ranks	11			34

weaker, or lower-ranking, students. Thus, if the two populations of students were identical in terms of academic potential, we would expect the value of T to be near the average of the above two values, or $(10 + 30)/2 = 20$.

Critical values of the MWW T statistic are provided in Table 10 of Appendix B for cases where both sample sizes are less than or equal to 10.* In these tables, n_1 refers to the sample size corresponding to the sample whose rank sum is being used in the test. The value of T_L is read directly from the tables, and the value of T_U is computed from equation (17.5).

$$T_U = n_1(n_1 + n_2 + 1) - T_L \qquad (17.5)$$

The null hypothesis of identical populations should be rejected if T is *strictly less than* T_L or *strictly greater than* T_U.

EXAMPLE 17.5
(CONTINUED)

Using Table 10 of Appendix B with a .05 level of significance, we see that the lower-tail critical value for the MWW statistic with $n_1 = 4$ (Garfield) and $n_2 = 5$ (Mulberry) is $T_L = 12$. The upper-tail critical value for the MWW statistic is computed using equation (17.5) as follows:

$$T_U = 4(4 + 5 + 1) - 12 = 28.$$

Thus, the MWW decision rule indicates that the null hypothesis of identical populations cannot be rejected if the sum of the ranks for the first sample (Garfield) is between 12 and 28, inclusively. The rejection rule can be written as follows:

Reject H_0 if $T < 12$ or if $T > 28$

Referring to Table 17.5, we see that $T = 11$. Thus, the null hypothesis H_0 is rejected, and we can conclude that the population of students at Garfield differs from the population of students at Mulberry in terms of academic potential. The higher class ranking obtained by the sample of Garfield students indicates that Garfield students appear to be better prepared for high school than the Mulberry students.

---

*A more complete table of critical values for the Mann-Whitney-Wilcoxon test can be found in *Practical Nonparametric Statistics,* by W. J. Conover.

LARGE-SAMPLE CASE

In the case where both sample sizes are greater than or equal to 10, a normal approximation of the distribution of T can be used to conduct the analysis for the MWW test. We show an example of the large-sample case by considering an application at the Third National Bank.

EXAMPLE 17.6▶ The Third National Bank has two branch offices. Data collected from two independent simple random samples, one from each branch, are shown in Table 17.6. Do the data indicate that the populations of checking account balances at the two branch banks are not identical?

The first step in the MWW test is to rank the *combined* data from the lowest to the highest values. Using the combined set of 22 observations shown in Table 17.6, we find the lowest data value of $750 (sixth item of sample 2) and assign to it a rank of 1. Continuing the ranking, we have the following results.

Account Balance	Item	Assigned Rank
$ 750	6th of sample 2	1
$ 800	5th of sample 2	2
$ 805	7th of sample 1	3
$ 850	2nd of sample 2	4
.	.	.
.	.	.
.	.	.
$1195	4th of sample 1	21
$1200	3rd of sample 1	22

TABLE 17.6   ACCOUNT BALANCES FOR TWO BRANCHES OF THE THIRD NATIONAL BANK

Branch 1		Branch 2	
Sampled Account	Account Balance ($)	Sample Account	Account Balance ($)
1	1095	1	885
2	955	2	850
3	1200	3	915
4	1195	4	950
5	925	5	800
6	950	6	750
7	805	7	865
8	945	8	1000
9	875	9	1050
10	1055	10	935
11	1025		
12	975		

In ranking the combined data, we may find that 2 or more data values are the same. In this case, these same values are given the *average* ranking of their positions in the combined data set. This situation of *ties* occurs with the ranking of the 22 account balances from the two branch banks. For example, the balance of $945 (eighth item of sample 1) will be assigned the rank of 11. However, the next two values in the data set are tied with values of $950 (see the sixth item of sample 1 and the fourth item of sample 2). Since these two values will be considered for assigned ranks of 12 and 13, they are both given the assigned rank of 12.5. At the next highest data value of $955, we continue the ranking process by assigning $955 the rank of 14. Table 17.7 shows the entire data set with the assigned rank of each observation.

The next step in the MWW test is to sum the ranks for each sample. These sums are shown in Table 17.7. The test procedure can be based on the sum of the ranks for either sample. We use the sum of the ranks for the sample from branch 1. Thus, for this example, $T = 169.5$.

Given that the sample sizes are $n_1 = 12$ and $n_2 = 10$, we can use the normal approximation to the sampling distribution of the rank sum, T. The appropriate sampling distribution is given by equations (17.6) and (17.7).

Sampling Distribution of T for Identical Populations

Mean: $\quad \mu_T = \frac{1}{2}n_1(n_1 + n_2 + 1)$ $\qquad$ **(17.6)**

Standard Deviation: $\quad \sigma_T = \sqrt{\frac{1}{12}n_1 n_2(n_1 + n_2 + 1)}$ $\qquad$ **(17.7)**

Distribution Form: $\quad$ Approximately normal, provided $n_1 \geq 10$ and $n_2 \geq 10$

TABLE 17.7   COMBINED RANKING OF THE DATA IN THE TWO SAMPLES FROM THE THIRD NATIONAL BANK

Branch 1			Branch 2		
Sampled Account	Account Balance ($)	Rank	**Sampled Account**	Account Balance ($)	Rank
1	1095	20	1	885	7
2	955	14	2	850	4
3	1200	22	3	915	8
4	1195	21	4	950	12.5
5	925	9	5	800	2
6	950	12.5	6	750	1
7	805	3	7	865	5
8	945	11	8	1000	16
9	875	6	9	1050	18
10	1055	19	10	935	10
11	1025	17		Sum of ranks	83.5
12	975	15			
	Sum of ranks	169.5			

For branch 1, we have

$$\mu_T = \tfrac{1}{2}12(12 + 10 + 1) = 138$$

$$\sigma_T = \sqrt{\tfrac{1}{12}12(10)(12 + 10 + 1)} = 15.17$$

The sampling distribution of T is shown in Figure 17.4. Following the usual hypothesis-testing procedure, we compute the standardized test statistic z to determine whether the observed value of T appears to be from the sampling distribution of Figure 17.4. If T appears to be from this distribution, we cannot reject the hypothesis that the two populations are identical. However, if T does not appear to be from this distribution, we will reject the null hypothesis and conclude that the populations are not identical. Computing the value of z, we have

$$z = \frac{T - \mu_T}{\sigma_T} = \frac{169.5 - 138}{15.17} = 2.08$$

At an $\alpha = .05$ level of significance, we know that we can reject H_0 if $z < -1.96$ or if $z > 1.96$. With $z = 2.08$, we reject H_0 and conclude that the two populations are not identical. That is, the populations of account balances at the two branches are not the same. ◀

In summary, the Mann-Whitney-Wilcoxon rank-sum test follows the steps outlined below to determine whether two independent random samples are selected from identical populations.

1. Rank the combined sample observations from lowest to highest, with tied values being assigned the average of the tied rankings.
2. Compute T, the sum of the ranks for the first sample.
3. In the large-sample case, make the test for significant differences between the two populations by using the observed value of T and comparing it to the sampling distribution of T for identical populations [see equations (17.6) and (17.7)]. The value of the standardized test statistic z will provide the basis for deciding whether or not to reject the null hypothesis of identical populations. In the small-sample case, use Table 10 in Appendix B to find the critical values for the test.

FIGURE 17.4   SAMPLING DISTRIBUTION OF T FOR THE THIRD NATIONAL BANK PROBLEM

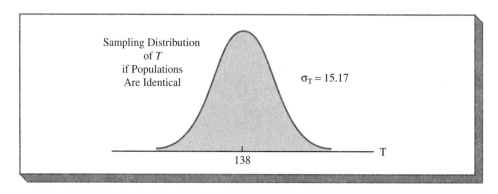

Sampling Distribution
of T
if Populations
Are Identical

$\sigma_T = 15.17$

138

T

NOTES AND
COMMENTS

The nonparametric test discussed in this section is used to determine whether two populations are identical. Parametric statistical tests, such as the *t* test described in Chapter 10, test the equality of two population means. When we reject the hypothesis that the means are equal, we conclude that the populations differ only in their means. When we reject the hypothesis that the populations are identical using the MWW test, we cannot state how they differ. The populations could have different means, different variances, and/or different forms. Nonetheless, if we believe that the populations are approximately the same in every way except for the means, a rejection of H_0 using the nonparametric method implies that the means differ. The major advantages of the MWW test, compared to the parametric *t* test, are that it does not require any assumptions about the form of the probability distribution from which the measurements come and the test may be used with ordinal data.

EXERCISES

APPLICATIONS

SELF TEST

18. Two fuel additives are being tested to determine their effect on gas mileage. Seven cars were tested using additive 1; another independent sample of nine cars was tested using additive 2. The data in the table below show the miles per gallon obtained using the two additives. Use $\alpha = .05$ and the MWW test to see if there is a significant difference in gasoline mileage.

Additive 1	Additive 2
17.3	18.7
18.4	17.8
19.1	21.3
16.7	21.0
18.2	22.1
18.6	18.7
17.5	19.8
	20.7
	20.2

SELF TEST

19. Starting-salary data for college graduates is reported by the College Placement Council (*USA Today,* April 6, 1992). Annual salaries (in $1000s) for a sample of accounting majors and a sample of finance majors are shown below.

Accounting	Finance	Accounting	Finance
28.8	26.3	28.1	29.0
25.3	23.6	24.7	27.4
26.2	25.0	25.2	23.5
27.9	23.0	29.2	26.9
27.0	27.9	29.7	26.2
26.2	24.5	29.3	24.0

a. Use a .05 level of significance to test the null hypothesis that there is no difference between the annual starting salaries for accounting majors and finance majors. What is your conclusion?

b. What are the sample means for accounting majors and finance majors?

20. The Anderson Company has sent two groups of employees to a privately run program providing word-processing training. One group was from the data-processing department; the other was from the typing pool. At the completion of the program, the Anderson Company received a report showing the class rank for each of its employees. Of the 70 persons finishing the program, the class ranks of the 13 employees of the Anderson Company are given in the table below. Use $\alpha = .10$, and test to see whether there is a performance difference between the two groups in the word-processing program.

Data-Processing Group	Typists
1	17
12	26
15	29
23	33
30	45
33	51
	62

21. Mileage performance tests were conducted for two models of automobiles. Twelve automobiles of each model were randomly selected, and a miles per gallon rating for each model was developed based on 1000 miles of highway driving. The data are shown below.

Model 1		Model 2	
Automobile	Miles per Gallon	Automobile	Miles per Gallon
1	20.6	1	21.3
2	19.9	2	17.6
3	18.6	3	17.4
4	18.9	4	18.5
5	18.8	5	19.7
6	20.2	6	21.1
7	21.0	7	17.3
8	20.5	8	18.8
9	19.8	9	17.8
10	19.8	10	16.9
11	19.2	11	18.0
12	20.5	12	20.1

Use $\alpha = .10$, and test for a significant difference in the populations of miles per gallon ratings for the two models.

22. Insurance costs for men and women were reported in *Newsweek,* May 8, 1989. Assume that the data in the table below show the annual cost of $100,000, 5-year term insurance policies for nonsmoking men and nonsmoking women. Use the MWW to test for a significant difference between the costs for men and women. Use $\alpha = .05$.

Men	Women	Men	Women
167	146	180	158
175	162	185	150
160	164	170	150
165	148	163	140
172	166	184	142

23. The following data from police records show the number of daily crime reports from a sample of days during the winter months and a sample of days during the summer months. Using a .05 level of significance, determine if there is a significant difference between the number of crime reports in the winter and summer months.

Winter	Summer	Winter	Summer
18	28	20	29
20	18	12	23
15	24	16	38
16	32	19	28
21	18	20	18

24. A certain brand of microwave oven was priced at 10 stores in Dallas and 13 stores in San Antonio. The data are shown in the table below. Use a .05 level of significance, and test whether prices for the microwave oven are the same in the two cities.

Dallas	San Antonio
445	460
489	451
405	435
485	479
439	475
449	445
436	429
420	434
430	410
405	422
	425
	459
	430

25. Miami University reported the starting salaries by major for graduates of the Richard T. Farmer School of Business Administration (*Miami University Class Profile,* 1990). The starting salaries shown below are reported in thousands of dollars. Use the Mann-Whitney-Wilcoxon test to see if there is evidence to conclude that the starting salaries for majors in management information systems differ from those for general business majors. Use a .05 level of significance.

Management Information Systems	General Business
27.2	25.6
29.0	29.1
28.0	21.7
27.0	24.0
27.5	25.2
30.5	25.0
24.3	25.0
32.5	23.8
26.0	

17.4 ▽ KRUSKAL-WALLIS TEST

The MWW test in Section 17.3 can be used to test whether or not two populations are identical. This test has been extended to the case of three or more populations by Kruskal and Wallis. The hypotheses for the *Kruskal-Wallis test* with $k \geq 3$ populations can be written as follows:

$$H_0: \text{ All } k \text{ populations are identical}$$

$$H_a: \text{ Not all populations are identical}$$

The Kruskal-Wallis test is based on the analysis of independent random samples from each of the k populations.

In Chapter 12 we introduced the completely randomized experimental design as a procedure that could be used to test for the equality of means among three or more populations. The parametric procedure known as analysis of variance (ANOVA) was used to analyze the data and conduct the test. The ANOVA procedure requires interval or ratio data and normally distributed populations with equal variances.

The nonparametric Kruskal-Wallis test can be used with ordinal data as well as with interval or ratio data. In addition, the Kruskal-Wallis test does not require the assumptions of normality and equal variances that are required by the parametric ANOVA procedure. Thus, whenever the data from $k \geq 3$ independent random samples is ordinal or whenever the assumptions of normality and equal variances are questionable, the Kruskal-Wallis test provides an alternative statistical procedure for testing whether or not the populations are identical. Let us demonstrate the Kruskal-Wallis test by using it in an employee-selection application.

EXAMPLE 17.7▶   Williams Manufacturing Company hires employees for its management staff from three area colleges. Recently, the company's personnel department has been collecting and reviewing annual performance ratings in an attempt to determine if there are differences in performance among the managers hired from the three colleges. Performance-rating data are available from independent samples of seven employees from college A, six employees from college B, and seven employees from college C. These data are summarized in Table 17.8; the overall performance rating of each manager is given on a 0 to 100 scale, with 100 being the highest possible performance evaluation. Suppose that we are interested in testing whether the three populations are identical with respect to performance evaluations.   ◀

TABLE 17.8   PERFORMANCE EVALUATION RATINGS FOR 20 EMPLOYEES

College A	College B	College C
25	60	50
70	20	70
60	30	60
85	15	80
95	40	90
90	35	70
80		75

The Kruskal-Wallis test statistic, which is based on the sum of ranks for each of the samples, can be computed as follows.

Kruskal-Wallis Test Statistic

$$W = \frac{12}{n_T(n_T + 1)} \left[\sum_{i=1}^{k} \frac{R_i^2}{n_i} \right] - 3(n_T + 1) \qquad \textbf{(17.8)}$$

where

$\quad k$ = the number of populations
$\quad n_i$ = the number of items in sample i
$\quad n_T = \Sigma n_i$ = total number of items in all samples
$\quad R_i$ = sum of the ranks for sample i

Kruskal and Wallis were able to show that under the null hypothesis that the populations are identical, the sampling distribution of W can be approximated by a chi-square distribution with $k - 1$ degrees of freedom. This approximation is generally acceptable if each of the sample sizes is greater than or equal to 5.

EXAMPLE 17.7
(CONTINUED)

In order to compute the W statistic, we must first rank-order all 20 data items. The lowest data value of 15 from the college B sample receives a rank of 1, whereas the highest data value of 95 from the college A sample receives a rank of 20. The data values, their associated ranks, and the sum of the ranks for the three samples are shown in Table 17.9. Note that we assign the average rank to tied items;* for example, the data values of 60, 70, 80, and 90 had ties.

The sample sizes are

$$n_1 = 7 \quad n_2 = 6 \quad n_3 = 7$$

and

$$n_T = \Sigma_{n_i} = 7 + 6 + 7 = 20$$

Using equation (17.8), the W statistic is computed as follows:

---

*If numerous tied ranks are observed, equation (17.8) must be modified; the modified formula can be found in *Practical Nonparametric Statistics*, by W. J. Conover.

TABLE 17.9   RANKINGS FOR THE 20 EMPLOYEES

College A	Rank	College B	Rank	College C	Rank
25	3	60	9	50	7
70	12	20	2	70	12
60	9	30	4	60	9
85	17	15	1	80	15.5
95	20	40	6	90	18.5
90	18.5	35	5	70	12
80	15.5			75	14
Sum of ranks	95		27		88

$$W = \frac{12}{20(21)} \left[\frac{(95)^2}{7} + \frac{(27)^2}{6} + \frac{(88)^2}{7} \right] - 3(20 + 1) = 8.92$$

The chi-square distribution table (Table 3 of Appendix B) shows that with $k - 1 = 2$ degrees of freedom and $\alpha = .05$ in the upper tail of the distribution, the critical chi-square value is $\chi^2 = 5.99147$. Since the test statistic $W = 8.92$ is greater than 5.99147, we reject the null hypothesis that the three populations are identical. As a result, we conclude that manager performance differs significantly depending on the college attended. Furthermore, since the performance ratings were the lowest for college B, it would appear reasonable for the company to either cut back on its recruiting from college B or at least do a more thorough evaluation of graduates from this college. ◀

NOTES AND COMMENTS

The Kruskal-Wallis procedure illustrated in Example 17.7 began with the collection of interval-scaled data showing employee-performance evaluation ratings. However, the procedure would have also worked had the data been the ordinal rankings of the 20 employees. In this case, the Kruskal-Wallis test could have been applied directly to the original data; the step of constructing the rank orderings from the performance evaluation ratings would have been omitted.

EXERCISES

METHODS

SELF TEST

26. Three products received the following performance ratings by a panel of 15 consumers.

	Product	
A	B	C
50	80	60
62	95	45
75	98	30
48	87	58
65	90	57

Use the Kruskal-Wallis test and $\alpha = .05$ to determine if there is a significant difference in the performance ratings for the products.

27. Three different admission-test-preparation programs are being evaluated. The scores obtained by a sample of 20 people utilizing the test-preparation programs yielded the results shown in the table below. Use the Kruskal-Wallis test to determine if there is a significant difference in the three test-preparation programs. Use $\alpha = .01$.

Program		
A	B	C
540	450	600
400	540	630
490	400	580
530	410	490
490	480	590
610	370	620
	550	570

APPLICATIONS

SELF TEST

28. An American Medical Association survey found that the average annual income for doctors is $155,000 (*St. Petersburg Times,* December 15, 1990). The data below show the annual income in thousands of dollars for samples of physicians specializing in surgery, radiology, and obstetrics. Do the data indicate differences in annual income for the following three specialties? Use a .05 level of significance. What is your conclusion?

Surgery	Radiology	Obstetrics
240	250	200
205	180	175
275	210	185
200	225	220
195	190	188
205	215	202

29. In Chapter 12 the ANOVA procedure was used to test for significant differences in gas mileage for three types of automobiles. The miles per gallon data obtained from tests on five automobiles of each type are shown in the table below.

Automobile		
A	B	C
19	19	24
21	20	26
20	22	23
19	21	25
21	23	27

a. Use the Kruskal-Wallis test with $\alpha = .05$ to determine if there is a significant difference in the gasoline mileage for the three automobiles.

b. What information available in the data is used by the ANOVA procedure and not the Kruskal-Wallis test?

30. A large corporation has been sending many of its first-level managers to an off-site supervisory skills course. Four different management-development centers offer this course, and the corporation is interested in determining if there are differences in the quality of training provided. A sample of 20 employees who have attended these programs has been chosen and the employees ranked with respect to supervisory skills. The results are shown below.

Course	Supervisory Skills Rank				
1	3	14	10	12	13
2	2	7	1	5	11
3	19	16	9	18	17
4	20	4	15	6	8

Note that the top-ranked supervisor attended course 2 and the lowest-ranked supervisor attended course 4. Use $\alpha = .05$, and test to see if there is a significant difference in the training provided by the four programs.

31. The better selling candies are higher in calories. Hershey's Milk Chocolate bar has an average of 240 calories; Twix has an average of 280 calories; and Reese's Peanut Butter Cups have an average of 250 calories (*USA Today,* April 7, 1992). Assume that the data in the table below show the calorie content from samples of M&Ms, Kit Kats, and Milky Way II's. Test for significant differences in the calorie content of these three candies. At a .05 level of significance, what is your conclusion?

M&Ms	Kit Kat	Milky Way II
230	225	200
210	205	208
240	245	202
250	235	190
230	220	180

17.5 SPEARMAN RANK CORRELATION

Correlation was introduced in Chapter 13 as a measure of the linear association between two variables for which interval or ratio data are available. In this section we consider measures of association between two variables when only rank-order (ordinal) data are available. The *Spearman rank-correlation coefficient,* r_s, has been developed for this purpose.

The formula for the Spearman rank-correlation coefficient is as follows.

Spearman Rank-Correlation Coefficient

$$r_s = 1 - \frac{6\Sigma d_i^2}{n(n^2 - 1)}$$ (17.9)

where

n = the number of items or individuals being ranked
x_i = the rank of item i with respect to one variable
y_i = the rank of item i with respect to a second variable
$d_i = x_i - y_i$

EXAMPLE 17.8 ▶ A company wants to determine whether individuals who were expected at the time of employment to be better salespersons actually turn out to have better sales records. To investigate this question, the vice president in charge of personnel has carefully reviewed the original job interview summaries, academic records, and letters of recommendation for 10 current members of the firm's sales force. Based on the review of this information, the vice president ranked the 10 individuals in terms of their potential for success, basing the assessment solely on the information available at the time of employment. Then a list was obtained of the number of units sold by each salesperson over the first 2 years. Then a second ranking of the 10 salespersons based on actual sales performance was carried out. Table 17.10 shows the relevant data and the two rankings. The statistical question involves determining whether or not there is agreement between the ranking of potential at the time of employment and the ranking based on the actual sales performance over the first 2 years. Let us compute the Spearman rank-correlation coefficient for the data in Table 17.10. The computations for the rank-correlation coefficient are summarized in Table 17.11. Here we see that the rank-correlation coefficient is a positive .73. The Spearman rank-correlation coefficient ranges from -1.0 to $+1.0$, with an interpretation similar to the sample correlation coefficient in that positive values near 1.0 indicate a strong positive association between the rankings; as one rank increases, the other rank

TABLE 17.10   **SALES POTENTIAL AND ACTUAL 2-YEAR SALES FOR 10 SALESPERSONS**

Salesperson	Ranking of Potential	2-Year Sales (Units)	Ranking According to 2-Year Sales
A	2	400	1
B	4	360	3
C	7	300	5
D	1	295	6
E	6	280	7
F	3	350	4
G	10	200	10
H	9	260	8
I	8	220	9
J	5	385	2

TABLE 17.11   **COMPUTATION OF THE SPEARMAN RANK-CORRELATION COEFFICIENT FOR SALES POTENTIAL AND SALES PERFORMANCE**

Salesperson	x_i = Ranking of Potential	y_i = Ranking of Sales Performance	$d_i = x_i - y_i$	d_i^2
A	2	1	1	1
B	4	3	1	1
C	7	5	2	4
D	1	6	−5	25
E	6	7	−1	1
F	3	4	−1	1
G	10	10	0	0
H	9	8	1	1
I	8	9	−1	1
J	5	2	3	9
				$\Sigma d_i^2 = 44$

$$r_s = 1 - \frac{6\Sigma d_i^2}{n(n^2-1)} = 1 - \frac{6(44)}{10(100-1)} = .73$$

increases. On the other hand, rank correlations near -1.0 indicate a strong negative association in the ranks (as one rank increases, the other rank decreases). The value $r_s = .73$ indicates a positive correlation between potential and actual performance. Individuals ranked high on potential tend to rank high on performance.

A TEST FOR SIGNIFICANT RANK CORRELATION

At this point, we have seen how sample results can be used to compute the sample rank-correlation coefficient. As with many other statistical procedures, we may wish to use the sample results to make an inference about the population rank correlation coefficient, ρ_s between two variables. In Example 17.8, the population rank-correlation coefficient could be obtained by making the rank-correlation coefficient computations for all members of the sales force. However, we would like to avoid all this data collection and make an inference about the population rank-correlation based on the sample rank-correlation coefficient, r_s. To make this inference, we test the following hypotheses:

$$H_0: \rho_s = 0$$

$$H_a: \rho_s \neq 0$$

Under the null hypothesis of no rank correlation ($\rho_s = 0$), the rankings are independent and the sampling distribution of r_s is as follows.

Sampling Distribution of r_s

Mean: $\mu_{r_s} = 0$ (17.10)

Standard deviation: $\sigma_{r_s} = \sqrt{\dfrac{1}{(n-1)}}$ (17.11)

Form: Approximately normal, provided $n \geq 10$

EXAMPLE 17.8
(CONTINUED)

The sample rank-correlation coefficient for sales potential and sales performance was $r_s = .73$. Use this value to test for a significant rank correlation.

From equation (17.10) we have $\mu_{r_s} = 0$, and from equation (17.11) we have $\sigma_{r_s} = \sqrt{1/(10 - 1)} = .33$. Thus, we have

$$z = \frac{r_s - \mu_{r_s}}{\sigma_{r_s}} = \frac{.73 - 0}{.33} = 2.21$$

Using a level of significance of $\alpha = .05$, the null hypothesis of no correlation can be rejected if $z < -1.96$ or if $z > 1.96$. Since $z = 2.21 > 1.96$, we reject the null hypothesis of no rank correlation. Thus, we can conclude $\rho_s \neq 0$ and that a significant rank correlation exists between sales potential and sales performance.

EXERCISES

METHODS

SELF TEST

32. Consider the set of rankings on a sample of 10 elements shown in the table below.

Element	x_i	y_i	Element	x_i	y_i
1	10	8	6	2	7
2	6	4	7	8	6
3	7	10	8	5	3
4	3	2	9	1	1
5	4	5	10	9	9

   a. Compute the Spearman rank-correlation coefficient for the data.
   b. Test for significant rank correlation using $\alpha = .05$, and state your conclusion.

33. Consider the following two sets of rankings for six items.

	Case One			Case Two	
Item	First Ranking	Second Ranking	Item	First Ranking	Second Ranking
A	1	1	A	1	6
B	2	2	B	2	5
C	3	3	C	3	4
D	4	4	D	4	3
E	5	5	E	5	2
F	6	6	F	6	1

Note that in the first case the rankings are identical, whereas in the second case the rankings are exactly opposite. What value should you expect for the Spearman rank-correlation coefficient for each of these cases? Explain. Calculate the rank-correlation coefficient for each case.

APPLICATIONS

SELF TEST ▷

34. *Financial World* (May 1992) presented a comparison of a number of educational statistics for the states. Shown below, for a sample of 11 states, are the ranks on pupil-teacher ratio (1 = lowest; 11 = highest) and the ranks on expenditure per pupil (1 = highest; 11 = lowest).

State	Rank Pupil-Teacher Ratio	Rank Expenditure per Pupil	State	Rank Pupil-Teacher Ratio	Rank Expenditure per Pupil
Arizona	10	9	Massachusetts	1	1
Colorado	8	5	Nebraska	2	7
Florida	6	4	North Dakota	7	8
Idaho	11	12	South Dakota	5	10
Iowa	4	6	Washington	9	3
Louisiana	3	11			

At the $\alpha = .05$ level, does there appear to be a relationship between expenditure per pupil and pupil-teacher ratio?

35. Airline passengers file complaints about lost, stolen, damaged, and delayed baggage. How do the airlines compare? According to the U.S. Department of Transportation (March 1989), 10 airlines can be ranked from fewest to most complaints per 1000 passengers. The table below shows the rankings of the airlines for December 1988 and January 1989.

December 1988	January 1989	December 1988	January 1989
Pan American	Pan American	USAir	Delta
Eastern	Northwest	American	American
United	United	Northwest	Continental
Continental	USAir	TWA	TWA
Delta	Eastern	Piedmont	Piedmont

a. Compute the rank correlation for the airlines for the two months of data.
b. Test for significant rank correlation using $\alpha = .05$. What is your conclusion?

36. In a poll of men and women television viewers, preferences for the top 10 shows led to the following rankings. Is there a relationship between the rankings by the two groups? Use $\alpha = .10$.

Television Show	Ranking by Men	Ranking by Women	Television Show	Ranking by Men	Ranking by Women
1	1	5	6	3	2
2	5	10	7	10	9
3	8	6	8	4	8
4	7	4	9	6	1
5	2	7	10	9	3

37. A student organization surveyed both recent graduates and current students in an attempt to obtain information on the quality of teaching at a particular university. An analysis of the responses provided the rankings shown in the table below for 10 professors on the basis of teaching ability. Do the rankings given by the current students agree with the rankings given by the recent graduates? Use $\alpha = .10$, and test for a significant rank correlation.

Professor	Ranking by Current Students	Ranking by Recent Graduates	Professor	Ranking by Current Students	Ranking by Recent Graduates
1	4	6	6	2	3
2	6	8	7	5	7
3	8	5	8	10	9
4	3	1	9	7	4
5	1	2	10	9	10

SUMMARY

In this chapter we have presented several statistical procedures that are classified as nonparametric methods. The parametric methods of the earlier chapters generally required interval or ratio data and were often based on assumptions such as the population having a normal population probability distribution.

Since nonparametric methods can be applied to nominal and ordinal data as well as to interval and ratio data and since nonparametric methods do not require population-distribution assumptions, nonparametric methods expand the class of problems that can be subjected to statistical analysis.

The sign test provides a nonparametric procedure for identifying differences between two populations. In the small-sample case, the binominal probability distribution can be used to determine the critical values for the sign test; in the large-sample case, a normal approximation may be used. The Wilcoxon signed-rank test provides a procedure for analyzing matched-sample data whenever interval- or ratio-scaled data are available for each matched pair. The Wilcoxon procedure tests the hypothesis that the two populations are identical.

The Mann-Whitney-Wilcoxon test provides a nonparametric method for testing for a difference between two populations based on two independent random samples. Tables of critical values were presented for the small-sample case, and a normal approximation was provided for the large-sample case. The Kruskal-Wallis test extended the Mann-Whitney-Wilcoxon test to the case of three or more populations. The Kruskal-Wallis test is the nonparametric version of the parametric ANOVA test for differences among population means.

In the last section of this chapter we introduced the Spearman rank-correlation coefficient as a measure of association for two rank-ordered sets of items.

GLOSSARY

Nonparametric methods A collection of statistical methods that generally require very few, if any, assumptions about the population distributions and the level of measurement. These methods may be applied when nominal or ordinal data are available.

Distribution-free methods Another name for nonparametric statistical methods suggested by the lack of assumptions required concerning the population distribution.

Sign test A nonparametric statistical test for identifying differences between two populations based on the analysis of two matched or paired samples.

Wilcoxon signed-rank test A nonparametric statistical test for identifying differences between two populations based on the analysis of two matched or paired samples.

Mann-Whitney-Wilcoxon (MWW) test A nonparametric statistical test for identifying differences between two populations based on the analysis of two independent samples.

Kruskal-Wallis test A nonparametric test for identifying differences among three or more populations based on an analysis of independent samples.

Spearman rank-correlation coefficient A correlation measure based on rank-order data for two variables.

KEY FORMULAS

Sign Test (Large-Sample Case—Normal Approximation)

$$\text{Mean: } \mu = .50n \tag{17.1}$$

$$\text{Standard Deviation: } \sigma = \sqrt{.25n} \tag{17.2}$$

Wilcoxon Signed-Rank Test

$$\text{Mean: } \mu_T = 0 \tag{17.3}$$

$$\text{Standard Deviation: } \sigma_T = \sqrt{\frac{n(n+1)(2n+1)}{6}} \tag{17.4}$$

Mann-Whitney-Wilcoxon (Normal Approximation)

$$\text{Mean: } \mu_T = \tfrac{1}{2}n_1(n_1 + n_2 + 1) \tag{17.6}$$

$$\text{Standard Deviation: } \sigma_T = \sqrt{\tfrac{1}{12}n_1 n_2 (n_1 + n_2 + 1)} \tag{17.7}$$

Kruskal-Wallis Test Statistic

$$W = \frac{12}{n_T(n_T + 1)}\left[\sum_{i=1}^{k}\frac{R_i^2}{n_i}\right] - 3(n_T + 1) \tag{17.8}$$

Spearman Rank-Correlation Coefficient

$$r_s = 1 - \frac{6\Sigma d_i^2}{n(n^2 - 1)} \tag{17.9}$$

REVIEW QUIZ

TRUE/FALSE

1. The sign test can be used to test whether individuals prefer one item over another.
2. Nonparametric methods are often referred to as distribution-free methods.
3. Nonparametric statistical methods are not applicable to ordinal data.
4. With the Wilcoxon signed-rank test, we must assume the populations sampled from are normal.
5. If ties occur in ranking the data from the two samples, the Mann-Whitney-Wilcoxon test cannot be used.
6. With the Mann-Whitney-Wilcoxon test, we need to know only the two sample sizes to compute the standard deviation of the rank sum.

7. With the Mann-Whitney-Wilcoxon test, we need to know only the two sample sizes to compute the expected value of the rank sum.

8. The Wilcoxon signed-rank test uses paired differences to test whether two populations are identical.

9. With the Wilcoxon signed-rank test, the expected value of the sum of signed ranks is equal to the sample size.

10. The Kruskal-Wallis test cannot be used unless the populations sampled from are approximately normal.

11. The Spearman rank-correlation coefficient can be computed only if quantitative data are available for the variables.

MULTIPLE CHOICE

12. A municipal transit system has collected data on which of two seating configurations is preferred by the passengers on its buses. In a sign test to determine whether one seating arrangement is significantly preferred, the null hypothesis would be
    a. H_0: $\mu = 0$ b. H_0: $\mu = .5$
    c. H_0: $p = 0$ d. H_0: $p = .5$

13. A nonparametric test for the equivalence of two populations would be used instead of a parametric test for the equivalence of the population parameters if
    a. the samples are very large
    b. the samples are not independent
    c. no information about the populations is available
    d. the parametric test is always used in this situation

14. A variable assumes the following values in a sample.

    $$10, 12, 15, 15, 16, 18.$$

    What rank is assigned to 15 in a rank-sum test?
    a. 3.5 for both values of 15
    b. 3 for one 15 and 4 for the other 15
    c. only one value of 15 should be used and given the rank 3
    d. no rank needed because ties should be omitted from the rankings

15. For a Mann-Whitney-Wilcoxon test, the first and second samples have sizes 15 and 24, respectively. The expected value of T is
    a. 180
    b. 300
    c. 0
    d. not enough information given

16. In a Wilcoxon signed-rank test, the two samples each have $n = 17$. The expected value of T is
    a. 0   b. 145
    c. 298   d. 300

SUPPLEMENTARY EXERCISES

38. Mueller Beverage Products of Milwaukee, Wisconsin, has conducted a market research study designed to determine if there is a consumer preference for Mueller's Old Brew Beer over the individual consumer's usual beer. Each individual participating in the test was provided with a glass of his or her usual beer and a glass of Mueller's Old Brew. The two glasses were not labeled; the individuals had no way of knowing beforehand which of the two glasses was Mueller's Old Brew and which was the individual's usual brand. The glass that each individual tasted first was randomly selected. After tasting the beer in each glass, the individuals were asked to indicate their preferred beer. The test results from a sample of 24 individuals are shown at the top of page 750.

Individual	Brand Preferred	Value Recorded	Individual	Brand Preferred	Value Recorded
1	Old Brew	+	13	Usual Brand	−
2	Old Brew	+	14	Usual Brand	−
3	Usual Brand	−	15	Old Brew	+
4	Old Brew	+	16	Usual Brand	−
5	Usual Brand	−	17	Old Brew	+
6	Old Brew	+	18	Old Brew	+
7	Usual Brand	−	19	Old Brew	+
8	Old Brew	+	20	Usual Brand	−
9	Old Brew	+	21	Old Brew	+
10	Usual Brand	−	22	Old Brew	+
11	Old Brew	+	23	Usual Brand	−
12	Usual Brand	−	24	Old Brew	+

If an individual selected Mueller's Old Brew as the preferred beer, a + sign was recorded. On the other hand, if the individual stated a preference for his or her usual brand, a − sign was recorded. Do the data for the 24 individuals indicate a significant difference in the preferences for the beers? Use $\alpha = .05$.

39. Two pilots for a prime-time television show (a western and a mystery show) are being tested. Both have been shown to a group of 12 viewers. The viewer preferences are shown.

Viewer	Preference	Viewer	Preference
1	Mystery	7	Mystery
2	Mystery	8	Western
3	Mystery	9	Mystery
4	Western	10	Mystery
5	Mystery	11	Mystery
6	Western	12	Mystery

Using $\alpha = .05$, test to see if there is a significant difference in preferences.

40. In a soft-drink taste test, 48 individuals stated a preference for one of two well-known brands. Results showed 28 favoring brand A, 16 favoring brand B, and 4 undecided. Use the sign test with $\alpha = .10$, and determine whether there is a significant difference in preferences for the two brands of soft-drinks.

41. Use the sign test, and perform the statistical analysis that will help us determine whether the task-completion times for two production methods differ. The data are shown in the table below. Use $\alpha = .05$.

Worker	Method 1 (minutes)	Method 2 (minutes)	Worker	Method 1 (minutes)	Method 2 (minutes)
1	10.2	9.5	7	10.6	10.5
2	9.6	9.8	8	10.0	10.0
3	9.2	8.8	9	11.2	10.6
4	10.6	10.1	10	10.7	10.2
5	9.9	10.3	11	10.6	9.8
6	10.2	9.3			

42. The national median price of new homes for 1988 was $123,500 (*U.S. News & World Report,* September 19, 1988). Assume that data on the prices of new homes were obtained from samples of loans recorded in Chicago and Dallas-Fort Worth. Use those data to test the hypothesis that the median price of homes in each of the two cities is the same as the national median price. Use $\alpha = .05$. State your conclusion for each city.

City	Greater Than $123,500	Equal to $123,500	Less Than $123,500
Chicago	55	6	28
Dallas-Fort Worth	42	3	36

43. Mayfield Products, Inc., has collected data on preferences of 12 individuals concerning the cleaning power of two brands of detergent. The individuals and their preferences are shown below. A + indicates a preference for brand A.

Individual	Brand A Versus Brand B	Individual	Brand A Versus Brand B
1	−	7	−
2	+	8	+
3	+	9	+
4	+	10	−
5	−	11	+
6	+	12	+

With $\alpha = .10$, test for a significant difference in the preference for the two brands.

44. Twelve homemakers were asked to estimate the retail selling price of two models of refrigerators. The estimates of selling price provided by the homemakers are shown in the table below. Use these data, and test at the .05 level of significance to determine if there is a difference in the homemakers' perception of selling price for the two models.

Homemaker	Model 1	Model 2
1	$650	$ 900
2	760	720
3	740	690
4	700	850
5	590	920
6	620	800
7	$700	$ 890
8	690	920
9	900	1000
10	500	690
11	610	700
12	720	700

45. A study was designed to evaluate the weight-gain potential of a new poultry feed. A sample of 12 chickens was used in a 6-week study. The weight of each chicken was recorded before and after the 6-week test period. The difference between the before and after weights of each chicken are as follows: 1.5, 1.2, −.2, .0, .5, .7, .8, 1.0, .0, .6, .2, −.01. A negative value indicates a weight loss during the test period, whereas .0 indicates no weight change over the period. Use a .05 level of significance to determine if the new feed appears to provide a weight gain for the chickens.

46. The data in the table below show product weights for items produced on two production lines. Test for a difference between the product weights for the two lines. Use $\alpha = .10$.

Production Line 1	Production Line 2
13.6	13.7
13.8	14.1
14.0	14.2
13.9	14.0
13.4	14.6
13.2	13.5
13.3	14.4
13.6	14.8
12.9	14.5
14.4	14.3
	15.0
	14.9

47. A client desires to determine if there is a significant difference in the time required to complete a program evaluation with the three different methods that are in common use. The times (in hours) required for each of 18 evaluators to conduct a program evaluation are given below.

Method 1	Method 2	Method 3
68	62	58
74	73	67
65	75	69
76	68	57
77	72	59
72	70	62

Use $\alpha = .05$, and test to see if there is a significant difference in the time required by the three methods.

48. A sample of 20 engineers, who have been with a company for 3 years, has been taken, and they have been rank-ordered with respect to managerial potential. Some of the engineers have attended the company's management-development course, others have attended an off-site management-development program at a local university, and the remainder have not attended any program. Use the rankings given in the table below and $\alpha = .025$ to test for a significant difference in the managerial potential of the three groups.

No Program	Company Program	Off-Site Program
16	12	7
9	20	1
10	17	4
15	19	2
11	6	3
13	18	8
	14	5

49. Shown below are course evaluation ratings for four instructors. Use $\alpha = .05$ and the Kruskal-Wallis procedure to test for a significant difference in teaching abilities.

Instructor	Course-Evaluation Rating								
Black	88	80	79	68	96	69			
Jennings	87	78	82	85	99	99	85	94	
Swanson	88	76	68	82	85	82	84	83	81
Wilson	80	85	56	71	89	87			

50. Wisman investment analysts ranked 12 companies, first with respect to book value and then with respect to growth potential.

Company	Ranking of Book Value	Ranking of Growth Potential
1	12	2
2	2	9
3	8	6
4	1	11
5	9	4
6	7	5
7	3	12
8	11	1
9	4	7
10	5	10
11	6	8
12	10	3

For these data, does a relationship exist between the companies' book values and growth potentials? Use $\alpha = .05$.

51. Two individuals provided the preference rankings of seven soft drinks shown in the table below. Compute the rank correlation for the two individuals.

Soft Drink	Ranking by Individual 1	Ranking by Individual 2	Soft Drink	Ranking by Individual 1	Ranking by Individual 2
A	1	3	E	7	6
B	3	2	F	4	1
C	5	5	G	2	4
D	6	7			

52. A sample of 15 students obtained the following rankings on midterm and final examinations in a statistics course.

Rank Midterm	Final	Rank Midterm	Final	Rank Midterm	Final
1	4	6	2	11	14
2	7	7	5	12	15
3	1	8	12	13	11
4	3	9	6	14	10
5	8	10	9	15	13

Compute the Spearman rank-correlation coefficient for the data, and test for a significant correlation with $\alpha = .10$.

CHAPTER 18

SAMPLE SURVEY

WHAT YOU WILL LEARN IN THIS CHAPTER

- What survey methods are commonly used
- The types of surveys conducted
- The difference between probabilistic and nonprobabilistic sampling
- The difference between sampling and nonsampling error
- How point and interval estimates of the population mean, total, and proportion are made when using simple random sampling, stratified simple random sampling, and cluster sampling
- How to choose a sample size that will provide the level of precision desired

CONTENTS

CG&E Surveys Commercial Customers

The Cincinnati Gas & Electric Company (CG&E) is a public utility that provides gas and electric power to customers in the Greater Cincinnati area. To improve service to its customers, CG&E continually strives to stay up-to-date with its customers' needs. In 1991, CG&E undertook a sample survey, the Building Characteristics Survey, to learn about the energy requirements of commercial buildings in its service area.

A variety of information concerning commercial buildings was sought, such as the floor space, number of employees, energy end-use, age of the building, type of building materials, and energy conservation measures. During preparations for the survey, CG&E analysts found that there were approximately 27,000 commercial buildings in the CG&E service area. Based on available funds and the precision desired in the results, they recommended that a sample of 616 commercial buildings be surveyed.

The sample design chosen was stratified simple random sampling. Total electrical usage over the past year for each commercial building in the service area was available from company records, and because many of the building characteristics of interest (size, number of employees, etc.) were related to usage, it was the criterion used to divide the population of buildings into six strata.

The first stratum contained the commercial buildings for the 100 largest energy users; each building in this stratum was included in the sample. Although these buildings constituted only .2% of the population, they accounted for 14.4% of the total electrical usage. For the other strata, the number of buildings sampled was determined on the basis of obtaining the greatest precision possible per unit cost.

A questionnaire was carefully developed and pretested before the actual survey was conducted. Data were collected through personal interviews. Completed surveys totaled 526 out of the sample of 616 commercial buildings. This response rate of 85.4% was considered to be excellent. Currently, CG&E is using the survey results to improve the forecasts of energy demand and to improve service to its commercial customers.

The authors are indebted to Mr. Jim Riddle of CG&E for providing this *Statistics in Practice.*

A *survey* is a process designed to collect data, or facts, about a situation. A survey of a complete population is called a *census;* a survey of a subset of a population is called a *sample survey.* The purpose of a sample survey is to collect data that can be used to develop inferences about population parameters such as the mean, total, and proportion. If properly designed, a sample survey can be quite precise and cost effective, especially when compared to a census. The purpose of this chapter is to provide an overview of several types of sample surveys and to describe the sampling plans that are most commonly employed. We also comment on some of the nonsampling issues associated with sample surveys.

18.1 TERMINOLOGY USED IN SAMPLE SURVEYS

In Chapter 1 we defined an element, a population, and a sample as follows:

- An *element* is the entity on which data are collected.
- A *population* is the collection of all the elements of interest.
- A *sample* is a subset of the population.

EXAMPLE 18.1 ▶ To illustrate these concepts, consider the following situation. Dunning Microsystems, Inc. (DMI), a manufacturer of personal computers and peripherals, would like to collect data about the characteristics of individuals who have purchased a DMI personal computer. To obtain these data, a sample survey of DMI personal computer owners could be conducted. The *elements* in this sample survey would be individuals who have purchased a DMI personal computer. The *population* would be the collection of all people who have purchased a DMI personal computer, and the *sample* would be the subset of DMI personal computer owners that we survey.   ◀

In sample surveys, it is necessary to distinguish between the *target population* and the *sampled population*. The target population is the population we want to make inferences about, while the sampled population is the population from which the sample is actually selected. It is important to understand that these two populations are not always the same.

EXAMPLE 18.1 ▶ In the DMI survey, the target population consists of all people who have purchased a DMI
(CONTINUED) personal computer. The sampled population, however, might be all owners who had sent warranty registration cards back to DMI. Since every person who buys a DMI personal computer does not send in the warranty card, the sampled population would differ from the target population.   ◀

Conclusions drawn from a sample survey apply only to the sampled population. Whether these conclusions can be extended to the target population depends on the judgment of the analyst. The key issue is whether the correspondence between the sampled population and the target population on the elements of interest is close enough to allow this extension.

Before sampling, the population must be divided into *sampling units*. In some cases, the sampling units are simply the elements. In other cases, the sampling units are groups of the elements.

EXAMPLE 18.2 ▶ Suppose we want to survey certified professional engineers who are involved in the design of heating and air conditioning systems for commercial buildings. If a list of all professional engineers involved in this work were available, the sampling units would be the professional engineers we want to survey. If such a list were not available, we would need an alternative approach. A business telephone directory might provide a list of all engineering firms involved in the design of heating and air conditioning systems. Given this list, we could select a sample of the engineering firms to survey; then, for each firm surveyed, we might interview all the professional engineers. In this case, the engineering firms would be the sampling units and the engineers interviewed would be the elements.   ◀

A list of the sampling units for a particular study is called a *frame*. In the sample survey of professional engineers, the frame is defined as all engineering firms listed in the business telephone directory; the frame is not a list of all professional engineers because this information is not available. The choice of a particular frame and hence the definition of the sampling units is often determined by the availability and reliability of a list. In practice, the development of the frame can be one of the most difficult and important parts of conducting a sample survey.

18.2 ▽ TYPES OF SURVEYS AND SAMPLING METHODS

The three most common types of surveys are mail, telephone, and personal interview surveys; each of these involves the design and administration of a questionnaire. There are also other types of surveys used to collect data that do not involve questionnaires. For example, auditing firms are often hired to sample a company's inventory of goods to estimate the value of inventory on the company's balance sheet. In such surveys, there is no questionnaire; someone simply counts the items and records the results.

In surveys that use questionnaires, the design of the questionnaire is critical. The designer must resist the temptation to add questions that *might* be of interest, since every question adds to the length of the questionnaire. Long questionnaires lead not only to interviewer fatigue, but also to respondent fatigue; this is especially true when mail and telephone surveys are conducted. However, if personal interviews are to be used, a longer and more complex questionnaire can be used. A large body of knowledge exists concerning the phrasing, sequencing, and grouping of questions for a questionnaire. These issues are discussed in more advanced books devoted to survey sampling; several good sources for this type of information are listed in the bibliography.

Sample surveys can also be classified in terms of the sampling method used. With a *probabilistic sampling method,* the probability of obtaining each possible sample can be computed; with a *nonprobabilistic sampling method,* this probability is unknown. Nonprobabilistic sampling methods should not be used if the researcher wants to make statements about the precision of the estimates. In contrast, probabilistic sampling methods can be used to develop confidence intervals that provide bounds on the sampling error. In the following sections, four of the most popular probabilistic sampling methods are discussed: simple random sampling, stratified simple random sampling, cluster sampling, and systematic sampling.

Although statisticians prefer to use a probabilistic sampling method, nonprobabilistic sampling methods often have to be used. The advantages of nonprobabilistic sampling methods are their low expense and ease of implementation. The disadvantage is that statistically valid statements cannot be made about the precision of the estimates. Two commonly used nonprobabilistic methods are convenience sampling and judgment sampling.

With *convenience sampling,* the units included in the sample are chosen because of accessibility. For example, a professor conducting a research study at a university may ask student volunteers to participate in a study simply because they are in the professor's class; in this case, the sample of students is referred to as a convenience sample. In some situations, convenience sampling is the only practical approach. For example, a shipment of oranges might be sampled by an inspector selecting oranges haphazardly from several crates since labeling each orange in the entire shipment to create a frame and employ a probabilistic method of sampling would be impractical. Wildlife captures and volunteer panels for consumer research are other examples of convenience samples.

Although convenience samples provide a relatively easy approach to sample selection and data gathering, it is impossible to evaluate the "goodness" of the sample statistics obtained in terms of their ability to estimate the population parameters of interest. A convenience sample may provide good results or it may not; there is no statistically justified procedure for making any statistical inferences from the sample results. Nevertheless, at times some researchers apply a statistical method designed for a probability sample to the data gathered from a convenience sample. In doing so, the researcher may argue that the convenience sample can be treated as if it were a random sample in the sense that it is representative of the population. However, this argument should be questioned; one should be very cautious in using a convenience sample to make statistical inferences about population parameters.

In using the nonprobability sampling technique referred to as *judgment sampling,* a person knowledgeable on the subject of the study selects sampling units that he or she feels are most representative of the population. Although judgment sampling is often a relatively easy way to select samples, users of the survey results must recognize that the quality of the results is dependent on the judgment of the person selecting the sample. Consequently, caution must be exercised when using judgment samples to make statistical inferences about a population parameter. In general, no statistical statements concerning the precision of the results from a judgment sample should be made.

Both probabilistic and nonprobabilistic sampling methods can be used to select a sample. The advantage of nonprobabilistic methods is that they are usually inexpensive and easy to use. However, if it is necessary to provide statements about the precision of the estimates, a probabilistic sampling method must be used. Almost all large sample surveys employ probabilistic sampling methods.

18.3 SURVEY ERRORS

Two types of errors can occur when conducting a survey. One type, *sampling error,* is defined as the magnitude of the difference between the point estimator and the population parameter. In other words, sampling error is the error that occurs because not every element in the population is surveyed. The second type, *nonsampling error,* refers to all other types of errors that can occur when a survey is conducted such as measurement error, interviewer error, and processing error. Although sampling error can only occur in a sample survey, nonsampling errors can occur in either a census or a sample survey.

NONSAMPLING ERROR

One of the most common types of nonsampling error occurs whenever we incorrectly measure the characteristic of interest. Measurement error can occur in a census or a sample survey. For either type of survey, the researcher must exercise care to ensure that any measuring instruments (e.g., the questionnaire) are properly calibrated and that the people who take the measurements are properly trained. Attention to detail is the best precaution in most situations.

Errors due to nonresponse are a concern to both the statistician responsible for the design of the survey and the manager using the results. This type of nonsampling error occurs when data cannot be obtained for some of the units surveyed or when only partial data are obtained. This problem is most serious when a bias is created. For example, if interviews were conducted to assess women's attitudes toward working outside the home, housecalls only during the daytime would create an obvious bias since women who work outside the home would be excluded from the sample.

Nonsampling errors due to lack of respondent knowledge occur frequently with technical surveys. For example, suppose building managers were surveyed to obtain detailed information about the types of ventilation systems used in office buildings. Managers of large office buildings may be very knowledgeable about such systems because they may attend training seminars and have support staff to help keep them current. In contrast, managers of smaller office buildings may be less knowledgeable about these systems because of the wide variety of duties they must perform. This difference in knowledge can significantly affect the survey results.

Two other types of nonsampling errors are selection errors and processing errors. Selection errors occur when an inappropriate item is included in the survey. Suppose a sample survey was designed to develop a profile of men with beards; if some interviewers interpreted the statement "men with beards" to include men with mustaches while other interviewers did not, the resulting data would be flawed. Processing errors occur whenever data are incorrectly recorded or incorrectly transferred from recording forms such as from questionnaires to computer files.

Although some nonsampling errors will occur in most surveys, they can be minimized by careful planning. Care should be taken to ensure that the sampled population corresponds closely to the target population, good questionnaire design principles should be followed, interviewers should be well trained, and so on. In the final report on a survey, there should be some discussion of the likely impact of nonsampling errors on the results.

SAMPLING ERROR

Recall the Dunning Microsystems (DMI) sample survey (Example 18.1). Suppose that DMI wanted to estimate the mean age of people who have purchased a DMI personal computer. If the entire population of DMI personal computer owners could be surveyed (a census), and nonsampling errors were not present, we would know the mean age exactly. On the other hand, suppose that less than 100% of the population of DMI owners can be surveyed. In this case, there will most likely be a difference between the sample mean and the population mean; the absolute value of this difference is the sampling error. In practice, it is not possible to know what the sampling error will be for any one particular sample because the population mean is unknown; however, it is possible to provide probability statements regarding the size of the sampling error.

As stated, sampling error occurs because a sample, and not the entire population, is surveyed. Even though sampling error cannot be avoided, it can be controlled. Selecting an appropriate sampling method or design is one important way to control this type of error. In the following sections we will discuss four probabilistic sampling methods: simple random sampling, stratified simple random sampling, cluster sampling, and systemic sampling.

18.4 SIMPLE RANDOM SAMPLING

Recall the definition of simple random sampling from Chapter 7:

> A simple random sample of size n from a finite population of size N is a sample selected such that every possible sample of size n has the same probability of being selected.

To conduct a sample survey using simple random sampling, we begin by developing a frame or list of all elements in the sampled population. Then a selection procedure, based on the use of random numbers, is used to ensure that each element in the sampled population has the same probability of being selected. In this section we show how estimates of a population mean, total, and proportion are made for sample surveys that use simple random sampling.

POPULATION MEAN

In most sample surveys, the form of the probability distribution for the population is unknown. In the DMI sample survey (Example 18.1), management wanted to estimate μ, the mean age of people who had purchased a DMI personal computer. Most likely, DMI would not know what the probability distribution of age is for the population of all DMI owners. In most cases, this is not a problem because the properties of the sampling distribution of $\bar{x}$, the point estimator of μ, mostly depend on the choice of the sample design.

In Chapter 7 we indicated that if a *large* ($n \geq 30$) *simple random sample* is selected, the central limit theorem enables us to conclude that the sampling distribution of $\bar{x}$ can be approximated by a normal probability distribution. In Chapter 8 we showed that for cases where the sampling distribution of $\bar{x}$ can be approximated by a normal probability distribution, an interval estimate of μ is given by

$$\bar{x} \pm z_{\alpha/2}\sigma_{\bar{x}} \tag{18.1}$$

where

$$\sigma_{\bar{x}} = \text{standard error of the mean}$$

Recall that $1 - \alpha$ is the confidence coefficient and $z_{\alpha/2}$ is the z value providing an area of $\alpha/2$ in the upper tail of the standard normal probability distribution. For a 95% confidence interval, $z_{.025} = 1.96$. Note also that the standard error of the mean, $\sigma_{\bar{x}}$, is just the standard deviation of the sampling distribution of $\bar{x}$. In general, whenever we use the term *standard error* in this chapter, we will be referring to the standard deviation of the sampling distribution for the point estimator being considered.

When a simple random sample of size n is selected from a finite population of size N, an estimate of the standard error of the mean is

$$s_{\bar{x}} = \sqrt{\frac{N-n}{N}}\left(\frac{s}{\sqrt{n}}\right) \tag{18.2}$$

In this case, the interval estimate of the population mean becomes

$$\bar{x} \pm z_{\alpha/2}s_{\bar{x}} \tag{18.3}$$

In a sample survey, it is common practice to use a value of $z = 2$ when developing interval estimates. Thus, when simple random sampling is used, an approximate 95% confidence interval estimate of the population mean is given by the following expression.

Approximate 95% Confidence Interval Estimate of the Population Mean

$$\bar{x} \pm 2s_{\bar{x}} \tag{18.4}$$

EXAMPLE 18.3 ▶ Consider the situation of the publisher of Great Lakes Recreation, a regional magazine specializing in articles on boating and fishing. The magazine currently has $N = 8000$ subscribers. A simple random sample of $n = 484$ subscribers provided a mean annual income of $30,500 with a standard deviation of $7040. An unbiased estimate of the mean annual income of all subscribers is given by $\bar{x} = \$30,500$. Using the sample results and equation (18.2), an estimate of the standard error of the mean is

$$s_{\bar{x}} = \sqrt{\frac{8000 - 484}{8000}} \left(\frac{7040}{\sqrt{484}} \right) = \$310$$

Therefore, using expression (18.4), an approximate 95% confidence interval estimate of the mean annual income for the magazine subscribers is

$$30,500 \pm 2(310) = 30,500 \pm 620$$

or $29,880 to $31,120.   ◀

The approach just described can be used to compute an interval estimate for other population parameters such as the population total or the population proportion. In cases where the sampling distribution of the point estimator can be approximated by a normal probability distribution, the approximate 95% confidence interval can be written as

Point Estimator $\pm$ 2(Estimate of the Standard Error of the Point Estimator)

The value of 2 times the estimate of the standard error of the point estimator is often referred to in sample surveys as the *bound on the error of the estimate* (B).

EXAMPLE 18.3 ▶
(CONTINUED)

For instance, in the Great Lakes Recreation sample survey, an estimate of the standard error of the point estimator is $s_{\bar{x}} = \$310$, and the bound on the error of the estimate is $B = 2(\$310) = \$620$.   ◀

POPULATION TOTAL

EXAMPLE 18.4 ▶

Consider the problem facing Northeast Electric and Gas (NEG). As part of an energy usage study, NEG needs to estimate the *total* square footage for the 500 public schools in its service area. We will denote the total square footage for the 500 schools as X; in other words, X denotes the population total.   ◀

The population total could be easily computed if we knew μ, the mean for the individual elements in the population. If μ were known, the value of X could be computed by multiplying N times μ. However, since μ is unknown, a point estimate of X is obtained by multiplying N times $\bar{x}$. We will denote the point estimator of X as $\hat{X}$.

Point Estimator of a Population Total
$$\hat{X} = N\bar{x} \qquad\qquad (18.5)$$

An estimate of the standard error of $\hat{X}$ is given by

$$s_{\hat{X}} = N s_{\bar{x}} \qquad\qquad (18.6)$$

where

$$s_{\bar{x}} = \sqrt{\frac{N - n}{N}} \left(\frac{s}{\sqrt{n}} \right) \qquad\qquad (18.7)$$

Note that equation (18.7) is just the formula for the estimated standard error of the mean. Using the above standard error, an approximate 95% confidence interval for the population total is given by the following expression.

Approximate 95% Confidence Interval Estimate of the Population Total

$$N\bar{x} \pm 2s_{\hat{X}} \qquad \text{(18.8)}$$

EXAMPLE 18.4
(CONTINUED)

Suppose that in the NEG study a simple random sample of $n = 50$ public schools was selected from the population of $N = 500$ schools; the sample mean was $\bar{x} = 22,000$ square feet, and the sample standard deviation was $s = 4000$ square feet. Using equation (18.5), the point estimator of the population total is

$$\hat{X} = (500)(22,000) = 11,000,000$$

Equation (18.7) can be used to compute an estimate of the standard error of the mean.

$$s_{\bar{x}} = \sqrt{\frac{500 - 50}{500}}\left(\frac{4000}{\sqrt{50}}\right) = 536.66$$

Then, using equation (18.6), an estimate of the standard error of $\hat{X}$ is

$$s_{\hat{X}} = (500)(536.66) = 268,330$$

Therefore, using equation (18.8), an approximate 95% confidence interval estimate of the total square footage for the 500 public schools in NEG's service area is

$$11,000,000 \pm 2(268,330) = 11,000,000 \pm 536,660$$

or 10,463,340 to 11,536,660 square feet.

POPULATION PROPORTION

The population proportion p is the fraction of the elements in the population with some characteristic of interest. In a market research study, for example, one might be interested in the proportion of consumers preferring a certain brand of product. The sample proportion $\bar{p}$ is an unbiased point estimator of the population proportion. An estimate of the standard error of the proportion is given by

$$s_{\bar{p}} = \sqrt{\left(\frac{N - n}{N}\right)\left(\frac{\bar{p}(1 - \bar{p})}{n - 1}\right)} \qquad \text{(18.9)}$$

An approximate 95% confidence interval estimate of the population proportion is given as follows.

Approximate 95% Confidence Interval Estimate of the Population Proportion

$$\bar{p} \pm 2s_{\bar{p}} \qquad \text{(18.10)}$$

EXAMPLE 18.4 (**CONTINUED**) ▶

Suppose that in the Northeast Electric and Gas sample survey, NEG would also like to estimate the proportion of the 500 public schools in its service area that use natural gas as fuel for heating. If 35 of the 50 sampled schools indicated that they use natural gas, the point estimate of the proportion of the 500 schools in the population that use natural gas is $\bar{p} = 35/50 = .70$. Using equation (18.9), we can compute an estimate of the standard error of the proportion.

$$s_{\bar{p}} = \sqrt{\left(\frac{500 - 50}{500}\right)\left(\frac{.7(1 - .7)}{50 - 1}\right)} = .062$$

Therefore, using equation (18.10), an approximate 95% confidence interval for the population proportion is

$$.7 \pm 2(.062) = .7 \pm .124$$

or .576 to .824. ◀

As the above example has shown, the width of the confidence interval can be rather large (.576 to .824) when estimating a population proportion. In general, large sample sizes are usually needed to obtain precise estimates of population proportions.

DETERMINING THE SAMPLE SIZE

An important consideration in sample design is the choice of sample size. The best choice usually involves a trade-off between cost and precision. Larger samples provide greater precision (tighter bounds on the sampling error), but are more costly. The budget for a study will often dictate how large the sample can be. In other cases, the size of the sample must be large enough to provide a specified level of precision.

A common approach to choosing the sample size is to first specify the precision desired and then determine the smallest sample size providing that precision. In this context, the term *precision* refers to the size of the approximate confidence interval; smaller confidence intervals provide more precision. Since the size of the approximate confidence interval depends on the bound B on the sampling error, choosing a level of precision amounts to choosing a value for B. Let us see how this approach works in choosing the sample size necessary to estimate the population mean.

Equation (18.2) showed that the estimate of the standard error of the mean is

$$s_{\bar{x}} = \sqrt{\frac{N - n}{N}}\left(\frac{s}{\sqrt{n}}\right)$$

Recall that the *bound on the error of the estimate* is "2 times the estimate of the standard error of the point estimator." Thus,

$$B = 2\sqrt{\frac{N - n}{N}}\left(\frac{s}{\sqrt{n}}\right) \tag{18.11}$$

Solving equation (18.11) for n will provide a bound on the sampling error equal to B. Doing so yields

$$n = \frac{Ns^2}{N\left(\dfrac{B^2}{4}\right) + s^2} \tag{18.12}$$

Once a desired level of precision has been selected (by choosing a value for B), equation (18.12) can be used to find the value of n that will provide the desired level of precision. Using equation (18.12) to choose n for a practical study presents problems, however. In addition to specifying the desired bound on the sampling error B, one must know the sample variance s^2. But, s^2 will not be known until the sample is actually taken.

Cochran* suggests several ways to develop an estimate of s^2 in practice. Three of these are stated below.

1. Take the sample in two stages. Use the value of s^2 found in stage 1 in equation (18.12); the resulting value of n is what the size of the total sample must be. Then, select the number of additional units needed at stage 2 to provide the total sample size determined in stage 1.
2. Use the results of a pilot survey or pretest to estimate s^2.
3. Use information from a previous sample.

EXAMPLE 18.5 ▶

Suppose we want an estimate of the population mean for starting salaries of college graduates at a particular university. We know that there are $N = 5000$ graduates, and we want to develop an approximate 95% confidence interval with a width of at most $1000. To provide such a confidence interval, $B = 500$. Before using equation (18.12) to determine the sample size, we need an estimate of s^2. Suppose a study of starting salaries was also conducted last year, and it was found that $s = \$3000$. We can use the data from this previous sample to estimate s^2. Using $B = 500$, $s = 3000$, and $N = 5000$, we can now use equation (18.12) to determine the sample size.

$$n = \frac{5000(3000)^2}{5000\left[\dfrac{(500)^2}{4}\right] + (3000)^2}$$

$$= 139.9689$$

Rounding up, we see that a sample size of 140 will provide an approximate 95% confidence interval of width $1000. Keep in mind, however, that this calculation is based on our initial estimate of $s = \$3000$. If s turns out to be larger in this year's sample survey, the resulting approximate confidence interval will have a width greater than $1000. Consequently, if cost considerations permit, the designer of the survey might choose a sample size of, say, 150 to provide added assurance that the final approximate 95% confidence interval will have a width less than $1000. ◀

The formula for determining the sample size necessary for estimating a population total with a bound B on the error of estimate is similar to that for the sample mean.

$$n = \frac{Ns^2}{\dfrac{B^2}{4N} + s^2} \tag{18.13}$$

EXAMPLE 18.5 ▶
(CONTINUED)

In Example 18.5 we wanted to estimate the mean starting salary with a bound on the sampling error of $B = 500$. Suppose we were also interested in estimating the total salary of all 5000 graduates with a bound of $2 million. Using equation (18.13) with

---

*William G. Cochran, *Sampling Techniques*, 3rd ed., Wiley, 1977.

$B = 2,000,000$, we see that the sample size needed to provide such a bound on the population total is

$$n = \frac{5000(3000)^2}{\dfrac{(2,000,000)^2}{4(5000)} + (3000)^2}$$

$$= 215.311$$

Rounding up, we see that a sample size of 216 is necessary to provide an approximate 95% confidence interval with a bound of $2 million. We note here that if the same survey is expected to provide a bound of $500 on the population mean and a bound of $2 million on the population total, a sample size of at least 216 must be used. This will provide a tighter bound than necessary on the population mean, while providing the minimum desired precision for the population total. ◄

To choose the sample size for estimating a population proportion, we use a formula very similar to the one for the population mean. We simply substitute $\bar{p}(1 - \bar{p})$ for s^2 in equation (18.12) to obtain

$$n = \frac{N\bar{p}(1 - \bar{p})}{N\left(\dfrac{B^2}{4}\right) + \bar{p}(1 - \bar{p})} \tag{18.14}$$

To use equation (18.14), the desired bound B must be specified and an estimate of $\bar{p}$ must be provided. If a good estimate of $\bar{p}$ is not available, we can use $\bar{p} = .5$; this will ensure that the resulting approximate confidence interval will have a bound on the sampling error at least as small as desired.

EXERCISES

METHODS

SELF TEST

1. Simple random sampling has been used to obtain a sample of $n = 50$ elements from a population of $N = 800$. The sample mean was $\bar{x} = 215$, and the sample standard deviation was found to be $s = 20$.
   a. Estimate the population mean.
   b. Estimate the standard error of the mean.
   c. Develop an approximate 95% confidence interval for the population mean.

2. Simple random sampling has been used to obtain a sample of $n = 80$ elements from a population of $N = 400$. The sample mean was $\bar{x} = 75$, and the sample standard deviation was found to be $s = 8$.
   a. Estimate the population total.
   b. Estimate the standard error of the population total.
   c. Develop an approximate 95% confidence interval for the population total.

3. Simple random sampling has been used to obtain a sample of $n = 100$ elements from a population of $N = 1000$. The sample proportion was $\bar{p} = .30$.
   a. Estimate the population proportion.
   b. Estimate the standard error of the proportion.
   c. Develop an approximate 95% confidence interval for the population proportion.

4. A sample is to be taken to develop an approximate 95% confidence interval estimate of the population mean. The population consists of 450 elements, and a pilot study has resulted in

$s = 70$. How large must the sample be if we want to develop an approximate 95% confidence interval with a width of 30?

APPLICATIONS

SELF TEST

5. In 1987, 153 U.S. banks operated foreign branches (*Statistical Abstract of the United States, 1989*). A sample of 30 of those banks showed average assets of $1.8 billion with a standard deviation of $.4 billion. The mean number of foreign branches for the banks in the sample was 6 with a standard deviation of 1.4.

   a. Develop an approximate 95% confidence interval for the population mean of the assets for the 153 banks.

   b. Develop an approximate 95% confidence interval for the total number of foreign branches of these banks.

   c. Suppose that 25% of the banks in the sample had one or more branches in the United Kingdom. Develop an approximate 95% confidence interval for the population proportion of banks having one or more branches in the United Kingdom.

6. A county in California had 724 corporate tax returns filed. The mean annual income reported was $161.22 thousand with a standard deviation of $31.3 thousand. How large a sample will be necessary next year to develop an approximate 95% confidence interval on mean annual corporate earnings? The precision required is an interval width of no more than $5000.

18.5 STRATIFIED SIMPLE RANDOM SAMPLING

To use *stratified simple random sampling,* the population must be divided into H groups, called strata. Then for stratum h, a simple random sample of size n_h is selected; the data from the H simple random samples are then combined to develop an estimate of a population parameter such as the population mean, total, or proportion.

If the variability within each stratum is smaller than the variability across the strata, a stratified simple random sample can lead to greater precision (narrower interval estimates of the population parameters). The basis for forming the various strata depends on the judgment of the designer of the sample. Depending on the application, a population can be stratified by department, location, age, product type, industry type, sales levels, and so on.

EXAMPLE 18.6 Suppose that the College of Business at Lakeland College wants to conduct a survey of this year's graduating class to learn about their starting salaries. There are five majors in the college: accounting, finance, information systems, marketing, and operations management. Of the $N = 1500$ students who graduated this year, there were $N_1 = 500$ accounting majors, $N_2 = 350$ finance majors, $N_3 = 200$ information systems majors, $N_4 = 300$ marketing majors, and $N_5 = 150$ operations management majors. Based on the analysis of previous salary data, it was believed that there would be more variability in starting salaries across majors than within each major. As a result, a stratified simple random sample of $n = 180$ students was selected; 45 of the 180 students that were sampled majored in accounting ($n_1 = 45$), 40 majored in finance ($n_2 = 40$), 30 majored in information systems ($n_3 = 30$), 35 majored in marketing ($n_4 = 35$), and 30 majored in operations management ($n_5 = 30$).

POPULATION MEAN

In stratified sampling, an unbiased estimate of the population mean is obtained by computing a weighted average of the sample means for each stratum. The weights used are the fraction of the population in each stratum. The resulting point estimator, denoted $\bar{x}_{st}$, is defined as follows.

Point Estimator of the Population Mean

$$\bar{x}_{st} = \sum_{h=1}^{H}\left(\frac{N_h}{N}\right)\bar{x}_h \qquad (18.15)$$

where

H = number of strata
$\bar{x}_h$ = sample mean for stratum h
N_h = number of elements in the population in stratum h
N = total number of elements in the population; $N = N_1 + N_2 + \cdots + N_H$

For stratified simple random sampling, the formula for computing an estimate of the standard error of the mean is

$$s_{\bar{x}_{st}} = \sqrt{\frac{1}{N^2}\sum_{h=1}^{H}N_h(N_h - n_h)\frac{s_h^2}{n_h}} \qquad (18.16)$$

Using the above results, an approximate 95% confidence interval estimate of the population mean is as follows.

Approximate 95% Confidence Interval Estimate of the Population Mean

$$\bar{x}_{st} \pm 2s_{\bar{x}_{st}} \qquad (18.17)$$

EXAMPLE 18.6 ▶
(CONTINUED)

The survey of 180 graduates of the College of Business at Lakeland College provided the sample results shown in Table 18.1. The sample mean for each major, or stratum, is as follows: $24,000 for accounting, $22,500 for finance, $25,500 for information systems, $21,000 for marketing, and $25,000 for operations management. Using these results and equation (18.15), we can compute a point estimate of the population mean.

TABLE 18.1   LAKELAND COLLEGE SAMPLE SURVEY OF STARTING SALARIES FOR COLLEGE SENIORS

Major	h	$\bar{x}_h$	s_h	N_h	n_h
Accounting	1	$24,000	2,000	500	45
Finance	2	$22,500	1,700	350	40
Information systems	3	$25,500	2,300	200	30
Marketing	4	$21,000	1,600	300	35
Operations management	5	$25,000	2,250	150	30

$$\bar{x}_{st} = \left(\frac{500}{1500}\right)(24,000) + \left(\frac{350}{1500}\right)(22,500) + \left(\frac{200}{1500}\right)(25,500)$$

$$+ \left(\frac{300}{1500}\right)(21,000) + \left(\frac{150}{1500}\right)(25,000) = \$23,350$$

In Table 18.2 we show a portion of the calculations needed to estimate the standard error; note that

$$\sum_{h=1}^{5} N_h(N_h - n_h)\frac{s_h^2}{n_h} = 42,909,037,698$$

Thus,

$$s_{\bar{x}_{st}} = \sqrt{\left[\frac{1}{(1500)^2}\right](42,909,037,698)} = \sqrt{19,070.68} = 138$$

Hence, using equation (18.17), an approximate 95% confidence interval estimate of the population mean is $23,350 \pm 2(138) = 23,350 \pm 276$, or \$23,074 to \$23,626.

POPULATION TOTAL

The point estimator of the population total (X) is obtained by multiplying N times $\bar{x}_{st}$.

Point Estimator of the Population Total

$$\hat{X} = N\bar{x}_{st} \qquad\qquad (18.18)$$

TABLE 18.2 **PARTIAL CALCULATIONS FOR THE ESTIMATE OF THE STANDARD ERROR OF THE MEAN FOR THE LAKELAND COLLEGE SAMPLE SURVEY**

Major	h	$N_h(N_h - n_h)\dfrac{s_h^2}{n_h}$
Accounting	1	$500(500 - 45)\dfrac{(2,000)^2}{45} = 20,222,222,222$
Finance	2	$350(350 - 40)\dfrac{(1,700)^2}{40} = 7,839,125,000$
Information systems	3	$200(200 - 30)\dfrac{(2,300)^2}{30} = 5,995,333,333$
Marketing	4	$300(300 - 35)\dfrac{(1,600)^2}{35} = 5,814,857,143$
Operations management	5	$150(150 - 30)\dfrac{(2,250)^2}{30} = \underline{3,037,500,000}$
		$42,909,037,698$

$$\sum_{h=1}^{5} N_h(N_h - n_h)\frac{s_h^2}{n_h}$$

An estimate of the standard error of $\hat{X}$ is

$$s_{\hat{X}} = N s_{\bar{x}_{st}} \qquad (18.19)$$

Thus, an approximate 95% confidence interval for the population total is given by the following.

Approximate 95% Confidence Interval Estimate of the Population Total

$$N\bar{x}_{st} \pm 2s_{\hat{X}} \qquad (18.20)$$

EXAMPLE 18.6 ▶
(CONTINUED)

The College of Business at Lakeland College would also like to estimate the total earnings of the 1500 business graduates in order to estimate their impact on the economy. Using equation (18.18), an unbiased estimate of the total earnings is

$$\hat{X} = (1500)23,350 = \$35,025,000$$

Using equation (18.19), an estimate of the standard error of the population total is

$$s_{\hat{X}} = 1500(138) = \$207,000$$

Thus, using equation (18.20) an approximate 95% confidence interval estimate of the total earnings of the 1500 graduates is $35,025,000 \pm 2(207,000) = 35,025,000 \pm 414,000$ or \$34,611,000 to \$35,439,000. ◀

POPULATION PROPORTION

An unbiased estimate of the population proportion p for stratified simple random sampling is provided by a weighted average of the proportions for each stratum. The weights used are the fraction of the population in each stratum. The resulting point estimator, denoted $\bar{p}_{st}$, is defined as follows.

Point Estimator of the Population Proportion

$$\bar{p}_{st} = \sum_{h=1}^{H} \left(\frac{N_h}{N} \right) \bar{p}_h \qquad (18.21)$$

where

H = number of strata
$\bar{p}_h$ = sample proportion for stratum h
N_h = number of elements in the population in stratum h
N = total number of elements in the population; $N = N_1 + N_2 + \cdots + N_H$

An estimate of the standard error of $\bar{p}_{st}$ is given by

$$s_{\bar{p}_{st}} \sqrt{\frac{1}{N^2} \sum_{h=1}^{H} N_h(N_h - n_h) \left[\frac{\bar{p}_h(1 - \bar{p}_h)}{n_h - 1} \right]} \qquad (18.22)$$

Thus, an approximate 95% confidence interval estimate of the population proportion is as follows.

Approximate 95% Confidence Interval Estimate of the Population Proportion

$$\bar{p}_{st} \pm 2s_{\bar{p}_{st}} \qquad (18.23)$$

EXAMPLE 18.6
(CONTINUED)

In the Lakeland College survey, the college wants to know the proportion of graduates receiving a starting salary of $30,000 or more. The results of the sample survey of 180 graduates show that 20 received starting salaries of $30,000 or more and that 4 of the 20 majored in accounting, 2 majored in finance, 7 majored in information systems, 1 majored in marketing, and 6 majored in operations management.

Using equation (18.21), the point estimate of the proportion receiving starting salaries of $30,000 or more is computed as follows:

$$\bar{p}_{st} = \left(\frac{500}{1500}\right)\left(\frac{4}{45}\right) + \left(\frac{350}{1500}\right)\left(\frac{2}{40}\right) + \left(\frac{200}{1500}\right)\left(\frac{7}{30}\right) + \left(\frac{300}{1500}\right)\left(\frac{1}{35}\right) + \left(\frac{150}{1500}\right)\left(\frac{6}{30}\right)$$

$$= .0981$$

In Table 18.3 we show a portion of the calculations needed to estimate the standard error; note that

$$\sum_{h=1}^{5} N_h(N_h - n_h)\left[\frac{\bar{p}_h(1 - \bar{p}_h)}{n_h - 1}\right] = 924.8305$$

TABLE 18.3   **PARTIAL CALCULATIONS FOR THE ESTIMATE OF THE STANDARD ERROR OF $\bar{p}_{st}$ FOR THE LAKELAND COLLEGE SAMPLE SURVEY**

Major	h	$N_h(N_h - n_h)\left[\dfrac{\bar{p}_h(1 - \bar{p}_h)}{n_h - 1}\right]$
Accounting	1	$500(500 - 45)\left[\dfrac{(4/45)(41/45)}{45 - 1}\right] = 418.7430$
Finance	2	$350(350 - 40)\left[\dfrac{(2/40)(38/40)}{40 - 1}\right] = 132.1474$
Information systems	3	$200(200 - 30)\left[\dfrac{(7/30)(23/30)}{30 - 1}\right] = 209.7318$
Marketing	4	$300(300 - 35)\left[\dfrac{(1/35)(34/35)}{35 - 1}\right] = 64.8980$
Operations management	5	$150(150 - 30)\left[\dfrac{(6/30)(24/30)}{30 - 1}\right] = \underline{99.3103}$
		924.8305

$$\sum_{h=1}^{5} N_h(N_h - n_h)\left[\frac{\bar{p}_h(1 - \bar{p}_h)}{n_h - 1}\right]$$

Thus,

$$s_{\bar{p}_{st}} = \sqrt{\frac{1}{(1500)^2}(924.8305)}$$

$$= .0203$$

Using equation (18.23), an approximate 95% confidence interval for the proportion of graduates receiving starting salaries of \$30,000 or more is .0981 ± 2(.0203) = .0981 ± .0406, or .0575 to .1387.

DETERMINING THE SAMPLE SIZE

With stratified simple random sampling, we can think of choosing a sample size as a two-step process. First, a total sample size n must be chosen. Second, we must decide on how to assign the sampled units to the various strata. Alternately, we could first decide how large a sample to take in each stratum and then sum the stratum sample sizes to obtain the total sample size. Since it is often of interest to develop estimates of the mean, total, and proportion for the individual strata, a combination of these two approaches is often employed. An overall sample size n and allocation that will provide the necessary precision for the overall population parameter of interest is found. Then, if the sample sizes in some of the strata are not large enough to provide the precision necessary for the estimates within the strata, the sample sizes for those strata are adjusted upward as necessary. In this subsection we discuss some of the issues pertinent to allocating the total sample to the various strata and present a method for choosing the total sample size and making the allocation.

The allocation issue has to do with deciding what fraction of the total sample should be assigned to each stratum. This determines how large the simple random sample will be in each stratum. The factors considered most important in making the allocation are

1. The number of elements in each stratum for the population
2. The variance of the elements within each stratum
3. The cost of selecting elements within each stratum

Generally speaking, larger samples should be assigned to the larger strata and to the strata with larger variances. Conversely, to get the most information for a given cost, smaller samples should be allocated to the strata where the cost per unit of sampling is greatest.

The individual stratum variances often differ greatly. For example, suppose that in a particular study we were interested in determining the mean number of employees per building; since there is greater variability in a stratum with larger buildings than there is in one with smaller buildings, a proportionately larger sample should be taken in such a stratum. The cost of selection can be an important consideration when significant travel is required of the interviewer or survey-taker between sampled units in some of the strata but not in others; this frequently occurs when some of the strata involve rural areas and others involve cities.

In many surveys, the cost per unit of sampling is approximately the same for each stratum (e.g., mail and telephone surveys); in such cases, the cost of sampling can be ignored in making the allocation. We present here the appropriate formulas for choosing the sample size and making the allocation in such cases. More advanced texts on sampling provide formulas for the case when sampling costs vary significantly across strata. The formulas we present in this section will minimize the total sampling cost for a given level of precision. This method, known as *Neyman allocation,* allocates the total sample *n* to the various strata as follows:

$$n_h = n\left(\frac{N_h s_h}{\displaystyle\sum_{h=1}^{H} N_h s_h}\right)$$ (18.24)

Equation (18.24) shows that the number of units allocated to a stratum increases with the stratum's size and variance. Note that to make this allocation, we need to first determine the total sample size *n*. Given a specified level of precision *B*, the formulas for choosing the total sample size when estimating the population mean and the population total are given below.

Sample Size for the Estimate of the Population Mean

$$n = \frac{\left(\displaystyle\sum_{h=1}^{H} N_h s_h\right)^2}{N^2\left(\dfrac{B^2}{4}\right) + \displaystyle\sum_{h=1}^{H} N_h s_h^2}$$ (18.25)

Sample Size for the Estimate of the Population Total

$$n = \frac{\left(\displaystyle\sum_{h=1}^{H} N_h s_h\right)^2}{\dfrac{B^2}{4} + \displaystyle\sum_{h=1}^{H} N_h s_h^2}$$ (18.26)

EXAMPLE 18.7 ▶ Suppose a Chevrolet dealer wants to survey the customers who have purchased a Corvette, Geo Prizm, or a Cavalier to obtain information the dealer feels will be helpful in determining future advertising. In particular, suppose the dealer wants to estimate the mean monthly income for these customers with a bound on the sampling error of $100. The dealer's 600 Corvette, Geo Prizm, and Cavalier customers have been divided into three strata: 100 Corvette owners, 200 Geo Prizm owners, and 300 Cavalier owners. A pilot survey was used to estimate the standard deviation in each stratum; the results are $s_1 = \$1300$, $s_2 = \$900$, and $s_3 = \$500$ for the Corvette, Geo Prizm, and Cavalier owners, respectively.

The first step in choosing a sample size for this survey is to use equation (18.25) to determine the total sample size necessary to provide a bound of $B = \$100$ on the estimate of the population mean. First, we compute

$$\sum_{h=1}^{3} N_h s_h = 100(1300) + 200(900) + 300(500) = 460,000$$

Next, we compute

$$\sum_{h=1}^{3} N_h s_h^2 = 100(1300)^2 + 200(900)^2 + 300(500)^2 = 406{,}000{,}000$$

Substituting these values into equation (18.25), we can determine the total sample size needed to provide a bound on the sampling error of $B = \$100$.

$$n = \frac{(460{,}000)^2}{\dfrac{(600)^2(100)^2}{4} + 406{,}000{,}000} = 162$$

Thus, a total sample size of 162 will provide the precision desired. To allocate the total sample to the three strata, we use equation (18.24).

$$n_1 = 162 \left[\frac{100(1300)}{460{,}000} \right] = 46$$

$$n_2 = 162 \left[\frac{200(900)}{460{,}000} \right] = 63$$

$$n_3 = 162 \left[\frac{300(500)}{460{,}000} \right] = 53$$

We would therefore recommend sampling 46 Corvette owners, 63 Geo Prizm owners, and 53 Cavalier owners for a total sample size of 162 customers.

To determine the sample size when estimating a population proportion, we simply substitute $\sqrt{\bar{p}_h(1 - \bar{p}_h)}$ for s_h in equation (18.25); the result is

$$n = \frac{\left(\sum_{h=1}^{H} N_h \sqrt{\bar{p}_h(1 - \bar{p}_h)} \right)^2}{N^2\left(\dfrac{B^2}{4}\right) + \sum_{h=1}^{H} N_h \bar{p}_h(1 - \bar{p}_h)} \tag{18.27}$$

Once the total sample size for the population proportion estimate has been determined, allocation to the various strata is again made using equation (18.24) with $\sqrt{\bar{p}_h(1 - \bar{p}_h)}$ substituted for s_h.

NOTES AND COMMENTS

1. An advantage of stratified simple random sampling is that estimates of population parameters for each stratum are automatically available as a byproduct of the sampling procedure. For example, besides obtaining an estimate of the average starting salary for all graduates in the Lakeland College sampling problem (Example 18.6), we also obtained an estimate of the average starting salary for each major. Since each of the starting-salary estimates was based on a simple random sample from each stratum, the procedure for developing an approximate confidence interval estimate when a simple random sample is selected [see equation (18.4)] can be used to compute an approximate 95% confidence interval estimate for the mean in each stratum. In a similar manner, interval estimates for the population total

—Continues on next page

—Continued from previous page

and the population proportion for each stratum can be developed using equations (18.8) and (18.10), respectively.

2. Another type of allocation that is sometimes used with stratified simple random sampling is called *proportional allocation*. Using this approach, the sample size allocated to each stratum is given by

$$n_h = n\left(\frac{N_h}{N}\right) \tag{18.28}$$

Proportional allocation is appropriate when the stratum variances are approximately equal and the cost per unit of sampling is about the same across strata. In the case where the stratum variances are equal, proportional allocation and the Neyman procedure result in the same allocation.

EXERCISES

METHODS

SELF TEST

7. A stratified simple random sample has been taken with the following results.

Stratum (h)	$\bar{x}_h$	s_h	$\bar{p}_h$	N_h	n_h
1	138	30	.50	200	20
2	103	25	.78	250	30
3	210	50	.21	100	25

   a. Develop an estimate of the population mean for each stratum.
   b. Develop an approximate 95% confidence interval for the population mean of each stratum.
   c. Develop an approximate 95% confidence interval for the overall population mean.

8. Reconsider the sample results in Exercise 7.
   a. Develop an estimate of the population total for each stratum.
   b. Develop a point estimate of the total for all 550 elements in the population.
   c. Develop an approximate 95% confidence interval for the population total.

9. Reconsider the sample results in Exercise 7.
   a. Develop an approximate 95% confidence interval for the proportion in each stratum.
   b. Develop a point estimate of the population proportion for the 550 elements in the population.
   c. Estimate the standard error of the population proportion.
   d. Develop an approximate 95% confidence interval for the population proportion.

10. A population has been divided into three strata with $N_1 = 300$, $N_2 = 600$, and $N_3 = 500$. From a past survey, the following estimates for the standard deviations in the three strata are available: $s_1 = 150$, $s_2 = 75$, $s_3 = 100$.
   a. Suppose an estimate of the population mean with a bound on the error of estimate of $B = 20$ is required. How large must the sample be? How many elements should be allocated to each stratum?
   b. Suppose that a bound of $B = 10$ is desired. How large must the sample be? How many elements should be allocated to each stratum?

c. Suppose an estimate of the population total with a bound of $B = 15,000$ is requested. How large must the sample be? How many elements should be allocated to each stratum?

APPLICATIONS

11. A drug store chain has stores located in four cities; there are 38 stores in Indianapolis, 45 in Louisville, 80 in St. Louis, and 70 in Memphis. Pharmacy sales in the four cities vary considerably due to the competition. The following sales data (in $1000s) are available from a sample survey. Each of the cities was considered a separate stratum, and a stratified simple random sample of stores was taken.

Pharmacy Sales			
Indianapolis	Louisville	St. Louis	Memphis
50.3	48.7	16.7	14.7
41.2	59.8	38.4	88.3
15.7	28.9	51.6	94.2
22.5	36.5	42.7	76.8
26.7	89.8	45.0	35.1
20.8	96.0	59.7	48.2
	77.2	80.0	57.9
	81.3	27.6	18.8
			22.0
			74.3

a. Estimate the mean sales for each city (stratum).
b. Develop an approximate 95% confidence interval for the mean sales in each city.
c. Estimate the proportion of stores with sales of $50,000 or more.
d. Develop an approximate 95% confidence interval for the proportion of stores with sales of $50,000 or more.

12. Reconsider the sample survey results in Exercise 11.
a. Estimate the population total for St. Louis.
b. Estimate the population total for Indianapolis.
c. Develop an approximate 95% confidence interval for mean pharmacy sales for the drug store chain.
d. Develop an approximate 95% confidence interval for total pharmacy sales for the drug store chain.

13. An accounting firm has a number of clients in the banking, insurance, and brokerage industries. There are $N_1 = 50$ banks, $N_2 = 38$ insurance companies, and $N_3 = 35$ brokerage firms. A marketing research firm has been hired to survey the accounting firm's clients in these three industries. The survey will ask a variety of questions about both the clients' businesses and their satisfaction with services provided by the accounting firm. Suppose an approximate 95% confidence interval is requested for the mean number of employees for the 123 clients with a bound on the error of estimation of $B = 30$.
a. Suppose a pilot study finds $s_1 = 80$, $s_2 = 150$, and $s_3 = 45$. Choose a total sample size, and explain how the sample size should be allocated to the three strata.
b. Suppose the pilot test is called into question and it is decided to assume the stratum standard deviations are all equal to 100 in choosing the sample size. Choose a total sample size, and determine how many elements should be sampled in each stratum.

18.6 ▽ CLUSTER SAMPLING

Cluster sampling requires that the population be divided into N groups of elements called clusters such that each element in the population belongs to one and only one cluster.

EXAMPLE 18.8 ▶

Suppose that we wanted to survey registered voters in the state of Ohio. One approach would be to develop a frame consisting of all registered voters in the state of Ohio and then select a simple random sample of voters from this frame. Alternatively, in cluster sampling, we might choose to define the frame as the list of the $N = 88$ counties in the state. In this approach, each county or cluster would consist of a group of registered voters, and each registered voter in the state would belong to one and only one cluster. A map of Ohio showing the 88 counties or clusters is provided in Figure 18.1.

Suppose that we selected a simple random sample of $n = 12$ of the 88 counties. At this point, we could collect data for *all* registered voters in each of the 12 sampled clusters, an approach referred to as *single-stage cluster sampling,* or we could select a simple random sample of registered voters from each of the 12 sampled clusters, an approach referred to as *two-stage cluster sampling.* ◀

Formulas are available for using the sample results obtained by cluster sampling to develop point and interval estimates of population parameters such as the population mean, total, or proportion. In this chapter, we only consider single-stage cluster sampling; more advanced texts on sampling present results for two-stage cluster sampling.

In the sense that both stratified and cluster sampling divide the population into groups of elements, the two sampling procedures are similar. The reasons for choosing cluster sampling, however, differ from the reasons for choosing stratified sampling. Cluster sampling tends to provide better results when the elements within the clusters are heterogeneous (not alike). In the ideal case, each cluster would be a small-scale version of the entire population. In this case, sampling a small number of clusters would provide good information about the characteristics of the entire population.

One of the primary applications of cluster sampling involves *area sampling,* where the clusters are counties, townships, city blocks, or other well-defined geographic sections of the population. Since data are collected from only a sample of the total geographic areas or clusters available, and since the elements within the clusters are typically close to one another, significant savings in time and cost can be realized when a data collector or interviewer is sent to a sampled unit. As a result, even if a larger total sample size is required, cluster sampling may be less costly than either simple random sampling or stratified simple random sampling. In addition, cluster sampling can minimize the time and cost associated with developing the frame or list of elements to be sampled, since cluster sampling does not require that a list of every element in the population be developed. One only needs a list of the elements in the clusters sampled.

EXAMPLE 18.9 ▶

Consider a survey conducted by the CPA (certified public accountant) Society of the 12,000 practicing CPAs in a particular state. As part of the survey, the CPA Society collected information on income, sex, and factors related to the CPA's life style. Because personal interviews were needed to obtain all the desired information, the CPA Society used a cluster sample to minimize the total travel and interviewing cost. The frame consisted of all CPA firms that were registered to practice accounting in the state. Suppose that there are $N = 1000$ clusters or CPA firms registered to practice accounting in the state, and that a simple random sample of $n = 100$ CPA firms is to be selected. ◀

FIGURE 18.1   COUNTIES OF THE STATE OF OHIO USED AS CLUSTERS OF REGISTERED VOTERS

In presenting the formulas for cluster sampling that are needed to develop approximate 95% confidence interval estimates of the population mean, total, and proportion, we will use the following notation:

N = number of clusters in the population
n = number of clusters selected in the sample
M_i = number of elements in cluster i
M = number of elements in the population; $M = M_1 + M_2 + \cdots + M_N$
$\bar{M} = M/N$ = average number of elements in a cluster
x_i = total of all observations in cluster i
a_i = number of observations in cluster i with a certain characteristic

For the CPA Society sample survey, we have

$$N = 1000$$

$$n = 10$$

$$M = 12,000$$

$$\bar{M} = 12,000/1000 = 12$$

Note that Table 18.4 shows the values of M_i and x_i for each of the sampled clusters as well as the number of female CPAs in the sampled firms (a_i). ◀

POPULATION MEAN

The point estimator of the population mean obtained from cluster sampling is given by the following formula.

Point Estimator of the Population Mean

$$\bar{x}_c = \frac{\displaystyle\sum_{i=1}^{n} x_i}{\displaystyle\sum_{i=1}^{n} M_i} \qquad (18.29)$$

An estimate of the standard error of $\bar{x}_c$ is

$$s_{\bar{x}_c} = \sqrt{\left(\frac{N-n}{Nn\bar{M}^2}\right)\frac{\displaystyle\sum_{i=1}^{n}(x_i - \bar{x}_c M_i)^2}{n-1}} \qquad (18.30)$$

TABLE 18.4   **RESULTS OF CPA SAMPLE SURVEY**

Firm (i)	CPAs (M_i)	Total Salary ($1000s) for Sample ($x_i$)	Females (a_i)
1	8	320	2
2	25	1125	8
3	4	115	0
4	17	714	6
5	7	247	1
6	3	94	2
7	15	634	2
8	4	147	0
9	12	481	5
10	33	1567	9
Totals	128	5444	35

Thus, an approximate 95% confidence interval estimate of the population mean is as follows.

> **Approximate 95% Confidence Interval Estimate of the Population Mean**
>
> $$\bar{x}_c \pm 2s_{\bar{x}_c} \qquad\qquad (18.31)$$

EXAMPLE 18.9
(CONTINUED)

Using the data in Table 18.4, an estimate of the mean salary for practicing certified public accountants is

$$\bar{x}_c = \frac{5444}{128} = 42.531$$

Since the salary data in Table 18.4 are in thousands of dollars, an estimate of the mean salary for practicing certified public accountants in the state is $42,531.

In Table 18.5, we show a portion of the calculations needed to estimate the standard error; note that

$$\sum_{i=1}^{n} (x_i - \bar{x}_c M_i)^2 = 39{,}178.688$$

Thus,

$$s_{\bar{x}_c} = \sqrt{\left[\frac{1000 - 10}{(1000)(10)(12)^2}\right]\frac{39{,}178.688}{10 - 1}} = 1.730$$

Hence, the standard error is $1730. Using equation (18.31), an approximate 95% confidence interval estimate for the mean annual salary is $42{,}531 \pm 2(1730) = 42{,}531 \pm 3460$ or $39,071 to $45,991. ◀

TABLE 18.5   **PARTIAL CALCULATIONS FOR THE ESTIMATE OF THE STANDARD ERROR FOR THE MEAN FOR CPA SAMPLE SURVEY WHERE $\bar{x}_c = 42.531$**

Firm (i)	M_i	x_i	$(x_i - 42.531M_i)^2$	
1	8	320	$[320 - 42.531(8)]^2 =$	409.982
2	25	1125	$[1125 - 42.531(25)]^2 =$	3,809.976
3	4	115	$[115 - 42.531(4)]^2 =$	3,038.655
4	17	714	$[714 - 42.531(17)]^2 =$	81.487
5	7	247	$[247 - 42.531(7)]^2 =$	2,572.214
6	3	94	$[94 - 42.531(3)]^2 =$	1,128.490
7	15	634	$[634 - 42.531(15)]^2 =$	15.721
8	4	147	$[147 - 42.531(4)]^2 =$	534.719
9	12	481	$[481 - 42.531(12)]^2 =$	862.714
10	33	1567	$[1567 - 42.531(33)]^2 =$	26,724.730
Totals	128	5444		39,178.688

$$\sum_{i=1}^{n} (x_i - \bar{x}_c M_i)^2$$

POPULATION TOTAL

The point estimator of the population total X is obtained for cluster sampling by multiplying M times $\bar{x}_c$.

Point Estimator of the Population Total

$$\hat{X} = M\bar{x}_c \tag{18.32}$$

An estimate of the standard error of $\hat{X}$ *is*

$$s_{\hat{X}} = Ms_{\bar{x}_c} \tag{18.33}$$

Thus, an approximate 95% confidence interval estimate for the population total is given by the following expression.

Approximate 95% Confidence Interval Estimate of the Population Total

$$M\bar{x}_c \pm 2s_{\hat{X}} \tag{18.34}$$

EXAMPLE 18.9
(CONTINUED)

For the CPA sample survey,

$$\hat{X} = M\bar{x}_c = 12{,}000(42{,}531) = \$510{,}372{,}000$$

$$s_{\hat{X}} = Ms_{\bar{x}_c} = 12{,}000(1730) = \$20{,}760{,}000$$

Thus, using equation (18.34), an approximate 95% confidence interval is $510,372,000 ± 2($20,760,000) = $510,372,000 ± $41,520,000 or $468,852,000 to $551,892,000.

POPULATION PROPORTION

The point estimator of the population proportion obtained from cluster sampling is as follows.

Point Estimator of the Population Proportion

$$\bar{p}_c = \frac{\displaystyle\sum_{i=1}^{n} a_i}{\displaystyle\sum_{i=1}^{n} M_i} \tag{18.35}$$

where

a_i = number of observations in cluster i with a certain characteristic

---

An estimate of the standard error of $\bar{p}_c$ is

$$s_{\bar{p}_c} = \sqrt{\left(\frac{N-n}{Nn\bar{M}^2}\right)\frac{\sum_{i=1}^{n}(a_i - \bar{p}_c M_i)^2}{n-1}} \qquad (18.36)$$

Thus, an approximate 95% confidence interval estimate for the population proportion is given by the following expression.

> **Approximate 95% Confidence Interval Estimate of the Population Proportion**
>
> $$\bar{p}_c \pm 2s_{\bar{p}_c} \qquad (18.37)$$

EXAMPLE 18.9 ▶
(CONTINUED)

For the CPA sample survey, we can use equation (18.35) and the data in Table 18.4 to develop an estimate of the number of practicing certified public accountants who are women.

$$\bar{p}_c = \frac{2 + 8 + \cdots + 9}{8 + 25 + \cdots + 33} = \frac{35}{128} = .2734$$

In Table 18.6 we show a portion of the calculations needed to estimate the standard error; note that

$$\sum_{i=1}^{n}(a_i - \bar{p}_c M_i)^2 = 15.2098$$

TABLE 18.6   **PARTIAL CALCULATIONS FOR THE ESTIMATE OF THE STANDARD ERROR OF $\bar{p}_c$ FOR THE CPA SAMPLE SURVEY WHERE $\bar{p}_c = .2734$**

Firm (i)	M_i	a_i	$(a_i - .2734M_i)^2$	
1	8	2	$[2 - .2734(8)]^2 =$	.0350
2	25	8	$[8 - .2734(25)]^2 =$	1.3572
3	4	0	$[0 - .2734(4)]^2 =$	1.1960
4	17	6	$[6 - .2734(17)]^2 =$	1.8284
5	7	1	$[1 - .2734(7)]^2 =$	.8350
6	3	2	$[2 - .2734(3)]^2 =$	1.3919
7	15	2	$[2 - .2734(15)]^2 =$	4.4142
8	4	0	$[0 - .2734(4)]^2 =$	1.1960
9	12	5	$[5 - .2734(12)]^2 =$	2.9556
10	33	9	$[9 - .2734(33)]^2 =$	.0005
Totals	128	35		15.2098

$$\sum_{i=1}^{n}(a_i - \bar{p}_c M_i)^2$$

Thus,

$$s_{\bar{p}_c} = \sqrt{\left[\frac{1000-10}{(1000)(10)(12)^2}\right]\frac{15.2098}{10-1}} = .0341$$

Hence, using equation (18.37), an approximate 95% confidence interval for the proportion of practicing CPAs who are women is $.2734 \pm 2(.0341) = .2734 \pm .0682$ or .2052 to .3416.   ◀

DETERMINING THE SAMPLE SIZE

Once the clusters have been formed, the primary issue in choosing a sample size is selecting the number of clusters n. The procedure for cluster sampling is similar to that for other methods of sampling. An acceptable level of precision is specified by choosing a value for B, the bound on the sampling error. Then a formula is developed for finding the value of n that will provide the desired precision.

The average cluster size and the variance between clusters are key factors in deciding how many clusters to include in the sample. If the clusters are similar, there will be a small variance between them, and the number of clusters sampled can be smaller. Also, if the average number of elements per cluster is larger, the number of clusters sampled can be smaller. The formulas for making an exact determination of sample size are included in more advanced texts on sampling.

▷

EXERCISES

METHODS

SELF TEST ▷

14. A sample of four clusters is to be taken from a population with $N = 25$ clusters and $M = 300$ elements. The values of M_i, x_i, and a_i for each cluster in the sample are given in the table below.

Cluster (i)	M_i	x_i	a_i
1	7	95	1
2	18	325	6
3	15	190	6
4	10	140	2
Totals	50	750	15

a. Develop point estimates of the population mean, total, and proportion.
b. Estimate the standard error for the estimates of part (a).
c. Develop an approximate 95% confidence interval for the population mean.
d. Develop an approximate 95% confidence interval for the population total.
e. Develop an approximate 95% confidence interval for the population proportion.

15. A sample of six clusters is to be taken from a population with $N = 30$ clusters and $M = 600$ elements in the population. The values of M_i, x_i, and a_i for each cluster in the sample are given at the top of page 785.

Cluster (i)	M_i	x_i	a_i
1	35	3,500	3
2	15	965	0
3	12	960	1
4	23	2,070	4
5	20	1,100	3
6	25	1,805	2
Totals	130	10,400	13

a. Develop point estimates of the population mean, total, and proportion.
b. Develop an approximate 95% confidence interval for the population mean.
c. Develop an approximate 95% confidence interval for the population total.
d. Develop an approximate 95% confidence interval for the population proportion.

APPLICATIONS

16. A public utility is conducting a survey of mechanical engineers to learn more about the factors influencing the choice of heating, ventilation, and air conditioning (HVAC) equipment for new commercial buildings. There are a total of 120 firms in the utility's service area engaged in designing HVAC systems. The sampling plan is to use cluster sampling with each firm representing a cluster. For each firm in the sample, all of the mechanical engineers will be interviewed. It is believed that there are approximately 500 mechanical engineers employed by the 120 firms.

A sample of 10 firms was taken. Among other things, the age of each respondent was recorded as well as whether or not the respondent had attended the local university.

Cluster (i)	M_i	Total of Respondents' Ages	Number Attending Local University
1	12	520	8
2	1	33	0
3	2	70	1
4	1	29	1
5	6	270	3
6	3	129	2
7	2	102	0
8	1	48	1
9	9	337	7
10	13	462	12
Totals	50	2000	35

a. Estimate the mean age of mechanical engineers engaged in this type of work.
b. Estimate the proportion of mechanical engineers in the utility's service area who attended the local university.
c. Develop an approximate 95% confidence interval for the mean age of mechanical engineers designing HVAC systems for commercial buildings.
d. Develop an approximate 95% confidence interval for the proportion of mechanical engineers in the utility's service area who attended the local university.

17. A national real estate company has just acquired a smaller firm that has 150 offices and 6000 agents in Los Angeles and other parts of southern California. The national firm has conducted a sample survey to learn about attitudes and other characteristics of its new employees. A sample of eight offices has been taken, and all of the agents at these offices have completed the questionnaire. Results of the survey for the eight offices are given below.

Office	Agents	Average Age	College Graduates	Males
1	17	37	3	4
2	35	32	14	12
3	26	36	8	7
4	66	30	38	28
5	43	41	18	12
6	12	52	2	6
7	48	35	20	17
8	57	44	25	26

  a. Estimate the mean age of the agents.

  b. Estimate the proportion of agents who are college graduates and the proportion who are male.

  c. Develop an approximate 95% confidence interval for the mean age of the agents.

  d. Develop an approximate 95% confidence interval for the proportion of agents who are college graduates.

  e. Develop an approximate 95% confidence interval for the proportion of agents who are male.

18.7 SYSTEMATIC SAMPLING

Systematic sampling is often used as an alternative to simple random sampling. In some sampling situations, especially those with large populations, it can be time-consuming to select a simple random sample by first finding a random number and then counting or searching through the frame until the corresponding element is found. Systematic sampling offers an alternative to simple random sampling in such cases. For example, if a sample size of 50 from a population containing 5000 elements is desired, we might sample 1 element for every 5000/50 = 100 elements in the population. A systematic sample for this case would involve randomly selecting 1 of the first 100 elements from the frame. The remaining sample elements are identified by starting with the first sampled element and then selecting every 100th element that follows in the frame. In effect, the sample of 50 is identified by moving systematically through the population and identifying every 100th element after the first randomly selected element. The sample of 50 will often be easier to select in this manner than would be the case if simple random sampling were used. Since the first element selected is a random choice, the assumption that a systematic sample has the properties of a simple random sample is often made. This assumption is usually appropriate when the frame is a random ordering of the elements in the population.

SUMMARY

We have provided a brief introduction to the field of survey sampling in this chapter. The purpose of survey sampling is to collect existing data for the purpose of making estimates of population parameters such as the population mean, total, or proportion. Survey sampling as a method of data collection can be contrasted with conducting experiments to generate data. When survey sampling is used, the design of the sampling plan is of critical importance in determining which existing data are collected. When experiments are employed, experimental design issues are of critical importance in determining which data are generated, or created.

There are two types of errors that can occur with sample surveys: sampling error and nonsampling error. Sampling error is the error that occurs because a sample, and not the entire population, is used to estimate a population parameter. Nonsampling error refers to all the other types of errors that can occur, such as measurement, interviewer, nonresponse, and processing error. Nonsampling errors are controlled by proper questionnaire design, thorough training of interviewers, careful verification of data, and so on. Sampling errors are minimized by a proper choice of sample design and by selecting an appropriate sample size.

We discussed four commonly used sample designs in this chapter: simple random sampling, stratified simple random sampling, cluster sampling, and systematic sampling. The objective of sample design is to get the most precise estimates for the least cost. When the population can be divided into strata so that the elements within each stratum are relatively homogeneous, stratified simple random sampling will provide more precision (smaller approximate confidence intervals) than simple random sampling. When the elements can be grouped in clusters so that all the elements in a cluster are close together geographically, cluster sampling often reduces interviewer cost; in these situations, cluster sampling will often provide the most precision per dollar. Systematic random sampling was presented as an alternative to simple random sampling.

GLOSSARY

Element The entity on which data are collected.

Population The collection of all elements of interest.

Sample A subset of the population.

Sampled population The population from which the sample is taken.

Target population The population about which inferences are made.

Sampling unit The units selected for sampling. A sampling unit may include several elements.

Frame A list of the sampling units for a study. The sample is drawn by selecting units from the frame.

Probabilistic sampling Any method of sampling in which the probability of each possible sample can be computed.

Nonprobabilistic sampling Any method of sampling for which the probability of selecting a sample cannot be computed.

Convenience sampling A nonprobabilistic method of sampling whereby elements are selected on the basis of convenience.

Judgment sampling A nonprobabilistic method of sampling whereby element selection is based on the judgment of the person doing the study.

Sampling error The error that occurs because a sample, and not the entire population, is used to estimate a population parameter.

Nonsampling error All types of errors other than sampling error, such as measurement error, interviewer error, and processing error.

Simple random sample A sample selected in such a manner that each sample of size n has the same probability of being selected.

Bound on error of the estimate A number added to and subtracted from a point estimate to create an approximate 95% confidence interval. It is given by 2 times the standard error of the point estimator.

Stratified simple random sampling A probabilistic method of selecting a sample in which the population is first divided into strata and a simple random sample is then taken from each stratum.

Cluster sampling A probabilistic method of sampling in which the population is first divided into clusters and then one or more clusters are selected for sampling. In single-stage cluster sampling, every element in each selected cluster is sampled; in two-stage cluster sampling, a sample of the elements in each selected cluster is collected.

Area sampling A version of cluster sampling in which the elements are formed into clusters on the basis of their geographic proximity.

Systematic sampling A method of choosing a sample by randomly selecting the first element and then selecting every kth element thereafter.

KEY FORMULAS

Simple Random Sampling

Interval Estimate of the Population Mean (σ Known)

$$\bar{x} \pm z_{\alpha/2}\sigma_{\bar{x}} \tag{18.1}$$

Estimate of the Standard Error of the Population Mean

$$s_{\bar{x}} = \sqrt{\frac{N-n}{N}}\left(\frac{s}{\sqrt{n}}\right) \tag{18.2}$$

Interval Estimate of the Population Mean

$$\bar{x} \pm z_{\alpha/2}s_{\bar{x}} \tag{18.3}$$

Approximate 95% Confidence Interval Estimate of the Population Mean

$$\bar{x} \pm 2s_{\bar{x}} \tag{18.4}$$

Point Estimator of a Population Total

$$\hat{X} = N\bar{x} \tag{18.5}$$

Estimate of the Standard Error of $\hat{X}$

$$s_{\hat{X}} = Ns_{\bar{x}} \tag{18.6}$$

Approximate 95% Confidence Interval Estimate of the Population Total

$$N\bar{x} \pm 2s_{\hat{X}} \tag{18.8}$$

Estimate of the Standard Error of the Population Proportion

$$s_{\bar{p}} = \sqrt{\left(\frac{N-n}{N}\right)\left(\frac{\bar{p}(1-\bar{p})}{n-1}\right)} \tag{18.9}$$

Approximate 95% Confidence Interval Estimate of the Population Proportion

$$\bar{p} \pm 2s_{\bar{p}} \tag{18.10}$$

Sample Size for an Estimate of the Population Mean

$$n = \frac{Ns^2}{N\left(\dfrac{B^2}{4}\right) + s^2} \qquad \textbf{(18.12)}$$

Sample Size for an Estimate of the Population Total

$$n = \frac{Ns^2}{\dfrac{B^2}{4N} + s^2} \qquad \textbf{(18.13)}$$

Sample Size for an Estimate of the Population Proportion

$$n = \frac{N\bar{p}(1 - \bar{p})}{N\left(\dfrac{B^2}{4}\right) + \bar{p}(1 - \bar{p})} \qquad \textbf{(18.14)}$$

Stratified Simple Random Sampling

Point Estimator of the Population Mean

$$\bar{x}_{\text{st}} = \sum_{h=1}^{H} \left(\frac{N_h}{N}\right)\bar{x}_h \qquad \textbf{(18.15)}$$

Estimate of the Standard Error of the Population Mean

$$s_{\bar{x}_{\text{st}}} = \sqrt{\frac{1}{N^2} \sum_{h=1}^{H} N_h(N_h - n_h)\frac{s_h^2}{n_h}} \qquad \textbf{(18.16)}$$

Approximate 95% Confidence Interval Estimate of the Population Mean

$$\bar{x}_{\text{st}} \pm 2s_{\bar{x}_{\text{st}}} \qquad \textbf{(18.17)}$$

Point Estimator of the Population Total

$$\hat{X} = N\bar{x}_{\text{st}} \qquad \textbf{(18.18)}$$

Estimate of the Standard Error of $\hat{X}$

$$s_{\hat{X}} = Ns_{\bar{x}_{\text{st}}} \qquad \textbf{(18.19)}$$

Approximate 95% Confidence Interval Estimate of the Population Total

$$N\bar{x}_{\text{st}} \pm 2s_{\hat{X}} \qquad \textbf{(18.20)}$$

Point Estimator of the Population Proportion

$$\bar{p}_{\text{st}} = \sum_{h=1}^{H} \left(\frac{N_h}{N}\right)\bar{p}_h \qquad \textbf{(18.21)}$$

Estimate of the Standard Error of $\bar{p}_{\text{st}}$

$$s_{\bar{p}_{\text{st}}} = \sqrt{\frac{1}{N^2} \sum_{h=1}^{H} N_h(N_h - n_h)\left[\frac{\bar{p}_h(1 - \bar{p}_h)}{n_h - 1}\right]} \qquad \textbf{(18.22)}$$

Approximate 95% Confidence Interval Estimate of the Population Proportion

$$\bar{p}_{\text{st}} \pm 2s_{\bar{p}_{\text{st}}} \tag{18.23}$$

Allocating the Total Sample n to the Strata: Neyman Allocation

$$n_h = n \left(\frac{N_h s_h}{\sum\limits_{h=1}^{H} N_h s_h} \right) \tag{18.24}$$

Sample Size for the Estimate of the Population Mean

$$n = \frac{\left(\sum\limits_{h=1}^{H} N_h s_h \right)^2}{N^2 \left(\dfrac{B^2}{4} \right) + \sum\limits_{h=1}^{H} N_h s_h^2} \tag{18.25}$$

Sample Size for the Estimate of the Population Total

$$n = \frac{\left(\sum\limits_{h=1}^{H} N_h s_h \right)^2}{\dfrac{B^2}{4} + \sum\limits_{h=1}^{H} N_h s_h^2} \tag{18.26}$$

Sample Size for the Estimate of the Population Proportion

$$n = \frac{\left(\sum\limits_{h=1}^{H} N_h \sqrt{\bar{p}_h(1 - \bar{p}_h)} \right)^2}{N^2 \left(\dfrac{B^2}{4} \right) + \sum\limits_{h=1}^{H} N_h \bar{p}_h(1 - \bar{p}_h)} \tag{18.27}$$

Proportional Allocation of Total Sample n to the Strata

$$n_h = n \left(\frac{N_h}{N} \right) \tag{18.28}$$

Cluster Sampling

Point Estimator of the Population Mean

$$\bar{x}_{\text{c}} = \frac{\sum\limits_{i=1}^{n} x_i}{\sum\limits_{i=1}^{n} M_i} \tag{18.29}$$

Estimate of the Standard Error of the Population Mean

$$s_{\bar{x}_{\text{c}}} = \sqrt{\left(\frac{N - n}{Nn\bar{M}^2} \right) \frac{\sum\limits_{i=1}^{n} (x_i - \bar{x}_{\text{c}} M_i)^2}{n - 1}} \tag{18.30}$$

Approximate 95% Confidence Interval Estimate of the Population Mean

$$\bar{x}_c \pm 2s_{\bar{x}_c} \qquad (18.31)$$

Point Estimator of the Population Total

$$\hat{X} = M\bar{x}_c \qquad (18.32)$$

Estimate of the Standard Error of $\hat{X}$

$$s_{\hat{X}} = Ms_{\bar{x}_c} \qquad (18.33)$$

Approximate 95% Confidence Interval Estimate of the Population Total

$$M\bar{x}_c \pm 2s_{\hat{X}} \qquad (18.34)$$

Point Estimator of the Population Proportion

$$\bar{p}_c = \frac{\sum_{i=1}^{n} a_i}{\sum_{i=1}^{n} M_i} \qquad (18.35)$$

Estimate of the Standard Error of $\bar{p}_c$

$$s_{\bar{p}_c} = \sqrt{\left(\frac{N-n}{Nn\bar{M}^2}\right)\frac{\sum_{i=1}^{n}(a_i - \bar{p}_c M_i)^2}{n-1}} \qquad (18.36)$$

Approximate 95% Confidence Interval Estimate of the Population Proportion

$$\bar{p}_c \pm 2s_{\bar{p}_c} \qquad (18.37)$$

REVIEW QUIZ

TRUE/FALSE

1. The nonsampling error is always zero when a census is taken.
2. The bound on the sampling error decreases as the sample size increases.
3. The frame for a sample survey is a list of the units making up the sample.
4. The precision in a sample survey refers to the care exercised in designing the questionnaire.
5. Judgment sampling is a type of nonprobabilistic sampling.
6. Simple random sampling is a form of probabilistic sampling.
7. Stratified random sampling works well in situations where the variance between strata is large relative to the variance within strata.
8. Cluster sampling works well when each cluster is representative of the population.
9. Cluster sampling is often used when controlling travel costs is an important consideration in administering a survey.
10. Prior to using systematic sampling, the population must be stratified.

MULTIPLE CHOICE

Use the following data for Questions 11–13. A simple random sample of 36 elements from a population of 100 elements yielded $\bar{x} = 720$ and $s = 24$.

11. An estimate of the standard error of the mean is
  a. 4.00   **b.** 3.20
  c. 10.24   **d.** 2.56

12. An approximate 95% confidence interval estimate for the population mean is
  a. 713.60 to 726.40
  b. 696 to 744
  c. 672 to 768
  d. 716.80 to 723.20

13. An approximate 95% confidence interval estimate for the population total is
  a. 70,000 to 74,000
  b. 71,200 to 72,800
  c. 71,680 to 72,320
  d. 71,600 to 72,320

Use the following data for Questions 14–16. A stratified simple random sample provided these results: $N_1 = 100$, $n_1 = 30$, $\bar{p}_1 = .2$; $N_2 = 150$, $n_2 = 40$, $\bar{p}_2 = .5$; $N_3 = 180$, $n_3 = 45$, $\bar{p}_3 = .4$.

14. The number of elements in the population is
  a. 115   **b.** 450
  c. 545   **d.** 430

15. A point estimate of the population proportion is
  a. .200   **b.** .500
  c. .293   **d.** .305

16. An estimate of the standard error of $\bar{p}_{st}$ is closest to
  a. .02   **b.** .03
  c. .04   **d.** .10

Use the following data for Questions 17 and 18. A cluster sample provided the following: $x_1 = 90$, $a_1 = 12$, $M_1 = 20$; $x_2 = 90$, $a_2 = 15$, $M_2 = 30$; $x_3 = 60$, $a_3 = 13$, $M_3 = 30$.

17. A point estimate of the population mean is given by
  a. .50   **b.** 3.00
  c. .33   **d.** 2.00

18. A point estimate of the population proportion is given by
  a. .50   **b.** 3.00
  c. .33   **d.** 2.00

SUPPLEMENTARY EXERCISES

18. A *USA Today*/CNN/Gallup telephone poll of 1421 adults nationwide (*USA Today*, January 16, 1992) concerned issues that were likely to influence voters. The following questions are based on that survey. (*Note:* Since the survey sampled a very small fraction of total adults, assume $(N - n)/N = 1$ in any formulas involving standard error.)

  a. Eighty percent indicated they would be more likely to vote for a candidate in favor of "requiring all able-bodied people on welfare to do work for their welfare checks." Develop an approximate 95% confidence interval for the population proportion.

  b. Sixty percent indicated they would be more likely to vote for a candidate in favor of "a national health care system paid for by new taxes." Develop an approximate 95% confidence interval for the population proportion.

  c. Forty-four percent indicated they would be more likely to vote for a candidate in favor of "giving families a tax credit for each child attending a private or parochial school" while 40% were less likely to vote for such a candidate and 16% did not express a preference. Develop an approximate 95% confidence interval for the proportion of adults more likely to vote for such a candidate.

d. *USA Today* reported that the "margin of sampling error is plus or minus 3 percentage points" for the survey. What does this mean, and how do you think they arrived at this number?

e. How might nonsampling error bias the results of such a survey? The actual polling was done by the Gallup organization.

19. The National Association of Children's Hospitals and Related Institutions and the new Coalition for America's Children (*Cincinnati Enquirer*, January 8, 1992) commissioned a nationwide telephone survey of 1083 registered voters concerning children's issues.

a. Seventy-one percent of respondents worried that "the quality of education is declining for children in public schools." Develop an approximate 95% confidence interval for the population proportion.

b. Sixty-two percent worried that "children are not provided with the basics that they need in health care, food and education." Develop an approximate 95% confidence interval for the population proportion.

c. Thirty-four percent worried that "pregnant mothers aren't getting the health care they need to make sure their child is healthy." Develop an approximate 95% confidence interval for this estimate.

d. It was reported that the "percent of error is plus or minus 3.1%." Interpret this statement.

20. A quality of life survey was conducted with employees of a manufacturing firm. Of the firm's 3000 employees, a sample of 300 were sent questionnaires. Two hundred usable questionnaires were obtained for a response rate of 67%.

a. The mean annual salary for the sample was $\bar{x} = \$23,200$ with $s = 3000$. Develop an approximate 95% confidence interval for the mean annual salary of the population.

b. Using the information in part (a), develop an approximate 95% confidence interval for the total salary of all 3000 employees.

c. Seventy-three percent of respondents reported that they were "generally satisfied" with their job. Develop an approximate 95% confidence interval for the population proportion.

d. Comment on whether or not you think the results in part (c) might be biased. Would your opinion change if you knew the respondents were guaranteed anonymity?

21. A U.S. Senate Judiciary Committee report showed the number of homicides state by state for 1991. In the states of Indiana, Ohio, and Kentucky, the number of homicides were, respectively, 380, 760, and 260. Suppose a stratified random sample with the following results were taken to learn more about the victims and the cause of death.

Stratum	Sample Size	Shootings	Beatings	Black Victims
Indiana	30	10	9	21
Ohio	45	19	12	34
Kentucky	25	7	11	15

a. Develop an approximate 95% confidence interval for the proportion of shooting deaths in Indiana.

b. Develop an estimate for the total number of shooting deaths in Ohio.

c. Develop an approximate 95% confidence interval for the proportion of shooting deaths in Ohio.

d. Develop an approximate 95% confidence interval for the proportion of shooting deaths across all three states.

22. Refer to the data in Exercise 21.
   a. Develop an estimate of the total number of deaths (in the three states) due to beatings.
   b. Develop an approximate 95% confidence interval for the proportion of deaths across all three states due to beatings.
   c. Develop an approximate 95% confidence interval for the proportion of victims who are black.
   d. Develop an estimate of the total number of victims who are black.

23. A stratified simple random sample is to be taken of a bank's customers to learn about a variety of attitudinal and demographic issues. The stratification is to be based on savings account balances as of June 30, 1992. A frequency distribution showing the number of accounts in each stratum, together with the standard deviation of account balances by stratum, is shown below.

Stratum ($)	Accounts	Standard Deviation of Account Balances
0.00– 1,000.00	3,000	80
1,000.01– 2,000.00	600	150
2,000.01– 5,000.00	250	220
5,000.01–10,000.00	100	700
over 10,000.00	50	3000

   a. Assuming the cost per unit sampled is approximately equal across strata, determine the total number of persons to include in the sample. Assume we would like a bound on the error of estimate of the population mean for savings account balances of $B = \$20$.
   b. Use the Neyman allocation procedure to determine the number to be sampled for each stratum.

24. A public agency is interested in learning more about the persons living in nursing homes in a particular city. There are a total of 100 nursing homes caring for 4800 people in the city and a cluster sample of six homes has been taken. Each person in the six homes has been interviewed. A portion of the sample results are shown below.

Home	Residents	Average Age of Residents	Disabled Residents
1	14	61	12
2	7	74	2
3	96	78	30
4	23	69	8
5	71	73	10
6	29	84	22

   a. Develop an estimate of the mean age of nursing home residents in this city.
   b. Develop an approximate 95% confidence interval for the proportion of disabled persons in the city's nursing homes.
   c. Estimate the total number of disabled persons residing in nursing homes in this city.

APPENDIXES

A
REFERENCES AND BIBLIOGRAPHY

B
TABLES

C
SUMMATION NOTATION

D
THE DATA DISK AND ITS USE

E
ANSWERS TO REVIEW QUIZZES

F
ANSWERS TO EVEN-NUMBERED EXERCISES

G
SOLUTIONS TO SELF-TEST EXERCISES

APPENDIX A

REFERENCES AND BIBLIOGRAPHY

GENERAL

Freedman, D., R. Pisani, and R. Purves, *Statistics,* New York, W. W. Norton, 1978.

Freund, J. E., and R. E. Walpole, *Mathematical Statistics,* 4th ed., Englewood Cliffs, N.J., Prentice-Hall, 1987.

Hoaglin, D. C., F. Mosteller, and J. W. Tukey, *Understanding Robust and Exploratory Data Analysis,* New York, Wiley, 1983.

Hogg, R. V., and A. T. Craig, *Introduction to Mathematical Statistics,* 4th ed., New York, Macmillian, 1978.

Mood, A. M., F. A. Graybill, and D. C. Boes, *Introduction to the Theory of Statistics,* 3rd ed., New York, McGraw-Hill, 1974.

Neter, J., W. Wasserman, and G. A. Whitmore, *Applied Statistics,* 3rd ed., Boston, Allyn & Bacon, 1987.

Roberts, H., *Data Analysis for Managers,* Redwood City, Calif., Scientific Press, 1988.

Ryan, T. A., B. L. Joiner, and B. F. Ryan, *Minitab Handbook,* 2nd ed., Boston, PWS-Kent, 1985.

Tanur, J. M., et al., *Statistics: A Guide to the Unknown,* 2nd ed., San Francisco, Holden-Day, 1978.

Tukey, J. W., *Exploratory Data Analysis,* Reading, Mass., Addison-Wesley, 1977.

Winkler, R. L., and W. L. Hays, *Statistics: Probability, Inference, and Decision,* 2nd ed., New York, Holt, Rinehart & Winston, 1975.

PROBABILITY

Barr, D. R., and P. W. Zehna, *Probability: Modeling Uncertainty,* Reading, Mass., Addison-Wesley, 1983.

Feller, W., *An Introduction to Probability Theory and Its Applications,* Vol. I, 3rd ed., New York, Wiley, 1968.

Feller, W., *An Introduction to Probability Theory and Its Applications,* Vol. II, 2nd ed., New York, Wiley, 1971.

Hoel, P. G., S. C. Port, and C. J. Stone, *Introduction to Probability Theory,* Boston, Houghton Mifflin, 1971.

Mendenhall, W., R. L. Scheaffer, and D. Wackerly, *Mathematical Statistics with Applications,* 4th ed., Boston, PWS-Kent, 1990.

Parzen, E., *Modern Probability Theory and Its Applications,* New York, Wiley, 1960.

Wadsworth, G. P., and J. G. Bryan, *Applications of Probability and Random Variables,* 2nd ed., New York, McGraw-Hill, 1974.

Zehna, P. W., *Probability Distributions and Statistics,* Boston, Allyn & Bacon, 1970.

EXPERIMENTAL DESIGN

Anderson, V. L., and R. A. McLean, *Design of Experiments: A Realistic Approach,* New York, Marcel Dekker, 1974.

Box, G. E. P., W. G. Hunter, and J. S. Hunter, *Statistics for Experimenters,* New York, Wiley, 1978.

Cochran, W. G., and G. M. Cox, *Experimental Designs,* 2nd ed., New York, Wiley, 1957.

Hicks, C. R., *Fundamental Concepts in the Design of Experiments,* 3rd ed., New York, Holt, Rinehart & Winston 1982.

Maxwell, S. E. and H. D. Delaney, *Designing Experiments and Analyzing Data,* Belmont, CA, Wadsworth, 1990.

Mendenhall, W., *Introduction to Linear Models and the Design and Analysis of Experiments,* Belmont, Calif., Wadsworth, 1968.

Montgomery, D. C., *Design and Analysis of Experiments,* New York, Wiley, 1976.

Winer, B. J., *Statistical Principles in Experimental Design,* 2nd ed., New York, McGraw-Hill, 1971.

REGRESSION ANALYSIS

Belsley, D. A., E. Kuh, and R. Welsch, *Regression Diagnostics: Identifying Influential Data and Sources of Collinearity,* New York, Wiley, 1980.

Chatterjee, S., and B. Price, *Regression Analysis by Example,* New York, Wiley, 1978.

Cook, R. D., and S. Weisberg, *Residuals and Influence in Regression,* New York, Chapman and Hall, 1982.

Daniel, C., and F. Wood, *Fitting Equations to Data,* 2nd ed., New York, Wiley, 1980.

Draper, N. R., and H. Smith, *Applied Regression Analysis,* 2nd ed., New York, Wiley, 1981.

Gunst, R. F., and R. L. Mason, *Regression Analysis and Its Application: A Data-Oriented Approach,* New York, Marcel Dekker, 1980.

Kleinbaum, D. G., and L. L. Kupper, *Applied Regression Analysis and Other Multivariable Methods,* North Scituate, Mass., Duxbury Press, 1978.

Mendenhall, W., *Introduction to Linear Models and the Design and Analysis of Experiments,* Belmont, Calif., Wadsworth, 1968.

Mosteller, F., and J. W. Tukey, *Data Analysis and Regression: A Second Course in Statistics,* Reading, Mass., Addison-Wesley, 1977.

Neter, J., W. Wasserman, and M. H. Kutner, *Applied Linear Statistical Models,* 2nd ed., Homewood, Ill., Richard D. Irwin, 1985.

Weisberg, S., *Applied Linear Regression,* 2nd ed., New York, Wiley, 1985.

Wesolowsky, G. O., *Multiple Regression and Analysis of Variance,* New York, Wiley, 1976.

Wonnacott, T. H., and R. J. Wonnacott, *Regression: A Second Course in Statistics,* New York, Wiley, 1981.

NONPARAMETRIC METHODS

Conover, W. J., *Practical Nonparametric Statistics,* 2nd ed., New York, Wiley, 1980.

Daniel, W. W., *Applied Nonparametric Statistics,* Boston, Houghton Mifflin, 1978.

Gibbons, J. D., *Nonparametric Statistical Inference,* New York, McGraw-Hill, 1971.

Gibbons, J. D., I. Olkin, and M. Sobel, *Selecting and Ordering Populations: A New Statistical Methodology,* New York, Wiley, 1977.

Hollander, M., and D. A. Wolfe, *Nonparametric Statistical Methods,* New York, Wiley, 1973.

Lehmann, E. L., *Nonparametrics: Statistical Methods Based on Ranks,* San Francisco, Holden-Day, 1975.

Mosteller, F., and R. E. K. Rourke, *Sturdy Statistics,* Reading, Mass., Addison-Wesley, 1973.

Siegel, S., *Nonparametric Statistics for the Behavioral Sciences,* New York, McGraw-Hill, 1956.

SAMPLING METHODS

Cochran, W. G., *Sampling Techniques,* 3rd ed., New York, Wiley, 1977.

Deming, W. E., *Sample Design in Business Research,* New York, Wiley, 1960.

Kish, L., *Survey Sampling,* New York, Wiley, 1965.

Levy, P. S., and S. Lemeshow, *Sampling of Populations: Methods and Applications,* New York, Wiley, 1991.

Scheaffer, R. L., W. Mendenhall, and L. Ott, *Elementary Survey Sampling,* 4th ed., Boston, PWS-Kent, 1990.

Williams, B., *A Sampler on Sampling,* New York, Wiley, 1978.

Warwick, D. P., and C. Lininger, *The Sample Survey: Theory and Practice,* New York, McGraw-Hill, 1975.

APPENDIX B

TABLES

TABLE 1

Standard Normal
Distribution

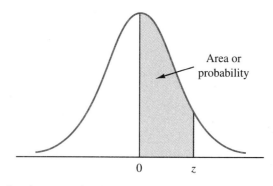

Area or
probability

0   z

Entries in the table give the area under the curve between the mean and z standard deviations above the mean. For example, for $z1.25$ the area under the curve between the mean and z is .3944.

z	.00	.01	.02	.03	.04	.05	.06	.07	.08	.09
.0	.0000	.0040	.0080	.0120	.0160	.0199	.0239	.0279	.0319	.0359
.1	.0398	.0438	.0478	.0517	.0557	.0596	.0636	.0675	.0714	.0753
.2	.0793	.0832	.0871	.0910	.0948	.0987	.1026	.1064	.1103	.1141
.3	.1179	.1217	.1255	.1293	.1331	.1368	.1406	.1443	.1480	.1517
.4	.1554	.1591	.1628	.1664	.1700	.1736	.1772	.1808	.1844	.1879
.5	.1915	.1950	.1985	.2019	.2054	.2088	.2123	.2157	.2190	.2224
.6	.2257	.2291	.2324	.2357	.2389	.2422	.2454	.2486	.2518	.2549
.7	.2580	.2612	.2642	.2673	.2704	.2734	.2764	.2794	.2823	.2852
.8	.2881	.2910	.2939	.2967	.2995	.3023	.3051	.3078	.3106	.3133
.9	.3159	.3186	.3212	.3238	.3264	.3289	.3315	.3340	.3365	.3389
1.0	.3413	.3438	.3461	.3485	.3508	.3531	.3554	.3577	.3599	.3621
1.1	.3643	.3665	.3686	.3708	.3729	.3749	.3770	.3790	.3810	.3830
1.2	.3849	.3869	.3888	.3907	.3925	.3944	.3962	.3980	.3997	.4015
1.3	.4032	.4049	.4066	.4082	.4099	.4115	.4131	.4147	.4162	.4177
1.4	.4192	.4207	.4222	.4236	.4251	.4265	.4279	.4292	.4306	.4319
1.5	.4332	.4345	.4357	.4370	.4382	.4394	.4406	.4418	.4429	.4441
1.6	.4452	.4463	.4474	.4484	.4495	.4505	.4515	.4525	.4535	.4545
1.7	.4554	.4564	.4573	.4582	.4591	.4599	.4608	.4616	.4625	.4633
1.8	.4641	.4649	.4656	.4664	.4671	.4678	.4686	.4693	.4699	.4706
1.9	.4713	.4719	.4726	.4732	.4738	.4744	.4750	.4756	.4761	.4767
2.0	.4772	.4778	.4783	.4788	.4793	.4798	.4803	.4808	.4812	.4817
2.1	.4821	.4826	.4830	.4834	.4838	.4842	.4846	.4850	.4854	.4857
2.2	.4861	.4864	.4868	.4871	.4875	.4878	.4881	.4884	.4887	.4890
2.3	.4893	.4896	.4898	.4901	.4904	.4906	.4909	.4911	.4913	.4916
2.4	.4918	.4920	.4922	.4925	.4927	.4929	.4931	.4932	.4934	.4936
2.5	.4938	.4940	.4941	.4943	.4945	.4946	.4948	.4949	.4951	.4952
2.6	.4953	.4955	.4956	.4957	.4959	.4960	.4961	.4962	.4963	.4964
2.7	.4965	.4966	.4967	.4968	.4969	.4970	.4971	.4972	.4973	.4974
2.8	.4974	.4975	.4976	.4977	.4977	.4978	.4979	.4979	.4980	.4981
2.9	.4981	.4982	.4982	.4983	.4984	.4984	.4985	.4985	.4986	.4986
3.0	.4986	.4987	.4987	.4988	.4988	.4989	.4989	.4989	.4990	.4990

TABLE 2

t Distribution

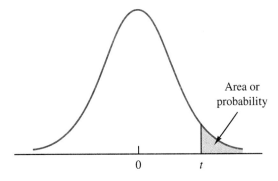

Area or probability

0   *t*

Entries in the table give *t* values for an area or probability in the upper tail of the *t* distribution. For example, with 10 degrees of freedom and a .05 area in the upper tail, $t_{.05}1.812$.

Degrees of Freedom	Area in Upper Tail				
	.10	.05	.025	.01	.005
1	3.078	6.314	12.706	31.821	63.657
2	1.886	2.920	4.303	6.965	9.925
3	1.638	2.353	3.182	4.541	5.841
4	1.533	2.132	2.776	3.747	4.604
5	1.476	2.015	2.571	3.365	4.032
6	1.440	1.943	2.447	3.143	3.707
7	1.415	1.895	2.365	2.998	3.499
8	1.397	1.860	2.306	2.896	3.355
9	1.383	1.833	2.262	2.821	3.250
10	1.372	1.812	2.228	2.764	3.169
11	1.363	1.796	2.201	2.718	3.106
12	1.356	1.782	2.179	2.681	3.055
13	1.350	1.771	2.160	2.650	3.012
14	1.345	1.761	2.145	2.624	2.977
15	1.341	1.753	2.131	2.602	2.947
16	1.337	1.746	2.120	2.583	2.921
17	1.333	1.740	2.110	2.567	2.898
18	1.330	1.734	2.101	2.552	2.878
19	1.328	1.729	2.093	2.539	2.861
20	1.325	1.725	2.086	2.528	2.845
21	1.323	1.721	2.080	2.518	2.831
22	1.321	1.717	2.074	2.508	2.819
23	1.319	1.714	2.069	2.500	2.807
24	1.318	1.711	2.064	2.492	2.797
25	1.316	1.708	2.060	2.485	2.787
26	1.315	1.706	2.056	2.479	2.779
27	1.314	1.703	2.052	2.473	2.771
28	1.313	1.701	2.048	2.467	2.763
29	1.311	1.699	2.045	2.462	2.756
30	1.310	1.697	2.042	2.457	2.750
40	1.303	1.684	2.021	2.423	2.704
60	1.296	1.671	2.000	2.390	2.660
120	1.289	1.658	1.980	2.358	2.617
∞	1.282	1.645	1.960	2.326	2.576

This table is reprinted by permission of Biometrika Trustees from Table 12, Percentage Points of the *t* Distribution, 3rd Edition, 1966. E. S. Pearson and H. O. Hartley, *Biometrika Tables for Statisticians,* Vol. I.

TABLE 3   Chi-Square Distribution

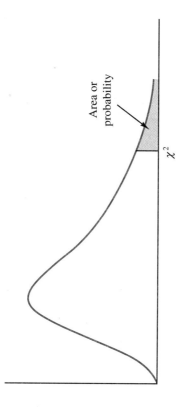

Area or probability

χ^2

Entries in the table give χ^2 values, for an area or probability in the upper tail of the chi-square distribution. For example, with 10 degrees of freedom and a .01 area in the upper tail, $\chi^2_{.01} = 23.2093$.

Area in Upper Tail

Degrees of Freedom	.995	.99	.975	.95	.90	.10	.05	.025	.01	.005
1	$392,704 \times 10^{-10}$	$157,088 \times 10^{-9}$	$982,069 \times 10^{-9}$	$393,214 \times 10^{-8}$	.0157908	2.70554	3.84146	5.02389	6.63490	7.87944
2	.0100251	.0201007	.0506356	.102587	.210720	4.60517	5.99147	7.37776	9.21034	10.5966
3	.0717212	.114832	.215795	.351846	.584375	6.25139	7.81473	9.34840	11.3449	12.8381
4	.206990	.297110	.484419	.710721	1.063623	7.77944	9.48773	11.1433	13.2767	14.8602
5	.411740	.554300	.831211	1.145476	1.61031	9.23635	11.0705	12.8325	15.0863	16.7496
6	.675727	.872085	1.237347	1.63539	2.20413	10.6446	12.5916	14.4494	16.8119	18.5476
7	.989265	1.239043	1.68987	2.16735	2.83311	12.0170	14.0671	16.0128	18.4753	20.2777
8	1.344419	1.646482	2.17973	2.73264	3.48954	13.3616	15.5073	17.5346	20.0902	21.9550
9	1.734926	2.087912	2.70039	3.32511	4.16816	14.6837	16.9190	19.0228	21.6660	23.5893

(TABLE CONTINUES)

10	2.15585	2.55821	3.24697	3.94030	15.9871	18.3070	20.4831	23.2093	25.1882
11	2.60321	3.05347	3.81575	4.57481	17.2750	19.6751	21.9200	24.7250	26.7569
12	3.07382	3.57056	4.40379	5.22603	18.5494	21.0261	23.3367	26.2170	28.2995
13	3.56503	4.10691	5.00874	5.89186	19.8119	22.3621	24.7356	27.6883	29.8194
14	4.07468	4.66043	5.62872	6.57063	21.0642	23.6848	26.1190	29.1413	31.3193
15	4.60094	5.22935	6.26214	7.26094	22.3072	24.9958	27.4884	30.5779	32.8013
16	5.14224	5.81221	6.90766	7.96164	23.5418	26.2962	28.8454	31.9999	34.2672
17	5.69724	6.40776	7.56418	8.67176	24.7690	27.5871	30.1910	33.4087	35.7185
18	6.26481	7.01491	8.23075	9.39046	25.9894	28.8693	31.5264	34.8053	37.1564
19	6.84398	7.63273	8.90655	10.1170	27.2036	30.1435	32.8523	36.1908	38.5822
20	7.43386	8.26040	9.59083	10.8508	28.4120	31.4104	34.1696	37.5662	39.9968
21	8.03366	8.89720	10.28293	11.5913	29.6151	32.6705	35.4789	38.9321	41.4010
22	8.64272	9.54249	10.9823	12.3380	30.8133	33.9244	36.7807	40.2894	42.7958
23	9.26042	10.19567	11.6885	13.0905	32.0069	35.1725	38.0757	41.6384	44.1813
24	9.88623	10.8564	12.4011	13.8484	33.1963	36.4151	39.3641	42.9798	45.5585
25	10.5197	11.5240	13.1197	14.6114	34.3816	37.6525	40.6465	44.3141	46.9278
26	11.1603	12.1981	13.8439	15.3791	35.5631	38.8852	41.9232	45.6417	48.2899
27	11.8076	12.8786	14.5733	16.1513	36.7412	40.1133	43.1944	46.9630	49.6449
28	12.4613	13.5648	15.3079	16.9279	37.9159	41.3372	44.4607	48.2782	50.9933
29	13.1211	14.2565	16.0471	17.7083	39.0875	42.5569	45.7222	49.5879	52.3356
30	13.7867	14.9535	16.7908	18.4926	40.2560	43.7729	46.9792	50.8922	53.6720
40	20.7065	22.1643	24.4331	26.5093	51.8050	55.7585	59.3417	63.6907	66.7659
50	27.9907	29.7067	32.3574	34.7642	63.1671	67.5048	71.4202	76.1539	79.4900
60	35.5346	37.4848	40.4817	43.1879	74.3970	79.0819	83.2976	88.3794	91.9517
70	43.2752	45.4418	48.7576	51.7393	85.5271	90.5312	95.0231	100.425	104.215
80	51.1720	53.5400	57.1532	60.3915	96.5782	101.879	106.629	112.329	116.321
90	59.1963	61.7541	65.6466	69.1260	107.565	113.145	118.136	124.116	128.299
100	67.3276	70.0648	74.2219	77.9295	118.498	124.342	129.561	135.807	140.169

This table is reprinted by permission of Biometrika Trustees from Table 8, Percentage Points of the χ^2 Distribution, by E. S. Pearson and H. O. Hartley, *Biometrika Tables for Statisticians*, Vol. I, 3rd Edition, 1966.

TABLE 4 F Distribution

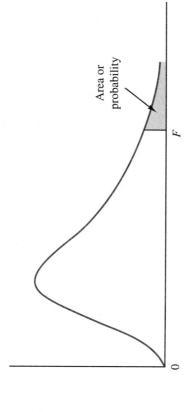

Area or probability

F

Entries in the table give F values, for an area or probability in the upper tail of the F distribution. For example, with 12 numerator degrees of freedom, 15 denominator degrees of freedom, and a .05 area in the upper tail, $F_{.05} = 2.48$.

Table of $F_{.05}$ Values

Denominator Degrees of Freedom	Numerator Degrees of Freedom																		
	1	2	3	4	5	6	7	8	9	10	12	15	20	24	30	40	60	120	∞
1	161.4	199.5	215.7	224.6	230.2	234.0	236.8	238.9	240.5	241.9	243.9	245.9	248.0	249.1	250.1	251.1	252.2	253.3	254.3
2	18.51	19.00	19.16	19.25	19.30	19.33	19.35	19.37	19.38	19.40	19.41	19.43	19.45	19.45	19.46	19.47	19.48	19.49	19.50
3	10.13	9.55	9.28	9.12	9.01	8.94	8.89	8.85	8.81	8.79	8.74	8.70	8.66	8.64	8.62	8.59	8.57	8.55	8.53
4	7.71	6.94	6.59	6.39	6.26	6.16	6.09	6.04	6.00	5.96	5.91	5.86	5.80	5.77	5.75	5.72	5.69	5.66	5.63
5	6.61	5.79	5.41	5.19	5.05	4.95	4.88	4.82	4.77	4.74	4.68	4.62	4.56	4.53	4.50	4.46	4.43	4.40	4.36

(TABLE CONTINUES)

df																			
6	5.99	5.14	4.76	4.53	4.39	4.28	4.21	4.15	4.10	4.06	4.00	3.94	3.87	3.84	3.81	3.77	3.74	3.70	3.67
7	5.59	4.74	4.35	4.12	3.97	3.87	3.79	3.73	3.68	3.64	3.57	3.51	3.44	3.41	3.38	3.34	3.30	3.27	3.23
8	5.32	4.46	4.07	3.84	3.69	3.58	3.50	3.44	3.39	3.35	3.28	3.22	3.15	3.12	3.08	3.04	3.01	2.97	2.93
9	5.12	4.26	3.86	3.63	3.48	3.37	3.29	3.23	3.18	3.14	3.07	3.01	2.94	2.90	2.86	2.83	2.79	2.75	2.71
10	4.96	4.10	3.71	3.48	3.33	3.22	3.14	3.07	3.02	2.98	2.91	2.85	2.77	2.74	2.70	2.66	2.62	2.58	2.54
11	4.84	3.98	3.59	3.36	3.20	3.09	3.01	2.95	2.90	2.85	2.79	2.72	2.65	2.61	2.57	2.53	2.49	2.45	2.40
12	4.75	3.89	3.49	3.26	3.11	3.00	2.91	2.85	2.80	2.75	2.69	2.62	2.54	2.51	2.47	2.43	2.38	2.34	2.30
13	4.67	3.81	3.41	3.18	3.03	2.92	2.83	2.77	2.71	2.67	2.60	2.53	2.46	2.42	2.38	2.34	2.30	2.25	2.21
14	4.60	3.74	3.34	3.11	2.96	2.85	2.76	2.70	2.65	2.60	2.53	2.46	2.39	2.35	2.31	2.27	2.22	2.18	2.13
15	4.54	3.68	3.29	3.06	2.90	2.79	2.71	2.64	2.59	2.54	2.48	2.40	2.33	2.29	2.25	2.20	2.16	2.11	2.07
16	4.49	3.63	3.24	3.01	2.85	2.74	2.66	2.59	2.54	2.49	2.42	2.35	2.28	2.24	2.19	2.15	2.11	2.06	2.01
17	4.45	3.59	3.20	2.96	2.81	2.70	2.61	2.55	2.49	2.45	2.38	2.31	2.23	2.19	2.15	2.10	2.06	2.01	1.96
18	4.41	3.55	3.16	2.93	2.77	2.66	2.58	2.51	2.46	2.41	2.34	2.27	2.19	2.15	2.11	2.06	2.02	1.97	1.92
19	4.38	3.52	3.13	2.90	2.74	2.63	2.54	2.48	2.42	2.38	2.31	2.23	2.16	2.11	2.07	2.03	1.98	1.93	1.88
20	4.35	3.49	3.10	2.87	2.71	2.60	2.51	2.45	2.39	2.35	2.28	2.20	2.12	2.08	2.04	1.99	1.95	1.90	1.84
21	4.32	3.47	3.07	2.84	2.68	2.57	2.49	2.42	2.37	2.32	2.25	2.18	2.10	2.05	2.01	1.96	1.92	1.87	1.81
22	4.30	3.44	3.05	2.82	2.66	2.55	2.46	2.40	2.34	2.30	2.23	2.15	2.07	2.03	1.98	1.94	1.89	1.84	1.78
23	4.28	3.42	3.03	2.80	2.64	2.53	2.44	2.37	2.32	2.27	2.20	2.13	2.05	2.01	1.96	1.91	1.86	1.81	1.76
24	4.26	3.40	3.01	2.78	2.62	2.51	2.42	2.36	2.30	2.25	2.18	2.11	2.03	1.98	1.94	1.89	1.84	1.79	1.73
25	4.24	3.39	2.99	2.76	2.60	2.49	2.40	2.34	2.28	2.24	2.16	2.09	2.01	1.96	1.92	1.87	1.82	1.77	1.71
26	4.23	3.37	2.98	2.74	2.59	2.47	2.39	2.32	2.27	2.22	2.15	2.07	1.99	1.95	1.90	1.85	1.80	1.75	1.69
27	4.21	3.35	2.96	2.73	2.57	2.46	2.37	2.31	2.25	2.20	2.13	2.06	1.97	1.93	1.88	1.84	1.79	1.73	1.67
28	4.20	3.34	2.95	2.71	2.56	2.45	2.36	2.29	2.24	2.19	2.12	2.04	1.96	1.91	1.87	1.82	1.77	1.71	1.65
29	4.18	3.33	2.93	2.70	2.55	2.43	2.35	2.28	2.22	2.18	2.10	2.03	1.94	1.90	1.85	1.81	1.75	1.70	1.64
30	4.17	3.32	2.92	2.69	2.53	2.42	2.33	2.27	2.21	2.16	2.09	2.01	1.93	1.89	1.84	1.79	1.74	1.68	1.62
40	4.08	3.23	2.84	2.61	2.45	2.34	2.25	2.18	2.12	2.08	2.00	1.92	1.84	1.79	1.74	1.69	1.64	1.58	1.51
60	4.00	3.15	2.76	2.53	2.37	2.25	2.17	2.10	2.04	1.99	1.92	1.84	1.75	1.70	1.65	1.59	1.53	1.47	1.39
120	3.92	3.07	2.68	2.45	2.29	2.17	2.09	2.02	1.96	1.91	1.83	1.75	1.66	1.61	1.55	1.50	1.43	1.35	1.25
∞	3.84	3.00	2.60	2.37	2.21	2.10	2.01	1.94	1.88	1.83	1.75	1.67	1.57	1.52	1.46	1.39	1.32	1.22	1.00

(**TABLE CONTINUES ON NEXT PAGE**)

This table is reprinted by permission of the Biometrika Trustees from Table 18, Percentage Points of the *F* Distribution, by E. S. Pearson and H. O. Hartley, *Biometrika Tables for Statisticians*, Vol. I, 3rd Edition, 1966.

Table of $F_{.01}$ Values

TABLE 4 (Continued)

Denominator Degrees of Freedom	Numerator Degrees of Freedom																		
	1	2	3	4	5	6	7	8	9	10	12	15	20	24	30	40	60	120	∞
1	4,052	4,999.5	5,403	5,625	5,764	5,859	5,928	5,982	6,022	6,056	6,106	6,157	6,209	6,235	6,261	6,287	6,313	6,339	6,366
2	98.50	99.00	99.17	99.25	99.30	99.33	99.36	99.37	99.39	99.40	99.42	99.43	99.45	99.46	99.47	99.47	99.48	99.49	99.50
3	34.12	30.82	29.46	28.71	28.24	27.91	27.67	27.49	27.35	27.23	27.05	26.87	26.69	26.60	26.50	26.41	26.32	26.22	26.13
4	21.20	18.00	16.69	15.98	15.52	15.21	14.98	14.80	14.66	14.55	14.37	14.20	14.02	13.93	13.84	13.75	13.65	13.56	13.46
5	16.26	13.27	12.06	11.39	10.97	10.67	10.46	10.29	10.16	10.05	9.89	9.72	9.55	9.47	9.38	9.29	9.20	9.11	9.06
6	13.75	10.92	9.78	9.15	8.75	8.47	8.26	8.10	7.98	7.87	7.72	7.56	7.40	7.31	7.23	7.14	7.06	6.97	6.88
7	12.25	9.55	8.45	7.85	7.46	7.19	6.99	6.84	6.72	6.62	6.47	6.31	6.16	6.07	5.99	5.91	5.82	5.74	5.65
8	11.26	8.65	7.59	7.01	6.63	6.37	6.18	6.03	5.91	5.81	5.67	5.52	5.36	5.28	5.20	5.12	5.03	4.95	4.86
9	10.56	8.02	6.99	6.42	6.06	5.80	5.61	5.47	5.35	5.26	5.11	4.96	4.81	4.73	4.65	4.57	4.48	4.40	4.31
10	10.04	7.56	6.55	5.99	5.64	5.39	5.20	5.06	4.94	4.85	4.71	4.56	4.41	4.33	4.25	4.17	4.08	4.00	3.91
11	9.65	7.21	6.22	5.67	5.32	5.07	4.89	4.74	4.63	4.54	4.40	4.25	4.10	4.02	3.94	3.86	3.78	3.69	3.60
12	9.33	6.93	5.95	5.41	5.06	4.82	4.64	4.50	4.39	4.30	4.16	4.01	3.86	3.78	3.70	3.62	3.54	3.45	3.36
13	9.07	6.70	5.74	5.21	4.86	4.62	4.44	4.30	4.19	4.10	3.96	3.82	3.66	3.59	3.51	3.43	3.34	3.25	3.17
14	8.86	6.51	5.56	5.04	4.69	4.46	4.28	4.14	4.03	3.94	3.80	3.66	3.51	3.43	3.35	3.27	3.18	3.09	3.00
15	8.68	6.36	5.42	4.89	4.56	4.32	4.14	4.00	3.89	3.80	3.67	3.52	3.37	3.29	3.21	3.13	3.05	2.96	2.87
16	8.53	6.23	5.29	4.77	4.44	4.20	4.03	3.89	3.78	3.69	3.55	3.41	3.26	3.18	3.10	3.02	2.93	2.84	2.75
17	8.40	6.11	5.18	4.67	4.34	4.10	3.93	3.79	3.68	3.59	3.46	3.31	3.16	3.08	3.00	2.92	2.83	2.75	2.65
18	8.29	6.01	5.09	4.58	4.25	4.01	3.84	3.71	3.60	3.51	3.37	3.23	3.08	3.00	2.92	2.84	2.75	2.66	2.57
19	8.18	5.93	5.01	4.50	4.17	3.94	3.77	3.63	3.52	3.43	3.30	3.15	3.00	2.92	2.84	2.76	2.67	2.58	2.49
20	8.10	5.85	4.94	4.43	4.10	3.87	3.70	3.56	3.46	3.37	3.23	3.09	2.94	2.86	2.78	2.69	2.61	2.52	2.42
21	8.02	5.78	4.87	4.37	4.04	3.81	3.64	3.51	3.40	3.31	3.17	3.03	2.88	2.80	2.72	2.64	2.55	2.46	2.36
22	7.95	5.72	4.82	4.31	3.99	3.76	3.59	3.45	3.35	3.26	3.12	2.98	2.83	2.75	2.67	2.58	2.50	2.40	2.31
23	7.88	5.66	4.76	4.26	3.94	3.71	3.54	3.41	3.30	3.21	3.07	2.93	2.78	2.70	2.62	2.54	2.45	2.35	2.26
24	7.82	5.61	4.72	4.22	3.90	3.67	3.50	3.36	3.26	3.17	3.03	2.89	2.74	2.66	2.58	2.49	2.40	2.31	2.21
25	7.77	5.57	4.68	4.18	3.85	3.63	3.46	3.32	3.22	3.13	2.99	2.85	2.70	2.62	2.54	2.45	2.36	2.27	2.17
26	7.72	5.53	4.64	4.14	3.82	3.59	3.42	3.29	3.18	3.09	2.96	2.81	2.66	2.58	2.50	2.42	2.33	2.23	2.13
27	7.68	5.49	4.60	4.11	3.78	3.56	3.39	3.26	3.15	3.06	2.93	2.78	2.63	2.55	2.47	2.38	2.29	2.20	2.10
28	7.64	5.45	4.57	4.07	3.75	3.53	3.36	3.23	3.12	3.03	2.90	2.75	2.60	2.52	2.44	2.35	2.26	2.17	2.06
29	7.60	5.42	4.54	4.04	3.73	3.50	3.33	3.20	3.09	3.00	2.87	2.73	2.57	2.49	2.41	2.33	2.23	2.14	2.03
30	7.56	5.39	4.51	4.02	3.70	3.47	3.30	3.17	3.07	2.98	2.84	2.70	2.55	2.47	2.39	2.30	2.21	2.11	2.01
40	7.31	5.18	4.31	3.83	3.51	3.29	3.12	2.99	2.89	2.80	2.66	2.52	2.37	2.29	2.20	2.11	2.02	1.92	1.80
60	7.08	4.98	4.13	3.65	3.34	3.12	2.95	2.82	2.72	2.63	2.50	2.35	2.20	2.12	2.03	1.94	1.84	1.73	1.60
120	6.85	4.79	3.95	3.48	3.17	2.96	2.79	2.66	2.56	2.47	2.34	2.19	2.03	1.95	1.86	1.76	1.66	1.53	1.38
∞	6.63	4.61	3.78	3.32	3.02	2.80	2.64	2.51	2.41	2.32	2.18	2.04	1.88	1.79	1.70	1.59	1.47	1.32	1.00

(TABLE CONTINUES)

TABLE 4 (Continued)

Table of $F_{.025}$ Values

Denominator Degrees of Freedom	Numerator Degrees of Freedom																		
	1	2	3	4	5	6	7	8	9	10	12	15	20	24	30	40	60	120	∞
1	647.8	799.5	864.2	899.6	921.8	937.1	348.2	956.7	963.3	968.6	976.7	984.9	993.1	997.2	1,001	1,006	1,010	1,014	1,018
2	38.51	39.00	39.17	39.25	39.30	39.33	39.36	39.37	39.39	39.40	39.41	39.43	39.45	39.46	39.46	39.47	39.48	39.49	39.50
3	17.44	16.04	15.44	15.10	14.88	14.73	14.62	14.54	14.47	14.42	14.34	14.25	14.17	14.12	14.08	14.04	13.99	13.95	13.90
4	12.22	10.65	9.98	9.60	9.36	9.20	9.07	8.98	8.90	8.84	8.75	8.66	8.56	8.51	8.46	8.41	8.36	8.31	8.26
5	10.01	8.43	7.76	7.39	7.15	6.98	6.85	6.76	6.68	6.62	6.52	6.43	6.33	6.28	6.23	6.18	6.12	6.07	6.02
6	8.81	7.26	6.60	6.23	5.99	5.82	5.70	5.60	5.52	5.46	5.37	5.27	5.17	5.12	5.07	5.01	4.96	4.90	4.85
7	8.07	6.54	5.89	5.52	5.29	5.21	4.99	4.90	4.82	4.76	4.67	4.57	4.47	4.42	4.36	4.31	4.25	4.20	4.14
8	7.57	6.06	5.42	5.05	4.82	4.65	4.53	4.43	4.36	4.30	4.20	4.10	4.00	3.95	3.89	3.84	3.78	3.73	3.67
9	7.21	5.71	5.08	4.72	4.48	4.32	4.20	4.10	4.03	3.96	3.87	3.77	3.67	3.61	3.56	3.51	3.45	3.39	3.33
10	6.94	5.46	4.83	4.47	4.24	4.07	3.95	3.85	3.78	3.72	3.62	3.52	3.42	3.37	3.31	3.26	3.20	3.14	3.08
11	6.72	5.26	4.63	4.28	4.04	3.88	3.76	3.66	3.59	3.53	3.43	3.33	3.23	3.17	3.12	3.06	3.00	2.94	2.88
12	6.55	5.10	4.47	4.12	3.89	3.73	3.61	3.51	3.44	3.37	3.28	3.18	3.07	3.02	2.96	2.91	2.85	2.79	2.72
13	6.41	4.97	4.35	4.00	3.77	3.60	3.48	3.39	3.31	3.25	3.15	3.05	2.95	2.89	2.84	2.78	2.72	2.66	2.60
14	6.30	4.86	4.24	3.89	3.66	3.50	3.38	3.29	3.21	3.15	3.05	2.95	2.84	2.79	2.73	2.67	2.61	2.55	2.49
15	6.20	4.77	4.15	3.80	3.58	3.41	3.29	3.20	3.12	3.06	2.96	2.86	2.76	2.70	2.64	2.59	2.52	2.46	2.40
16	6.12	4.69	4.08	3.73	3.50	3.34	3.22	3.12	3.05	2.99	2.89	2.79	2.68	2.63	2.57	2.51	2.45	2.38	2.32
17	6.04	4.62	4.01	3.66	3.44	3.28	3.16	3.06	2.98	2.92	2.82	2.72	2.62	2.56	2.50	2.44	2.38	2.32	2.25
18	5.98	4.56	3.95	3.61	3.38	3.22	3.10	3.01	2.93	2.87	2.77	2.67	2.56	2.50	2.44	2.38	2.32	2.26	2.19
19	5.92	4.51	3.90	3.56	3.33	3.17	3.05	2.96	2.88	2.82	2.72	2.62	2.51	2.45	2.39	2.33	2.27	2.20	2.13
20	5.87	4.46	3.86	3.51	3.29	3.13	3.01	2.91	2.84	2.77	2.68	2.57	2.46	2.41	2.35	2.29	2.22	2.16	2.09
21	5.83	4.42	3.82	3.48	3.25	3.09	2.97	2.87	2.80	2.73	2.64	2.53	2.42	2.37	2.31	2.25	2.18	2.11	2.04
22	5.79	4.38	3.78	3.44	3.22	3.05	2.93	2.84	2.76	2.70	2.60	2.50	2.39	2.33	2.27	2.21	2.14	2.08	2.00
23	5.75	4.35	3.75	3.41	3.18	3.02	2.90	2.81	2.73	2.67	2.57	2.47	2.36	2.30	2.24	2.18	2.11	2.04	1.97
24	5.72	4.32	3.72	3.38	3.15	2.99	2.87	2.78	2.70	2.64	2.54	2.44	2.33	2.27	2.21	2.15	2.08	2.01	1.94
25	5.69	4.29	3.69	3.35	3.13	2.97	2.85	2.75	2.68	2.61	2.51	2.41	2.30	2.24	2.18	2.12	2.05	1.98	1.91
26	5.66	4.27	3.67	3.33	3.10	2.94	2.82	2.73	2.65	2.59	2.49	2.39	2.28	2.22	2.16	2.09	2.03	1.95	1.88
27	5.63	4.24	3.65	3.31	3.08	2.92	2.80	2.71	2.63	2.57	2.47	2.36	2.25	2.19	2.13	2.07	2.00	1.93	1.85
28	5.61	4.22	3.63	3.29	3.06	2.90	2.78	2.69	2.61	2.55	2.45	2.34	2.23	2.17	2.11	2.05	1.98	1.91	1.83
29	5.59	4.20	3.61	3.27	3.04	2.88	2.76	2.67	2.59	2.53	2.43	2.32	2.21	2.15	2.09	2.03	1.96	1.89	1.81
30	5.57	4.18	3.59	3.25	3.03	2.87	2.75	2.65	2.57	2.51	2.41	2.31	2.20	2.14	2.07	2.01	1.94	1.87	1.79
40	5.42	4.05	3.46	3.13	2.90	2.74	2.62	2.53	2.45	2.39	2.29	2.18	2.07	2.01	1.94	1.88	1.80	1.72	1.64
60	5.29	3.93	3.34	3.01	2.79	2.63	2.51	2.41	2.33	2.27	2.17	2.06	1.94	1.88	1.82	1.74	1.67	1.58	1.48
120	5.15	3.80	3.23	2.89	2.67	2.52	2.39	2.30	2.22	2.16	2.05	1.94	1.82	1.76	1.69	1.61	1.53	1.43	1.31
∞	5.02	3.69	3.12	2.79	2.57	2.41	2.29	2.19	2.11	2.05	1.94	1.83	1.71	1.64	1.57	1.48	1.39	1.27	1.00

TABLE 5
Binomial Probabilities

Entries in the table give the probability of x successes in n trials of a binomial experiment, where p is the probability of a success on one trial. For example, with six trials and $p = .05$, the probability of two successes is .0305.

						p				
n	x	.01	.02	.03	.04	.05	.06	.07	.08	.09
2	0	.9801	.9604	.9409	.9216	.9025	.8836	.8649	.8464	.8281
	1	.0198	.0392	.0582	.0768	.0950	.1128	.1302	.1472	.1638
	2	.0001	.0004	.0009	.0016	.0025	.0036	.0049	.0064	.0081
3	0	.9703	.9412	.9127	.8847	.8574	.8306	.8044	.7787	.7536
	1	.0294	.0576	.0847	.1106	.1354	.1590	.1816	.2031	.2236
	2	.0003	.0012	.0026	.0046	.0071	.0102	.0137	.0177	.0221
	3	.0000	.0000	.0000	.0001	.0001	.0002	.0003	.0005	.0007
4	0	.9606	.9224	.8853	.8493	.8145	.7807	.7481	.7164	.6857
	1	.0388	.0753	.1095	.1416	.1715	.1993	.2252	.2492	.2713
	2	.0006	.0023	.0051	.0088	.0135	.0191	.0254	.0325	.0402
	3	.0000	.0000	.0001	.0002	.0005	.0008	.0013	.0019	.0027
	4	.0000	.0000	.0000	.0000	.0000	.0000	.0000	.0000	.0001
5	0	.9510	.9039	.8587	.8154	.7738	.7339	.6957	.6591	.6240
	1	.0480	.0922	.1328	.1699	.2036	.2342	.2618	.2866	.3086
	2	.0010	.0038	.0082	.0142	.0214	.0299	.0394	.0498	.0610
	3	.0000	.0001	.0003	.0006	.0011	.0019	.0030	.0043	.0060
	4	.0000	.0000	.0000	.0000	.0000	.0001	.0001	.0002	.0003
	5	.0000	.0000	.0000	.0000	.0000	.0000	.0000	.0000	.0000
6	0	.9415	.8858	.8330	.7828	.7351	.6899	.6470	.6064	.5679
	1	.0571	.1085	.1546	.1957	.2321	.2642	.2922	.3164	.3370
	2	.0014	.0055	.0120	.0204	.0305	.0422	.0550	.0688	.0833
	3	.0000	.0002	.0005	.0011	.0021	.0036	.0055	.0080	.0110
	4	.0000	.0000	.0000	.0000	.0001	.0002	.0003	.0005	.0008
	5	.0000	.0000	.0000	.0000	.0000	.0000	.0000	.0000	.0000
	6	.0000	.0000	.0000	.0000	.0000	.0000	.0000	.0000	.0000
7	0	.9321	.8681	.8080	.7514	.6983	.6485	.6017	.5578	.5168
	1	.0659	.1240	.1749	.2192	.2573	.2897	.3170	.3396	.3578
	2	.0020	.0076	.0162	.0274	.0406	.0555	.0716	.0886	.1061
	3	.0000	.0003	.0008	.0019	.0036	.0059	.0090	.0128	.0175
	4	.0000	.0000	.0000	.0001	.0002	.0004	.0007	.0011	.0017
	5	.0000	.0000	.0000	.0000	.0000	.0000	.0000	.0001	.0001
	6	.0000	.0000	.0000	.0000	.0000	.0000	.0000	.0000	.0000
	7	.0000	.0000	.0000	.0000	.0000	.0000	.0000	.0000	.0000
8	0	.9227	.8508	.7837	.7214	.6634	.6096	.5596	.5132	.4703
	1	.0746	.1389	.1939	.2405	.2793	.3113	.3370	.3570	.3721
	2	.0026	.0099	.0210	.0351	.0515	.0695	.0888	.1087	.1288
	3	.0001	.0004	.0013	.0029	.0054	.0089	.0134	.0189	.0255
	4	.0000	.0000	.0001	.0002	.0004	.0007	.0013	.0021	.0031
	5	.0000	.0000	.0000	.0000	.0000	.0000	.0001	.0001	.0002
	6	.0000	.0000	.0000	.0000	.0000	.0000	.0000	.0000	.0000
	7	.0000	.0000	.0000	.0000	.0000	.0000	.0000	.0000	.0000
	8	.0000	.0000	.0000	.0000	.0000	.0000	.0000	.0000	.0000

(TABLE CONTINUES)

TABLE 5

(Continued)

n	x	.01	.02	.03	.04	p .05	.06	.07	.08	.09
9	0	.9135	.8337	.7602	.6925	.6302	.5730	.5204	.4722	.4279
	1	.0830	.1531	.2116	.2597	.2985	.3292	.3525	.3695	.3809
	2	.0034	.0125	.0262	.0433	.0629	.0840	.1061	.1285	.1507
	3	.0001	.0006	.0019	.0042	.0077	.0125	.0186	.0261	.0348
	4	.0000	.0000	.0001	.0003	.0006	.0012	.0021	.0034	.0052
	5	.0000	.0000	.0000	.0000	.0000	.0001	.0002	.0003	.0005
	6	.0000	.0000	.0000	.0000	.0000	.0000	.0000	.0000	.0000
	7	.0000	.0000	.0000	.0000	.0000	.0000	.0000	.0000	.0000
	8	.0000	.0000	.0000	.0000	.0000	.0000	.0000	.0000	.0000
	9	.0000	.0000	.0000	.0000	.0000	.0000	.0000	.0000	.0000
10	0	.9044	.8171	.7374	.6648	.5987	.5386	.4840	.4344	.3894
	1	.0914	.1667	.2281	.2770	.3151	.3438	.3643	.3777	.3851
	2	.0042	.0153	.0317	.0519	.0746	.0988	.1234	.1478	.1714
	3	.0001	.0008	.0026	.0058	.0105	.0168	.0248	.0343	.0452
	4	.0000	.0000	.0001	.0004	.0010	.0019	.0033	.0052	.0078
	5	.0000	.0000	.0000	.0000	.0001	.0001	.0003	.0005	.0009
	6	.0000	.0000	.0000	.0000	.0000	.0000	.0000	.0000	.0001
	7	.0000	.0000	.0000	.0000	.0000	.0000	.0000	.0000	.0000
	8	.0000	.0000	.0000	.0000	.0000	.0000	.0000	.0000	.0001
	9	.0000	.0000	.0000	.0000	.0000	.0000	.0000	.0000	.0001
	10	.0000	.0000	.0000	.0000	.0000	.0000	.0000	.0000	.0001
12	0	.8864	.7847	.6938	.6127	.5404	.4759	.4186	.3677	.3225
	1	.1074	.1922	.2575	.3064	.3413	.3645	.3781	.3837	.3827
	2	.0060	.0216	.0438	.0702	.0988	.1280	.1565	.1835	.2082
	3	.0002	.0015	.0045	.0098	.0173	.0272	.0393	.0532	.0686
	4	.0000	.0001	.0003	.0009	.0021	.0039	.0067	.0104	.0153
	5	.0000	.0000	.0000	.0001	.0002	.0004	.0008	.0014	.0024
	6	.0000	.0000	.0000	.0000	.0000	.0000	.0001	.0001	.0003
	7	.0000	.0000	.0000	.0000	.0000	.0000	.0000	.0000	.0000
	8	.0000	.0000	.0000	.0000	.0000	.0000	.0000	.0000	.0000
	9	.0000	.0000	.0000	.0000	.0000	.0000	.0000	.0000	.0000
	10	.0000	.0000	.0000	.0000	.0000	.0000	.0000	.0000	.0000
	11	.0000	.0000	.0000	.0000	.0000	.0000	.0000	.0000	.0000
	12	.0000	.0000	.0000	.0000	.0000	.0000	.0000	.0000	.0000
15	0	.8601	.7386	.6333	.5421	.4633	.3953	.3367	.2863	.2430
	1	.1303	.2261	.2938	.3388	.3658	.3785	.3801	.3734	.3605
	2	.0092	.0323	.0636	.0988	.1348	.1691	.2003	.2273	.2496
	3	.0004	.0029	.0085	.0178	.0307	.0468	.0653	.0857	.1070
	4	.0000	.0002	.0008	.0022	.0049	.0090	.0148	.0223	.0317
	5	.0000	.0000	.0001	.0002	.0006	.0013	.0024	.0043	.0069
	6	.0000	.0000	.0000	.0000	.0000	.0001	.0003	.0006	.0011
	7	.0000	.0000	.0000	.0000	.0000	.0000	.0000	.0001	.0001
	8	.0000	.0000	.0000	.0000	.0000	.0000	.0000	.0000	.0000
	9	.0000	.0000	.0000	.0000	.0000	.0000	.0000	.0000	.0000
	10	.0000	.0000	.0000	.0000	.0000	.0000	.0000	.0000	.0000
	11	.0000	.0000	.0000	.0000	.0000	.0000	.0000	.0000	.0000
	12	.0000	.0000	.0000	.0000	.0000	.0000	.0000	.0000	.0000
	13	.0000	.0000	.0000	.0000	.0000	.0000	.0000	.0000	.0000
	14	.0000	.0000	.0000	.0000	.0000	.0000	.0000	.0000	.0000
	15	.0000	.0000	.0000	.0000	.0000	.0000	.0000	.0000	.0000

(TABLE CONTINUES ON NEXT PAGE)

TABLE 5
(Continued)

n	x	.01	.02	.03	.04	.05	.06	.07	.08	.09
18	0	.8345	.6951	.5780	.4796	.3972	.3283	.2708	.2229	.1831
	1	.1517	.2554	.3217	.3597	.3763	.3772	.3669	.3489	.3260
	2	.0130	.0443	.0846	.1274	.1683	.2047	.2348	.2579	.2741
	3	.0007	.0048	.0140	.0283	.0473	.0697	.0942	.1196	.1446
	4	.0000	.0004	.0016	.0044	.0093	.0167	.0266	.0390	.0536
	5	.0000	.0000	.0001	.0005	.0014	.0030	.0056	.0095	.0148
	6	.0000	.0000	.0000	.0000	.0002	.0004	.0009	.0018	.0032
	7	.0000	.0000	.0000	.0000	.0000	.0000	.0001	.0003	.0005
	8	.0000	.0000	.0000	.0000	.0000	.0000	.0000	.0000	.0001
	9	.0000	.0000	.0000	.0000	.0000	.0000	.0000	.0000	.0000
	10	.0000	.0000	.0000	.0000	.0000	.0000	.0000	.0000	.0000
	11	.0000	.0000	.0000	.0000	.0000	.0000	.0000	.0000	.0000
	12	.0000	.0000	.0000	.0000	.0000	.0000	.0000	.0000	.0000
	13	.0000	.0000	.0000	.0000	.0000	.0000	.0000	.0000	.0000
	14	.0000	.0000	.0000	.0000	.0000	.0000	.0000	.0000	.0000
	15	.0000	.0000	.0000	.0000	.0000	.0000	.0000	.0000	.0000
	16	.0000	.0000	.0000	.0000	.0000	.0000	.0000	.0000	.0000
	17	.0000	.0000	.0000	.0000	.0000	.0000	.0000	.0000	.0000
	18	.0000	.0000	.0000	.0000	.0000	.0000	.0000	.0000	.0000
20	0	.8179	.6676	.5438	.4420	.3585	.2901	.2342	.1887	.1516
	1	.1652	.2725	.3364	.3683	.3774	.3703	.3526	.3282	.3000
	2	.0159	.0528	.0988	.1458	.1887	.2246	.2521	.2711	.2818
	3	.0010	.0065	.0183	.0364	.0596	.0860	.1139	.1414	.1672
	4	.0000	.0006	.0024	.0065	.0133	.0233	.0364	.0523	.0703
	5	.0000	.0000	.0002	.0009	.0022	.0048	.0088	.0145	.0222
	6	.0000	.0000	.0000	.0001	.0003	.0008	.0017	.0032	.0055
	7	.0000	.0000	.0000	.0000	.0000	.0001	.0002	.0005	.0011
	8	.0000	.0000	.0000	.0000	.0000	.0000	.0000	.0001	.0002
	9	.0000	.0000	.0000	.0000	.0000	.0000	.0000	.0000	.0000
	10	.0000	.0000	.0000	.0000	.0000	.0000	.0000	.0000	.0000
	11	.0000	.0000	.0000	.0000	.0000	.0000	.0000	.0000	.0000
	12	.0000	.0000	.0000	.0000	.0000	.0000	.0000	.0000	.0000
	13	.0000	.0000	.0000	.0000	.0000	.0000	.0000	.0000	.0000
	14	.0000	.0000	.0000	.0000	.0000	.0000	.0000	.0000	.0000
	15	.0000	.0000	.0000	.0000	.0000	.0000	.0000	.0000	.0000
	16	.0000	.0000	.0000	.0000	.0000	.0000	.0000	.0000	.0000
	17	.0000	.0000	.0000	.0000	.0000	.0000	.0000	.0000	.0000
	18	.0000	.0000	.0000	.0000	.0000	.0000	.0000	.0000	.0000
	19	.0000	.0000	.0000	.0000	.0000	.0000	.0000	.0000	.0000
	20	.0000	.0000	.0000	.0000	.0000	.0000	.0000	.0000	.0000

(TABLE CONTINUES)

TABLE 5
(Continued)

						p				
n	x	.10	.15	.20	.25	.30	.35	.40	.45	.50
2	0	.8100	.7225	.6400	.5625	.4900	.4225	.3600	.3025	.2500
	1	.1800	.2550	.3200	.3750	.4200	.4550	.4800	.4950	.5000
	2	.0100	.0225	.0400	.0625	.0900	.1225	.1600	.2025	.2500
3	0	.7290	.6141	.5120	.4219	.3430	.2746	.2160	.1664	.1250
	1	.2430	.3251	.3840	.4219	.4410	.4436	.4320	.4084	.3750
	2	.0270	.0574	.0960	.1406	.1890	.2389	.2880	.3341	.3750
	3	.0010	.0034	.0080	.0156	.0270	.0429	.0640	.0911	.1250
4	0	.6561	.5220	.4096	.3164	.2401	.1785	.1296	.0915	.0625
	1	.2916	.3685	.4096	.4219	.4116	.3845	.3456	.2995	.2500
	2	.0486	.0975	.1536	.2109	.2646	.3105	.3456	.3675	.3750
	3	.0036	.0115	.0256	.0469	.0756	.1115	.1536	.2005	.2500
	4	.0001	.0005	.0016	.0039	.0081	.0150	.0256	.0410	.0625
5	0	.5905	.4437	.3277	.2373	.1681	.1160	.0778	.0503	.0312
	1	.3280	.3915	.4096	.3955	.3602	.3124	.2592	.2059	.1562
	2	.0729	.1382	.2048	.2637	.3087	.3364	.3456	.3369	.3125
	3	.0081	.0244	.0512	.0879	.1323	.1811	.2304	.2757	.3125
	4	.0004	.0022	.0064	.0146	.0284	.0488	.0768	.1128	.1562
	5	.0000	.0001	.0003	.0010	.0024	.0053	.0102	.0185	.0312
6	0	.5314	.3771	.2621	.1780	.1176	.0754	.0467	.0277	.0156
	1	.3543	.3993	.3932	.3560	.3025	.2437	.1866	.1359	.0938
	2	.0984	.1762	.2458	.2966	.3241	.3280	.3110	.2780	.2344
	3	.0146	.0415	.0819	.1318	.1852	.2355	.2765	.3032	.3125
	4	.0012	.0055	.0154	.0330	.0595	.0951	.1382	.1861	.2344
	5	.0001	.0004	.0015	.0044	.0102	.0205	.0369	.0609	.0938
	6	.0000	.0000	.0001	.0002	.0007	.0018	.0041	.0083	.0156
7	0	.4783	.3206	.2097	.1335	.0824	.0490	.0280	.0152	.0078
	1	.3720	.3960	.3670	.3115	.2471	.1848	.1306	.0872	.0547
	2	.1240	.2097	.2753	.3115	.3177	.2985	.2613	.2140	.1641
	3	.0230	.0617	.1147	.1730	.2269	.2679	.2903	.2918	.2734
	4	.0026	.0109	.0287	.0577	.0972	.1442	.1935	.2388	.2734
	5	.0002	.0012	.0043	.0115	.0250	.0466	.0774	.1172	.1641
	6	.0000	.0001	.0004	.0013	.0036	.0084	.0172	.0320	.0547
	7	.0000	.0000	.0000	.0001	.0002	.0006	.0016	.0037	.0078
8	0	.4305	.2725	.1678	.1001	.0576	.0319	.0168	.0084	.0039
	1	.3826	.3847	.3355	.2670	.1977	.1373	.0896	.0548	.0312
	2	.1488	.2376	.2936	.3115	.2965	.2587	.2090	.1569	.1094
	3	.0331	.0839	.1468	.2076	.2541	.2786	.2787	.2568	.2188
	4	.0046	.0185	.0459	.0865	.1361	.1875	.2322	.2627	.2734
	5	.0004	.0026	.0092	.0231	.0467	.0808	.1239	.1719	.2188
	6	.0000	.0002	.0011	.0038	.0100	.0217	.0413	.0703	.1094
	7	.0000	.0000	.0001	.0004	.0012	.0033	.0079	.0164	.0312
	8	.0000	.0000	.0000	.0000	.0001	.0002	.0007	.0017	.0039

(TABLE CONTINUES ON NEXT PAGE)

Table 5

(Continued)

n	x	.10	.15	.20	.25	p .30	.35	.40	.45	.50
9	0	.3874	.2316	.1342	.0751	.0404	.0207	.0101	.0046	.0020
	1	.3874	.3679	.3020	.2253	.1556	.1004	.0605	.0339	.0176
	2	.1722	.2597	.3020	.3003	.2668	.2162	.1612	.1110	.0703
	3	.0446	.1069	.1762	.2336	.2668	.2716	.2508	.2119	.1641
	4	.0074	.0283	.0661	.1168	.1715	.2194	.2508	.2600	.2461
	5	.0008	.0050	.0165	.0389	.0735	.1181	.1672	.2128	.2461
	6	.0001	.0006	.0028	.0087	.0210	.0424	.0743	.1160	.1641
	7	.0000	.0000	.0003	.0012	.0039	.0098	.0212	.0407	.0703
	8	.0000	.0000	.0000	.0001	.0004	.0013	.0035	.0083	.0176
	9	.0000	.0000	.0000	.0000	.0000	.0001	.0003	.0008	.0020
10	0	.3487	.1969	.1074	.0563	.0282	.0135	.0060	.0025	.0010
	1	.3874	.3474	.2684	.1877	.1211	.0725	.0403	.0207	.0098
	2	.1937	.2759	.3020	.2816	.2335	.1757	.1209	.0763	.0439
	3	.0574	.1298	.2013	.2503	.2668	.2522	.2150	.1665	.1172
	4	.0112	.0401	.0881	.1460	.2001	.2377	.2508	.2384	.2051
	5	.0015	.0085	.0264	.0584	.1029	.1536	.2007	.2340	.2461
	6	.0001	.0012	.0055	.0162	.0368	.0689	.1115	.1596	.2051
	7	.0000	.0001	.0008	.0031	.0090	.0212	.0425	.0746	.1172
	8	.0000	.0000	.0001	.0004	.0014	.0043	.0106	.0229	.0439
	9	.0000	.0000	.0000	.0000	.0001	.0005	.0016	.0042	.0098
	10	.0000	.0000	.0000	.0000	.0000	.0000	.0001	.0003	.0010
12	0	.2824	.1422	.0687	.0317	.0138	.0057	.0022	.0008	.0002
	1	.3766	.3012	.2062	.1267	.0712	.0368	.0174	.0075	.0029
	2	.2301	.2924	.2835	.2323	.1678	.1088	.0639	.0339	.0161
	3	.0853	.1720	.2362	.2581	.2397	.1954	.1419	.0923	.0537
	4	.0213	.0683	.1329	.1936	.2311	.2367	.2128	.1700	.1208
	5	.0038	.0193	.0532	.1032	.1585	.2039	.2270	.2225	.1934
	6	.0005	.0040	.0155	.0401	.0792	.1281	.1766	.2124	.2256
	7	.0000	.0006	.0033	.0115	.0291	.0591	.1009	.1489	.1934
	8	.0000	.0001	.0005	.0024	.0078	.0199	.0420	.0762	.1208
	9	.0000	.0000	.0001	.0004	.0015	.0048	.0125	.0277	.0537
	10	.0000	.0000	.0000	.0000	.0002	.0008	.0025	.0068	.0161
	11	.0000	.0000	.0000	.0000	.0000	.0001	.0003	.0010	.0029
	12	.0000	.0000	.0000	.0000	.0000	.0000	.0000	.0001	.0002
15	0	.2059	.0874	.0352	.0134	.0047	.0016	.0005	.0001	.0000
	1	.3432	.2312	.1319	.0668	.0305	.0126	.0047	.0016	.0005
	2	.2669	.2856	.2309	.1559	.0916	.0476	.0219	.0090	.0032
	3	.1285	.2184	.2501	.2252	.1700	.1110	.0634	.0318	.0139
	4	.0428	.1156	.1876	.2252	.2186	.1792	.1268	.0780	.0417
	5	.0105	.0449	.1032	.1651	.2061	.2123	.1859	.1404	.0916
	6	.0019	.0132	.0430	.0917	.1472	.1906	.2066	.1914	.1527
	7	.0003	.0030	.0138	.0393	.0811	.1319	.1771	.2013	.1964
	8	.0000	.0005	.0035	.0131	.0348	.0710	.1181	.1647	.1964
	9	.0000	.0001	.0007	.0034	.0116	.0298	.0612	.1048	.1527
	10	.0000	.0000	.0001	.0007	.0030	.0096	.0245	.0515	.0916
	11	.0000	.0000	.0000	.0001	.0006	.0024	.0074	.0191	.0417
	12	.0000	.0000	.0000	.0000	.0001	.0004	.0016	.0052	.0139
	13	.0000	.0000	.0000	.0000	.0000	.0001	.0003	.0010	.0032
	14	.0000	.0000	.0000	.0000	.0000	.0000	.0000	.0001	.0005
	15	.0000	.0000	.0000	.0000	.0000	.0000	.0000	.0000	.0000

(Table Continues)

TABLE 5
(Continued)

n	x	.10	.15	.20	.25	p .30	.35	.40	.45	.50
18	0	.1501	.0536	.0180	.0056	.0016	.0004	.0001	.0000	.0000
	1	.3002	.1704	.0811	.0338	.0126	.0042	.0012	.0003	.0001
	2	.2835	.2556	.1723	.0958	.0458	.0190	.0069	.0022	.0006
	3	.1680	.2406	.2297	.1704	.1046	.0547	.0246	.0095	.0031
	4	.0700	.1592	.2153	.2130	.1681	.1104	.0614	.0291	.0117
	5	.0218	.0787	.1507	.1988	.2017	.1664	.1146	.0666	.0327
	6	.0052	.0301	.0816	.1436	.1873	.1941	.1655	.1181	.0708
	7	.0010	.0091	.0350	.0820	.1376	.1792	.1892	.1657	.1214
	8	.0002	.0022	.0120	.0376	.0811	.1327	.1734	.1864	.1669
	9	.0000	.0004	.0033	.0139	.0386	.0794	.1284	.1694	.1855
	10	.0000	.0001	.0008	.0042	.0149	.0385	.0771	.1248	.1669
	11	.0000	.0000	.0001	.0010	.0046	.0151	.0374	.0742	.1214
	12	.0000	.0000	.0000	.0002	.0012	.0047	.0145	.0354	.0708
	13	.0000	.0000	.0000	.0000	.0002	.0012	.0045	.0134	.0327
	14	.0000	.0000	.0000	.0000	.0000	.0002	.0011	.0039	.0117
	15	.0000	.0000	.0000	.0000	.0000	.0000	.0002	.0009	.0031
	16	.0000	.0000	.0000	.0000	.0000	.0000	.0000	.0001	.0006
	17	.0000	.0000	.0000	.0000	.0000	.0000	.0000	.0000	.0001
	18	.0000	.0000	.0000	.0000	.0000	.0000	.0000	.0000	.0000
20	0	.1216	.0388	.0115	.0032	.0008	.0002	.0000	.0000	.0000
	1	.2702	.1368	.0576	.0211	.0068	.0020	.0005	.0001	.0000
	2	.2852	.2293	.1369	.0669	.0278	.0100	.0031	.0008	.0002
	3	.1901	.2428	.2054	.1339	.0716	.0323	.0123	.0040	.0011
	4	.0898	.1821	.2182	.1897	.1304	.0738	.0350	.0139	.0046
	5	.0319	.1028	.1746	.2023	.1789	.1272	.0746	.0365	.0148
	6	.0089	.0454	.1091	.1686	.1916	.1712	.1244	.0746	.0370
	7	.0020	.0160	.0545	.1124	.1643	.1844	.1659	.1221	.0739
	8	.0004	.0046	.0222	.0609	.1144	.1614	.1797	.1623	.1201
	9	.0001	.0011	.0074	.0271	.0654	.1158	.1597	.1771	.1602
	10	.0000	.0002	.0020	.0099	.0308	.0686	.1171	.1593	.1762
	11	.0000	.0000	.0005	.0030	.0120	.0336	.0710	.1185	.1602
	12	.0000	.0000	.0001	.0008	.0039	.0136	.0355	.0727	.1201
	13	.0000	.0000	.0000	.0002	.0010	.0045	.0146	.0366	.0739
	14	.0000	.0000	.0000	.0000	.0002	.0012	.0049	.0150	.0370
	15	.0000	.0000	.0000	.0000	.0000	.0003	.0013	.0049	.0148
	16	.0000	.0000	.0000	.0000	.0000	.0000	.0003	.0013	.0046
	17	.0000	.0000	.0000	.0000	.0000	.0000	.0000	.0002	.0011
	18	.0000	.0000	.0000	.0000	.0000	.0000	.0000	.0000	.0002
	19	.0000	.0000	.0000	.0000	.0000	.0000	.0000	.0000	.0000
	20	.0000	.0000	.0000	.0000	.0000	.0000	.0000	.0000	.0000

(TABLE CONTINUES ON NEXT PAGE)

TABLE 5
(Continued)

n	x	.55	.60	.65	.70	.75	.80	.85	.90	.95
2	0	.2025	.1600	.1225	.0900	.0625	.0400	.0225	.0100	.0025
	1	.4950	.4800	.4550	.4200	.3750	.3200	.2550	.1800	.0950
	2	.3025	.3600	.4225	.4900	.5625	.6400	.7225	.8100	.9025
3	0	.0911	.0640	.0429	.0270	.0156	.0080	.0034	.0010	.0001
	1	.3341	.2880	.2389	.1890	.1406	.0960	.0574	.0270	.0071
	2	.4084	.4320	.4436	.4410	.4219	.3840	.3251	.2430	.1354
	3	.1664	.2160	.2746	.3430	.4219	.5120	.6141	.7290	.8574
4	0	.0410	.0256	.0150	.0081	.0039	.0016	.0005	.0001	.0000
	1	.2005	.1536	.1115	.0756	.0469	.0256	.0115	.0036	.0005
	2	.3675	.3456	.3105	.2646	.2109	.1536	.0975	.0486	.0135
	3	.2995	.3456	.3845	.4116	.4219	.4096	.3685	.2916	.1715
	4	.0915	.1296	.1785	.2401	.3164	.4096	.5220	.6561	.8145
5	0	.0185	.0102	.0053	.0024	.0010	.0003	.0001	.0000	.0000
	1	.1128	.0768	.0488	.0284	.0146	.0064	.0022	.0005	.0000
	2	.2757	.2304	.1811	.1323	.0879	.0512	.0244	.0081	.0011
	3	.3369	.3456	.3364	.3087	.2637	.2048	.1382	.0729	.0214
	4	.2059	.2592	.3124	.3601	.3955	.4096	.3915	.3281	.2036
	5	.0503	.0778	.1160	.1681	.2373	.3277	.4437	.5905	.7738
6	0	.0083	.0041	.0018	.0007	.0002	.0001	.0000	.0000	.0000
	1	.0609	.0369	.0205	.0102	.0044	.0015	.0004	.0001	.0000
	2	.1861	.1382	.0951	.0595	.0330	.0154	.0055	.0012	.0001
	3	.3032	.2765	.2355	.1852	.1318	.0819	.0415	.0146	.0021
	4	.2780	.3110	.3280	.3241	.2966	.2458	.1762	.0984	.0305
	5	.1359	.1866	.2437	.3025	.3560	.3932	.3993	.3543	.2321
	6	.0277	.0467	.0754	.1176	.1780	.2621	.3771	.5314	.7351
7	0	.0037	.0016	.0006	.0002	.0001	.0000	.0000	.0000	.0000
	1	.0320	.0172	.0084	.0036	.0013	.0004	.0001	.0000	.0000
	2	.1172	.0774	.0466	.0250	.0115	.0043	.0012	.0002	.0000
	3	.2388	.1935	.1442	.0972	.0577	.0287	.0109	.0026	.0002
	4	.2918	.2903	.2679	.2269	.1730	.1147	.0617	.0230	.0036
	5	.2140	.2613	.2985	.3177	.3115	.2753	.2097	.1240	.0406
	6	.0872	.1306	.1848	.2471	.3115	.3670	.3960	.3720	.2573
	7	.0152	.0280	.0490	.0824	.1335	.2097	.3206	.4783	.6983
8	0	.0017	.0007	.0002	.0001	.0000	.0000	.0000	.0000	.0000
	1	.0164	.0079	.0033	.0012	.0004	.0001	.0000	.0000	.0000
	2	.0703	.0413	.0217	.0100	.0038	.0011	.0002	.0000	.0000
	3	.1719	.1239	.0808	.0467	.0231	.0092	.0026	.0004	.0000
	4	.2627	.2322	.1875	.1361	.0865	.0459	.0185	.0046	.0004
	5	.2568	.2787	.2786	.2541	.2076	.1468	.0839	.0331	.0054
	6	.1569	.2090	.2587	.2965	.3115	.2936	.2376	.1488	.0515
	7	.0548	.0896	.1373	.1977	.2670	.3355	.3847	.3826	.2793
	8	.0084	.0168	.0319	.0576	.1001	.1678	.2725	.4305	.6634

(TABLE CONTINUES)

TABLE 5
(Continued)

n	x	.55	.60	.65	.70	.75	.80	.85	.90	.95
						p				
9	0	.0008	.0003	.0001	.0000	.0000	.0000	.0000	.0000	.0000
	1	.0083	.0035	.0013	.0004	.0001	.0000	.0000	.0000	.0000
	2	.0407	.0212	.0098	.0039	.0012	.0003	.0000	.0000	.0000
	3	.1160	.0743	.0424	.0210	.0087	.0028	.0006	.0001	.0000
	4	.2128	.1672	.1181	.0735	.0389	.0165	.0050	.0008	.0000
	5	.2600	.2508	.2194	.1715	.1168	.0661	.0283	.0074	.0006
	6	.2119	.2508	.2716	.2668	.2336	.1762	.1069	.0446	.0077
	7	.1110	.1612	.2162	.2668	.3003	.3020	.2597	.1722	.0629
	8	.0339	.0605	.1004	.1556	.2253	.3020	.3679	.3874	.2985
	9	.0046	.0101	.0207	.0404	.0751	.1342	.2316	.3874	.6302
10	0	.0003	.0001	.0000	.0000	.0000	.0000	.0000	.0000	.0000
	1	.0042	.0016	.0005	.0001	.0000	.0000	.0000	.0000	.0000
	2	.0229	.0106	.0043	.0014	.0004	.0001	.0000	.0000	.0000
	3	.0746	.0425	.0212	.0090	.0031	.0008	.0001	.0000	.0000
	4	.1596	.1115	.0689	.0368	.0162	.0055	.0012	.0001	.0000
	5	.2340	.2007	.1536	.1029	.0584	.0264	.0085	.0015	.0001
	6	.2384	.2508	.2377	.2001	.1460	.0881	.0401	.0112	.0010
	7	.1665	.2150	.2522	.2668	.2503	.2013	.1298	.0574	.0105
	8	.0763	.1209	.1757	.2335	.2816	.3020	.2759	.1937	.0746
	9	.0207	.0403	.0725	.1211	.1877	.2684	.3474	.3874	.3151
	10	.0025	.0060	.0135	.0282	.0563	.1074	.1969	.3487	.5987
12	0	.0001	.0000	.0000	.0000	.0000	.0000	.0000	.0000	.0000
	1	.0010	.0003	.0001	.0000	.0000	.0000	.0000	.0000	.0000
	2	.0068	.0025	.0008	.0002	.0000	.0000	.0000	.0000	.0000
	3	.0277	.0125	.0048	.0015	.0004	.0001	.0000	.0000	.0000
	4	.0762	.0420	.0199	.0078	.0024	.0005	.0001	.0000	.0000
	5	.1489	.1009	.0591	.0291	.0115	.0033	.0006	.0000	.0000
	6	.2124	.1766	.1281	.0792	.0401	.0155	.0040	.0005	.0000
	7	.2225	.2270	.2039	.1585	.1032	.0532	.0193	.0038	.0002
	8	.1700	.2128	.2367	.2311	.1936	.1329	.0683	.0213	.0021
	9	.0923	.1419	.1954	.2397	.2581	.2362	.1720	.0852	.0173
	10	.0339	.0639	.1088	.1678	.2323	.2835	.2924	.2301	.0988
	11	.0075	.0174	.0368	.0712	.1267	.2062	.3012	.3766	.3413
	12	.0008	.0022	.0057	.0138	.0317	.0687	.1422	.2824	.5404
15	0	.0000	.0000	.0000	.0000	.0000	.0000	.0000	.0000	.0000
	1	.0001	.0000	.0000	.0000	.0000	.0000	.0000	.0000	.0000
	2	.0010	.0003	.0001	.0000	.0000	.0000	.0000	.0000	.0000
	3	.0052	.0016	.0004	.0001	.0000	.0000	.0000	.0000	.0000
	4	.0191	.0074	.0024	.0006	.0001	.0000	.0000	.0000	.0000
	5	.0515	.0245	.0096	.0030	.0007	.0001	.0000	.0000	.0000
	6	.1048	.0612	.0298	.0116	.0034	.0007	.0001	.0000	.0000
	7	.1647	.1181	.0710	.0348	.0131	.0035	.0005	.0000	.0000
	8	.2013	.1771	.1319	.0811	.0393	.0138	.0030	.0003	.0000
	9	.1914	.2066	.1906	.1472	.0917	.0430	.0132	.0019	.0000
	10	.1404	.1859	.2123	.2061	.1651	.1032	.0449	.0105	.0006
	11	.0780	.1268	.1792	.2186	.2252	.1876	.1156	.0428	.0049
	12	.0318	.0634	.1110	.1700	.2252	.2501	.2184	.1285	.0307
	13	.0090	.0219	.0476	.0916	.1559	.2309	.2856	.2669	.1348
	14	.0016	.0047	.0126	.0305	.0668	.1319	.2312	.3432	.3658
	15	.0001	.0005	.0016	.0047	.0134	.0352	.0874	.2059	.4633

(TABLE CONTINUES ON NEXT PAGE)

TABLE 5
(Continued)

n	x	.55	.60	.65	.70	.75	.80	.85	.90	.95
18	0	.0000	.0000	.0000	.0000	.0000	.0000	.0000	.0000	.0000
	1	.0000	.0000	.0000	.0000	.0000	.0000	.0000	.0000	.0000
	2	.0001	.0000	.0000	.0000	.0000	.0000	.0000	.0000	.0000
	3	.0009	.0002	.0000	.0000	.0000	.0000	.0000	.0000	.0000
	4	.0039	.0011	.0002	.0000	.0000	.0000	.0000	.0000	.0000
	5	.0134	.0045	.0012	.0002	.0000	.0000	.0000	.0000	.0000
	6	.0354	.0145	.0047	.0012	.0002	.0000	.0000	.0000	.0000
	7	.0742	.0374	.0151	.0046	.0010	.0001	.0000	.0000	.0000
	8	.1248	.0771	.0385	.0149	.0042	.0008	.0001	.0000	.0000
	9	.1694	.1284	.0794	.0386	.0139	.0033	.0004	.0000	.0000
	10	.1864	.1734	.1327	.0811	.0376	.0120	.0022	.0002	.0000
	11	.1657	.1892	.1792	.1376	.0820	.0350	.0091	.0010	.0000
	12	.1181	.1655	.1941	.1873	.1436	.0816	.0301	.0052	.0002
	13	.0666	.1146	.1664	.2017	.1988	.1507	.0787	.0218	.0014
	14	.0291	.0614	.1104	.1681	.2130	.2153	.1592	.0700	.0093
	15	.0095	.0246	.0547	.1046	.1704	.2297	.2406	.1680	.0473
	16	.0022	.0069	.0190	.0458	.0958	.1723	.2556	.2835	.1683
	17	.0003	.0012	.0042	.0126	.0338	.0811	.1704	.3002	.3763
	18	.0000	.0001	.0004	.0016	.0056	.0180	.0536	.1501	.3972
20	0	.0000	.0000	.0000	.0000	.0000	.0000	.0000	.0000	.0000
	1	.0000	.0000	.0000	.0000	.0000	.0000	.0000	.0000	.0000
	2	.0000	.0000	.0000	.0000	.0000	.0000	.0000	.0000	.0000
	3	.0002	.0000	.0000	.0000	.0000	.0000	.0000	.0000	.0000
	4	.0013	.0003	.0000	.0000	.0000	.0000	.0000	.0000	.0000
	5	.0049	.0013	.0003	.0000	.0000	.0000	.0000	.0000	.0000
	6	.0150	.0049	.0012	.0002	.0000	.0000	.0000	.0000	.0000
	7	.0366	.0146	.0045	.0010	.0002	.0000	.0000	.0000	.0000
	8	.0727	.0355	.0136	.0039	.0008	.0001	.0000	.0000	.0000
	9	.1185	.0710	.0336	.0120	.0030	.0005	.0000	.0000	.0000
	10	.1593	.1171	.0686	.0308	.0099	.0020	.0002	.0000	.0000
	11	.1771	.1597	.1158	.0654	.0271	.0074	.0011	.0001	.0000
	12	.1623	.1797	.1614	.1144	.0609	.0222	.0046	.0004	.0000
	13	.1221	.1659	.1844	.1643	.1124	.0545	.0160	.0020	.0000
	14	.0746	.1244	.1712	.1916	.1686	.1091	.0454	.0089	.0003
	15	.0365	.0746	.1272	.1789	.2023	.1746	.1028	.0319	.0022
	16	.0139	.0350	.0738	.1304	.1897	.2182	.1821	.0898	.0133
	17	.0040	.0123	.0323	.0716	.1339	.2054	.2428	.1901	.0596
	18	.0008	.0031	.0100	.0278	.0669	.1369	.2293	.2852	.1887
	19	.0001	.0005	.0020	.0068	.0211	.0576	.1368	.2702	.3774
	20	.0000	.0000	.0002	.0008	.0032	.0115	.0388	.1216	.3585

TABLE 6

Values of $e^{-\mu}$

μ	$e^{-\mu}$	μ	$e^{-\mu}$	μ	$e^{-\mu}$
.00	1.0000				
.05	.9512	2.05	.1287	4.05	.0174
.10	.9048	2.10	.1225	4.10	.0166
.15	.8607	2.15	.1165	4.15	.0158
.20	.8187	2.20	.1108	4.20	.0150
.25	.7788	2.25	.1054	4.25	.0143
.30	.7408	2.30	.1003	4.30	.0136
.35	.7047	2.35	.0954	4.35	.0129
.40	.6703	2.40	.0907	4.40	.0123
.45	.6376	2.45	.0863	4.45	.0117
.50	.6065	2.50	.0821	4.50	.0111
.55	.5769	2.55	.0781	4.55	.0106
.60	.5488	2.60	.0743	4.60	.0101
.65	.5220	2.65	.0707	4.65	.0096
.70	.4966	2.70	.0672	4.70	.0091
.75	.4724	2.75	.0639	4.75	.0087
.80	.4493	2.80	.0608	4.80	.0082
.85	.4274	2.85	.0578	4.85	.0078
.90	.4066	2.90	.0550	4.90	.0074
.95	.3867	2.95	.0523	4.95	.0071
1.00	.3679	3.00	.0498	5.00	.0067
1.05	.3499	3.05	.0474	5.05	.0064
1.10	.3329	3.10	.0450	5.10	.0061
1.15	.3166	3.15	.0429	5.15	.0058
1.20	.3012	3.20	.0408	5.20	.0055
1.25	.2865	3.25	.0388	5.25	.0052
1.30	.2725	3.30	.0369	5.30	.0050
1.35	.2592	3.35	.0351	5.35	.0047
1.40	.2466	3.40	.0334	5.40	.0045
1.45	.2346	3.45	.0317	5.45	.0043
1.50	.2231	3.50	.0302	5.50	.0041
1.55	.2122	3.55	.0287	5.55	.0039
1.60	.2019	3.60	.0273	5.60	.0037
1.65	.1920	3.65	.0260	5.65	.0035
1.70	.1827	3.70	.0247	5.70	.0033
1.75	.1738	3.75	.0235	5.75	.0032
1.80	.1653	3.80	.0224	5.80	.0030
1.85	.1572	3.85	.0213	5.85	.0029
1.90	.1496	3.90	.0202	5.90	.0027
1.95	.1423	3.95	.0193	5.95	.0026
2.00	.1353	4.00	.0183	6.00	.0025
				7.00	.0009
				8.00	.000335
				9.00	.000123
				10.00	.000045

TABLE 7
Poisson
Probabilities

Entries in the table give the probability of x occurrences for a Poisson process with a mean μ.
For example, when $\mu = 2.5$, the probability of four occurrences is .1336.

					μ					
x	0.1	0.2	0.3	0.4	0.5	0.6	0.7	0.8	0.9	1.0
0	.9048	.8187	.7408	.6703	.6065	.5488	.4966	.4493	.4066	.3679
1	.0905	.1637	.2222	.2681	.3033	.3293	.3476	.3595	.3659	.3679
2	.0045	.0164	.0333	.0536	.0758	.0988	.1217	.1438	.1647	.1839
3	.0002	.0011	.0033	.0072	.0126	.0198	.0284	.0383	.0494	.0613
4	.0000	.0001	.0002	.0007	.0016	.0030	.0050	.0077	.0111	.0153
5	.0000	.0000	.0000	.0001	.0002	.0004	.0007	.0012	.0020	.0031
6	.0000	.0000	.0000	.0000	.0000	.0000	.0001	.0002	.0003	.0005
7	.0000	.0000	.0000	.0000	.0000	.0000	.0000	.0000	.0000	.0001

					μ					
x	1.1	1.2	1.3	1.4	1.5	1.6	1.7	1.8	1.9	2.0
0	.3329	.3012	.2725	.2466	.2231	.2019	.1827	.1653	.1496	.1353
1	.3662	.3614	.3543	.3452	.3347	.3230	.3106	.2975	.2842	.2707
2	.2014	.2169	.2303	.2417	.2510	.2584	.2640	.2678	.2700	.2707
3	.0738	.0867	.0998	.1128	.1255	.1378	.1496	.1607	.1710	.1804
4	.0203	.0260	.0324	.0395	.0471	.0551	.0636	.0723	.0812	.0902
5	.0045	.0062	.0084	.0111	.0141	.0176	.0216	.0260	.0309	.0361
6	.0008	.0012	.0018	.0026	.0035	.0047	.0061	.0078	.0098	.0120
7	.0001	.0002	.0003	.0005	.0008	.0011	.0015	.0020	.0027	.0034
8	.0000	.0000	.0001	.0001	.0001	.0002	.0003	.0005	.0006	.0009
9	.0000	.0000	.0000	.0000	.0000	.0000	.0001	.0001	.0001	.0002

					μ					
x	2.1	2.2	2.3	2.4	2.5	2.6	2.7	2.8	2.9	3.0
0	.1225	.1108	.1003	.0907	.0821	.0743	.0672	.0608	.0550	.0498
1	.2572	.2438	.2306	.2177	.2052	.1931	.1815	.1703	.1596	.1494
2	.2700	.2681	.2652	.2613	.2565	.2510	.2450	.2384	.2314	.2240
3	.1890	.1966	.2033	.2090	.2138	.2176	.2205	.2225	.2237	.2240
4	.0992	.1082	.1169	.1254	.1336	.1414	.1488	.1557	.1622	.1680

(TABLE CONTINUES)

TABLE 7
(Continued)

x	μ									
	2.1	2.2	2.3	2.4	2.5	2.6	2.7	2.8	2.9	3.0
5	.0417	.0476	.0538	.0602	.0668	.0735	.0804	.0872	.0940	.1008
6	.0146	.0174	.0206	.0241	.0278	.0319	.0362	.0407	.0455	.0504
7	.0044	.0055	.0068	.0083	.0099	.0118	.0139	.0163	.0188	.0216
8	.0011	.0015	.0019	.0025	.0031	.0038	.0047	.0057	.0068	.0081
9	.0003	.0004	.0005	.0007	.0009	.0011	.0014	.0018	.0022	.0027
10	.0001	.0001	.0001	.0002	.0002	.0003	.0004	.0005	.0006	.0008
11	.0000	.0000	.0000	.0000	.0000	.0001	.0001	.0001	.0002	.0002
12	.0000	.0000	.0000	.0000	.0000	.0000	.0000	.0000	.0000	.0001

x	μ									
	3.1	3.2	3.3	3.4	3.5	3.6	3.7	3.8	3.9	4.0
0	.0450	.0408	.0369	.0344	.0302	.0273	.0247	.0224	.0202	.0183
1	.1397	.1304	.1217	.1135	.1057	.0984	.0915	.0850	.0789	.0733
2	.2165	.2087	.2008	.1929	.1850	.1771	.1692	.1615	.1539	.1465
3	.2237	.2226	.2209	.2186	.2158	.2125	.2087	.2046	.2001	.1954
4	.1734	.1781	.1823	.1858	.1888	.1912	.1931	.1944	.1951	.1954
5	.1075	.1140	.1203	.1264	.1322	.1377	.1429	.1477	.1522	.1563
6	.0555	.0608	.0662	.0716	.0771	.0826	.0881	.0936	.0989	.1042
7	.0246	.0278	.0312	.0348	.0385	.0425	.0466	.0508	.0551	.0595
8	.0095	.0111	.0129	.0148	.0169	.0191	.0215	.0241	.0269	.0298
9	.0033	.0040	.0047	.0056	.0066	.0076	.0089	.0102	.0116	.0132
10	.0010	.0013	.0016	.0019	.0023	.0028	.0033	.0039	.0045	.0053
11	.0003	.0004	.0005	.0006	.0007	.0009	.0011	.0013	.0016	.0019
12	.0001	.0001	.0001	.0002	.0002	.0003	.0003	.0004	.0005	.0006
13	.0000	.0000	.0000	.0000	.0001	.0001	.0001	.0001	.0002	.0002
14	.0000	.0000	.0000	.0000	.0000	.0000	.0000	.0000	.0000	.0001

x	μ									
	4.1	4.2	4.3	4.4	4.5	4.6	4.7	4.8	4.9	5.0
0	.0166	.0150	.0136	.0123	.0111	.0101	.0091	.0082	.0074	.0067
1	.0679	.0630	.0583	.0540	.0500	.0462	.0427	.0395	.0365	.0337
2	.1393	.1323	.1254	.1188	.1125	.1063	.1005	.0948	.0894	.0842
3	.1904	.1852	.1798	.1743	.1687	.1631	.1574	.1517	.1460	.1404
4	.1951	.1944	.1933	.1917	.1898	.1875	.1849	.1820	.1789	.1755
5	.1600	.1633	.1662	.1687	.1708	.1725	.1738	.1747	.1753	.1755
6	.1093	.1143	.1191	.1237	.1281	.1323	.1362	.1398	.1432	.1462
7	.0640	.0686	.0732	.0778	.0824	.0869	.0914	.0959	.1002	.1044
8	.0328	.0360	.0393	.0428	.0463	.0500	.0537	.0575	.0614	.0653
9	.0150	.0168	.0188	.0209	.0232	.0255	.0280	.0307	.0334	.0363

(TABLE CONTINUES ON NEXT PAGE)

TABLE 7
(Continued)

					μ					
x	4.1	4.2	4.3	4.4	4.5	4.6	4.7	4.8	4.9	5.0
10	.0061	.0071	.0081	.0092	.0104	.0118	.0132	.0147	.0164	.0181
11	.0023	.0027	.0032	.0037	.0043	.0049	.0056	.0064	.0073	.0082
12	.0008	.0009	.0011	.0014	.0016	.0019	.0022	.0026	.0030	.0034
13	.0002	.0003	.0004	.0005	.0006	.0007	.0008	.0009	.0011	.0013
14	.0001	.0001	.0001	.0001	.0002	.0002	.0003	.0003	.0004	.0005
15	.0000	.0000	.0000	.0000	.0001	.0001	.0001	.0001	.0001	.0002

					μ					
x	5.1	5.2	5.3	5.4	5.5	5.6	5.7	5.8	5.9	6.0
0	.0061	.0055	.0050	.0045	.0041	.0037	.0033	.0030	.0027	.0025
1	.0311	.0287	.0265	.0244	.0225	.0207	.0191	.0176	.0162	.0149
2	.0793	.0746	.0701	.0659	.0618	.0580	.0544	.0509	.0477	.0446
3	.1348	.1293	.1239	.1185	.1133	.1082	.1033	.0985	.0938	.0892
4	.1719	.1681	.1641	.1600	.1558	.1515	.1472	.1428	.1383	.1339
5	.1753	.1748	.1740	.1728	.1714	.1697	.1678	.1656	.1632	.1606
6	.1490	.1515	.1537	.1555	.1571	.1584	.1594	.1601	.1605	.1606
7	.1086	.1125	.1163	.1200	.1234	.1267	.1298	.1326	.1353	.1377
8	.0692	.0731	.0771	.0810	.0849	.0887	.0925	.0962	.0998	.1033
9	.0392	.0423	.0454	.0486	.0519	.0552	.0586	.0620	.0654	.0688
10	.0200	.0220	.0241	.0262	.0285	.0309	.0334	.0359	.0386	.0413
11	.0093	.0104	.0116	.0129	.0143	.0157	.0173	.0190	.0207	.0225
12	.0039	.0045	.0051	.0058	.0065	.0073	.0082	.0092	.0102	.0113
13	.0015	.0018	.0021	.0024	.0028	.0032	.0036	.0041	.0046	.0052
14	.0006	.0007	.0008	.0009	.0011	.0013	.0015	.0017	.0019	.0022
15	.0002	.0002	.0003	.0003	.0004	.0005	.0006	.0007	.0008	.0009
16	.0001	.0001	.0001	.0001	.0001	.0002	.0002	.0002	.0003	.0003
17	.0000	.0000	.0000	.0000	.0000	.0001	.0001	.0001	.0001	.0001

					μ					
x	6.1	6.2	6.3	6.4	6.5	6.6	6.7	6.8	6.9	7.0
0	.0022	.0020	.0018	.0017	.0015	.0014	.0012	.0011	.0010	.0009
1	.0137	.0126	.0116	.0106	.0098	.0090	.0082	.0076	.0070	.0064
2	.0417	.0390	.0364	.0340	.0318	.0296	.0276	.0258	.0240	.0223
3	.0848	.0806	.0765	.0726	.0688	.0652	.0617	.0584	.0552	.0521
4	.1294	.1249	.1205	.1162	.1118	.1076	.1034	.0992	.0952	.0912
5	.1579	.1549	.1519	.1487	.1454	.1420	.1385	.1349	.1314	.1277
6	.1605	.1601	.1595	.1586	.1575	.1562	.1546	.1529	.1511	.1490
7	.1399	.1418	.1435	.1450	.1462	.1472	.1480	.1486	.1489	.1490
8	.1066	.1099	.1130	.1160	.1188	.1215	.1240	.1263	.1284	.1304
9	.0723	.0757	.0791	.0825	.0858	.0891	.0923	.0954	.0985	.1014

(TABLE CONTINUES)

TABLE 7
(Continued)

	μ									
x	6.1	6.2	6.3	6.4	6.5	6.6	6.7	6.8	6.9	7.0
10	.0441	.0469	.0498	.0528	.0558	.0588	.0618	.0649	.0679	.0710
11	.0245	.0265	.0285	.0307	.0330	.0353	.0377	.0401	.0426	.0452
12	.0124	.0137	.0150	.0164	.0179	.0194	.0210	.0227	.0245	.0264
13	.0058	.0065	.0073	.0081	.0089	.0098	.0108	.0119	.0130	.0142
14	.0025	.0029	.0033	.0037	.0041	.0046	.0052	.0058	.0064	.0071
15	.0010	.0012	.0014	.0016	.0018	.0020	.0023	.0026	.0029	.0033
16	.0004	.0005	.0005	.0006	.0007	.0008	.0010	.0011	.0013	.0014
17	.0001	.0002	.0002	.0002	.0003	.0003	.0004	.0004	.0005	.0006
18	.0000	.0001	.0001	.0001	.0001	.0001	.0001	.0002	.0002	.0002
19	.0000	.0000	.0000	.0000	.0000	.0000	.0000	.0001	.0001	.0001

	μ									
x	7.1	7.2	7.3	7.4	7.5	7.6	7.7	7.8	7.9	8.0
0	.0008	.0007	.0007	.0006	.0006	.0005	.0005	.0004	.0004	.0003
1	.0059	.0054	.0049	.0045	.0041	.0038	.0035	.0032	.0029	.0027
2	.0208	.0194	.0180	.0167	.0156	.0145	.0134	.0125	.0116	.0107
3	.0492	.0464	.0438	.0413	.0389	.0366	.0345	.0324	.0305	.0286
4	.0874	.0836	.0799	.0764	.0729	.0696	.0663	.0632	.0602	.0573
5	.1241	.1204	.1167	.1130	.1094	.1057	.1021	.0986	.0951	.0916
6	.1468	.1445	.1420	.1394	.1367	.1339	.1311	.1282	.1252	.1221
7	.1489	.1486	.1481	.1474	.1465	.1454	.1442	.1428	.1413	.1396
8	.1321	.1337	.1351	.1363	.1373	.1382	.1388	.1392	.1395	.1396
9	.1042	.1070	.1096	.1121	.1144	.1167	.1187	.1207	.1224	.1241
10	.0740	.0770	.0800	.0829	.0858	.0887	.0914	.0941	.0967	.0993
11	.0478	.0504	.0531	.0558	.0585	.0613	.0640	.0667	.0695	.0722
12	.0283	.0303	.0323	.0344	.0366	.0388	.0411	.0434	.0457	.0481
13	.0154	.0168	.0181	.0196	.0211	.0227	.0243	.0260	.0278	.0296
14	.0078	.0086	.0095	.0104	.0113	.0123	.0134	.0145	.0157	.0169
15	.0037	.0041	.0046	.0051	.0057	.0062	.0069	.0075	.0083	.0090
16	.0016	.0019	.0021	.0024	.0026	.0030	.0033	.0037	.0041	.0045
17	.0007	.0008	.0009	.0010	.0012	.0013	.0015	.0017	.0019	.0021
18	.0003	.0003	.0004	.0004	.0005	.0006	.0006	.0007	.0008	.0009
19	.0001	.0001	.0001	.0002	.0002	.0002	.0003	.0003	.0003	.0004
20	.0000	.0000	.0001	.0001	.0001	.0001	.0001	.0001	.0001	.0002
21	.0000	.0000	.0000	.0000	.0000	.0000	.0000	.0000	.0001	.0001

(TABLE CONTINUES ON NEXT PAGE)

TABLE 7

(Continued)

x	8.1	8.2	8.3	8.4	8.5	8.6	8.7	8.8	8.9	9.0
0	.0003	.0003	.0002	.0002	.0002	.0002	.0002	.0002	.0001	.0001
1	.0025	.0023	.0021	.0019	.0017	.0016	.0014	.0013	.0012	.0011
2	.0100	.0092	.0086	.0079	.0074	.0068	.0063	.0058	.0054	.0050
3	.0269	.0252	.0237	.0222	.0208	.0195	.0183	.0171	.0160	.0150
4	.0544	.0517	.0491	.0466	.0443	.0420	.0398	.0377	.0357	.0337
5	.0882	.0849	.0816	.0784	.0752	.0722	.0692	.0663	.0635	.0607
6	.1191	.1160	.1128	.1097	.1066	.1034	.1003	.0972	.0941	.0911
7	.1378	.1358	.1338	.1317	.1294	.1271	.1247	.1222	.1197	.1171
8	.1395	.1392	.1388	.1382	.1375	.1366	.1356	.1344	.1332	.1318
9	.1256	.1269	.1280	.1290	.1299	.1306	.1311	.1315	.1317	.1318
10	.1017	.1040	.1063	.1084	.1104	.1123	.1140	.1157	.1172	.1186
11	.0749	.0776	.0802	.0828	.0853	.0878	.0902	.0925	.0948	.0970
12	.0505	.0530	.0555	.0579	.0604	.0629	.0654	.0679	.0703	.0728
13	.0315	.0334	.0354	.0374	.0395	.0416	.0438	.0459	.0481	.0504
14	.0182	.0196	.0210	.0225	.0240	.0256	.0272	.0289	.0306	.0324
15	.0098	.0107	.0116	.0126	.0136	.0147	.0158	.0169	.0182	.0194
16	.0050	.0055	.0060	.0066	.0072	.0079	.0086	.0093	.0101	.0109
17	.0024	.0026	.0029	.0033	.0036	.0040	.0044	.0048	.0053	.0058
18	.0011	.0012	.0014	.0015	.0017	.0019	.0021	.0024	.0026	.0029
19	.0005	.0005	.0006	.0007	.0008	.0009	.0010	.0011	.0012	.0014
20	.0002	.0002	.0002	.0003	.0003	.0004	.0004	.0005	.0005	.0006
21	.0001	.0001	.0001	.0001	.0001	.0002	.0002	.0002	.0002	.0003
22	.0000	.0000	.0000	.0000	.0001	.0001	.0001	.0001	.0001	.0001

x	9.1	9.2	9.3	9.4	9.5	9.6	9.7	9.8	9.9	10
0	.0001	.0001	.0001	.0001	.0001	.0001	.0001	.0001	.0001	.0000
1	.0010	.0009	.0009	.0008	.0007	.0007	.0006	.0005	.0005	.0005
2	.0046	.0043	.0040	.0037	.0034	.0031	.0029	.0027	.0025	.0023
3	.0140	.0131	.0123	.0115	.0107	.0100	.0093	.0087	.0081	.0076
4	.0319	.0302	.0285	.0269	.0254	.0240	.0226	.0213	.0201	.0189
5	.0581	.0555	.0530	.0506	.0483	.0460	.0439	.0418	.0398	.0378
6	.0881	.0851	.0822	.0793	.0764	.0736	.0709	.0682	.0656	.0631
7	.1145	.1118	.1091	.1064	.1037	.1010	.0982	.0955	.0928	.0901
8	.1302	.1286	.1269	.1251	.1232	.1212	.1191	.1170	.1148	.1126
9	.1317	.1315	.1311	.1306	.1300	.1293	.1284	.1274	.1263	.1251

(TABLE CONTINUES)

TABLE 7
(Continued)

	μ									
x	9.1	9.2	9.3	9.4	9.5	9.6	9.7	9.8	9.9	10.0
10	.1198	.1210	.1219	.1228	.1235	.1241	.1245	.1249	.1250	.1251
11	.0991	.1012	.1031	.1049	.1067	.1083	.1098	.1112	.1125	.1137
12	.0752	.0776	.0799	.0822	.0844	.0866	.0888	.0908	.0928	.0948
13	.0526	.0549	.0572	.0594	.0617	.0640	.0662	.0685	.0707	.0729
14	.0342	.0361	.0380	.0399	.0419	.0439	.0459	.0479	.0500	.0521
15	.0208	.0221	.0235	.0250	.0265	.0281	.0297	.0313	.0330	.0347
16	.0118	.0127	.0137	.0147	.0157	.0168	.0180	.0192	.0204	.0217
17	.0063	.0069	.0075	.0081	.0088	.0095	.0103	.0111	.0119	.0128
18	.0032	.0035	.0039	.0042	.0046	.0051	.0055	.0060	.0065	.0071
19	.0015	.0017	.0019	.0021	.0023	.0026	.0028	.0031	.0034	.0037
20	.0007	.0008	.0009	.0010	.0011	.0012	.0014	.0015	.0017	.0019
21	.0003	.0003	.0004	.0004	.0005	.0006	.0006	.0007	.0008	.0009
22	.0001	.0001	.0002	.0002	.0002	.0002	.0003	.0003	.0004	.0004
23	.0000	.0001	.0001	.0001	.0001	.0001	.0001	.0001	.0002	.0002
24	.0000	.0000	.0000	.0000	.0000	.0000	.0000	.0001	.0001	.0001

	μ									
x	11	12	13	14	15	16	17	18	19	20
0	.0000	.0000	.0000	.0000	.0000	.0000	.0000	.0000	.0000	.0000
1	.0002	.0001	.0000	.0000	.0000	.0000	.0000	.0000	.0000	.0000
2	.0010	.0004	.0002	.0001	.0000	.0000	.0000	.0000	.0000	.0000
3	.0037	.0018	.0008	.0004	.0002	.0001	.0000	.0000	.0000	.0000
4	.0102	.0053	.0027	.0013	.0006	.0003	.0001	.0001	.0000	.0000
5	.0224	.0127	.0070	.0037	.0019	.0010	.0005	.0002	.0001	.0001
6	.0411	.0255	.0152	.0087	.0048	.0026	.0014	.0007	.0004	.0002
7	.0646	.0437	.0281	.0174	.0104	.0060	.0034	.0018	.0010	.0005
8	.0888	.0655	.0457	.0304	.0194	.0120	.0072	.0042	.0024	.0013
9	.1085	.0874	.0661	.0473	.0324	.0213	.0135	.0083	.0050	.0029
10	.1194	.1048	.0859	.0663	.0486	.0341	.0230	.0150	.0095	.0058
11	.1194	.1144	.1015	.0844	.0663	.0496	.0355	.0245	.0164	.0106
12	.1094	.1144	.1099	.0984	.0829	.0661	.0504	.0368	.0259	.0176
13	.0926	.1056	.1099	.1060	.0956	.0814	.0658	.0509	.0378	.0271
14	.0728	.0905	.1021	.1060	.1024	.0930	.0800	.0655	.0514	.0387
15	.0534	.0724	.0885	.0989	.1024	.0992	.0906	.0786	.0650	.0516
16	.0367	.0543	.0719	.0866	.0960	.0992	.0963	.0884	.0772	.0646
17	.0237	.0383	.0550	.0713	.0847	.0934	.0963	.0936	.0863	.0760
18	.0145	.0256	.0397	.0554	.0706	.0830	.0909	.0936	.0911	.0844
19	.0084	.0161	.0272	.0409	.0557	.0699	.0814	.0887	.0911	.0888

(TABLE CONTINUES ON NEXT PAGE)

TABLE 7
(Continued)

x	μ									
	11	12	13	14	15	16	17	18	19	20
20	.0046	.0097	.0177	.0286	.0418	.0559	.0692	.0798	.0866	.0888
21	.0024	.0055	.0109	.0191	.0299	.0426	.0560	.0684	.0783	.0846
22	.0012	.0030	.0065	.0121	.0204	.0310	.0433	.0560	.0676	.0769
23	.0006	.0016	.0037	.0074	.0133	.0216	.0320	.0438	.0559	.0669
24	.0003	.0008	.0020	.0043	.0083	.0144	.0226	.0328	.0442	.0557
25	.0001	.0004	.0010	.0024	.0050	.0092	.0154	.0237	.0336	.0446
26	.0000	.0002	.0005	.0013	.0029	.0057	.0101	.0164	.0246	.0343
27	.0000	.0001	.0002	.0007	.0016	.0034	.0063	.0109	.0173	.0254
28	.0000	.0000	.0001	.0003	.0009	.0019	.0038	.0070	.0117	.0181
29	.0000	.0000	.0001	.0002	.0004	.0011	.0023	.0044	.0077	.0125
30	.0000	.0000	.0000	.0001	.0002	.0006	.0013	.0026	.0049	.0083
31	.0000	.0000	.0000	.0000	.0001	.0003	.0007	.0015	.0030	.0054
32	.0000	.0000	.0000	.0000	.0001	.0001	.0004	.0009	.0018	.0034
33	.0000	.0000	.0000	.0000	.0000	.0001	.0002	.0005	.0010	.0020
34	.0000	.0000	.0000	.0000	.0000	.0000	.0001	.0002	.0006	.0012
35	.0000	.0000	.0000	.0000	.0000	.0000	.0000	.0001	.0003	.0007
36	.0000	.0000	.0000	.0000	.0000	.0000	.0000	.0001	.0002	.0004
37	.0000	.0000	.0000	.0000	.0000	.0000	.0000	.0000	.0001	.0002
38	.0000	.0000	.0000	.0000	.0000	.0000	.0000	.0000	.0000	.0001
39	.0000	.0000	.0000	.0000	.0000	.0000	.0000	.0000	.0000	.0001

TABLE 8

Random Numbers

63271	59986	71744	51102	15141	80714	58683	93108	13554	79945
88547	09896	95436	79115	08303	01041	20030	63754	08459	28364
55957	57243	83865	09911	19761	66535	40102	26646	60147	15702
46276	87453	44790	67122	45573	84358	21625	16999	13385	22782
55363	07449	34835	15290	76616	67191	12777	21861	68689	03263
69393	92785	49902	58447	42048	30378	87618	26933	40640	16281
13186	29431	88190	04588	38733	81290	89541	70290	40113	08243
17726	28652	56836	78351	47327	18518	92222	55201	27340	10493
36520	64465	05550	30157	82242	29520	69753	72602	23756	54935
81628	36100	39254	56835	37636	02421	98063	89641	64953	99337
84649	48968	75215	75498	49539	74240	03466	49292	36401	45525
63291	11618	12613	75055	43915	26488	41116	64531	56827	30825
70502	53225	03655	05915	37140	57051	48393	91322	25653	06543
06426	24771	59935	49801	11082	66762	94477	02494	88215	27191
20711	55609	29430	70165	45406	78484	31639	52009	18873	96927
41990	70538	77191	25860	55204	73417	83920	69468	74972	38712
72452	36618	76298	26678	89334	33938	95567	29380	75906	91807
37042	40318	57099	10528	09925	89773	41335	96244	29002	46453
53766	52875	15987	46962	67342	77592	57651	95508	80033	69828
90585	58955	53122	16025	84299	53310	67380	84249	25348	04332
32001	96293	37203	64516	51530	37069	40261	61374	05815	06714
62606	64324	46354	72157	67248	20135	49804	09226	64419	29457
10078	28073	85389	50324	14500	15562	64165	06125	71353	77669
91561	46145	24177	15294	10061	98124	75732	00815	83452	97355
13091	98112	53959	79607	52244	63303	10413	63839	74762	50289
73864	83014	72457	22682	03033	61714	88173	90835	00634	85169
66668	25467	48894	51043	02365	91726	09365	63167	95264	45643
84745	41042	29493	01836	09044	51926	43630	63470	76508	14194
48068	26805	94595	47907	13357	38412	33318	26098	82782	42851
54310	96175	97594	88616	42035	38093	36745	56702	40644	83514
14877	33095	10924	58013	61439	21882	42059	24177	58739	60170
78295	23179	02771	43464	59061	71411	05697	67194	30495	21157
67524	02865	39593	54278	04237	92441	26602	63835	38032	94770
58268	57219	68124	73455	83236	08710	04284	55005	84171	42596
97158	28672	50685	01181	24262	19427	52106	34308	73685	74246
04230	16831	69085	30802	65559	09205	71829	06489	85650	38707
94879	56606	30401	02602	57658	70091	54986	41394	60437	03195
71446	15232	66715	26385	91518	70566	02888	79941	39684	54315
32886	05644	79316	09819	00813	88407	17461	73925	53037	91904
62048	33711	25290	21526	02223	75947	66466	06232	10913	75336
84534	42351	21628	53669	81352	95152	08107	98814	72743	12849
84707	15885	84710	35866	06446	86311	32648	88141	73902	69981
19409	40868	64220	80861	13860	68493	52908	26374	63297	45052
57978	48015	25973	66777	45924	56144	24742	96702	88200	66162
57295	98298	11199	96510	75228	41600	47192	43267	35973	23152
94044	83785	93388	07833	38216	31413	70555	03023	54147	06647
30014	25879	71763	96679	90603	99396	74557	74224	18211	91637
07265	69563	64268	88802	72264	66540	01782	08396	19251	83613
84404	88642	30263	80310	11522	57810	27627	78376	36240	48952
21778	02085	27762	46097	43324	34354	09369	14966	10158	76089

This table is reprinted from page 44 of *A Million Random Digits with 100,000 Normal Deviates* by The Rand Corporation (New York: The Free Press, 1955). Copyright 1955 and 1983 by The Rand Corporation. Used by permission.

TABLE 9
Critical Values for
the Durbin-Watson
Test for
Autocorrelation

Entries in the table give the *critical* values for a one-tailed Durbin-Watson test for autocorrelation. For a two-tailed test, the level of significance is doubled.

Significance Points of d_L and d_U: $\alpha = .05$
Number of Independent Variables

k	1		2		3		4		5	
n	d_L	d_U	d_L	d_U	d_L	d_U	d_L	d_U	d_L	d_U
15	1.08	1.36	0.95	1.54	0.82	1.75	0.69	1.97	0.56	2.21
16	1.10	1.37	0.98	1.54	0.86	1.73	0.74	1.93	0.62	2.15
17	1.13	1.38	1.02	1.54	0.90	1.71	0.78	1.90	0.67	2.10
18	1.16	1.39	1.05	1.53	0.93	1.69	0.82	1.87	0.71	2.06
19	1.18	1.40	1.08	1.53	0.97	1.68	0.86	1.85	0.75	2.02
20	1.20	1.41	1.10	1.54	1.00	1.68	0.90	1.83	0.79	1.99
21	1.22	1.42	1.13	1.54	1.03	1.67	0.93	1.81	0.83	1.96
22	1.24	1.43	1.15	1.54	1.05	1.66	0.96	1.80	0.86	1.94
23	1.26	1.44	1.17	1.54	1.08	1.66	0.99	1.79	0.90	1.92
24	1.27	1.45	1.19	1.55	1.10	1.66	1.01	1.78	0.93	1.90
25	1.29	1.45	1.21	1.55	1.12	1.66	1.04	1.77	0.95	1.89
26	1.30	1.46	1.22	1.55	1.14	1.65	1.06	1.76	0.98	1.88
27	1.32	1.47	1.24	1.56	1.16	1.65	1.08	1.76	1.01	1.86
28	1.33	1.48	1.26	1.56	1.18	1.65	1.10	1.75	1.03	1.85
29	1.34	1.48	1.27	1.56	1.20	1.65	1.12	1.74	1.05	1.84
30	1.35	1.49	1.28	1.57	1.21	1.65	1.14	1.74	1.07	1.83
31	1.36	1.50	1.30	1.57	1.23	1.65	1.16	1.74	1.09	1.83
32	1.37	1.50	1.31	1.57	1.24	1.65	1.18	1.73	1.11	1.82
33	1.38	1.51	1.32	1.58	1.26	1.65	1.19	1.73	1.13	1.81
34	1.39	1.51	1.33	1.58	1.27	1.65	1.21	1.73	1.15	1.81
35	1.40	1.52	1.34	1.58	1.28	1.65	1.22	1.73	1.16	1.80
36	1.41	1.52	1.35	1.59	1.29	1.65	1.24	1.73	1.18	1.80
37	1.42	1.53	1.36	1.59	1.31	1.66	1.25	1.72	1.19	1.80
38	1.43	1.54	1.37	1.59	1.32	1.66	1.26	1.72	1.21	1.79
39	1.43	1.54	1.38	1.60	1.33	1.66	1.27	1.72	1.22	1.79
40	1.44	1.54	1.39	1.60	1.34	1.66	1.29	1.72	1.23	1.79
45	1.48	1.57	1.43	1.62	1.38	1.67	1.34	1.72	1.29	1.78
50	1.50	1.59	1.46	1.63	1.42	1.67	1.38	1.72	1.34	1.77
55	1.53	1.60	1.49	1.64	1.45	1.68	1.41	1.72	1.38	1.77
60	1.55	1.62	1.51	1.65	1.48	1.69	1.44	1.73	1.41	1.77
65	1.57	1.63	1.54	1.66	1.50	1.70	1.47	1.73	1.44	1.77
70	1.58	1.64	1.55	1.67	1.52	1.70	1.49	1.74	1.46	1.77
75	1.60	1.65	1.57	1.68	1.54	1.71	1.51	1.74	1.49	1.77
80	1.61	1.66	1.59	1.69	1.56	1.72	1.53	1.74	1.51	1.77
85	1.62	1.67	1.60	1.70	1.57	1.72	1.55	1.75	1.52	1.77
90	1.63	1.68	1.61	1.70	1.59	1.73	1.57	1.75	1.54	1.78
95	1.64	1.69	1.62	1.71	1.60	1.73	1.58	1.75	1.56	1.78
100	1.65	1.69	1.63	1.72	1.61	1.74	1.59	1.76	1.57	1.78

(TABLE CONTINUES)

This table comes from J. Durbin and G. S. Watson, "Testing for serial correlation in least square regression II," *Biometrika,* 38, 1951, 159–178.

TABLE 9
(Continued)

Significance Points of d_L and d_U: $\alpha.025$
Number of Independent Variables

k	1		2		3		4		5	
n	d_L	d_U	d_L	d_U	d_L	d_U	d_L	d_U	d_L	d_U
15	0.95	1.23	0.83	1.40	0.71	1.61	0.59	1.84	0.48	2.09
16	0.98	1.24	0.86	1.40	0.75	1.59	0.64	1.80	0.53	2.03
17	1.01	1.25	0.90	1.40	0.79	1.58	0.68	1.77	0.57	1.98
18	1.03	1.26	0.93	1.40	0.82	1.56	0.72	1.74	0.62	1.93
19	1.06	1.28	0.96	1.41	0.86	1.55	0.76	1.72	0.66	1.90
20	1.08	1.28	0.99	1.41	0.89	1.55	0.79	1.70	0.70	1.87
21	1.10	1.30	1.01	1.41	0.92	1.54	0.83	1.69	0.73	1.84
22	1.12	1.31	1.04	1.42	0.95	1.54	0.86	1.68	0.77	1.82
23	1.14	1.32	1.06	1.42	0.97	1.54	0.89	1.67	0.80	1.80
24	1.16	1.33	1.08	1.43	1.00	1.54	0.91	1.66	0.83	1.79
25	1.18	1.34	1.10	1.43	1.02	1.54	0.94	1.65	0.86	1.77
26	1.19	1.35	1.12	1.44	1.04	1.54	0.96	1.65	0.88	1.76
27	1.21	1.36	1.13	1.44	1.06	1.54	0.99	1.64	0.91	1.75
28	1.22	1.37	1.15	1.45	1.08	1.54	1.01	1.64	0.93	1.74
29	1.24	1.38	1.17	1.45	1.10	1.54	1.03	1.63	0.96	1.73
30	1.25	1.38	1.18	1.46	1.12	1.54	1.05	1.63	0.98	1.73
31	1.26	1.39	1.20	1.47	1.13	1.55	1.07	1.63	1.00	1.72
32	1.27	1.40	1.21	1.47	1.15	1.55	1.08	1.63	1.02	1.71
33	1.28	1.41	1.22	1.48	1.16	1.55	1.10	1.63	1.04	1.71
34	1.29	1.41	1.24	1.48	1.17	1.55	1.12	1.63	1.06	1.70
35	1.30	1.42	1.25	1.48	1.19	1.55	1.13	1.63	1.07	1.70
36	1.31	1.43	1.26	1.49	1.20	1.56	1.15	1.63	1.09	1.70
37	1.32	1.43	1.27	1.49	1.21	1.56	1.16	1.62	1.10	1.70
38	1.33	1.44	1.28	1.50	1.23	1.56	1.17	1.62	1.12	1.70
39	1.34	1.44	1.29	1.50	1.24	1.56	1.19	1.63	1.13	1.69
40	1.35	1.45	1.30	1.51	1.25	1.57	1.20	1.63	1.15	1.69
45	1.39	1.48	1.34	1.53	1.30	1.58	1.25	1.63	1.21	1.69
50	1.42	1.50	1.38	1.54	1.34	1.59	1.30	1.64	1.26	1.69
55	1.45	1.52	1.41	1.56	1.37	1.60	1.33	1.64	1.30	1.69
60	1.47	1.54	1.44	1.57	1.40	1.61	1.37	1.65	1.33	1.69
65	1.49	1.55	1.46	1.59	1.43	1.62	1.40	1.66	1.36	1.69
70	1.51	1.57	1.48	1.60	1.45	1.63	1.42	1.66	1.39	1.70
75	1.53	1.58	1.50	1.61	1.47	1.64	1.45	1.67	1.42	1.70
80	1.54	1.59	1.52	1.62	1.49	1.65	1.47	1.67	1.44	1.70
85	1.56	1.60	1.53	1.63	1.51	1.65	1.49	1.68	1.46	1.71
90	1.57	1.61	1.55	1.64	1.53	1.66	1.50	1.69	1.48	1.71
95	1.58	1.62	1.56	1.65	1.54	1.67	1.52	1.69	1.50	1.71
100	1.59	1.63	1.57	1.65	1.55	1.67	1.53	1.70	1.51	1.72

(**TABLE CONTINUES ON NEXT PAGE**)

TABLE 9
(Continued)

Significance Points of d_L and d_U: $\alpha.01$
Number of Independent Variables

k	1		2		3		4		5	
n	d_L	d_U	d_L	d_U	d_L	d_U	d_L	d_U	d_L	d_U
15	0.81	1.07	0.70	1.25	0.59	1.46	0.49	1.70	0.39	1.96
16	0.84	1.09	0.74	1.25	0.63	1.44	0.53	1.66	0.44	1.90
17	0.87	1.10	0.77	1.25	0.67	1.43	0.57	1.63	0.48	1.85
18	0.90	1.12	0.80	1.26	0.71	1.42	0.61	1.60	0.52	1.80
19	0.93	1.13	0.83	1.26	0.74	1.41	0.65	1.58	0.56	1.77
20	0.95	1.15	0.86	1.27	0.77	1.41	0.68	1.57	0.60	1.74
21	0.97	1.16	0.89	1.27	0.80	1.41	0.72	1.55	0.63	1.71
22	1.00	1.17	0.91	1.28	0.83	1.40	0.75	1.54	0.66	1.69
23	1.02	1.19	0.94	1.29	0.86	1.40	0.77	1.53	0.70	1.67
24	1.04	1.20	0.96	1.30	0.88	1.41	0.80	1.53	0.72	1.66
25	1.05	1.21	0.98	1.30	0.90	1.41	0.83	1.52	0.75	1.65
26	1.07	1.22	1.00	1.31	0.93	1.41	0.85	1.52	0.78	1.64
27	1.09	1.23	1.02	1.32	0.95	1.41	0.88	1.51	0.81	1.63
28	1.10	1.24	1.04	1.32	0.97	1.41	0.90	1.51	0.83	1.62
29	1.12	1.25	1.05	1.33	0.99	1.42	0.92	1.51	0.85	1.61
30	1.13	1.26	1.07	1.34	1.01	1.42	0.94	1.51	0.88	1.61
31	1.15	1.27	1.08	1.34	1.02	1.42	0.96	1.51	0.90	1.60
32	1.16	1.28	1.10	1.35	1.04	1.43	0.98	1.51	0.92	1.60
33	1.17	1.29	1.11	1.36	1.05	1.43	1.00	1.51	0.94	1.59
34	1.18	1.30	1.13	1.36	1.07	1.43	1.01	1.51	0.95	1.59
35	1.19	1.31	1.14	1.37	1.08	1.44	1.03	1.51	0.97	1.59
36	1.21	1.32	1.15	1.38	1.10	1.44	1.04	1.51	0.99	1.59
37	1.22	1.32	1.16	1.38	1.11	1.45	1.06	1.51	1.00	1.59
38	1.23	1.33	1.18	1.39	1.12	1.45	1.07	1.52	1.02	1.58
39	1.24	1.34	1.19	1.39	1.14	1.45	1.09	1.52	1.03	1.58
40	1.25	1.34	1.20	1.40	1.15	1.46	1.10	1.52	1.05	1.58
45	1.29	1.38	1.24	1.42	1.20	1.48	1.16	1.53	1.11	1.58
50	1.32	1.40	1.28	1.45	1.24	1.49	1.20	1.54	1.16	1.59
55	1.36	1.43	1.32	1.47	1.28	1.51	1.25	1.55	1.21	1.59
60	1.38	1.45	1.35	1.48	1.32	1.52	1.28	1.56	1.25	1.60
65	1.41	1.47	1.38	1.50	1.35	1.53	1.31	1.57	1.28	1.61
70	1.43	1.49	1.40	1.52	1.37	1.55	1.34	1.58	1.31	1.61
75	1.45	1.50	1.42	1.53	1.39	1.56	1.37	1.59	1.34	1.62
80	1.47	1.52	1.44	1.54	1.42	1.57	1.39	1.60	1.36	1.62
85	1.48	1.53	1.46	1.55	1.43	1.58	1.41	1.60	1.39	1.63
90	1.50	1.54	1.47	1.56	1.45	1.59	1.43	1.61	1.41	1.64
95	1.51	1.55	1.49	1.57	1.47	1.60	1.45	1.62	1.42	1.64
100	1.52	1.56	1.50	1.58	1.48	1.60	1.46	1.63	1.44	1.65

TABLE 10

T_L Values for the Mann-Whitney-Wilcoxon Test

Reject the hypothesis of identical populations if the sum of the ranks for the n_1 items is *less than* the value T_L shown in the following table or if the sum of the ranks for the n_1 items is *greater than* the value T_U where

$$T_U = n_1(n_1 + n_2 + 1) - T_L$$

						n_2				
$\alpha = .05$		2	3	4	5	6	7	8	9	10
	2	3	3	3	3	3	3	4	4	4
	3	6	6	6	7	8	8	9	9	10
	4	10	10	11	12	13	14	15	15	16
	5	15	16	17	18	19	21	22	23	24
n_1	6	21	23	24	25	27	28	30	32	33
	7	28	30	32	34	35	37	39	41	43
	8	37	39	41	43	45	47	50	52	54
	9	46	48	50	53	56	58	61	63	66
	10	56	59	61	64	67	70	73	76	79

						n_2				
$\alpha = .10$		2	3	4	5	6	7	8	9	10
	2	3	3	3	4	4	4	5	5	5
	3	6	7	7	8	9	9	10	11	11
	4	10	11	12	13	14	15	16	17	18
	5	16	17	18	20	21	22	24	25	27
n_1	6	22	24	25	27	29	30	32	34	36
	7	29	31	33	35	37	40	42	44	46
	8	38	40	42	45	47	50	52	55	57
	9	47	50	52	55	58	61	64	67	70
	10	57	60	63	67	70	73	76	80	83

TABLE 11 Critical Values of the Studentized Range Distribution

α.05

Number of Populations

Degrees of Freedom	2	3	4	5	6	7	8	9	10	11	12	13	14	15	16	17	18	19	20
1	18.0	27.0	32.8	37.1	40.4	43.1	45.4	47.4	49.1	50.6	52.0	53.2	54.3	55.4	56.3	57.2	58.0	58.8	59.6
2	6.08	8.33	9.80	10.9	11.7	12.4	13.0	13.5	14.0	14.4	14.7	15.1	15.4	15.7	15.9	16.1	16.4	16.6	16.8
3	4.50	5.91	6.82	7.50	8.04	8.48	8.85	9.18	9.46	9.72	9.95	10.2	10.3	10.5	10.7	10.8	11.0	11.1	11.2
4	3.93	5.04	5.76	6.29	6.71	7.05	7.35	7.60	7.83	8.03	8.21	8.37	8.52	8.66	8.79	8.91	9.03	9.13	9.23
5	3.64	4.60	5.22	5.67	6.03	6.33	6.58	6.80	6.99	7.17	7.32	7.47	7.60	7.72	7.83	7.93	8.03	8.12	8.21
6	3.46	4.34	4.90	5.30	5.63	5.90	6.12	6.32	6.49	6.65	6.79	6.92	7.03	7.14	7.24	7.34	7.43	7.51	7.59
7	3.34	4.16	4.68	5.06	5.36	5.61	5.82	6.00	6.16	6.30	6.43	6.55	6.66	6.76	6.85	6.94	7.02	7.10	7.17
8	3.26	4.04	4.53	4.89	5.17	5.40	5.60	5.77	5.92	6.05	6.18	6.29	6.39	6.48	6.57	6.65	6.73	6.80	6.87
9	3.20	3.95	4.41	4.76	5.02	5.24	5.43	5.59	5.74	5.87	5.98	6.09	6.19	6.28	6.36	6.44	6.51	6.58	6.64
10	3.15	3.88	4.33	4.65	4.91	5.12	5.30	5.46	5.60	5.72	5.83	5.93	6.03	6.11	6.19	6.27	6.34	6.40	6.47
11	3.11	3.82	4.26	4.57	4.82	5.03	5.20	5.35	5.49	5.61	5.71	5.81	5.90	5.98	6.06	6.13	6.20	6.27	6.33
12	3.08	3.77	4.20	4.51	4.75	4.95	5.12	5.27	5.39	5.51	5.61	5.71	5.80	5.88	5.95	6.02	6.09	6.15	6.21
13	3.06	3.73	4.15	4.45	4.69	4.88	5.05	5.19	5.32	5.43	5.53	5.63	5.71	5.79	5.86	5.93	5.99	6.05	6.11
14	3.03	3.70	4.11	4.41	4.64	4.83	4.99	5.13	5.25	5.36	5.46	5.55	5.64	5.71	5.79	5.85	5.91	5.97	6.03
15	3.01	3.67	4.08	4.37	4.59	4.78	4.94	5.08	5.20	5.31	5.40	5.49	5.57	5.65	5.72	5.78	5.85	5.90	5.96
16	3.00	3.65	4.05	4.33	4.56	4.74	4.90	5.03	5.15	5.26	5.35	5.44	5.52	5.59	5.66	5.73	5.79	5.84	5.90
17	2.98	3.63	4.02	4.30	4.52	4.70	4.86	4.99	5.11	5.21	5.31	5.39	5.47	5.54	5.61	5.67	5.73	5.79	5.84
18	2.97	3.61	4.00	4.28	4.49	4.67	4.82	4.96	5.07	5.17	5.27	5.35	5.43	5.50	5.57	5.63	5.69	5.74	5.79
19	2.96	3.59	3.98	4.25	4.47	4.65	4.79	4.92	5.04	5.14	5.23	5.31	5.39	5.46	5.53	5.59	5.65	5.70	5.75
20	2.95	3.58	3.96	4.23	4.45	4.62	4.77	4.90	5.01	5.11	5.20	5.28	5.36	5.43	5.49	5.55	5.61	5.66	5.71
24	2.92	3.53	3.90	4.17	4.37	4.54	4.68	4.81	4.92	5.01	5.10	5.18	5.25	5.32	5.38	5.44	5.49	5.55	5.59
30	2.89	3.49	3.85	4.10	4.30	4.46	4.60	4.72	4.82	4.92	5.00	5.08	5.15	5.21	5.27	5.33	5.38	5.43	5.47
40	2.86	3.44	3.79	4.04	4.23	4.39	4.52	4.63	4.73	4.82	4.90	4.98	5.04	5.11	5.16	5.22	5.27	5.31	5.36
60	2.83	3.40	3.74	3.98	4.16	4.31	4.44	4.55	4.65	4.73	4.81	4.88	4.94	5.00	5.06	5.11	5.15	5.20	5.24
120	2.80	3.36	3.68	3.92	4.10	4.24	4.36	4.47	4.56	4.64	4.71	4.78	4.84	4.90	4.95	5.00	5.04	5.09	5.13
∞	2.77	3.31	3.63	3.86	4.03	4.17	4.29	4.39	4.47	4.55	4.62	4.68	4.74	4.80	4.85	4.89	4.93	4.97	5.01

(TABLE CONTINUES)

Reprinted by permission of the Biometrika Trustees from *Biometrika Tables for Statisticians*, E. S. Pearson and H. O. Hartley, Vol. 1, 176–77.

TABLE 11 (Continued)

$\alpha = .01$

Degrees of Freedom	Number of Populations																		
	2	3	4	5	6	7	8	9	10	11	12	13	14	15	16	17	18	19	20
1	90.0	135.	164.	186.	202.	216.	227.	237.	246.	253.	260.	266.	272.	277.	282.	286.	290.	294.	298.
2	14.0	19.0	22.3	24.7	26.6	28.2	29.5	30.7	31.7	32.6	33.4	34.1	34.8	35.4	36.0	36.5	37.0	37.5	37.9
3	8.26	10.6	12.2	13.3	14.2	15.0	15.6	16.2	16.7	17.1	17.5	17.9	18.2	18.5	18.8	19.1	19.3	19.5	19.8
4	6.51	8.12	9.17	9.96	10.6	11.1	11.5	11.9	12.3	12.6	12.8	13.1	13.3	13.5	13.7	13.9	14.1	14.2	14.4
5	5.70	6.97	7.80	8.42	8.91	9.32	9.67	9.97	10.2	10.5	10.7	10.9	11.1	11.2	11.4	11.6	11.7	11.8	11.9
6	5.24	6.33	7.03	7.56	7.97	8.32	8.61	8.87	9.10	9.30	9.49	9.65	9.81	9.95	10.1	10.2	10.3	10.4	10.5
7	4.95	5.92	6.54	7.01	7.37	7.68	7.94	8.17	8.37	8.55	8.71	8.86	9.00	9.12	9.24	9.35	9.46	9.55	9.65
8	4.74	5.63	6.20	6.63	6.96	7.24	7.47	7.68	7.87	8.03	8.18	8.31	8.44	8.55	8.66	8.76	8.85	8.94	9.03
9	4.60	5.43	5.96	6.35	6.66	6.91	7.13	7.32	7.49	7.65	7.78	7.91	8.03	8.13	8.23	8.32	8.41	8.49	8.57
10	4.48	5.27	5.77	6.14	6.43	6.67	6.87	7.05	7.21	7.36	7.48	7.60	7.71	7.81	7.91	7.99	8.07	8.15	8.22
11	4.39	5.14	5.62	5.97	6.25	6.48	6.67	6.84	6.99	7.13	7.25	7.36	7.46	7.56	7.65	7.73	7.81	7.88	7.95
12	4.32	5.04	5.50	5.84	6.10	6.32	6.51	6.67	6.81	6.94	7.06	7.17	7.26	7.36	7.44	7.52	7.59	7.66	7.73
13	4.26	4.96	5.40	5.73	5.98	6.19	6.37	6.53	6.67	6.79	6.90	7.01	7.10	7.19	7.27	7.34	7.42	7.48	7.55
14	4.21	4.89	5.32	5.63	5.88	6.08	6.26	6.41	6.54	6.66	6.77	6.87	6.96	7.05	7.12	7.20	7.27	7.33	7.39
15	4.17	4.83	5.25	5.56	5.80	5.99	6.16	6.31	6.44	6.55	6.66	6.76	6.84	6.93	7.00	7.07	7.14	7.20	7.26
16	4.13	4.78	5.19	5.49	5.72	5.92	6.08	6.22	6.35	6.46	6.56	6.66	6.74	6.82	6.90	6.97	7.03	7.09	7.15
17	4.10	4.74	5.14	5.43	5.66	5.85	6.01	6.15	6.27	6.38	6.48	6.57	6.66	6.73	6.80	6.87	6.94	7.00	7.05
18	4.07	4.70	5.09	5.38	5.60	5.79	5.94	6.08	6.20	6.31	6.41	6.50	6.58	6.65	6.72	6.79	6.85	6.91	6.96
19	4.05	4.67	5.05	5.33	5.55	5.73	5.89	6.02	6.14	6.25	6.34	6.43	6.51	6.58	6.65	6.72	6.78	6.84	6.89
20	4.02	4.64	5.02	5.29	5.51	5.69	5.84	5.97	6.09	6.19	6.29	6.37	6.45	6.52	6.59	6.65	6.71	6.76	6.82
24	3.96	4.54	4.91	5.17	5.37	5.54	5.69	5.81	5.92	6.02	6.11	6.19	6.26	6.33	6.39	6.45	6.51	6.56	6.61
30	3.89	4.45	4.80	5.05	5.24	5.40	5.54	5.65	5.76	5.85	5.93	6.01	6.08	6.14	6.20	6.26	6.31	6.36	6.41
40	3.82	4.37	4.70	4.93	5.11	5.27	5.39	5.50	5.60	5.69	5.77	5.84	5.90	5.96	6.02	6.07	6.12	6.17	6.21
60	3.76	4.28	4.60	4.82	4.99	5.13	5.25	5.36	5.45	5.53	5.60	5.67	5.73	5.79	5.84	5.89	5.93	5.98	6.02
120	3.70	4.20	4.50	4.71	4.87	5.01	5.12	5.21	5.30	5.38	5.44	5.51	5.56	5.61	5.66	5.71	5.75	5.79	5.83
∞	3.64	4.12	4.40	4.60	4.76	4.88	4.99	5.08	5.16	5.23	5.29	5.35	5.40	5.45	5.49	5.54	5.57	5.61	5.65

APPENDIX C

SUMMATION NOTATION

SUMMATIONS

Definition

$$\sum_{i=1}^{n} x_i = x_1 + x_2 + \cdots + x_n \tag{C.1}$$

Example for $x_1 = 5, x_2 = 8, x_3 = 14$:

$$\sum_{i=1}^{3} x_i = x_1 + x_2 + x_3$$
$$= 5 + 8 + 14$$
$$= 27$$

Result 1

For a constant c:

$$\sum_{i=1}^{n} c = (c + c + \ldots + c) = nc \tag{C.2}$$

n times

Example for $c = 5, n = 10$:

$$\sum_{i=1}^{10} 5 = 10(5) = 50$$

Example for $c = \bar{x}$:

$$\sum_{i=1}^{n} \bar{x} = n\bar{x}$$

Result 2

$$\sum_{i=1}^{n} cx_i = cx_1 + cx_2 + \cdots + cx_n$$

$$= c(x_1 + x_2 + \cdots + x_n) = c \sum_{i=1}^{n} x_i \tag{C.3}$$

Example for $x_1 = 5, x_2 = 8, x_3 = 14, c = 2$:

$$\sum_{i=1}^{3} 2x_i = 2 \sum_{i=1}^{3} x_i = 2(27) = 54$$

Result 3

$$\sum_{i=1}^{n} (ax_i + by_i) = a \sum_{i=1}^{n} x_i + b \sum_{i=1}^{n} y_i \qquad \textbf{(C.4)}$$

Example for $x_1 = 5, x_2 = 8, x_3 = 14, a = 2, y_1 = 7, y_2 = 3, y_3 = 8, b = 4$:

$$\sum_{i=1}^{3} (2x_i + 4y_i) = 2 \sum_{i=1}^{3} x_i + 4 \sum_{i=1}^{3} y_i$$
$$= 2(27) + 4(18)$$
$$= 54 + 72$$
$$= 126$$

DOUBLE SUMMATIONS

Consider the following data involving the variable x_{ij}, where i is the subscript denoting the row position and j is the subscript denoting the column position:

		Column		
		1	2	3
Row	1	$x_{11} = 10$	$x_{12} = 8$	$x_{13} = 6$
	2	$x_{21} = 7$	$x_{22} = 4$	$x_{23} = 12$

Definition

$$\sum_{i=1}^{n} \sum_{j=1}^{m} x_{ij} = (x_{11} + x_{12} + \cdots + x_{1m}) + (x_{21} + x_{22} + \cdots + x_{2m})$$
$$+ (x_{31} + x_{32} + \cdots + x_{3m}) + \cdots + (x_{n1} + x_{n2} + \cdots + x_{nm}) \qquad \textbf{(C.5)}$$

Example:

$$\sum_{i=1}^{2} \sum_{j=1}^{3} x_{ij} = x_{11} + x_{12} + x_{13} + x_{21} + x_{22} + x_{23}$$
$$= 10 + 8 + 6 + 7 + 4 + 12$$
$$= 47$$

Definition

$$\sum_{i=1}^{n} x_{ij} = x_{1j} + x_{2j} + \cdots + x_{nj} \qquad \textbf{(C.6)}$$

Example:

$$\sum_{i=1}^{2} x_{i2} = x_{12} + x_{22}$$
$$= 8 + 4$$
$$= 12$$

SHORTHAND NOTATION

Sometimes when a summation is for all values of the subscript, we use the following shorthand notations:

$$\sum_{i=1}^{n} x_i = \Sigma x_i \tag{C.7}$$

$$\sum_{i=1}^{n} \sum_{j=1}^{m} x_{ij} = \Sigma\Sigma x_{ij} \tag{C.8}$$

$$\sum_{i=1}^{n} x_{ij} = \sum_{i} x_{ij} \tag{C.9}$$

THE DATA DISK AND ITS USE

The Data Disk contains most of the larger data sets presented in the text. A list of the data set filenames is shown below. To retrieve a data set using Minitab or The Data Analyst software package, you only need to know the filename of the data set. For example, the data set corresponding to Exercise 16 in Chapter 2 is named AUTOCOST; thus, if you are using Minitab, you would access these data by entering the Minitab command Retrieve 'AUTOCOST'. If you are using The Data Analyst, a list of each data set filename is displayed whenever you select Choice 2, Retrieve a Previously Saved Data Set, from the Data Set Selection Menu; in this case, you simply select the filename corresponding to the data set that you want to retrieve. Note that each data set on The Data Disk is also identified in the text with a logo that appears in the margin.

Chapter 2
AUTOCOST	Exercise 16
EXSALARY	Exercise 18
MPGDATA	Exercise 20
COMPUTER	Exercise 21
JOBSAT	Exercise 27
LAWAGES	Exercise 36
COMSTOCK	Exercise 38
GRADEAVE	Exercise 39
ROOMCOST	Exercise 40
STATES	Exercise 42
CITIES	Exercise 44
CONSOLID	Computer Case

Chapter 3
STARTSAL	Exercise 5
LAWAGES	Exercise 26
GROWTH	Exercise 44
INJURY	Exercise 47
WORLD	Exercise 48
PRICES	Exercise 56
MORTGAGE	Exercise 59
EXAM	Exercise 65
DUKE	Exercise 70
CONSOLID	Computer Case 1
HEALTH-1	Computer Case 2
HEALTH-2	Computer Case 2

Chapter 8
AUTO	Computer Case
QUALITY	Computer Case

Chapter 10
GOLF	Computer Case

Chapter 11
TRAINING	Computer Case

Chapter 12
INFO	Exercise 8
MACHINES	Exercise 10
JUDGMENT	Exercise 29
PAINT	Exercise 30
SERVICE	Exercise 54
ASSEMBLY	Exercise 55
DOWJONES	Exercise 58
BROWSE	Exercise 61
MEDICAL1	Computer Case
MEDICAL2	Computer Case

Chapter 13
PRESCRIP	Exercise 45
HOME1	Exercise 46
HOME2	Exercise 55

TEMPSC	Exercise 56			
HIGHLOW	Exercise 65		**FOOTBALL**	Exercise 52
SAFETY	Computer Case		**CONSUMER**	Computer Case

Chapter 14

BUTLER	Table 14.2		**Chapter 15**	
MOVIE	Exercise 6		**REYNOLDS**	Table 15.1
MOWER	Exercise 7		**TYLER**	Table 15.2
PHARMACY	Exercise 8		**MPG**	Table 15.4
SCHOOLS1	Exercise 9		**POLAROID**	Exercise 8
HOUSING	Exercise 10		**HOUSING**	Exercise 9
FORBES1	Exercise 23		**FOOTBALL**	Exercise 12
REPAIR	Exercise 33		**STROKE**	Exercise 14
FORBES2	Exercise 35		**CRAVENS**	Table 15.5
STROKE	Exercise 36		**SCHOOLS2**	Exercise 15
MOVIE	Exercise 39		**FORBES2**	Exercise 19
JOBS	Exercise 50		**AUDIT**	Exercise 31
			JOBS	Exercise 32
			LAYOFFS	Computer Case

NOTE REGARDING THE FILES ON THE DATA DISK

If you look at the directory of the files on The Data Disk you will find that there are two files corresponding to each data set. Files that can be accessed by Minitab have the filename extension MTW (for example, AUTOCOST.MTW), and files that can be accessed by The Data Analyst have the filename extension ASW (e.g., AUTOCOST.ASW). You *do not* have to use these filename extensions when using Minitab or The Data Analyst.

APPENDIX E

ANSWERS TO REVIEW QUIZZES

CHAPTER 1

1. F	5. T	9. F	13. b	17. b
2. T	6. F	10. T	14. b	18. b
3. F	7. T	11. b	15. a	
4. T	8. F	12. b	16. b	

CHAPTER 2

1. F	4. T	6. T	8. d	10. c
2. F	5. T	7. c	9. a	11. a
3. F				

CHAPTER 3

1. F	5. F	8. F	11. b	14. d
2. F	6. F	9. T	12. b	15. a
3. T	7. F	10. T	13. a	16. b
4. T				

CHAPTER 4

1. T	5. F	9. T	13. a	17. d
2. T	6. T	10. F	14. c	18. a
3. F	7. T	11. b	15. b	19. d
4. F	8. F	12. a	16. d	

CHAPTER 5

1. T	5. F	9. T	13. c	16. c
2. F	6. T	10. F	14. a	17. d
3. T	7. T	11. F	15. b	18. c
4. F	8. T	12. F		

CHAPTER 6

1. F	5. F	9. T	12. a	15. b
2. F	6. T	10. T	13. a	16. d
3. T	7. T	11. d	14. c	17. c
4. F	8. F			

CHAPTER 7

1. T	5. T	8. F	11. a	14. b
2. T	6. F	9. c	12. b	15. c
3. F	7. F	10. c	13. b	16. d
4. T				

CHAPTER 8

1. F	6. F	11. T	15. b	19. a
2. T	7. T	12. F	16. b	20. b
3. T	8. F	13. F	17. a	21. b
4. T	9. F	14. c	18. c	22. c
5. T	10. T			

CHAPTER 9

1. F	3. T	5. F	7. d	9. d
2. F	4. T	6. T	8. c	10. b

CHAPTER 10

1. T	5. F	8. T	11. d	14. b
2. F	6. F	9. F	12. d	15. a
3. T	7. T	10. T	13. d	16. a
4. F				

CHAPTER 11

1. F	4. F	7. F	10. c	13. c
2. F	5. T	8. T	11. b	14. d
3. T	6. T	9. a	12. b	15. b

CHAPTER 12

1. T	6. T	10. F	14. c	18. d
2. T	7. F	11. F	15. d	19. b
3. F	8. T	12. F	16. a	20. c
4. F	9. F	13. b	17. b	21. d
5. F				

CHAPTER 13

1. T	6. T	11. T	15. c	19. d
2. F	7. T	12. T	16. a	20. b
3. F	8. T	13. a	17. a	21. c
4. T	9. F	14. b	18. a	22. a
5. F	10. F			

CHAPTER 14

1. F	4. T	7. F	10. d	13. c
2. F	5. T	8. T	11. a	14. b
3. F	6. T	9. b	12. b	15. b

CHAPTER 15

1. F	5. F	9. F	13. d
2. T	6. T	10. T	14. d
3. F	7. F	11. T	15. d
4. F	8. F	12. c	16. b

CHAPTER 16

1. T	5. T	9. T	12. c	15. b
2. F	6. T	10. F	13. c	16. b
3. F	7. F	11. a	14. a	17. d
4. F	8. T			

CHAPTER 17

1. T	5. F	8. T	11. F	14. a
2. T	6. T	9. F	12. c	15. b
3. F	7. T	10. F	13. c	16. a
4. F				

CHAPTER 18

1. F	5. T	9. T	13. c	17. b
2. T	6. T	10. F	14. d	18. a
3. F	7. T	11. b	15. c	
4. F	8. T	12. a	16. c	

APPENDIX F

ANSWERS TO EVEN-NUMBERED EXERCISES

CHAPTER 1

2. a. 7
   b. 4
   c. 7
   d. Pay versus corporation profit rating is qualitative; others quantitative
   e. Pay versus corporation profit rating is ordinal; others are ratio
4. **a**, **c**, and **d** are quantitative; **b** and **e** are qualitative
6. a. 1000
   b. Nominal
   c. Percentages
   d. 17%
8. a. Quantitative, ratio
   b. Qualitative, nominal
   c. Qualitative, ordinal
   d. Qualitative, nominal
   e. Quantitative, ratio
10. a. Visitors to Hawaii
   b. Yes; vast majority travel to Hawaii by air
   c. 1 and 4 are quantitative; 2 and 3 are qualitative
12. a. Product taste and preference data
   b. Actual test data from young adults
14. a. 40% in the sample died of heart disease
   b. Qualitative; nominal scale
18. a. All adult viewers reached by the television station
   b. The viewers contacted in the telephone survey
   c. A sample requires less time and a lower cost compared to contacting all viewers in the population
20. a. Correct
   b. Challenge
   c. Correct
   d. Challenge
   e. Challenge

CHAPTER 2

2. a. .20
   b. 40
   c.

Class	Frequency
A	44
B	36
C	80
D	40

4. a. Scale is nominal
   b.

Beverage	Frequency	Relative Frequency
Milk	3	.10
Fruit juice	4	.13
Soft drink	13	.43
Beer	8	.27
Bottled water	2	.07
Totals	30	1.00

   d. Soft drink
6. a.

Video	Frequency	Relative Frequency
B	9	.150
E	13	.217
F	14	.233
H	8	.133
L	6	.100
T	10	.167
Totals	60	1.000

   b. Fantasia

8. a.

Position	Frequency	Relative Frequency
P	17	.309
H	4	.073
1	5	.091
2	4	.073
3	2	.036
S	5	.091
L	6	.109
C	5	.091
R	7	.127
Totals	55	1.000

b. Pitcher
c. 3rd base
d. Rightfield
e. Infielders 16 to outfielders 18

10. a. Ordinal
b.

Rating	Frequency	Relative Frequency
Poor	2	.03
Fair	4	.07
Good	12	.20
Very good	24	.40
Excellent	18	.30
Totals	60	1.00

12.

Class	Cumulative Frequency	Cumulative Relative Frequency
≤19	8	.200
≤29	20	.500
≤39	35	.875
≤49	40	1.000

14. b, c.

Class	Frequency	Relative Frequency
6.0–7.9	4	.20
8.0–9.9	2	.10
10.0–11.9	8	.40
12.0–13.9	3	.15
14.0–15.9	3	.15
Totals	20	1.00

16.

Prices	New Freq.	New Rel. Freq.	Used Freq.	Used Rel. Freq.
0.0–4.9	0	.000	3	.25
5.0–9.9	1	.071	6	.50
10.0–14.9	5	.357	3	.25
15.0–19.9	5	.357	0	.00
20.0–24.9	3	.214	0	.00
Totals	14		12	

18.

Salary ($1000s)	Freq.	Rel. Freq.	Cum. Freq.
0–499	1	.04	1
500–999	9	.36	10
1000–1499	11	.44	21
1500–1999	2	.08	23
2000–2499	2	.08	25
Totals	25	1.00	

20. a, b.

Miles per Gallon	Frequency	Relative Frequency
24.0–25.9	2	.07
26.0–27.9	5	.17
28.0–29.9	10	.33
30.0–31.9	9	.30
32.0–33.9	3	.10
34.0–35.9	1	.03
Totals	30	1.00

d. Civic

22.

5	7 8	
6	4 5 8	
7	0 2 2 5 5 6 8	
8	0 2 3 5	

24.

11	6
12	0 2
13	0 6 7
14	2 2 7
15	5
16	0 2 8
17	0 2 3

26.

```
-1 | 3  7
 0 | 1  8
 1 | 3  4  5  6
 2 | 0  1  4  5  5  6
 3 | 0  3  3
 4 | 5  9
 . | .
 . | .
 . | .
12 | 0
```

28. a.

```
1 | 5  6  8  9
2 | 1  2
3 | 4
4 | 1  6  8
5 |
6 | 5  6  8
7 | 1
8 | 9  9
9 | 2  8
```

30. Cell frequencies

A	B	C	Total
25	30	5	60
5	20	15	40

A appears best.

32. a, b

Industry	Frequency	Relative Frequency
Beverage	2	.10
Chemicals	3	.15
Electronics	6	.30
Food	7	.35
Aerospace	2	.10
Totals	20	1.00

34. a.

Party Affiliation	Frequency	Relative Frequency
Democrat	17	.425
Republican	17	.425
Independent	6	.150
Totals	40	1.000

36.

Hourly Wage	Freq.	Rel. Freq.	Cum. Freq.	Cum. Rel. Freq.
4.00–5.99	1	.04	1	.04
6.00–7.99	3	.12	4	.16
8.00–9.99	8	.32	12	.48
10.00–11.99	6	.24	18	.72
12.00–13.99	5	.20	23	.92
14.00–15.99	2	.08	25	1.00
Totals	25	1.00		

38.

Closing Price	Freq.	Rel. Freq.	Cum. Freq.	Cum. Rel. Freq.
0–9⅞	9	.225	9	.225
10–19⅞	10	.250	19	.475
20–29⅞	5	.125	24	.600
30–39⅞	11	.275	35	.875
40–49⅞	2	.050	37	.925
50–59⅞	2	.050	39	.975
60–69⅞	0	.000	39	.975
70–79⅞	1	.025	40	1.000
Totals	40	1.000		

40. a, b.

Cost	Frequency	Relative Frequency
2000–2499	2	.04
2500–2999	7	.14
3000–3499	16	.32
3500–3999	11	.22
4000–4499	5	.10
4500–4999	4	.08
5000–5499	2	.04
5500–5999	3	.06
Totals	50	1.00

42.

Income	Frequency	Relative Frequency
12,000–13,999	1	.020
14,000–15,999	10	.196
16,000–17,999	17	.333
18,000–19,999	11	.216
20,000–21,999	6	.118
22,000–23,999	3	.059
24,000–25,999	3	.059
Totals	51	1.000

44. a.

```
4 | 4
4 | 7
5 | 0  0  0  2  2
5 | 5  6  6  8
6 | 1  1  2  3
6 | 6  7  8  9
7 | 1  3
7 | 9
8 | 0  0
```

b.

```
2 | 5  8  8  9  9  9
3 | 2
3 | 5  5  5  5  5  7  9
4 | 0  1  1
4 | 9
5 | 0  0  3
5 | 6  7  8
```

d. 7

e.

Temperature	Frequency	
	High Temp.	Low Temp.
20–29	0	6
30–39	0	8
40–49	2	4
50–59	9	6
60–69	8	0
70–79	3	0
80–89	2	0

CHAPTER 3

2. 16, 16.5

4. 2 from each end

6. a. 171.25, 175, 145

   b. 145, 202.5

8. a. 38.75, 29

   b. 38.61, 38.38

   c. 38.5

d. 29.5, 47.5

e. 31

10. a. 178

   b. 178

   c. Do not report a mode

   d. 184

12. a. 48.33, 49; do not report a mode

   b. 45, 55

   c. 45, 55

14.   *City*: mean = 15.58, median = 15.9, mode = 15.3

   Country: mean = 18.92, median = 18.7, mode = 18.6 and 19.4

16. a. 38, 39.5, 42

   b. 27, 43.5

   c. $Q_1 = 28$, $Q_3 = 42$

     L. Hinge = 28, U. Hinge = 42

18. 16, 4

20. a. *New:* \$16.05, \$16.40

     Used: \$7.53, \$7.35

   b. *New:* 14.9, 4.0, 4.21

     Used: 7.5, 5.15, 2.80

   c. New cars are \$8000–9000 more expensive

22. a. Range = 32, IQR = 10

   b. 92.75, 9.63

24. *Dawson:* range = 2, s = .67

     Clark: range = 8, s = 2.58

26. a. 10.40

   b. 10.05

   c. 9.50

   d. 3.60

   e. 6.75

   f. 2.60

28. *Quarter milers:* s = .056, Coef. of Var. = 5.8

     Milers: s = .130, Coef. of Var. = 2.9

30. .20, 1.50, 0, −.50, −2.20

32. a. 95%

   b. Almost all

   c. 68%

34. a. $\bar{x} = 77.5$, s = 9.86

   b. z = 3.30, an outlier

   c. 16%, 2.5%

36. a. At least 75%

   b. At least 89%

   c. 95%; almost all

38. a. 75%, 84%, 89%

   b. 95%; almost all

40. 15, 22.5, 26, 29, 34

42. 5, 8, 10, 15, 18

44. a. 1.9, 10.0, 14.5, 20.4, 55.9

   b. *Inner fences:* −5.6, 36.0

     Outer fences: −21.2, 51.6

c. 46.1 and 49.9 are mild outliers;
   55.9 is an extreme outlier

46. a. 79.31, 78.5
   b. 76.5, 80.5
   c. 72, 76.5, 78.5, 80.5, 90
   d. Camcorders have higher variation
   e. Yes; Hitachi and Mitsubishi are mild outliers

48. a. 9.32, 9.8
   b. 6.5, 14.6
   c. *Inner fences:* −5.65, 26.75
   Outer fences: −17.8, 38.9
   Six mild outliers plus
   Hong Kong is an extreme outlier

50. 25, 5

52. 74.02, 158.78, 12.6

54. 10.74, 25.63, 5.06

56. a. 3.19, 3.30, 3.70
   b. 2.90, 3.50
   c. 2.80, .60
   d. .30, .54
   e. *Inner fences:* 2.00, 4.40
   Two mild outliers

58. a. $\bar{x} = 1028.18$, median = 1000, no mode
   b. Range = 510, IQR = 220
   c. $s^2 = 24,256.36$, $s = 155.74$
   d. No outliers

60. a. 12.33
   b. 12
   c. 18
   d. 10
   e. 17
   f. 30.75
   g. 5.55

62. a. 68%
   b. 95%
   c. $z = 4.56$, yes

64. a. Public: 32; auto: 32
   b. Public: 4.64; auto: 1.83
   c. Auto has less variability

66. a. 400, 624, 836, 999, 1278
   c. *Inner fences:* 61.5, 1561.5
   No outliers

68. 51.50, 227.37, 15.08

70. a. *Duke:* 87.97, 88.50
   Opponent: 72.64, 71.50
   b. *Duke:* 50, 21.5
   Opponent: 52, 15.5

CHAPTER 4

2. a. 1/4
   b. 3/4

4. 24

6. 6

8. a. 36
   c. 6

10. a. 9

12. a. 9
   c. 5
   d. 1

14. 230, 300

16. 1106/2038, 826/2038, 106/2038

18. a. 6
   b. Relative frequency
   c. .12, .24, .30, .20, .10, .04

20. No; $P(0) = $ 1/4, $P(1) = $ 1/2, $P(2) = $ 1/4

22. a. 1/4
   b. 1/2
   c. 3/4

24. a. 36
   c. 1/6
   d. 5/18
   e. No; $P(\text{Odd}) = $ 1/2, $P(\text{Even}) = $ 1/2
   f. Classical

26. a. Yes
   b. .58
   c. .12

28. a. .84
   b. .20
   c. .24

30. a. .54
   b. .34

32. a. .75
   b. .80
   c. .55

34. a. No
   b. .73
   c. .82
   d. .12
   e. .68

36. .26

38. .50

40. a. .33
   b. .28
   c. .21

42. a. A = first car starts, B = second car starts
   $P(A) = .80$, $P(B) = .40$,
   $P(A \cap B) = .30$
   b. .90
   c. .10

44. a. No
   b. .54
   c. No

46. a. .75
   b. .667

c. .50

d. No

e. No

48. a.

	Single	Married	Total
Under 30	.55	.10	.65
30 or over	.20	.15	.35
Total	.75	.25	1.00

b. Higher probability of under 30

c. Higher probability of single

d. .55

e. .8462

f. No

50. a.

		Own U.S. Car		
		Yes	*No*	**Total**
Own	*Yes*	.15	.05	.20
Foreign				
Car	*No*	.75	.05	.80
Total		.90	.10	1.00

b. 20% own a foreign car; 90% own a U.S. car

c. .15

d. .95

e. .17

f. .75

g. No

52. c. .72

d. .40

e. No

54. a. .5541

c. .67

56. a. .197

b. .121

c. No

58. a. .7

b. 1

c. .375

d. .625

e. .56

60. a. .21

b. Yes

62. .6007

64. a. Three

b. .10, .50, .40

66. a. $\{E_1, E_2, E_3, E_4\}$

b. $\{E_1, E_5, E_6, E_7, E_8\}$

c. $\{E_2\}$

d. $\{E_7, E_8\}$

e. None

f. $\{E_4, E_5, E_6, E_7, E_8\}$

g. $\{E_1, E_2, E_3, E_4\}$

h. $\{E_1, E_2, E_3, E_4\}$

i. $\{E_1, E_2, E_3\}$

j. No

k. Yes

68. a. .02

b. .64

c. .93

d. Yes

70. a. .745

b. .192

c. .475

d. Democrat

e. .391

f. No

72. b. .2022

c. .4618

d. .4005

74. a. .90, 0

b. No

76. a.

	Smoker	Non-smoker	Total
Record of Heart Disease	.10	.08	.18
No Record of Heart Disease	.20	.62	.82
Total	.30	.70	1.00

b. .10

c. 30% smokers; 70% nonsmokers; 18% with record of heart disease; 82% with no record of heart disease

d. .333

e. .114

f. No

g. Smokers have a higher probability of having a record of heart disease

78. a. .25, .40, .10

b. .25

c. B and S are independent; program appears to have no effect

80. a. .20

b. .35

c. 65%

82. Call back since $P(\text{sale} \mid \text{call back}) = .21$

84. 3.44%

86. a. .12

b. .625

c. .305

88. a. .0625
b. .0132
c. Three

CHAPTER 5

2. a. Discrete
b. Discrete
c. Discrete
d. Continuous
e. Continuous
f. Discrete
4. 0, 1, 2, 3, 4, 5
6. a. 0, 1, 2, . . . , 20; discrete
b. 0, 1, 2, . . . ; discrete
c. 0, 1, 2, . . . , 50; discrete
d. $0 \leq x \leq 8$; continuous
e. $x > 0$; continuous
8. a. Yes
b. .25
c. .35
d. .40
10. a. _____

x	$f(x)$
1	.298
2	.092
3	.066
4	.035
5	.066
6	.019
7	.424
	1.000

12. b. .20
c. .70
14. a. Yes
b. ⅓
c. ⅚
16. a. 6
b. 4.50
c. 2.12
18. a. 1.25
b. 1.2875
c. 1.13
20. a. 13.474%
b. $\sigma^2 = 71.79$, $\sigma = 8.47$
c. 6.964, 1.7401
22. a. $166
b. −94; concern is to protect against the expense of a big accident
24. a. 445
b. 1250 loss

26. a. 145, 140; medium preferred
b. 2725, 12,400; medium preferred
28. a. .5220
b. .3685
c. .1095
30. a. .2348
b. .0936
c. 0
32. a. $f(0) = .3487$
b. $f(2) = .1937$
c. .9298
d. .6513
e. 1
f. $\sigma^2 = .9000$, $\sigma = .9487$
34. a. .2301
b. .3410
c. .8784
36. a. Probability of a defective part must be .03 for each trial; trials must be independent
c. 2
d.

Number of defects	0	1	2
Probability	.9409	.0582	.0009

38. a. $P(x \geq 8) = .2402$
b. $f(1) = .2488$
40. a. .90
b. .99
c. .999
d. Yes
42. a. .5905
b. .1937
c. .6082
44. a. Three trials
b. 2; boy or girl
c. .5; they are the same for each trial
d. Trials are independent
e. $x =$ number of girls; discrete; 0, 1, 2, 3
46. a. .2340
b. .0995
48. 212.5, 31.88
50. a. $f(x) = \dfrac{3^x e^{-3}}{x!}$
b. .2241
c. .1494
d. .8008
52. a. .0183
b. .0733
c. .1465
54. a. .1952
b. .1048
c. 4; .0183
d. .0907
56. a. .0821
b. .0653
c. .3840

58. a. .000045
   b. .010245
   c. .0821
   d. .9179
60. .1249
62. .0189
64. a. .50
   b. .067
   c. .4667
   d. .30
66. .00582
68. a. .01
   b. .07
   c. .92
   d. .07
70. a. .2013
   b. .1244
   c. .1323
   d. .2541
72. a. 1.6
   b. $120
74. a. 2.2
   b. 1.16
76. a. $f(x) \geq 0$ and $\Sigma f(x) = 1$
   b. $17.25
   c. $1.25, 7.8%
   d. 1.3875
   e. The first stock
78. a. 3 hours
   b. 1
80. a. .1712
   b. .2529
82. a. .2785
   b. .3417
84. a. .9510
   b. .0480
   c. .0490

Chapter 6

2. b. .50
   c. .60
   d. 15
   e. 8.33
4. b. .50
   c. .30
   d. .40
6. a. .40
   b. .64
   c. .68
10. a. .3413
   b. .4332

   c. .4772
   d. .4938
12. a. .2967
   b. .4418
   c. .3300
   d. .5910
   e. .8849
   f. .2388
14. a. $z = 1.96$
   b. $z = .61$
   c. $z = 1.12$
   d. $z = .44$
16. a. $z = 2.33$
   b. $z = 1.96$
   c. $z = 1.645$
   d. $z = 1.28$
18. a. .1814
   b. .9656
   c. $12,816 or more
20. a. 50.77%
   b. 15.87%
   c. 23.88%
22. a. .7745
   b. 36.32
   c. 19%
24. a. 2.865 years
   b. .6247
26. a. .1151
   b. .2852
   c. 49 or less
28. a. .1685
   b. .8261
   c. .0033
   d. .2148
30. a. 11
   b. .5780
   c. .1170
   d. .6950
32. a. Approximately 1
   b. .0228
   c. .0228
34. a. .5276
   b. .3935
   c. .4724
   d. .1341
36. a. $1 - e^{-x_0/3}$
   b. .4866
   c. .3679
   d. .8111
   e. .3245
38. b. .6321
   c. .3935
   d. .0821

40. a. 4 hours
   b. $1/4\ e^{-x/4}$
   c. .7788
   d. .1353
42. b. .30
   c. .15
   d. .40
   e. 2.50 minutes
44. b. .25
46. a. .8106
   b. .0833
   c. .6955
   d. .9500
48. a. .3174, 317.4 defects
   b. .0028, 2.8 defects
50. .0062
52. a. .5899
   b. 30 or more
54. a. 47.06%
   b. .0475
   c. 42,480
56. a. .0228
   b. Do not use the die
   c. 3.36733
58. a. 5.16%
   b. 57.87%
   c. 99.55%
   d. Approximately 0

Chapter 7

2. a. All college students receiving federal aid
   b. Students whose records were studied
   c. Less time and lower cost
4. a. All United flights arriving at O'Hare
   b. The 104 flights studied
   c. 2.88%
   d. Prefer FAA
6. 22, 147, 229, 289
8. 20
10. a. Finite
   b. Infinite
12. a. 10
   b. 1/10
14. a. B and A
   b. E and C
16. a. L.A. Gear, New Balance, Adidas, Avia, Reebok
   b. 0
   c. 252

18. 2782, 493, 825, 1807, 289
20. 447, 348, 499, 568, 055, 392
   126, 036, 599, 294, 570, 159
22. a. Randomly select a page (1–853), and then randomly
   select a line (1–400) on the sampled page
   b. Skip or ignore inappropriate lines, and repeat the
   sampling procedure of part (a)
24. a. 2.175
   b. 2.191
   c. Numerical values are approximately the same
   d. 0.9927
26. 3.54, 2.50, 2.04, 1.77
   $\sigma_{\bar{x}}$ decreases as n increases
28.

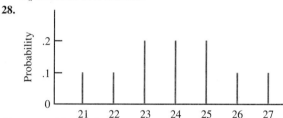

30. a. $\mu = 14$, $\sigma = 4$
   b. 17, 13, 11, 15, 16, 14, 18, 10, 14, 12
   c.

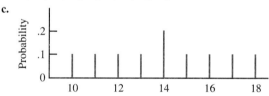

   d. 14, 2.45
   e. 14, 2.45
   f. Approach in part (e) is easier
32. 170, 4.43
34. a. 100
   b. 2
   c. Normal
   d. Normal curve with mean 100 and standard deviation 2
36. a. 200
   b. 5
   c. Normal with $E(\bar{x}) = 200$ and $\sigma_{\bar{x}} = 5$
   d. The probability distribution of $\bar{x}$
38. a. 105
   b. 0.55
   c. Normal
   d. Normal curve with mean 105 and standard deviation
   .55
40. a. Only for $n = 30$ and $n = 40$
   b. $n = 30$; normal with $E(\bar{x}) = 400$ and $\sigma_{\bar{x}} = 9.13$
   $n = 40$; normal with $E(\bar{x}) = 400$ and $\sigma_{\bar{x}} = 7.91$

42. Normal with $E(\bar{x}) = 68$ and $\sigma_{\bar{x}} = 1.58$

44. 0.4772

46. a. Normal with $E(\bar{x}) = 12.55$ and $\sigma_{\bar{x}} = .6325$

    b. .8858

    c. .5704

48. a. .6528

    b. .3616

50. a. Normal with $E(\bar{x}) = 16{,}012$ and $\sigma_{\bar{x}} = 420$

    b. .9826

    c. .7660, .4514, .1896

    d. Increase the sample size

52. a. Normal with $E(\bar{x}) = 14.25$ and $\sigma_{\bar{x}} = .2828$

    b. .9441

    c. .6212

    d. .9878, .7888

54. a. Assume population has a normal distribution

    b. .9266

    c. Increase n to at least 30

56. 4324, 2875, 318, 538, 4771

58. a. $\mu = 4.5$, $\sigma = 0.9574$

    b. 2 at 3.5, 3 at 4.0, 5 at 4.5, 3 at 5.0, and 2 at 5.5

    d. 4.5, 0.606

    e. 4.5, 0.606

60. a. .4714

    b. .7198

    c. 693

62. a. Normal with $E(\bar{x}) = 49{,}000$ and $\sigma_{\bar{x}} = 1200$

    b. .5934

    c. .7620

    d. .9050

    e. 553

64. a. 67

    b. 1.5

    c. Normal with $E(\bar{x}) = 67$ and $\sigma_{\bar{x}} = 1.5$

    d. .9082

    e. .4972

66. a. No, since $n/N = .01$

    b. Use $\sigma_{\bar{x}} = .0566$

    c. .9232

68. a. 625

    b. .7888

CHAPTER 8

2. a. 30.60 to 33.40

    b. 30.34 to 33.66

    c. 29.82 to 34.18

4. 62

6. 397.60 to 322.40

8. a. 23,480 to 24,520

    b. 23,380 to 24,620

    c. 23,186 to 24,814

10. a. 46.15 to 47.85

    b. 42.53 to 45.47

    c. Men, due to larger n

12. a. 6

    b. 96

14. a. 24.31 to 26.15

    b. 24.14 to 26.32

16. a. 1.734

    b. -1.321

    c. 3.365

    d. -1.761 and 1.761

    e. -2.048 and 2.048

18. a. 15.97 to 18.53

    b. 15.71 to 18.79

    c. 15.14 to 19.36

20. a. 13.2

    b. 7.8

    c. 7.62 to 18.78

    d. Wide interval; larger sample desirable

22. 6.28 to 6.78

24. a. 21.15 to 23.65

    b. 21.12 to 23.68

    c. Intervals are essentially the same

26. 4.51 to 6.59

28. a. 9

    b. 35

    c. 78

30. a. 50, 89, 200

    b. Only if $E = 1$ is essential

32. 384

34. 59

36. a. 49

    b. 30 to 34

38. a. H_0: $\mu \leq 14$

       H_a: $\mu > 14$

40. a. H_0: $\mu \geq 220$

       H_a: $\mu < 220$

42. a. H_0: $\mu \leq 1$

       H_a: $\mu > 1$

    b. Claiming $\mu > 1$ when it is not true

    c. Claiming $\mu \leq 1$ when it is not true

44. a. H_0: $\mu \geq 220$

       H_a: $\mu < 220$

    b. Claiming $\mu < 220$ when it is not true

    c. Claiming $\mu \geq 220$ when it is not true

46. a. Reject H_0 if $z > 2.05$

    b. 1.36

    c. .0869

    d. Do not reject H_0

48. a. .0344; reject H_0

    b. .3264; do not reject H_0

    c. .0668; do not reject H_0

    d. Approximately 0; reject H_0

e. .8413; do not reject H_0

50. a. $H_0: \mu \le 6.5; H_a: \mu > 6.5$
  b. $z = 5.91$; reject H_0
  c. Current cars being driven longer
52. a. $z = 1.77$; do not reject H_0
54. $z = -2.74$; reject H_0
  p-value $= .0031$
56. a. Reject H_0 if $z < -1.645$
  b. $z = -1.98$; reject H_0
  c. .0239
58. a. Reject H_0 if $z < -2.33$ or $z > 2.33$
  b. -1.13
  c. .2584
  d. Do not reject H_0
60. a. .0718; do not reject H_0
  b. .6528; do not reject H_0
  c. .0404; reject H_0
  d. Approximately 0; reject H_0
  e. .3174; do not reject H_0
62. a. $z = -1.06$; do not reject H_0
  b. .2892
64. $z = 6.37$; reject H_0
66. a. $50,367 to $53,633
  b. Reject H_0
68. a. $8.66 or less
  b. Reject H_0
70. a. 18
  b. 1.41
  c. Reject H_0 if $t < -2.571$ or $t > 2.571$
  d. -3.47
  e. Reject H_0
72. a. .01; reject H_0
  b. .10; do not reject H_0
  c. Between .025 and .05; reject H_0
  d. Greater than .10; do not reject H_0
  e. Approximately 0; reject H_0
74. $t = 1.20$; do not reject H_0
76. a. $t = -3.33$; reject H_0
  b. p-value is less than .005
78. a. .2912
  b. Type II error
  c. .0031
80. a. Concluding $\mu \le 15$ when it is not true
  b. .2676
  c. .0179
82. a. Concluding $\mu = 28$ when it is not true
  b. .0853, .6179, .6179, .0853
  c. .9147
84. .1151, .0015
  Increasing n reduces β
86. 214
88. 109
90. 324

92. 2.11 to 2.39
94. 55.15 to 56.05
96. a. 1483.1
  b. 374.9
  c. 1136.4 to 1829.8
98. 9.20 to 14.80
100. 710
102. 37
104. 54
106. $z = 2.26$; reject H_0
108. a. $z = 1.80$; do not reject H_0
  b. .0718
  c. 349 to 375
110. a. .0143
  b. Reject H_0
112. 219

CHAPTER 9

2. a. .6156
  b. .8530
4. a. .6156
  b. .7814
  c. .9488
  d. .9942
  e. Higher probability with larger n
6. a. 3 with $\bar{p} = .25$, 9 with $\bar{p} = .50$ and 3 with $\bar{p} = .75$
  b.

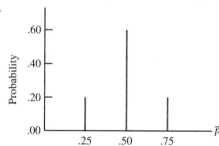

8. a. .9742
  b. .7372
10. a. .7062
  b. .1469
  c. .0025
12. a. Normal with $E(\bar{p}) = .80$ and $\sigma_{\bar{p}} = .02$
  b. .8664
  c. .9596
14. a. Normal with $E(\bar{p}) = .15$ and $\sigma_{\bar{p}} = .0505$
  b. .4448
  c. .8389
16. a. .25
  b. .0217
  c. .2075 to .2925
18. 350
20. .7825 to .8175

22. .3233 to .3767

24. .6007 to .7393

26. a. .2271 to .3529

  b. .2537 to .3263

  c. .2619 to .3181

  d. The interval width decreases, indicating a better precision

28. a. .4081 to .5319

  b. 383

30. a. 72

  b. 129

  c. 289

  d. 801

  e. n becomes larger

32. a. .4811 to .6189

  b. 381

34. a. Reject H_0 if $z > 1.645$

  b. $z = 1.69$; reject H_0

36. a. $z = -2.80$; .0026; reject H_0

  b. $z = -1.20$; .1151; do not reject H_0

  c. $z = -2.00$; .0228; reject H_0

  d. $z = .80$; .7881; do not reject H_0

38. $z = 1.43$; do not reject H_0

40. $z = -1.31$; do not reject H_0

  p-value $= .1902$

42. $z = -1.20$; do not reject H_0

44. $z = 1.37$; do not reject H_0

46. $z = -.72$; do not reject H_0

  p-value $= .4716$

48. a. Many different samples of 80 provide different values for $\bar{p}$

  b. Normal with $E(\bar{p}) = .35$ and $\sigma_{\bar{p}} = .0533$

  c. $\sigma_{\bar{p}}$ decreases to .0337

50. .8461

52. a. Normal with $E(\bar{p}) = .15$ and $\sigma_{\bar{p}} = .0292$

  b. .6970

54. .0947 to .1253

56. .0438 to .2362

58. .5165 to .6035

  .2216 to .2984

  A greater proportion use recognition

60. a. 1267

  b. 1508

62. a. 406

  b. .7356 to .8044

  c. .6008 to .6792

  d. Yes

64. $z = 2.52$; reject H_0

  p-value $= .0059$

66. a. Show $p < .50$

  b. $z = -6.62$; reject H_0

68. $z = 1.07$; do not reject H_0

  p-value $= .1423$

70. $z = -1.66$; do not reject H_0

  p-value $= .0485$

Chapter 10

2. a. 2.4

  b. 5.27

  c. .09 to 4.71

4. a. $4800

  b. $3651 to $5949

6. 2548.73 to 2951.27

8. a. 2

  b. 1.022 to 2.9878

  c. Populations normal with equal variances

10. a. 1.63

  b. 1.48 to 3.52

12. a. $z = -1.53$; do not reject H_0

  b. .1260

14. .1446; do not reject H_0

16. $z = 2.66$; reject H_0

18. a. $t = 2.22$; reject H_0

  b. .11 to 3.27 (thousands)

20. a. 3, -1, 3, 5, 3, 0, 1

  b. 2

  c. 2.082

  d. 2

  e. .07 to 3.93

22. a. $t = 2.38$; reject H_0

  b. .03 to 1.39

24. $\bar{d} = 3$; $t = 2.23$; reject H_0

26. a. $t = 5.88$; reject H_0

  b. 1.4 to 3.0

28. a. .12

  b. .0586 to .1814

  c. .0469 to .1931

30. .2122 to .2879

32. a. $z = 2.42$; reject H_0

  p-value $= .0078$

  b. Be aware of potential for less spending

  c. .0156 to .1444

34. $z = 5.67$; reject H_0

36. a. $z = 2.33$; reject H_0

  b. .02 to .22

38. 1354 to 2646

40. $t = -1.69$; do not reject H_0

42. $t = 2.29$; reject H_0

44. a. H_0: $p_1 - p_2 \leq 0$

    H_a: $p_1 - p_2 > 0$

  b. $z = 1.80$; reject H_0

    p-value $= .0359$

46. .0174; reject H_0

CHAPTER 11

2. a. 15.76 to 46.95
   b. 14.46 to 53.33
   c. 3.8 to 7.3
4. a. .22 to .71
   b. .47 to .84
6. a. 7.56 to 12.77
   b. 7.95 to 11.81
   c. 8.35 to 11.03
   d. The estimate is more precise for larger n
8. a. $\chi^2 = 32.39$; do not reject H_0
   b. 40.83 to 133.02
   c. 6.39 to 11.53
10. $\chi^2 = 31.5$; reject H_0
12. $\chi^2 = 10.16$; do not reject H_0
14. $F = 2.42$; reject H_0
16. $F = 1.19$; do not reject H_0
18. $F = 1.13$; do not reject H_0
20. $F = 2.20$; reject H_0
22. a. $F = 4$; reject H_0
   b. Drive carefully on wet pavement
24. 10.72 to 24.68
26. a. $\chi^2 = 27.44$; reject H_0
   b. .00012 to .00042
28. a. 15
   b. 6.25 to 11.13
30. $F = 1.39$; do not reject H_0
32. $F = 2.08$; reject H_0

CHAPTER 12

2. a. MSB = 268
   b. MSW = 92.04
   c. Cannot reject H_0 since $F = 2.91 < F_{.05} = 4.26$
   d.

Source of Variation	Sum of Squares	Degrees of Freedom	Mean Square	F
Between	536	2	268	2.91
Within	828.39	9	92.04	
Total	1364.39	11		

4. b. Reject H_0 since $F = 80 > F_{.05} = 2.76$
6. Reject H_0 since $F = 10.63 > F_{.05} = 4.26$
8. Source of information does not significantly affect the dissemination of the information; $F = .27 < F_{.05} = 3.47$
10. Significant difference; $F = 19.86 > F_{.05} = 3.10$
12. a. Significant difference; $F = 7.87 > F_{.05} = 4.26$
   b. 1 and 2; 2 and 3

   c. 1 and 2
   d. 1 and 2; 2 and 3
14. -9.37 to $-.63$
16. Reject the hypothesis that the means are equal
18. 2 and 3
20. a.

Source of Variation	Sum of Squares	Degrees of Freedom	Mean Square	F
Treatment	1488	2	744	5.50
Error	2029.55	15	135.3	
Total	3517.55	17		

   b. Significant difference between A and C
22. a. H_0: $\mu_1 = \mu_2 = \mu_3 = \mu_4 = \mu_5$
      H_a: Not all the population means are equal
   b. Reject H_0 since $F = 14.07 > 2.69$
24. Significant difference; $F = 43.99$ exceeds the critical value, which is between 3.15 and 3.23
26. b. Significant difference; $F = 9.87 > F_{.05} = 3.35$
28. Not significant; $F = 1.78 < F_{.05} = 3.89$
30. Not significant; $F = 2.54 < F_{.05} = 3.24$
32. Means are all different
34. Significant; $F = 6.60 > F_{.05} = 4.46$
36. Significant; $F = 12.60 > F_{.05} = 3.07$
38. Significant; $F = 7.12 > F_{.05} = 3.26$
40. Factor A is not significant since $F = 2.05 < F_{.05} = 5.99$
   Factor B is not significant since $F = 4.06 < F_{.05} = 5.14$
   Interaction is significant since $F = 7.66 > F_{.05} = 5.14$
42. Factor A is significant since $F = 10.75 > F_{.05} = 5.14$
   Factor B is not significant since $F = 3 < F_{.05} = 5.99$
   Interaction is not significant since $F = 1.75 < F_{.05} = 5.14$
44. Size is not significant since $F = 1.17 < F_{.05} = 5.99$
   Price is significant since $F = 6.15 > F_{.05} = 5.14$
   Interaction is not significant since $F = 4.71 < F_{.05} = 5.14$
46. The means of the k populations have to be equal
48. MSTR is based on the variation between sample means whereas MSE is computed based on the variation within each sample
50. a. Not significant since $t = .42 < t_{.025} = 2.477$
   b. Not significant since $F = .18 < F_{.05} = 5.99$
52. Significant; $F = 10.59$ exceeds the critical value, which is between 2.84 and 2.92
54. Significant; $F = 11.65 > F_{.05} = 3.13$
56. Significant; $F = 7.23 > F_{.05} = 4.26$
58. Not significant; $F = 1.66 < F_{.05} = 3.01$
60. Significant; $F = 5.19 > F_{.05} = 4.26$
62. Significant; $F = 6.99 > F_{.05} = 4.46$
64. Not significant; $F = 1.67 < F_{.01} = 10.92$
66. Type of machine is significant; type of loading system and interaction are not significant

CHAPTER 13

2. b. There appears to be a linear relationship between x and y
   d. $\hat{y} = 30.33 - 1.88x$
   e. 19.05
4. b. There appears to be a linear relationship between x and y
   d. $\hat{y} = -240.5 + 5.5x$
   e. 106 pounds
6. c. $\hat{y} = .38 + .35x$
   d. 9.13
8. a. $\hat{y} = 49.5 + 75.5x$
   b. 276
10. d. $\hat{y} = 6.38 + 45.68x$, where x and y are measured in 1000s
   e. $129,716
12. c. $\hat{y} = -55.84 + 1.67x$
   d. 44.36%
   e. $\hat{y} = 36.01$; predicted value is almost the same as the observed value
14. b. $\hat{y} = 192.77 + 59.07x$
   c. 783.47; estimated value is much greater than the observed value
16. a. SSE = 6.3325, SST = 114.80, SSR = 108.47
   b. $r^2 = .945$
18. a. SSE = 85,135.14, SST = 335,000, SSR = 249,864.86
   b. $r^2 = .746$
20. a. $\hat{y} = -54.85 + 98.80x$
   b. $r^2 = .992$
   c. 38.13 mgs
22. a. 4.133
   b. 2.033
   c. .643
   d. Reject H_0 since $t = 4.04 > t_{.05} = 3.182$
   e. Reject H_0 since $F = 16.36 > F_{.05} = 10.13$
24. Not significant since $t = 1.82 < t_{.05} = 3.182$
26. They are related since $F = 48.17 > F_{.01} = 34.12$
28. They are related since $F = 106.92 > F_{.05} = 5.32$
30. a. 1.11
   b. 7.07 to 14.13
   c. 2.32
   d. 3.22 to 17.98
32. *Confidence interval:* $-.4$ to 4.98
   Prediction interval: -2.27 to 7.31
34. $1443.51 to $1824.05
36. $1186.32 to $2081.24
38. a. $\hat{y} = 39.0162 - .2861x$
   b. They are related since $F = 38.72 > F_{.05} = 5.32$
   c. $r^2 = .829$; a good fit

d. 23.10 to 26.32
e. 20.10 to 29.32

40.

Source of Variation	Sum of Squares	Degrees of Freedom	Mean Square	F
Regression	108.47	1	108.47	51.41
Error	6.33	3	2.11	
Total	114.80	5		

42. a. 9
   b. $\hat{y} = 20.0 + 7.21x$
   c. 1.3626
   d. Significant relationship since $F = 28 > F_{.05} = 5.59$
   e. $380,500
44. a. $\hat{y} = 80.0 + 50.0x$
   b. 30
   c. Significant relationship since $F = 83.17 > F_{.05} = 4.20$
   d. $680,000
46. b. Yes
   c. $\hat{y} = 17.3 + 1.32x$
   d. Significant relationship; p-value = 0.000
   e. $r^2 = .537$; not a very good fit
   f. $60,940 to $65,903
   g. $53,493 to $65,903
48. a. $\hat{y} = 2.32 + .64x$
   b. No; the variance does not appear to be the same for all values of x
50. b. Yes
52. a. Yes; $x = 135$, $y = 145$ may be an outlier
   b. Yes
   c. Yes
54. a. $\hat{y} = -.23 + .049x$
   b. Minitab identifies observation 10 as an influential observation; the standardized residual plot shows an unusual trend in the residuals
56. a. $\hat{y} = -10.9 + 1.04x$
   b. Observation 10 is a possible outlier
58. b. There appears to be a linear relationship between x and y
   c. $s_{xy} = 26.5$
   d. $r_{xy} = .69$
60. b. Yes; linear or even a possible curvilinear relationship
   c. $s_{xy} = 36.8$
   d. $r_{xy} = .92$
62. b. Yes
   c. $r_{xy} = .87$
64. a. Yes
   b. $r_{xy} = .85$
   c. Repeat H_0 since $4.56 > t_{.005} = 3.355$
70. a. $\hat{y} = 22.173 - .1478x$

    b. Significant relationship since $F = 11.33 > F_{.05} = 7.71$

    c. $r^2 = .739$; a reasonably good fit

    d. 12.296 to 17.270

72. a. $\hat{y} = 10.5 + .953x$

    b. Significant relationship; p-value $= 0.000$

    c. \$2874 to \$4952

    d. Yes

74. a. $\hat{y} = 3767 - 322x$

    b. Significant relationship; p-value $= 0.000$

    c. $r^2 = .944$; an excellent fit

    d. \$2115.90 to \$2839.70

76. a. $\hat{y} = 20.7 - .234x$

    b. Not significant; p-value $= .096$

    c. Linear relationship is not appropriate

78. a. $\hat{y} = 220 + 132x$

    b. Significant relationship; p-value $= 0.000$

    c. $r^2 = .873$; a very good fit

    d. \$559.50 to \$933.90

80. a. $\hat{y} = 5.85 + .830x$

    b. Significant relationship; p-value $= 0.000$

    c. 84.65

    d. 65.35 to 103.96

CHAPTER 14

2. a. $\hat{y} = 45.06 + 1.94x_1$; $\hat{y} = 132.36$

    b. $\hat{y} = 85.22 + 4.32x_2$; $\hat{y} = 150.02$

    c. $\hat{y} = -18.37 + 2.01x_1 + 4.74x_2$; $\hat{y} = 143.18$

4. a. \$255,000

6. a. $\hat{y} = 83.2 + 2.29$ TVADV $+ 1.30$ NEWSADV

    b. No

8. a. $\hat{y} = 2.1 + .0138x_1 + .00584x_2$

    b. Advertising expenditure, per capita income, store size

10. a. $\hat{y} = -5.7 + 1.54$ STARTS $+ 1.81$ INCOME

    b. 96.96

12. a. .926

    b. .905

    c. Yes

14. a. .75

    b. .68

16. a. $R^2 = .377$, $R_a^2 = .282$

    b. The fit is not very good

18. a. Significant; p-value $= .000$

    b. Significant; p-value $= .000$

    c. Significant; p-value $= .000$

20. a. SSE $= 4000$, $s^2 = 571.43$, MSR $= 6000$

    b. Significant; $F = 10.50 > F_{.05} = 4.74$

22. Significant; $F = 6.58 > F_{.05} = 4.74$

24. a. Significant; p-value $= .046 < \alpha = .05$

    b. Not significant; p-value $= .219$

    c. Not significant; p-value $= .099$

26. a. 132.16 to 154.15

    b. 111.15 to 175.17

28. a. 71,130 to 112,350

    b. 42,870 to 140,610

30. a. $x_2 = 0$ if level 1; $x_2 = 1$ if level 2

    $E(y) = \beta_0 + \beta_1 x_1 + \beta_2 x_2$

    b. $E(y) = \beta_0 + \beta_1 x_1$

    c. $E(y) = \beta_0 + \beta_1 x_1 + \beta_2$

    d. $\beta_2 = E(y \mid \text{level 2}) - E(y \mid \text{level 1})$

32. a. \$15,300

    b. \$56,100

    c. \$41,600

34. a. $\beta_2 = E(y \mid \text{electrical problem}) - E(y \mid \text{mechanical problem})$

    b. $\hat{y} = .930 + .388x_1 + 1.26x_2$

    c. Significant; p-value $= .001 < \alpha = .05$

    d. Yes, $R_a^2 = .819$

    e. $b_2 = 1.26$ hours

36. a. $\hat{y} = -91.8 + 1.08$ AGE $+ .252$ PRESSURE $+ 8.74$ SMOKER

    b. Significant; p-value $= .01 < \alpha = .05$

    c. 95% prediction interval is 21.35 to 47.19 or a probability of .2135 to .4719; quit smoking and begin some type of treatment to reduce his blood pressure

38. a. $\hat{y} = -53.3 + 3.11x$

    b. -1.40, $-.15$, 1.36, $.47$, -1.39; no

    c. .38, .28, .22, .20, .98; no

    d. .60, .00, .26, .03, 11.09; yes, the fifth observation

40. a. Shape of the plot supports the assumptions

    b. Observation 13 is an outlier

    c. No

42. b. 3.19

44. a. $R^2 = .95$, $R_a^2 = .93$

    b. Significant; $F = 57.84 > F_{.01} = 5.21$

46. a. 3.04, 3.61, 5.08

    b. Both are significant

    d. .91

48. b. Significant; $F = 22.79 > F_{.05} = 5.79$

    c. $R_a^2 = .861$; good fit

    d. Both are significant

50. a. $\hat{y} = -16.7 + 1.02$ %WOMEN

    b. $R^2 = .868$; very good fit

c.

D_1	D_2	D_3	D_4	D_5	Industry
0	0	0	0	0	Industrial
1	0	0	0	0	Technology
0	1	0	0	0	Consumer
0	0	1	0	0	Retailing
0	0	0	1	0	Media
0	0	0	0	1	Financial

$\hat{y} = 11.5 + 7.75D_1 + 7.75D_2 + 26.7D_3 + 24.8D_4 + 33.5D_5$

d. %WOMEN is a better predictor than the type of industry

e. $\hat{y} = -20.0 + 1.18$ %WOMEN $- 1.96D_1 - 5.49D_2 - 8.20D_3 + 3.74D_4 - 7.99D_5$

f. Adding type of industry to the model involving %WOMEN is not significant

52. a. POINTS $= 170 + 6.61$ TEAMINT

b. Pattern is acceptable; observation 13 looks unusual

c. No outliers

d. Yes; observation 13

Chapter 15

2. a. $\hat{y} = 9.32 + .424x$; p-value $= .117$ indicates a weak relationship between x and y

b. $\hat{y} = -8.10 + 2.41x - .0480x^2$; $R_a^2 = .932$, a very good fit

c. 20.965

4. a. $\hat{y} = 943 + 8.71x$

b. Significant; p-value $= .005 < \alpha = .01$

6. b. No, the relationship appears to be curvilinear

c. Several possible models;
e.g., $\hat{y} = 2.90 - .185x + .00351x^2$

8. b. Simple linear model may be appropriate; however, there may be some curvilinear effect in the data

c. $\hat{y} = -6.60 - 14.2x + .508x^2$

10. a. Significant; $t = 4.34 > 2.069$

b. Not significant; $t = 1.48 < 2.069$

c. Significant; $t = -4.46 < -2.069$

d. x_2 can be dropped

12. a. $\hat{y} = 170 + 6.61$ TEAMINT

b. $\hat{y} = 280 + 5.18$ TEAMINT $- .0037$ RUSHING $- 3.92$ OPPONINT

c. Addition of the two independent variables is not significant

14. a. $\hat{y} = -111 + 1.32$ AGE $+ .296$ PRESSURE

b. $\hat{y} = -123 + 1.51$ AGE $+ .448$ PRESSURE $+ 8.87$ SMOKER $- .00276$ AGEXEXPRES

c. Significant

16. a. $\hat{y} = 780 + 3.06$ %COLLEGE

b. $\hat{y} = 780 + 3.06$ %COLLEGE

18. a. $\hat{y} = 383 - 12.2$ WINS

b. Use WINS, PASSING, and OPPONINT

20. $\hat{y} = -91.8 + 1.08$ AGE $+ .252$ PRESSURE $+ 8.74$ SMOKER

22. $d = 1.60$; test is inconclusive

24.

x_1	x_2	Treatment
0	0	1
1	0	2
0	1	3

$x_3 = 0$ if block 1; $x_3 = 1$ if block 2
$E(y) = \beta_0 + \beta_1 x_1 + \beta_2 x_2 + \beta_3 x_3$

26. a.

D_1	D_2	Manufacturer
0	0	1
1	0	2
0	1	3

$E(y) = \beta_0 + \beta_1 D_1 + \beta_2 D_2$

b. $\hat{y} = 23.0 + 5.00D_1 - 2.00D_2$

c. H_0: $\beta_1 = \beta_2 = 0$

d. Mean time is not the same for each manufacturer; p-value $= .004$

28. Significant difference between the two analyzers

30. b. There may be some curvilinear effect

c. $\hat{y} = 2081 + 21.9$ TIME $- .0417$ TIMESQ

32. a. $\hat{y} = -16.7 + 1.02$ %WOMEN

b. Use %WOMEN and D4, where $D_4 = 1$ if media, $D_4 = 0$ otherwise

34. a. $\hat{y} = 70.6 + 12.7$ INDUS $- 2.92$ ICQUAL

b. No obvious pattern

c. Test is inconclusive

36. Significant differences between comfort levels for the three types of browsers

Chapter 16

2. $\chi^2 = 15.33$, $\chi^2_{.05} = 7.81473$, reject H_0

4. $\chi^2 = 12.51$, $\chi^2_{.05} = 7.81473$; reject H_0

6. $\chi^2 = 6.24$, $\chi^2_{.10} = 4.60517$; reject H_0

8. $\chi^2 = 8.89$, $\chi^2_{.05} = 9.48773$; do not reject H_0

10. $\chi^2 = .49$, $\chi^2_{.10} = 7.799$; do not reject H_0

12. $\chi^2 = 19.78$, $\chi^2_{.05} = 9.48773$; reject H_0

14. $\chi^2 = 6.31$, $\chi^2_{.05} = 9.48773$; do not reject H_0

16. $\chi^2 = 14.72$, $\chi^2_{.05} = 5.99147$; reject H_0

18. $\chi^2 = 37.17$, $\chi^2_{.01} = 9.21034$; reject H_0

20. $\chi^2 = 7.96$ $\chi^2_{.05} = 9.48773$; do not reject H_0

22. a. Do not reject, $\chi^2 = .39$

   b. Do not reject, $\chi^2 = 1.57$

   c. 9 males and 3 females, $\chi^2 = 3.53$

24. $\chi^2 = 7.44$, $\chi^2_{.05} = 9.48773$; do not reject assumption

26. $\chi^2 = 31.43$, $\chi^2_{.05} = 7.81473$; do not plan on same number each day.

28. $\chi^2 = 22.87$, $\chi^2_{.05} = 9.48773$; reject H_0

30. $\chi^2 = 6.20$, $\chi^2_{.05} = 12.5916$; do not reject H_0

32. $\chi^2 = 99.42$, $\chi^2_{.05} = 7.81473$; conclude that opinions differ

CHAPTER 17

2. $z = 3.27$; reject H_0

4. $z = 3.15$; reject H_0

6. $z = 1.90$; do not reject H_0

8. $z = 3.76$; reject H_0

10. $z = 1.27$; do not reject H_0

12. $z = 2.43$; reject H_0

14. $z = 1.89$; reject H_0

16. $z = -2.05$; reject H_0

18. $T = 34$; reject H_0

20. $T = 28.5$; reject H_0

22. $z = 3.33$; reject H_0

24. $z = -.25$; do not reject H_0

26. $W = 10.22$; reject H_0

28. $W = 3.06$; do not reject H_0

30. $W = 8.03$; reject H_0

32. a. .68

   b. $z = 2.06$; reject H_0

34. $z = .72$; do not reject H_0

36. $r_s = .04$; $z = .12$; do not reject H_0

38. $z = .82$; do not reject H_0

40. $z = 1.81$; reject H_0

42. $z = 2.96$; reject H_0

   $z = .68$; do not reject H_0

44. $z = -2.59$; reject H_0

46. $z = -2.97$; reject H_0

48. $W = 12.61$; reject H_0

50. $r_s = -.92$; $z = -3.05$; reject H_0

52. $r_s = .76$; $z = 2.84$; reject H_0

CHAPTER 18

2. a. 30,000

   b. 320

   c. 29,360 to 30,640

4. 73

6. 337

8. a. *Stratum 1:* 27,600

   Stratum 2: 25,750

   Stratum 3: 21,000

   b. 74,350

   c. 70,599.88 to 78,100.12

10. a. n = 93, $n_1 = 30$, $n_2 = 30$, $n_3 = 33$

   b. $n = 306$, $n_1 = 98$, $n_2 = 98$, $n_3 = 109$

   c. $n = 275$, $n_1 = 88$, $n_2 = 88$, $n_3 = 98$

12. a. \$3,617,000

   b. \$1,122,265

   c. \$41,066 to \$56,499

   d. \$9,568,261 to \$13,164,197

14. a. 15, 4500, .30

   b. 1.4708, 441.24, .0484

   c. 12.0584 to 17.9416

   d. 3617.52 to 5382.48

   e. .2032 to .3968

16. a. 40

   b. .70

   c. 35.8634 to 44.1366

   d. .5234 to .8766

18. a. .7788 to .8212

   b. .5740 to .6260

   c. .4136 to .4664

20. a. \$22,790 to \$23,610

   b. \$68,370,366 to \$70,829,634

   c. .6692 to .7908

22. a. 431

   b. .2175 to .3983

   c. .6230 to .8002

   d. 996

24. a. 75.275

   b. .198 to .502

   c. 1680

APPENDIX G

SOLUTIONS TO SELF-TEST EXERCISES

CHAPTER 1

3. a. Five variables: sales, rank, profit, assets, and industry
  b. Rank and industry are qualitative; sales, profit, and assets are quantitative
  c. Sales—ratio; rank—ordinal; profit—ratio; assets—ratio; industry—nominal
13. a. All adults in the United States
  b. Favorite way to spend time in the evening
  c. Qualitative since the variable is nominal
  d. 1500
  e. 70% of the adults in the sample selected "staying at home with the family"
  f. Statistical inference would be the inference that 70% of the population of adults prefer "staying at home with the family" as the favorite way to spend time in the evening

CHAPTER 2

3. a. $360° \times 58/120 = 174°$
  b. $360° \times 42/120 = 126°$
  c.

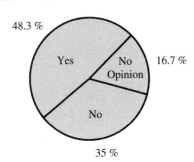

48.3 % Yes   No Opinion 16.7 % No 35 %

d.

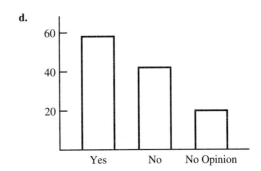

7.

Rating	Frequency	Relative Frequency
Outstanding	19	.38
Very good	13	.26
Good	10	.20
Average	6	.12
Poor	2	.04

Management should be pleased with these results: 64% of the ratings are very good to outstanding, and 84% of the ratings are good or better; comparing these ratings to previous results will show whether or not the restaurant is making improvements in its customers' ratings of food quality

12.

Class	Cumulative Frequency	Cumulative Relative Frequency
≤19	8	.200
≤29	20	.500
≤39	35	.925
≤49	40	1.000

15. a, b.

Waiting Time	Frequency	Relative Frequency
0–4	4	.20
5–9	8	.40
10–14	5	.25
15–19	2	.10
20–24	1	.05
Totals	20	1.00

c, d.

Waiting Time	Cumulative Frequency	Cumulative Relative Frequency
≤4	4	.20
≤9	12	.60
≤14	17	.85
≤19	19	.95
≤24	20	1.00

e. $^{12}/_{20} = .60$

23.
```
 6 | 3
 7 | 5  5  7
 8 | 1  3  4  8
 9 | 3  6
10 | 0  4  5
11 | 3
```

25.
```
 9 | 8  9
10 | 2  4  6  6
11 | 4  5  7  8  8  9
12 | 2  4  5  7
13 | 1  2
14 | 4
15 | 1
```

29.

Response	Distance (Miles) 0–24	25–99	100 or more	Total
Satisfied	7	2	3	12
	(.35)	(.10)	(.15)	(.60)
Not Satisfied	3	2	3	8
	(.15)	(.10)	(.15)	(.40)
Total	10	4	6	20
	(.50)	(.20)	(.30)	(1.00)

Some Preliminary Conclusions:

- People most satisfied traveled the fewest miles
- More people are satisfied than not satisfied
- The greatest percentage traveled less than 25 miles

CHAPTER 3

3. Arrange data in order: 15, 20, 25, 25, 27, 28, 30, 34

$i = \dfrac{20}{100}(8) = 1.6$; round up to position 2

20th percentile = 20

$i = \dfrac{25}{100}(8) = 2$; use positions 2 and 3

25th percentile $= \dfrac{20 + 25}{2} = 22.5$

$i = \dfrac{65}{100}(8) = 5.2$; round up to position 6

65th percentile = 28

$i = \dfrac{75}{100}(8) = 6$; use positions 6 and 7

75th percentile $= \dfrac{28 + 30}{2} = 29$

8. a. $\bar{x} = \dfrac{\Sigma x_i}{n} = \dfrac{775}{20} = 38.75$

Mode = 29 (appears three times)

b. .05(20) = 1; trim lowest (22) and highest (58) values

5% trimmed mean $= \dfrac{695}{18} = 38.61$

.10(20) = 2; trim two lowest (22, 24) and two highest (57, 58) values

10% trimmed mean $= \dfrac{614}{16} = 38.38$

c. Data in order: 22, 24, 29, 29, 29, 30, 31, 31, 32, 37, 40, 41, 44, 44, 46, 49, 50, 52, 57, 58

Median (10th and 11th positions)

$\dfrac{37 + 40}{2} = 38.5$

At home workers are slightly younger

d. $i = \dfrac{25}{100}(20) = 5$; use positions 5 and 6

$Q_1 = \dfrac{29 + 30}{2} = 29.5$

$i = \dfrac{75}{100}(20) = 15$; use positions 15 and 16

$Q_3 = \dfrac{4649}{2} = 47.5$

e. $i = \dfrac{32}{100}(20) = 6.4$; round up to position 7

32nd percentile = 31

At least 32% of the people are 31 or younger

19. Range = 34 − 15 = 19

Arrange data in order: 15, 20, 25, 25, 27, 28, 30, 34

$i = \dfrac{25}{100}(8) = 2; Q_1 = \dfrac{20 + 25}{2} = 22.5$

$i = \dfrac{75}{100}(8) = 6; Q_3 = \dfrac{28 + 30}{2} = 29$

$IQR = Q_3 - Q_1 = 29 - 22.5 = 6.5$

$\bar{x} = \dfrac{\Sigma x_i}{n} = \dfrac{204}{8} = 25.5$

x_i	$(x_i - \bar{x})$	$(x_i - \bar{x})^2$
27	1.5	2.25
25	−.5	.25
20	−5.5	30.25
15	−10.5	110.25
30	4.5	20.25
34	8.5	72.25
28	2.5	6.25
25	−.5	.25
		242.00

$s^2 = \dfrac{\Sigma(x_i - \bar{x})^2}{n - 1} = \dfrac{242}{8 - 1} = 34.57$

$s = \sqrt{34.57} = 5.88$

25. a. Range $= 190 - 168 = 22$

b. $\bar{x} = \dfrac{\Sigma x_i}{n} = \dfrac{1068}{6} = 178$

$s^2 = \dfrac{\Sigma(x_i - \bar{x})^2}{n - 1}$

$= \dfrac{4^2 + (-10)^2 + 6^2 + 12^2 + (-8)^2 + (-4)^2}{6 - 1}$

$= \dfrac{376}{5} = 75.2$

c. $s = \sqrt{75.2} = 8.67$

d. $\dfrac{s}{\bar{x}}(100) = \dfrac{8.67}{178}(100) = 4.87$

31. Chebyshev's theorem: *at least* $(1 - 1/k^2)$

a. $k = \dfrac{40 - 30}{5} = 2; (1 - \frac{1}{2}^2) = .75$

b. $k = \dfrac{45 - 30}{5} = 3; (1 - \frac{1}{3}^2) = .89$

c. $k = \dfrac{38 - 30}{5} = 1.6; (1 - \frac{1}{1.6}^2) = .61$

d. $k = \dfrac{42 - 30}{5} = 2.4; (1 - \frac{1}{2.4}^2) = .83$

e. $k = \dfrac{48 - 30}{5} = 3.6; (1 - \frac{1}{3.6}^2) = .92$

34. a. $\bar{x} = \dfrac{\Sigma x_i}{n} = 77.5$

$s = \sqrt{\dfrac{\Sigma(x_i - \bar{x})^2}{n - 1}} = \sqrt{\dfrac{874.5}{10 - 1}} = 9.86$

b. $z = \dfrac{x_i - \bar{x}}{s} = \dfrac{110 - 77.5}{9.86} = 3.30$; yes

c. $z = \dfrac{87 - 77.5}{9.86} = .96$ or approximately 1

68% are within ± 1;
32% are outside ± 1 (half above $+ 1$ and half below -1);
therefore, 32%/2 = 16% should be 87 or above

$z = \dfrac{58 - 77.5}{9.86} = -1.98$ or approximately -2

95% are within ± 2; 5% are outside ± 2 (half above $+2$ and half below -2); therefore, 5%/2 = 2.5% should be 58 or below

42. Arrange data in order: 5, 6, 8, 10, 10, 12, 15, 16, 18

$i = \dfrac{25}{100}(9) = 2.25$; round up to position 3

$Q_1 = 8$

Median (5th position) $= 10$

$i = \dfrac{75}{100}(9) = 6.75$; round up to position 7

$Q_3 = 15$

5-number summary: 5, 8, 10, 15, 18

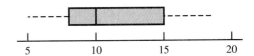

45. a. Arrange data in order: 484, 598, 636, 1049, 1061, 1220, 2188, 2366, 2472, 2731, 3122, 3261, 4514, 8030, 8258, 15,980, 32,249

$i = \dfrac{25}{100}(17) = 4.25$; round up to 5th position

$Q_1 = 1061$

Median (9th position) $= 2472$

$i = \dfrac{75}{100}(17) = 12.75$; round up to 13th position

$Q_3 = 4514$

5-number summary: 484, 1061, 2472, 4514, 32,249

b. $IQR = Q_3 - Q_1 = 4514 - 1061 = 3453$

Inner Fences: $Q_1 - 1.5IQR = 1061 - 1.5(3453) = -4118$

$\qquad\qquad Q_3 + 1.5IQR = 4514 + 1.5(3453) = 9694$

Outer Fences: $Q_1 - 3IQR = 1061 - 3(3453) = -9298$

$\qquad\qquad Q_3 + 3IQR = 4514 + 3(3453) = 14873$

c. Dow Chemical and DuPont are extreme outliers with very large sales compared to the rest of the industry.

d.

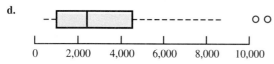

0	2,000	4,000	6,000	8,000	10,000

49.

f_i	M_i	f_iM_i
4	5	20
7	10	70
9	15	135
5	20	100
25		325

$$\bar{x} = \frac{\Sigma f_iM_i}{n} = \frac{325}{25} = 13$$

50.

f_i	M_i	$(M_i - \bar{x})$	$(M_i - \bar{x})^2$	$f_i(M_i - \bar{x})^2$
4	5	−8	64	256
7	10	−3	9	63
9	15	2	4	36
5	20	7	49	245
25				600

$$s^2 = \frac{\Sigma f_i(M_i - \bar{x})^2}{n-1} = \frac{600}{25-1} = 25$$

$$s = \sqrt{25} = 5$$

CHAPTER 4

5. $\binom{6}{3} = \frac{6!}{3!3!} = \frac{6 \cdot 5 \cdot 4 \cdot 3 \cdot 2 \cdot 1}{(3 \cdot 2 \cdot 1)(3 \cdot 2 \cdot 1)} = 20$

ABC	ACE	BCD	BEF
ABD	ACF	BCE	CDE
ABE	ADE	BCF	CDF
ABF	ADF	BDE	CEF
ACD	AEF	BDF	DEF

14. $\binom{50}{4} = \frac{50!}{4!46!} = \frac{50 \cdot 49 \cdot 48 \cdot 47}{4 \cdot 3 \cdot 2 \cdot 1} = 230,300$

16. $P(\text{never married}) = \frac{1106}{2038}$

$P(\text{married}) = \frac{826}{2038}$

$P(\text{other}) = \frac{106}{2038}$

Note that the sum of the probabilities equals 1

23. a. S = {ace of clubs, ace of diamonds, ace of hearts, ace of spades}

b. S = {2 of clubs, 3 of clubs, ..., 10 of clubs, J of clubs, Q of clubs, K of clubs, A of clubs}

c. There are 12; jack, queen, or king in each of the four suits

d. For (a): 4/52 = 1/13 = .08

For (b): 13/52 = 1/4 = .25

For (c): 12/52 = .23

33. a. $P(A) = P(E_1) + P(E_4) + P(E_6)$
$= .05 + .25 + .10 = .40$
$P(B) = P(E_2) + P(E_4) + P(E_7)$
$= .20 + .25 + .05 = .50$
$P(C) = P(E_2) + P(E_3) + P(E_5) + P(E_7)$
$= .20 + .20 + .15 + .05 = .60$

b. $A \cup B = \{E_1, E_2, E_4, E_6, E_7\}$
$P(A \cup B) = P(E_1) + P(E_2) + P(E_4) + P(E_6) + P(E_7)$
$= .05 + .20 + .25 + .10 + .05$
$= .65$

c. $A \cap B = \{E_4\}, P(A \cap B) = P(E_4) = .25$

d. Yes, they are mutually exclusive

e. $B^c = \{E_1, E_3, E_5, E_6\}$

$P(B^c) = P(E_1) + P(E_3) + P(E_5) + P(E_6)$
$= .05 + .20 + .15 + .10$
$= .50$

40. Let: B = rented a car for business reasons
P = rented a car for personal reasons

a. $P(B \cap P) = P(B)P(P) - P(B \cup P)$
$= .54 + .51 - .72$
$= .33$

b. $P(\text{did not rent}) = 1 - P(B \cup P)$
$= 1 - .72$
$= .28$

c. $P(\text{business only}) = P(B \cap P^c)$
$= P(B) - P(B \cap P)$
$= .54 - .33$
$= .21$

45. a. $P(A \mid B) = \frac{P(A \cap B)}{P(B)} = \frac{.40}{.60} = .6667$

b. $P(B \mid A) = \frac{P(A \cap B)}{P(A)} = \frac{.40}{.50} = .80$

c. No, because $P(A \mid B) \neq P(A)$

49. a.

	Reason for Applying			
	Quality	*Cost/Convenience*	*Other*	**Total**
Full-time	.218	.204	.039	.461
Part-time	.208	.307	.024	.539
Total	.426	.511	.063	1.00

b. It is most likely a student will cite cost or convenience as the first reason (probability = .511); school quality is the first reason cited by the second largest number of students (probability = .426)

c. P(quality | full-time) = .218/.461 = .473

d. P(quality | part-time) = .208/.539 = .386

e. For independence, we must have $P(A)P(B) = P(A \cap B)$; from the table,

$$P(A \cap B) = .218, P(A) = .461, P(B) = .426$$
$$P(A)P(B) = (.461)(.426) = .196$$

Since $P(A)P(B) \neq P(A \cap B)$, the events are not independent

57. a. Yes, since $P(A_1 \cap A_2) = 0$

b. $P(A_1 \cap B) = P(A_1)P(B | A_1) = .40(.20) = .08$

$P(A_2 \cap B) = P(A_2)P(B | A_2) = .60(.05) = .03$

c. $P(B) = P(A_1 \cap B) + P(A_2 \cap B) = .08 + .03 = .11$

d. $P(A_1 | B) = \dfrac{.08}{.11} = .7273$

$P(A_2 | B) = \dfrac{.03}{.11} = .2727$

60. M = missed payment
D_1 = customer defaults
D_2 = customer does not default
$P(D_1) = .05, P(D_2) = .95, P(M | D_2) = .2, P(M | D_1) = 1$

a. $P(D_1 | M) = \dfrac{P(D_1)P(M | D_1)}{P(D_1)P(M | D_1) + P(D_2)P(M | D_2)}$

$= \dfrac{(.05)(1)}{(.05)(1) + (.95)(.2)}$

$= \dfrac{.05}{.24} = .21$

b. Yes, the probability of default is greater than .20

CHAPTER 5

1. a. (H, H, H) (H, H, T) (H, T, H) (H, T, T)
(T, H, H) (T, H, T) (T, T, H) (T, T, T)

b. x = number of heads on three coin tosses

c.

Outcome	(H, H, H)	(H, H, T)	(H, T, H)	(H, T, T)	(T, H, H)	(T, H, T)	(T, T, H)	(T, T, T)
Value of x	3	2	2	1	2	1	1	0

d. Discrete; it may assume four values; 0, 1, 2, and 3.

3. Let: Y = position is offered
N = position is not offered

a. $S = \{(Y, Y, Y), (Y, Y, N), (Y, N, Y), (Y, N, N),$
$(N, Y, Y), (N, Y, N), (N, N, Y), (N, N, N)\}$

b. Let N = number of offers made; N is a discrete random variable

c.

Experimental Outcome	(Y, Y, Y)	(Y, Y, N)	(Y, N, Y)	(Y, N, N)	(N, Y, Y)	(N, Y, N)	(N, N, Y)	(N, N, N)
Value of N	3	2	2	1	2	1	1	0

8. a. $f(x) \geq 0$ for all values of x

$\Sigma f(x) = 1$; therefore, it is a proper probability distribution

b. Probability $x = 30$ is $f(30) = .25$

c. Probability $x < 25$ is $f(20) + f(25) = .20 + .15 = .35$

d. Probability $x > 30$ is $f(35) = .40$

9. a.

x	$f(x)$
1	3/20 = .15
2	5/20 = .25
3	8/20 = .40
4	4/20 = .20
	Total 1.00

b.

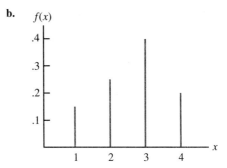

c. $f(x) \geq 0$ for $x = 1, 2, 3, 4$
$\Sigma f(x) = 1$

17. a.

y	f(y)	yf(y)
2	.20	.40
4	.30	1.20
7	.40	2.80
8	.10	.80
Totals	1.00	5.20

$$E(y) = \mu = 5.20$$

b.

y	$y - \mu$	$(y - \mu)^2$	f(y)	$(y - \mu)^2 f(y)$
2	−3.20	10.24	.20	2.048
4	−1.20	1.44	.30	.432
7	1.80	3.24	.40	1.296
8	2.80	7.84	.10	.784
			Total	4.560

$$Var(y) = 4.56$$

$$\sigma = \sqrt{4.56} = 2.14$$

20. a. $E(x) = 26.0(.31) + 8.6(.23) + 7.7(.45) - 2.9(.01)$
$= 13.474$

The expected return on \$1.00 for an individual investor is 13.474%

b.

x − 13.474	$(x - 13.474)^2$	f(x)	$(x - 13.474)^2 f(x)$
12.526	156.900680	.31	48.639211
−4.874	23.755876	.23	5.463851
−5.774	33.339076	.45	15.002584
16.374	268.107880	.01	2.681079
		Total	71.786725

$$Var(x) = 71.7867$$

$$\sigma = 8.4727$$

c.

x	f(x)	xf(x)	$(x - \mu)$	$(x - \mu)^2$	$(x - \mu)^2 f(x)$
5.0	.31	1.550	−1.964	3.857296	1.195762
8.6	.23	1.978	1.636	2.676496	.615594
7.7	.45	3.465	.736	.541696	.243763
−2.9	.01	−.029	−9.864	97.298496	.972985
	Totals	6.964			3.028104

$$E(x) = 6.964$$

$$Var(x) = 3.028104$$

$$\sigma = 1.7401$$

27. a.

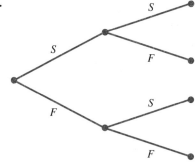

b. $f(1) = \binom{2}{1}(.4)^1(.6)^1 = \frac{2!}{1!1!}(.4)(.6) = .48$

c. $f(0) = \binom{2}{0}(.4)^0(.6)^2 = \frac{2!}{0!2!}(1)(.36) = .36$

d. $f(2) = \binom{2}{2}(.4)^2(.6)^0 = \frac{2!}{2!0!}(.16)(1) = .16$

e. $P(x \geq 1) = f(1) + f(2) = .48 + .16 = .64$

f. $E(x) = np = 2(.4) = .8$
$Var(x) = np(1 - p) = 2(.4)(.6) = .48$
$\sigma = \sqrt{.48} = .6928$

36. a. Probability of a defective part being produced must be .03 for each trial; trials must be independent

b. Let: D = defective
G = not defective

1st part	2nd part	Experimental Outcome	Number Defective
	D	(D, D)	2
D	G	(D, G)	1
G	D	(G, D)	1
	G	(G, G)	0

c. Two outcomes result in exactly one defect

d. $P(\text{no defects}) = (.97)(.97) = .9409$
$P(1 \text{ defect}) = 2(.03)(.97) = .0582$
$P(2 \text{ defects}) = (.03)(.03) = .0009$

51. a. $f(x) = \dfrac{2^x e^{-2}}{x!}$

b. $\mu = 6$ for 3 time periods

c. $f(x) = \dfrac{6^x e^{-6}}{x!}$

d. $f(2) = \dfrac{2^2 e^{-2}}{2!} = \dfrac{4(.1353)}{2} = .2706$

e. $f(6) = \dfrac{6^6 e^{-6}}{6!} = .1606$

f. $f(5) = \dfrac{4^5 e^{-4}}{5!} = .1563$

54. a. $\mu = 48(5/60) = 4$

$$f(3) = \dfrac{4^3 e^{-4}}{3!} = \dfrac{(64)(.0183)}{6} = .1952$$

b. $\mu = 48(15/60) = 12$

$$f(10) = \dfrac{12^{10} e^{-12}}{10!} = .1048$$

c. $\mu = 48(5/60) = 4$; one can expect 4 callers to be waiting after 5 minutes

$$f(0) = \dfrac{4^0 e^{-4}}{0!} = .0183;$$ the probability none will be waiting after 5 minutes is .0183

d. $\mu = 48(3/60) = 2.4$

$$f(0) = \dfrac{2.4^0 e^{-2.4}}{0!} = .0907;$$ the probability of no interruptions in 3 minutes is .0907

64. a. $f(1) = \dfrac{\binom{3}{1}\binom{10-3}{4-1}}{\binom{10}{4}} = \dfrac{\left(\dfrac{3!}{1!2!}\right)\left(\dfrac{7!}{3!4!}\right)}{\dfrac{10!}{4!6!}}$

$$= \dfrac{(3)(35)}{210} = .50$$

b. $f(2) = \dfrac{\binom{3}{2}\binom{10-3}{2-2}}{\binom{10}{2}} = \dfrac{(3)(1)}{45} = .067$

c. $f(0) = \dfrac{\binom{3}{0}\binom{10-3}{2-0}}{\binom{10}{2}} = \dfrac{(1)(21)}{45} = .4667$

d. $f(2) = \dfrac{\binom{3}{2}\binom{10-3}{4-2}}{\binom{10}{4}} = \dfrac{(3)(21)}{210} = .30$

68. $N = 60, n = 10$

a. $r = 20, x = 0$

$$f(0) = \dfrac{\binom{20}{0}\binom{40}{10}}{\binom{60}{10}} = \dfrac{(1)\left(\dfrac{40!}{10!30!}\right)}{\dfrac{60!}{10!50!}} = \left(\dfrac{40!}{10!30!}\right)\left(\dfrac{10!50!}{60!}\right)$$

$$= \dfrac{40 \cdot 39 \cdot 38 \cdot 37 \cdot 36 \cdot 35 \cdot 34 \cdot 33 \cdot 32 \cdot 31}{60 \cdot 59 \cdot 58 \cdot 57 \cdot 56 \cdot 55 \cdot 54 \cdot 53 \cdot 52 \cdot 51}$$

$$\approx .01$$

b. $r = 20, x = 1$

$$f(1) = \dfrac{\binom{20}{1}\binom{40}{9}}{\binom{60}{10}} = 20\left(\dfrac{40!}{9!31!}\right)\left(\dfrac{10!50!}{60!}\right)$$

$$\approx .07$$

c. $1 - f(0) - f(1) = 1 - .08 = .92$

d. Same as the probability one will be from Hawaii; in part (b) that was found to equal approximately .07

Chapter 6

1. a.

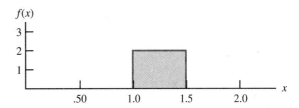

b. $P(x = 1.25) = 0$; the probability of any single point is zero since the area under the curve above any single point is zero

c. $P(1.0 \le x \le 1.25) = 2(.25) = .50$

d. $P(1.20 < x < 1.5) = 2(.30) = .60$

4. a.

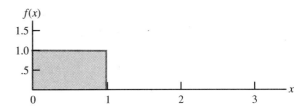

b. $P(.25 < x < .75) = 1(.50) = .50$

c. $P(x \le .30) = 1(.30) = .30$

d. $P(x > .60) = 1(.40) = .40$

13. a. $.4761 + .1879 = .6640$
  b. $.3888 - .1985 = .1903$
  c. $.4599 - .3508 = .1091$

15. a. Look in the table for an area of $.5000 - .2119 = .2881$; since the value we are seeking is below the mean, the z value must be negative; thus, for an area of $.2881$, $z = -.80$

  b. Look in the table for an area of $.9030/2 = .4515$; $z = 1.66$

  c. Look in the table for an area of $.2052/2 = .1026$; $z = .26$

  d. Look in the table for an area of $.9948 - .5000 = .4948$; $z = 2.56$

  e. Look in the table for an area of $.6915 - .5000 = .1915$; since the value we are seeking is below the mean, the z value must be negative; thus, $z = -.50$

18. a. With $z = \dfrac{(12{,}000 - 10{,}000)}{2200} = .91$,

$$P(x \le 12{,}000) = P(z \le .91) = .5000 + .3186$$
$$= .8186$$

$$P(x \ge 12{,}000) = 1 - P(x \le 12{,}000) = 1 - .8186$$
$$= .1814$$

So, the probability that a rehabilitation program will cost at least $12{,}000 is .1814

  b. With $z = \dfrac{(6000 - 10{,}000)}{2200} = -1.82$,

$$P(x \ge 6000) = P(z \ge -1.82) = .4656 + .5000$$
$$= .9656$$

The probability that a rehabilitation program will cost at least $6000 is .9656

  c. First, find the value of z that cuts off an area of .10 in the upper tail of the standard normal distribution; a value of $z = 1.28$ does this.
  Now find the value of x corresponding to $z = 1.28$

$$\frac{x - 10{,}000}{2200} = 1.28$$

$$x = 10{,}000 + 1.28(2200) = 12{,}816$$

The cost range for the 10% most expensive programs is $12{,}816 or more

27. $\mu = np = (20)(.4) = 8$

$$\sigma = \sqrt{np(1 - p)} = \sqrt{(20)(.4)(.6)} = 2.19$$

a. At $x = 3.5$: $z = \dfrac{3.5 - 8}{2.19} = -2.05$ Area $= .4798$

At $x = 2.5$: $z = \dfrac{2.5 - 8}{2.19} = -2.51$ Area $= .4940$

$P(x = 3) = .4940 - .4798 = .0142$

Binomial probability from table $= .0123$

  b. At $x = 4.5$: $z = \dfrac{4.5 - 8}{2.19} = -1.60$ Area $= .4452$

At $x = 3.5$: $z = \dfrac{3.5 - 8}{2.19} = -2.05$ Area $= .4798$

$P(x = 4) = .4798 - .4452 = .0346$

Binomial probability from table $= .0350$

  c. At $x = 7.5$: $z = \dfrac{7.5 - 8.0}{2.19} = \dfrac{-.5}{2.19} = -.23$

Area $= .0910$

$P(x \ge 8) = .5000 + .0910 = .5910$

Binomial probability from table $= .5841$

  d. At $x = 5.5$: $z = \dfrac{5.5 - 8}{2.19} = -1.14$ Area $= .3729$

$P(x < 6) = .5000 - .3729 = .1271$

Binomial probability from table $= .1255$

Note that this is the binomial probability of 5 or less.

29. a. Use $\mu = np = (50)(.3) = 15$

$$\sigma = \sqrt{np(1 - p)}$$
$$= \sqrt{15(.7)}$$
$$= 3.24$$

Find probability of $9.5 \le x \le 10.5$

At $x = 9.5$: $z = \dfrac{9.5 - 15}{3.24} = \dfrac{-5.5}{3.24} = -1.70$

At $x = 10.5$: $z = \dfrac{10.5 - 15}{3.24} = \dfrac{-4.5}{3.24} = -1.39$

Probability is $.4554 - .4177 = .0377$

  b. At $x = 19.5$: $z = \dfrac{19.5 - 15}{3.24} = \dfrac{4.5}{3.24} = 1.39$

Probability is $.5000 - .4177 = .0823$

c. At $x = 9.5$: $z = \dfrac{9.5 - 15}{3.24} = \dfrac{-5.5}{3.24} = -1.70$

At $x = 20.5$: $z = \dfrac{20.5 - 15}{3.24} = \dfrac{5.5}{3.24} = 1.70$

Probability of $(9.5 \leq x \leq 20.5)$ is $2(.4554) = .9108$

36. a. $P(x \leq x_0) = 1 - e^{-x_0/3}$

b. $P(x \leq 2) = 1 - e^{-2/3} = 1 - .5134 = .4866$

c. $P(x \geq 3) = 1 - P(x \leq 3) = 1 - (1 - e^{-3/3})$
$= e^{-1} = .3679$

d. $P(x \leq 5) = 1 - e^{-5/3} = 1 - .1889 = .8111$

e. $P(2 \leq x \leq 5) = P(x \leq 5) - P(x \leq 2)$
$= .8111 - .4866 = .3245$

38. a.

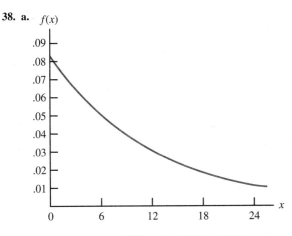

b. $P(x \leq 12) = 1 - e^{-12/12} = 1 - .3679 = .6321$

c. $P(x \leq 6) = 1 - e^{-6/12} = 1 - .6065 = .3935$

d. $P(x \geq 30) = 1 - P(x < 30)$
$= 1 - (1 - e^{-30/12})$
$= .0821$

CHAPTER 7

3. a. The sample consists of the 163 students who participated in the study

b. Yes, there could be problems here: the sample of students came from high schools located in three small cities; there is no statistical support for the sample being representative of all high school students in the United States

5. a. AB, AC, AD, AE, BC, BD, BE, CD, CE, DE

b. With 10 samples, each has a $\frac{1}{10}$ probability

c. E and C because 8 and 0 do not apply; 5 identifies E; 7 does not apply; 5 is skipped since E is already in the sample; 3 identifies C; 2 is not needed since the sample of size 2 is complete

15. 554, 459, 147, 385, 689, 640, 113, 340, 756, 953, 403, 827

23. a.

Sample	Sample Mean
1–2	1.5
1–3	1.0
1–4	2.0
1–5	2.5
2–3	0.5
2–4	1.5
2–5	2.0
3–4	1.0
3–5	1.5
4–5	2.5

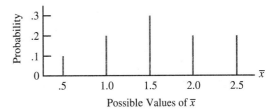

b.

Sample	Sample Mean
1–2–3	1.00
1–2–4	1.67
1–2–5	2.00
1–3–4	1.33
1–3–5	1.67
1–4–5	2.33
2–3–4	1.00
2–3–5	1.33
2–4–5	2.00
3–4–5	1.67

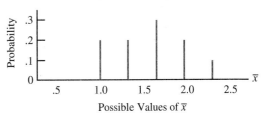

29. a. 6; AB, AC, AD, BC, BD, CD where A is Albert, B is Becky, C is Cindy, and D is David

b.

Sample	Sample Mean
AB	16
AC	17
AD	16
BC	18
BD	17
CD	18

c.

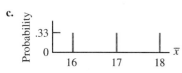

Possible Values of $\bar{x}$

32. $E(\bar{x}) = \mu = 170$ $\sigma_{\bar{x}} = \dfrac{\sigma}{\sqrt{n}} = \dfrac{28}{\sqrt{40}} = 4.43$

36. a. $E(\bar{x}) = \mu = 200$

    b. $\sigma_{\bar{x}} = \dfrac{\sigma}{\sqrt{n}} = \dfrac{50}{\sqrt{100}} = 5$

    c. Normal with $E(\bar{x}) = 200$ and $\sigma_{\bar{x}} = 5$

    d. It shows the probability distribution of all possible sample means that can be observed with random samples of size 100; this distribution can be used to compute the probability that $\bar{x}$ is within a specified $\pm$ from μ

39. Normal curve with $E(\bar{x}) = 18$ and standard deviation

$$\sigma_{\bar{x}} = \frac{\sigma}{\sqrt{n}} = \frac{4}{\sqrt{50}} = .566$$

43. $E(\bar{x}) = 100$ $\sigma_{\bar{x}} = \dfrac{\sigma}{\sqrt{n}} = \dfrac{12}{\sqrt{36}} = 2$

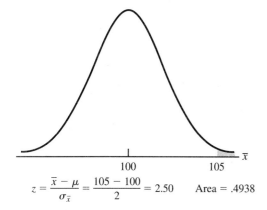

$z = \dfrac{\bar{x} - \mu}{\sigma_{\bar{x}}} = \dfrac{105 - 100}{2} = 2.50$ Area $= .4938$

$P(\bar{x} \geq 105) = .5000 - .4938 = .0062$

46. a.

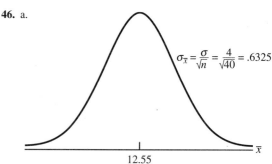

$\sigma_{\bar{x}} = \dfrac{\sigma}{\sqrt{n}} = \dfrac{4}{\sqrt{40}} = .6325$

b. $P(11.55 \leq \bar{x} \leq 13.55) = ?$

$$z = \frac{13.55 - 12.55}{.6325} = 1.58$$

Area $= .4429 \times 2 = .8858$

c. $P(12.05 \leq \bar{x} \leq 13.05) = ?$

$$z = \frac{13.05 - 12.55}{.6325} = .79$$

Area $= .2852 \times 2 = .5704$

CHAPTER 8

2. Use $\bar{x} \pm z_{\alpha/2}(\sigma/\sqrt{n})$ with the sample standard deviation s used to estimate σ

    a. $32 \pm 1.645(6/\sqrt{50})$
        32 ± 1.4; (30.6 to 33.4)

    b. $32 \pm 1.96(6/\sqrt{50})$
        32 ± 1.66; (30.34 to 33.66)

    c. $32 \pm 2.575(6/\sqrt{50})$
        32 ± 2.18; (29.82 to 34.18)

5. a. $\sigma_{\bar{x}} = \sigma/\sqrt{n} = 2.50/\sqrt{49} = .3571$

    b. Sampling error less than or equal to $1.96\sigma_{\bar{x}} = .70$

    c. $12.60 \pm .70$ or (11.90 to 13.30)

17. a. $\bar{x} = \dfrac{\Sigma x_i}{n} = \dfrac{80}{8} = 10$

    b. $s = \sqrt{\dfrac{\Sigma(x_i - \bar{x})^2}{n - 1}} = \sqrt{\dfrac{84}{8 - 1}} = 3.46$

    c. With 7 degrees of freedom, $t_{.025} = 2.365$

$$\bar{x} \pm t_{.025}\frac{s}{\sqrt{n}}$$

$$10 \pm 2.365\frac{3.46}{\sqrt{8}}$$

10 ± 2.89; (7.11 to 12.89)

19. At 90%, $80 \pm t_{.05}(s/\sqrt{n})$

with df = 17, $t_{.05} = 1.740$

$$80 \pm 1.740(10/\sqrt{18})$$
$$80 \pm 4.10; \ (75.90 \text{ to } 84.10)$$

At 95%, $80 \pm t_{.025}(10/\sqrt{18})$

with df = 17, $t_{.025} = 2.110$

$$80 \pm 4.97; \ (75.03 \text{ to } 84.97)$$

28. a. Planning value of $\sigma = \dfrac{\text{Range}}{4} = \dfrac{36}{4} = 9$

b. $n = \dfrac{z_{.025}^2 \sigma^2}{E^2} = \dfrac{(1.96)^2(9)^2}{(3)^2} = 34.6 \approx 35$

c. $n = \dfrac{(1.96)^2(9)^2}{(2)^2} = 77.8 \approx 78$

29. Use $n = \dfrac{z_{\alpha/2}^2 \sigma^2}{E^2}$

$$n = \dfrac{(1.96)^2(6.82)^2}{(1.5)^2} = 79.4 \text{ or } 80$$

$$n = \dfrac{(1.645)^2(6.82)^2}{(2)^2} = 31.5 \text{ or } 32$$

38. a. $H_0: \mu \leq 14$

$H_a: \mu > 14$; research hypothesis

b. There is no statistical evidence that the new bonus plan increases sales volume

c. The research hypothesis that $\mu > 14$ is supported; we can conclude that the new bonus plan increases the mean sales volume

41. a. Claiming $\mu > 6.08$ when it is not; the researcher would claim an annual rate higher than the national average when this is not the case

b. Concluding $\mu \leq 6.08$ when it is not; the researcher would not detect the fact that adults in Des Moines have a higher annual rate

46. a. $z = 2.05$; reject H_0 if $z > 2.05$

b. $z = \dfrac{\bar{x} - \mu}{\sigma/\sqrt{n}} = \dfrac{16.5 - 15}{7/\sqrt{40}} = 1.36$

c. Area at $z = 1.36 = .4131$

p-value $= .5000 - .4131 = .0869$

d. Do not reject H_0

49. a. $H_0: \mu \geq 26{,}100$

$H_a: \mu < 26{,}100$; research hypothesis

b. Reject H_0 if $z < -1.645$

$$z = \dfrac{\bar{x} - \mu_0}{\sigma/\sqrt{n}} = \dfrac{25{,}000 - 26{,}100}{2400/\sqrt{36}} = -2.75$$

Reject H_0 and conclude that the mean cost is less than $26,100

c. p-value $= .5000 - .4970 = .0030$

58. a. Reject H_0 if $z < -2.33$ or $z > 2.33$

b. $z = \dfrac{\bar{x} - \mu}{\sigma/\sqrt{n}} = \dfrac{14.2 - 15}{5/\sqrt{50}} = -1.13$

c. p-value $= 2(.5000 - .3708) = .2584$

d. Do not reject H_0

61. a. $H_0: \mu = 9.70$

$H_a: \mu \neq 9.70$

Reject H_0 if $z < -1.96$ or if $z > 1.96$

$$z = \dfrac{\bar{x} - \mu_0}{\sigma/\sqrt{n}} = \dfrac{9.30 - 9.70}{1.05/\sqrt{49}} = -2.67$$

Reject H_0; conclude that wage rates in the city are not $\mu = 9.70$

b. p-value $= 2(.5000 - .4962) = .0076$

70. a. $\bar{x} = \dfrac{\Sigma x_i}{n} = \dfrac{108}{6} = 18$

b. $s = \sqrt{\dfrac{\Sigma(x_i - \bar{x})}{n-1}} = \sqrt{\dfrac{10}{6-1}} = 1.41$

c. Reject H_0 if $t < -2.571$ or $t > 2.571$

d. $t = \dfrac{\bar{x} - \mu}{s/\sqrt{n}} = \dfrac{18 - 20}{1.41/\sqrt{6}} = -3.46$

e. Reject H_0; conclude H_a is true

73. a. $H_0: \mu \leq 200$

$H_a: \mu > 200$

With 9 degrees of freedom, reject H_0 if $t > 1.833$

$$\bar{x} = \dfrac{\Sigma x_i}{n} = 218$$

$$s = \sqrt{\dfrac{\Sigma(x_i - \bar{x})^2}{n-1}} = \sqrt{\dfrac{6210}{9}} = 26.27$$

$$t = \dfrac{\bar{x} - \mu_0}{s/\sqrt{n}} = \dfrac{218 - 200}{26.27/\sqrt{10}} = 2.17$$

Reject H_0; conclude the population mean rental rate exceeds the $200 per month rate in Baltimore

b. Using the t distribution table with 9 degrees of freedom, we find

$$t_{.05} = 1.833 \quad \text{and} \quad t_{.025} = 2.262$$

The p-value is between .025 and .05; interpolation will show the p-value associated with $t = 2.17$ is approximately:

$$p\text{-value} = .05 - .025\dfrac{(2.17 - 1.833)}{(2.262 - 1.833)} = .03$$

78. $\sigma_{\bar{x}} = \dfrac{\sigma}{\sqrt{n}} = \dfrac{5}{\sqrt{120}} = .46$

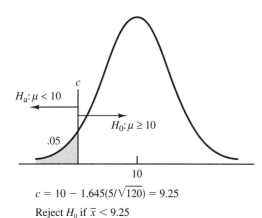

$c = 10 - 1.645(5/\sqrt{120}) = 9.25$

Reject H_0 if $\bar{x} < 9.25$

a. When $\mu = 9$,

$$z = \dfrac{9.25 - 9}{5/\sqrt{120}} = .55$$

$$P(H_0) = (.5000 - .2088) = .2912$$

b. Type II error

c. When $\mu = 8$,

$$z = \dfrac{9.25 - 8}{5/\sqrt{120}} = 2.74$$

$$\beta = (.5000 - .4969) = .0031$$

81. a. H_0: $\mu \geq 25$
H_a: $\mu < 25$
Reject H_0 if $z < -2.05$

$$z = \dfrac{\bar{x} - \mu_0}{\sigma/\sqrt{n}} = \dfrac{\bar{x} - 25}{3/\sqrt{30}} = -2.05$$

Solve for $\bar{x} = 23.88$
Decision Rule: Accept H_0 if $\bar{x} \geq 23.88$
Reject H_0 if $\bar{x} < 23.88$

b. For $\mu = 23$,

$$z = \dfrac{23.88 - 23}{3/\sqrt{30}} = 1.61$$

$$\beta = .5000 - .4463 = .0537$$

c. For $\mu = 24$,

$$z = \dfrac{23.88 - 24}{3/\sqrt{30}} = -.22$$

$$\beta = .5000 + .0871 = .5871$$

d. The Type II error cannot be made in this case; note that when $\mu = 25.5$, H_0 is true. The Type II error can only be made when H_0 is false

86. $n = \dfrac{(z_\alpha + z_\beta)\sigma^2}{(\mu_0 - \mu_a)^2} = \dfrac{(1.645 + 1.28)^2(5)^2}{(10 - 9)^2} = 214$

89. At $\mu_0 = 400$, $\alpha = .02$; $z_{.02} = 2.05$
At $\mu_a = 385$, $\beta = .10$; $z_{.10} = 1.28$
With $\sigma = 30$,

$$n = \dfrac{(z_\alpha + z_\beta)^2\sigma^2}{(\mu_0 - \mu_a)^2} = \dfrac{(2.05 + 1.28)^2(30)^2}{(400 - 385)^2} = 44.4 \text{ or } 45$$

CHAPTER 9

2. a. $E(\bar{p}) = .40$

$$\sigma_{\bar{p}} = \sqrt{\dfrac{p(1-p)}{n}} = \sqrt{\dfrac{(.40)(.60)}{200}} = .0346$$

$$z = \dfrac{\bar{p} - p}{\sigma_{\bar{p}}} = \dfrac{.03}{.0346} = .87$$

Area $= .3078 \times 2 = .6156$

b. $z = \dfrac{\bar{p} - p}{\sigma_{\bar{p}}} = \dfrac{.05}{.0346} = 1.45$

Area $= .4265 \times 2 = .8530$

5. a.

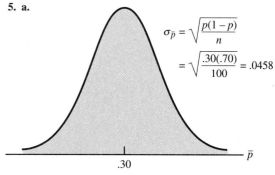

$$\sigma_{\bar{p}} = \sqrt{\dfrac{p(1-p)}{n}}$$

$$= \sqrt{\dfrac{.30(.70)}{100}} = .0458$$

The normal distribution is appropriate because $np = 100(.30) = 30$ and $n(1-p) = 100(.70) = 70$ are both greater than 5

b. $P(.20 \leq \bar{p} \leq .40) = ?$

$$z = \dfrac{.40 - .30}{.0458} = 2.18$$

Area $= .4854 \times 2 = .9708$

c. $P(.25 \leq \bar{p} \leq .35) = ?$

$$z = \dfrac{.35 - .30}{.0458} = 1.09$$

Area $= .3621 \times 2 = .7242$

16. a. $\bar{p} = \dfrac{100}{400} = .25$

b. $\sqrt{\dfrac{\bar{p}(1-\bar{p})}{n}} = \sqrt{\dfrac{.25(.75)}{400}} = .0217$

c. $\bar{p} \pm z_{.025} \sqrt{\dfrac{\bar{p}(1-\bar{p})}{n}}$

$.25 \pm 1.96(.0217)$

$.25 \pm .0425; \ (.2075 \text{ to } .2925)$

21. a. $\bar{p} = 248/400 = .62$

b. $\bar{p} \pm 1.645 \sqrt{\dfrac{.62(.38)}{400}}$

$.62 \pm .04; \ (.58 \text{ to } .66)$

35. a. Reject H_0 if $z < -1.96$ or $z > 1.96$

b. $\sigma_{\bar{p}} = \sqrt{\dfrac{.20(.80)}{400}} = .02$

$z = \dfrac{\bar{p} - p}{\sigma_{\bar{p}}} = \dfrac{.175 - .20}{.02} = -1.25$

c. p-value $= 2(.5000 - .3944) = .2112$

d. Do not reject H_0

38. $H_0\colon p \le .113$

$H_a\colon p > .113$

Reject H_0 if $z > 1.645$

$\sigma_{\bar{p}} = \sqrt{\dfrac{p(1-p)}{n}} = \sqrt{\dfrac{.113(.887)}{200}} = .0224$

$\bar{p} = \dfrac{29}{200} = .145$

$z = \dfrac{\bar{p} - p_0}{\sigma_{\bar{p}}} = \dfrac{.145 - .113}{.0224} = 1.43$

Do not reject H_0; there is no statistical evidence to justify concluding $p > .113$

CHAPTER 10

1. a. $\bar{x}_1 - \bar{x}_2 = 13.6 - 11.6 = 2$

b. $s_{\bar{x}_1 - \bar{x}_2} = \sqrt{\dfrac{s_1^2}{n_1} + \dfrac{s_2^2}{n_2}} = \sqrt{\dfrac{(2.2)^2}{50} + \dfrac{3^2}{35}} = .595$

$2 \pm 1.645(.595)$

$2 \pm .98; \ (1.02 \text{ to } 2.98)$

c. $2 \pm 1.96(.595)$

$2 \pm 1.17; \ (.83 \text{ to } 3.17)$

8. a. $\bar{x}_1 - \bar{x}_2 = 10 - 8 = 2$ hours

b. $s^2 = \dfrac{9(1.2)^2 + 11(1)^2}{10 + 2 - 2} = 1.198$

$s_{\bar{x}_1 - \bar{x}_2} = \sqrt{1.198\left(\dfrac{1}{10} + \dfrac{1}{12}\right)} = 0.469$

$2 \pm 2.186(0.469)$

$2 \pm 0.978, \quad (1.022 \text{ to } 2.9878)$

c. Populations are normally distributed with equal variances

11. a. $s_{\bar{x}_1 - \bar{x}_2} = \sqrt{\dfrac{s_1^2}{n_1} + \dfrac{s_2^2}{n_2}} = \sqrt{\dfrac{(5.2)^2}{40} + \dfrac{6^2}{50}} = 1.18$

$z = \dfrac{(\bar{x}_1 - \bar{x}_2) - (\mu_1 - \mu_2)}{s_{\bar{x}_1 - \bar{x}_2}} = \dfrac{(25.2 - 22.8)}{1.18} = 2.03$

Reject H_0 if $z > 1.645$; therefore reject H_0; conclude H_a is true and $\mu_1 > 0$

b. p-value $= .5000 - .4788 = .0212$

15. Population 1 is supplier A.

Population 2 is supplier B.

$H_0\colon \mu_1 - \mu_2 < 0$ \quad Stay with supplier A

$H_a\colon \mu_1 - \mu_2 > 0$ \quad Change to supplier B

Reject H_0 if $2 < 1.645$

$z = \dfrac{(\bar{x}_1 - \bar{x}_2) - (\mu_1 - \mu_2)}{\sqrt{\sigma_1^2/n_1 + \sigma_2^2/n_2}} = \dfrac{(14 - 12.5) - 0}{\sqrt{(3)^2/50 + (2)^2/30}}$

$= \dfrac{5}{2.07} = 2.42$

Reject H_0; change to supplier B

19. a. 1, 2, 0, 0, 2

b. $\bar{d} = \dfrac{\Sigma d_i}{n} = \dfrac{5}{5} = 1$

c. $s_d = \sqrt{\dfrac{\Sigma(d_i - \bar{d})^2}{n-1}} = \sqrt{\dfrac{4}{5-1}} = 1$

d. With 4 degrees of freedom, $t_{.05} = 2.132$; reject H_0 if $t > 2.132$

$t = \dfrac{\bar{d} - \mu_d}{s_d/\sqrt{n}} = \dfrac{1 - 0}{1/\sqrt{5}} = 2.24$

Reject H_0; conclude $\mu_d > 0$

21. Difference = rating after − rating before

$H_0\colon \mu_d \le 0$

$H_a\colon \mu_d > 0$

With 7 degrees of freedom, reject H_0 if $t > 1.895$; when $\bar{d} = .625$ and $s_d = 1.302$,

$$t = \frac{\bar{d} - \mu_d}{s_d/\sqrt{n}} = \frac{.63 - 0}{1.302/\sqrt{8}} = 1.36$$

Do not reject H_0; we cannot conclude that seeing the commercial improves the potential to purchase

29. a. $\bar{p} = \dfrac{n_1\bar{p}_1 + n_2\bar{p}_2}{n_1 + n_2} = \dfrac{200(.22) + 300(.16)}{200 + 300} = .184$

$s_{\bar{p}_1 - \bar{p}_2} = \sqrt{(.184)(.816)\left(\dfrac{1}{200} + \dfrac{1}{300}\right)} = .0354$

Reject H_0 if $z > 1.645$

$$z = \frac{(.22 - .16) - 0}{.0354} = 1.69$$

Reject H_0

b. p-value $= (.5000 - .4545) = .0455$

31. $\bar{p}_1 = 270/500 = .54;\ \bar{p}_2 = 162/360 = .45$

$\bar{p}_1 - \bar{p}_2 = .54 - .45 = .09$

$$.09 \pm 1.96\sqrt{\frac{.54(.46)}{500} + \frac{.45(.55)}{360}}$$

$.09 \pm .0675$, or $.0225$ to $.1575$

CHAPTER 11

2. $s^2 = 25$

a. With 19 degrees of freedom, $\chi^2_{.05} = 30.1435$ and $\chi^2_{.95} = 10.1170$

$$\frac{19(25)}{30.1435} \le \sigma^2 \le \frac{19(25)}{10.1170}$$

$$15.76 \le \sigma^2 \le 46.95$$

b. With 19 degrees of freedom, $\chi^2_{.025} = 32.8523$ and $\chi^2_{.975} = 8.90655$

$$\frac{19(25)}{32.8523} \le \sigma^2 \le \frac{19(25)}{8.90655}$$

$$14.46 \le \sigma^2 \le 53.33$$

c. $3.8 \le \sigma \le 7.3$

9. $H_0: \sigma^2 \le .0004$

$H_a: \sigma^2 > .0004$

$n = 30$

$\chi^2_{.05} = 42.5569$ (29 degrees of freedom)

$$\chi^2 = \frac{(29)(.0005)}{.0004} = 36.25$$

Do not reject H_0; the product specification does not appear to be violated

15. We recommend placing the larger sample variance in the numerator; with $\alpha = .05$, $F_{.025,20,24} = 2.33$: reject H_0 if $F > 2.33$

$$F = \frac{8.2}{4.0} = 2.05;\ \text{do not reject } H_0$$

Or, had we used the lower tail F value,

$$F_{.025,20,24} = \frac{1}{F_{.025,24,20}} = \frac{1}{2.46} = .41$$

$$F = \frac{4.0}{8.2} = .49$$

$F > .41$; do not reject H_0

17. a. Let: σ_1^2 = variance in repair costs (4-year-old automobiles)

σ_2^2 = variance in repair costs (2-year-old automobiles)

$H_0: \sigma_1^2 \le \sigma_2^2$

$H_a: \sigma_1^2 > \sigma_2^2$

b. $s_1^2 = (170)^2 = 28{,}900$

$s_2^2 = (100)^2 = 10{,}000$

$$F = \frac{s_1^2}{s_2^2} = \frac{28{,}900}{10{,}000} = 2.89$$

$F_{.01,24,24} = 2.66$

Reject H_0; conclude that automobiles 4 years old have a larger variance in annual repair costs compared to automobiles 2 years old; this is expected due to the fact that older automobiles are more likely to have some very expensive repairs that lead to greater variance in the annual repair costs

CHAPTER 12

1. a. See Figure G12.1a

b. $\bar{\bar{x}} = (30 + 45 + 36)/3 = 37$

$$\text{SSB} = \sum_{j=1}^{k} n_j(\bar{x}_j - \bar{\bar{x}})^2$$

$$= 5(30 - 37)^2 + 5(45 - 37)^2$$

$$+ 5(36 - 37)^2 = 570$$

$$\text{MSB} = \frac{\text{SSB}}{k-1} = \frac{570}{2} = 285$$

FIGURE G12.1a

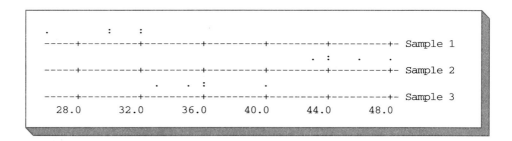

c.

$$SSW = \sum_{j=1}^{k} (n_j - 1)s_j^2$$

$$= 4(6) + 4(4) + 4(6.5) = 66$$

$$MSW = \frac{SSW}{n_T - k} = \frac{66}{15 - 3} = 5.5$$

d. $F = \dfrac{MSB}{MSW} = \dfrac{285}{5.5} = 51.82$

$F_{.05} = 3.89$ (2 degrees of freedom numerator and 12 denominator)

Since $F = 51.82 > F_{.05} = 3.89$, we reject the null hypothesis that the means of the three populations are equal

e.

Source of Variation	Sum of Squares	Degrees of Freedom	Mean Square	F
Between	570	2	285	51.82
Within	66	12	5.5	
Total	636	14		

6.

	Manufacturer 1	Manufacturer 2	Manufacturer 3
Sample mean	23	28	21
Sample variance	6.67	4.67	3.33

$$\bar{\bar{x}} = (23 + 28 + 21)/3 = 24$$

$$SSB = \sum_{j=1}^{k} n_j(\bar{x}_j - \bar{\bar{x}})^2$$

$$= 4(23 - 24)^2 + 4(28 - 24)^2$$

$$+ 4(21 - 24)^2 = 104$$

$$MSB = \frac{SSB}{k - 1} = \frac{104}{2} = 52$$

$$SSW = \sum_{j=1}^{k} (n_j - 1)s_j^2$$

$$= 3(6.67) + 3(4.67) + 3(3.33) = 44.01$$

$$MSW = \frac{SSW}{n_T - k} = \frac{44.01}{12 - 3} = 4.89$$

$$F = \frac{MSB}{MSW} = \frac{52}{4.89} = 10.63$$

$F_{.05} = 4.26$ (2 degrees of freedom numerator and 9 denominator)

Since $F = 10.63 > F_{.05} = 4.26$, we reject the null hypothesis that the mean time needed to mix a batch of material is the same for each manufacturer

11. a. $LSD = t_{\alpha/2}\sqrt{MSW\left(\dfrac{1}{n_i} + \dfrac{1}{n_j}\right)}$

$$= t_{.025}\sqrt{5.5\left(\frac{1}{5} + \frac{1}{5}\right)}$$

$$= 2.776\sqrt{2.2} = 4.12$$

$|\bar{x}_1 - \bar{x}_2| = |30 - 45| = 15 > LSD$; significant difference

$|\bar{x}_1 - \bar{x}_3| = |30 - 36| = 6 > LSD$; significant difference

$|\bar{x}_2 - \bar{x}_3| = |45 - 36| = 9 > LSD$; significant difference

b. $\bar{x}_1 - \bar{x}_2 \pm t_{\alpha/2}\sqrt{MSW\left(\dfrac{1}{n_1} + \dfrac{1}{n_2}\right)}$

$$(30 - 45) \pm 2.776\sqrt{5.5\left(\frac{1}{n_1} + \frac{1}{n_2}\right)}$$

$$-15 \pm 4.12 = -19.12 \text{ to } -10.88$$

c. $\alpha = .05/3 = .017$

$t_{.017/2} = t_{.0085}$, which is approximately $t_{.01} = 3.747$

$$BSD = 3.747\sqrt{5.5\left(\frac{1}{5} + \frac{1}{5}\right)} = 5.56$$

Thus, if the absolute value of the difference between any two sample means exceeds 5.56, there is sufficient evidence to reject the hypothesis that the corresponding population means are equal; since the differences in absolute value are each greater than 5.56, all three means appear to be different

d. $\text{TSD} = q\sqrt{\dfrac{\text{MSW}}{n}} = 5.04\sqrt{\dfrac{5.5}{5}} = 5.29$

Since the absolute value of the differences between any two sample means exceeds 5.29, there is sufficient evidence to conclude that the three means are different

e. $\bar{x}_i - \bar{x}_j \pm q\sqrt{\dfrac{\text{MSW}}{n}}$

$\bar{x}_i - \bar{x}_j \pm 5.04\sqrt{\dfrac{5.5}{5}}$

$\bar{x}_i - \bar{x}_j \pm 5.29$

Populations 1 and 2:

$\qquad 30 - 45 \pm 5.29 = -20.29 \text{ to } -9.71$

Populations 1 and 3:

$\qquad 30 - 36 \pm 5.29 = -11.29 \text{ to } -.71$

Populations 2 and 3: $45 - 36 \pm 5.29 = 3.71 \text{ to } 14.29$

13. $\text{LSD} = t_{\alpha/2}\sqrt{\text{MSW}\left(\dfrac{1}{n_1} + \dfrac{1}{n_3}\right)}$

$\qquad = t_{.025}\sqrt{4.89\left(\dfrac{1}{4} + \dfrac{1}{4}\right)}$

$\qquad = 2.262\sqrt{2.45} = 3.54$

Since $|\bar{x}_1 - \bar{x}_3| = |23 - 21| = 2 < 3.54$, there does not appear to be any significant difference between the means of populations 1 and 2

14. $\bar{x}_1 - \bar{x}_2 \pm q\sqrt{\dfrac{\text{MSW}}{n}}$

$23 - 28 \pm 3.95\sqrt{\dfrac{4.89}{4}}$

$-5 \pm 4.37 = -9.37 \text{ to } -.63$

19. a. $\bar{\bar{x}} = (156 + 142 + 134)/3 = 144$

$\text{SSTR} = \displaystyle\sum_{j=1}^{k} n_j(\bar{x}_j - \bar{\bar{x}})^2$

$\qquad = 6(156 - 144)^2 + 6(142 - 144)^2$

$\qquad + 6(134 - 144)^2 = 1488$

b. $\text{MSTR} = \dfrac{\text{SSTR}}{k-1} = \dfrac{1488}{2} = 744$

c. $s_1^2 = 164.35, \quad s_2^2 = 131.10, \quad s_3^2 = 110.46$

$\text{SSE} = \displaystyle\sum_{j=1}^{k} (n_j - 1)s_j^2$

$\qquad = 5(164.35) + 5(131.10) + 5(110.46)$

$\qquad = 2029.55$

d. $\text{MSE} = \dfrac{\text{SSE}}{n_T - k} = \dfrac{2029.55}{18 - 3} = 135.3$

e. $F = \dfrac{\text{MSTR}}{\text{MSE}} = \dfrac{744}{135.3} = 5.50$

$F_{.05} = 3.68$ (2 degrees of freedom numerator and 15 denominator)

Since $F = 5.50 > F_{.05} = 3.68$, we reject the hypothesis that the means for the three treatments are equal

34. *Treatment Means*

$\bar{x}_{.1} = 13.6 \qquad \bar{x}_{.2} = 11.0 \qquad \bar{x}_{.3} = 10.6$

Block Means

$\bar{x}_{1.} = 9 \qquad \bar{x}_{2.} = 7.67 \qquad \bar{x}_{3.} = 15.67$

$\bar{x}_{4.} = 18.67 \qquad \bar{x}_{5.} = 7.67$

Overall Mean

$\bar{\bar{x}} = 176/15 = 11.73$

Step 1

$\text{SST} = \displaystyle\sum_i \sum_j (x_{ij} - \bar{\bar{x}})^2$

$\qquad = (10 - 11.73)^2 + (9 - 11.73)^2 + \cdots + (8 - 11.73)^2$

$\qquad = 354.93$

Step 2

$\text{SSTR} = b\displaystyle\sum_j (\bar{x}_{.j} - \bar{\bar{x}})^2$

$\qquad = 5[(13.6 - 11.73)^2 + (11.0 - 11.73)^2$

$\qquad + (10.6 - 11.73)^2] = 26.53$

Step 3

$\text{SSBL} = k\displaystyle\sum_i (\bar{x}_{i.} - \bar{\bar{x}})^2$

$\qquad = 3[(9 - 11.73)^2 + (7.67 - 11.73)^2 + (15.67 - 11.73)^2$

$\qquad + (18.67 - 11.73)^2 + (7.67 - 11.73)^2] = 312.32$

Step 4

$\text{SSE} = \text{SST} - \text{SSTR} - \text{SSBL}$

$\qquad = 354.93 - 26.53 - 312.32 = 16.08$

Source of Variation	Sum of Squares	Degrees of Freedom	Mean Square	F
Treatments	26.53	2	13.27	6.60
Blocks	312.32	4	78.08	
Error	16.08	8	2.01	
Total	354.93	14		

$F_{.05} = 4.46$ (2 numerator degrees of freedom and 8 denominator)

Since $F = 6.60 > F_{.05} = 4.46$, we reject the null hypothesis that the means of the three treatments are equal

TABLE G12.40

		Factor B			Factor A
		Level 1	Level 2	Level 3	Means
Factor A	Level 1	$\bar{x}_{11} = 150$	$\bar{x}_{12} = 78$	$\bar{x}_{13} = 84$	$\bar{x}_{1\cdot} = 104$
	Level 2	$\bar{x}_{21} = 110$	$\bar{x}_{22} = 116$	$\bar{x}_{23} = 128$	$\bar{x}_{2\cdot} = 118$
Factor B Means		$\bar{x}_{\cdot1} = 130$	$\bar{x}_{\cdot2} = 97$	$\bar{x}_{\cdot3} = 106$	$\bar{\bar{x}} = 111$

40. See Table G12.40

Step 1

$$SST = \sum_i \sum_j \sum_k (x_{ijk} - \bar{\bar{x}})^2$$

$$= (135 - 111)^2 + (165 - 111)^2 + \cdots$$
$$+ (136 - 111)^2 = 9028$$

Step 2

$$SSA = br \sum_i (\bar{x}_{i\cdot} - \bar{\bar{x}})^2$$

$$= 3(2)[(104 - 111)^2 + (118 - 111)^2] = 588$$

Step 3

$$SSB = ar \sum_j (\bar{x}_{\cdot j} - \bar{\bar{x}})^2$$

$$= 2(2)[(130 - 111)^2 + (97 - 111)^2$$
$$+ (106 - 111)^2] = 2328$$

Step 4

$$SSAB = r \sum_i \sum_j (\bar{x}_{ij} - \bar{x}_{i\cdot} - \bar{x}_{\cdot j} + \bar{\bar{x}})^2$$

$$= 2[(150 - 104 - 130 + 111)^2$$
$$+ (78 - 104 - 97 + 111)^2 + \cdots$$
$$+ (128 - 118 - 106 + 111)^2] = 4392$$

Step 5

$$SSE = SST - SSA - SSB - SSAB$$

$$= 9028 - 588 - 2328 - 4392 = 1720$$

Source of Variation	Sum of Squares	Degrees of Freedom	Mean Square	F
Factor A	588	1	588	2.05
Factor B	2328	2	1164	4.06
Interaction	4392	2	2196	7.66
Error	1720	6	286.67	
Total	9028	11		

$F_{.05} = 5.99$ (1 degree of freedom numerator and 6 denominator)

$F_{.05} = 5.14$ (2 degrees of freedom numerator and 6 denominator)

Since $F = 2.05 < F_{.05} = 5.99$, factor A is not significant
Since $F = 4.06 < F_{.05} = 5.14$, factor B is not significant
Since $F = 7.66 > F_{.05} = 5.14$, interaction is significant

CHAPTER 13

1. a.

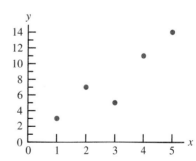

b. There appears to be a linear relationship between x and y

c. Many different straight lines can be drawn to provide a linear approximation of the relationship between x and y; in part (d) we will determine the equation of a straight line that "best" represents the relationship according to the least squares criterion

d. $\Sigma x_i = 15$, $\Sigma y_i = 40$, $\Sigma x_i y_i = 146$, $\Sigma x_i^2 = 55$

$$b_1 = \frac{\Sigma x_i y_i - (\Sigma x_i \Sigma y_i)/n}{\Sigma x_i^2 - (\Sigma x_i)^2/n}$$

$$= \frac{146 - (15)(40)/5}{55 - (15)^2/5} = 2.6$$

$$b_0 = \bar{y} - b_1 \bar{x}$$

$$= 8 - 2.6(3) = .2$$

$$\hat{y} = .2 + 2.6x$$

e. $\hat{y} = .2 + 2.6x = .2 + 2.6(4) = 10.6$

4. a.

b. It indicates there may be a linear relationship between the variables

c. Many different straight lines can be drawn to provide a linear approximation of the relationship between x and y; in part (d) we will determine the equation of a straight line that "best" represents the relationship according to the least squares criterion

d. $\Sigma x_i = 325$, $\Sigma y_i = 585$; $\Sigma x_i y_i = 38,135$, $\Sigma x_i^2 = 21,145$

$$b_1 = \frac{\Sigma x_i y_i - (\Sigma x_i \, \Sigma y_i)/n}{\Sigma x_i^2 - (\Sigma x_i)^2/n}$$

$$= \frac{38,135 - (325)(585)/5}{21,145 - (325)^2/5} = 5.5$$

$$b_0 = \bar{y} - b_1 \bar{x}$$

$$= 117 - 5.5(65) = -240.5$$

$$\hat{y} = -240.5 + 5.5x$$

e. $\hat{y} = -240.55.5(63) = 106$

The estimate of weight is 106 pounds

---

15. a. $\hat{y}_i = .2 + 2.6x$ and $\bar{y} = 8$

x_i	y_i	$\hat{y}_i$	$y_i - \hat{y}_i$	$(y_i - \hat{y}_i)^2$	$y_i - \bar{y}$	$(y_i - \bar{y})^2$
1	3	2.8	.2	.04	−5	25
2	7	5.4	1.6	2.56	−1	1
3	5	8.0	−3.0	9.00	−3	9
4	11	10.6	.4	.16	3	9
5	14	13.2	.8	.64	6	36
				SSE = 12.40		SST = 80

$$\text{SSR} = \text{SST} - \text{SSE} = 80 - 12.4 = 67.6$$

b. $r^2 = \dfrac{\text{SSR}}{\text{SST}} = \dfrac{67.6}{80} = .845$

The least squares line provided a very good fit; 84.5% of the variability in y has been explained by the least squares line

c. $\Sigma x_i = 15$, $\Sigma y_i = 40$, $\Sigma x_i y_i = 146$,
$$\Sigma x_i^2 = 55, \Sigma y_i^2 = 400$$

$$\text{SSR} = \frac{[\Sigma x_i y_i - (\Sigma x_i \Sigma y_i)/n]^2}{\Sigma x_i^2 - (\Sigma x_i)^2/n}$$

$$= \frac{[146 - (15)(40)/5]^2}{55 - (15)^2/5} = 67.6$$

$$\text{SST} = \Sigma y_i^2 - \frac{(\Sigma y_i)^2}{n}$$

$$= 400 - \frac{(40)^2}{5} = 80$$

Note that these are the same values shown in part (a)

18. a. $\Sigma x_i = 19.2$, $\Sigma y_i = 10,500$, $\Sigma x_i y_i = 34,030$
$$\Sigma x_i^2 = 62.18, \Sigma y_i^2 = 18,710,000$$

$$\text{SSR} = \frac{[\Sigma x_i y_i - (\Sigma x_i \Sigma y_i)/n]^2}{\Sigma x_i^2 - (\Sigma x_i)^2/n}$$

$$= \frac{[34,030 - (19.2)(10,500)/6]^2}{62.18 - (19.2)^2/6} = 249,864.86$$

$$\text{SST} = \Sigma y_i^2 - \frac{(\Sigma y_i)^2}{n}$$

$$= 18,710,000 - \frac{(10,500)^2}{6} = 335,000$$

$$\text{SSE} = \text{SST} - \text{SSR} = 335,000 - 249,864.86$$

$$= 85,135.14$$

b. $r^2 = \dfrac{\text{SSR}}{\text{SST}} = \dfrac{249{,}864.86}{335{,}000} = .746$

The least squares line accounted for 74.6% of the total sum of squares

22. a. $s^2 = \text{MSE} = \dfrac{\text{SSE}}{n-2} = \dfrac{12.4}{3} = 4.133$

b. $s = \sqrt{\text{MSE}} = \sqrt{4.133} = 2.033$

c. $\Sigma x_i = 15,\ \Sigma x_i^2 = 55$

$$s_{b_1} = \dfrac{s}{\sqrt{\Sigma x_i^2 - (\Sigma x_i)^2/n}}$$

$$= \dfrac{2.033}{\sqrt{55 - (15)^2/5}} = .643$$

d. $t = \dfrac{b_1 - \beta_1}{s_{b_1}} = \dfrac{2.6 - 0}{.643} = 4.04$

$t_{.025} = 3.182$ (3 degrees of freedom)

Since $t = 4.04 > t_{.05} = 3.182$, we reject $H_0\colon \beta_1 = 0$

e. $\text{MSR} = \dfrac{\text{SSR}}{1} = 67.6$

$$F = \dfrac{\text{MSR}}{\text{MSE}} = \dfrac{67.6}{4.133} = 16.36$$

$F_{.05} = 10.13$ (1 degree of freedom numerator and 3 denominator)

Since $F = 16.36 > F_{.05} = 10.13$, we reject $H_0\colon \beta_1 = 0$

25. a. $s^2 = \text{MSE} = \dfrac{\text{SSE}}{n-2} = \dfrac{85{,}135.14}{4} = 21{,}283.79$

$s = \sqrt{\text{MSE}} = \sqrt{21{,}283.79} = 145.89$

$\Sigma x_i = 19.2,\ \Sigma x_i^2 = 62.18$

$$s_{b_1} = \dfrac{s}{\sqrt{\Sigma x_i^2 - (\Sigma x_i)^2/n}}$$

$$= \dfrac{145.89}{\sqrt{62.18 - (19.2)^2/6}} = 169.59$$

$t = \dfrac{b_1 - \beta_1}{s_{b_1}} = \dfrac{581.08 - 0}{169.59} = 3.43$

$t_{.025} = 2.776$ (4 degrees of freedom)

Since $t = 3.43 > t_{.025} = 2.776$, we reject $H_0\colon \beta_1 = 0$

b. $\text{MSR} = \dfrac{\text{SSR}}{1} = \dfrac{249{,}864.86}{1} = 249{,}864.86$

$$F = \dfrac{\text{MSR}}{\text{MSE}} = \dfrac{249{,}864.86}{21{,}283.79} = 11.74$$

$F_{.05} = 7.71$ (1 degree of freedom numerator and 4 denominator)

Since $F = 11.74 > F_{.05} = 7.71$, we reject $H_0\colon \beta_1 = 0$

30. a. $s = 2.033$

$\Sigma x_i = 15,\quad \Sigma x_i^2 = 55$

$$s_{\hat{y}_p} = s\sqrt{\dfrac{1}{n} + \dfrac{(x_p - \bar{x})^2}{\Sigma x_i^2 - (\Sigma x_i)^2/n}}$$

$$= 2.033\sqrt{\dfrac{1}{5} + \dfrac{(4 - 3)^2}{55 - (15)^2/5}} = 1.11$$

b. $\hat{y} = .2 + 2.6x = .2 + 2.6(4) = 10.6$

$\hat{y}_p \pm t_{\alpha/2}s_{\hat{y}_p}$

$10.6 \pm 3.182(1.11)$

10.6 ± 3.53 or 7.07 to 14.13

c. $s_{\text{ind}} = s\sqrt{1 + \dfrac{1}{n} + \dfrac{(x_p - \bar{x})^2}{\Sigma x_i^2 - (\Sigma x_i)^2/n}}$

$$= 2.033\sqrt{1 + \dfrac{1}{5} + \dfrac{(4 - 3)^2}{55 - (15)^2/5}} = 2.32$$

d. $\hat{y}_p \pm t_{\alpha/2}s_{\hat{y}_p}$

$10.6 \pm 3.182(2.32)$

10.6 ± 7.38 or 3.22 to 17.98

34. $s = 145.89,\quad \Sigma x_i = 19.2,\quad \Sigma x_i^2 = 62.18$

$\hat{y} = -109.46 + 581.08x = -109.46 + 581.08(3)$

$\quad = 1633.78$

$$s_{\hat{y}_p} = s\sqrt{\dfrac{1}{n} + \dfrac{(x_p - \bar{x})^2}{\Sigma x_i^2 - (\Sigma x_i)^2/n}}$$

$$= 145.89\sqrt{\dfrac{1}{6} + \dfrac{(3 - 3.2)^2}{62.18 - (19.2)^2/6}} = 68.54$$

$\quad = 68.54$

$\hat{y}_p \pm t_{\alpha/2}s_{\hat{y}_p}$

$1633.78 \pm 2.776(68.54)$

1633.78 ± 190.27 or $\$1443.51$ to $\$1824.05$

36. $s = 145.89,\quad \Sigma x_i = 19.2,\quad \Sigma x_i^2 = 62.18$

$\hat{y} = -109.46 + 581.08x = -109.46 + 581.08(3)$

$\quad = 1633.78$

$$s_{\text{ind}} = s\sqrt{1 + \dfrac{1}{n} + \dfrac{(x_p - \bar{x})^2}{\Sigma x_i^2 - (\Sigma x_i)^2/n}}$$

$$= 145.89\sqrt{1 + \dfrac{1}{6} + \dfrac{(3 - 3.2)^2}{62.18 - (19.2)^2/6}} = 161.19$$

$\hat{y}_p \pm t_{\alpha/2}s_{\text{ind}}$

$1633.78 \pm 2.776(161.19)$

1633.78 ± 447.46 or $\$1186.32$ to $\$2081.24$

39.

Source of Variation	Sum of Squares	Degrees of Freedom	Mean Square	F
Regression	67.6	1	67.6	16.36
Error	12.4	3	4.133	
Total	80.0	4		

42. a. 9

b. $\hat{y} = 20.0 + 7.21x$

c. 1.3626

d. $\text{SSE} = \text{SST} - \text{SSR}$
$$= 51{,}984.1 - 41{,}587.3 = 10{,}396.8$$
$$\text{MSE} = 10{,}396.8/7 = 1485.3$$
$$F = \frac{\text{MSR}}{\text{MSE}} = \frac{41{,}587.3}{1485.3} = 28.0$$
$F_{.05} = 5.59$ (1 degree of freedom numerator and 7 denominator)
Since $F = 28 > F_{.05} = 5.59$, we reject $H_0: \beta_1 = 0$

e. $\hat{y} = 20.0 + 7.21(50) = 380.5$ or $\$380{,}500$

47. a. $\Sigma x_i = 70$, $\Sigma y_i = 76$, $\Sigma x_i y_i = 1264$, $\Sigma x_i^2 = 1106$

$$b_1 = \frac{\Sigma x_i y_i - (\Sigma x_i \, \Sigma y_i)/n}{\Sigma x_i^2 - (\Sigma x_i)^2/n}$$

$$= \frac{1264 - (70)(65)/5}{1106 - (70)^2/5} = 1.5873$$

$$b_0 = \bar{y} - b_1 \bar{x}$$
$$= 15.2 - 1.5873(14) = -7.0222$$

$$\hat{y} = -7.02 + 1.59x$$

b.

x_i	y_i	$\hat{y}_i$	$y_i - \hat{y}_i$
6	6	2.52	3.48
11	8	10.47	-2.47
15	12	16.83	-4.83
18	20	21.60	-1.60
20	30	24.78	5.22

c.

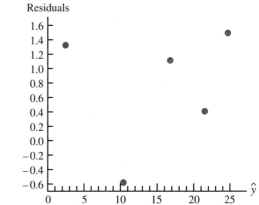

With only five observations, it is difficult to determine whether the assumptions are satisfied; however, the plot does suggest curvature in the residuals, which would indicate that the error term assumptions are not satisfied; the scatter diagram for these data also indicates that the underlying relationship between x and y may be curvilinear

d. $s^2 = 23.78$

$$h_i = \frac{1}{n} + \frac{(x_i - \bar{x})^2}{\Sigma x_i^2 - (\Sigma x_i)^2/n}$$

$$= \frac{1}{5} + \frac{(x_i - 14)^2}{1106 - (70)^2/5} = \frac{1}{5} + \frac{(x_i - 14)^2}{126}$$

x_i	h_i	$s_{y_i - \hat{y}_i}$	$y_i - \hat{y}_i$	Standardized Residuals
6	.7079	2.64	3.48	1.32
11	.2714	4.16	-2.47	-.59
15	.2079	4.34	-4.83	-1.11
18	.3270	4.00	-1.60	-.40
20	.4857	3.50	5.22	1.49

e. Standardized Residuals

The plot of the standardized residuals against $\hat{y}$ has the same shape as the original residual plot; as stated in part (c), the curvature observed indicates that the assumptions regarding the error term may not be satisfied

49. $\Sigma x_i = 57$, $\Sigma y_i = 294$, $\Sigma x_i y_i = 2841$,

$\Sigma x_i^2 = 753$, $\Sigma y_i^2 = 13{,}350$

$$b_1 = \frac{\Sigma x_i y_i - (\Sigma x_i \Sigma y_i)/n}{\Sigma x_i^2 - (\Sigma x_i)^2/n}$$

$$= \frac{2841 - (57)(294)/7}{753 - (57)^2/7} = 1.5475$$

$b_0 = \bar{y} - b_1\bar{x}$

$= 42 - (1.5475)(8.1492) = 29.3989$

$\hat{y} = 29.40 + 1.55x$

b. From Exercise 19 we have SSR = 691.72 and SST = 1002; therefore SSE = 1002 − 691.72 = 310.28

$$F = \frac{MSR}{MSE} = \frac{691.72}{310.28/5} = 11.15$$

$F_{.05} = 6.61$ (1 degree of freedom numerator and 5 denominator)

Since $F = 11.47 > F_{.05} = 6.61$, we reject H_0: $\beta_1 = 0$; the relationship is significant at the .05 level

c.

x_i	y_i	$\hat{y}_i = 29.40 + 1.55x_i$	$y_i - \hat{y}_i$
1	19	30.95	−11.95
2	32	32.50	−.50
4	44	35.60	8.40
6	40	38.70	1.30
10	52	44.90	7.10
14	53	51.10	1.90
20	54	60.40	−6.40

d. The residuals plot shown below leads us to question the assumption of a linear relationship between x and y; even though the relationship is significant at the $\alpha = .05$ level, it would be extremely dangerous to extrapolate beyond the range of the data (e.g., $x > 20$)

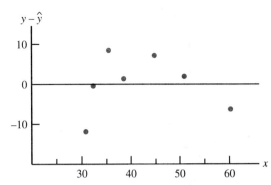

52. a. Using Minitab, we obtained the estimated regression equation $\hat{y} = 66.1 + .4023x$; a portion of the Minitab output is shown in Figure G13.52a

The fitted values and standardized residuals are shown below:

x_i	y_i	$\hat{y}_i$	Standardized Residuals
135	145	120.41	2.11
110	100	110.35	−1.08
130	120	118.40	.14
145	120	124.43	−.38
175	130	136.50	−.78
160	130	130.47	−.04
120	110	114.38	−.41

FIGURE G13.52a

```
Predictor      Coef      Stdev     t-ratio        p
Constant       66.10     32.06        2.06    0.094
X              0.4023    0.2276       1.77    0.137

s = 12.62      R-sq = 38.5%     R-sq(adj) = 26.1%

Analysis of Variance

SOURCE         DF          SS          MS        F        p
Regression     1        497.2       497.2     3.12    0.137
Error          5        795.7       159.1
Total          6       1292.9

Unusual Observations
Obs.        X          Y      Fit Stdev.Fit  Residual   St.Resid
  1       135     145.00   120.42      4.87     24.58      2.11R
```

b. Standardized Residuals

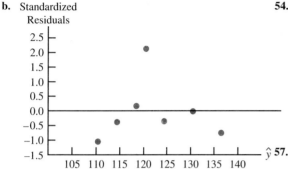

The standardized residual plot indicates that the observation $x = 135$, $y = 145$ may be an outlier; note that this observation has a standardized residual of 2.11

c. The scatter diagram is shown below:

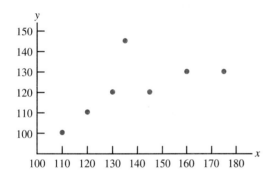

The scatter diagram also indicates that the observation $x = 135$, $y = 145$ may be an outlier; the implication is that for simple linear regression outliers can be identified by looking at the scatter diagram

54. a. Using Minitab, we obtained the estimated regression equation $\hat{y} = -.23 + .049x$

b. A portion of the Minitab output is shown in Figure G13.54b.

Note that Minitab identifies observation 10 as an influential observation; the standardized residual plot for these data also shows a very unusual trend in the residuals, an indication that the assumptions for ϵ may not be satisfied for these data

57. a.

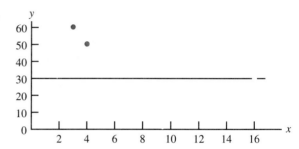

b. There appears to be a negative linear relationship between x and y

c.

x_i	y_i	$x_i - \bar{x}$	$y_i - \bar{y}$	$(x_i - \bar{x})(y_i - \bar{y})$
4	50	-4	4	-16
6	50	-2	4	-8
11	40	3	-6	-18
3	60	-5	14	-70
16	30	8	-16	-128
40	230	0	0	-240

$\bar{x} = 8$; $\bar{y} = 46$

$$s_{xy} = \frac{\Sigma(x_i - \bar{x})(y_i - \bar{y})}{n - 1} = \frac{-240}{4} = -60$$

FIGURE **G13.54b**

```
Predictor      Coef       Stdev     t-ratio        p
Constant     -0.2303     0.1138      -2.02      0.078
X           0.048987   0.002769      17.69      0.000

s = 0.1086      R-sq = 97.5%      R-sq(adj) = 97.2%

Analysis of Variance

SOURCE          DF         SS          MS          F         p
Regression       1      3.6946      3.6946     313.04     0.000
Error            8      0.0944      0.0118
Total            9      3.7890

Unusual Observations
Obs.      X          Y       Fit Stdev.Fit   Residual    St.Resid
  10    67.0     3.0000    3.0518    0.0843    -0.0518     -0.76 X
```

The sample covariance indicates a negative linear association between x and y

d. $r_{xy} = \dfrac{s_{xy}}{s_x s_y} = \dfrac{-60}{(5.43)(11.40)} = -.97$

The sample correlation coefficient of $-.97$ is indicative of a reasonably strong negative linear relationship

CHAPTER 14

1. a. $b_1 = .5906$ is an estimate of the change in y corresponding to a 1-unit change in x_1 when x_2 is held constant

   $b_2 = .4980$ is an estimate of the change in y corresponding to a 1-unit change in x_2 when x_1 is held constant

6. a. The Minitab output is shown in Figure G14.6a

   b. No, it is 1.60 in 5a and in Figure G14.6a; in this exercise it represents the marginal change in revenue due to an increase in television advertising with newspaper advertising held constant

12. a. $R^2 = \dfrac{SSR}{SST} = \dfrac{14,052.2}{15,182.9} = .926$

    b. $R_a^2 = 1 - (1 - R^2)\dfrac{n-1}{n-p-1}$

    $= 1 - (1 - .926)\dfrac{10-1}{10-2-1} = .905$

    c. Yes; after adjusting for the number of independent variables in the model, we see that 90.5% of the variability in y has been accounted for

14. a. $R^2 = \dfrac{SSR}{SST} = \dfrac{12,000}{16,000} = .75$

b. $R_a^2 = 1 - (1 - R^2)\left(\dfrac{n-1}{n-p-1}\right)$

   $= 1 - (.25)\left(\dfrac{9}{7}\right)$

   $= .68$

c. The adjusted coefficient of determination shows that 68% of the variability has been explained by the two independent variables; thus, we conclude that the model does explain a large amount of the variability

17. a. $MSR = \dfrac{SSR}{p} = \dfrac{6216.375}{2} = 3108.188$

    $MSE = \dfrac{SSE}{n-p-1} = \dfrac{507.75}{10-2-1} = 72.536$

    b. $F = \dfrac{MSR}{MSE} = \dfrac{3108.188}{72.536} = 42.85$

    $F_{.05} = 4.74$ (2 degrees of freedom numerator and 7 denominator)

    Since $F = 42.85 > F_{.05} = 4.74$, the overall model is significant

    c. $t = \dfrac{b_1}{s_{b_1}} = \dfrac{.5906}{.0813} = 7.26$

    $t_{.025} = 2.365$ (7 degrees of freedom)
    Since $t = 7.26 > t_{.025} = 2.365$, β_1 is significant

    d. $t = \dfrac{b_2}{s_{b_2}} = \dfrac{.4980}{.0567} = 8.78$

    Since $t = 8.78 > t_{.025} = 2.365$, β_2 is significant

20. a. $SSE = SST - SSR = 16000 - 12000 = 4000$

    $s^2 = \dfrac{SSE}{n-p-1} = \dfrac{4000}{7} = 571.43$

    $MSR = \dfrac{SSR}{p} = \dfrac{12,000}{2} = 6000$

FIGURE G14.6a

```
The regression equation is
REVENUE = 83.2 + 2.29 TVADV + 1.30 NEWSADV


Predictor      Coef       Stdev     t-ratio        p
Constant     83.230       1.574       52.88    0.000
TVADV        2.2902       0.3041        7.53    0.001
NEWSADV      1.3010       0.3207        4.06    0.010


s = 0.6426      R-sq = 91.9%      R-sq(adj) = 88.7%


Analysis of Variance


SOURCE          DF          SS          MS         F        p
Regression       2      23.435      11.718     28.38    0.002
Error            5       2.065       0.413
Total            7      25.500
```

b. $F = \dfrac{MSR}{MSE} = \dfrac{6000}{571.43} = 10.50$

$F_{.05} = 4.74$ (2 degrees of freedom numerator and 7 denominator)

Since $F = 10.50 > F_{.05} = 4.74$, reject H_0; there is a significant relationship among the variables

26. a. Using Minitab, the 95% confidence interval is 132.16 to 154.15

b. Using Minitab, the 95% prediction interval is 111.15 to 175.17

28. a. Using Minitab, the 95% confidence interval is 71.13 to 112.35; in terms of units: 71,130 to 112,350

b. Using Minitab, the 95% prediction interval is 42.87 to 140.61; in terms of units: 42,870 to 140,610

This prediction interval estimate would probably not be of interest to Heller since they are a manufacturer of mowers and not a retailer

30. a. $E(y) = \beta_0 + \beta_1 x_1 + \beta_2 x_2$

where $x_2 = \begin{cases} 0 \text{ if level 1} \\ 1 \text{ if level 2} \end{cases}$

b. $E(y) = \beta_0 + \beta_1 x_1 + \beta_2(0) = \beta_0 + \beta_1 x_1$

c. $E(y) = \beta_0 + \beta_1 x_1 + \beta_2(1) = \beta_0 + \beta_1 x_1 + \beta_2$

d. $\beta_2 = E(y \mid \text{level 2}) - E(y \mid \text{level 1})$

β_1 is the change in $E(y)$ for a 1-unit change in x_1 holding x_2 constant

32. a. $15,300, since $b_3 = 15.3$

b. $\hat{y} = 10.1 - 4.2(2) + 6.8(8) + 15.3(0)$

$= 10.1 - 8.4 + 54.4$

$= 56.1$

Sales prediction: $56,100

c. $\hat{y} = 10.1 - 4.2(1) + 6.8(3) + 15.3(1)$

$= 10.1 - 4.2 + 20.4 + 15.3$

$= 41.6$

Sales prediction: $41,600

37. a. The Minitab output is shown in Figure G14.37a

b. Using Minitab, we obtained the following values:

x_i	y_i	$\hat{y}_i$	Standardized Residual
1	3	2.8	.16
2	7	5.4	.94
3	5	8.0	−1.65
4	11	10.6	.24
5	14	13.2	.62

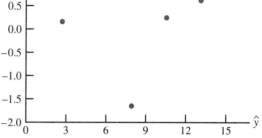

The point (3,5) does not appear to follow the trend of the remaining data; however, the value of the standardized residual for this point, −1.65, is not large enough for us to conclude that (3,5) is an outlier

FIGURE G14.37a

```
The regression equation is
Y = 0.20 + 2.60 X

Predictor       Coef      Stdev     t-ratio       p
Constant       0.200      2.132       0.09     0.931
X             2.6000     0.6429       4.04     0.027

s = 2.033      R-sq = 84.5%     R-sq(adj) = 79.3%

Analysis of Variance

SOURCE         DF         SS         MS        F        p
Regression      1     67.600     67.600    16.35    0.027
Error           3     12.400      4.133
Total           4     80.000
```

c. Using Minitab, we obtained the following values:

x_i	y_i	Studentized Deleted Residual
1	3	.13
2	7	.92
3	5	−4.42
4	11	.19
5	14	.54

$t_{.025} = 4.303 \ (n - p - 2 = 5 - 1 - 2 =$

2 degrees of freedom)

Since the studentized deleted residual for (3,5) is −4.42 < −4.303, we conclude that the 3rd observation is an outlier

39. a. The Minitab output appears in Figure G14.6a; the estimated regression equation is

$$\text{REVENUE} = 83.2 + 2.29 \ \text{TVADV}$$
$$+ \ 1.30 \ \text{NEWSADV}$$

b. Using Minitab, we obtained the following values:

$\hat{y}_i$	Standardized Residual	$\hat{y}_i$	Standardized Residual
96.63	−1.62	94.39	1.10
90.41	−1.08	94.24	−.40
94.34	1.22	94.42	−1.12
92.21	−.37	93.35	1.08

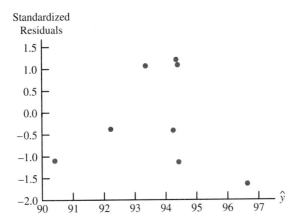

Standardized Residuals

With relatively few observations, it is difficult to determine if any of the assumptions regarding ϵ have been violated; for instance, an argument could be made that there does not appear to be any pattern in the plot; alternatively, an argument could be made that there is a curvilinear pattern in the plot

c. The values of the standardized residuals are greater than −2 and less than +2; thus, using this test, there are no outliers

As a further check for outliers, we used Minitab to compute the following studentized deleted residuals:

Observation	Studentized Deleted Residual	Observation	Studentized Deleted Residual
1	−2.11	5	1.13
2	−1.10	6	−.36
3	1.31	7	−1.16
4	−.33	8	1.10

$t_{.025} = 2.776 \ (n - p - 2 = 8 - 2 - 2 =$

4 degrees of freedom)

Since none of the studentized deleted residuals is less than −2.776 or greater than 2.776, we conclude that there are no outliers in the data

d. Using Minitab, we obtained the following values:

Observation	h_i	D_i
1	.63	1.52
2	.65	.70
3	.30	.22
4	.23	.01
5	.26	.14
6	.14	.01
7	.66	.81
8	.13	.06

The critical leverage value is

$$\frac{3(p + 1)}{n} = \frac{3(2 + 1)}{8} = 1.125$$

Since none of the values exceed 1.125, we conclude that there are no influential observations

However, using Cook's distance measure, we see that $D_1 > 1$ (rule of thumb critical value); thus, we conclude the first observation is influential

Final conclusion: observation 1 is an influential observation

CHAPTER 15

1. a. The Minitab output is shown in Figure G15.1a
b. Since the p-value corresponding to $F = 6.85$ is .059 > $\alpha = .05$, the relationship is not significant

FIGURE G15.1a

```
The regression equation is
Y = - 6.8 + 1.23 X

Predictor        Coef        Stdev      t-ratio        p
Constant        -6.77        14.17        -0.48      0.658
X              1.2296        0.4697        2.62      0.059

s = 7.269        R-sq = 63.1%      R-sq(adj) = 53.9%

Analysis of Variance

SOURCE         DF          SS          MS          F          p
Regression      1       362.13      362.13       6.85      0.059
Error           4       211.37       52.84
Total           5       573.50
```

c.

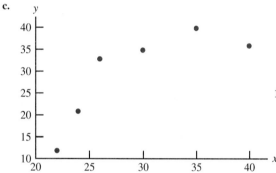

The scatter diagram suggests that a curvilinear relationship may be appropriate

d. The Minitab output is shown in Figure F15.1d

e. Since the p-value corresponding to $F = 25.68$ is $.013 < \alpha = .05$, the relationship is significant

f. $\hat{y} = -168.88 + 12.187(25) - .17704(25)^2 = 25.145$

5. a. The Minitab output is shown in Figure F15.5a

b. Since the relationship between y and x was significant in Exercise 4, the relationship which includes x^2 must also be significant; looking at the p-value, we see that $.003 < .01$, and thus we would reject $H_0: \beta_1 = \beta_2 = 0$

c. See Figure G15.5c

11. a. $R^2 = \dfrac{\text{SSR}}{\text{SST}} = \dfrac{1760}{1805} = .975$

b. $R_a^2 = 1 - (1 - R^2)\left(\dfrac{n - 1}{n - p - 1}\right)$

$= 1 - (.025)\left(\dfrac{29}{25}\right)$

$= .971$

c. SSE $= 1805 - 1760 = 45$

$F = \dfrac{\text{MSR}}{\text{MSE}} = \left(\dfrac{1760/4}{45/25}\right) = 244.44$

$F_{.05} = 2.76$ (4 degrees of freedom numerator and 25 denominator)

Since $244.44 > 2.76$, reject H_0, and conclude the relationship is significant

12. a. The Minitab output is shown in Figure G15.12a

b. The Minitab output is shown in Figure G15.12b

c. $F = \dfrac{[\text{SSE(reduced)} - \text{SSE(full)}]/(\text{# extra terms})}{\text{MSE(full)}}$

$= \dfrac{(23,157 - 14,317)/2}{1432} = 3.09$

$F_{.05} = 4.10$ (2 degrees of freedom numerator and 10 denominator)

Since $F = 3.09 < F_{.05} = 4.10$, the addition of the two independent variables is not significant

Note: Suppose that we consider adding only the number of interceptions made by the opponents; the corresponding Minitab output is shown in Figure G15.12c; in this case,

$F = \dfrac{(23,157 - 14,335)/1}{1303} = 6.77$

$F_{.05} = 4.84$ (1 degree of freedom numerator and 11 denominator)

Since $F = 6.77 > F_{.05} = 4.84$, the addition of the number of interceptions made by the opponents is significant

15. a. The Minitab output is shown in Figure G15.15a

b. Stepwise procedure (see Figure G15.15b)

c. Backward elimination procedure (see Figure G15.15c)

d. Best subsets regression of %COLLEGE (see Figure G15.15d)

FIGURE G15.1d

```
The regression equation is
Y = - 169 + 12.2 X - 0.177 XSQ

Predictor        Coef        Stdev      t-ratio          p
Constant       -168.88       39.79       -4.24       0.024
X               12.187        2.663        4.58       0.020
XSQ            -0.17704      0.04290      -4.13       0.026

s = 3.248        R-sq = 94.5%      R-sq(adj) = 90.8%

Analysis of Variance

SOURCE         DF          SS          MS          F          p
Regression      2        541.85      270.92      25.68      0.013
Error           3         31.65       10.55
Total           5        573.50
```

FIGURE G15.5a

```
The regression equation is
Y = 433 + 37.4 X -0.383 XSQ

Predictor        Coef        Stdev      t-ratio          p
Constant        432.6        141.2        3.06       0.055
X               37.429        7.807        4.79       0.017
XSQ            -0.3829       0.1036       -3.70       0.034

s = 15.83        R-sq = 98.0%      R-sq(adj) = 96.7%

Analysis of Variance

SOURCE         DF          SS          MS          F          p
Regression      2        36643       18322       73.15      0.003
Error           3         751         250
Total           5        37395
```

FIGURE G15.5c

```
    Fit    Stdev.Fit            95% C.I.                95% P.I.
 1302.01        9.93      (1270.41, 1333.61)     (1242.55, 1361.47)
```

FIGURE G15.12a

```
The regression equation is
POINTS = 170 + 6.61 TEAMINT

Predictor        Coef        Stdev      t-ratio         p
Constant       170.13       44.02         3.86      0.002
TEAMINT          6.613       2.258         2.93      0.013

s = 43.93        R-sq = 41.7%     R-sq(adj) = 36.8%

Analysis of Variance

SOURCE         DF           SS           MS          F         p
Regression      1         16546        16546       8.57     0.013
Error          12         23157         1930
Total          13         39703

Unusual Observations
Obs. TEAMINT    POINTS       Fit Stdev.Fit  Residual   St.Resid
 13     33.0     340.0     388.4      34.2     -48.4      -1.75 X

X denotes an obs. whose X value gives it large influence.
```

FIGURE G15.12b

```
The regression equation is
POINTS = 280 + 5.18 TEAMINT - 0.0037 RUSHING - 3.92 OPPONINT

Predictor        Coef        Stdev      t-ratio         p
Constant       280.34       81.42         3.44      0.006
TEAMINT          5.176       2.073         2.50      0.032
RUSHING       -0.00373     0.03336        -0.11      0.913
OPPONINT        -3.918       1.651        -2.37      0.039

s = 37.84        R-sq = 63.9%     R-sq(adj) = 53.1%

Analysis of Variance

SOURCE         DF           SS           MS          F         p
Regression      3         25386         8462       5.91     0.014
Error          10         14317         1432
Total          13         39703

SOURCE         DF       SEQ SS
Regression      1        16546
Error           1          776
Total           1         8064
```

```
The regression equation is
POINTS = 274 + 5.23 TEAMINT - 3.96 OPPONINT

Predictor         Coef        Stdev      t-ratio          p
Constant        273.77        53.81         5.09      0.000
TEAMINT          5.227        1.931         2.71      0.020
OPPONINT        -3.965        1.524        -2.60      0.025

s = 36.10       R-sq = 63.9%     R-sq(adj) = 57.3%

Analysis of Variance

SOURCE          DF          SS           MS          F          p
Regression       2        25386        12684       9.73      0.004
Error           11        14335         1303
Total           13        39703

SOURCE          DF       SEQ SS
TEAMINT          1        16546
OPPONINT         1         8822
```

```
The regression equation is
%COLLEGE = -26.6 + 0.0970 SATSCORE

Predictor         Coef        Stdev      t-ratio          p
Constant        -26.61        37.22        -0.72      0.485
SATSCORE       0.09703      0.03734         2.60      0.019

s = 12.83       R-sq = 29.7%     R-sq(adj) = 25.3%

Analysis of Variance

SOURCE          DF          SS           MS          F          p
Regression       1       1110.8       1110.8       6.75      0.019
Error           16       2632.3        164.5
Total           17       3743.1
```

```
STEP                   1            2
CONSTANT          -26.61       -26.93

SATSCORE           0.097        0.084
t-RATIO             2.60         2.46

%TAKESAT                        0.204
t-RATIO                          2.21

s                   12.8         11.5
R-sq               29.68        46.93
```

FIGURE G15.15c

```
STEP                   1          2          3          4
CONSTANT             33.71      17.46     -32.47     -26.93

SIZE                 -1.56      -1.39
t-RATIO              -1.43      -1.42

STUDENT$           -0.0024    -0.0026    -0.0019
t-RATIO              -1.47      -1.75      -1.31

SALARY            -0.00026
t-RATIO              -0.40

SATSCORE             0.077      0.081      0.095      0.084
t-RATIO               2.06       2.36       2.77       2.46

%TAKESAT             0.285      0.274      0.291      0.204
t-RATIO               2.47       2.53       2.60       2.21

s                     11.2       10.9       11.2       11.5
R-sq                 59.65      59.10      52.71      46.93
```

FIGURE G15.15d

```
                                  S   S %
                                  T   A T
                                  U   S T A
                                  D   A S K
                              S   E   L C E
                              I   N   A O S
                  Adj.        Z   T   R R A
Vars   R-sq   R-sq      S     E   $   Y E T

  1    29.7   25.3   12.826            X
  1    25.5   20.8   13.203              X
  2    46.9   39.9   11.508            X X
  2    38.2   30.0   12.417    X       X
  3    52.7   42.6   11.244        X   X X
  3    49.5   38.7   11.618    X       X X
  4    59.1   46.5   10.852    X X     X X
  4    52.8   38.3   11.660        X X X X
  5    59.6   42.8   11.219    X X X X X
```

FIGURE **G15.21a**

```
The regression equation is
P/E = 6.51 + 0.569 %PROFIT

Predictor        Coef        Stdev      t-ratio         p
Constant        6.507        1.509        4.31       0.000
%PROFIT        0.5691       0.1281        4.44       0.000

s = 2.580        R-sq = 53.7%      R-sq(adj) = 51.0%

Analysis of Variance

SOURCE          DF          SS          MS          F          p
Regression       1        131.40      131.40      19.74      0.000
Error           17        113.14        6.66
Total           18        244.54
```

21. a. The Minitab output is shown in Figure F15.21a
  b. Residual plot as a function of the order in which the data are presented is shown below; there does not appear to be any pattern indicative of positive autocorrelation

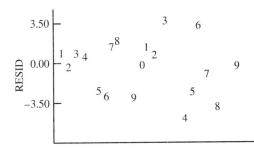

  c. The Durbin-Watson statistic (obtained from Minitab) is $d = 2.34$; at $\alpha = .05$, $d_L = 1.18$ and $d_U = 1.39$; since $d > d_U$, there is no significant positive autocorrelation

23.

x_1	x_2	x_3	Treatment
0	0	0	A
1	0	0	B
0	1	0	C
0	0	1	D

$$E(y) = \beta_0 + \beta_1 x_1 + \beta_2 x_2 + \beta_3 x_3$$

26. a.

D_1	D_2	Manufacturer
0	0	1
1	0	2
0	1	3

$$E(y) = \beta_0 + \beta_1 D_1 + \beta_2 D_2$$

  b. See Figure G15.26b
  c. H_0: $\beta_1 = \beta_2 = 0$
  d. Since the p-value is .004 $< \alpha = .05$, we conclude that the mean time to mix a batch of material is not the same for each manufacturer

CHAPTER 16

1. *Expected frequencies:* $e_1 = 200(.40) = 80$,
$$e_2 = 200(.40) = 80,$$
$$e_3 = 200(.20) = 40$$
Actual frequencies: $f_1 = 60, f_2 = 120, f_3 = 20$

$$\chi^2 = \frac{(60 - 80)^2}{80} + \frac{(120 - 80)^2}{80} + \frac{(20 - 40)^2}{40}$$
$$= \frac{400}{80} + \frac{1600}{80} + \frac{400}{40}$$
$$= 5 + 20 + 10 = 35$$

$\chi^2_{.01} = 9.21034$ with $k - 1 = 3 - 1 = 2$ degrees of freedom

Since $\chi^2 = 35 > 9.21034$, reject the null hypothesis; that is, the population proportions are not as stated in the null hypothesis

FIGURE **G15.26b**

```
The regression equation is
Time = 23.0 + 5.00 D1 - 2.00 D2

Predictor        Coef       Stdev      t-ratio         p
Constant       23.000      1.106        20.80      0.000
D1              5.000      1.563         3.20      0.011
D2             -2.000      1.563        -1.28      0.233

s = 2.211        R-sq = 70.3%      R-sq(adj) = 63.7%

Analysis of Variance

SOURCE          DF           SS           MS          F         p
Regression       2      104.000       52.000      10.64     0.004
Error            9       44.000        4.889
Total           11      148.000
```

3. *Expected frequencies:* $300(.29) = 87, 300(.28) = 84,$
$\qquad\qquad 300(.25) = 75, 300(.18) = 54$

$e_1 = 87, e_2 = 84, e_3 = 75, e_4 = 54$

Actual frequencies: $f_1 = 95, f_2 = 70, f_3 = 89, f_4 = 46$

$\chi^2_{.05} = 7.81$ (3 degrees of freedom)

$$\chi^2 = \frac{(95-87)^2}{87} + \frac{(70-84)^2}{84}$$

$$+ \frac{(89-75)^2}{75} + \frac{(46-54)^2}{54} = 6.87$$

Do not reject H_0; there is no significant change in the viewing audience proportions

11. H_0: The column factor is independent of the row factor
H_a: The column factor is not independent of the row factor

Expected frequencies:

	A	B	C
P	28.5	39.9	45.6
Q	21.5	30.1	34.4

$$\chi^2 = \frac{(20-28.5)^2}{28.5} + \frac{(44-39.9)^2}{39.9} + \frac{(50-45.6)^2}{45.6}$$

$$+ \frac{(30-21.5)^2}{21.5} + \frac{(26-30.1)^2}{30.1} + \frac{(30-34.4)^2}{34.4}$$

$$= 7.86$$

$\chi^2_{.025} = 7.37776$ with $(2-1)(3-1) = 2$ degrees of freedom

Since $\chi^2 = 7.86 > 7.37776$, reject H_0; that is, conclude that the column factor is not independent of the row factor

13. H_0: There is no difference in shooting percentage among the teams
H_a: There is a difference in shooting percentage among the teams

Row 1 total = 629; row 2 total = 988
Column 1 total = 374; column 2 total = 341
Column 3 total = 369; column 4 total = 533

Overall total = 1617

Using these totals, we compute the expected frequencies

Expected frequencies:

	Duke	**Mich.**	**Ind.**	**Cin.**
Made	145.4830	132.6463	143.5380	207.3327
Missed	228.5170	208.3537	225.4620	325.6673

$$\chi^2 = \frac{(160-145.4830)^2}{145.4830} + \frac{(113-132.6463)^2}{132.6463} + \cdots$$

$$+ \frac{(331-325.6673)^2}{325.6673}$$

$$= 1.4486 + 2.9098 + .7625 + .1372 + .9222$$

$$+ 1.8525 + .4855 + .0873 = 8.6056$$

$\chi^2_{.05} = 7.81473$ with 3 degrees of freedom

Since $\chi^2 = 8.6056 > 7.81473$, reject H_0; that is, conclude that there is a difference in 3-point shooting ability for the teams

CHAPTER 17

1. Binomial probabilities for $n = 10, p = .50$

x	Probability	x	Probability
0	.0010	6	.2051
1	.0098	7	.1172
2	.0439	8	.0439
3	.1172	9	.0098
4	.2051	10	.0010
5	.2461		

$P(0)P(1) = .0108$; Adding $P(2)$, exceeds .025 required in the tail; therefore, reject H_0 if the number of plus signs is less than 2 or greater than 8;

Number of plus signs is 7

Do not reject H_0; conclude that there is no indication that a difference exists

2. $n = 27$ cases in which a value different from 150 is obtained

Use normal approximation with $\mu = np = .5(27) = 13.5$ and $\sigma = \sqrt{.25n} = \sqrt{.25(27)} = 2.6$

Use $x = 22$ as the number of plus signs and obtain the following test statistic:

$$z = \frac{x - \mu}{\sigma} = \frac{22 - 13.5}{2.6} = 3.27$$

With $\alpha = .01$, we reject if $z > 2.33$;

Since $z = 3.27 > 2.33$, reject H_0 and, conclude the median is greater than 150

4. We need to determine the number of "better" responses and the number of "worse" responses; the sum of the two is the sample size used for the study

$$n = .34(1253) + .29(1253) = 789.4$$

Use the large-sample test using the normal distribution; this means the value of $n(n = 789.4$ above) need not be integer. Use

$$\mu = .5n = .5(789.4) = 394.7$$

$$\sigma = \sqrt{.25n} = \sqrt{.25(394.7)} = 9.93$$

Let: $p =$ proportion of adults who feel children will have a better future

$H_0: p \leq .50$

$H_a: p > .50$

$$x = .34(1253) = 426.0$$

$$z = \frac{x - \mu}{\sigma} = \frac{426.0 - 394.7}{9.93} = 3.15$$

With $\alpha = .05$, we reject if $z > 1.645$;

Since $z = 3.15 > 1.645$, reject H_0 and, conclude that more than half of the adults feel their children will have a better future

12. H_0: The populations are identical

H_a: The populations are not identical

Additive			Absolute		Signed
1	*2*	Difference	Value	Rank	Rank
20.12	18.05	2.07	2.07	9	+9
23.56	21.77	1.79	1.79	7	+7
22.03	22.57	−.54	.54	3	−3
19.15	17.06	2.09	2.09	10	+10
21.23	21.22	.01	.01	1	+1
24.77	23.80	.97	.97	4	+4
16.16	17.20	−1.04	1.04	5	−5
18.55	14.98	3.57	3.57	12	+12
21.87	20.03	1.84	1.84	8	+8
24.23	21.15	3.08	3.08	11	+11
23.21	22.78	.43	.43	2	+2
25.02	23.70	1.32	1.32	6	+6
					$T = 62$

$$\mu_T = 0$$

$$\sigma_T = \sqrt{\frac{n(n+1)(2n+1)}{6}} = \sqrt{\frac{12(13)(25)}{6}} = 25.5$$

$$z = \frac{T - \mu_T}{\sigma_T} = \frac{62 - 0}{25.5} = 2.43$$

Two-tailed test; reject H_0 if $z < -1.96$ or $z > 1.96$

Since $z = 2.43 > 1.96$, reject H_0 and, conclude that there is a significant difference in the additives

13.

Without Relaxant	With Relaxant	Difference	Rank of Absolute Difference	Signed Rank
15	10	5	9	9
12	10	2	3	3
22	12	10	10	10
8	11	−3	6.5	−6.5
10	9	1	1	1
7	5	2	3	3
8	10	−2	3	−3
10	7	3	6.5	6.5
14	11	3	6.5	6.5
9	6	3	6.5	6.5
				$T = 36$

$\mu_T = 0$

$$\sigma_T = \sqrt{\frac{n(n+1)(2n+1)}{6}} = \sqrt{\frac{10(11)(21)}{6}} = 19.62$$

$$z = \frac{T - \mu_T}{\sigma_T} = \frac{36}{19.62} = 1.83$$

One-tailed test; reject H_0 if $z > 1.645$

Reject H_0; there is a significant difference in favor of the relaxant

18. Rank the combined samples and find rank sum for each sample; this is a small-sample test since $n_1 = 7$ and $n_2 = 9$

Additive 1		Additive 2	
MPG	Rank	MPG	Rank
17.3	2	18.7	8.5
18.4	6	17.8	4
19.1	10	21.3	15
16.7	1	21.0	14
18.2	5	22.1	16
18.6	7	18.7	8.5
17.5	3	19.8	11
	34	20.7	13
		20.2	12
			102

$T = 34$

With $\alpha = .05$, $n_1 = 7$, and $n_2 = 9$

$$T_L = 41 \text{ and } T_U = 7(7 + 9 + 1) - 41 = 78$$

Since $T = 34 < 41$, reject H_0; and, conclude that there is a significant difference in gasoline mileage

19. a. Rank the combined samples, and find rank sum for each sample; with $n_1 = 12$ and $n_2 = 12$, this is a large-sample case

H_0: There are no differences in the distribution of starting salaries

H_a: There is a difference between the distributions of starting salaries

We reject H_0 if $z < -1.96$ or $z > 1.96$

	Accounting		Finance	
	Salary	Rank	Salary	Rank
	28.8	20	26.3	13
	25.3	9	23.6	3
	26.2	11	25.0	7
	27.9	17.5	23.0	1
	27.0	15	27.9	17.5
	26.2	11	24.5	5
	28.1	19	29.0	21
	24.7	6	27.4	16
	25.2	8	23.5	2
	29.2	22	26.9	14
	29.7	24	26.2	11
	29.3	23	24.0	4
Totals	327.6	185.5	307.3	114.5

$$\mu_T = \tfrac{1}{2}n_1(n_1 + n_2 + 1) = \tfrac{1}{2}12(12 + 12 + 1) = 150$$

$$\sigma_T = \sqrt{\tfrac{1}{12}n_1 n_2(n_1 + n_2 + 1)} = \sqrt{\tfrac{1}{12}(12)(12)(25)} = 17.32$$

$T = 185.5$

$$z = \frac{T - \mu_T}{\sigma_T} = \frac{185.5 - 150}{17.32} = 2.05$$

Since $z = 2.05 > 1.96$, reject H_0; and, conclude that there is a difference in starting salaries

26. Rankings:

Product A	Product B	Product C
4	11	7
8	14	2
10	15	1
3	12	6
9	13	5
Sums 34	65	21

$$W = \frac{12}{(15)(16)}\left[\frac{(34)^2}{5} + \frac{(65)^2}{5} + \frac{(21)^2}{5}\right] - 3(16)$$

$$= 58.22 - 48 = 10.22$$

$\chi^2_{.05} = 5.99147$ (2 degrees of freedom)

Reject H_0; and conclude the ratings for the products differ

28. Specialty rankings:

	Surgery	Radiology	Obstetrics
	16	17	7.5
	10.5	2	1
	18	12	3
	7.5	15	14
	6	5	4
	10.5	13	9
Totals	68.5	64	38.5

$$W = \frac{12}{(18)(19)}\left[\frac{(68.5)^2}{6} + \frac{(64)^2}{6} + \frac{(38.5)^2}{6}\right] - 3(19)$$

$$= 60.06 - 57 = 3.06$$

$$\chi^2_{.05} = 5.99147 \text{ (2 degrees of freedom)}$$

Since $3.06 \leq 5.99147$, do not reject H_0; for these three specialties, we cannot conclude that there are significant differences in salary

32. a. $\Sigma d_i^2 = 52$

$$r_s = 1 - \frac{6\Sigma d_i^2}{n(n^2 - 1)} = 1 - \frac{6(52)}{10(99)} = .68$$

b. $\sigma_{r_s} = \sqrt{\frac{1}{n-1}} = \sqrt{\frac{1}{9}} = .33$

$$z = \frac{r_s - 0}{\sigma_{r_s}} = \frac{.68}{.33} = 2.06$$

Reject H_0 if $z < -1.96 \text{ or } z > 1.96$;
Since $z = 2.06 > 1.96$, reject H_0 and conclude that significant rank correlation exists

34. $\Sigma d_i^2 = 170$

$$r_s = 1 - \frac{6\Sigma d_i^2}{n(n^2 - 1)} = 1 - \frac{6(170)}{11(120)} = .23$$

$$\sigma_{r_s} = \sqrt{\frac{1}{n-1}} = \sqrt{\frac{1}{10}} = .32$$

$$z = \frac{r_s - 0}{\sigma_{r_s}} = \frac{.23}{.32} = .72$$

Reject H_0 if $z < -1.96 \text{ or } z > 1.96$; since $z = .72$, do not reject H_0; we cannot conclude that there is a significant relationship between the rankings

CHAPTER 18

1. a. $\bar{x} = 215$ is an estimate of the population mean

b. $s_{\bar{x}} = \frac{20}{\sqrt{50}} \sqrt{\frac{800 - 50}{800}} = 2.7386$

c. $215 \pm 2(2.7386)$ or 209.5228 to 220.4772

5. a. $\bar{x} = 1.8$ and $s = .4$

$$s_{\bar{x}} = \left(\frac{.4}{\sqrt{30}}\right) \sqrt{\frac{153 - 30}{153}} = .0655$$

Approximate 95% confidence interval:

$$1.8 \pm 2(.0655)$$

or $\$1.669$ to $\$1.931$ billion

b. $\bar{x} = 6$ and $s = 1.4$

$$N\bar{x} = 153(6) = 918$$

$$Ns_{\bar{x}} = 153\left(\frac{1.4}{\sqrt{30}}\right) \sqrt{\frac{153 - 30}{153}} = 35.0643$$

Approximate 95% confidence interval:

$$918 \pm 2(35.0643)$$

or 847.8714 to 988.1286

c. $\bar{p} = .25$

$$s_{\bar{p}} = \sqrt{\left(\frac{153 - 30}{153}\right)\frac{(.25)(.75)}{29}} = .0721$$

Approximate 95% confidence interval:

$$.25 \pm 2(.0721)$$

or $.1058$ to $.3942$

This is a rather large interval; sample sizes must be rather large to obtain tight confidence intervals on a population proportion

7. a. *Stratum 1:* $\bar{x}_1 = 138$
Stratum 2: $\bar{x}_2 = 103$
Stratum 3: $\bar{x}_3 = 210$

b. *Stratum 1*

$$\bar{x}_1 = 138; s_{\bar{x}_1} = \left(\frac{30}{\sqrt{20}}\right)\sqrt{\frac{200 - 20}{200}} = 6.3640$$

Approximate 95% confidence interval:

$$138 \pm 2(6.3640)$$

or 125.272 to 150.728

Stratum 2

$$\bar{x}_2 = 103; s_{\bar{x}_2} = \left(\frac{25}{\sqrt{30}}\right)\sqrt{\frac{250 - 30}{250}} = 4.2817$$

Approximate 95% confidence interval:

$$103 \pm 2(4.2817)$$

or 94.4366 to 111.5634

Stratum 3

$$\bar{x}_3 = 210; \; s_{\bar{x}_3} = \left(\frac{50}{\sqrt{25}}\right)\sqrt{\frac{100-25}{100}} = 8.6603$$

Approximate 95% confidence interval:

$$210 \pm 2(8.6603)$$

or 192.6794 to 227.3206

c. $\bar{x}_{st} = \left(\frac{200}{550}\right)138 + \left(\frac{250}{550}\right)103 + \left(\frac{100}{550}\right)210$

$= 50.1818 + 46.8182 + 38.1818 = 135.1818$

$$s_{\bar{x}_{st}} = \sqrt{\left(\frac{1}{(550)^2}\right)\left(200(180)\frac{(30)^2}{20} + 250(220)\frac{(25)^2}{30} + 100(75)\frac{(50)^2}{25}\right)}$$

$$= \sqrt{\left(\frac{1}{(550)^2}\right)3{,}515{,}833.3} = 3.4092$$

Approximate 95% confidence interval:

$$135.1818 \pm 2(3.4092)$$

or 128.3634 to 142.0002

14. a. $\bar{x}_c = \dfrac{\Sigma x_i}{\Sigma M_i} = \dfrac{750}{50} = 15$

$\hat{X} = M\bar{x}_c = 300(15) = 4500$

$\bar{p}_c = \dfrac{\sum a_i}{\sum M_i} = \dfrac{15}{50} = .30$

b. $\sum(x_i - \bar{x}_c M_i)^2 = [95 - 15(7)]^2 + [325 - 15(18)]^2$

$\qquad + [190 - 15(15)]^2 + [140 - 15(10)]^2$

$\qquad = (-10)^2 + (55)^2 + (-35)^2$

$\qquad + (-10)^2 = 4450$

$$s_{\bar{x}_c} = \sqrt{\left(\frac{25-4}{(25)(4)(12)^2}\right)\left(\frac{4450}{3}\right)} = 1.4708$$

$s_{\hat{X}} = Ms_{\bar{x}_c} = 300(1.4708) = 441.24$

$\sum(a_i - \bar{p}_c M_i)^2 = [1 - .3(7)]^2 + [6 - .3(18)]^2$

$\qquad + [6 - .3(15)]^2 + [2 - .3(10)]^2$

$\qquad = (-1.1)^2 + (.6)^2 + (1.5)^2$

$\qquad + (-1)^2 = 4.82$

$$s_{\bar{p}_c} = \sqrt{\left(\frac{25-4}{(25)(4)(12)^2}\right)\left(\frac{4.82}{3}\right)} = .0484$$

**c. Approximate 95% confidence interval
for population mean:**

$$15 \pm 2(1.4708)$$

or 12.0584 to 17.9416

**d. Approximate 95% confidence interval
for population total:**

$$4500 \pm 2(441.24)$$

or 3617.52 to 5382.48

**e. Approximate 95% confidence interval for
population proportion:**

$$.30 \pm 2(.0484)$$

or .2032 to .3968

INDEX

Numbers that are followed by an *f* are references to a figure.
Numbers that are followed by a *t* are references to a table.

PHOTO CREDITS

Page 21, © Jean-Claude Lejeune/Stock Boston
Page 61, © Martha Bates/Stock Boston
Page 111, © Jean-Claude Lejeune/Stock Boston
Page 157, top © Cincinnati Reds, 1985
Page 157, bottom © National Baseball Library and Archive,
                 Cooperstown, N.Y.
Page 199, © 1991, PhotoResource, Mpls, MN
Page 233, © 1992, PhotoResource, Mpls, MN
Page 267, © 1990, Jacobs/Custom Medical Stock Photo, Inc.
Page 341, © 1990, Mike Mazzaschi/Stock Boston
Page 411, © Steve McCutcheon/Visuals Unlimited
Page 435, © 1991, Elizabeth Crews/Stock Boston
Page 583, © Tony Stone Worldwide/Howard Grey
Page 641, © The Stock Market/Paul Barton, 1991
Page 695, © 1992, PhotoResource, Mpls, MN.
Page 715, © 1989, Elizabeth Crews/Stock Boston

Please remember that this is a library book,
and that it belongs only temporarily to each
person who uses it. Be considerate. Do
not write in this, or any, library book.

519.5 A546i 1994

Anderson, David Ray,
 1941-
Introduction to
 statistics : concepts
      c1994.